The Truth About Energy

The transition to renewable energy is vital and fast-paced, but how do we choose which technologies to drive this energy transition? This timely book provides everyone interested in the renewable energy transition with an introduction to and technical foundation for understanding modern energy technology. It traces everyday power generation through history, from the Industrial Revolution to today. It examines the use of wood, coal, oil, natural gas, hydro, and nuclear to produce energy, before discussing renewable energy sources such as biomass, photovoltaics, concentrated solar power, wind, wave, and geothermal. The book examines to what extent and how each technology can contribute to a clean, green infrastructure. *The Truth About Energy* explains the science and engineering of energy to help everyone understand and compare current and future advances in renewable energy, providing the context to critically examine the different technologies that are competing in a fast-evolving engineering, political, and economic landscape.

JOHN K. WHITE is a physicist, writer, and educator, who has worked in the engineering, science, and education fields in Canada, the Netherlands, Ireland, and Spain. He is the editor of the website E21NS (Energy in the 21st Century News Service) and author of *Do the Math! On Growth, Greed, and Strategic Thinking* (Sage). He was a lecturer in the School of Physics, University College Dublin, and the Department of Education, University of Oviedo, where he taught courses in atomic physics, optics, and science education.

"White's writing convincingly glides between scientific and technological descriptions of energy and the social world in which those descriptions both form, and are formed by, that same science and technology. Part exposé, part explanatory guide, the book provides an excellent and balanced foundation from which to understand the transitions we must undergo in our troubled relationship to energy. The analysis is thoroughly researched and not shy about naming names as it builds upon specifics. Now thrown into crisis mode, we face extremely consequential choices – digesting this book enables the reader to understand the social consequences of energy choices."

Professor Kirk W. Junker, Vice-Rector for Sustainability,
University of Cologne

"Written by a physicist who has done his homework – on chemistry, biology, and even some social science – this is a well-informed and engaging book that conveys the excitement the author believes we should all feel about renewable energy. For those with the interest, there is ample technological and historical background to understand the complex technologies on offer, both renewable and non-renewable. But there are also personal asides and charming juxtapositions, such as when a fusion reactor and Vincent van Gogh share space within a single paragraph. The author's interests are wide-ranging and, despite my own background in the topics he covers, I learned a lot. I can readily see incorporating it as a supplementary text in a course in environmental studies, energy systems, or technology."

Dr. Eric Kemp-Benedict, Associate Professor of Ecological Economics,
University of Leeds

The Truth About Energy

Our Fossil-Fuel Addiction and the Transition to Renewables

JOHN K. WHITE

formerly University College Dublin

CAMBRIDGE
UNIVERSITY PRESS

Shaftesbury Road, Cambridge CB2 8EA, United Kingdom

One Liberty Plaza, 20th Floor, New York, NY 10006, USA

477 Williamstown Road, Port Melbourne, VIC 3207, Australia

314–321, 3rd Floor, Plot 3, Splendor Forum, Jasola District Centre,
New Delhi – 110025, India

103 Penang Road, #05–06/07, Visioncrest Commercial, Singapore 238467

Cambridge University Press is part of Cambridge University Press & Assessment,
a department of the University of Cambridge.

We share the University's mission to contribute to society through the pursuit of
education, learning and research at the highest international levels of excellence.

www.cambridge.org
Information on this title: www.cambridge.org/9781009433198

DOI: 10.1017/9781009433181

First published 2024

A catalogue record for this publication is available from the British Library.

Library of Congress Cataloging-in-Publication Data
Names: White, John K. (John Kingston), 1959–, author.
Title: The truth about energy : our fossil-fuel addiction and the transition to renewables / John
K. White.
Description: Cambridge ; New York, NY : Cambridge University Press, [2023] | Includes
bibliographical references and index.
Identifiers: LCCN 2023026500 (print) | LCCN 2023026501 (ebook) | ISBN 9781009433198
(paperback) | ISBN 9781009433181 (ebook)
Subjects: LCSH: Power resources – History.
Classification: LCC TJ163.2 .W475 2023 (print) | LCC TJ163.2 (ebook) | DDC 333.7909–
dc23/eng/20230908
LC record available at https://lccn.loc.gov/2023026500
LC ebook record available at https://lccn.loc.gov/2023026501

ISBN 978-1-009-43319-8 Paperback

For Belén

Contents

Preface

The word *revolution* was first employed to mean "great change" after the 1543 publication of *On the Revolutions of the Heavenly Spheres* by the Polish mathematician, astronomer, and clergyman Nicolaus Copernicus. Positing a central sun to explain the precise workings of a seemingly complicated celestial sky, the simpler heliocentric solar system shook the Middle Ages from its long intellectual slumber, correcting an outdated model of the heavens that had become overly cumbersome and unwieldy. At the same time, antiquated political systems run by hereditary royals and aristocratic cliques began toppling to differing degrees in the newly emerging nation states across the globe.

But the powers-that-be don't cede control easily, keen to maintain order and protect their vast amounts of tribute wealth. The new thinking was subjected to scorn and ridicule, although ultimately no one could stop what had become clear: that the Earth was not the center of the universe (nor even the center of its own solar system) and that the rights of man were more important than the wealth of a few elites and their families. And there were more changes to come. From the 1830s on, the Industrial Revolution began remaking a labor-intensive, agrarian society into an energy-based, consumer-led world, where machines and fossil-fuel energy replaced animal and human muscle, while the Information Revolution of the 1960s overhauled how we create, store, and exchange information with the advent of the transistor, integrated circuit, and microprocessor, revamping an entire global economy and the lives of billions. The ramifications of each revolution are still felt today, while disagreements abound over how to implement the changes without undoing the very fabric of society.

All transitions take time before the revolutionary becomes the everyday. Newton's *Principia* was written 144 years after *De revolutionibus*, itself only published after Copernicus's death for fear of contradicting the Church. What's more, Copernicus wasn't entirely modern, still harboring Ptolemaic artifacts such as equants and perfect circles, literally wheels inside of wheels to describe planetary motion. Often considered the last of the old rather than the first of the new, Copernicus nonetheless aimed to rework a mistaken orbital geometry by putting the Sun at the center, laboring for four decades in his spare time as a devout Polish clergyman. By doing away with a 14-centuries-old, Earth-centered solar system, despite including numerous Ptolemaic fixes, Copernicus paved the way for Galileo, Kepler, and Newton, who refined what he could not to secure the revolution. Galileo would also suffer for his beliefs, incarcerated by the Church and subjected to condemnation, but Newton would finally put paid to Aristotle, Ptolemy, and the Church's belief that the Earth didn't move. Progress is slow: 1,400 years for a Polish clergyman and astronomy hobbyist to overturn Ptolemy (2,000 if we go back to Aristotle). And 400 more for the Church to apologize to Galileo, unwilling to see what he (Jovian moons), Kepler (elliptical orbits), and Newton (inverse-square gravitation) showed them.

Today, we are beginning another revolution, just as profound yet entirely different than any that came before. Renewable energy and connectivity are remaking power generation in a thoroughly novel way, combining rather than destroying the industrial and technological foundations of the past. In the coming decades, nothing will be the same. Hundreds of millions of jobs will be created, supplanting economic and political hierarchies of the past with laterally organized and interconnected nodal power structures. Author and social critic Jeremy Rifkin has called the coming upheaval the "Third Industrial Revolution" (TIR), a post-carbon, distributed capitalism, where renewable energy and internet technologies "merge to create a powerful new infrastructure."

Others have called the ongoing transition a fourth industrial revolution that includes disruptive technologies such as the Internet of things, robotics, artificial intelligence, and virtual reality. Whatever the number or however far we are on the road to a seemingly inevitable future, the new energy systems are fundamentally changing long-established relationships between producers and consumers, impacting everyday business, governmental oversight, and even our own civic responsibilities. Faced with increased global warming, others are also calling for a systematic dismantling of our oil-addicted lives before it is too late to make a difference, a.k.a. the "fossil endgame."

The next revolution isn't advancing smoothly, however, because of a still dominant, multitrillion-dollar, petroleum-based economy at the forefront of everyday living as well as our many entrenched habits, despite diminishing oil supplies and increasingly harmful consequences to health and the atmosphere from burnt carbon emissions. As with all revolutions, we begin the new amid the old until we can no longer ignore the obvious. Institutional change comes slowly. Revolutions are disruptive. They change everything.

Renewable energy isn't well understood either, despite much of the underlying technology having been around for ages, whether generating electrons from photons in a solar cell (first developed in the 1950s), electric propulsion in an induction motor (initially demonstrated in the 1830s), or the chemistry of charge storage in a battery (a device with us since 1799). And while the age of oil produced an unparalleled explosion of people and power – global population more than quadrupling to almost eight billion in the last century – the major disruptions of the twenty-first century will be renewable energy, electric vehicles, and storage batteries. Each will become cheaper, longer lasting, and ubiquitous, just as power stations, gasoline-powered vehicles (gasmobiles), and electric light did over a century ago, thanks in part to the ingenuity and inspiration of more than a few dedicated inventors and innovators. Some are household names, such as Thomas Edison, Nikola Tesla, and George Westinghouse, while scores of others are not as well known, such as Charles Parson, William Stanley, and Samuel Insull.

In Part I, I examine coal, oil, and nuclear power, technologies that created our modern industrial world and will continue to produce energy for decades to come, increasingly so in some countries. Knowing more about fossil fuels is essential to understand the feedstock of modernity as well as global supply chains, some that have become increasingly more dangerous amid shifting geopolitical borders and tenuous international alliances. Armed with the knowledge of an existing infrastructure, we can fix our past mistakes and step boldly into a cleaner future to generate the power we want and need, ensuring a safer and more secure world. In some cases, old-world infrastructure can be retooled to accommodate renewables, including transmission lines, liquid-fuel transportation, and long-haul delivery.

In Part II, I examine the technologies of renewable energy (solar, wind, water) and distributed power systems that are remaking our twenty-first-century lives. After years of inaction, policy makers are beginning to see the benefits of modernizing a petroleum-based economy, no longer citing a drag on growth but an economic opportunity to implement proven solutions on a wider scale to curb pollution and greenhouse gas emissions. No one should doubt, however, the power of those seeking to slow the change or co-opt new ways as their own. Led by Big Oil and indeed Big Gas, organized climate denialism and climate delayism continue to report falsehoods and made-up controversies as in the past with pesticides, acid rain, and tobacco addiction, while war in Ukraine has especially upended the production and distribution of new energy.

At the same time, each of us can make a difference with simple, smart decisions about our own energy needs and consumptive lifestyles (Part III). As we saw during the COVID-19 pandemic, where energy use dropped significantly across the globe, many of us can live with less, lowering consumption and carbon emissions. Conservation need not be a dirty word practiced only by a few, but by everyone seeking to reduce consumption and save money.

Wobbling through the extraordinary changes that lie ahead, we must become better informed about the fundamentals of power generation as we transition from brown to green. We can no longer burn whatever we please, ignoring the obvious consequences of our actions. Instead, we must rebuild, rewire, and rework our failed past, careful not to add more layers of sophistication or increase an already widening digital divide. Decentralized green energy sources also mean less conflict over supply chains that can produce devastating conditions in low-income and developing countries across the globe and even on the edges of Europe.

The mechanics of motive power, electricity, and networks are examined in the context of new industries competing for our daily needs and their value in a modern sustainable world. Connectivity is especially remaking the way we create and share power – soon we will borrow a cup of charge from a neighbor as easily as a cup of sugar. The means to implement an economically viable and sustainable future is discussed, building on a needed "energy literacy" in a fast-changing world as we look under the hood of electric vehicles, rechargeable batteries, and distributed resources, defining the origin and meaning of common units along the way (e.g., watts, joules, kilowatt-hours) and the latest new-energy terminology.

We must find more efficient ways to generate power for electricity, heat, and transportation, recognizing that clean, sustainable energy is the only way forward for a warming planet with finite resources. We have no choice but to leave the world better than how we found it. What an opportunity to be present during the change from old to new, from brown to green, from a dirty, old-world, fossil-fuel-burning past to a modern, renewables-powered, clean reality. Unlike in the days of Copernicus, Thomas Jefferson, or James Watt, this time many of us can both see and be a part of the revolution.

Acknowledgments

Writing *The Truth About Energy* has been an extraordinary experience, thanks largely to those who helped explain ideas and direct me to new vistas along the way. I would especially like to thank my godparents, Tom and Pat Lawson, who were Green long before we capitalized the word; my mother Daisy, who always supported me and shared with me her unwavering enthusiasm for life; and Belén Roza, the love of my life and constant companion, without whom this book would not be possible.

I have been asking why things work for as long as I can remember. Fortunately, many generous people have been there with a ready answer. I would like to thank Stefan Bäumer, Tom Clynes, Ian Fairlie, Andrew Fitzpatrick, Brian Gallagher, Charles Gasparovic, John German, Robin Marjoribanks, Lisa MacKenzie, Molly Mulloy, Paz Roza, Fernando Santirso, Jeffrey St. Clair, Avelino Valle, Fernando Vaquero, David White, Deirdre White, Peter Whitney, and V. Zajac for their kind sharing. I would also like to thank Brian Gallagher, Charles Gasparovic, and Belén Roza, who read parts or all of the manuscript and gave me essential feedback. Thanks also to Avelino Valle, who created a number of original illustrations. He has excellently illuminated in pictures what I could not in words.

Thanks also to everyone at Cambridge University Press for turning the draft into a finished book. In particular, I would like to thank Matt Lloyd who started the ball rolling and for his overall supervision, Maya Zakrzewska-Pim, whose constant help and guidance was invaluable throughout, Sapphire Duveau and Jenny van der Meijden for their excellent content management, Daniel Jebarajan, Geethanjali Rangaraj, and the design team at Integra for their dedicated organization, and Sara Brunton who edited the book to the highest standard and attention to detail. I am indebted to each of them for their professionalism, hard work, and kindness throughout.

Charles Gasparovic was especially helpful, talking me through new ideas, sending me source material, and encouraging me from the start. He was the first to read the entire manuscript from beginning to end, and see the strengths among the weaknesses. He suggested many invaluable additions, unique thoughts, and pertinent calculations, without which this book would be incomplete. I am indebted to him for his ever helpful guidance.

My thanks to the many dedicated writers who have recorded their observations for all to see, especially from *The New York Times*, *The Guardian*, and *CounterPunch*, from which I have précised a few already perfect analyses. Try as I might, I can't possibly read and digest all. Thanks also to *National Geographic* for bravely showing the truth about our beautiful and changing world. Thanks to all those who kindly gave permission to use their images: the City of Toronto, the M. King Hubbert Collection, Vaisala, Nokia Bell Labs, NASA, Solar AquaGrid, UNDP, IBIS Power, and Archives ESB (Ireland).

I have had to put a number of different hats on to understand the subject more deeply. As a physicist, I am used to thinking in terms of relationships, essentially A leads to B, though watch out for C. For example, sunlight converted to electricity can be explained in the language of material science and atomic physics, but needs a bit of atmospheric science and spherical geometry to get the details right. I was surprised how much of the fundamentals involved biology, geology, and chemistry, areas where my footing is not as sure. I have learned everything from photosynthesis and soil strata to the physiology of cow stomachs and the smell of a biofuel plant. I am never surprised at how much there is to learn or how there is always more to grasp. Long may the wonder last.

I have used publicly accessible data wherever possible, into which one can delve further for more. All links to external websites were correct at the time of publication. You can check many of the sources in a companion website to this book entitled E21NS. I also provide adjunct educational material on my website at www.johnkwhite.ie.

There is no end to a book on energy. The changes are breathtaking from all sectors. Hopefully, I have helped to clarify the material and cleared a way to further one's own journey. Thank you to all. Onwards to a happier and cleaner future.

A Word about Numbers

Normally, one puts reference numbers in an appendix, stuffed away at the back of a book, which a reader may not even realize is there until the end. But I think a few essential numbers up front make a good overview to the material, even if they appear out of context at this point. It doesn't matter if one doesn't fully understand their meaning yet or the accompanying units. A quick look now may help with the explanation and context later.

Note that some of the numbers are approximate. It is easy to get bogged down in exact measures and lose sight of the order of magnitude of a quantity. But with a rough idea at one's fingertips, one can more easily compare the myriad different amounts of things. Relative numbers and percentages can be of much more use.

For example, globally we emit about 50 billion tons of greenhouse gases (GHGs) per year, an amount that likely doesn't mean much to most people. Broken into percentages by country or industry, however, the numbers may mean more, where China emits roughly 30%, the United States 15%, and the European Union 10%, while the rest of the world combined emits 45%. Furthermore, the amount has been growing by about 500 million tons per year over the last decade (~1%/year), although not at the same rate everywhere. The more industrialized nations are responsible for most of the emissions, three-quarters of which is carbon dioxide, the GHG everyone is talking about. Historically since 1750, Europe has emitted 37%, the United States 26%, Asia 15%, China 14%, and the rest of the world 9%.

As we "decarbonize" energy, we must also correspondingly increase grid power for electric vehicles, home heating, and storage batteries. To get an idea of the scale of the challenge, the global electrical power grid has a generating capacity of about 8 trillion watts (8 TW), 15% in the United States. If we plan to replace petroleum with electricity to run our vehicles, the amount will need to increase by anywhere from 10% to 50%. Add electrical heating and industrial processes and the global grid could more than double. Time to get started!

Chapter		
1	horsepower	1 horsepower (hp) = 746 watts (W)
1	global oil use	~100 million barrels per day (mbpd)
1	global natural gas use	~10 billion cubic meters per day (bcmpd)
2	greenhouse gas emissions	~50 billion tons per year (75% carbon dioxide)
3	U-235 half-life	~700 million years
3	number of nuclear weapons	~12,000 (~6,000 US and Russia each)
4	earth–sun average distance	150 million km (1 AU)
4	visible energy band	~380 nm (blue) to ~780 nm (red)
4	solar irradiance (on the ground)	~1,000 W/m^2
4	earth atmosphere	~100 km (the Kármán line)
4	atmospheric carbon dioxide	400 ppm (2014), 419 ppm (2023)
5	wind turbine coverage	1 MW = ~$1 million, ~1,000 people
5	global/US primary energy	~600 EJ/100 EJ (2022)
5	global/US grid power capacity	~8,000 GW (8 TW)/1,200 GW (1.2 TW)
6	Li-ion battery range	~30 kWh = 1 hour = 100 miles (160 km)
6	Li-ion battery energy density, cost	~250 Wh/kg, ~$100/kWh
7	home electrical use (per couple)	~30 kWh/day (~$100/month @ 10 cents/kWh)
7	home electrical use (per person)	~15 kWh/day or 5,000 kWh/year per person

Introduction

There is much to do to create a modern energy paradigm, one that is clean, sustainable, and economically viable, but the changes are coming as overall efficiencies improve and manufacturing costs decrease for today's renewable technologies. In 2000, 0.6% of total global energy production was generated either by wind or solar, a 50% increase in a decade; by 2010, the amount had doubled.[1] By 2013, Spain had achieved a global first as wind-generated power became its main source of energy (21% of total demand, enough to run 7 million homes[2]), while both Portugal and Denmark now regularly produce days powered 100% by wind.

Historically cloudy England scored a first, as solar became the largest source of grid energy during an especially sunny 2018 spring bank-holiday weekend,[3] while in the midst of high winds from Storm Bella on Boxing Day in 2020, the UK was more than half powered by wind, a new record. On April 30, 2022, 100% of California's electricity was generated by renewables for a short period on the weekend, two-thirds from solar panels, an enviable milestone in the world's fifth largest economy.[4] Even in rural Africa, simple solar mobile phone rechargers now service hundreds of thousands of homes,[5] while in developing countries rural solar hook-ups provide basic lighting that reduces harmful kerosene burning as "micro-grids" bypass the need for expensive grid tie-ins.

Quoting the nineteenth-century philosopher Arthur Schopenhauer, the German parliamentarian Hermann Scheer stated, "A new idea will firstly become denounced as ridiculous, secondly there are many fights against it, and finally all people were in favor of it from the early beginning."[6] Scheer was responsible for the 2000 German Renewable Energy Act, which spurred on an avant-garde approach to energy technology via consumer subsidies and grid buybacks, transforming Germany into a world leader in solar power, all in a country with a mean latitude of 51.5° and a daily average of 4.1 hours of sunshine. As noted in a 2012 *New York Times* op-ed, "More than one million

Germans have installed solar panels on their roofs, enough to provide close to 50 percent of the nation's power, even though Germany averages the same amount of sunlight as Alaska."[7]

Other countries are looking to Germany's example as renewable energy technologies become more viable year on year. In particular, China has seen remarkable double-digit growth over the past 20 years and is racing ahead in photovoltaic (PV) technology, producing solar panels cheaper than anywhere else and bringing the cost of solar infrastructure into the realm of everyday possibility. A Chinese-made solar cell today is 200 times cheaper than in the 1970s, spurring on the rise of gigawatt-sized power plants. As if announcing their coming dominance in globalized markets, some Chinese solar power plants are even being designed in the shape of a giant panda bear that employ three different types of modules (mono-c, bifacial dual-glazed, and thin-film). China also boasts more wind farms than any other country under its Wind Base program, having already installed over 200 GW of power with plans to reach 400 GW by 2030 and 1000 GW (1 terawatt!) by 2050, for a total national penetration covering two-thirds of its existing grid from wind power alone.

And yet in much of the developed world, green markets are still advancing slowly, while governments continue to support pollution-spewing coal, oil, and natural-gas (methane) industries that have alarming consequences for our health and well-being in an increasingly warming planet. If temperatures continue to rise at current rates, we may not be able to stop the damage.

Rather than encouraging cleaner technologies, the politics of the day hinders progress, including the United States Supreme Court overriding the ability of the Environmental Protection Agency (EPA) to regulate emissions and the European Union labeling natural gas and nuclear power as green in its taxonomy of environmentally sustainable economic activities. To be sure, a powerful, anti-environmental lobby is keeping carbon in business, preferring monopolistic, central-metered power over local, autonomous, grid independence. Some even claim global warming is a hoax to disadvantage leading industrialized nations.

Resistance is especially strong in the United States, where at the height of the 1970s oil crisis Jimmy Carter had 32 solar-thermal panels installed on the White House roof that were promptly dismantled by his successor Ronald Reagan, who alarmingly claimed that trees caused more pollution than automobiles, while working to maintain a gluttonous, unquestioning dependence on oil. In a strikingly similar anti-green reboot, Donald Trump actively promoted "clean coal," endorsed lower vehicle-emissions standards, and pulled the US out of the 2015 Paris Agreement, which had outlined new-energy transition initiatives and sought to coordinate worldwide climate action. Relishing his role as a green Grinch, he even had a bike-sharing hub outside the White House removed.

But despite an apparent institutional resistance to change, the United States has pioneered many ground-breaking renewable energy sector (RES) projects, such as the first megawatt (MW) capacity photovoltaic solar-powered plant in Hesperia, California, in 1982, followed two years later at Rancho Seco, now run as part of the Sacramento Municipal Utility District power grid that generates 3.2 MW, enough to electrify 2,200 homes.[8] Another pioneering technology has been operating since 1989 at Kramer Junction in the Mojave Desert – concentrated solar power, ideal for a location averaging 340 days of sunshine per year.[9] Modern solar power plants are slowly being added piece-meal across the country, scaled to size as needed, while wind power is surging in the vast oil-laden regions of West Texas.

When it comes to making fundamental changes, however, the United States is notoriously hands off, preferring coal-, oil-, and natural-gas-fired power generation with generous government subsidies. There is no rationale for the resistance other than to maintain the status quo in a presumed market economy, odd given the many paradigm-changing technologies developed there in the past, such as in aviation, satellites, and communication, and a well-established entrepreneurial ethos. As NYU physicist Martin Hoffert noted, "Most of the modern technology that has been driving the US economy did not come spontaneously from market forces. The Internet was supported for 20 years by the military and for 10 more years by the National Science Foundation before Wall Street found it."[10] To be sure, change comes slowly, especially when faced with a relentless and well-healed opposition.

Hoping to kick-start changes in a country that emits 15% of emissions from only 4% of the world's population, Barack Obama announced a new Clean Power Plan in 2015, letting states implement their own means to reduce carbon emissions by 20% by 2020 and providing $1 billion in guaranteed loans for household solar installation. In early 2016, Obama and Canadian Prime Minister Justin Trudeau also agreed to reduce methane gas emissions in their countries by up to 45% below 2012 levels by 2025, while discussing ways to transition to a low-carbon economy. Alas, many of the changes were undone by Obama's successor, who characterized regulations as a restrictive bureaucratic killjoy. In 2022, Joe Biden's Build Back Better clean-energy plan was also gutted, stuck in debate between those who want to incentivize new technolo-gies (clean energy) and those beholden to the past (subsidized fossil fuels). Highlighting the extent of the challenge, annual government subsidies for fossil fuels in 2022 doubled across the globe from the previous year to almost $700 billion.

The role of international agreements seems even more fruitless. In December 2015, delegates from 195 countries gathered in Paris to hammer out a deal to reduce greenhouse gas emissions and curb increased global warming. The meeting helped to overcome previously intransigent positions, such as reporting on emissions, maintaining and updating nationally determined contributions (NDCs) every 5 years, and setting an ambitious goal of reaching "net-zero" emissions by 2050, although the final agreement was not legally binding. There were also problems with the report language, such as "shall" versus "should," while various high-carbon subgroups cut advantageous side deals. As with the earlier unratified 1997 Kyoto Protocol, the US pulled out before rejoining at the start of the Biden administration.

Agreement is an exasperatingly slow process as politicians first think about their own regions, unwilling to prod the well-established, fossil-fuel industries that propel much of their gross domestic products (GDPs). Scheer even wondered if international conferences have done more harm than good, noting that "While the delegates have been debating over the past decade, emissions have been rising by an unprecedented 30 per cent." Sadly, he surmised that "The effect of the climate change negotiations has thus been to preserve the status quo."[11] One meme doing the rounds at the 2021 COP26 meeting in Glasgow summed up the growing frustration: "Number of years leaders have been coming to COP – 26, number of years GHG emissions have dropped – 0." Or as the teenage Swedish climate crusader and youth activist Greta Thunberg put it, "three decades of 'blah blah blah'."[12] Green meetings may come with grand themes, but change is never easy. Unfortunately, as we argue about our differences and the consequences of our inaction, the stakes keep getting higher.

***

It is hard to imagine life without energy. We burn fuel to heat and electrify our homes, light our streets, and run our cars. Modern industry demands cheap electricity to manufacture the basics of everyday life, while oil is refined to make numerous household products from plastics to lubricants and tires to ink, as well as keeping us moving from A to B. There is no turning back the modern, energy-based lifestyle we all enjoy. Alas, an economy that produces energy by burning carbon-containing fuel is responsible for millions of deaths per year, creating unwanted by-products in the process, primarily carbon dioxide and methane, but also nitrous oxide, benzene, and other carcinogenic aromatic hydrofluorocarbons.

The carbon economy is also limited. Deposited in the earth 300 million years ago during the Carboniferous Period, when high pressures converted decaying plants and animals into so-called "fossil fuels," the supply will eventually run out.

In 1956, the American geologist M. King Hubbert calculated how much was left in the ground, noting that oil output had doubled in the United States every 8.7 years from 1880 to 1930.[13] In his seminal report to the American Petroleum Institute Hubbert predicted that the US would reach a maximum production of conventional oil (a.k.a. "peak oil") in the 1970s, which it did, further predicting a world peak oil for 2000, a date still much debated and scrutinized.

Today, some analysts believe there may be only a few decades left given our current annual consumption of 35 billion barrels per year, and even less if demand from China, India, and other developing countries continues to increase. Newer methods to secure oil and gas are postponing the end, such as offshore drilling, fracking, and oil sands, but at some point in this century fossil fuels will diminish beyond easy extraction and eventually run out. Regardless the exact end day, there is no such thing as a fossil-fuel free lunch – when it comes to consumption, one always has to pay the bill.

Despite the declining supplies, worsening pollution, and increased greenhouse gas emissions, global energy consumption continues, however, to be powered primarily by fossil fuels – oil (30%), coal (27%), and natural gas (23%) – with biomass (10%), nuclear (4.9%), hydroelectric (2.6%), and renewables (2.5%) accounting for the rest. To be sure, the old-world infrastructure is massive with more than 60,000 power plants,[14] generating roughly four-fifths of all power, including almost 20,000 oil plants generating at least 1 MW in the United States alone. Globally, there are about 7,500 coal plants, more than a third in China, and 440 nuclear plants. Currently, wind and solar provide about 2.5% of the total primary energy supply (10% of electrical power), while renewable technologies (wind, water, sun – WWS) are beginning to account for more of the energy mix and in some places are already at parity or cheaper than traditional power production (Figure 1 and Table 1). No one can argue anymore about the cost.

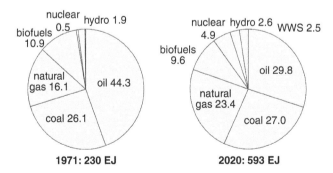

Figure 0.1 Global energy supply by fuel source: (a) 1971 percentage and (b) 2020 percentage. (*Source*: International Energy Agency).

Table 0.1 *Global energy supply by fuel source: 1971 and 2020 (exajoules, percentage)*

Fuel source	1971 (EJ)	%	2020 (EJ)	%
Oil	102	44.3	177	29.8
Coal	60	26.1	160	27.0
Natural gas	37	16.2	139	23.4
Biofuels	25	10.8	57	9.6
Nuclear	1.2	0.5	29	4.9
Hydro	4.4	1.9	16	2.6
Renewables	0.5	0.2	15	2.5
Total	**230**	**100.0**	**593**	**100.0**

Sources: "Total primary energy supply by fuel, 1971 and 2019," International Energy Agency, August 6, 2021. www.iea.org/data-and-statistics/charts/total-pri mary-energy-supply-by-fuel-1971-and–2019 (1971) and "Total energy supply (TES) by source," World 1990–2020, International Energy Agency, October 26, 2022. www.iea.org/data-and-statistics/data-tools/energy-statistics-data-browser (2020).

The four great capital-accumulating world powers of history were the Genoese-backed Spanish, the Dutch – both nations powered by wind and the limits of intermittent favor – a coal-fueled industrialized Britain – vastly more advanced and wealthier than any before, yet as dirty as it was rich – and the oil-powered United States, whose brokered petroleum politics and never-ending thirst for more has created unparalleled convenience, yet put our planet on a perilous path of uncertain change. In the relentless quest for more, modern capital slows for no one, even as its seeds its own destruction.

Indeed, a modern renewable-energy paradigm will fundamentally change how we all live, creating abundance, independence, and security. As noted by Scheer, "Making the groundbreaking transition to an economy based on solar energy and solar resources will do more to safeguard our common future than any other economic development since the Industrial Revolution."[15]

It has been almost 60 years since the publication of *Silent Spring*, Rachel Carson's groundbreaking account of the effects of chemical pesticides on the environment, which helped forge new attitudes about our relationship with the Earth, ultimately leading to the creation of the Environmental Protection Agency (EPA) in the United States during the presidency of Richard Nixon and a better understanding of industrial neglect and waste. Here is the start of her final chapter, entitled "The Other Road." The case for change can't be made any better.

We stand now where two roads diverge. But unlike the roads in Robert Frost's familiar poem, they are not equally fair. The road we have long been traveling is deceptively easy, a smooth superhighway on which we progress with great speed, but at its end lies disaster. The other fork of the road – the one "less traveled by" – offers our last, our only chance to reach a destination that assures the preservation of our earth.[16]

***

Despite the difficult choices ahead, there is cause for optimism. As if awakened to the challenges of remaking a failing industrial past with a revitalized green future, newly elected US president Joe Biden announced a host of measures to combat the growing "existential dangers" of climate change. In the preamble to his January 21, 2021, "Executive Order on Tackling the Climate Crisis at Home and Abroad,"[17] he announced plans to double offshore wind by 2030, build 500,000 roadside electric-vehicle charging stations, and update the entire 650,000-plus fleet of federal vehicles from gasoline to electric, initiating an American electric-vehicle "arms race" that soon saw $85 billion in promised spending by GM and Ford. The measures also included a pause on new oil-and-gas leases on federal lands and waters and signaled his intention to ask Congress to repeal $40 billion in annual oil-industry subsidies, while encouraging private-sector investment in modern renewables, increased international financing, and more directed research and development.

Ambitious in scope, yet full of practical plans, Biden, John Kerry (special presidential envoy for climate), and Gina McCarthy (the first-ever US national climate adviser and a former EPA head) outlined a way to fight climate change at home and across the globe. Rejoining the Paris Agreement, the United States also re-engaged with others to implement planet-wide actions on global greenhouse gas emissions, especially with China and Europe. The new president also stressed job creation in the burgeoning green sector, capping a million abandoned oil and gas wells, revitalizing dying fossil-fuel communities (particularly coal), and building 1.5 million new energy-efficient homes. Although late to the game, the US sought to lead with its enormous economic might, hoping to recapture the spirit of past ideals.

On the heels of Biden's clean-energy initiatives, the International Energy Agency (IEA) announced its own groundbreaking recommendations to transition to renewable-energy sources in a 2021 report entitled *"Net Zero by 2050: A Roadmap for the Global Energy Sector."* Recognizing that the emission-reduction goals pledged by national governments were insufficient to keep global warming below a targeted 1.5°C in the 2015 Paris Agreement, the IEA report listed 400 steps to achieve net zero emissions, in particular an immediate end to investing in new fossil-fuel supply projects or "unabated" coal plants, no sales of new internal combustion engine passenger cars by 2035, and a net-zero global electricity sector by 2040. The report also called for "a major global push to accelerate innovation" by adding a gigawatt of solar photovoltaics almost every day, a gigawatt of wind power every other day, and energy efficiency improvements of 4% per year until 2030. Intermittency, cyber security, and critical mineral resources were also cited as potential areas of concern.

Net Zero by 2050 outlined how different the energy landscape could look in 2050 if almost 90% of electricity generation came from renewable-energy sources (70% wind power and PV solar), while at the same time reducing global energy demand by almost 8% despite 2 billion more people. The report further emphasized how most emission reductions could come from already available technologies, requiring only increased investment to roll out advanced batteries, hydrogen electrolyzers, and carbon-capture systems, all of which exist in various stages of development. The IEA Executive Director, Fatih Birol, also noted that transitioning to net zero is a huge economic opportunity for national economies, but that "The transition must be fair and inclusive, leaving nobody behind. We have to ensure that developing economies receive the financing and technological know-how they need to build out their energy systems to meet the needs of their expanding populations and economies in a sustainable way."[18] The IEA also reported that $4 trillion was needed to fund the change to renewables over the next decade.

Other countries chimed in with their own plans, the UK announcing its goal to reach net zero by 2050 by revamping home heating (insulation and replacing boilers with heat pumps), decarbonized travel (electric-vehicle incentives, supply chains, and charging stations), and increased green-energy production (wind, solar, clean electricity by 2035). Tree planting, peat restoration, carbon capture, and hydrogen gas incentives for heavy industry were also cited, although with only £2 billion in investment and 440,000 new jobs promised, the plan was dismissed as insufficient to tackle the formidable problems that lay ahead. Policy reversals in a new government also impacted the proposed plans.

Biden's so-called Build Back Better "reset," which included provisions for home-energy efficiency, grid enhancements, electric ferries, and more electrified mass public transit, also hit a roadblock when $150 billion in clean-energy programs from his initial $3.5 trillion bill were scrapped after objections from within his own party, in particular by Joe Manchin, a coal-sponsored Democrat senator from West Virginia. After months of wrangling, the overall package was gutted because of presumed fears about inflation. The politics of jobs and fossil-fuel profits is still very much the norm, despite the devastating dangers of business as usual.

What's more, despite the change in green language at the government level, more drilling permits were approved in the first year of Biden's presidency than in the first three years of the previous Trump administration. Biden's climate advisor, Gina McCarthy, also announced she was leaving her post, upset at the slow pace of change. The reclassification of natural gas and nuclear power in the EU taxonomy – a green investment label to help consumers, energy providers, and governments implement renewable technology in a structural fashion – was also called out by environmentalists as "greenwashing."

Hoping to inscribe into law the tenuous agreements reached in Paris, delegates from around the world also met again in November 2021, in Glasgow for COP26, where a number of significant measures were passed, including pledges by some countries for an 85% reduction in deforestation, a 30% reduction in methane emissions, and an end to the financing of overseas fossil-fuel projects. The pact failed to call for an end to coal, however, because of disagreements between developing countries whose economies rely on cheap, accessible coal and developed countries who want to oversee and usher in their plans for a green-led economy. Nonetheless, the pact was signed by 197 countries, despite a last-minute change of wording in the final declaration to a phase *down* from a phase *out* of coal. Prompted by India and China, the change in wording was about more "targeted support" by the developed "carbon colonialist" nations. In other words, money.

No fossil-fuel "non-proliferation treaty" or ban on fossil-fuel subsidies were included in the text, but coal and fossil fuels were mentioned for the first time, to some a step in the right direction after 25 years of meetings. The chief EU delegate Frans Timmermans eloquently reminded everyone that "European wealth was built on coal and if we don't get rid of coal, European death will also be built on coal." Of course, a large emissions gap of 50 billion tons per year of GHGs still remains, while a week later the US proceeded with the largest-ever total acreage auction of oil-and-gas leases in the Gulf of Mexico at almost 7,000 square miles, whose extraction and burning will last beyond a decade and continue to pollute and increase global warming. Urging others to quit coal while ramping up oil, natural gas, and nuclear power may be more about ensuring that leading industrial nations maintain their competitive advantage over weaker countries than helping them to leapfrog fossil fuels.

The issue is particularly thorny for developing countries whose newly built coal plants will be rendered useless in a globally mandated closing. The 2022 outbreak of war in Ukraine also severely altered the energy landscape for European countries reliant on Russian natural gas, who then backtracked on their coal-reduction promises, even restarting previously shuttered plants. Winter warmth today trumps future concerns about the admittedly abstract nature of global warming via unseen greenhouse gases. Alas, with each passing year, business as usual becomes less of an option.

And then in the summer of 2022, some hope appeared in the guise of a remade US federal bill, the Inflation Reduction Act (IRA), which provided $369 billion in green spending over a decade, especially for wind, solar, and battery technologies, although other measures for carbon capture also meant more funding for petroleum and pipelines as did the usual side deals. Our long-traveled road continues to be trodden upon by those with their hands on the steering wheel. To be sure, transitioning to clean, renewable energy is an enormous undertaking.

Given the extent of the resistance to change, however, one wonders if the lofty rhetoric is more about controlling finances than transitioning to renewables. At the start of 2023 at the World Economic Forum in Davos, Switzerland, the UN Secretary General António Guterres called out the continued extraction of fossil fuels as "inconsistent with human survival." Yet at the same time, the CEO of the Abu Dhabi National Oil Company was appointed COP president; in other words, more of the same for oil companies seeking to maintain control of world energy supplies.

The truth about energy is that we use too much, are insufficiently concerned about deleterious effects (pollution, GHG emissions, global supply chains), and aren't in any hurry to implement cleaner, safer alternatives. After more than two centuries of unchecked growth on the back of cheap fossil fuels, there is no urgency to mend our ways or corral those who would willfully destroy or damage Mother Earth. Although we can see the dangers, economic activity is primarily designed to make money without considering the consequences, while governments won't regulate or punish those who care nothing for the environment.

It may be simplistic to pass history off as a set of energy dominoes, each better constructed for consumption and growth than the previous, but the next revolution is creating rather than destroying and managing rather than exploiting resources to reinvent a cleaner future after centuries of neglect. A new energy era has begun, one that can be implemented over time, similar to how hydro, thermal, and other power-generation technologies were added to an expanding grid in the twentieth century. The answers are right in front of us, including new ideas about growth itself, there for all to see, just as Galileo showed us if we would only look: an empire of sun, wind, water, and earth.

It is time to make the change and plug into a better way – we simply can't keep burning away our future. It is time to learn more about how energy and power works and put into practice new ways of consuming in our everyday lives. It's time to learn the truth about energy.

PART I

Out with the Old

I have not failed. I've just found 10,000 ways that won't work.

Thomas Edison

The meek shall inherit the Earth, but not its mineral rights.

J. Paul Getty

Nothing in life is to be feared, it is only to be understood.

Marie Curie

.

1

Wood to Coal

A Short History of the Industrial Revolution

1.1 Fire and Water Become Heat and Steam

We have been warming ourselves and cooking with fire since prehistory, but not until the eighteenth century did we begin to understand its properties. Fire was one of the four ancient elements, along with earth, water, and air, and was considered to be hot and dry, in contrast to water which was cold and wet. Aristotle described fire as "finite" and its locomotion as "up," compared to earth which was "down." Not much was known of its material properties, however, other than in contrast to something else, although Aristotle did note that "Heraclitus says that at some time all things become fire" (*Physics*, Book 3, Part 5).

By the early eighteenth century, fire was being tamed in England in large furnaces at high temperatures to turn clay into pottery, for example, in the kilns of the early industrialist Josiah Wedgwood, "potter to her majesty."[1] More importantly, fire was being used to heat water to create steam, which could move a piston to lift a load many times that of man or horse. When the steam was condensed, a vacuum was created that lowered the piston to begin another cycle. For a burgeoning mining industry, mechanical power was essential to pump water from flooded mines, considered "the great engineering problem of the age," employing simple boilers fueled by wood and then coal, which was more than twice as efficient, easy to mine, and seemingly limitless in the English Midlands.

Wedgwood's need for more coal to fire the shaped clay in his growing pottery factory turned his mind to seek better access to local mines, helping to spur on an era of canal building. Distinctively long and slender to accommodate two-way traffic, canal barges were first pulled by horses walking alongside on adjacent towpaths, but would soon be powered by a revolutionary new engine.

In today's technology-filled world, units can often be confusing, but one is essential, the watt (W), named after Scotsman James Watt, who is credited with inventing the first "general-purpose" steam engine, forever transforming the landscape of a once pastoral England. The modern unit of electrical power, the watt (W), or ability to "do work" or "use energy" in a prescribed time, was named in his honor, and replaced the "horsepower" (hp), a unit Watt himself had created to help customers quantify the importance of his invention. One horse lifting 550 pounds (roughly the weight of three average men) a distance of one foot in one second equals 1 hp or 746 watts.[2]

First operated in 1776, Watt sufficiently improved upon the atmospheric water pump of the English ironmonger Thomas Newcomen, who had reworked an earlier crude design of Denis Papin, which in turn was based on the vacuum air pump of the Anglo-Irish chemist Robert Boyle. Papin was a young French Huguenot who had fled Paris for London, becoming an assistant to Boyle, but was unable to finance a working model of his "double-acting" steam engine, despite having explained the theory in his earlier 1695 publication *Collection of Various Letters Concerning Some New Machines*.[3]

Before Watt, various rudimentary machines pumped water and wound elevator cables in deep coalmines, all of which employed a boiler to make the steam to raise the piston, which was then lowered by atmospheric pressure over a vacuum created when the steam was condensed, providing a powerful motive energy to ease man's labor.[4] But such "engines" were severely limited because of engineering constraints, poor fuel efficiency, and the inability to convert up-and-down reciprocal motion to the axial rotation of a wheel.

Watt improved the condensation process by adding a separate condenser to *externally* cool the steam, keeping the piston cylinder as hot as possible. The separate condenser saved energy by not heating, cooling, and reheating the cylinder, which was hugely wasteful, ensuring more useful work was done moving steam instead of boiling water. After tinkering and experimenting for years on the demonstration models of Newcomen's pump (which sprayed cold water directly into the cylinder) and Papin's digester (a type of pressure cooker), the idea came to him while out walking in Glasgow Green one spring Sunday in 1765.[5] Watt also incorporated an automatic steam inlet to regulate the speed via a "fly-ball" governor – an early self-regulating or negative feedback control system – and a crankshaft gear system to translate up–down motion to rotary motion. Most importantly, his engine was much smaller and used one-quarter as much coal.

With business partner Matthew Boulton, who acquired a two-thirds share of net profits while paying all expenses, Watt developed his steam engine full time, building an improved version to work in the copper and tin mines of

Cornwall, where limited coal supplies made the Newcomen water pump expensive to operate. Adapted for smooth operation via gears and line shafts, their new steam engine was soon powering the cloth mills of Birmingham, initiating the industrial transformation of Britain. Controlled steam had created mechanical work, indispensable in the retooled textile factories sprouting up across England. The model Newcomen atmospheric steam engine that Watt had improved on still resides in the University of Glasgow, where he studied and worked as a mathematical instrument maker when he began experimenting on Newcomen's cumbersome and inefficient water pump.

By the end of their partnership, Boulton and Watt had built almost 500 steam engines in their Soho Manufactory near Birmingham, the world's first great machine-making plant, both to pump water and to turn the spindles of industry. As noted by historian Richard Rhodes, the general-purpose steam engine not only pumped water ("the miner's great enemy"), but "came to blow smelting furnaces, turn cotton mills, grind grain, strike medals and coins, and free factories of the energetic and geographic constraints of animal or water power."[6] Following the earlier replacement of charcoal (partially burnt wood) with coke (baked or "purified" coal) to smelt iron, which "lay at the heart of the Industrial Revolution,"[7] a truly modern age emerged after Watt's invention, along with a vast increase in the demand for coal.

With subsequent improvements made by Cornish engineer Richard Trevithick, such as an internal fire box in a cylindrical boiler, smaller-volume, high-pressure steam known as "strong steam" – thus no need of an external condenser, the piston returning by its own force – exhaust blasts, and a double-acting, horizontal mount, the first *locomotion* engine was ready for testing. Dubbed the "puffing devil," Trevithick's steam carriage was the world's first self-propelled "automobile," initially tested on Christmas Eve in 1801 on the country roads of Cornwall. Alas, his curious contraption was too large, crude, and sputtered unevenly. In the autumn of 1803, he demonstrated an improved version on the streets of London that fared much better despite the bumpy cobblestone ride. Called "Trevithick's Dragon," his three-wheeled "horseless carriage" managed 6 miles per hour, although no investors were interested in further developing his unique, steam-powered, road vehicle.[8]

It didn't take long, however, for the steam engine to replace horse-drawn carriages in purpose-built "iron roads" and on canals and rivers more suited for safe and smooth travel, which would soon come to epitomize the might of the Industrial Revolution. The expiration in 1800 of Watt's original 1769 patent[9] helped stimulate innovation, the first successful demonstration of a steam "locomotive" on rails occurring on February 21, 1804, as part of a wager between two Welsh ironmasters. Trevithick's "Tram Waggon" logged almost

20 miles on a return trip from Penydarren to the canal wharf in Abercynon, north of Cardiff in south Wales. The locomotive was fully loaded with 10 tons of iron and 70 men in 5 wagons, reaching almost 5 mph, triumphantly besting horse power and showing that flanged iron wheels wouldn't slip on iron rails although the rails themselves were damaged.[10] With improved engineering, there would be enough power to propel the mightiest of trains and ships as the changes came fast and furious, beginning a century-long makeover from animal and human to machine power.

Across the Atlantic in the United States, Robert Fulton made the steamship a viable commercial venture, the engine power transmitted by iconic vertical paddle wheels, after modifying the important though fruitless work of three unheralded American inventors – Oliver Evans (whose Orukter Amphibolos dredger was the first steam-powered land and water vehicle, albeit hardly functional) and James Rumsey and John Fitch (who both tried to run motorized steam ships upstream with little success). Cutting his business teeth on the windy and tidal waters of the Hudson River, navigable as far as Albany, and having taken the *New Orleans* on its maiden voyage downriver from Pittsburgh to its namesake town in 82 days – interrupted by fits of odd winter weather throughout and a major earthquake – Fulton acquired the exclusive rights in 1811 to run a river boat *up* the Mississippi River.[11]

Having established steam and continental travel in the psyche of a young American nation, the steamboat greatly fueled westward expansion and the need for immigrant workers. Refashioning the low-pressure, reciprocating steam engine into a more powerful, high-pressure, motive steam engine, the *Enterprise* made the first ever return journey along the Mississippi in 1815, a model that would become standard for generations,[12] while two years later the *Washington* went as far as Louisville in a record 25 days, both riverboats designed by Henry M. Shreve after whom Shreveport is named.

The American west would never be the same, steam power opening up new lands for commerce and habitation. In 1817, only 17 steamboats operated on western rivers; by 1860 there were 735.[13] Prior to the success of the railroads from the 1840s on, steamboat travel accounted for 60% of all US steam power.[14] As noted by Thomas Crump in *The Age of Steam*, "large-scale settlement of the Mississippi river system – unknown before the nineteenth century – was part and parcel of steamboat history."[15] (For the most part, because of continental geography, boat traffic in the USA is north–south while railways go east–west.)

It would be in the smaller, coal-rich areas of northern England, however, where steam was initially exploited for travel on land, the first steam-powered locomotive rail route operating between the Teesside towns of Stockton and

Darlington. The massive contraption was designed by George Stephenson, a self-educated Tyneside mining engineer, who improved the engine efficiency via a multiple-fire-tube boiler that increased the amount of water heated to steam. Primarily employed to haul coal to a nearby navigable waterway, be it river, canal, or sea, Stephenson also used wrought iron rails that were stronger and lighter than brittle cast iron to support the heavier loads, implemented a standard track gauge at 4 foot 8½ inches,[16] and introduced the whistle blast to vent exhaust steam via the chimney to increase the draft.[17]

After numerous tests and an 1826 Act of Parliament to secure funds, the first official passenger train service opened on September 15, 1830, bringing excited travelers from Liverpool to Manchester, including the leading dignitaries of the day such as the prime minister, the Duke of Wellington, and the home secretary Robert Peel. The 30-mile line passed through a mile-long tunnel, across a viaduct, and over the 4-mile-long Chat Moss peat bog west of Manchester. Driven by Stephenson, the service was powered by his *Northumbrian* locomotive steam engine, and was meant to cover the journey at 14 miles per hour, although a fatal accident on the day disrupted the festivities.[18]

Steam-powered trains soon brought travelers from across Britain to new destinations, most of whom had never been further from their home than a day's return travel by foot or horse, such as day trips to the shore and other "railway tourism" spots. Transporting a curious Victorian public here and there in a now rapidly shrinking world, the confines of the past were loosed as locomotion improved. Expanding the boundaries of travel as never before, markets around the globe flourished with increased transportation speed and distance.

In 1838, the SS (Steam Ship) *Sirius* made the first transatlantic crossing faster than sail, from Cork to New York, ushering in the global village as well as standardizing schedules rather than relying on capricious winds (in an early example of "range anxiety" the first steamships were too small to carry enough coal for an ocean crossing). High-pressure compound engines, twin screw propellers,[19] tin alloy bearings, and steel cladding followed, much of the new technology tested for battle or in battle.[20] By 1900, a voyage from Liverpool to New York took less than 2 weeks instead of 12 as immigrants arrived in record numbers, peaking between 1890 and 1914 when 200,000 made the crossing per year.[21] A disproportionately smaller number of travelers crossed in the opposite direction, covered by shipping back more consumer goods, including for the first time coveted American perishables.

The race was on to exploit the many industrial applications of steam power as coal-powered factories and mills manufactured goods faster and cheaper – textiles, ironmongery, pottery, foodstuffs – spreading from the 1840s across

Europe and beyond. By the year of the great Crystal Palace Exhibition in 1851 in Hyde Park, London – the world's first international fair (a.k.a. The Great Exhibition of the Works of Industry of all Nations) that brought hundreds of thousands of tourists from all over the UK to the capital for the first time by rail – Victorian Britain ruled the world, producing two-thirds of its coal and half its iron.[22]

Railways in the United States also expanded rapidly, in part because of the 1849 California gold rush as tens of thousands of fortune seekers went west. Government grants offered $16,000 per mile to lay tracks on level ground and $48,000 per mile in mountain passes.[23] By 1860 there were 30,000 miles of track – three times that of the UK – greatly increasing farm imports from west to east.[24] More track opened the frontier to migrant populations as new territory was added to the federal fold, increasingly so after the Civil War. While the steamboat opened up the American west, the railroad cemented its credentials.

On May 10, 1869, the Union Pacific met the Central Pacific at Promontory Summit in Utah, just north of Great Salt Lake, triumphantly uniting west and east and cutting a 30-day journey by stagecoach from New York to San Francisco to just 7 days by train "for those willing to endure the constant jouncing, vagaries of weather, and dangers of travelling through hostile territory,"[25] beginning at least in principle the nascent creed of Manifest Destiny over the North American continent. As if to anoint the American industrial era, when the trains met, the eastbound Jupiter engine burned wood and the westbound Locomotive No. 119 coal.[26]

Signaling the importance of rail travel to the birth of a nation and modern identity, British Columbia joined a newly confederated Canada in 1871, amid the fear of more continental growth by its rapidly expanding neighbor to the south, in part because of a promised rail connection to the east. The "last spike" was hammered at Craigellachie, BC, on November 7, 1885, inaugurating at the time the world's longest railroad at almost 5,000 km.

Born of fire and steam, the world's first technological revolution advanced at a prodigious pace as the world outgrew its horse-, water-, and wind-powered limits. Engines improved in efficiency and power, from the 2.9-ton axel load of Stephenson's 1830 *Northumbrian* to the 46,000 hp of the White Star Line's *Titanic* 80 years on. What we call the Industrial Revolution was in fact an *energy* revolution, replacing the inconsistent labor of men and beasts of burden with the organized and predictable workings of steam-powered machines.

In his biography of Andrew Carnegie, one of the most successful indus-trialists to master the financial power of a rapidly changing mechanized world via his "vertically integrated" Pittsburgh steel mills, Harold Livesay noted that "steam boilers replaced windmills, waterwheels, and men as

power sources for manufacturing" and that the new machinery "could be
run almost continuously on coal, which in England was abundant and
accessible in all weathers."[27] As transportation networks grew, the need
for more coal increased, as did the need for more steel that required an
enormous amount of coking coal to manufacture. Soon, a modern, coal- and
steel-powered United States would rule the world.

If not for government assistance, however, the revolution would have
been severely hampered. Subsidies made possible the fortunes of early
industrialists such as Wedgwood and Thomas Telford in Britain and
Carnegie and Cornelius Vanderbilt in the United States, while railway
construction across the world would have been impossible without secure
government funding, much of it abused by those at the helm:

> Governments assumed much of the financial risk involved in building the American
> railway network. States made loans to railway companies, bought equipment for
> them and guaranteed their bonds. The federal government gave them free grants of
> land. However, promoters and construction companies, often controlled by
> American or British bankers, retained the profits from building and operating the
> railways. Thus the railways, too, played a part in the transformation of the industrial
> and corporation pattern that was sharply changing the class structure of the United
> States.[28]

To be sure, governments enlisted and encouraged development, although
many early railway builders ensured routes passed through territories
they themselves had purchased prior to public listing, which were then
resold at great profit. Some even moved the proposed routes after dumping the soon-to-be-worthless land. Money was the main by-product of the
Industrial Revolution in a nascent political world, systematically supporting a new owner class that demanded control of a vast, newly mechanized infrastructure and a pliant workforce. Harnessing steam power was
the product of many inquisitive minds seeking to make man's labor
easier, although never before had the masters of business owned so
much.

Of course, building infrastructure takes time and funds. The first half
of the rail networks that now service Britain, Europe, and the eastern US
took four decades to construct from 1830 to 1870, by which time the
most critical operational problems were solved.[29] But despite legal challenges, labor shortages, safety (fatal accidents a regular occurrence), and
the politics of the day, the Industrial Revolution prevailed because of
those who demanded change. By the sweat and ingenuity of many, fire
and water was converted into motive and industrial energy.

1.2 The Energy Content of Fuel: Wood, Peat, Coal, Biomass, etc.

As more improved lifting and propulsion devices were developed, using what Watt called "fire engines," mechanical power would become understood in terms of the work done by a horse in a given amount of time. The heat generated in the process, either as hot vapor or friction, however, was still as mysterious as fire, thought of as either a form of self-repellent "caloric" that flowed like blood between a hot and a cold object or as a weightless gas that filled the interstices of matter. Heat is *calor* in Latin, also known as *phlogiston*, Greek for "combustible." What heat *was*, however, was still unknown.

In the seventeenth century, Robert Boyle had investigated the properties of air and gases, formulating one of the three laws to relate the volume, pressure, and temperature of a vapor or gas, important concepts that explain the properties of heated and expanding chambers of steam.[30] But not until the middle of the nineteenth century was a workable understanding of heat made clear, essential to improve the efficiency of moving steam in a working engine. In 1798, the American-born autodidact and physicist Benjamin Thompson published the paper, "An Experimental Enquiry Concerning the Source of the Heat which is Excited by Friction" that disputed the idea of a caloric fluid, and instead described heat as a form of motion.

As a British loyalist during the American Revolution, Thompson moved to England and then to Bavaria, where he continued his experiments on heat, studying the connection between friction and the boring of cannons, a process he observed could boil water. As Count Rumford – a title bestowed on Thompson by the Elector of Bavaria for designing the grounds of the vast English Garden in Munich and reorganizing the German military – he went on to pioneer a new field of science called *thermodynamics*, that is, motion from heat, which could be used to power machines and eventually generate electricity.

Building on Thompson's work, James Prescott Joule, who came from a long line of English brewers and was well versed in the scientific method, would ultimately devise various experiments that showed how mechanical energy was converted to heat, by which heat would become understood as a *transfer of energy* from one object to another. Joule's work would lead to the formulation of the idea of conservation of energy (a.k.a. the first law of thermodynamics), a cornerstone of modern science and engineering. Essentially, Joule let a falling weight turn a paddle enclosed in a liquid and noted the temperature increase of the liquid. Knowing the mechanical work done by the falling weight, Joule was able to measure the equivalence of heat and energy, which he initially called *vis-viva*, from which one could calculate the efficiency of a heat engine. Notably for the time, Joule measured temperature to within 1/200th of a degree Fahrenheit, essential to the precision of his results.[31]

Joule would also establish the relationship between the flow of current in a wire and the heat dissipated, now known as Joule's Law, still important today regarding efficiency and power loss during electrical transmission. To understand Joule heating, feel the warm connector cable as a phone charges, the result of electrons bumping into copper atoms.[32] In his honor, the joule (J) was made the System International (SI) unit of work and its energy equivalent.

As we have seen, the watt is the SI unit of power, equivalent to doing 1 joule of work in 1 second (1 W = 1 J/s). The watt and joule both come in metric multiples, for example, the kilowatt (kW), megawatt (MW), gigawatt (GW), and terawatt (TW). The first commercial electricity-generating power plant in 1882 produced about 600,000 watts or 600 kW (via Edison's six coal-fired Pearl Street dynamos in Lower Manhattan as we shall see), the largest power plant in the UK produces 4,000 MW (4,000 million watts or 4 GW in the Drax power station in north Yorkshire, generating almost 7% of the UK electrical supply), the installed global photovoltaic power is now over 100 billion watts per year (100 GW/year), while the amount of energy from the Sun reaching the Earth per second is about 10,000 TW (over 1,000 times current electric grid capacity). As for the joule, the total primary energy produced annually across the globe is about 600×10^{18} joules, that is, 600 billion billion, 600 quintillion, or 600 *exa*joules (EJ).

To measure the power generated or consumed in a period of time, that is, the *energy* expressed in joules, you will often see another unit, the kilowatt-hour (kWh), which is more common than joules when discussing electrical power. The British thermal unit (or BTU) is also used, as are other modern measures – for example, tonnes of oil equivalent (TOE) and even gasoline gallon equivalent (GGE) – but we don't want to get bogged down in conversions between different units, so let's stick to joules (J) for energy and watts (W) for power wherever possible. The important thing to remember is that power (in watts) is the *rate* of using energy (in joules) or the *rate* of doing work (in joules) – a horse lifting three average men *twice* as fast exerts *twice* the power.

We also have to consider the quantity of fuel burned that creates the heat to boil water to make steam and generate electric power, which also comes in legacy units such as cords, tons, or barrels. To compare the energy released from different materials, Table 1.1 lists the standard theoretical energy content in megajoules per kilogram (MJ/kg). A kilogram of wood produces less energy when burned than a kilogram of coal, which has less energy than a kilogram of oil. Indeed, coal has

Table 1.1 *Energy density (or specific energy) of various fuels*

Fuel	Energy content (MJ/kg)
Wood (green/air dry)	10.9/15.5
Peat	14.6
Coal (lignite, a.k.a. "brown")	14.7–19.3
Coal (sub-bituminous)	19.3–24.4
Coal (bituminous, a.k.a. "black")	24.4–32.6
Coal (anthracite)	>32.6
Biomass (ethanol)	29.7
Biomass (biodiesel)	45.3
Crude oil	41.9
Gasoline	45.8
Methane	55.5
Hydrogen gas (H_2)	142

Source: Elert, G., "Chemical potential energy," The Physics Hyper Textbook. http://physics.info/energy-chemical/.

about three times the energy content of wood and was thus better suited to exploit the motive force of steam-powered machines, especially black coal with over twice the energy of brown coal (32.6/14.7). We see also that dry wood has 40% more energy than green wood (15.5/10.9), which has more moisture and thus generates more smoke when burned, while gasoline (45.8) gives more bang for your buck than coal, peat, or wood, and is especially valued as a liquid.

Some fuels also have many varieties and thus the efficiency depends on the type. Hardwoods (e.g., maple and oak) contain more stored energy than softwoods (e.g., pine and spruce) and will burn longer with greater heat when combusted in a fireplace (hardwoods are better during colder outdoor temperatures). The efficiency also depends on the density, moisture content, and age. Furthermore, how we burn the wood – in a simple living-room fireplace, wood-burning stove, cast-iron insert, or as pressed pellets in a modern biomass power station – affects the efficiency, but the energy content is still the same no matter how the fuel is burnt. For example, a traditional, open fireplace is eight times less efficient than a modern, clean-burning stove or fan-assisted, cast-iron insert, because much of the energy is lost up the chimney due to poor air circulation (oxygen is essential for combustion). Similar constraints follow for all fuels, whether wood, peat, coal, biomass, oil, gasoline, methane, or hydrogen.

Two of the fuels listed in Table 1.1 have been around forever – wood and peat – considered now by some as a renewable-energy source along with a very modern biomass. Remade as a presumed clean-energy miracle and labeled "renewable" in a modern industrial makeover, we will see if they are as green as claimed. As usual, the devil is in the detail.

***

Wood was our first major energy source, requiring little technology to provide usable heat for cooking or warmth in a dwelling. As a sustainable combustion fuel, however, wood requires large tracts of land and lacks sufficient energy content and supply, unless managed in a controlled way, for example, as practiced for millennia in Scandinavia. With the advent of coal in land-poor Britain, wood as an energy source fell from one-third in the time of Elizabeth I in the sixteenth century to 0.1% by the time of Victoria three centuries later.[33] By 1820, the amount of woodland needed to equal coal consumption in the UK was 10 times the national forest cover, highlighting the limitations of wood as a viable fuel source.[34] The transition also helped save wood for other essential industries rather than inefficient burning.

The modern practice of sustainability in forest management can be traced to the early eighteenth century in Germany, where replanting homogeneous species – mostly for timber – followed clear cutting, which also led to a ban on collecting deadwood and acorns in previously common lands. The German economist and philosopher Karl Marx would begin to form his paradigm-changing ideas about class in Bavaria in the 1840s from "the privatization of forests and exclusion of communal uses,"[35] setting off the battle over public and private control of natural resources and the ongoing debate about unregulated capitalism.

Today, a cooperative and sustainable "tree culture" exists in many northern countries, especially in Scandinavia, where roughly 350 kg of wood per person per year is burned.[36] Wood has also become part of a controversial new energy business, converted to pellets and burned in refitted power plants as a renewable fuel replacement for coal (which we'll look at later).

Formed mostly in the last 10,000 years, peat is partially decomposed plant material found on the surface layer of mires and bogs (cool, acidic, oxygen-depleted wetlands), where temperatures are high enough for plant growth but not for microbial breakdown. Peat is also found in areas as diverse as the temperate sub-arctic and in rainforests. According to the World Energy Council, 3% of the world's land surface is peat, 85% in Russia, Canada, the USA, and Indonesia, although less than 0.1% is used as energy.[37] The three largest consumers of peat in Europe are Finland (59%), Ireland (29%), and Sweden (11%), comprising 99% of total consumption.

Peat (or *turf* as it's known in Ireland) has been burnt for centuries in Irish fireplaces, mostly threshed from plentiful midland bogs. Burned in three national power stations, peat contributes 8.5% of Irish electrical needs, and is also processed into milled briquettes by the state-run turf harvester Bord na Móna for household fireplaces, readily purchased in local corner shops and petrol stations across the country. For years, Ireland's three peat-fired power plants provided roughly 300 MW of electrical power to a national grid of about 3.2 GW, although in 2018 Bord na Móna announced it would close all 62 "active bogs" and discontinue burning peat by 2025, citing increased global warming and the need to decarbonize energy (small-scale threshing is still allowed for local domestic use via so-called "turbary rights").[38] Elsewhere, peat is still burned for warmth and electric power. Globally, 125 peat-fired power stations (some co-fired with biomass) provide electricity to almost two million people.

To combat increased greenhouse gases, however, peatlands have become strategically important as carbon sinks. The World Energy Council estimates that additional atmospheric carbon dioxide from the deforestation, drainage, degradation, and conversion to palm oil and paper pulp tree plantations – particularly in the farmed peatlands of Southeast Asia – is equivalent to 30% of global carbon dioxide emissions from fossil fuels. Peat contains roughly one-third of all soil carbon – twice that stored in forests (~2,000 billion tons of CO_2) – while about 20% of the world's peatlands have already been cleared, drained, or used for fuel, releasing centuries of stored carbon into the atmosphere.[39]

Holding one-third of the earth's tropical peatlands, Indonesia has been particularly degraded from decades of land-use change for palm-oil plantations, while icy northern bogs are becoming increasingly vulnerable as more permafrost melts from increased global warming (the peat resides on top and within the permafrost that holds about 80% of all stored peatland carbon). As Russia expands petroleum exploration in the north, more Arctic tundra is also being destroyed, a double whammy for the environment via increased GHG emissions and depleted carbon stores. Although trees and plants can be grown and turned into biomass for more efficient burning (as seen in Table 1.1), bogs are not as resilient and care is needed to avoid depleting a diminishing resource. Today, the rate of extraction is greater than the rate of growth, prompting peat to be reclassified as a "non-renewable fossil fuel." In countries with a historic connection, conservation is more important than as any viable energy source.[40]

The transformation from old to new continues as we grapple with increasing energy needs. Whether we use steam to move a piston in a Watt "fire engine" or a giant turbine in a modern power station that burns peat, coal, biomass, oil, or natural gas, the efficiency still depends on the fuel used to boil the water to make the steam. The demand for fuel with better energy content changed the global economy, firstly after coal replaced wood (at least twice as efficient), which we look at now, and again as oil replaced coal (roughly twice as efficient again), which we'll look at in Chapter 2. A transition simplifies or improves the efficiency of old ways (see Table 1.1), turning intellect into industry with increased capital – when both transpire, change becomes unstoppable.

1.3 The Original Black Gold: Coal

Found in wide seams up to 3 km below the ground, coal is a "fossil fuel" created about 300 million years ago during the Carboniferous Period as heat and pressure compressed the peat that had formed from pools of decaying plant matter, storing carbon throughout the Earth's crust. Aristotle called it "the rock that burns." As far back as the thirteenth century, sea coals that had washed up on the shores and beaches of northeast England and Scotland were collected for cooking and heating in small home fires as well as for export by ship to London and abroad.

By the seventeenth century, coal had been found throughout England, burned to provide heat at home and in small manufacturing applications. Supplies were plentiful, initially dug out by "miners" lowered into bell pits by rope, before deeper mines were excavated by carving horizontally through the ground. Coal seams as wide as 10 feet were worked with hammers, picks, and shovels, the lengthening tunnels supported by wooden props. The broken coal bits hewn from the coal face were removed by basket or pushed and pulled in wagons along a flat "gallery" or slightly inclined "drift" before being mechanically hoisted from above through vertical shafts. Animals and railcars eventually replaced men to remove the coal from within the sunless pits. To help evacuate the deepening, water-laden mines, mechanical pumps were employed, inefficiently fueled with coal from the very same mines they serviced.

As we have seen, manufacturing industries evolved to incorporate newly invented coal-powered steam engines: spindles turned imported American cotton into cloth, large mills greatly increased grinding capacities of wheat between two millstones to make bread flour as well as other processed foodstuffs, and transportation was transformed across Britain and beyond. The Boulton & Watt Soho Foundry made cast iron and wrought iron parts to manufacture bigger and better engines. Glass, concrete, paper, and steel plants all prospered thanks to highly mechanized, coal-fueled, high-temperature manufacturing and steam-driven technology. By the beginning of the nineteenth century, "town gas" formed from the gasification of coal also lit the lamps of many cities, in particular London.

A few critics wondered about the social costs and safety of mining and burning coal. In his poem "Jerusalem," the English Romantic artist and poet William Blake asked "And was Jerusalem builded here / Among these dark Satanic Mills?" The Victorian writer Charles Dickens questioned the inhumane treatment in large mechanized factories, where more than 80% of the work was done by children, many of whom suffered from tuberculosis, while child labor was still rampant in the mines. Adults did most of the extraction, but "boys as young as six to eight years were employed for lighter tasks" and "some of the heaviest work was done by women and teenage girls" carrying coal up steep ladders in "baskets on their back fastened with straps to the forehead."[41] Pit ponies eventually replaced women and children as beasts of burden, some permanently stabled underground and equally mistreated.

The 1819 Cotton Mills and Factories Act in Britain limited to 11 hours per day the work of children aged 5–11 years, but in reality did little to aid their horrible plight. In *Hard Times*, an account of life in an 1840s Lancashire mill town, Dickens wrote about the abuse at the hands of seemingly well-intended Christian minders, "So many hundred Hands in this Mill; so many hundred horse Steam Power. It is known, to the force of a single pound weight, what the engine will do; but, not all the calculators of the National Debt can tell me the capacity for good or evil."[42] The 1842 Mines and Collieries Act finally outlawed women and children under 10 from working in the mines, often sidestepped by uncaring owners.

By the mid-nineteenth century, 500 gigantic chimneys spewed out toxins over the streets of Manchester, the center of early British coal-powered manufacturing. Known as "the chimney of the world," Manchester had the highest mortality rate for respiratory diseases and rickets in the UK, countless dead trees from acid rain, and an ashen daylight reduced almost by half.[43] As seen in J. M. W. Turner's ephemeral paintings, depicting the soot, smoke, and grime of nineteenth-century living, black was the dominant color of a growing industrialization, expelled from an increasing number of steam-powered trains and towering factory smokestacks. In a letter home to his brother Theo in 1879, Vincent van Gogh noted the grim reality of the Borinage coal regions in southern Belgium, where he came to work as a lay priest prior to becoming a painter: "It's a sombre place, and at first sight everything around it has something dismal and deathly about it. The workers there are usually people, emaciated and pale owing to fever, who look exhausted and haggard, weather-beaten and prematurely old, the women generally sallow and withered."[44]

Health problems, low life expectancy, and regular accidents were the reality of pit life as miners worked in oxygen-depleted and dangerous dust-filled mines, subject to pneumoconiosis (black lung or miner's asthma), nystagmus (dancing eyes), photophobia, black spit, roof cave-ins, runaway cars, cage crashes, and coal-dust and methane gas explosions (fire damp). Prior to battery-operated filament lamps, the Davy lamp fastened an iron gauze mesh around an open flame to lower the point temperature and reduce mine gas ignitions, although economic output mattered more than safety as the world reveled in a vast newfound wealth. A 1914 report calculated that "a miner was severely injured every two hours, and one killed every six hours."[45]

Mining is not for the faint-hearted nor weak-bodied. Even seasoned miners cricked their backs, crawling doubled-over through four-foot-high, dimly lit gallery tunnels to and from the coal face, traveling a mile or more both ways to work each day (for which they were not paid). In *The Road to Wigan Pier*, the English novelist and journalist George Orwell likened the mines with their roaring machines and blackened air to hell because of the "heat, noise, confusion, darkness, foul air, and above all, unbearably cramped space."[46] Published in 1937, Orwell described the miners as "blackened to the eyes, with their throats full of coal dust, driving their shovels forward with arms and belly muscles of steel" as they labored in "hideous" slag-heaps and lived amid "frightful" landscapes.[47] Writing at the outset of his depressingly frank report on the invisible drudgery of working-class miners in Yorkshire and Lancashire, Orwell nonetheless considered the coal miner "second in importance only to the man who ploughs the soil," while the whole of modern civilization was "founded on coal."[48]

Almost a century after Dickens had written about the horrid state of working-class life in the factories of industrial England, the smell and occupational status of a person was still the unwritten measure of a man. Bleak industrial towns remained full of "warped lives" and "ailing children," while the ugliness of industrialism was evident everywhere amidst mass unemployment at its worst. Coal may have been the essential ingredient of the Industrial Revolution, fortuitously located where many displaced laborers lived, yet did little to improve the lives of those who had no choice but to take whatever work they could get.

At the height of the British coal industry in the 1920s, 1.2 million workers were employed,[49] each producing on average about one ton of coal per day, one-third of which was exported.[50] With the conversion of a labor-intensive, coal-fired steam to more easily maintained diesel-electric trains; however, 400,000 jobs were lost.[51] By the 1950s, only 165,000 men worked 300 pits in the Northumberland and Durham coalfields, dubbed the Great Northern Coalfield.

North Sea gas in the 1980s also replaced coal gas in home heating and cooking, reducing the demand for coal even further. Mostly comprised of methane, "natural" gas is so-called because the gas isn't manufactured like synthetic coal gas, but comes straight from the ground. Unable to compete with oil and gas finds from the North Sea, coal was finally phased out in the UK from 1984, after a prolonged battle between politicians and workers despite the National Coal Board claiming 400 years' worth still left to mine. By 1991 only a few deep mines remained in operation with just 1,000 employees.[52] Having once shipped 10,000 tons a day from the northeast to London in the mid-nineteenth century until being upended by the railways, by 2004 the docks of Newcastle had become a net *importer* of coal.[53] The last deep UK mine – the Kellingley Colliery in North Yorkshire – shut in 2015.

Just as coal and steam had taken away jobs from low-output, rural handicraft workers two centuries earlier, oil and diesel made coal workers redundant, their labor-intensive output no longer required. The transition would engulf Britain, pitting the power of technology and capital against long-standing working communities, emboldening the prime minister of the day Margaret Thatcher and the trade-union leader Arthur Scargill, who battled over ongoing pit closures. Coal mining was dying in the very place it had begun as Ewan MacColl sang in "Come all ye Gallant Colliers" (sung to the tune of Morrissey and the Russian Sailor):

> Come all ye gallant colliers and listen to me tale
> How they closed the Aberaman Pit in Aberdale, South Wales
> It was in 1842 that coal there was first won
> She's yielded 40 million tons but now her days are done

By contrast, coal remained a thriving industry in the United States – even into the twenty-first century, the lust for abundant, cheap energy unabated. Prior to 1812, the USA imported most of its coal from Britain, but after the War of 1812 deposits were mined in the plentiful seams of the Appalachian Mountains, comprising parts of six eastern states (Pennsylvania, Maryland, Ohio, Kentucky, Virginia, and West Virginia). After the Civil War, coal production grew rapidly – almost 10-fold in 30 years – especially needed for steam engines on trains as coal replaced wood toward the end of the nineteenth century. The United States grew rapidly, building more energy-intensive infrastructure to secure its vast, ever-expanding borders.

As in Britain, coal brought many health and safety problems, as well as conflicts between workers and owners. From 1877 to 1879, labor unrest in

Pennsylvania led to the hanging of 20 supposed saboteurs known as the Molly
Maguires, who sought better conditions and basic workers' rights. In 1920,
after attempting to organize as part of the United Mine Workers of America
(UMWA) union in the company coal town of Matewan, West Virginia, miners
were repeatedly manhandled by private guards leading to a shoot-out on the
main street and 10 deaths. With an abundant and inexpensive resource, dug out
of remote mountainous regions, lawlessness prevailed; even the wives of
injured miners were forced into prostitution to cover their husbands' down
time.[54] Disputes continued well into the twentieth century as profits were
valued above health, safety, and security.

In his Pulitzer Prize-winning novel *King Coal*, Upton Sinclair described the
wretched conditions faced by miners and their families during the 1910s in his
telling account of working for the fictional General Fuel Company. Drudgery,
grime, headaches, accidents, company debt in a company town, and uncertain
futures were the harsh reality of the early coal industry in a "don't ask" world of
corporate paternalism. Hal Warner is the protagonist, who learns first-hand about
the grim existence of a coal worker's life, many recent immigrants with families
in search of a better life, yet destined instead to a life ever on the periphery:

> There was a part of the camp called "shanty-town," where, amid miniature
> mountains of slag, some of the lowest of the newly arrived foreigners had been
> permitted to build themselves shacks out of old boards, tin, and sheets of tar-paper.
> These homes were beneath the dignity of chicken-houses, yet in some of them
> a dozen people were crowded, men and women sleeping on old rags and blankets on
> a cinder floor. Here the babies swarmed like maggots.[55]

Country singer Loretta Lynn sang about her experiences growing up in
poverty in a Kentucky mining town as a coal miner's daughter "in a cabin on
a hill in Butcher Holler," where her daddy shoveled coal "to make a poor man's
dollar." The drudgery was endless and "everything would start all over come
break of morn."

Before basic safety measures were introduced in the 1930s, 2,000 miners
died on average each year in the United States. Methane gas monitors, ventilation systems, double shafts, roof bolts, safety glasses, steel-toed boots, battery-
operated lamps, and emergency breathing packs helped improve conditions,
while continuous mining replaced hazardous blasting, employing remote-
control cutting heads to break up the coalface and hydraulic roof supports.
Various innovations transformed the coal industry, many invented by American
engineer Joseph Francis Joy, such as mechanized loading devices, shuttle cars,
and long-wall machines that did the work of a thousand men, turning hewers
and haulers into machine operators.[56]

Naturally found in coalmines, outgassing is particularly dangerous, responsible for one of the strangest detection methods ever: a caged canary that dies in the presence of methane or carbon monoxide, signaling impending peril to nearby miners. Today, coal mining is highly mechanized and much safer – modern mining operations include stabilizing old tunnels with coal paste and 5G monitoring to increase safety – although accidents are still very much a part of a miner's life despite the more modern measures. Sadly, thousands of workers die each year in the mining business, while every country has its own horrid story of lost miners and devastated families. Not all accidents can be explained as bad luck.

In 2018, a methane gas explosion in a black-coal mine in the Czech Republic killed 13 miners working about 1 km down, where the level of methane was almost five times the allowed level, while in 2021, 52 Russian miners died after a methane explosion ripped through a Siberian coal mine that had previously been certified safe without inspection. Killed in an explosion at the Pike River mine in New Zealand, the remains of 29 workers went undiscovered for over a decade until improved bore-hole drilling technology revealed the location of some of the men in 2021.[57] In 2022, 41 miners died after a methane explosion in a coastal Black Sea mine, the worst mining disaster in Turkey since 2014 when 301 miners lost their lives. Risk warnings had been given prior to the accident about outgassing below 300 m.

No amount of protection can make mining entirely safe, but open-cast or surface mining avoids many of the problems of underground mining, and is now more prevalent. Accounting for about one-fifth of electrical power generation in the United States, two-thirds of American coal is surface-mined in open pits, where the "overburden" is first cleared to access the giant seams, some as wide as 80 feet. Giant draglines and tipper shovels load the coal into 400-ton trucks for removal. Nonetheless, a 2023 accident at an open-pit mine in northern China killed five miners instantly when a mine wall collapsed, while 47 more were buried under tons of rubble and presumed dead.

Mechanization, lower operating costs, and better management in the USA has helped to counter much of the conflict that engulfed the UK, where mine owners and miners battled for decades over low wages, working conditions, and long hours amid mounting inefficiency. Mediated by beholden governments, industry control in the UK regularly changed hands, epitomized by the 1926 General Strike, the 1938 Coal Act, and eventual nationalization of Britain's premier industry eight years later. Prior to World War I, Britain had astoundingly supplied two-thirds of the world's coal, but unregulated profits, lowered living standards, and blind loyalty to a national cause was unsustainable.

Table 1.2 *Global coal-fired power plants by country: number, power, CO_2 emissions, 2022 (*>30 MW)*

#	Country	Number*	Power (GW)	%	CO_2 (Gtons/year)	%
1	China	1,118	1,074	52.0	4.7	49.7
2	India	285	233	11.3	1.0	11.1
3	United States	225	218	10.5	1.1	11.3
4	Japan	92	51	2.4	0.2	2.5
5	South Africa	19	44	2.1	0.2	2.2
6	Indonesia	87	41	2.0	0.2	2.0
7	Russia	71	40	1.9	0.2	2.3
8	South Korea	23	38	1.8	0.2	1.7
9	Germany	63	38	1.8	0.2	1.9
10	Poland	44	30	1.5	0.2	1.7
	Rest of world	261	261	12.6	1.3	3.6
	Total	2,439	2,067	100.0	9.4	100.0

Source: "Summary Tables: By Country," *Global Coal Plant Tracker, Global Energy Monitor*, July 2022. https://globalenergymonitor.org/projects/global-coal-plant-tracker/summary-tables/.

Today, China generates more than half and India and the United States about 10% each, burning roughly three-quarters of an 8.5 billion ton annual production (Table 1.2). Coal is available in vast quantities with deposits in over 75 countries with enough reserves to last 100 years at current rates, generating electricity from over 6,000 units worldwide. Despite what may appear to be the dying days of coal, the original Industrial Revolution fuel still accounts for almost 40% of globally produced electric power. What's more, even if the percentage of coal power were to decrease from 40% to 25% in the next two decades as has been called for to limit global warming to under 1.5°C, more coal will still be burnt if production increases as predicted by 2% per year.[58] Coal is also burned to produce the high temperatures to smelt steel, make cement, and produce coal-to-liquid fuel. Any proposed "phase down" let alone "phase out" won't happen easily or without a fight.

***

There are five coal types, generally ranked by age, based on the percentage of carbon and thus energy content or "calorific value." From lowest to highest rank: peat, lignite (a.k.a. brown), sub-bituminous, bituminous (a.k.a. black), and anthracite. Younger coal has less carbon and more moisture, and is less efficient when burned (low calorific value). Peat is included as the youngest, least-efficient coal; lignite is high in moisture (up to 40%) and lower in energy (20 MJ/kg); whereas anthracite has less moisture (as low as 3%) and more energy (24 MJ/kg). The older "mature" coal is darker, harder, shinier, and produces more carbon dioxide when burned. (The ratio of carbon to hydrogen and oxygen atoms increases in higher-ranked coals.)

In the United States, the older, dirtier coal is typically found in the east in the anthracite-filled Appalachian regions along with bituminous coal. The relatively cleaner sub-bituminous coal is mostly found in the west, while the cleanest, lignite, is in the south. Wyoming produces over 200 million tons each year, more than 40% of American supplies, primarily in the North Antelope Rochelle and Black Thunder surface mines of the Powder River Basin, at about one-fifth the cost of east-coast anthracite (Figure 1.1). Mile-long trains transport the cheaper, easy-to-access coal to power plants as far away as New York. Underscoring its importance to the American economy, almost half of all US train freight by tonnage is coal.[59]

According to the US Energy Information Administration, the Appalachian Region – the historical heartland of American coal mining – still accounts for almost one-third of US production, mostly in West Virginia (14%), Pennsylvania (7%), Illinois (6%), and Kentucky (5%).[60] Much of the coal is anthracite – older, dirtier, and hard-to-access. As in other coal regions around the world, long-standing communities and livelihoods depend on mining. Nevertheless, 40,000 jobs have been shed in the past decade as the American coal industry declines because of

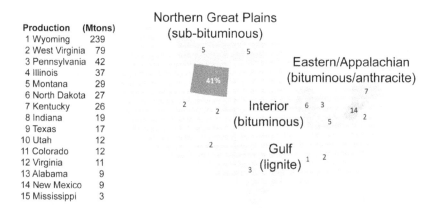

Figure 1.1 Coal production in the United States (million tons and percent, 2021). *Source:* US Energy Information Administration, "Coal Data: By state and mine type," October 18, 2022. eia.gov/coal/data.php#production.

higher costs, worsening pollution, and cheaper, cleaner alternatives despite attempts by some to reinvent a past glory or claim a supposed "war on coal."

Since its peak in 2007, US coal production has dropped almost 50%, while more coal plants were retired in the first two years of the Trump administration than in the first four years of his predecessor.[61] Two long-standing coal plants on the Monongahela River south of Pittsburgh were shuttered on the same day in 2013 – the 370-MW Mitchell Power Station and 1.7-GW Hatfield's Ferry Power Station – undercut by cheaper natural gas and renewables. Just as diesel and natural gas spelled the end of coal in the UK, igniting an intense worker–owner battle, cheap natural gas from fracking has undercut American coal, widening an already fierce political divide.

Of course, consumption in the USA is now dwarfed by China, which has remade its economy on the back of coal-fired power, helping to raise hundreds of millions from poverty. Having accelerated the building of power plants since liberalizing state controls in the late 1970s, China now burns more coal than the rest of the world combined (50.6%), providing more than two-thirds of its domestic energy needs, albeit with poorer emissions standards. Life expectancy in China has been reduced by 5 years because of coal pollution with over one million premature deaths per year compared to 650,000 in India and 25,000 in the USA.[62]

Unfortunately, coal has dramatically increased air pollution as seen in regular public-health alerts. Pollution levels in some Chinese cities are so unbearable that inhabitants routinely wear masks, don't eat outdoors, and run air purifiers in their homes. Chinese journalist Chai Jing notes that smokestacks were once considered a sign of progress in a developing China, yet she now worries about the link between air pollution and cancer: "In the past 30 years, the death rate from lung cancer has increased by 465%."[63] Xiao Lijia, a Chinese electrical engineer who began his career at the second-largest coal plant in Heilongjiang Province northeast of Beijing, adds "It is still a great challenge to improve the surrounding environment."[64]

While developing countries continue adding coal power (although at a lesser rate than in the past two decades), many developed countries are cutting back or even phasing out coal in the face of rising global temperatures and increased pollution. After failing to meet its 2015 Paris Agreement emissions targets, Germany – the world's fourth-largest economy and sixth-largest consumer of coal – announced in 2019 that it would end coal-fired electrical power generation by 2038. Highly subsidized, deep "black-coal" mining in the Ruhr had already closed in 2018, but under Germany's new plan 34 GW of coal power will be halved by 2030 before being completely shut down.

The affected regions will be offered €40 billion in aid, in particular the open-pit "brown-coal" mining communities of Lusatia near the Polish border. A member of both the overseeing commission and executive board of Germany's second-largest trade union offered an ironic though pragmatic analysis: "It's a compromise that hurts everyone. That's always a good sign."[65] The plan includes converting Germany's largest open-pit mine at Hambach, North Rhine-Westphalia, into a lake, the deepest artificially created space in the country, as well as "rehabilitating" other disused mining areas.

In a bid to tackle air pollution and desertification, China has been planting trees at a rate of more than a billion a year since 1978 as part of the Great Green Wall forestation project. In 2019, 60,000 soldiers were reassigned to plant even more, primarily in coal-heavy Hebei Province, the main source of pollution to neighboring Beijing, where pollution levels are 10 times safe limits. The plan is to increase woodland area by more than 300,000 km^2 (from 21.7% to 23%). The Chinese government has also pledged to reduce its carbon intensity and "peak" emissions by 2030 (and become overall carbon neutral or "net zero" by 2060), but to achieve any real reductions China must significantly limit coal consumption.

As more jobs in the renewable-energy sector supplant mining jobs, coal becomes less viable by the day. The International Renewable Energy Agency (IRENA) counted 11.5 million global renewable-energy jobs in 2019, including 3.8 million in solar, 2.5 million in liquid biofuels, 2 million in hydropower, and 1.2 million in wind power.[66] California now has more clean-energy jobs than all coal-related jobs in the rest of the United States,[67] while Wall Street continues to dump coal investments. In communities that have worked the mines for generations, King Coal is dying after almost two centuries of rule. And yet the damage continues, wherever brown jobs and old thinking is championed over cleaner alternatives, while in the developing world coal still thrives. In 2021, the amount of coal-powered electrical power generation actually increased despite an acknowledged recognition of global warming at the annual COP meeting that year in Glasgow.

Discarded in the land of its birth because of poor economics and increased health problems, and in decline in most developed countries around the world, without viable alternatives countries will still generate power with "the rock that burns" no matter the obvious problems to health and increased global warming. Who would have thought we would be burning the black stuff three centuries after the start of the Industrial Revolution? Who would have thought we would ignore all the consequences?

1.4 Coal Combustion, Health, and the Forgotten Environment

Steam-powered technology has brought numerous changes to the way we live, especially a thirst for *more*, the rally cry of the Industrial Revolution. The most onerous by-product, however, is pollution. In *The Ascent of Man*, author and philosopher Jacob Bronowski describes the reality quite simply: "We think of pollution as a modern blight, but it's not. It's another expression of the squalid indifference to health and decency that for centuries had made the plague a yearly visitation."[68]

We no longer have an annual plague, but we do have increased occurrences of bronchitis, asthma, lung cancer, and other respiratory illnesses, all attributable in part to burning wood and coal. To be sure, we have reaped numerous benefits from the Industrial Revolution, but we also suffer unwanted side effects from burning carbon-containing fuels: pollution, smog, acid rain, increased cancer rates, and now a rising global temperature caused by a secondary absorption effect of the main combustion by-product, carbon dioxide (1 kg of coal produces about 2 kg of CO_2).

We didn't use to care about the by-products, some of which contain valuable nutrients that can be ploughed back into the ground as fertilizer (for example, nitrogen-rich wood ash), but after almost two centuries of ramped-up industrial use, our energy waste products have become problematic. The main culprits are carbon dioxide (CO_2) and methane (CH_4), atmospheric gases that absorb infrared heat reradiated from a sun-heated earth, which enhances the natural greenhouse effect, turning the earth into an increasingly warming hothouse. Many other combustion by-products are also extremely harmful to the environment and our health.

Combustion is the process of burning fuel in the presence of oxygen, which produces energy and waste; for example, simple charcoal and ash from wood, coke and ash from coal, or a witches' brew of by-products from various long-winded, hard-to-pronounce hydrocarbons.[69] Any material can be burned given sufficient heat and oxygen, and thus the question with all fuels is: What can we burn that gives us enough energy to heat our homes, run our power plants, and fuel our cars, but doesn't produce so much of the bad stuff?

Not all burning is the same: as more coal replaced wood in urban fireplaces, the waste became a particularly toxic combination of coal-creosote and sulfuric acid, causing numerous health issues. In London, smoke and river fog would combine in a dense "smog," forever labeling the first great industrial capital with its infamous moniker "The Smoke" and permanently dividing the city into a wealthier, upwind west end and a poorer, downwind east end, based solely on proximity to the "aerial sewage."

Fueled by numerous power plants, industrial sites, and household hearths that burned more coal along the Thames and elsewhere, the smog became unbearable after an especially bad cold spell during the winter of 1952. Visibility was almost impossible throughout London as football games and greyhound races were cancelled (the dogs couldn't see what they were chasing), people drowned in canals they hadn't seen, and buildings and monuments rotted in a sulfurous bath, while one airplane pilot got lost taxiing to the terminal after landing, requiring a search party that promptly got lost too.[70] "Fog cough" was rampant as more people perished in Greater London in a single month than "died on the entire country's roads in the whole year."[71] After the so-called "Great Smog" of 1952 that killed almost 4,000 people in a few days[72] – a level of death unseen since the Blitz – the horror was eventually addressed, prompting a switch from solid fuels such as coal to liquid fuel: firstly coal-derived town gas, followed by imported propane, and finally "natural" gas (methane) after the 1965 North Sea finds.[73]

To counter the smog, the Clean Air Act of 1956 mandated "smoke-free" zones, where only smokeless fuel could be burned, although others were slow to heed the warnings. Not until 1989 did Dublin introduce similar restrictions, following an especially bad winter in 1981–82,[74] adding an extra 33 deaths per 100,000 people from the increase in smog caused by burning coal in open grates.[75] Following the ban, air pollution mortalities dropped by half.[76] The move from coal-powered to diesel locomotives in many industrialized countries also reduced the rampant railway exhaust that had gone unchecked for over a century.

In poorer countries, however, it is harder to convince people to burn expensive, imported natural gas rather than cheap and readily available coal or give up their coal jobs without alternatives. In Ulan Bator, Mongolia, the world's coldest capital, particulate matter levels have been measured at more than 100 times WHO limits because of inefficient coal burning, leading to higher rates of pneumonia, a 40% reduction in lung function for children, and a 270% spike in respiratory diseases.[77]

Even in Europe, which has vowed to end coal, countries such as Poland still heat their homes primarily with coal (87%). Talk of "phasing out" coal or switching to a "bridge" fuel such as natural gas seems cynical at best in the face of a shivering population. Even in the United States, one in 25 homes is still heated by wood, much of it in substandard stoves where particulate matter is four times that of coal, contributing to 40,000 early deaths per year.[78] Warmth always comes at a hefty price when inefficient wood- or coal-burning is the source.

In December 2015, just as national delegates were convening in Paris for the COP21 climate change meetings, China raised its first-ever four-color warning system to a maximum "red alert." The government recommended that "everyone should avoid all outdoor activity," while the high-range Air Quality Index (AQI) almost doubled, alarming officials and posing a "serious risk of respiratory effects in the general population."[79] As Edward Wong noted, "Sales of masks and air purifiers soared, and parents kept their children indoors during mandatory school closings."[80]

As noted in a 2019 WHO study, 8.7 million people across the globe die prematurely each year from bad air – even more than from smoking – while air pollution in Europe has lowered life expectancy by more than 2 years,[81] where all forms of pollution have been linked to one in eight deaths.[82] The economic cost of an increasingly toxic environment has been estimated at almost $3 trillion per year from premature deaths, diminished health, and lost work.[83]

To explain the toxic blight in countries that have scrimped on safety, a new global lexicon has arisen, including "airpocalypse" and "PM2.5" indicating the size in microns of the particulate matter (PM) released to the air, which can travel deep into the respiratory tract and enter the bloodstream through the lungs. Another modern measure is the "social cost of carbon dioxide" (SCCO2), a 2021 report pegging SCCO2 at between $51/tonne and $310/tonne, that is, annual costs ranging from 2 to 11 trillion dollars (the upper estimate includes future uncertainty and feedback loops[84]). Developing countries hoping to expand their energy needs by burning coal do so at a huge cost to public health and climate change.

***

Despite the danger, some remedies are available. In early conventional, coal-fired plants, "lump" coal was burned on a grate to create steam, but is now milled to a fine powder (<200 microns) and automatically stoked to increase the surface area for faster, more efficient burning. In such a pulverized coal combustion (PCC) system, powdered coal is blown into a combustion chamber to burn at a higher temperature, combusting about 25 tons per minute.[85] Although coal combustion is more efficient than in the past, two-thirds of the energy is nonetheless lost as heat if no thermal capture systems are employed, for example, collecting waste heat for district heating in a combined heat and power (CHP) plant.

Washing coal before combustion helps reduce toxic emissions in older, dirtier ranked coal, while precipitators can be installed to filter fly ash after burning.[86] New carbon storage methods are also being sought to dispose of emissions, such as burying high-pressure carbon dioxide thousands of feet underground in disused oil and gas reservoirs. In a typical coal plant, carbon dioxide accounts for roughly 15% of the exhaust gas, but capturing the outflow is prohibitively expensive and is more about extending outdated systems than improving bad air (as we will see later).

But despite the improvements and PR about "clean coal" or "carbon capture," the pollutants are still the same, the amount depending on the elemental makeup of the coal type. Coal is not pure carbon, but is mostly carbon (70–95%) along with hydrogen (2–6%) and oxygen (2–20%), plus various levels of sulfur, nitrogen, mercury, other metals, and moisture, as well as an inorganic, incombustible mineral part left behind as ash that is full of rare-earth elements (REEs), especially in lignite. The particulate-matter emissions include numerous pollutants: carbon monoxide, sulfur dioxide, nitrogen oxide, mercury, lead, cadmium, arsenic, and various hydrocarbons, some of which are carcinogenic.

According to the Physicians for Social Responsibility, coal contributes to four of the five top causes of US deaths: heart disease, cancer, stroke, and chronic lower respiratory diseases. Ill effects include asthma, lung disease, lung cancer, arterial occlusion, infarct formation, cardiac arrhythmias, congestive heart failure, stroke, and diminished intellectual capacity, while "between 317,000 and 631,000 children are born in the US each year with blood mercury levels high enough to reduce IQ scores and cause lifelong loss of intelligence."[87]

The higher the sulfur content, the more sulfur dioxide, perhaps the nastiest coal-burning by-product, which creates acid rain (the sulfur reacts with oxygen during burning and is absorbed by water vapor in the atmosphere). Sulfur content varies by coal rank, but bituminous coal has as much as nine times more sulfur than sub-bituminous coal, creating more health problems in the eastern United States.[88] Flue-gas desulfurization (FGD) systems, better known as "scrubbers," have been available in the USA since 1971 to reduce sulfur emissions and acid rain, but are not required in every state. According to the US Energy Information Administration, plants fitted with FGD equipment produced more than half of all electricity generated from coal yet only a quarter of the sulfur dioxide emissions.[89]

Since Richard Nixon signed the 1970 Clean Air Act, the US government has mandated emission controls for both carbon and sulfur emissions in coal-fired power plants, including the use of scrubbers, although emissions are still higher than anywhere else in the world besides China. In 2015, Barack Obama signed a new Clean Air Act, allowing states to set their own means to reduce emissions by 20% by 2020, realistically achievable only by shuttering coal-powered plants, especially older ones fueled by the dirtier, darker coal found in the Illinois Basin, Central Appalachia, and Northern Appalachia.

Efficient coal plants make a difference. According to the World Energy Council, "A one percentage point improvement in the efficiency of a conventional pulverized coal combustion plant results in a 2–3% reduction in CO_2 emissions. Highly efficient modern supercritical and ultra-supercritical coal plants emit almost 40% less CO_2 than subcritical plants."[90] Unfortunately, despite almost half of all new coal-fired power plants fitted with high-efficiency, low-emission (HELE) coal technologies, about three-quarters still do not avail of the latest emission-reduction technologies.

Stiff opposition continues by lobbyists and politicians who routinely down-play the dangers, especially where coal accounts for much of the economy. Long-time Republican senate leader Mitch McConnell from Kentucky stated that "carbon-emission regulations are creating havoc in my state and other states,"[91] while Democrat senator Joe Manchin from coal-rich West Virginia has made millions from coal, derailing legislation curbing its use. It isn't easy to turn off the coal switch in a country so reliant on a readily accessible energy source when vested interests dictate policy, and indeed regulations were rolled back in 2017 under Donald Trump. As always, jobs are at stake, but the emissions stake is higher, while the whole coal-use cycle must be costed – increased temperatures, air pollution, and expensive cleanups in sensitive water-table areas affected by mining run-off.

As the world wrestles with how to transition from dirty to clean energy, including how to pay the bills, the ongoing problem of coal was highlighted at the 2018 COP24 summit in southwest Poland, the largest coal-mining region in Europe that provides 80% of Poland's electric power. The meeting took place in the Silesian capital of Katowice, where the smog hangs in the open air. Christiana Figueres, the former UN climate executive secretary and one of the architects of the Paris Agreement, acknowledged that Katowice was on the frontline of industrial change amid the increased perils of global warming, "a microcosm of everything that's happening in the world."[92]

Alas, even developed countries are finding coal hard to quit amid the lofty rhetoric about "net zero" and a "carbon-neutral" future, which is now part of a regular discourse on transitioning to cleaner energy. At the same time as the 2021 COP26 meeting in Glasgow, the UK was considering reopening coalmine operations on the Cumbrian coast to supply coking coal for steel, citing critical local jobs, despite cleaner alternatives such as Sweden's pioneering green steel (which has 5% the carbon footprint of conventional steel) or recycling via electric arc furnaces.[93] Rather than "consign coal to history" as called for at the meeting by the then UK prime minister and contravening renewed ideals about the right to clean air, a new license was also granted at the Aberpergwm Colliery in south Wales to mine 40 million tonnes of high-grade anthracite, most of which will be used to make steel at the nearby Port Talbot steelworks.[94] The war in Ukraine also pushed other European countries to reconsider shuttering coal-powered plants as more Russian natural gas was curtailed by Western sanctions and Russian retaliation. One can hardly expect the developing world to ramp down coal use when the rich, developed countries won't abide by their own rules.

Despite the dangers, turning off coal is no simple fix. And indeed, coal consumption has continued to rise, 25% over the last decade in Asia, which accounts for 77% of all coal use, two-thirds from China.[95] Once thought of as "the fuel of the future" in Asia, more than 75% of planned coal plants have at least been shelved since the 2015 COP21 in Paris.[96] Nonetheless, without effective global policy agreements, coal will continue to provide more power in poorer countries. In late 2021, Poland was fined €500,000 per day rather than close an open-cast lignite mine in Turów, Poland's prime minister noting "it would deprive millions of Polish families of electricity."[97] Soon after, however, China announced it would no longer build new coal plants abroad, although more domestic plants would continue to be built.

Converting from brown to green is a huge challenge when so much of our generated power depends on the rock that burns. New investment strategies for stagnate regions and economic support for displaced workers – a.k.a. a "just transition" – is essential if we want to manage the social dangers and political fallout, both in the richer developed West and in the poorer, coal-reliant developing world. The goal is to maintain the supply of readily available, "dispatchable" power without continuing to pollute or add to global warming.

1.5 The Economic By-Products of Coal: Good, Bad, Ugly, and Indifferent

Coal and steam have brought many advantages to the world. Global trade expanded as raw materials such as cotton were shipped in greater quantities from Charleston, South Carolina, to Liverpool, ferried by canal barge and train to Manchester, and turned into finished goods in the burgeoning textile factories of "Cottonopolis," before being shipped around the world at great profit.

But more than just locomotion and industry changed with steam technology. In 1866, the first transatlantic cable was laid between Valencia Island, Kerry, and Heart's Content, Newfoundland, transmitting news in minutes instead of weeks via telegraph between the stock floors and investment houses of New York and London.[98] In 1873, the French writer, Jules Verne, wrote about circumnavigating the globe in 80 days as the world shrank more each year, forever changing the concept of distance and time, and accelerating the transformation to global village we know today.

If we take the start of the Industrial Revolution[99] as the year Thomas Newcomen invented the atmospheric water pump in 1712 for the Conygree Coalworks near Birmingham and the end as the sinking of the Titanic in 1912, we can try to gauge the vast changes in a 200-year period from old to new. Some mark 1776 or 1830 as the start, while others think the revolution is still rolling on today, but these dates or thereabouts are only a convenient benchmark. No single date can be ascribed, while the extent of industrialization has varied widely across the globe.

From 1700 to 1900, the population of England increased almost six-fold, from 5.8 to 32.5 million, whereas Europe's population more than trebled from 86.3 to 284 million. In the same period, the population of the world more than doubled from almost 750 million to 1.65 billion (passing one billion around 1800). Towns and cities flourished with industrialization despite worrying a few concerned souls such as the English cleric and political economist Thomas Malthus, who thought that a linearly increasing food supply couldn't keep pace with a doubling population. But double it did, again and again, to more than eight billion today, although the overall rate of increase has slowed of late.

Urbanization also increased as more people moved from the country to seek work and to avail of better goods and services in the newly expanding towns and cities. From 1717, Manchester grew from a pastoral town of 10,000 to more than 2.3 million by 1911.[100] In 1800, 20% of Britons lived in towns of at least 10,000, while by 1912, over half did. Urbanization ultimately led to suburbanization and city sprawl as trains went underground, making modern commuting possible as well as speeding the change from dirty, coal-powered steam to clean, electric locomotion. By 1950, 79% of the population of the United Kingdom lived in cities, a number expected to rise to over 92% by 2030.[101]

Urbanization is the main sociological change of the Industrial Revolution. Indeed, everyone has neighbors now. By comparison, in China, more than a quarter of a billion people are expected to move from the country to the city by 2030, dwarfing in a decade Europe's century-long industrial transformation.

The number of farms and people working the land dropped accordingly as newly mechanized tools reduced the need for farm labor and an urbanized manufacturing economy replaced a centuries-old, rural, agrarian way of life, referred to by Jacob Bronowski as being "paid in coin not kind."[102] From 1700 to 1913, Britain's GDP increased more than 20-fold (from £9 billion to £195 billion),[103] establishing Britain as the greatest capital-accumulating empire in history, the third such after Spain and the Netherlands, and the most vast prior to the rise of the United States.

By 1847, at the peak of the second rail mania, 4% of male UK laborers were employed in rail construction.[104] In 1877, there were 322 million train passengers, increasing to 797 million by 1890, including on the luxury 2,000-km Orient Express, which by 1883 left daily from Paris to Constantinople, epitomizing the newly minted power and wealth of the industrial age.[105] In the midst of all the progress, old jobs were supplanted, and boom–bust cycles became the norm as demand periodically lagged supply and vice versa.

Industrialization created specialization and the need for organized economic planning, about which Adam Smith wrote in detail in *The Wealth of Nations*, his seminal treatise on the causes and consequences of a rapidly industrializing world. In the midst of the Industrial Revolution, workers were displaced and opposition to the loss of jobs intensified, some a way of life for generations.

A group of artisan textile workers from the English Midlands, the Luddites (1811–1813), were the first to strike out, literally fighting for their livelihoods as they smashed loom frames in a number of nocturnal attacks, pitting man versus machine in a new kind of war that sought to include more than just price in the emerging doctrine of quantity over quality. Conditions had been deteriorating for years as unskilled outside workers, including those fleeing the Enclosures, sliced into their business via a cheaper, mechanized and steam-powered factory system.

The stakes were high, the government passing the 1812 Frame Breaking Act, which instigated the death penalty for those found guilty. In defense of the beleaguered workers, the poet and politician Lord Byron declared in a speech to the House of Lords that the outrages had "arisen from circumstances of the most unparalleled distress. The perseverance of these miserable men in their proceedings tends to prove that nothing but absolute want could have driven a large and once honest and industrious body of the people into the commission of excesses so hazardous to themselves, their families, and the community."[106]

Byron's assertion was not well received in the corridors of power, but the cause was lost to the unstoppable pace of industrialization. There could be no returning to making pins, lace, and linen from home as politicians began to measure the success of a nation on its economic output rather than the health and well-being of its citizens. The same fight happens with all disruptive technologies, including today between ride-hailing companies and taxi drivers, checkout scanners versus cashiers, and as auto-drive cars and drones begin to replace delivery workers.

Worst was the treatment of children, essentially forced to work as slaves. The utopian socialist and mill owner Robert Owen recognized the darker side to unregulated industry, especially child labor. In his New Lanark mills in Lanarkshire, Scotland, Owen reduced the workload of children from 13 hours to 12 hours a day and excused all children under 10 from work – hardly enlightened by today's standards, but a respite nonetheless from a constant working existence and much better than practiced elsewhere.[107]

Others came to champion workers' rights against the onslaught of technology that was shaping modern life. The most influential critic of "oppressive" capitalism was the transplanted German philosopher and economist Karl Marx, who wrote extensively about the consequences of unregulated industrialization in Britain and elsewhere and how laborers had become part of the machinery, a "commodity, like every other article of commerce, and are consequently exposed to all the vicissitudes of competition, to all the fluctuations of the market."[108]

One thinks of propriety and prurience as the predominant characteristics of Victorian Britain, but the truth was much dirtier and plain for all to see in the grime, filth, and poverty that followed industrialization. Scarred by the combustion of coal and a relentless thirst for more, a new world was rising from the ashes of feudalism. The Industrial Revolution created more jobs than ever, some little more than bonded servitude, yet also brought misery to those who could neither take advantage of a changing economy nor lobby for better treatment, and were ultimately left behind by governments more interested in work than workers. In Britain, the creation of the Labour Party in 1900 sought to restore the balance, mustering support from trade unions and disenfranchised workers.

Many examples of unbridled capitalism can be found in *The Ragged Trousered Philanthropists*, Robert Tressell's 1914 book about a group of English housepainters living and working under the screws of an uncaring ownership. The "philanthropists" of Tressell's title are the workers who make money for their Edwardian "betters" without complaint, beholden to a trickle-*up* system where unorganized labor pays for the riches of a politically protected upper class. Tressell's book helped to elect the first British Labour government in 1924, headed by Scottish socialist Ramsay MacDonald.

By 1945, in response to the promised rewards for sacrifices made during World War II, modern reforms were established to mitigate the excesses of unregulated capitalism, such as an expanded national health service, social insurance, minimum wage, and, as if in homage to the first great progenitor of change, a public train system. Coal, steam, motive power, and mechanized manufacturing created the modern world, one that rarely stops to measure the consequences of burning carbon-containing fuel or the endless cry for more. The battle continues today to keep the bounty of industrial progress in place for all.

1.6 A Very Old Very New Fuel: Biomass

In an attempt to curtail the worst of coal's excesses, an old staple has been refashioned, engineered to increase efficiency. Declared as cleaner and labeled "carbon neutral," both "biomass" and "biofuel" are being hailed as substitutes for coal, to continue to generate electricity via steam-powered combustion plants and augment the liquid fuel needs of an ever-growing transportation sector. The truth, however, is never quite the same as the PR.

Biomass was our first fuel – perhaps straw, dried leaves, or wood – as early humans looked to burn what they found nearby for warmth, light, and more chewable protein. Wood was also the first major pre-industrial fuel, long before anyone called it biomass, helping to keep alive those who lived in inhospitable winter climates. Today, biomass refers to a number of materials of biological origin from forestry, agriculture, and the biodegradable part of industrial and residential waste, and is rightly called a "solar resource"[109] to highlight the connection to the Sun – photosynthesis has been providing energy for plants since the beginning of time to turn sunlight into oxygen and sugar. More exotic biomass now includes "biofuels" for transport and "bioliquids" for nontransport purposes (heat and electricity generation), which we look at in this and the next section. The main types of biomass and biofuels are listed in Figure 1.2.

In Europe, half of all renewable energy comes from wood, as much as 80% in Poland and Finland.[110] If grown in properly managed forests, wood is considered "carbon neutral," where the carbon emitted during burning is absorbed by new growth, although some trees can take up to a century to mature enough

BIOMASS	BIOFUELS	
forestry residue	ETHANOL	BIODIESEL
pressed pellets	sugarcane (Brazil)	soybean (USA)
		rapeseed (China)
agricultural waste	corn (USA)	canola (Canada)
		sunflower
industrial/residential waste	cellulosic	cooking oil
		algae

Figure 1.2 (a) Power-generation biomass and (b) transportation biofuels (ethanol/ biodiesel).

to offset the lost carbon, while biomass combustion still contributes to increased air pollution. The rush to use wood as a renewable energy source was spurred on by countries trying to reach an ambitious goal of obtaining 20% of all energy needs from renewables by 2020.

The list of materials that classify as biomass is extensive, but some of the more notable – besides wood – are plants, plant parts, wood waste products, forestry residue, manure, sewage sludge, landfill gas, flotsam, paper, cardboard, pasteboard, tires (from natural rubber and fibers), vegetable oil, and a variety of farm crops, such as sugarcane and corn (for ethanol) and soybeans and rapeseed/canola (for biodiesel). If you've ever wondered where your household "paper waste" goes, some of it is turned into biomass.

Subsidies have played a major role in making wood a modern industrial fuel. The 4-GW Drax Power Station in north Yorkshire that burns both coal and biomass – the UK's largest power provider and world's largest biomass power plant at 2.6 GW – can receive more than half a billion pounds per year burning wood pellets, almost triple profits.[111] Although other fuels are also imported from afar, much of the source material is imported from forests in North America, a carbon-intensive process, especially because wood has a much lower energy content than coal or oil and is much less dense, requiring more space to transport. Some of the pellets have even been manufactured from illegally logged, "high-carbon" primary forests. What's more, converting wood to pellets for fuel reduces the availability of wood as an essential ingredient in the pulp and paper, construction, and furniture industries, while making pellets is also carbon-intensive, negating the idea of a carbon-neutral source:

> The process of making pellets out of wood involves grinding it up, turning it into a dough, and putting it under pressure. That, plus the shipping, requires energy and produces carbon: 200 kg of CO_2 for the amount of wood needed to provide 1 MWh of electricity.[112]

A major problem in chopping down and replanting trees to produce biomass is the different rates of carbon emission and absorption – a fast, one-time, carbon release during burning versus a slow, long-term absorption during growth that can take as long as 100 years to offset, ill-advised when global temperatures are rising today.[113] Indeed, there is no such thing as instantaneous carbon neutrality, although as much as five times the growth can be stimulated by "coppicing" – growing a new tree in the root system of an older felled tree or stool – but requires a rigorously controlled replanting system.[114] Managed planting is also worse when hardwood trees are replaced by faster-growing species such as pine, because natural forests equilibrate carbon better than plantations, while repeated harvesting and replanting reduces soil productivity and increases nitrogen dioxide emissions.[115] Called "fake forests" by those who consider natural forest regeneration the smarter policy, monoculture tree plantations are also more susceptible to single pests and thus not as resilient.

Biomass power generation is still small compared to coal, oil, and natural gas, with about 4,500 power stations generating a total of 75 GW across the globe.[116] The world's largest all-biomass power plant was the UK's Ironbridge station in the Severn Gorge near Birmingham, a 740-MW converted coal-power plant that burnt over one million tons of wood pellets a year, followed by the 450-MW Lynemouth plant in Northumberland, another coal-conversion in the heart of old coal country. The partly converted Drax coal plant generates 2.6 GW from burning seven million tons of wood pellets per year.

Other biomass power stations burn forest residue, wood chips, "black liquor" (a pulp-and-paper industry waste product), sugarcane, peat, and various other solid recovered fuels such as unclean plastic, paper, and cardboard.[117] As with peat, biomass can be burned to generate heat or electrical power on its own or co-fired with other fuels. Bagasse (leftover fiber after the juice is squeezed from pulverized sugarcane) and recycled urban wood is burned in the largest American biomass power plant, the 75-MW Okeelanta Biomass Cogeneration plant in South Bay, Florida.[118]

In Europe, wood-burning emissions (a.k.a. "forest energy") are not included in the carbon budget, an odd accounting fudge given that biomass produces much the same or even more carbon dioxide as coal (roughly 2 kg of CO_2 per 1 kg of wood). Classified as renewable (although hardly green), the European Union is making global warming worse by offering subsidies for large-scale, "carbon-neutral" wood burning. David Carr of the US Southern Environmental Law Center noted that "cutting down trees and burning them as wood pellets in power plants is a disaster for climate policy, not a solution,"[119] while increased deforestation annually clears five million hectares of forest.[120]

The EU's 2009 classification of biomass as renewable has, in fact, turned forestry into a callous extraction industry without concern for natural habitats. To balance its carbon budget, emissions are also measured where a tree is cut down rather than where a processed pellet is burned; that is, a tree farm in North Carolina rather than a European biomass power plant. Instead of managing waste wood in a viable organic recycling program, industrial biomass promotes the wholesale cutting down of vast areas of woodland. As noted by a former Tuft University environmental policy professor, "I can't think of anything that harms nature more than cutting down trees and burning them," while Ed Markey, US senator for Massachusetts, stated "Biomass is categorically incompatible with our climate, justice and health goals."[121] Indeed, if one wanted to reduce global warming, the easiest solution is to plant *more* trees, not cut more down.

Indigenous woodland species are particularly endangered by industrialized forestry that aims to produce commercial timber and biomass feedstock as fast as possible. A 2022 *Nature* study reported a more than 30% reduction of birds over a decade and a 66% loss in breeding habitat in the past 35 years, caused by human-induced changes to forest composition, in particular thinning and replanting with single species trees for merchantable timber that is fueling a biodiversity crisis in our woodlands.[122] Procuring a constant supply of renewable cellulose is hardly sustainable if it doesn't protect native bird and wildlife populations from a constant makeover of their habitats.

In the developing world, however, there is often little choice other than to burn what is available, albeit very low-tech. According to the World Energy Council, "The predominant use of biomass today consists of fuel wood used in non-commercial applications, in simple inefficient stoves for domestic heating and cooking in developing countries," and accounts for about one-fifth of total primary energy (that is, raw energy found in nature, not derived or refined from other sources).[123] Globally, burning biomass provides about 3% of total primary energy, mostly from heat and power applications.

Fortunately, burning wood has become much more efficient since clean-burning stoves were introduced in the 1980s, although few in the developing world have access to such technology. As noted by Lars Mytting, author of *Norwegian Wood: Chopping, stacking, and drying wood the Scandinavian way*, "The emissions from good-quality wood burned in a modern, clean-burning stove using the proper techniques amount to less than 5 percent of the emissions from the old stoves . . . . They combust more efficiently too – in some models as much as 92 percent of the energy potential in the wood is used."[124] Despite the improvements, however, burning wood still accounts for half of atmospheric pollution, especially via incomplete combustion and poorly managed fires (for example, to keep a slow burn at night).

Across the Scandinavian countries, small-scale "intelligent wood burning" in clean-burning stoves ensures a sustainable source that has kept the home fires burning for thousands of years in some of the coldest winter worlds. Alas, not everyone has access to expert methods, or the finances to change how they burn our most ancient fuel.

***

Transportation biomass (a.k.a. biofuel) is also on the rise, lowering emissions and reducing petroleum dependence, while garnering plenty of press and attention, not all positive. Produced from common farm crops to make ethanol and biodiesel for use as "drop-in" fuels in automobile engines, biofuels comprise 1.5% of all transport fuel (~7% in the US).[125] Tellingly, Rudolf Diesel, inventor of the compression-ignition internal combustion engine, experimented with a number of biological and fossil fuels while developing his more powerful "diesel" engine from 1893 to 1897 in Augsburg, Germany, trying both peanut oil and coal dust before choosing petroleum. When it comes to burning, energy content is king.

Ethanol is produced from sugarcane, corn, and waste cellulose, while biodiesel is produced from various vegetable oils (primarily soybean, rapeseed/canola, and sunflower), as well as recycled cooking grease and algae. Pure ethanol (E100) has 67% the energy content of gasoline (a.k.a. petrol in the UK), whereas pure biodiesel (B100) is about 96% as efficient. Easily blended with gasoline, almost all of the 150 billion gallons of gasoline annually consumed in the United States is already at least 10% ethanol (E10), including on the racetrack where Daytona 500 drivers mix 15% corn ethanol into their fuel (E15).[126] With more than 250 million vehicles on American roads, a cleaner, hybrid "flex-fuel" helps reduce excessive dependence on foreign oil as well as lowering nasty emissions. The director of the US National Bioenergy Center noted that "Biofuels are the easiest fuels to slot into the existing fuel system."[127]

Highlighting how easily an internal combustion engine can run on either gasoline or alcohol, Henry Ford provided a flex-fuel option for his famous Model T until 1931, before alcohol prices rose and distribution was curtailed during Prohibition.[128] Neither punitive US taxes on alcohol nor Standard Oil, which held a monopoly on the global oil market, helped the fledgling grain-based fuel industry.[129] Consumer options are only as good as the price at the pump, yet hardly viable without a pump.

The increased need for fossil-fuel alternatives, however, has rekindled interest in farm-based fuels, in particular ethanol (ethyl alcohol) and biodiesel. Sugarcane ethanol is an alcohol-based fuel, produced by the fermentation of sugarcane juice (about 10% of pulverized sugarcane) and molasses.[130] Pioneered in Brazil in the 1920s (mostly in the state of São Paulo) and expanded during the global oil crisis of the 1970s, sugarcane ethanol is burned in all Brazilian flex-fuel vehicles and accounts for 50% of the country's auto-fuel needs, with every gas station having at least one ethanol pump. Sugarcane provides over 16% of Brazil's total energy supply, taking up 1.5% of arable land for production. As a transportation fuel, sugarcane ethanol also lowers greenhouse gas emissions by as much as 90% compared to gasoline.[131] Subsequently, Brazil produces much fewer transport emissions than elsewhere.

Corn ethanol is similar, using yeast to ferment starch in ground corn kernels as in traditional grain-based alcohol production (starch to sugar to alcohol), but is only about 20% as efficient as sugarcane ethanol. Corn ethanol is also more expensive to produce than sugarcane ethanol, which is already a sugar and is thus more easily fermented without the need of extensive preprocessing. Nonetheless, corn is the number one US crop and ethanol production has grown rapidly throughout the midwestern "Corn Belt" states where the vast farming infrastructure helped to facilitate the change from growing corn for food to growing corn for fuel.

Targets originally cited in the 2005 US Energy Policy Act were expanded in the 2007 Energy Independence and Security Act to foster growth and reduce American reliance on foreign oil. In the midst of rising prices after the Russian invasion of Ukraine at the start of 2022, the Biden administration announced an increase to E15 to lower gasoline consumption and prices, although the volume of blended renewable fuel had already been mandated to quadruple to 36 billion gallons by 2022 via the Renewable Fuel Standard (RFS) program.[132]

Corn ethanol production has numerous detractors, however, primarily because of large government subsidies, combustion pollution, and the amount of petroleum in the production process, perhaps equal in energy to the petroleum replaced. Economist C. Ford Runge itemized ethanol's failings, especially increasing the "blend wall" from 10% (E10) to 30% (E30): "Higher-ethanol blends still produce significant levels of air pollution, reduce fuel efficiency, jack up corn and other food prices, and have been treated with skepticism by some car manufacturers for the damage they do to engines."[133] Runge further noted that increased production of corn ethanol is not about economics, but special interests and the federally mandated RFS program. What's more, besides toxic, ethanol-based particulate matter impacting humans, "more recent scientific analysis links corn for ethanol to declining bee populations, with potentially catastrophic implications for many other high-value agricultural crops (almonds, apples) that depend on these insects for pollination."[134]

The large-scale conversion has also prompted a vigorous food-versus-fuel debate (a.k.a. "eat or heat"), while excessive, profit-only mono-farming poses risks to ecosystems and biodiversity. Biofuels are not without politics, especially corn ethanol that pits farmers against the petroleum industry. In Iowa and other Corn Belt states, where one out of every two bushels of corn goes toward biofuels, the required amount of blending is also debated. Initiated by the Trump administration, smaller oil refineries easily bypassed existing biofuel quotas, citing dubious competitive hardship via so-called small refinery exemptions (SREs), lowering ethanol consumption by more than 10% and angering farmers.[135]

Part of the problem is the extent of the conversion as energy-industry analyst James Conca noted, "In 2000, over 90% of the US corn crop went to feed people and livestock, many in undeveloped countries, with less than 5% used to produce ethanol. In 2013, however, 40% went to produce ethanol, 45% was used to feed livestock, and only 15% was used for food and beverage."[136] Corn prices have also risen to satisfy the need for more fuel, although increased meat consumption has upped prices too because more corn is needed to produce more animal feed. There is no magic bullet, one-system-fits-all solution to global energy needs, and a complete conversion of all available biomass farmland to biofuel-producing crops would displace about 25% of the gasoline annually consumed in the United States. Clearly, other options are needed.

Cellulosic ethanol – a so-called "second-generation" biofuel – is less controversial, using microbes (thermochemical gasification) or enzymes (saccharification) to break down cellulose in waste agriculture, such as woody fibers, trees, and grasses. The leftover, nonedible parts of corn plants on existing farmlands and land not suitable for food also support cellulosic ethanol production. The earth's most abundant carbon-based molecule, half of all modern biomass is cellulose, $(C_6H_{10}O_5)_n$, the result of photosynthesis-generated glucose that bonds together to form the cell wall of green plants and algae.[137] Cellulosic ethanol doesn't use as much fossil fuels for fertilizers or cultivation, increase food prices, or cause as much environmental damage as grain-based ethanol, while burning cellulosic ethanol also produces 60% less greenhouse gas emissions than burning petroleum.[138]

One novel power plant – the world's largest upon completion – was built by Dupont in 2015 in Nevada, Iowa, annually producing 30 million gallons of cellulosic ethanol from 375,000 tons of corn stover (stalk, leaf, husk, and cob leftover in the corn harvest), before being sold to a German biofuels company in 2019 and retooled to produce natural gas.[139] Unfortunately, preprocessing cellulose from energy crop waste is still hugely expensive, although the costs could decrease with improved biomass yield via faster-growing crops and plants or from using winter cover crops.

***

Americans also annually consume about 50 billion gallons of diesel fuel. Most trucks, buses, and tractors run on the more robust diesel engine, and thus biodiesel has become increasingly popular since commercial production began in the 1990s. A diesel engine is more powerful than a gasoline engine – combusting fuel by compression instead of spark ignition – and is better suited for larger vehicles. As with ethanol/gasoline, blends are common, B20 (20% biodiesel by volume), as well as pure B100. Biodiesel is clean, extremely efficient, doesn't require a retrofit to existing diesel engines, and produces less particulate matter and toxic emissions than petroleum-based diesel fuel.[140] Biodiesel also causes less engine wear and increases lubricity and engine efficiency, while emitting 60% less carbon dioxide than petroleum diesel.[141]

Most biodiesel is made from soybeans, rapeseed/canola, and "yellow grease" (recycled cooking oil). First planted in the United States in 1765, soybeans were farmed for livestock feed before becoming a major oilseed crop in the 1940s. The biodiesel is created by pressing soybeans to extract the oil (about 18% of the seed) that is then refined through a process called transesterification, reacting methanol with sodium hydroxide (lye) as a catalyst to produce methyl esters (the biodiesel) and glycerine.

Today, the United States is the leading producer of soybeans followed by Brazil, together accounting for 60% of global production, although Iowa soybean output was significantly reduced after higher tariffs were implemented in the 2018 USA–China trade war. According to an analysis by the Department of Biological Systems Engineering at the University of Nebraska–Lincoln, using all soybean crops for biodiesel production could produce about five billion gallons per year, although two billion gallons is more realistic, ultimately providing 4% of American diesel fuel needs.[142]

Rapeseed can also produce biodiesel and other traditional petroleum products such as ink, and is readily seen in the vast fields of yellow found alongside many country highways. First employed as a steam engine lubricant in Canada during World War II, before being bred in the 1970s to remove toxins such as erucic acid for use as an edible cooking oil and livestock feed,[143] rapeseed oil was given the trademark *Canola* from "Canadian oil low acid," losing its capitalization to become regular *canola*, a popular North American cooking oil. Like all biodiesels, canola weighs less and causes less engine wear than petroleum-based diesel.

In various other countries, barley, castor beans, sunflowers, jatropha, yellow grease, and coconut and palm oils have also been converted to biodiesel, although much less than from soybean and rapeseed plants. Wheat, rice, and sugar beet are also viable feedstocks to make ethanol along with sugarcane, molasses, and corn.

Biomass makes up roughly 10% of world energy production, including almost three million barrels per day of biofuels (~2/3 ethanol, 1/3 biodiesel), a more than five-fold increase in 20 years (Figure 1.3). As with fossil fuels, production is linked to available sources, North America producing almost 50% of all biofuels (corn and soybeans), followed by South America 30% (sugarcane and corn) and Europe 20% (corn and soybeans). Asia kicks in with under 5% (mostly Indonesian palm oil).

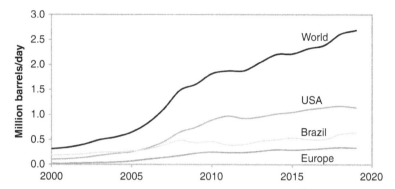

Figure 1.3 World biofuel production 2000 to 2019 (million barrels/day). *Source*:
US Energy Administration Information, "Biofuels Production," International
Energy Statistics, www.eia.gov/international/data/world/biofuels/biofuels-
production.

Highlighting the large agricultural footprint of such "energy crops," US corn
ethanol cropland comprises almost 20% of 84 million total acres and soybean
biodiesel 25% of 300 million acres. To increase production, more farmland will
be needed, while to run all gasoline vehicles in the USA on ethanol would
require an area twice the size of Texas, an unlikely proposition. In the UK,
which consumes about one-sixth the amount of fuel as in the USA, an area
equal to the whole of the UK would be needed to run its petrol-fueled vehicles.

Despite the lower carbon footprint compared to burning petroleum, there are
serious concerns about biofuels, including reduced food crops, higher food
prices, water shortages, and deforestation (especially in Brazil and Indonesia),
while the energy required to produce biofuels is greater than the energy
generated in an internal combustion engine, underscoring the inefficiency.
Some Indonesian forests have even been cut down to grow sugarcane for "low-
carbon" biofuels, a perverse attempt to make "green" fuel.[144]

It is also unlikely there will ever be enough land or plant-based feedstock to
make the liquid fuel needed to satisfy existing transportation needs if petroleum
fuels are completely phased out. As a simple comparison, the amount of
farmland to make industrial biofuels is much greater than via wind turbines
or solar panels (50 times for wheat to 110 times for sugar beet[145]), never mind
the pollution.

Although the debate continues about the value of growing crops for fuel and
power, obtained either from fermented carbohydrate crops (sugarcane and corn
ethanol), pressed oil-bearing plants (soybean and rapeseed/canola biodiesel),
or complex second-generation biofuels that require vast amounts of organic

feedstock (cellulosic ethanol) – whether on their own, in co-fired plants, or as a liquid-fuel alternative to gasoline and diesel – biomass and biofuels offer a unique, modern, farm-engineered energy source. Alas, given the vast infrastructure already in place for liquid fuels, biofuels are about extending the lifetime of the internal combustion engine and liquid-fuel infrastructure via pretend green additives. And while the synthetic fuels (synfuels) derived from organic feedstocks such as sugarcane, corn, soybeans, and rapeseed/canola can reduce dependence on petroleum, they still produce carbon dioxide when burned and are a net energy loss if more energy is required to produce them in the first place than is regained during combustion. Indeed, despite the more modern methods to turn food into fuel, the technology has not overcome the inherent inefficiency of biomass, the reason we changed to fossil fuels in the first place.

We can, however, reduce the massive amount of biological waste that would otherwise go straight to a landfill by managing the organic output from farms, businesses, and homes. Fortunately, much of what we waste can be put to use, turning seemingly unimportant organic matter into energy as a bioliquid, which we look at next.

1.7 The Closed Loop: Making Waste Profitable

Redefining what we mean by "waste," a modern biogas plant helps dispose of the heaps of agricultural by-products, industry food waste, and household garbage (for example, kitchen leftovers) that we create every day, converting waste to energy (WTE) rather than just throwing everything out in one fell swoop. Today's industrial systems can convert "waste to watts" or waste to biogas in under two months.

The heart of a biogas plant is the anaerobic digester (AD) that converts organic matter to methane, which is then burned for cooking, to generate electricity, or even to run a retrofitted bio-burner car. Food waste, recycled cooking oils, animal manure, and water-treatment sludge produced during industrial manufacturing and in homes can all be "digested" inside an airtight holding tank by bacteria in the absence of oxygen (anaerobic) that produces methane (~60%), carbon dioxide (~40%), and a leftover liquid-fertilizer "digestate."

Methane is already generated in many industrial sewage treatment works: the bacteria introduced into heated, airless digester tanks that react with the protein and carbohydrates in the settled sludge. The nitrogen- and phosphorous-rich leftover solids also make perfectly good soil-conditioning compost. The waste plant heat can also be captured in a combined heat-and-power (CHP) unit instead of being lost, improving the conversion efficiency by over 80% and cutting greenhouse gas emissions in half.[146]

The liquid digestate functions as an organic alternative to chemical fertilizers, fulfilling the sought-after environmental "closed loop" described as the basis of farming by British agricultural scientist Albert Howard in the Law of Return, that is, "the faithful return to the soil of all available vegetable, animal, and human wastes."[147] Essentially, the digestate (waste product of our own waste via industrial-scale composting) helps produce more food while reducing production costs. Either aerobically (with oxygen) or typically anaerobically (without oxygen) as in a backyard composter or in today's industrial-scale anaerobic digester, composting produces essential humus to enrich soil and limits household food waste that might otherwise end up causing emission problems down the road.

Numerous cities and communities have started to collect household organic matter to generate methane and compost, a better practice than just dumping the waste into a landfill and then boring through the dirt-covered garbage mounds after the fact (if at all) to collect the seeping landfill gas. Some are even fueling the very same trucks that pick up the waste as in an organic collection system begun in Toronto in 2002 that has been converting the waste into biogas since 2022 to provide end-use heat, electricity, and compressed natural gas to run the vehicles (Figure 1.4).

Seville's famous bitter marmalade oranges are now being collected in a novel methane-gas pilot program in the Andalusian capital to clear the streets of over five million kilograms of fallen fruit from almost 50,000 trees. Creating a nuisance for citizens and city cleaners alike, the formerly discarded organic mess is instead fermented in an existing water company's organic waste facility to generate electricity (50 kWh from 1,000 kg). The head of the company's environmental department noted, "It's not just about saving money. The oranges are a problem for the city and we're producing added value from waste."[148]

Cows provide the ideal source for an anaerobic digester, where some farms already produce methane to generate electricity in what could be called "poop power" or "manure management." Along with a daily supply of milk from 700 cows, Brett Reinford's Pennsylvania dairy farm collects 7,000 gallons of manure, which is shoveled into a hole with some local food waste, ground up into a slurry, and dumped into the digester, where microorganisms break down the smoothed-out mixture to produce solid waste and gas (60% methane) that is then converted to electricity and heat. As Reinford noted, the generated energy "heats our home, it heats our workshop, it heats all the hot water in our barn,

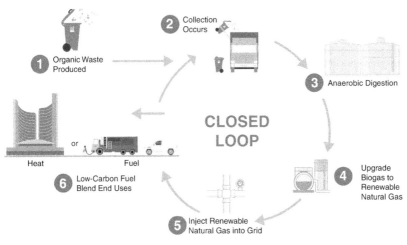

Figure 1.4 Closed loop organic recycling from door to vehicle. *Source*: City of Toronto, "Turning waste into renewable natural gas – Circular model showing how waste can ultimately be used to create green energy," 2023. www.toronto.ca/ services-payments/recycling-organics-garbage/solid-waste-facilities/renewable-natural-gas/.

and then, in the fall, we also use the heat that comes off of the radiator to dry all of our corn here on our farm. So we're not using any fossil fuels on this farm for the things that we need to heat."[149]

Having previously spent $2,500 per month on hot water, Reinford's heating bill was reduced to zero while the excess electricity was sold on to the grid. Food-industry critic and author Tristram Stuart noted that "Methane from farm manures was recently found to be among the most environmentally beneficial biofuels currently available. Increasingly these systems are also being touted as one of the best ways of treating food waste."[150]

Benefitting from its 2000 Renewable Energy Act, Germany has over 9,000 biodigesters, while by 2008 China operated more than 3,500 biogas plants on commercial farms to treat animal waste as well as 18 million domestic units.[151] In the United States, roughly 300 farms operate a digester, although the EPA estimates more than 8,000 farms could benefit from better manure management.[152]

One must be careful with the smell, however, as residents in Lowell, Michigan, (pop.: 4,000) can attest, after a biogas plant was built there in 2015, the first of its kind in the USA. Operated by Sustainable Partners (Spart), the 25,000 square foot, $6 million digester annually converts about

8,000 tons of manure from a local farm and 1.5 million gallons of wastewater from a nearby food company to generate 800 kW, while also treating wastewater previously hauled away by truck and dumped. Unfortunately, fats, oils, and greases (FOG) from area restaurants were added in the process, creating a foul odor, especially unbearable on humid days. Forced to close within 2 years, the plant reopened after enclosing the open-air tanks and fitting carbon filters to keep out the nasty smells. As Spart president Greg Northrup noted, "A biogas plant is a natural fit for baseload distributed generation. We can count on the cows for a constant supply of feedstock, which allows the plant to produce energy 24/7/365."[153] Baseload, as in 24-hour, year-round, poop power.[154]

Nonetheless, subsidies and incentive programs are still needed to run an AD, despite providing a viable organic waste-management system. Furthermore, it may seem environmentally friendly to run a car on cow dung – the manure from one cow per year can produce enough compressed natural gas (CNG) to run a car for 9,000 km or from 2 tons of processed straw – but in the end we are still burning fuel. What's more, less than 1% of the energy content is recovered, while burning the output of our food waste from an anaerobic digester might replace 1% of the domestic electrical supply.[155] Synthetic methane gas production can also leak up to 15%, adding to harmful GHG emissions, problematic for increasing biomethane facilities. It's also better to grow less food in the first place or waste less rather than burn everything after the fact (as we will see in Chapter 7).

For now, at least, if we can eat something we can create energy and digestate from what we throw away, burning the by-product and using the residue as feed or fertilizer. Biomass is easy to grow: all you need is seed, earth, water, and sun, all available in abundance (and nitrogen-rich fertilizers for large-scale production), but is a highly controversial way to grow fuel for transport or to generate power. As part of a closed-loop food cycle incorporated into our daily lives, however, we can certainly make better use of the massive amounts of organic waste we create, be it agricultural, industrial, or homemade. If we want to eat we can't avoid producing waste, although we can be much smarter after the fact to reduce the global warming hit.

In *The Ascent of Man*, Jacob Bronowski noted that 10,000 years ago seed plantation underwent an important change in the Middle East, when by a genetic accident wild wheat crossed with goat grass to form a fertile hybrid that again crossed with another goat grass to give bread wheat, a staple of the human diet. At that moment, we changed from nomadic living to settled farming, initiating an agricultural revolution, what in the wider sense Bronowski called a *biological* revolution, "the largest single step in the ascent of man."[156] All of civilization followed from the start of human settlement and sustainable agriculture.

Today, that revolution is being enhanced to broaden the scope of farming and include what we used to think of as food as energy. As we look at the origins of a revolution in *energy*, the second largest step was to tame fire for other than warmth and cooking, engineering heat to move objects our bodies couldn't, machine-aided by burning another fuel created from ancient plant remains. With coal, there was more than enough energy to burn.

1.8 Moving Heat Becomes Electrical Power

The science of heat engines applies to all fuels as the French engineer and scientist Sadi Carnot discovered while helping to improve the industrial capacity of post-Napoleonic France, which had fallen behind its Enlightenment rivals. Carnot's insight was to show that a heat engine is a cyclical system, whose theoretical efficiency depends on the temperature *difference* of its hot "source" and cold "sink." Water is boiled to steam, which enters a cylinder to raise a piston. The hot steam is then vented, for example, to an external condenser in Watt's steam-condenser engine, leaving behind a vacuum that lowers the piston. The cycle is repeated over and over. The hotter the cylinder and colder the condenser, the more efficient the engine.

One also sees in Carnot's great insight of engine efficiency resulting from a temperature *difference* that heat is not matter, for example, the soon-to-be disproved caloric fluid as many thought including Carnot himself, but rather a *transfer of energy* from a high-temperature area to a low-temperature area. In an open fire we see the characteristic crackling of air currents moving in all directions, the lighter hot air rising everywhere to be replaced by cold air. A heat engine is the organized movement of heat from a hot source (T_H) to a cold sink (T_C), where the greater the temperature difference ($T_H - T_C$) the faster the transfer of energy and the higher the engine efficiency.

In fact, as Carnot proposed, a heat engine is improved by *preheating* the steam, thus increasing the temperature of the source. Engine efficiency would later be codified by the simple relation $1 - T_C/T_H$, which shows that increasing T_H or decreasing T_C increases the efficiency. Note, 1 is a perfectly efficient heat engine where all the heat is extracted, theoretically achievable only with a cold sink at zero. Here, zero is *absolute* zero or 0 K, where all motion ceases. The K is for Kelvin in honor of William Thomson, a.k.a. Lord Kelvin, the Belfast-born and University of Glasgow natural philosophy professor who first determined the value of absolute zero (0 K = $-273\,°C$ or $-460\,°F$).

We can also make a source hotter by *pressurizing* the steam, a dangerous reality of early engines. Pioneered by Richard Trevithick, improved boiler technology could move a piston at a higher pressure, the "strong steam" increasing performance in smaller locomotive engines as the increased pressure created higher temperatures and thus greater efficiency ($P \propto T$). But the advent of strong steam also increased the danger of mechanical failure. From 1800 to 1870, over 5,000 people died in 1,600 boiler explosions resulting from high-pressure steam. In the USA, more than 2,000 people died from boiler explosions in the first 30 years of steamboat operations.[157]

As engineering theory and manufacturing standards improved, steam would be employed to move the largest of ocean liners. Steam's next great achievement, however, would be to move the smallest of things, the electron. In 1882, the first commercial electrical power station was built to convert the motion of a piston into a continuous electric current, transmitting electricity through a wire for use at remote distances. For the first great modern innovation that most characterizes the world we know today and which for many of us as ordinary citizens has impacted daily life the most, we turn to a self-taught, telegraph operator cum inventor, Thomas Alva Edison, who built a power station operated by a coal-fueled "dynamo" to light an entire neighborhood and then the world. While steam locomotion expanded the limits of everyday travel and enabled a global population to cohabit for the first time, electrical power transmission expanded time itself, turning the world on round the clock and literally illuminating our existence 24/7.

But to understand the dynamo and Edison's novel engineering achievements, we must first look at a few scientists who paved the way: Italian Alessandro Volta, Dane Hans Christian Ørsted, Englishman Michael Faraday, and Anglo-Irish engineer Charles Algernon Parson. Many more were involved, but Volta, Ørsted, Faraday, Parson, and Edison designed the essential components of today's electric power-generating station. One can also acknowledge the contributions of the American founding father and jack-of-all trades Benjamin Franklin, who courageously tamed the power of atmospheric charge in his famous thunderstorm experiments, showing that lightning was in fact electrical, and the English chemist and quintessential scientific showman Humphry Davy, who as president of the Royal Society of London[158] helped popularize the wonders of an emerging new science, in particular, his brilliant display of artificial "arc" lighting. The Serbian electrical engineer Nikola Tesla also deserves credit as does the American engineer William Stanley, who developed the transformer for long-distance power transmission, but we'll cover their unique contributions later.

Alessandro Volta developed the first "battery" in 1799, the term coined by Franklin in the 1740s after he compared a battery of cannon to a series of charge jars (a.k.a. Leyden jars),[159] which he filled via a kite, thread, and key to store *static* electricity from the ambient charge of an electrical storm, but was now applied to a chemical source. Showing that electricity could be *chemically* generated for the first time, Volta's "wet" battery consisted of two metals with different electronegativities (χ) wrapped around a brine-soaked electrolyte (the ion/electron source) to create a movement of electrons from negative to positive electrode, that is, a continuous *current* rather than a brief jolt of static. The terms *positive* and *negative* charge were also coined by Franklin. To increase the voltage in his rudimentary metal–electrolyte–metal construction, Volta added more layers in series, mimicking the repeating chambers on the back of an electric torpedo fish, whose remarkable electric killing ability the English scientist Henry Cavendish had earlier analyzed.

Experimenting with different metals that he characterized from least to most electronegative (zinc, lead, tin, iron, copper, silver, gold, graphite[160]) and observing that a greater charge was created the further apart the two metals fell on his scale, Volta quantified electricity in the first-ever electric circuit. His initial "voltaic pile" was made of tin leaf and brass for the electrodes with salt-water-soaked pieces of cloth for the electrolyte, although Volta soon turned to zinc, copper, and diluted acid, essentially equivalent to today's chemical battery (Figure 1.5).[161]

One can see how a battery works by making one at home, sandwiching a series of aluminum washers (or pieces of aluminum foil, $\chi = 1.61$) and copper pennies ($\chi = 1.90$) around cut cardboard soaked in a fizzy soft drink as the electron/ion source and connecting a light bulb or LED to the ends (the electrodes). All batteries are based on the same principle to create a flow of electrons (current) from a negative anode to a positive cathode (the electrodes) through an ion-rich substance (the electrolyte), where at the same time positive ions diffuse in the opposite direction through the electrolyte from cathode to anode. Even a potato or lemon can be used as the acidic electrolyte between two different metals.[162]

In 1809, Humphry Davy utilized Volta's invention to demonstrate arc lighting, connecting the ends of a massive battery to two thin charcoal sticks that produced a spark when he touched the sticks together and a brilliant "blue–white" light when he moved them apart. To make his new chemical batteries – the largest yet assembled – Davy separated the highly reactive metals from their natural compounds by electrolysis, further isolating six new elements for the first time: potassium (from potash), sodium (from soda), and magnesium (from magnesia) as well as barium, strontium, and calcium.[163]

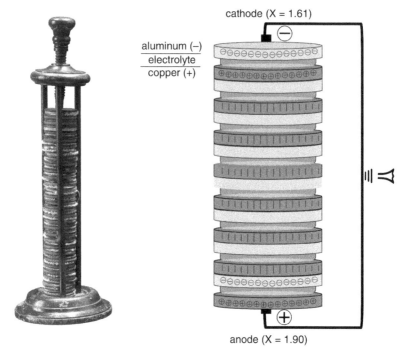

Figure 1.5 (a) Volta's original pile (*source:* Rama CC BY-SA 3.0 FR) and (b) a simple homemade battery using alternating strips of copper pennies, electrolyte (soda-soaked cardboard), and aluminum washers.

The scientific and commercial benefits of electricity were catching up to the wonders of its terrifying and unknown origins, isolating new elements and creating a controllable continuous electrical power source. With a chemical battery and two rudimentary carbon filaments, the mysteries of electricity were being made practical in a simple display that defines modernity like none before – artificial illumination. The telegraph, telephone, light bulb, power station, and all of modernity would follow.

The big step, however, was to produce a voltage that didn't need a permanently charged battery. Inspired by Ørsted's discovery that an electric current in a wire could move a magnetized needle – observed by chance in a University of Copenhagen science class he was teaching (the same effect is seen when a compass needle moves in an electrical storm)[164] – Michael Faraday wondered if a magnet could create an electric current in a wire, so-called "induction." Faraday demonstrated the principle in his lab on October 17, 1831, sharing the results a month later at the Royal Society of London.

Initially trained as a bookbinder, Faraday had been appointed assistant to Davy, the then Society president, after sitting in on Davy's public lectures and presenting him with a bound set of meticulously recorded notes. Appointed lab demonstrator and later manager, Faraday reveled in explaining electricity and magnetism without having to delve into the mathematical rigor, to which he was ill-equipped. In his paradigm-changing demonstration, Faraday moved a bar magnet in and out of a metal-wound iron ring and saw that an electric current was indeed induced – alternating current as his detector needle moved in opposite directions as he moved the magnet back and forth – showing that mechanical and electrical energy were interchangeable and could be exploited either in a generator or a motor.

Moving a "rotor" in and out of a stationary "stator" cuts the field lines and induces – that is, without touch – an electric current in the coiled wires wrapped around the stator. Interestingly, the generated magnetic force diminished with the inverse square of the separation ($1/r^2$), known as "action at a distance" in Newtonian parlance or "interacting fields" in Faraday's nonmathematical "lines of influence" description, seen in the picturesque contours of iron filings aligned between two poles of a magnet, indicating at each point the field strength.[165]

The Scottish mathematician and physicist James Clerk Maxwell would formulate Faraday's nonmathematical description of interacting electric and magnetic fields into four equations now at the core of electrical technology in a tour de force of theoretical mastery, which also showed how all electromagnetic (EM) radiation travels at the speed of light. Note, a discharging chemical battery produces a constant *direct* current (DC), while a rotating generator produces a sinusoidal varying *alternating* current (AC). Maxwell would also definitively show that heat and temperature resulted from moving atoms and molecules, and was thus not a flowing substance.

All that remained was a constant motion to create a constant electrical source, for which two inventors, Belgian Zénobe Gramme and American Charles F. Brush, would make viable with the creation of an electric generator that Gramme first demonstrated at the 1873 Vienna International Exhibition. The steam-powered *dynamo* was perfect for arc lighting, its name derived from the *dynamically* induced electric current via a rotating magnetic field. The easiest setup is a magnetized central rotor moving inside a fixed, wire-coiled outer stator, the basis of today's electric generator, where the current is generated in the stator wires via magnetic induction to power an external "load." Nikola Tesla would later perfect the AC generator at Niagara Falls (and AC motor) as we will see in Chapter 5.

***

Enter the genius of Edison, the aptly named "wizard of Menlo Park" and the world's most celebrated inventor, who seemingly worked around the clock from his own New Jersey "invention factory," eating at his desk and sleeping in short naps squirreled away in hidden parts of his laboratory. Having already invented the stock ticker, designed and built a commercially viable telephone, and wowing all with his 1877-patented tin-foil phonograph ("writer of sound") as he played back his recordings of "Mary Had a Little Lamb" to astonished audiences everywhere, the world's most famous "practical problem solver" wanted to light the world in a safe and controllable way, for which he would of course also charge a fee.

Outdoor street lighting had begun in London along Pall Mall, the world's first gas-lit street in 1807. Initially lit by "coal gas" – a mix of methane, hydrogen, and carbon monoxide produced by the anaerobic heating of coal – the Gas Light and Coke Company produced enough coal gas by 1820 to light almost 30,000 lamps over 120 miles of London's main streets.[166] Much of the initial technology to produce the gas, such as the retort heating of coal and the coal-gas distribution system, was developed and financed by Boulton, Watt and Co., who had first demonstrated gas lighting five years earlier at their Soho Manufactory, despite the dangers of frequent explosions and carcinogenic by-products in the coal-distillation process.[167] The process also produced a coke residue, perfectly suited for smelting. The name follows from the process of "cooking" the coal.

Glasgow, Liverpool, Dublin, and other European cities were soon illuminated by coal gas, followed by others along the east coast of the United States. The first commercial American gasworks was the Gas Light Company of Baltimore, lighting its first street in 1817 on the corner of Market and Lemon Streets, while other US cities would soon compete with the already well-established whale-oil and tallow industry.[168] The lit night immediately improved inner-city safety and the social habits of a growing urban life, but also lengthened the working day. Nor did the irony go unnoticed of burning coal gas for lighting, which produced the soot that hindered health and visibility, reducing sunlight by almost half in the worst-affected cities.[169] Some streets were also illuminated by a crude form of electric arc lighting, as we saw above, which although powerful was extremely dangerous, resulting in numerous deaths when wires broke or workmen were tangled.

Indoor lighting was also available via coal gas or kerosene, but was hot, smelly, and dangerous. Untreated coal gas contains toxic amounts of carbon monoxide, ammonia, and hydrogen sulfide with its characteristic rotten-egg smell, while in the 25 years prior to converting gas to electric light, hundreds of theaters burned down. In one famous disaster, a gas-lamp fire destroyed the original Harrods in Knightsbridge in the west end of London, rebuilt on the same spot to its current splendor (including the world's first escalator). Belfast's Crown Liquor Saloon is still lit by gas today, although the source is "natural" gas supplied via the same distribution lines. Both are, in fact, methane despite the different methods of production.

After numerous attempts in his large, purpose-built, Menlo Park lab, where he had tested 6,000 different "filament" materials, Edison finally improved on his earlier, rudimentary versions of electric illumination, producing in 1879 a safe, long-lasting glowing light, by electrically heating a high-resistance, carbonized cotton sewing thread in an evacuated glass bulb. *Incandescent* electric light had arrived, the filament glowing softly instead of rapidly burning out, because carbon absorbs more heat and oxidizes less in a vacuum, followed soon after by even longer-lasting carbonized cardboard, bamboo, and then drawn-wire metal filaments. The inefficient, unsafe, and highly toxic practice of *burning* carbon-containing fuel to produce light was replaced by a simpler, incandescent, *glowing* of a high-resistance wire thread, which was safer, cleaner, and cheaper, and hence much easier to exploit commercially. (An even higher-resistance coiled tungsten filament with higher melting point is still in use today.[170])

Creating power to light a whole building, street, or neighborhood, however, would require a stronger, more constant power source than the simple lab batteries in Edison's popular demonstrations. The illustrious practical problem solver needed a power station and a distribution system to deliver steady, *scalable*, electric power. That power station would ultimately include his own massive dynamos – steam-driven piston and coil assemblies powered by the continuous burning of coal. With the backing of New York banker J. P. Morgan, Edison started up the Edison Illuminating Company, and set out to light the world.

It took four years and half a million dollars to work out the details of his electric network – dynamo design, switching, metering, laying of insulated cables underneath existing streets (as modeled on the earlier setup of gaslight distribution) – the work much delayed by one-off, "isolated lighting" units built for eager customers and demanding financiers who wanted to show off their newly lit wealth. Morgan's 219 Madison Avenue mansion and William Vanderbilt's brownstone double house on Fifth Avenue both sported electric lighting powered by a basement dynamo, although Vanderbilt's system was removed when a smoldering fire appeared in the metallic-threaded wallpaper.

But after much labor and engineering innovation, Edison was finally ready to illuminate a part of downtown New York via an electric-powered, copper-wire transmission system:

> On the afternoon of 4 September 1882, [Edison], Bergmann, Kruesi, and a few other
> Edison Electric staff members went to the offices of J. P. Morgan in the Drexel

Building at Broad and Wall Streets. At 3:00 P.M., a switch was ceremoniously
thrown, and Edison's electric lights came to life.[171]

Six 27-ton, 100-kW "Jumbo" dynamos – named for P. T. Barnum's prized
elephant – generated the coal-fueled electric power, Edison's Pearl Street
Station transmitting 100 volts of direct-current electricity underground to
400 lights in a one-square-mile area of Lower Manhattan's First District
(bounded by Wall Street, Nassau Street, Spruce Street, and Ferry Street).
The area included the offices of *The New York Times*, which the
following day noted that "The light was soft, mellow and grateful to the
eye, and it seemed almost like writing by daylight to have a light without
a particle of flicker."[172]

London's Savoy Theatre was the first public building to be lit by its own
internally generated electric power system, displayed at a performance of
Gilbert and Sullivan's *Patience*, while Macy's in Midtown Manhattan was
the world's first electrified department store.[173] But with Edison's novel distribution system, *externally* transmitted electricity became all the rage – light
powered offsite from a *remote* power station.

Life changed utterly with Edison's invention as on-demand electricity created a new reality. Within six months Edison and Morgan's company sold more
than 330 isolated plants[174] and by 1888 had built 121 centrally operated power
stations across the United States. By 1902, there were over 3,500 central plants
and 50,000 one-offs that powered a factory or single home.[175] Gas (and arc)
lighting would ably try to compete with the wonders of incandescent lighting,
offering lower prices and improved technology without much success, before
switching altogether to cooking and heating.[176] The safe, clean, and aesthetic-
ally pleasing appeal of Edison's new light would wire the world with electric
power, invisibly generated from afar.

There were still wrinkles to iron out. Although Edison had created new ways
to transmit more electricity, such as devising a "feeder" and "mains" system, he
couldn't maintain sufficient power for more than about a mile because of large
transmission line losses, at least without laying an excessive amount of thick,
expensive, copper wire. The "try again" T.A. he wrote beside each of his failed
filament materials may have spurred him on to develop electric lighting, but
would fail him in his battle with the ever-pragmatic George Westinghouse, who
used AC to deliver more electricity at a higher voltage instead of having to build
numerous local dynamos as in Edison's DC system. Loath to admit his greatest
triumph had been one-upped by his fiercest competitor, Edison's transmission
method was better suited for smaller local installations and one-off sites,
especially in wealthier urban areas.

Although the engineering of electrical power was being developed on the fly, direct current is okay at a high voltage (V) because the current (I) remains low and thus less power (P = IV) is lost during transmission, but increasing a DC voltage is harder and more expensive than increasing an AC voltage. The key is the transformer (which we'll look at in Chapter 5), allowing the same amount of power to be transmitted farther at a high voltage (e.g., 60 kV) before being reduced back to a safe level for household use at the other end (e.g., 100 V).

Alternating current would eventually supplant direct current to provide more efficient, long-distance power transmission, although the brilliant but pig-headed Edison would persist with DC. His tawdry attempts to dissuade the world of AC included electrocuting elephants and other poor animals, burned to death by AC "electromort." Never one to be shackled by the truth, high-voltage direct current is just as lethal as alternating current.

***

Today, we use the same principles of electrical generation as in Edison's first DC power station: heat is converted to mechanical energy by burning fuel in a steam engine that is converted to electricity via an induction generator. One more innovation, however, was needed to smooth out the action of the cumbersome, up-and-down mechanical piston to make large-scale electric power feasible – the turbine, a device to convert fluid flow into mechanical energy, where the fluid can either be fast-moving steam or rushing water.

In 1884, Charles Parson, the youngest son of a celebrated Irish astronomer, designed a *rotating* steam-powered turbine for the Holborn Street dynamo in London, creating more energy with the same amount of steam by passing it through a modified set of windmill-style fans instead of jerkily moving a piston up and down. Parson's insight was to make the fan blades as small as possible in a compound turbine assembly – two rotors surrounding a middle stator – that rotated the assembly faster. The power increased as the revolutions per minute on his new turbine rose from 500 to 4,800.[177]

Niagara Falls was the game changer, falling water diverted from Lake Erie on its way to Lake Ontario to turn a turbine that generated alternating current in an AC induction generator, created by another electrical genius (as we will see in Chapter 5). The electricity was first used to power industries in the adjacent town of Niagara Falls, New York, before being transmitted 22 miles to the nearby city of Buffalo without any significant power losses and thereafter incorporated into other types of power stations whatever the fuel: coal, oil, natural gas, or water. Ironically, so much power was gobbled up by the industries set up along the banks of the Falls – "yoked to the cataract" – that they could have been powered by DC.

With electric power safely generated outside the city and transmitted on demand to any hooked-up industrial or residential site, the electrification of the world began for both lighting and power. Highlighting the original lighting application, other electrical appliances were initially screwed into a light socket before two- and then three-prong wall outlets were designed to connect any type of electrical device. The so-called "War of the Currents" would be won by Westinghouse and AC, Edison bought out by his own company in the aftermath. There would be no holding the great Wizard back, however, as he moved on to more inventions, including the Kinetoscope and moving pictures, while his Edison General Electric would merge with Thomson-Houston, an early maker of arc lighting, to become General Electric, one of the world's largest-ever companies, disparagingly called by some the "Electric Trust."

All fuel-burning power stations today are variations of Edison and Parson's creation – boiler, turbine, generator, transmission, wall outlet – the thermal efficiency increasing from 2.5% in Edison's Pearl Street Station to as much as 40% by the 1960s with improvements in turbine materials and increased steam temperatures and pressures.[178] Today, the 4-GW Drax power station, the largest power plant in the UK, generates enough power to satisfy the electrical needs of six million people. There is seemingly no end to the might of fire and steam, and its applied power at a distance, electricity. (Falling water also produces mechanical energy in a wheel or turbine to generate electric power directly as we will see later.)

Prior to the Industrial Revolution, science was about nature as found, and afterwards about what nature could become, using energy to transform the natural world instead of humbly accepting its offerings. In *The Ascent of Man*, Jacob Bronowski noted that "Energy had become the central concept in science, and its main concern was the unity of nature of which energy is the core."[179] With mechanized power, energy becomes something for everyone to enjoy, not just those with sufficient means, allowing more people to partake of its technological fruits – tea pots, durable fabrics, a trip to the sea on a train, and, in time, electricity, central heating, and air conditioning. The electric bonanza continues into the twenty-first century with our own modern gadgets and household appliances as we continually step into an ever-changing future. Building order from chaos, the sublime achievement of the Industrial Revolution.

We are in a similar place today as we continue to reap the benefits of the Industrial Revolution while seeking to remove the ill effects. Just as with the quest for new and better ways to produce power, we must apply our own best efforts to improve the energy equation, using only materials and methods that generate an overall net benefit. Innovation creates systematic change to everyday life, offering opportunities to anyone who can turn invention into industry as well as benefitting those who simply want to enjoy the fruits of progress without delving into the details.

Steam-driven road carriages would be the next big thing to follow from industrialization and would start to appear as commercial "automobiles" by the turn of the twentieth century. Alas, the next big thing was a chaotic contraption. It is one thing to run a mammoth train on a purpose-built railway or a giant ship on the high seas, ostensibly safe from accident, or a dynamo at a remote location to deliver electric power to the masses, but another to use a large and smoke-spewing steam engine on the open road. A new paradigm was needed, like when we changed from wood to coal. The change didn't have long to wait.

The great discoveries of the past 250 years – the steam engine, electromagnetic induction, the electric power grid, the internal combustion engine, the transistor, personal computers, the Internet – change not just the way we live, but an entire global economy. Nothing, however, would create more change or make more millionaires than one discovery. By the early 1900s, the iron carriage had made its appearance on the streets of our booming cities, but a new kind of engine and a new kind of fuel would be needed to make a "gasmobile" run. Oil.

2

Oil and Gas: Twentieth-Century Prosperity

2.1 The New Black Gold

Throughout history, oil has been used to embalm the dead, caulk ships, fasten jewelry, light lamps, and treat various ailments. Marco Polo observed camels being treated for mange by the Caspian Sea on his travels through Persia to the Orient. Christopher Columbus began caulking his ships with Trinidad pitch (bitumen) after he landed there on his third voyage in 1498,[1] while the Holy Roman Emperor Charles V used imported Venezuelan oil to treat gout.[2] The Native American Seneca tribe traded oil found in shallow pits and on the surface of lakes – collected by blankets and squeezed out – to be used as "salve, mosquito repellent, purge and tonic, as well as wigwam waterproofing, body paint, and for various religious practices."[3] Not until such "crude" oil was refined, however, was its modern use made apparent – first for kerosene lighting and then as fuel for a new kind of engine.

On August 27, 1859, in Titusville, Pennsylvania, a town of little more than 100 people, a 40-year-old former railroad conductor and jack-of-all-trades turned oil prospector Edwin Drake succeeded in tapping into an underground oil deposit beneath the bedrock, initiating the world's first commercial oil well. Armed with a favorable analysis of nearby "seep" oil by a Yale chemist[4] and the help of a blacksmith cum salt-well driller from nearby Tarentum known as Uncle Billy, "Colonel" Drake – the title added to provide credibility to the "mad" venture – had been drilling for months, almost giving up from lack of funds until oil was eventually found at 69.5 feet (21.2 m), brought to the surface in a hand-fashioned pail.[5] Of such determination is the modern world made.

Seepage onto adjacent farms from the banks of Oil Creek had brought Drake and his bosses in the first American oil company – reformed from the Pennsylvania Rock Oil Company as Seneca Oil (later referred to as "snake oil") – to northwest Pennsylvania in search of new sources for a burgeoning lamp-fuel industry. The newly discovered "rock oil" was little different than the oil found in nearby salt wells, seeps, or bubbling up on lakes and rivers, variably used as a machine lubricant, sawmill grease, healing ointment (known locally as Mustang Liniment), or distilled into lamp oil, but was more plentiful, seemingly present in vast subterranean fields.

The rock oil would decimate the once mighty whale-oil industry that had lit the night lamps of homes for centuries. President Thomas Jefferson noted that whale fishery on the whole was difficult for both merchants and sailors, but that sperm oil was unique – luminous, odorless, and didn't easily freeze as temperatures dropped.[6] In his epic 1851 tale *Moby-Dick*, Herman Melville called sperm oil "the sweetest of all oils," which filled "almost all the tapers, lamps, and candles that burn round the globe," noting that a "well reaped" whale harvest was caught mostly by American whalers, annually netting as much as 7 million dollars from a navy of more than 700 ships manned by 18,000 men.[7] Each sperm whale produced on average almost 2,000 gallons of oil (about 60 barrels), rendered onboard from the boiled blubber and spermaceti-filled head, delivering "a quality oil, which brought a higher price than the oil from any other species."[8]

Hunting off the coast of Long Island and Nantucket since the 1690s, the American whaling industry had expanded quickly as adventurous whalers sought oil in far-flung regions around the globe, sailing as far as the Arctic and Japan in trips lasting up to four years. The "strike rate" was less than one hit for every 10 throws, until explosive-head harpoon guns improved their catch. By its peak in 1846, during the "golden age" after the War of 1812, the whaling industry employed 70,000 people and serviced 735 ships. As James Coleman notes in a history of the American whaling industry, "After one four-year voyage, the *Charles W. Morgan* returned with 1,150 barrels, grossing $44,138.75 for the trip. This equated to $38/barrel from 19 whales."[9]

Coleman also noted that New England was the port of call of eight of every 10 whaling ships, providing a rich income to owners and captains, although "sailors earned only a scant living,"[10] often returning in hawk to the company store or purposely left ashore before the end of a journey to avoid payment. With the advent of industrially distilled kerosene, followed by electric light, the need for whale oil eventually declined after a few disastrous seasons. By 1890, decades after the first oil wells had begun to produce cheap and plentiful petroleum, only 200 whaling vessels remained.[11] The largest recorded whale-oil haul on a four-year voyage was equaled by the Drake well in only 82 days.[12]

Early "drilling" was more bashing than drilling, where a heavy cast-iron bit tied to a rope or a length of pipe was repeatedly winched up by hand or steam power and dropped to break through the rock and shale below, the drilling speed depending on the underlying strata. In the Drake well, 10-foot lengths of 3-inch-diameter, half-inch-thick pipe were inserted to keep the drill hole from collapsing or filling with water.[13] Today, in a not dissimilar procedure, where modern rigs are assembled like giant erector sets, the time to drill one foot is called the "penetration," although a rotary bit is now employed rather than a percussion weight. Acknowledging the laborious process, the name for petroleum is derived from the Latin for both rock "petra" and oil "oleum."

The amount of oil extracted from Drake's well averaged about 10–20 barrels a day in the first year, initially brought up in an improvised pail and then with the aid of a hand-operated pitcher pump and stored in available whiskey barrels, hence the modern measure. Petroleum geologist and author Parke Dickey noted that, "Oil at this time was worth about $1.25/gal, and the production of several hundred gallons per day impressed everyone with this astonishing new source of wealth."[14] Oil soon began to flow in abundance from underground fields around Oil Creek, including Drake's second well in 1860 at a depth of 480 feet (146 m) that produced 24 barrels per day.[15] By the end of the year, 74 wells produced 1,165 barrels per day. (Note, a standard oil barrel contains 42 US gallons.)

In 1861, Henry Rouse struck the first "gusher" at the 320-foot (98-m) deep Little and Merrick well, north of the newly founded Oil City, the underground gas pressure sufficient to raise an oil "fountain" as high as 40 feet in the air.[16] More than 3,000 barrels per day gushed out, although the oil unfortunately caught fire, killing 19 people including Rouse, who miraculously dictated his will onsite before he died. The fire spread to other nearby wells and took 70 hours to extinguish "by smothering the fire with dirt and manure."[17]

Without any blowout-preventer system, oil fires were common in early drilling, as highly flammable, associated gas is ignited in the large reserves. An oil fire can cause enormous damage as traditional methods fail to extinguish the flames, requiring a well-placed detonation to evacuate the oxygen. In his 1927 novel, *Oil!*, Upton Sinclair wrote:

> There was a tower of flame and the most amazing spectacle – the burning oil would hit the ground, and bounce up, and explode, and leap again and fall again, and great red masses of flame would unfold, and burst, and yield black masses of smoke, and these in turn red. Mountains of smoke rose to the sky, and mountains of flame came seething down to the earth; every jet that struck the ground turned into a volcano, and rose again, higher than before; the whole mass, boiling and bursting, became a river of fire, a lava flood that went streaming down the valley, turning everything it touched into flame, then swallowing it up and hiding the flames in a cloud of smoke.[18]

Figure 2.1 The 1860s oil rush in Venango County, Pennsylvania.

The dangers did not deter the hopeful masses arriving daily in search of fortune. More gushers were found on the banks of Oil Creek and along the Allegheny River and its tributaries, bringing workers in droves to western Pennsylvania, creating the first-ever "black-gold" oil rush (Figure 2.1). Oil was easy to extract, seemingly limitless, and by the 1870s cost less than drinking water. The barrels were initially loaded onto river barges and transported south to Pittsburgh, as much as two-thirds lost along the way from leakage, before a rail link was established in Titusville and a pipeline built from the well heads to the rail terminus.[19] At its peak, the fields around Oil City produced half the world's oil.[20]

While various oils from antiquity had been used as lamp fuel, the distillation of kerosene from the vast new "crude" oil reserves would create an entirely different market for illumination. Prior to Drake's first well, Samuel Kier operated a one-barrel, cast-iron kettle still in Pittsburgh, distilling both salt-well oil and seep oil into lamp fuel by traditional whiskey-still methods – essentially retort heating to produce vapor condensed out to a liquid. Having started in the family salt business in Tarentum on the Allegheny River north of Pittsburgh, Kier had originally discarded the oil that seeped into the salt wells, but when his wife took ill one day, he hit on the idea of selling the unwanted oil as a cure-all in half-pint bottles at 50 cents for "internal and external application, treating rheumatism, gout, neuralgia, coughs, sprains, bruises, and many other conditions."[21] In one brochure, Kier cited the medicinal powers of Rock Oil and cautioned against other "spurious" sellers:

A remedy of wonderful efficacy / Lame would walk, blind could see / Cure for rheumatism, gout, neuralgia / Put up as it flows from the bosom of the earth, without anything added or taken from it

At the peak of his cure-all business, Kier employed 50 traveling agents plying their trade in colorful Medicine Show wagons, although he would soon return to refining the increased output of Allegheny crude drilled from Drake's and other Venango County wells, barged from Oil City to Pittsburgh. After adding a five-barrel still, he began the first commercial oil refinery in the United States,[22] selling the refined crude oil (called camphene or "carbon oil") as an illuminant for a new lamp design, comprised of a four-pronged wick holder to remove smoke and odor. By double distillation, Kier produced a much-improved lighter liquid at "about six barrels of distillate per day, which sold for $1.50 per gallon."[23]

Although the English chemist James Young was the first to practically manufacture illuminating oil from minerals – distilling gas from heated coal and paraffin from oil shale – the Canadian geologist Abraham Gesner was the first to turn petroleum into kerosene, distilling asphalt from a 16-foot-thick vein of vertical bitumen in Albert County, New Brunswick.[24] Patenting the process in 1849 in his native country, Gesner called his new product "keroselain" from *keros* and *elaion*, Greek for wax and oil, later shortened to "kerosene."

Originally distilled from New Brunswick oil shale (known as cannel coal) and then boghead coal from Bathgate, Scotland, Gesner reported that his kerosene burned "with a brilliant white light without smoke or the naphthalous odor so offensive in so many hydrocarbons," and was 13 times brighter than burning lard, wood-pulp oil, or common whale oil, six times sperm oil, four times gaslight, and 2.5 times rapeseed or camphene (distilled pine tree resin or turpentine).[25] Efficiently and safely converting crude oil into usable lamp fuel would ultimately set the oil business on a path to riches.

Moving to the United States and acquiring an American patent in 1854 (which would be argued for years), Gesner's Kerosene Oil Company produced 5,000 gallons per day by 1859, burned in almost every US home before more competitors appeared, increasing the daily American consumption to 30,000 gallons, essentially ending the centuries-old, whale-oil business.[26] Petroleum-derived kerosene soon bested all other illuminants such as grain alcohol, whale oil, camphene, and shale-oil kerosene, in part because of excessive government taxes during the Civil War. As historian Richard Rhodes notes:

> Not for the last time, a petroleum product – kerosene – rode into the marketplace on a government subsidy – exclusion from a punitive tax – and crowded out other fuels. By 1870, camphene and alcohol had all but disappeared from the market, while petroleum-derived kerosene sales had reached 200 million gallons annually.[27]

Charles Lockhart and John Gracie improved the distillation process, described in their 1863 patent *Improvement in Stills for Petroleum, etc.*, which "relates, first, to a means for taking off vapor from the still at the same height from the surface of the oil and at different heights during the process of distillation; second, to a means for keeping the bottom of the still clean or free from incrustation." Distilling lighter gases from crude oil produced usable lamp fuel, but left a heavy residue, at the time a discarded nuisance.

We know today that crude oil is a mixture of a number of hydrocarbons (labeled C_1–C_{70+}), separated by heating, where smaller-molecule hydrocarbons with lower boiling points evaporate more quickly than larger ones, and are condensed out in turn into their various component parts, so-called "fractional distillation." From lightest to heaviest the refined hydrocarbons consist of petroleum gases (C_1–C_4), naphtha (chemicals), gasoline, kerosene, diesel, lubricants, waxes, fuel oil, and finally asphalt ($>C_{70}$).

A barrel of crude oil contains 42 US gallons (159 liters), producing on average about 19 gallons of gasoline (45%), 13 gallons of diesel (30%), four gallons of jet fuel (10%), two gallons of liquefied petroleum gases (5%), one gallon of heavy fuel oil, one gallon of heating oil, and various amounts of other petroleum products. The percentages vary widely for different crudes, depending on how long the "kerogen" was heated and pressurized in the ground. For example, Nigeria's Bonny Light contains more gasoline and less asphalt, while Venezuela's Orinoco Heavy has more diesel and fuel oil. Pennsylvania oil is a foul-smelling, wax-based paraffin crude, 20% of which is convertible to gasoline.

Today, we use kerosene for heating oil, jet fuel, and a solvent for insecticide sprays, as well as in first-stage rocket propellant, called RP-1, which utilizes an explosive mix of kerosene and liquid oxygen for short-term thrust in the high-gravity, dense-air lower atmosphere. Improved refining methods would ultimately exploit the rest of the crude, including the leftover C_{70+} tar residue.

An idealized distillation process is shown in Figure 2.2, where the lighter, low-boiling-point distillates are condensed at greater heights in the distillation tower, beside the average breakdown of one barrel of crude by percentage. The refinery yield is determined by the US Energy Information Administration (EIA) for petroleum and other liquids. The largest percentage is gasoline at 45%, followed by diesel (distillate fuel oil) at 30%, and kerosene-type jet fuel at 10%. Interestingly, crude oil on its own is of little use, but when refined into parts becomes the basis for numerous everyday products, most importantly gasoline, diesel, and jet fuel, but also propane, butane, motor oil, naphtha,[28] plastics, wax, and asphalt.

With an increased need for refined petroleum products, more refineries began to appear in and around the oil regions of western Pennsylvania. Having started ferrying oil from Oil City to Samuel Kier's Pittsburgh distillery,

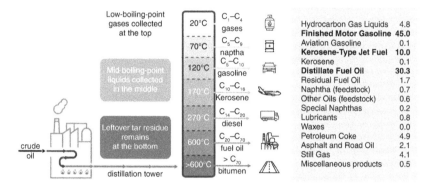

Figure 2.2 (a) Fractional distillation process and (b) EIA percentage refinery yield (*source*: "Refinery Yield" (January 2023), US Energy Information Agency. www .eia.gov/dnav/pet/pet_pnp_pct_dc_nus_pct_m.htm).

Charles Lockhart began his own refinery, the Brilliant, producing up to 250 barrels of kerosene per day: "The technology used at Brilliant involved distillation, probably followed by treatment with caustic soda, sulfuric acid and finally water washing."[29] Lockhart eventually joined up with a Cleveland merchant turned refiner, John D. Rockefeller, merging eight Pennsylvania refineries (Brilliant, National, Nonpareil, Standard, Lily, Crystal, and Model in Pittsburgh and Atlantic in Philadelphia) into Standard Oil, kick-starting the American oil industry. Prior to the advent of gasoline, kerosene was its major product. By 1875, the Standard Oil Company would control almost 90% of American crude-oil refining, transportation, and sales, 16 years after Drake had first discovered oil at Titusville.

In the highly productive although uncertain early exploration years, boom–bust cycles followed the ups and downs of supply and demand, where prior to proper scientific surveying "creekologists" sought oil by sight, smell, and sound, while "practical oil men" used their experience and "rule-of-thumb" methods, ready to "discourse learnedly on rocks, formations, strata, shales, sandstones,"[30] ever suspicious of professional geologists well-versed in the academic language of rocks and stratification. However the oil was got, each large new find further lowered prices as supply swamped demand – shades of modern gluts and shocks.

In Pennsylvania, for example, output decreased slightly in 1862 but increased annually thereafter for nine years. Production leveled off again in 1873 but soon began climbing and peaked at thirty million barrels annually in 1882. ... Continued production swamped existing transport and refining facilities and prompted businessmen to begin thinking about organizing their industry more efficiently.[31]

Amid the chaos of the early oil industry, one man sought to make sense of the confusion. Having formed the Standard Oil Company, Inc., of Ohio in 1870, John D. Rockefeller set out to "standardize" the various kerosene grades for safe use. In time, as refining methods improved, petroleum would become indispensable and much in demand for a new kind of engine fuel: gasoline. By the time John D. – himself the son of a former wandering snake-oil salesman – left his mark on the business, Standard Oil would be the world's largest oil company, and Rockefeller its first billionaire (as we will see next).

In Paul Thomas Anderson's film *There Will Be Blood*, based on Upton Sinclair's 1927 novel *Oil!*, the traveling oil prospector Daniel Plainview cum millionaire oil tycoon (played by Daniel Day-Lewis) explains the importance of securing the rights to underground resources, exclaiming to his pastor nemesis Eli, "If you have a milkshake and I have a milkshake and I have a straw and my straw reaches across the room starts to drink your milkshake, I drink your milkshake! I drink it up." Such was the way of the early oil business.

To the victors go the spoils, which by the advent of the internal combustion engine at the start of the twentieth century would encompass the entire globe. Oil created our modern world, and a global economy. New reserves were sought the world over, making oil not only the most valuable commodity ever, but its own currency. From humble beginnings, pulled out of a shallow well in a makeshift pail by a make-believe colonel drilled near a known seep on a small farm in northern Pennsylvania, oil would eventually come to run and rule the world.

2.2 The Standard Runs the World

As part of a 2014 United Nations climate change summit in New York, it was announced that the $860 million Rockefeller Brothers Fund would divest itself of all fossil-fuel holdings, signaling an end to Standard Oil billionaire John D. Rockefeller's vast petroleum legacy. No change in plans was announced for the companies created from Standard Oil, such as ExxonMobil, the world's largest publicly traded oil and gas company, although Steven Rockefeller (the son of John D.'s grandson Nelson A. Rockefeller) stated there was "a moral and economic dimension" to stockpiling climate change fuels. Fund chairperson Valerie Wayne (the granddaughter of John D.'s grandson

John D. Rockefeller III) also claimed they wanted "to use the fund to advance environmental issues."[32]

All a far cry from how *McClure's* magazine writer Ida Tarbell – who grew up in Venango County and knew first-hand how "tar and oil stained everything" – portrayed Rockefeller Sr.'s intentions in a series of "muckraking" articles from 1902 to 1904. In her scathing accounts that helped lead to the 1911 court-ordered break-up of the largest monopoly in history, Tarbell noted, "It seemed to be an intellectual necessity for him to be able to direct the course of any particular gallon of oil from the moment it gushed from the earth until it went into the lamp of a housewife."[33]

Born in 1839 to an impecunious traveling "snake-oil" salesman, John D. moved to Cleveland, Ohio, with his family at the age of 14, his bigamist father leaving for good soon after. Starting his professional life as a bookkeeper, the young entrepreneur became a wholesale produce merchant at the age of 20, forming his first company with Englishman Maurice Clark, 12 years his senior. Taking advantage of the Great Lakes boat trade and then furnishing army supplies to the Union during the Civil War, which John D. evaded by paying a proxy fighter, their small firm saw a golden opportunity to make good in petroleum, overflowing in the nascent oil regions between Cleveland and Pittsburgh.

By 1862, the kerosene industry was fast expanding, Cleveland competing with Pittsburgh as the center of crude-oil refining, ideally situated near the oil fields of western Pennsylvania and Lake Erie with easy access to the Erie Canal and two railway trunk lines to New York (the Erie and the Central). Rockefeller and Clark purchased their first Cleveland oil refinery in the Flats, adjacent to a railway, and with the help of another Englishman, Samuel Andrews, who would develop the process of fractional distillation, set out to improve the refining process and expand their business. As Tarbell noted, Rockefeller thrived as the financial brains of the operation:

> In the new firm Andrews attended to the manufacturing. The pushing of the business, the buying and the selling, fell to Rockefeller. From the start his effect was tremendous. He had the frugal man's hatred of waste and disorder, of middlemen and unnecessary manipulation, and he began a vigorous elimination of these from his business. The residuum that other refineries let run into the ground, he sold. Old iron found its way to the junk shop. He bought his oil directly from the wells. He made his own barrels. He watched and saved and contrived.[34]

Towards the end of the Civil War, Rockefeller sold his interests in his produce commission business to focus entirely on oil. Struggling at first and almost going bankrupt, the young oil merchant offered something he didn't yet have to the shipping tycoon Cornelius Vanderbilt: 60 trains full of oil to run daily on Vanderbilt's Central Line from Cleveland to New York, even demanding a rebate for doing so. Hoping to see business boom with a sweetheart deal, Rockefeller set out to corner the kerosene market.

Under the guise of the South Improvement Company, a series of shrewd deals enabled Rockefeller to gain favorable access to railway lines in and out of the oil regions, facilitated by the railway companies giving large rebates to high-volume shippers such as the newly formed Standard Oil, while raising the rates on smaller competitors. Claims of conspiracy echoed from Titusville to the state and federal capitals, a relatively unknown "John Rockefeller" listed as one of the seven directors of the secretive but far-reaching Southern Improvement Company, dubbed "The Anaconda" by outraged oil workers.

Breaking the backs of his competition through preferential transportation schemes, Rockefeller next organized the takeover of other refineries, a.k.a. the Cleveland Massacre, what Tarbell called "as dazzling an achievement as it was a hateful one."[35] In six weeks, Standard took over 22 of 26 refineries, positioning itself to dictate the terms of the entire oil trade. Daily shipping capacity rose from 1,500 to 10,000 barrels,[36] destroying the livelihood of many independent operators.

By 1872, the Cleveland refineries were annually refining more oil than anywhere else in the country, one refiner stating "he was informed by Rockefeller, of the Standard, that if he would not sell out he should be crushed out."[37] In a damning *Atlantic Monthly* exposé published a decade later, Henry Demarest Lloyd wrote about Standard's cut-throat practices after the takeovers:

> The great majority of these refineries, when bought by the Standard, were dismantled and the "junk" was hauled to other refineries. The Vesta and Cosmos refineries, which cost about $800,000, were sold at sheriff's sale to the Standard for $80,000, and are now run vigorously by that company. The Germania, which was run to its full capacity as long as the Pennsylvania Railroad gave its proprietor transportation, is now leased to the Standard, but stands idle, as that concern can make more money by limiting the production and maintaining an artificial price than by giving the people cheap light. The Standard became practically the only refiner of oil in Western Pennsylvania, and its rule was bankruptcy to all attempting to lead an independent existence.[38]

Having garnered control of the refineries from Pittsburgh to Cleveland, Rockefeller next played the railways against each other (the Erie, Central, and Pennsylvania), demanding and receiving continued favorable rates even after the disclosure of the South Improvement Company and various legal proceedings to limit Standard's control, which as always were easily dodged. In the early years of the oil industry, control of transportation was essential, shady dealings regularly made with the railway companies, often with the approval of pliant politicians. Ron Chernow, a modern Rockefeller biographer, noted that "John D. and his colleagues regarded government regulators as nuisances to be bypassed wherever possible."[39] In the early battle between private enterprise and public good, it was clear that businessmen not politicians ran the country.

When the railways tried to fight back, Rockefeller had a "pipe-line" network built, laying 1.5 miles of 3-inch pipe per day to circumvent their influence and that of other proposed distribution networks, such as the competing Tidewater Pipeline to Philadelphia, which was eventually bought out. Not subject to the same laws and regulations as the railway lines, Standard's new pipeline network devastated the railway industry, one-third of 360 railroad companies going bankrupt, precipitating a steep decline in stock prices and leading to the Panic of 1873 and the first great American depression, the longest on record with 65 months of contraction.[40]

Price fixing was common, unions brutally suppressed, and politicians openly corrupt.[41] Lloyd noted that "the money of the Standard was more powerful than the petition of business men who asked only for a fair chance."[42] Millions across the country lost jobs, while unemployment peaked in 1878 at 14%. The stranglehold was so complete that Lloyd even hoped a recent find in the Hanover petroleum district of Germany would redress the madness: "German oil wells, German refineries, and the Canadian canals may yet give the people of the interior of this continent what the American Standard and the American railroads have denied them – cheap light."[43]

After organizing the oil refineries and establishing a fixed mode of transportation, Rockefeller then went after the oil markets. With its ruthless business practices, trust companies devised to navigate thorny interstate commerce laws, and a vast network of agents, Standard Oil would eventually control 90% of the entire oil business – from well to pipe to wagon to home.

Considered one of the first investigative journalists, Ida Tarbell pulled no punches about Standard's involvement in politics to solicit favorable legislation and its "sordid methods of securing confidential information,"[44] writing in 1905 "There was no doubt that the investigation of 1876 and the first bill to regulate interstate commerce introduced at that time had been squelched largely through the efforts of two members of Congress, one of them directly and the other indirectly interested in the Standard."[45] It was also claimed that the Ohio 1884 senate seat of Henry Payne was bought by Standard money. When asked by the New York Senate in 1888 if he was part of a trust or if it enjoyed favorable freight rates, Rockefeller himself would lie under oath as would Vanderbilt about Payne's relationship with Standard.

Lloyd's 1881 article was also damning in its every detail: "There was apparently no trick the Standard would not play,"[46] he wrote. Bad deeds included price fixing, barrel famines, and selling below cost to kill off smaller interests or force them to join up, as well as refusing transportation that damaged communities, wells, and the environment as unmarketable oil was casually discarded into rivers, while thousands of ordinary workers lost their jobs. Contrary to the pioneering American myth, it was not what you knew or the sweat of your brow that made the man, but who you knew and what infrastructure they controlled, while the environment was routinely ignored. Standard's unrelenting quest for more had no bounds.

By the time Rockefeller was done, he had bested the independent suppliers in the oil regions around Titusville, the great shipping magnate Cornelius Vanderbilt (pioneer of the hostile takeover), refinery owners in Cleveland, Pittsburgh, and beyond, Jay Gould and Tom Scott of the Erie and Pennsylvania railroads, the Producer's Protective Association, as well as numerous state and federal officials, having paid off more than a few politicians to kill off restrictive legislation.

Rockefeller was the embodiment of a new America that had finally matured after more than a century of floundering in the wilderness since independence and the ravages of civil war – single-minded, relentless, and unapologetic, seeing strength only in what one owned or could control. His methods were merciless, "forcing companies to sell out or to join with him," while disrupting his competitors' operations with "strong-arm squads as well as continuous threats of economic destruction."[47] Rockefeller had learned well the lessons of his impoverished youth – "let the money be my slave and not make myself a slave to money."

In 1882, Rockefeller doubled down on his control over the business, forming Standard Oil Trust, a "corporation of corporations." The first of its kind, the Trust was centrally organized, yet comglomerated 40 smaller enterprises solely to circumvent state laws, its Agreement kept secret for six years. Dissolved in 1892 after much pressure and the passing of the 1890 Sherman Antitrust Act, Standard's legendary lawyer Samuel Dodd famously created the "holding company" in response, an English common-law entity designed to evade laws against combinations and maintain a monopoly. With numerous presumably independent Standards, the main Standard, headquartered at 26 Broadway in New York, could continue its ruthlessness unabated.

Rockefeller would claim he was only protecting his company against speculators, while trying to control prices that had dipped to 15 cents per gallon in 1875. Questioned under oath, however, one Standard Oil director stated the real objective of the Trust was "Simply to hold up the price of oil – to get all we can for it."[48] Indeed, between 1875 and 1904, Standard Oil controlled anywhere from 80% to 95% of the refining business, an economic grip Tarbell noted was "so complete that the price of oil, both crude and refined, is actually issued from its headquarters!"[49] The supposed free market of capitalism was in ruins.

Rockefeller saw oil as his own personal fiefdom and his way of business the only way. Like Vanderbilt in shipping and Carnegie in steel, Rockefeller demanded efficient administration, which to him meant establishing a monopoly to eliminate "wasteful conditions of competition." In the once ruggedly individualistic United States, independence was dead and a new corporate America had its first king. The real purpose of Standard Oil was to regulate the price of oil, vertically controlling all facets of operation, with a de-facto veto over any restrictions to its reach and a smug disregard for the rights of others.

Whether collusion among presumed competitors or closely supervised best practice down to the smallest bung – ever scrutinized by John D. himself – the production, refining, and distribution of oil expanded enormously, providing households in the United States and around the world with clean, safe, and affordable lighting fuel. Soon, kerosene for home use – what John D. called the "poor man's light" – was exported to the four corners of the globe under the Standard flag. Exports to Europe and the Far East increased from 80 million gallons in 1868 to over 400 million gallons by 1879.[50] In 1880, Americans consumed 220 million gallons for a population of 50 million.[51] Only cotton was more valuable. As Lloyd noted, Americans used "more kerosene lamps than Bibles."[52]

***

Standard's days of total control, however, were numbered as public opinion about monopolies started to change, in part due to Tarbell's reporting as well as populist politicians such as Theodore Roosevelt, who wanted to reign in the unlawful antics of large companies. In 1911, the United States Supreme Court found that Standard Oil had violated the Sherman Antitrust Act of 1890, which stated that "Every contract, combination in the form of trust or otherwise, or conspiracy, in restraint of trade or commerce among the several States, or with foreign nations, is declared to be illegal." In 1914, the Sherman Antitrust Act was amended to the Clayton Antitrust Act to include "unfair competition."

The Standard was broken up into 34 "Baby Standards," the largest including Standard Oil of New Jersey (renamed Esso as in "SO" and then Exxon), Standard Oil of New York (renamed Mobil), Standard Oil of California (renamed Socal and then Chevron), and Standard Oil of Indiana (renamed Amoco). Highlighting the friendly, good-old-boys nature of a presumed competitive business, the dismantling has reversed of late, in part to increase the production base in uncertain times. In 1960, Chevron acquired Kentucky Standard (joined with Gulf in 1984 and Texaco in 2000), BP acquired Standard Oil of Ohio in 1968 and Amoco in 1998, while Exxon and Mobil remerged in 1998 as ExxonMobil, the then largest merger in history at $74 billion. Today, talks are ongoing about an ExxonMobil and Chevron merger.

At the time of Standard's dissolution in 1911, the Baby Standards were worth $600 million, almost quintupling to $2.9 billion a decade later. Much of the growth was due to an increase in oil production prior to the Great Depression, the petroleum market exploding with the introduction of the internal combustion engine (using gasoline, a hitherto wasted crude oil portion), increased demand for home heating oil, and the building of oil-fired electrical power plants (604,000 barrels per day in 1911 to 2,760,000 barrels per day in 1929[53]).

Oil and all of its components were in ever more greater demand. Fueled by the previously unused gasoline portion, automobile sales increased from almost 17,000 in 1914 to nearly 1.7 million by 1919.[54] The introduction of the thermostat in the 1920s created on-demand home heating at easily controlled temperatures, spurring on heating-oil sales, while more electric power-generating plants were constructed after the first Niagara Falls hydroelectric plant, mostly fueled by diesel oil to run factories, electricity grids, and elevated city railways. Instigated by President Eisenhower, the Federal Aid Highway Act of 1956 greatly expanded the need for asphalt, the thickest petroleum distillate in the complex refining process.

New methods of refining would turn out hundreds of crude-oil products, which would create and come to define the modern world: propane cooking stoves, butane lighters, pen ink, vinyl records, plastic bags, shingles, asphalt, even pharmaceuticals and chemotherapy. Propane (C_3H_8) and butane (C_4H_{10}) are the two liquefied petroleum gases (LPG) distilled from crude oil at the lowest boiling points. Asphalt (C_{70+}) is the leftover part of the distillation residue after boiling out all the other lighter components. Today, the largest oil refinery in the United States is in Port Arthur, Texas, 14 miles southeast of the 1901 Spindletop gusher. Now owned by Saudi Aramco, the Motiva refinery can process over 600,000 barrels per day.[55]

The everyday uses of refined crude include acetylene (welding), propane and butane (cooking liquid, heating, and refrigeration), synthetic toluene (the second T in TNT), acetone (jet fuel antiknock), isopropyl alcohol (shampoo, perfume, cosmetics dehydrating agents, solvents, preservatives), plastics (furniture, milk cartons, food containers), golf balls, panty hose, lacquers, kerosene (aviation, tractor, and stove fuel), lubricating oil,[56] grease, gasoline (originally unwanted!), diesel fuel, boiler oil, solvents, wax, wax paper, candles, chewing gum, sealants, water proofing, transformer insulating oil, petroleum jelly, cosmetics (lipstick and eye liner), ointments, asphalt residuum (asphalt, shingles, roofing paper), and petroleum coke (almost pure carbon). The list is the story of the twentieth century.

Petroleum remade the world, our everyday lives unrecognizable from the early production days in western Pennsylvania, when half of all oil came from around Oil City. Although Pennsylvania now accounts for a paltry 0.2% of American production – 20,000 of roughly 10 million barrels a day – Rockefeller's legacy long remains. Today, 97% of US oil comes from seven states (Texas 42%, New Mexico 11%, North Dakota 10%, Alaska 4%, Colorado 4%, Oklahoma 3.5%, and California 3%) as well as offshore in the Gulf of Mexico (15%), as shown in Figure 2.3.[57]

In 1937, at the age of 97, the patriarch of one of the richest families in history died. His net worth at the time of his death was over $1 billion, not counting another half billion he gave away during his lifetime (the Rockefeller Foundation was the world's largest charity until the Ford Foundation). As the poster boy for the Gilded Age, Rockefeller created or endowed the University of Chicago, Spellman College for freed black slave women, the Museum of Modern Art, Radio City Music Hall, Lincoln Center, the Cloisters – European castles were dismantled after World War I and reconstructed in his New York City museum – and the Rockefeller Institute for Medical Research, although some organizations questioned the "moral issues" of accepting "tainted money."[58] Towards the end, he even saw to giving away dimes to strangers.

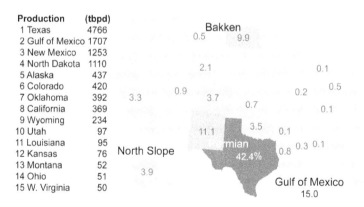

Figure 2.3 US oil production (thousand barrels per day and percent, 2021). *Source:* "Crude Oil Production," US Energy Information Agency, October 31, 2021. www.eia.gov/dnav/pet/pet_crd_crpdn_adc_mbblpd_a.htm. Note, Gulf of Mexico is Federal Offshore (PADD 3).

Some say Rockefeller was the fittest survivor in a Social Darwinist game of winners and losers, deemed right because of the size of his fortune. More than any other Gilded Age "fellow millionaire," many thought him a shrewd industry titan and a nation builder who helped pioneer a "vertically integrated" business model. Others remember Rockefeller as a ruthless businessman, one who cared little for competition or the environment, the embodiment of a new American win-at-all-cost ethos sanctioned by his Christian piety.

As the normally laconic Rockefeller himself remarked in an address at Brown University, where his son was a student, "The growth of large business is merely a survival of the fittest ... It is merely the working out of the law of nature and the law of God." His great grandson, West Virginia senator John D. Rockefeller IV, summed up his achievements with a modern addendum: "It's dazzlingly amazing that one individual was able to corner about 98% of all the kerosene and eventually oil production in the world. He was a brilliant business person. I can only give him credit for that. I don't have to give him credit for the way he became that powerful."[59]

At the height of his wealth, John D.'s financial worth was equal to 1.6% of American wealth, unprecedented and still unmatched today (see Table 2.1). But should one work backwards to anoint a victor if the winning was illegal? And who provides the infrastructure under which everyone can flourish or at least be given equal opportunity? Indeed, many have questioned the ethics of making money by destroying others while liberally availing of public resources. Perhaps by divesting their family wealth from fossil fuels, John D.'s descendants hope to redress past ills, adding a different sheen to the legacy of the world's first billionaire and progenitor of our oil-run world. Divestment won't repair the damage from burning petroleum without thought to the consequences, but can at least provide funds to help with the transition from fossil fuels to renewables.

Table 2.1 *Net worth of some of the richest men (*at time of death or highest percentage)*

Titan	Dates	Wealth ($billion)	US GDP ($billion)*	%
Cornelius Vanderbilt	1794–1877	0.105	9	1.2
Andrew Carnegie	1835–1919	0.475	77	0.6
John D. Rockefeller	1839–1937	1.500	92	1.6
Henry Ford	1863–1947	1.200	240	0.5
J. Paul Getty	1892–1976	2.000	1,800	0.1
Bill Gates	1955–	82.000	18,000 (2015)	0.5
Elon Musk	1971–	200.000	23,000 (2021)	0.9

2.3 The Engine Makers: Otto, Daimler, Maybach, Benz, and Diesel

New Yorkers made more than one million carriage trips in 1870, requiring over 150,000 horses. On average, each horse relieved itself of more than 20 pounds of manure a day,[60] totaling about 50,000 tons per month in droppings. Some of the manure was collected and sold on to farmers, turning a tidy profit for an unwanted waste product before industrial fertilizers, but as cities around the world grew, the manure kept piling up, raising concerns about disease and insurmountable dung piles at every turn, not to mention the urine. And then, as if in an answer to a modern prayer, city life changed with the rise of the electric tram and the automobile. The stench of rotting animal waste would be forever gone from our streets.

It is said that success has many parents while failure is an orphan. The internal combustion engine has both, including Swiss inventor François Isaac de Rivaz, who used a mixture of hydrogen and oxygen as the fuel in a rudimentary design; Belgian-French engineer Jean J. Lenoir, who built a double-spark, coal-gas engine; and three German engineers Nikolaus Otto, Gottlieb Daimler, and Wilhelm Maybach, who jointly developed a four-stroke engine that compressed coal gas prior to flamed ignition, forming the basis of today's liquid-fuel gasoline engine. To this list, we can add Thomas Newcomen, James Watt, Richard Trevithick, and George Stephenson, who all made important contributions to the earlier *external* combustion steam engine, as well as Karl Benz and Rudolf Diesel, who would each make essential design improvements. However, if we must choose only one as the inventor of a useful internal combustion engine, Nikolaus Otto can be so honored.

In 1864, Otto and the German sugar maker Eugen Langen designed a two-stroke, "vertical free piston atmospheric gas engine" based on the Lenoir engine, using piped-in coal gas from their Cologne factory lighting supply as a fuel source, winning the "Grand Prix" at the 1867 Paris Exposition. The bulky, stationary, 2-horsepower (hp) engine weighed about 2 tons and was ideally suited for industrial applications in workshops and warehouses.[61] Over the next decade, they built 35,000 stationary engines at the Deutz-AG-Gasmotorenfabrik, part-owned by Otto, and which for a short time was the world's largest manufacturer of engines.[62]

Working with Daimler (the technical director) and Maybach (the chief designer), Otto then built the first four-stroke engine in 1876, the coal-gas fuel introduced into a cylinder, *compressed*, burned, and expelled, thus moving a piston down and up in a resulting controlled explosion. The four strokes are (1) intake (down), (2) compression (up), (3) ignition (down), and (4) exhaust

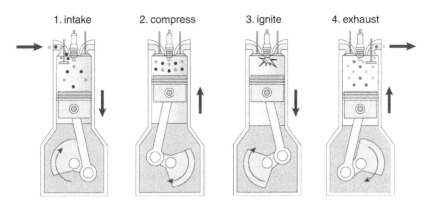

Figure 2.4 The four-stroke Otto Cycle: intake, compress, ignite, and exhaust.

(up), now called the Otto Cycle, where the piston transits one complete cycle (down, up, down, up) as the fuel–air mixture is repeatedly introduced, compressed, burnt, and expelled in the cylinder (Figure 2.4). To convert the reciprocating up–down motion into rotation, the piston is connected to a rotating crankshaft.

Putting a practical motor on the road, however, was still the holy grail of late-nineteenth-century engineering. There had been previous attempts to fix a steam engine to a road vehicle, notably the three-wheeled, 2.5-mph French military tractor of Nicolas Cugnot in 1769 that needed to stop every 15 minutes or so to build up enough steam,[63] and Trevithick's "puffing devil" in 1801 that terrorized the back roads of Cornwall. But despite a perfect fit for factories, trains, and ships, steam was unsuitable for the open road because the engine was too big, horribly inefficient, and burned too much fuel, while large exhaust puffs clouded the vision of both pedestrians and drivers following behind. Although the Stanley Steamer would enjoy limited success in the United States, steam power was impractical for automobiles.

Various other fuels had been imagined or tried, such as gunpowder (Dutch physicist Christiaan Huygens), hydrogen–oxygen (Isaac de Rivaz), coal gas (Lenoir), and kerosene (American engineer George Brayton), but were either too dangerous to run or the fuel too volatile to store. Otto also tried ethyl alcohol (ethanol), which lacked sufficient power, while Rudolf Diesel would even try peanut oil in one of his early designs. The solution was a safe, easily portable liquid, converted into highly combustible vapor as needed. Enter the gasoline-injection carburetor that introduced an atomized, air–gas mixture into the combustion chamber, patented in 1887 by Maybach. The idea may have come from a perfume sprayer.

Figure 2.5 (a) 1885 Daimler–Maybach Reitwagen (*source:* Wladyslaw CC BY-SA 2.0 DE) and (b) 1886 Benz Patent-Motorwagen driven by Bertha Benz from Mannheim to Pforzheim.

After starting up their own company in Bad Cannstatt, then just a village outside of Stuttgart, Daimler and Maybach improved the basics of the Otto engine (fuel vapor, hot-tube ignition, float carburetor, automatic intake valve), most importantly reducing the weight 10-fold, such that an engine could be fitted onto a vehicle. In 1885, they affixed their smaller 0.5-hp motor to a wooden bicycle, the world's first motorcycle with a top speed of 7 mph, which they called a "Reitwagen" (riding wagon; Figure 2.5a). A year later, in nearby Manheim, Karl Benz built the first three-wheeled motor vehicle, the 0.8-hp "Patent-Motorwagen" (Figure 2.5b) that had all the essential components of today's gasoline-fueled "car" (carriage), including igniting the fuel with a battery-powered, high-voltage spark current.[64]

Benz would receive the first patent for a "vehicle powered by a gas engine" in 1886, famously demonstrated two years later by his wife Bertha and their two sons on a 180-km round-trip journey from Mannheim to Pforzheim, requiring fuel top-ups along the way at pharmacies in Wiesloch, Langenbrücken, and Bruchsal as the highly flammable cleaning solvent fuel "Ligroin" repeatedly ran out (a C_5–C_8 mixture located in a 50-km-capacity tank).[65] With Daimler in Stuttgart and Benz in Mannheim, their two firms – Daimler-Motoren-Gesellschaft (DMG) and Benz & Cie. – worked to improve both the engine and the "automobile" (integrated engine and chassis, 2 V-slanted cylinders, four-speed, tooth-geared transmission) as they competed for customers.

We can lose the meaning of words with casual use, such as "internal combustion engine," but an internal combustion engine is just that – fuel burned inside a chamber to apply pressure to a reciprocating piston that drives a rotating crank, rather than boiling water for steam in a separate external chamber and condensing the motive steam to create a pressure difference. With the essential mechanics worked out, all that was needed was the right liquid fuel. As oil had been a nuisance in salt wells, so too was gasoline a nuisance to early oil refineries, burnt off during the production of kerosene. Not any more. Because of the higher volatility than kerosene, gasoline was better for both industrial and vehicle combustion engines.

Soon, the gasoline or "petrol" engine – so dubbed by Daimler to calm fears about burning highly flammable benzene (C_6H_6) – and its "motor spirit" fuel were in great demand as cars, boats, and any transport means possible were motorized in increasing numbers. By 1893, Benz & Cie. had built 1,000 gasoline engines, Daimler the same by 1895.[66] In 1890, Maybach designed the first four-cylinder, in-line boat engine, fitting Count Zeppelin's LZ 1 airship a decade later with two, four-cylinder, light-alloy engines for its maiden flight over Lake Constance on the Swiss–German border. The first successful engine-assisted airplane flight was in 1903 at Kitty Hawk, North Carolina, the Wright brothers' heavier-than-air Flyer staying aloft for 12 seconds, powered by a crude gasoline engine with a bicycle-chain drive to turn the propeller. The Wright brothers had started out as bicycle makers in Dayton, Ohio.

The Mercedes, however, would change everything, spurred on by Emil Jellinek, an Austrian entrepreneur and sportsman and one of Daimler's first customers. Having bought a belt-driven, 6-hp, two-cylinder, 24-km/h top-speed DMG car for himself in 1897, followed the next year with two Phoenix cars – the world's first four-cylinder road vehicle with a front-mounted, 8-hp engine, and high-efficiency, spray-nozzle carburetor that powered trucks (1896), taxis (1897), and buses (1898)[67] – Jellinek ordered 10 cars in 1899 and another 29 in 1900, which he mostly sold to his upper-crust business and sporting friends, becoming DMG's primary sales agent. As the market called for more innovation, Jellinek passed on ideas to Daimler, who would then build the redesigned cars to spec, one of which had a much improved engine – 6.6 kg/ hp, a top speed of almost 90 km/h, inflow air-cooling radiator – and a new name, that of Jellinek's daughter Mercedes.

Developed by Maybach, the 35-hp Mercedes racing car with "low center of gravity, pressed-steel frame, lightweight high-performance engine and honey-comb radiator"[68] was delivered on December 22, 1900, ushering in the twenti-eth century and a new age of motor-assisted transportation. A huge success, Jellinek changed his name to Jellinek-Mercedes, joking how he was the first man to take his daughter's name, while Daimler's name would be subsumed by his company's leading product in the eventual merger with his main competitor. In 1926, amid the economic travails of post-war Germany, DMG and Benz would be forced to merge to form Mercedes-Benz.

(a)   (b)

Figure 2.6 The diesel engine: (a) Rudolf Diesel's 1894 engine (13 kW) now at the Augsburg MAN Diesel Museum (*source:* Tiia Monto CC BY-SA 3.0) and (b) 1904 MAN diesel engine installation (1.2 MW) at the Kiev Municipal Transport Authority.

The automobile engineers continued to tinker, looking for performance advantages in ever newer designs. Inspired by the explosive compression of tinder in a Victorian cigar lighter, Rudolf Diesel hit on the idea of increasing fuel compression to eliminate the need for sparked ignition (recall that temperature is proportional to pressure) creating what would be called the "diesel" engine in his honor (Figure 2.6). More powerful than a gasoline engine, a diesel engine is also more efficient (27%) compared to gas (17%), gasoline (12%), oil (10%), or steam (6%).[69] Refined at a higher temperature than gasoline in the fractional distillation process, another previously wasted crude oil portion now known as diesel could be newly exploited. Initially run on peanut oil, Diesel's engine won the "Grand Prix" at the 1900 Paris Exhibition.

Also called a "compression-ignition" engine, a diesel engine is more powerful than a gasoline engine because of the greater compression ratio: $CR = V_{max}/V_{min}$, where V_{max} is the maximum chamber volume and V_{min} is the minimum chamber volume (at maximum compression), which ultimately determines piston power. V_{min} is much smaller in a diesel engine, making it more powerful (CR at ~20:1 compared to a midsize gasoline engine at ~8:1) and characteristically noisier as it operates at higher pressures.

While gasoline engines were being developed to power cars and motorcycles, diesel engines were built for more heavy lifting. After a diesel engine was successfully installed in the French submarine *L'Aigrette* in 1902, the Danish company Burmeister & Wain commissioned the 7,400-ton MS

(Motor Ship) *Selandia*, the world's first ocean-going, diesel-powered vessel. Launched in 1911, the *Selandia* drew much attention on its maiden voyage from Copenhagen to Bangkok because the 113-m ship had no smoke stacks – combusted diesel exhaust is vented directly into the ocean – and much less fuel and maintenance costs.

The higher-performance compression-ignition engines were built to power all kinds of factories, first installed in the United States in 1898 for bottling at the St. Louis Anheuser-Busch Brewery under license from Diesel. By 1907, American central power stations totaled 10,000 brake hp[70] (68% of all hp[71]), a 1914 catalogue listing the diesel engine's many advantages, including longer operating hours, longer life, uniform stress, less maintenance, simple start, as well as no need for a carburetor, vaporizer, hot-bulb flame, or electrical ignition as oil spray combustion was due solely to the heat generated by compression. What's more, if the compression failed, there would be no ignition and the engine would stop, improving safety and efficiency.[72]

Although coal-fired external steam engines had been used to run power stations since Edison's first Pearl Street Station dynamo in 1882 (a.k.a. prime movers), diesel units were soon put to greater commercial use, helping to create the industrialized world we know today. By 1913, every industrial sector had converted to internal combustion engines, mostly powered by diesel. The scale of adoption can be seen in an exhaustive list of applications cited that year in the USA and elsewhere, such as automobile works, battery charging, brewery, cement works, chocolate works, copper mining, electric light plant, fertilizer works, flour mill, gas works, glass works, gold mines, ice and cold storage, locomotive works, machine builders, municipal plants, paper mills, phosphate mining, printing ink, roller mills, street railway, structural steel and iron, waterworks, and woolen mills.[73]

Having simplified the means to power a maturing twentieth-century Industrial Revolution, the diesel engine also brought an end to coal-powered locomotion as diesel-powered ships and trains refueled faster, took less storage and boiler space, and required much less maintenance, the liquid fuel conveniently pumped rather than shoveled. Bunkering at sea was no longer a logistics nightmare for transportation companies and militaries needing to refuel to continue their journeys in far-flung regions around the world. Underscoring how change marks its own time, oil was initially sprayed onto the coals to increase engine efficiency before coal was completely replaced, while half as many men were needed to service the boiler rooms of the new navies and cruise ships.[74]

Launched in 1910, Shell's 650-hp *Vulcanus* was the first diesel-driven tanker, while the last coal-fired US battleship commissioned for active service was in 1914.[75] Powered by a Sulzer-Diesel engine, a Rivadavia-class battleship became the world's largest and swiftest dreadnought, starting a worrying arms race in South America prior to World War I. Although the *Titanic*, launched in Belfast in 1912, was still coal-fueled and only one of its three propellers turbine-driven, coal was effectively phased out as a transportation fuel by the end of World War I, no longer needed other than to generate electricity and in high-temperature industry manufacturing.[76] The great age of steam transportation ended with World War I, although steam is still used in power plants wherever water is boiled to turn a turbine to make electricity, whether coal-fired via "steam" coal, oil-fired, natural-gas-fired, or uranium-fueled.

Compression-ignition engines and their applications changed the world in every way, Rudolf Diesel himself noting a year before his death that "to diesel" had become a new verb in the lexicon of power. In New Hampshire, a diesel system was run in parallel with a hydroelectric station to generate 7 MW of electric power. Oil for American trains also increased 14-fold in the first 20 years of the twentieth century.[77] Today, diesel power plants are especially important for remote and emergency backup power, while diesel locomotives are rated as high as 3.5 MW and the largest supertankers at almost 100 MW.[78]

The diesel engine was also a boon to farming, helping to revolutionize the labor-intensive grain harvest. As in many modern global businesses, companies are bought, sold, merged, and split-up with such regularity it is hard to keep up, but having purchased Deutz-Allis in 1990, the Georgia-based Agco can trace its lineage to Nikolaus Otto and Eugen Langen's first great engine manufacturing company in Cologne, where Otto first demonstrated his four-stroke engine. Agco is one of the world's largest farm equipment manufacturers with a product line that includes tractors, combines, and engines.[79] Today, diesel engines in the USA "power more than two-thirds of all farm equipment, move 90 percent of its product and pump one-fifth of its water."[80] Market penetration is much the same elsewhere.

Diesel is synonymous with power and reliability, although before the inclusion of block heaters cold starts in vehicles were more difficult than gasoline engines. Because of pressure combustion, diesel has less knock – uneven burning of unstable hydrocarbons where fuel explodes prior to spark ignition rather than being evenly burnt – and gets better fuel economy especially when towing, typically up to 40% more miles per gallon.[81] For some consumers, however, the only concern is which nozzle to use at the pump. Fortunately for gasoline-fueled cars, a diesel nozzle is bigger and can only fit in a diesel tank, while diesel car drivers must be more careful because the smaller gas nozzle fits both.

Rudolf Diesel was full of ideas yet terrible at business, making a fortune from his invention before losing everything on bad investments. In the autumn of 1913, during a Channel crossing from Antwerp to Harwich aboard the *Dresden*, he mysteriously went missing. Some say he jumped, others that he was pushed, possibly by spies who didn't want his engine used for (or against) the German war effort, or that he had too many advanced social ideas such as worker-run factories, cooperatives, and decentralized industry. Diesel's enormous ingenuity, however, lives on. Today, marine vehicles are predominantly powered by diesel, freight trains rely almost exclusively on diesel, and more than half of buses are either diesel or diesel-hybrid, although the percentage is now decreasing with the advent of electric motors.

There were many innovations to come to reach today's modern internal combustion engine car, whether gasoline- or diesel-powered: modern assembly, electric starter (1912), injection timing (four-cylinder synchronizing to maximize the power stroke), angled V8 fuel injection (replacing the carburetor), electric fuel injection (first successfully applied to the 1940 Alfa Romeo), fuel additives (such as lead to remove gasoline knock although later discontinued), turbochargers (to increase air intake volume), and catalytic converters (to lower toxic exhaust emissions by reducing incomplete carbon burning and nitrogen oxides) to name but a few. There are over 100,000 automobile-related patents, but despite the many innovations the basic design of Otto's original four-stroke reciprocating engine has essentially remained the same.

No other internal combustion engine has enjoyed the success of the four-stroke, fueled either by gasoline or diesel, designed by two German engineers over a century ago.[82] That is, until the whole design was ripped out in favor of a much simpler propulsion system, one that no longer exploded hydrocarbons to turn a wheel: electromagnetic induction (which we'll look at in Chapter 6). Not only was the electric motor simpler and more efficient, one doesn't need to worry about any nasty combustion by-products.[83]

We've come a long way since the 1870s' horse-manure dilemma in our crowded cities, but emissions of a different kind would become a bigger concern by the turn of the twentieth century, thanks to industrial manufacturing, power plants, and transportation, all powered by burnt hydrocarbons. After World War II, with the growth of the modern consumerist society, car registrations shot up across the United States, more than doubling from roughly 26 million in 1945 to over 52 million by 1955.[84] As the car became an everyday way of life, however, the toxic by-products of combustion could no longer be ignored.

2.4 A New World Culture: Fast and Furious

As gasoline became more available – spurred on by new oil finds in 1901 in East Texas and around the world in Borneo and the East Indies – and automotive manufacturing continued to evolve, gasoline-powered cars became more affordable. Daimler, Maybach, Benz, Diesel, and others saw to the engineering needs of making better engines, their innovations applied to powering trains, ships, planes, trucks, buses, farm equipment, and electricity-generating power plants, but it was the success of the automobile that most changed the world. Nowhere was that success more apparent than in the United States.

By 1900, there were almost 8,000 American cars in operation,[85] powered either by steam (smooth acceleration, but slow-starting, bulky, and inefficient fuel), electric charge (quiet, but a heavy and limited battery), or gasoline (noisy and prone to breakdowns) – production numbers for 1900 are shown in Table 2.2. Driven mostly by wealthy enthusiasts and considered a "rich man's toy," one-off manufacturing, inadequate supply chains, and uneven dirt roads made "motoring" impossible for the average citizen, although the cycling craze of the 1890s encouraged the building of some new roads and better manufacturing techniques.

One innovator, however, would change the manufacturing process forever, introducing interchangeable parts and patterned assembly methods, and can be said to have created the modern automotive industry. Another expanded the process into an easily repeatable system and can be said to have inaugurated the modern world. As with other great engineering developments, such as steam power in northern England and the internal combustion engine in southern Germany, the two started working in close proximity and would go on to make their names in what would soon become known as Motor City, a.k.a. Motown.

Ransom E. Olds started out in Lansing, Michigan, making steam-powered cars ("steamers"), before moving to nearby Detroit, where he built a gasoline car with interchangeable parts that would soon become known as an

Table 2.2 *American production of steam-powered, electric, and internal-combustion automobiles in 1900 (number and average price)*

	Steam-powered	Electric	Internal combustion
Number of cars	1,681	1,575	936
Average price	$682	$1,822	$938

Source: Bakker, S., From Luxury to Necessity: What the railways, electricity and the automobile teach us about the IT revolution, p. 128, Boom, Amsterdam, 2017.

"Oldsmobile," although his initial attempts were too expensive for everyday workers. After a fire destroyed his Olds Motor Works factory in 1901, he began to "mass produce" the 7-hp Curved Dash runabout with standard parts manufactured locally and assembled in-house, employing subcontractors that included the Cadillac founder Henry Leyland and the Dodge brothers. Although the Benz Velo is considered the world's first "series-produced" automobile at almost 1,200 vehicles, making Benz & Cie. the largest car manufacturer prior to 1900,[86] Olds quickly caught up. By 1905, he had produced 18,500 cars at a more affordable price of $650,[87] establishing Detroit as a world-class car-making hub. In time, there would be more automobiles on the roads than horses.[88]

Henry Ford was born on his father's farm in Dearborn, Michigan, on the outskirts of Detroit. Starting out in the Edison Illuminating Co. as a young engineer with new ideas about practical invention, Ford would learn first-hand how Edison was, as he later described him, "the greatest inventive genius in the world," but one who "knows almost nothing about business."[89] In Detroit, Ford would engineer his own business acumen, availing of his lifelong wisdom of surrounding himself with smart people.

Revolutionizing the manufacturing process by building cars with standardized parts made on site instead of shipped in by external suppliers, Ford initiated the mechanized moving assembly line, an idea he adapted from the meatpacking industry, where an incomplete car moved past a stationary worker performing a specialized task. Letters were attached to each new design – A, B, AC, C, F, K, N, R, and S[90] – perhaps stealing a page from his one-time mentor's book, who initialed each of his early light-bulb filament attempts with T.A. for "Try Again" Edison.

Always looking to cut costs, Ford eventually focused on one model, a lightweight (1,200 pounds), four-passenger, 20-hp car with a top speed of 45 mph, but most importantly made of a simple, rugged design at a low price. At his Highland Park factory, newly built in a small community just north of Detroit, the Model T was churned out faster and cheaper, assembled around the clock in three 8-hour shifts, taking one-tenth the time as previous designs.[91] Having redesigned the assembly process to simplify construction and maximize throughput, Ford sold 10,000 Model Ts in 1909, the first year of production. Five years later, an entire Model T could be assembled in just 26 and a half minutes.[92]

After lowering the price from $950 to $490 – about half the average worker's annual salary – Ford would sell almost a quarter of a million Model Ts by 1914,[93] the new assembly-line system so rigidly automated he could joke, "Any customer can have a car painted any color that he wants so long as it is black." By 1921, half of all cars in the United States were Model Ts,[94] while by the time the model was finally discontinued in 1927 – as customers called for more innovation, comfort, and different colors – the Ford Motor Company had sold more than 15 million of its ubiquitous brand.[95]

Ford spearheaded many features found in today's car: the one-piece engine block, removable cylinder heads, a flexible suspension system (essential, given the dire condition of early American roads, less than 10% of which were "surfaced"), lightweight vanadium alloy steel (rather than carbon steel), and the left-side steering wheel to better gauge on-coming traffic, although he was reluctant to employ innovations by other manufacturers, such as "electric starters, hydraulic brakes, windshield wipers and more luxurious interiors."[96] In 1932 at the age of 69, he helped popularize the V8 engine, a boon for big-engine enthusiasts and the evolving American psyche (2 V-angled banks of four cylinders to ignite eight cylinders at different times for smooth, increased power). There had even been plans for a Ford–Edison electric car that alas didn't materialize because of battery limitations (as we'll see later).

More improvements in engine performance and driving comfort were regularly made by competing manufacturers in the hunt for greater sales in a now rapidly expanding global automobile industry (see Figure 2.7), initially fueled by success on the racetrack ("win on Sunday sell on Monday"), such as the

Figure 2.7 Changing car styles: (a) 1900 Mercedes 35, (b) 1910 Ford Model T, (c) 1953 Chevrolet Corvette, and (d) 1955 Chrysler Chevrolet.

closed body, safety glass, torsion-beam suspension, shock absorbers, stream-lined aerodynamics (essential as speeds increased), aluminum–copper alloy pistons, supercharger airflow, air-cooled engines, and hydraulic brakes. Power windows (first adapted to the Packard in 1940), the curb feeler (patented in 1952), and the intermittent windshield wiper (patented in 1964) became must-have accessories to make driving easier as well as a few savvy inventors rich. Innovation did not initially include better working conditions or unions, Ford a particularly virulent anti-union boss, but workforce management methods such as Taylorism kept the line running smoothly and efficiently.

Perhaps the greatest innovation of all was the installment plan, which allowed customers to buy a car without paying full price at the time of purchase, an idea the banks initially thought was too risky and Ford opposed, preferring customers to save the full amount in regular installments *prior* to delivery, a scheme with obvious drawbacks. The new car mortgage helped to "democratize the automobile," reinventing the American Dream as a place of freedom for intrepid wanderers, transforming the automotive industry into a money-making giant as carmakers cornered the market on transportation.

Founded by William Durant in 1908 after buying out Ransom Olds' pion-eering Olds Motor Works, General Motors introduced the one-third-down, easy-financing loan, helping GM to supplant the more cautious Ford Motor Company as the world's top car manufacturer. The automobile was no longer a novelty, affordable only to wealthy sportsmen, but a commodity available to all, which also helped to ratchet control over consumers via increased eco-nomic dependency on banks and lending institutions. In a single decade, the number of financing companies increased from 40 in 1917 to almost 2,000 in 1927, while by 1963 GM had financed loans for nearly 50 million customers through its GMAC lending arm.[97]

When Chrysler – founded in 1925 by Walter P. Chrysler – introduced the six-cylinder engine in 1929, the "Big Three" was born, bent on wooing the American public to the open road with slick marketing and easy financing. American capitalism had officially begun as production followed consumption via a newly *manufactured* need, Ford anointing the emerging creed of modern consumerism in a 1926 article entitled "Mass Production," writing that "The necessary, precedent condition of mass production is a capacity, latent or developed, of mass consumption, the ability to absorb large production. The two go together, and in the latter may be traced the reasons for the former."

In the process, workers were reduced to automatons, requiring little skill to perform a highly repetitive job, although Ford famously solved the problem of large turnover by doubling salaries and offering a $5-a-day minimum wage. With innovation and increased automation, car making would become monotonous and dehumanized drudgery for "unskilled" line workers, self-styled "shoprats" whose "every movement was a plodding replica of the one that had gone before," as noted by a third-generation veteran GM riveter in the 1970s.[98]

Ford had built the car he always wished, "so low in price that no man making a good salary will be unable to own one – and enjoy with his family the blessings of hours of pleasure in God's great open space."[99] The "road" – wherever and however one could find it – was opened to more than just the elite and their luxury cars, freeing the countryside and beyond for motorized adventure. By the time Henry Ford died, he had amassed one of the largest fortunes ever by selling everyday cars to the multitudes, upwards of $2 billion (more than Bill Gates today as a percentage of American GDP).

Further innovations would make the car a standard feature of modern life, including more roads and highways, paid for by happy drivers via gasoline taxes and spurred on by government policy. In 1919, starting from just south of the White House lawn, Lt. Col. Dwight D. Eisenhower traveled to San Francisco in a convoy of trucks, cars, motorcycles, and ambulances to highlight the need for better roads. In what he called "Through Darkest America with Truck and Tank," the coast-to-coast trip took 2 months, at times managing less than 6 miles per hour on roads Eisenhower rated as "average to nonexistent."[100] Returning 25 years later after World War II as Supreme Allied Commander and understanding the strategic importance of the German autobahn to move men and materials à la the earlier *route royale* of Louis XIV, the eventual President Eisenhower would set out to build an entire national highway system via an ambitious law – the Federal Aid Highway Act of 1956 – that earmarked $25 billion for the construction of 41,000 miles of interstate roads (over $1 trillion today as a percentage of GDP). Thanks to a government largesse unmatched in history, Route 66 became reality and the beaten path a way of life.

With the building up of an essential petroleum infrastructure during war, such as oil refineries and pipelines for military airfields, the post-war "petrolization" of the West exploded, fueling the success of the automobile from the '50s onwards, while the vast new interstate highway system created construction booms at exits and expanded individualized travel beyond one's locale.[101] Eisenhower's highway system – designed in part to ease military transport and the evacuation of cities in the event of nuclear attack – was called "the most ambitious and expansive public works project in all of human history" by economic theorist Jeremy Rifkin, making the USA the most prosperous country on Earth via its "unparalleled economic expansion" and "commensurate multiplier effect."[102] At the same time, the economy became permanently tied to the car and a continuous supply of plentiful and cheap oil.

More than any other invention, the car created our modern consumer world, one of fast-food outlets, box stores, hotel chains, and newly constructed satellite towns. The resultant suburban sprawl became full of drive-to amenities rather than centralized hubs where neighbors walked to for supplies and interacted with fellow citizens. With more cars came weekend traffic, cloverleaf interchanges, standardized signage, road maps, and service stations, as well as motels, malls, drive-in restaurants (1921), cinemas (1933), and churches (1955), turning a triumphant post-war United States into an obsessed and car-reliant culture. Ironically, the democratization of the car also splintered American society into distinct groups with disparate needs – urbanites seeking more government regulation and rural dwellers less interference.

No longer dictated by a fixed train or tram schedule, life soon became somewhere to go to rather than a place to be as drivers tested the limits of their newfound freedom. Tourist attractions became more accessible, such as a rebuilt London Bridge in Lake Havasu City, Arizona, near the Colorado River, south of Las Vegas via US Route 95, while daily life was transformed into a whirlwind adventure full of possibility and unknown potential. A new future could be seen outside the windshield of one's own wheels, the past a forgotten glance in the rear-view mirror as one churned through miles of landscape caring for little, least of all the exhaust. Unasked in the adventure is whether one is satisfied with the revealed future in all the motion.

Sal Paradise summed up the changed, post-war ethos in Jack Kerouac's quintessential American road trip and frenzied beat tale *On the Road*. Riding shotgun at the start of their epic, cross-country journey with his joyriding companion Dean Moriarty behind the wheel of a "sleek" and "spacious" 1949 Hudson,[103] Sal exclaims "And he hunched over the wheel and gunned her; he was back in his element, everybody could see that. We were all delighted, we all realized we were leaving confusion behind and performing our one and noble function of the time, *move*. And we moved!"[104]

From the 1950s on, the automobile was king – stylistic, sexy, powerful, an extension of the willed self – while the life of the city declined. Streetcar tracks were ripped up, sidewalks forgotten, and bicycle lanes discarded.[105] Although the commuter train in the more densely populated Europe would bring citizens closer together in a fast-changing urbanized landscape, the United States preferred isolated box communities that relied on the car, a full tank of gas, and the presumed freedom of being in charge of one's own transport. Having run out of room after years of moving west, the wide-open space could be tamed only by constant motion.

No other product has reinvented human culture so completely. At its peak in 1965, one out of six American jobs was in automobile manufacturing.[106] Today, Detroit carmakers crank out almost 20 million vehicles a year, iconic brands synonymous with the Motor City, made either by Ford, GM, or Chrysler (or companies bought out by the Big Three). For the first time in 2010, the number of cars topped one billion across the globe, 10,000 times more than in 1900. Laid end to end, today's global network of highways would reach the moon and back 20 times.

Foreign automakers such as Volkswagen, Toyota, and Hyundai would eventually make in-roads into the lucrative American market, offering better mileage and lower costs, leaving the Detroit carmakers behind, who like Ford were slow to recognize a changing mindset. Toyota improved efficiency with a U-shaped assembly line and multiple worker tasks, significantly cutting labor costs – the Corolla is the best-selling car ever at over 50 million. In a world predicated on motion, one must move with the times to stay competitive. Economic hardships also contributed to bad times for the Motor City and its famed carmakers, GM and Chrysler bailed out by the US government after facing bankruptcy from the 2007 global financial crisis. Detroit officially went broke July 18, 2013, although the Big Three would survive to compete again in a thoroughly international market thanks to federal aid.

The top 10 manufactures comprise about two-thirds of annual car sales (~$3 trillion worth!) as shown in Table 2.3. Subject to the economic vagaries of boom and bust, the list regularly changes as do the names whenever companies merge or split. GM and Ford are ranked 4 and 5, while Chrysler is number 8 as part of the larger FCA cum Stellantis conglomerate[107] (the year founded is when the company or original parent first started making vehicles).

Table 2.3 *Top 10 global car producers (2017) (*merged in 2020)*

#	Company	Year	HQ	Annual production (millions)
1	Toyota	1937	Aichi, Japan	10.5
2	Volkswagen	1937	Wolfsburg, Germany	10.4
3	Hyundai Kia	1967	Seoul, South Korea	7.2
4	General Motors	1908	Detroit, MI, USA	6.9
5	Ford	1903	Dearborn, MI, USA	6.4
6	Nissan	1911	Yokohama, Japan	5.8
7	Honda	1937	Tokyo, Japan	5.2
8*	Fiat/Chrysler (FCA)	1899	London, England	4.6
9	Renault	1899	Paris, France	4.2
10*	Groupe PSA	1896	Paris, France	3.6

Source: "World Motor Vehicle Production: OICA correspondents survey (2018)," OICA, 2018. www.oica.net/wp-content/uploads/World-Ranking-of-Manufacturers-1.pdf.

Today, almost 100 million vehicles are sold annually across the globe, compared to 66 million in 2005, as sales continue to climb in the developing world (roughly 30 million in China compared to six million in 2005[108]). In 2009, Chinese sales surpassed those in the USA for the first time (11 million) and Europe in 2012 (19 million), although Americans still enjoy the distinction of living in the only country with almost as many cars as people (1.25 people per car).[109] Oil and gas are the essential ingredients that keep the industry going, maintaining the flow of goods across a network of modern roads and a revamped individualism begun in the twinkling of American adventure. In homage to the past, GM continued manufacturing the Oldsmobile until 2004, more than 100 years after Ransom E. Olds first assembled his dream car.

Alas, common sense often lags behind progress in the rush to produce more, with safety and health an afterthought to sales. After the publication of his landmark 1965 book, *Unsafe at Any Speed*, American lawyer and consumer advocate Ralph Nader finally got the car companies to take notice, and basic safety features were installed, most famously the seat belt. In the USA, only New Hampshire still doesn't have an automatic buckle-up law. Thanks to Nader and improved safety standards, cars are much safer than 50 years ago, the Center for Auto Safety estimating that 3.5 million lives were saved from 1966 to 2014.[110] When first proposed in 1967, Henry Ford II warned they "would shut down the industry." Although not the future one might suppose in an efficiently mechanized world, Ford later admitted, "We wouldn't have the kinds of safety built into automobiles that we have had unless there had been a federal law."[111]

In the land of stylized freedom, fast and furious would become the norm, no matter the noise, fumes, or danger. The paved road was a fairytale come true, open to all who giddily roared by, such as Mr. Toad in Kenneth Grahame's satirical Edwardian story *The Wind in the Willows*, who crashes his motor car into a pond without a care in the world as he breathlessly exclaims his rush of excitement:

> Then he burst into song again, and chanted with uplifted voice –
> 'The motor-car went Poop-poop-poop,
>   As it raced along the road.
> Who was it steered it into a pond?
>   Ingenious Mr. Toad!
> O, how clever I am! How clever, how clever, how very clev –'

2.5 Hydrocarbons and the Darker Side to Oil

A funny thing happened on the way to the future, when the car was king and freedom was a given – we forgot about the exhaust. Each year the fumes grew, responsible for millions of deaths. By 1929, there were already 300,000 gas stations in the United States, each emblazoned with their own sleek company logo of flaming torches, yellow shells, arrow-pierced diamonds, white stars, or plain-old company names, a cornucopia of shiny happy symbols selling freedom. As noted by historian Daniel Yergin in *The Prize*, "If not quite a religion, the sale of gasoline at retail outlets had become . . . a big and very competitive business."[112] The road was laid wide open to smoke-spewing drivers, although as the number of cars and pollution rose, a few concerned citizens started to ask questions.

There is much technical jargon in the chemistry of petroleum combustion, but we can simplify the composition and burning process to understand the basics, not least to combat pollution or learn how Volkswagen, Fiat, and Renault gamed in-car emissions tests to cheat industry regulations. Petroleum contains thousands of chemical compounds (mostly hydrocarbons), along with nitrogen (0–1%), sulfur (0–10%), and oxygen (0–5%) – the so-called NSO compounds – as well as trace amounts of various metals (V, Ni, Fe, Al, Na, Ca, Cu, and U), the majority of which are vanadium and nickel.[113] Essentially, petroleum contains 80–87% carbon (C) and 10–15% hydrogen (H), while gasoline comprises up to 200 different hydrocarbons (HCs). The petroleum hydrogen/carbon mix is similar to plants and animals, which clearly points to a biological origin.

Recall that crude oil is composed of a range of C_1–C_{70+} hydrocarbons, from the more easily distilled, light, low-boiling-point C_1–C_4 range molecules – methane (CH_4), ethane (C_2H_6), and the liquid petroleum gases propane (C_3H_8) and butane (C_4H_{10}) – to the heavy tar and asphalt residuum C_{70+} molecules. Gasoline is composed of the "gasoline range" molecules, C_4–C_{12}, which are distilled as "straight-run" gasoline and mainly include butane, pentane (C_5H_{12}), hexane (C_6H_{14}), heptane (C_7H_{16}), and octane (C_8H_{18}) (a.k.a. straight-chain n-alkanes or saturated hydrocarbons with general formula C_nH_{2n+2}), as well as benzene (C_6H_6) and its derivatives (toluene, ethylene, and xylene, a.k.a. BTEX[114]), naphthalene ($C_{10}H_8$), and various branch-chain isoalkanes, cell-structure cycloalkanes, and alkenes (a.k.a. olefins), the exact percentages depending on the crude and refining process.[115] A mouthful to be sure.

Crudes are first defined by field of origin (we'll look at density and sulfur content later, the other main attributes). Table 2.4 shows the hydrocarbon composition of three crudes from Kuwait, Prudhoe Bay (Alaska), and South

Table 2.4 *Principal hydrocarbon concentration in three select crude oils by percent weight*

Hydrocarbon	Kuwait	Prudhoe Bay	South Louisiana
Paraffins	16.2	12.5	8.8
Naphthenes	4.1	7.4	7.7
Aromatics	2.4	3.2	2.1
Asphaltenes	77.3	76.8	81.4

Source: National Research Council, "Oil in the sea: Inputs, fates, and effects," p. 19.

Louisiana. The Kuwait crude is considered light, that is, more small-molecule alkanes (paraffins) and is more easily refined into gasoline, while the Louisiana crude is heavier, that is, more high-boiling, large-molecule "asphaltenes." For comparison, the relative amounts of hydrocarbons in three different fuel types are shown in Table 2.5 – premium gasoline, regular gasoline, and diesel – divided into lower-range aliphatics (alkanes, alkenes, and alkynes) and higher-range BTEX aromatics (sweet smelling yet highly toxic).[116] Here, you can see the greater number of higher-molecule hydrocarbons (middle distillates) in diesel (typically C_{10}–C_{21}) and the much greater amount of BTEX aromatics in premium and regular gasoline.

Of course, not all gasoline is completely burned during combustion, the major cause of tailpipe pollution, especially in low-performance cars. The idealized combustion process in a gasoline engine, assuming a *complete* burning of *only* octane, is $2C_8H_{18} + 25O_2 \rightarrow 16CO_2 + 18H_2O$, but as with wood and coal *incomplete* combustion also creates carbon (C) (soot) and carbon monoxide (CO), along with the carbon dioxide (CO_2) and water (H_2O) from complete combustion.[117]

Nitrogen oxides are also emitted during high-temperature incomplete combustion and released to the atmosphere, in particular nitric oxide (NO) and nitrogen dioxide (NO_2), collectively referred to as NOx, which contribute to smog, acid rain, and tropospheric ozone. Particulate matter (PM) is also produced – primarily a mix of nitrates and sulfates from fuels containing nitrogen and sulfur, aromatic hydrocarbon gases trapped in the balled-up solid phase constituents of PM, and trace metals. Emitted in various sizes, PM is divided into fine and course categories for convenience, labeled PM2.5 (less than 2.5 µm) and PM10 (less than 10 µm). The smaller the size the greater toxicity, while PM2.5 (a.k.a. fine particles) from diesel poses a particularly dangerous health risk to humans, all the more so with the increased numbers of

Table 2.5 *Relative hydrocarbon concentration in three sample Alaskan fuels (premium gasoline, regular gasoline, and diesel), divided by range into aliphatics and aromatics*

Hydrocarbon type	Hydrocarbon range	Premium gasoline	Regular gasoline	Diesel fuel
Aliphatic	C_5–C_{10}	40.6	49.0	5.7
	C_{10}–C_{21}	11.2	10.2	78.7
	C_{21}–C_{35}	0.0	0.0	0.8
	Total	51.9	59.3	85.3
Aromatic	C_9–C_{10}	2.4	2.7	0.0
	C_{10}–C_{21}	12.7	10.5	0.2
	C_{21}–C_{35}	2.9	2.4	0.1
	Benzene	14.0	11.1	0.8
	Toluene	8.6	7.2	0.1
	Ethylene	7.5	6.8	12.5
	Xylene	0.0	0.0	1.0
	Total	48.1	40.7	14.7
Total		100.0	100.0	100.0

Source: "Hydrocarbon characterization for use in the hydrocarbon risk calculator and example characterizations of selected Alaskan fuels," Geosphere, Inc., September 2006.

vehicles in urban areas. Unburned vapors also enter the atmosphere during refining, handling, and filling.

In the early years of production, the maximum amount of straight-run gasoline refined from crude via fractional distillation was about 20% (much less with a poorer Pennsylvania crude), which prior to the discovery of the internal combustion engine was typically discarded as worthless. After the 1911 court-ordered breakup of Standard Oil, a team of petroleum scientists working at one of the Baby Standards began to experiment with new methods to turn a larger percentage into gasoline. Led by William Burton, a Johns Hopkins PhD chemist, the Indiana Standard team started to "crack" leftover gas oil by heating it under very high pressure to over 650°F, hoping to break down the higher-molecule hydrocarbons into the shorter-length C_4–C_{12} gasoline range. The Burton thermal-cracking process was fraught with danger, but produced a synthetic gasoline that doubled the amount of usable fuel.[118]

The discovery resulted in part because of more independence at the spin-off Standards, removed from the "rigid and controlling grip of 26 Broadway"[119] that allowed new technological ideas to flourish despite the distractions of Standard of Indiana's chairman Colonel Robert Stewart (and friend of Calvin

Coolidge) being investigated for taking bribes in the private leasing of the newly created Teapot Dome federal oil reserve in Wyoming. The strategic military backup supply (a.k.a. Naval Petroleum Reserve No. 3) had been established in 1915 to ensure a secure flow of oil in the case of emergency.

"Thermal cracking" as the process became known was a boon for the nascent car industry, as gasoline started to outsell kerosene for the first time. The synthetic gasoline was even preferred to straight-run gas (a.k.a. white gas) because it had better antiknock properties, although its octane number (~50) would soon be too low for the newer, higher compression-ratio engines being built. Although the cracking process was further improved to increase the octane number to around 70, by the late 1930s other means were needed to produce even higher-octane gasoline for the constantly improving, more powerful engines. (As seen at any service station today, modern octane ratings range from 87 (regular) to 94 (premium).)

Note that the octane number does not equate to the amount of octane (C_8) in the gasoline, but is a measure of the burn efficiency, expressed as a percentage between 0 for heptane (C_5), which does not burn smoothly, and 100 for iso-octane (C_8), which does. Thus, an 87-rated gasoline burns with the same auto-ignition characteristics as a mixture of 13% heptane and 87% iso-octane. As for diesel engines, an octane number doesn't mean much, because diesel fuel is pressure-ignited – air is compressed and heated to ignite the fuel, by definition producing a more even burn and thus less knock.[120]

"Catalytic" or "cat cracking" was discovered prior to World War II by Eugene Houdry, increasing the octane rating even more. A French World War I veteran, steel company heir, and motor racer, Houdry found that high-octane gasoline could be made by introducing a clay catalyst into the cracking process, greatly increasing the amount of higher-performance gasoline available from crude up to 45% (a catalyst quickens but does not participate in a reaction). Today, a synthetic zeolite mineral is used as the catalyst to produce octane numbers from 85 to 100, the output depending on vehicle performance, region, and even altitude. The chemist and international coal expert Harold Schobert noted that some "consider the development of catalytic cracking to be the greatest triumph of chemical engineering."[121] Further polymerization (small to big) and alkylation (big to small) processes also transformed non-gasoline-range molecules into gasoline with suitably high octane numbers.

The hunt for higher-octane gasoline was essential to reduce engine "knock," the uneven burning caused by early fuel ignition as different hydrocarbon components ignite at different temperatures. From the mid-20s to the mid-80s, tetraethyl lead (TEL) was famously added to increase octane ratings and reduce knock, despite the obvious health hazards of using a known environmental poison and the availability of other, safer, antiknock additives such as alcohol. In 1924, five Standard Oil researchers went insane and died after working with lead additives, while 49 others were hospitalized. Lead was originally tried "purely by guess" in the search for a light, antiknock additive that also included foul-smelling contenders such as aniline (rotten fish), selenium (rotten horse radish), and tellurium ("satanic garlic").[122] The leaded mixture was labeled "ethyl" gasoline and dyed red to warn of its toxicity.

Sold to the public for the first time in 1923 at a filling station in Dayton, Ohio, 90% of American gasoline would be leaded by 1936 and more than 98% by 1963. Money was the motivation for using lead as an additive, Charles Kettering and Thomas Midgely – the two GM engineers in charge of the program – keen to reap "the potential profit from developing a patented additive rather than adding unpatentable alcohol to motor fuel."[123] Just 3 grams per leaded gallon (gplg) raised the octane number by 10–15,[124] but after 60 years lead was finally made illegal (>0.10 gplg[125]) as a precursor to introducing catalytic converters to reduce emissions, which were themselves contaminated by lead. A 1985 EPA study calculated that up to 5,000 Americans died annually from lead-related heart disease, while since the ban "the mean blood-lead level of the American population has declined more than 75 percent."[126]

Note "unleaded" at the pump doesn't mean lead has been removed, just that it is not *added*, although sadly lead is still used as a gasoline additive in countries with lax environmental regulations, including China, and is still added to boat and jet fuel. Today, high "research octane number" (RON) additives are mixed in, such as benzene (RON = 104), although benzene is a human carcinogen, one of almost two dozen "mobile source air toxics" (MSAT) and as such limited to between 0.62% and 1.3% by volume according to the second phase of the 2012 US Clean Air Act (MSAT-2). Ethanol is also blended to increase octane levels and engine performance and to conserve petroleum reserves (as we saw earlier). Alas, as lead content is reduced in high-performance cars, the amount of benzene and toluene increases.

Knowing the danger and doing something, however, is never simple with cars. In 1965, the US Congress called for national emission standards, while by the 1960s the smog in Los Angeles had become so bad that people started demanding controls on car emissions, much as they had a decade earlier with burning sulfur-rich coal in London's Big Smoke. Although the cause of the hazy brown LA air wasn't immediately understood, breathing was becoming more difficult, while eye irritation and crop damage increased. Asked to analyze a "dark-brown, vile-smelling" drop of liquid air condensed from the smog, Arie Haagen-Smit, a Dutch chemist working at Caltech, reported that the previously unknown smog was full of saturated and unsaturated hydrocarbons, which had originated in "all the materials lost at the oil fields, refineries, filling stations, automobiles, etc." [127] The massive amount of spewed exhaust from two million cars in the tangled mess of freeways was being oxidized by the Californian sunlight and ozone to create a new kind of danger – photochemical smog.

Further analysis by Haagen-Smit showed that "the combustion in the automobile engine was not as complete as the industry had assumed and that considerable quantities of carbon monoxide and unburned hydrocarbons were actually released."[128] Trained to analyze the bio-organic source of flavors in pineapples and wines, Haagen-Smit had constructed the first standard air-quality measurements for motor vehicles, becoming the first chair of the California Air Resources Board (CARB) in 1968 and a dedicated advocate for clean air. As Mary Nichols, another long-time CARB chair, noted "It wasn't until the Clean Air Act in 1970 that you had a law that said, 'we're going to set an air quality standard based on a public health measurement, and then the government will go out and take whatever action is needed to reach those limits'."[129] Getting from A to B would no longer be a thoughtless luxury.

Unfortunately, automobile pollution remains a major problem, despite emission levels being set in some progressive jurisdictions, while the effect of sunlight on reactive hydrocarbons and nitrogen oxides that linger in the atmosphere can still produce bad air episodes. In 1979, Los Angeles saw a 50% rise in the number of hospital patients with "chronic lung diseases such as emphysema and asthma."[130] In 2015, the smog in Paris was so bad the Eifel Tower and the Sacre Coeur were almost completely shrouded, prompting an emergency vehicle restriction – alternate-day, even–odd plate numbers – that reduced particulate matter by 40%. It seems in our race to get to where we want to go as fast as we can, we forgot about how we got there. Burning petroleum always comes at a price and not just at the pump – almost $6/gallon for diesel and $5/gallon for gasoline in the USA at its peak in 2022 (Figure 2.8).

2.6 Volkswagen, Emissions, and a Careless Car Culture

In the United States, one can easily check the rated emissions of any vehicle. As noted by Greenercars.org, "All vehicles have a mandatory under-the-hood label that identifies the emission standard(s) – so while you're standing on the dealer's lot, just pop the hood and have a look."[131] One can also calculate emissions from different vehicles with an online calculator for comparison as shown in Table 2.6 for two sample vehicles, a Ford Fiesta and a Volkswagen

Table 2.6 *Estimated tailpipe emissions, Ford Fiesta and Volkswagen Polo (gas/diesel engine, driven normal/fast)*

Vehicle	CO_2 (g/km)	CO_2 (tonnes)	NOx (mg/km)	NOx + PM (kg)	MPG
Fiesta 1.6 L (gas, normal)	138	3.67	18	2.99	37.4
Fiesta 1.5 L (diesel, normal)	94	2.65	51	6.47	61.3
Fiesta 1.6 L (gas, fast)		4.17		3.32	
Fiesta 1.5 L (diesel, fast)		3.00		7.61	
Polo 1.4 L (gas, normal)	110	3.18	37	3.16	43.6
Polo 1.4 L (diesel, normal)	93	2.83	52	6.63	57.7
Polo 1.4 L (gas, fast)		3.61		3.52	
Polo 1.4 L (diesel, fast)		3.21		7.80	

Source: Lane, B., "Car emissions calculator," Nextgreencar.com.

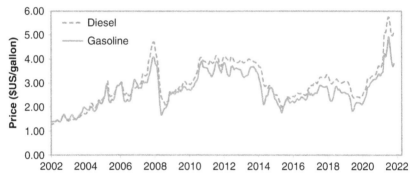

Figure 2.8 US gasoline and diesel prices per gallon (2002–2022). *Source*: "Gasoline and Diesel Fuel Update," US Energy Information Agency. www.eia .gov/petroleum/gasdiesel/gaspump_hist.php (gasoline) and www.eia.gov/petrol eum/gasdiesel/dieselpump_hist.php (diesel).

Polo running on gasoline or diesel and driven either normally or fast in real-world conditions for 10,000 miles. Note, PM is PM10 for direct tailpipe emissions plus indirect emissions generated during fuel and vehicle production.

There is plenty of information to digest here from just two cars, never mind the different ways to list emissions, yet we can see from the data that gasoline-fueled engines emit more CO_2, while diesel-fueled engines emit more NOx + PM (although they get better mileage). Driving aggressively also increases emissions by about 10% for both CO_2 and NOx + PM.

There are simple tips to improve fuel efficiency, reduce emissions, and save money: avoid aggressive driving, drive steadily at posted speed limits (efficiency drops above 60 mph), use the proper gear, engage the cruise control if possible on flat terrains, and cut down on air conditioning (open windows at low speeds).[132] One should also regularly service a vehicle, ensure the tires are properly inflated (1 psi under consumes 3% more fuel), burn the appropriate octane fuel (for example, 87/89/91/94), turn off power accessories before starting, tighten the gas cap (to avoid evaporation), use a block heater in winter, and shut off the engine instead of idling. An idling car releases more emissions because the engine is designed to run at speed.

Excessive speed also increases fuel consumption and thus emissions. Lowering the speed limit in the USA, for example, from 65 to 55 mph in 1975, saved an estimated 73 million barrels of oil in the first year and significantly reduced emissions. Even parking in shady areas, combining errands, and walking or cycling for short trips saves fuel and money. Fuel savings can be substantial from following just a few basic rules, while the health benefits are immeasurable. Trip length is also a factor as more than half of all fuel is consumed on trips within 3 miles of home.[133]

How a car is driven and the fuel type especially affects what comes out of the exhaust, essentially why Volkswagen was caught cheating in 2015. Once sold as a greener option to gasoline (less CO_2), diesel cars are now public enemy number 1, creating serious smog issues in Paris, Madrid, and other large cities, largely because of higher NOx. Spurred on by government subsidies, about one-third of vehicles in Europe are diesel, yet diesel has become a political hot potato over concerns about particulate matter emissions (clearly seen in the simple analysis above). Market penetration is much less in the USA (<1%) because diesel has always had a dirtier connotation, but with billions of dollars up for grabs, carmakers regularly use aggressive tactics to find new buyers for their supposed cleaner models.

The battle reached a head in October 2015, when the EPA charged Volkswagen with cheating on emission tests. As the findings showed, VW had installed a so-called "defeat device," initiated whenever a car's computer identified the standard conditions of an EPA test – smooth in-house rolling road, constant speed pattern, no steering – such that the on-board software turned on a "dyno calibration" to burn the fuel more completely by adjusting air–fuel ratios and exhaust flows, thus achieving lower emissions during the test.

The scandal was particularly disturbing because drivers had been tricked into thinking their diesel cars emitted fewer pollutants, when in fact under normal road conditions they were spewing NOx at levels 40 times the measured test results, possibly emitting up to one million extra tonnes per year,[134] while a 2017 study showed that almost 40,000 people per year were dying early "due to the failure of diesel vehicles to meet official limits in real driving conditions."[135]

The world's second-largest automaker with annual sales of almost 10 million vehicles, maker of "the people's" car, Herbie the Love Bug, and the brake-challenged family van in *Little Miss Sunshine*, Volkswagen had brazenly cheated both the regulatory bodies and the public. As reported by the US deputy attorney general in June 2016, after an extensive investigation into the carmaker's illegal actions, "Volkswagen's efforts to evade emissions rules [was] one of the most flagrant violations of environmental and consumer laws in our country's history. . . . By duping the regulators, Volkswagen turned nearly half a million American drivers into unwitting accomplices in an unprecedented assault on our environment."[136]

As part of the settlement, VW was ordered to buy back faulty cars – nearly 500,000 American 2.0-liter diesel vehicles sold from 2009 to 2015 (mostly VW Passat, VW Jetta, Audi, and Porsche) – pay compensation, and contribute $14.7 billion to green investments, the largest auto-industry class-action case in US history. The resulting fallout saw Volkswagen shares drop 35%, while claims are still pending in the USA and Europe, where more than 10 million VW diesel cars were equipped with the same defeat device.

There would seem to be an established culture of cheating in the auto industry. Fiat (turning off exhaust filters after a proscribed test time) and Renault (mismatched in-house and on-road levels) were also investigated, while Mitsubishi admitted to cheating on mileage ratings over 25 years and Mercedes-Benz paid almost $3 billion to settle its own charges in the diesel emissions scam. In 2019, the US Department of Justice announced that Ford was also being investigated for misleading consumers about emissions, citing false in-car EPA labels.

VW's CEO Martin Winterkorn, however, claimed the cheating at Volkswagen was due to "the grave errors of very few," despite the systematic gaming that would require knowledge of more than a few engineers in a 600,000-strong company. Ferdinand Piëch, VW chairman and grandson of Ferdinand Porsche – who built the first hybrid car in 1898, worked as an engineer for Mercedes-Benz, and later designed the iconic VW Beetle – was eventually forced to resign. Winterkorn would also resign amid allegations he knew about the defeat device. In 2019, German prosecutors further announced criminal charges against VW's next CEO Herbert Diess, chairman Hans Dieter Poetsch, and Winterkorn for stock market manipulation, while a 450,000-strong, diesel-owners' class-action suit rattled the German economy, where the auto industry accounts for 15% of the export market. Diess was finally ousted in 2022, while in 2023 the former Audi CEO Rupert Sadler eventually plead guilty to fraud and admitted others knew.

Despite the bravado and chummy government roots – the Lower Saxony government controls about 20% of VW – a previously unchallenged modus operandi is being called to task for its polluting and cheating ways. In a book about the scandal, *New York Times* reporter Jack Ewing noted that Piëch "created a company culture that allowed the diesel fraud to fester," where "people in charge were all too willing to resort to illegal behavior to fulfill the company's ambition."[137] A feared taskmaster, Piëch believed in "total dominance" and "impossible doesn't exist," personal creeds that precipitated the crisis. In a cruel attempt to show that NOx emissions were harmless, caged animals were even subjected to exhaust, while the company lied through its teeth about diesel's green cred in a highly successful American advertising campaign that saw sales shoot up across the more diesel-wary United States.[138]

Curiously, many of us seem unconcerned about tailpipe emissions, thinking unseen exhaust isn't a problem. But we know emissions are harmful – carbon monoxide is sufficiently toxic to cause death in a sealed-off garage, while those who have ever smoked understand the adverse effect of breathing in toxins. Second-hand smoke and first-hand toxins are killers, but what choice do ordinary consumers have when limited by price? An expensive hybrid Toyota Prius at $25,000 or an all-electric Tesla Model 3 at a minimum of $35,000 is a hard sell compared to a gasoline-fueled Ford Fiesta at $14,000 or the reduced cost of government-subsidized diesel cars in Europe. Electric vehicle prices have a long way to fall before the average buyer can fork out a large chunk of savings to help save the environment.

Fortunately, many carmakers provide buybacks on older, dirtier cars in exchange for newer, cleaner models, while some governments offer incentives to trade in older gasmobiles in various "cash-for-clunkers" programs. In 2016, the Government of Ontario announced a program of expanded rebates of more than $14,000 for electric vehicles, free overnight charging for 4 years, and mandatory electric garage outlets in new homes and condos. The program included incentives for truckers to switch from diesel to natural gas, propane, or gasoline mixed with ethanol.[139] The US government offers a tax credit up to $7,500 for new electric vehicle purchases, while other governments offer similar subsidies.

There are an estimated 1.5 billion vehicles worldwide, 97.5% of which run on gasoline or diesel in an internal combustion engine (ICE), although the number could rise to 2.5 billion by 2050, primarily because of continued growth in China where only 100 million vehicles were in operation in 2014, an amount that has more than tripled in just a decade to over 300 million.[140] Obviously, if the number of ICE vehicles increases so will emissions, expanding both our carbon and toxic footprint. Alas, little has been done to improve on an engine that wastes 80% of its energy as heat, while spewing out poisonous fumes in the process.

As we've seen, the main culprits are NOx, VOCs (volatile organic compounds such as benzene, toluene, 1,3-butadiene, and formaldehyde), PM2.5, and CO, although catalytic converters have reduced pollutants by 30% and reformed fuel has eliminated low-boiling-point aromatics. Diesel engines now emit less NOx and PM pollution because of exhaust gas recirculation, selective catalytic reduction via urea injection, and diesel particulate filters,[141] while newer models automatically shut off when stopped rather than idle (for example, at a traffic light), but the transportation sector still accounts for almost half of all NOx emissions. The contribution of road transport emissions in Europe is shown in Table 2.7 from a Thematic Strategy on Air Pollution (TSAP) report, highlighting the troubling amounts.

The history of the automobile shows that change is slow and incremental, wheels inside of wheels if you like, adding a fix here or a redesign there, typically making the engine perform better but not addressing the underlying problem of toxic exhaust filling our streets. Compared to exploding hydrocarbons, electromagnetic propulsion is a snap – an electric current produces a magnetic field in an outer coiled stator that repels an inner magnetized rotor forcing it to turn – all without having to pump, refine, or ship a drop of oil or jerry-rig an inefficient pollution-spewing engine to make the fuel burn better.

Table 2.7 *NOx, VOC, PM2.5, NH_3 CH_4, and SO_2 emissions in Europe (kilotons) and major contributors (kilotons, percent), 2010*

Emission	Kilotons	Major contributor	Amount (percent major contributor)	Amount (percent road transport)
NOx	8,805	Road transport	3,751 (42.6%)	3,751 (42.6%)
VOC	7,512	Solvent use	3,037 (40.4%)	1,100 (14.6%)
PM2.5	1,616	Domestic	695 (43.0%)	217 (13.4%)
NH_3	3,678	Waste treatment	174 (4.7%)	88 (2.4%)
CH_4	19,070	Agriculture	9,525 (50.0%)	84 (0.4%)
SO_2	4,837	Power generation	2,739 (56.6%)	7 (0.1%)

Source: Amann, M. "The Final Policy Scenarios of the EU Clean Air Policy Package," TSAP Report #11 Version 1.1a, International Institute for Applied Systems Analysis, February 2014. https://previous.iiasa.ac.at/web/home/research/researchPrograms/air/policy/TSAP_11-finalv1-1a.pdf.

Alas, changes to standard sales practices don't come easily – while the number of electric vehicles is increasing, the auto industry knows all the tricks since its inception over a century ago. Following the publication of *Unsafe at Any Speed*, which focused on GM's Corvair, General Motors sent detectives to spy on Ralph Nader and even set a supermarket honey trap to catch him out. As Nader noted: "Behind it all was, they didn't want the federal regulators telling them how to build a car in terms of safety. And then they said, 'If it's done in safety, it will be done in pollution control, and it will be done in fuel efficiency'."[142] Essentially, a free reign to dictate the free-wheeling ways of a permanently captive American motoring market. In a resulting congressional hearing into automobile safety, overseen by then attorney general Robert F. Kennedy, GM was forced to apologize for trying to besmirch Nader, turning *Unsafe at Any Speed* into a bestseller.

No apologies, however, came from the company that bought out trolley systems across the country in the 1930s and 1940s, tore out tram tracks to ensure cars ruled in an open American landscape, or backed the rise of the Greyhound bus line over inter-city electric trains, travesties Nader calls "one of the greatest economic conspiracies in American history." No apologies either for continuing to pollute the air we breathe and slowing the change to a cleaner future.

2.7 A New World Order: Money, Mecca, and the Middle East

Petroleum finds around the world have been creating new oil regions and new booms since the time of Colonel Drake, essential to an oil-run economy. One of the next major finds after Titusville was in Baku, Azerbaijan, on the western shores of the Caspian Sea, the world's largest lake, and near to where Marco Polo witnessed oil being applied as a horse ointment on his way to Xanadu and the court of the Mongol king. The oil there was initially turned into kerosene for Russian lamps, but was soon to be refined as gasoline for a growing automobile industry, bunker fuel for ships, and diesel to turn the engines of industry. The venture was backed by Rothschild family money and the technical expertise of three brothers: Robert, Ludvig, and Alfred Nobel.

Bigger tankers, however, were needed to move the increasing supplies around the world and reduce the high costs of transporting barrels over long distances. Middle brother Ludvig developed the bulk tanker, beginning service from Baku to the internal Russian market up the Volga, the longest river in Europe. Overcoming weight balance and safety issues, one ship captain noted,

"The difficulty was that the oil seemed to move quicker than water, and in rough weather when the vessel was pitched forward the oil would rush down and force the vessel into the waves."[143] Having solved the physics of a large, constantly shifting, ballast by installing vertical baffled compartments, the Nobel Brothers Petroleum Producing Company, a.k.a. Branobel, launched the first successful bulk tanker in 1878, capable of transporting 240 tons of kerosene.[144] The *Zoroaster* was deployed on the inland Caspian Sea, helping Branobel capture half of the Russian kerosene market.

More tankers were launched from the Black Sea, the Baku oil first transported through the Caucasus Mountains either by train to the Georgian port of Batumi via the Transcaucasus Railway or by pipeline. The route had been cleared with considerable quantities of a revolutionary new explosive, developed by the youngest Nobel brother, Alfred. Producing 30% of the world market, Branobel soon rivaled Rockefeller's Standard Oil.

Astonishingly, after Ludvig died in 1888, an obituary appeared in a French newspaper for the younger Alfred, announcing "*Le marchand de la morte est mort*" ("The merchant of death is dead") and that he "became rich by finding ways to kill more people faster than ever before."[145] Wanting to be remembered for more than dynamite and death, Alfred bequeathed most of his estate to establish prizes in Physics, Chemistry, Medicine, Literature, and Peace that bear his name and which have been awarded annually since 1901 almost without exception.

The English businessman, trader, and Shell founder Marcus Samuel further improved on the Nobels' tanker design to transport larger loads to markets in the Far East via the Black Sea and Suez Canal, which had opened in 1869 to connect the Mediterranean and the Red Sea. Launched in 1892 and filled with 4,000 tons of kerosene, the *Murex* was certified safe from fire and explosion, the tanks designed to withstand expanding and contracting volumes of kerosene at different temperatures. Having inherited his father's Far East trading business, Samuel had a fascination both with exotic shells and ferrying oil to market:

> By the end of 1893, Samuel had launched ten more ships, all of them named for seashells – the *Conch*, the *Clam*, the *Elax*, the *Cowrie*, and so on. By the end of 1895, sixty-nine tanker passages had been made through the Suez Canal, all but four ships owned or chartered by Samuel. By 1902, of all the oil to pass through the Suez Canal, 90 percent belonged to Samuel and his group.[146]

Thanks to plentiful reserves, the invention of more mechanized extraction methods, and a newly industrializing world, by the 1900s Azerbaijan was producing half of the world's oil. In a land where "Azer" means fire and the name of a god the locals worshiped for millennia, oil was worth more than gold. Unlike other fuels, petroleum is especially coveted because of its simple extraction methods (especially compared to coal), easy transportation, and versatility, whether to generate electricity, heat homes, run cars, trains, ships, and planes, or as a feedstock in various industrial processes such as making plastics, lubricants, and asphalt. The global petroleum market was inaugurated to exploit new sources and expanding markets. With each new find, the extraction of liquid gold increased and the world was soon awash in oil as the tankers and bankers filled and refilled their coffers.

In the United States, the familiar up-and-down "rocking horse" or "nodding donkey" pumpjacks could be seen dotting the plains of Ohio, Illinois, Oklahoma, California, and Texas, especially East Texas after the 100,000 barrel-per-day Spindletop strike in 1901 that tripled US production, spawning Texaco and the Mellon family's Gulf Oil. More finds in Sumatra (Royal Dutch), Borneo (Shell), and Persia (Anglo-Persia) created a growing international market that would challenge the dominance of American kerosene sales, led as ever by the monopolist giant Standard Oil (a.k.a. the Octopus). The global supply chain widened as the oil majors followed wherever the black gold flowed:

Mexico became a producer in 1901, followed by Argentina in 1907 and Trinidad in 1908. By 1910, world production had grown to 900,000 barrels per day, the bulk originating in the USA (560,000) and most of the remainder in Russia (200,000). Oil was found in Persia (Iran) in 1908, and exports there commenced in 1911, leading to the prominence of that region as a source of crude oil. Production in British Borneo and Venezuela began in 1911 and 1914, respectively.[147]

By the turn of the century, the "Great Game for Oil" was being played out across Asia and Europe, Russia controlling the kerosene markets of Austria–Hungary, the Balkans, and Turkey and the United States much of Germany, France, and England. Covert operations flourished with oil experts posing as agricultural workers, rubber contractors, and even diplomats to ensure continued access to the supplies and to keep the tap open whatever the circumstances.[148]

Although World War I and the Great Depression slowed exports, exploration continued across the globe, wherever oil's tell-tale signs were found and a suitable analysis supported further survey. Having heard about oil seepages in the Persian Gulf, a former World War I British army quartermaster found oil on May 31, 1932, at al-Hasa in Bahrain, a seemingly nondescript island off the east coast of Saudi Arabia. Major Frank Holmes would become known to the locals as "Abu al-Naft" (the "Father of Oil"), although full-scale extraction would lag amid the wrangling of competing interests and the coming winds of another world war.

***

In time, the biggest prize of all would be Saudi Arabia, known for its endless deserts and clan-centered Bedouin wanderers, living as best they could among inhospitable sands and oases, where temperatures can vary from a blazing midday 50°C to a frigid 0°C at night. The Arabian Peninsula's claim to fame – the two historic cities of Mecca and Medina – brought tens of thousands of pilgrims to the Muslim Holy Land during the Hajj, beginning 70 days after Ramadan, as well as financial tributes to whomever ruled. After decades of tribal war, Abdulaziz ibn Muhammad al Saud was proclaimed King of the Hejaz and Nejd in 1926, returning his family to dominance in the Holy Land, in charge of Mecca and, more importantly, the tribute money.

With a reduction of pilgrims during the Depression years, however, funds were scarce, and although wary of foreigners averse to strict Wahhabi laws traipsing around his kingdom, the Saudi king – known as Ibn Saud in the West – was eager to strike an exploration deal after the nearby Bahrain find. In 1933, Ibn Saud received enough gold to keep his restored kingdom afloat, while the American oil company Socal landed a 60-year concession over almost half of his lands, an area bigger than Texas. The Bedu may have initially thought the automobile was the "devil's invention," but the lure of black gold soon changed their thinking.

Joining with Texaco in 1936, Socal reformed as the American oil company Caltex, initially looking for oil in Saudi Arabia's Eastern Province at the site of a tell-tale, rocky, dome-shaped hill in the Dammam Zone. After minimal results from the first six holes, 10 million dollars spent, and tribute money scarce from a lingering global depression, however, the investors wanted to pull the plug, but fortunately Dammam No. 7 finally hit. After 5 years of hard slogging, just under a mile down on March 4, 1938, the field immediately began producing 1,500 barrels a day, upped to almost 4,000 by the end of the month.[149]

Building on the success of No. 7's "deep play," holes 2 and 4 were redrilled deeper, ultimately producing commercial quantities that would continue for almost five decades. Although the new fields wouldn't be effectively tapped until after World War II, the smell of oil was thick and Ibn Saud was safely ensconced in power over 90% of the Arabian Peninsula. Dammam No. 7 would ultimately produce over 32 *billion* barrels before being shut down in 1982.[150]

Managed by Caltex, the Dammam finds skirted the terms of the "Red Line" Agreement, a non-compete zone established in 1928 by the main Middle East players – Standard, Shell, Royal Dutch, Anglo-Persian, and Gulf – to keep a lid on costs and unnecessary competition. The deal had been successfully brokered by Calouste Gulbenkian, a.k.a. "Mr. Five Per Cent," an Armenian business-man, who had the wherewithal to keep digging in the sands long before the West came calling. Gulbenkian was the first great transnational oil wheeler dealer and founder of the Turkish Petroleum Company (later the Iraq Petroleum Company).

Deemed strategically unimportant during World War II, Saudi Arabia escaped the carnage wrought across much of Europe, primarily because of its still immature wells. Franklin Roosevelt referred to Saudi Arabia in 1941 as "a little far afield for us," but the United States soon saw the importance of keeping the Saudi king happy, authorizing assistance through the Lend–Lease policy initiated to keep the UK afloat during the Battle of Britain despite American neutrality. Although American interests were slow to materialize, FDR would ultimately reappraise the USA–Saudi relationship in a February 1943 statement: "I hereby find the defense of Saudi Arabia vital to the defense of the United States."[151] By the end of the hostilities, it was clear the Middle East would become a battlefield in another kind of war.

Returning from an oil reconnaissance trip to the Gulf in early 1944, the famed Texas oilman Everette Lee DeGolyer – pioneer of the seismograph and its use to search for oil, and who once discovered a well producing 110,000 barrels per day – knew that world politics was about to change, a member of his team quipping, "The oil in this region is the greatest single prize in all history."[152] That prize would come to comprise Saudi Arabia, Iraq, Iran, Kuwait, Qatar, and Bahrain, whose collective reserves potentially totaled as much as 300 *billion* barrels, a third located in Saudi Arabia. In the post-war rush to ensure supplies for the oil-run armies of a modern world, another conflict was looming – a war for oil.

After some bitter wrangling between the USA and the UK, who both thought the other was trying to expand on their presumed concession, the two agreed to cooperate, effectively carving up the Middle East between them: Iran to Great Britain, Saudi Arabia to the United States, and Iraq and Kuwait to be shared. An orderly production and price stability was the stated goal. As Daniel Yergin noted in *The Prize*, his epic tale about the global petroleum business, "A rising tide of cheap oil from the Persian Gulf after the war could be as destabilizing as the flood of oil from East Texas in the early 1930s. At the same time, many in the United States continued to fear the exhaustion of US reserves and wanted to reduce the call on American oil."[153]

Over time, Saudi Arabia was effectively groomed as the world's oil capital, outright controlled by the United States, who in 1940 had "almost single-handedly fuelled the entire Allied war effort,"[154] producing 63% of global oil, most of which was exported from the Gulf of Mexico. Recognizing that its domestic supply was insufficient in the event of another war, the USA initiated a special relationship with the Saudi royal family, personally cemented in 1944 during a secret meeting between Roosevelt and Ibn Saud aboard the USS *Quincy* on Great Bitter Lake in the Suez Canal after Roosevelt's return from the Big Three Yalta Conference: "For the Saudi King, this was perhaps only his second trip outside his kingdom since that day, forty-five years ago, when he had left exile in Kuwait to take his first step – the assault on Riyadh – toward regaining Arabia."[155] That time, Ibn Saud had been smuggled out of his native homeland in the saddlebag of a camel.

The two leaders hit it off during their long, intense discussions, where "the King declared that he was the 'twin' brother of the President because of their close ages, the responsibilities for their nations' well-being, their interests in farming, and their grave physical infirmities."[156] In an exchange of gifts, Roosevelt gave Ibn Saud his spare wheelchair, which the king – arthritic, his hands shaking from Parkinson's, and part-invalid from war wounds – proudly displayed thereafter at his home in Riyadh. The resultant arrangement to develop the vast Saudi Arabian oil fields would become known as "solidifica- tion," code for American aid in exchange for oil, establishing a Saudi– American quid pro quo relationship that prevails today. At the same time, British colonial power was being eroded across the region, the Brits forced to play second-fiddle to the United States, to which the United Kingdom owed its existence after the devastation of World War II.

By 1949, after the dust had settled on the travesties of war, the United States was responsible for 60% of global industrial production, beginning the domin- ance of American foreign policy and the US dollar as the de-facto world currency, not least to counter Soviet expansion as oil became a political weapon against growing labor unions and "left-wing European workers' movements tied to coal."[157] Oil is easier to produce, control, and much less labor-intensive than coal, helping to keep workers in line who are less able to organize, while unions are powerless to counter the globalized petroleum industry unchecked by a national governance that exists to keep the oil profits flowing.

***

In the immediate aftermath of World War II, securing the distribution of oil became paramount to Western interests, especially after Egyptian president and pan-Arab nationalist Gamal Abdel Nasser appropriated the French- and British-run Suez Canal in 1956, ultimately using the profits to build the Aswan Dam.[158] The move blocked transit through the main oil route from the Persian Gulf to the West, drastically reducing oil supplies to Europe during the winter of 1956–57. Although the crisis was resolved without a full-blown conflict, larger "supertankers" were built to ensure an economically viable passage around the Cape of Good Hope, avoiding further disruptions to the flow of oil from field to consumer.

Ironically, having established the Suez as a viable route to transport petroleum – ferrying kerosene to the Far East in the first open-sea bulk carrier *Murex* – Shell led the way, designing and building larger carriers up to 200,000 deadweight tons (40 times *Murex*) and installing improved safety features to reduce dangerous static and explosive gas build-up in the storage tanks. Although the distance was more than doubled – adding 6,000 miles to a 5,000-mile journey – bypassing Suez ensured a free flow of oil in case of more closures to the canal through which two-thirds of Europe's oil passed. Indeed, the canal would be closed again for 8 years after the 1967 Arab–Israeli Six-Day War (instead of 5 months as in 1956–57). Rather than share resources, new refineries were also built closer to consumers to lessen the impact of reduced output in war-torn, politically unstable, and unfriendly regions.

Today, most of the world's oil is transported in very large crude carriers (VLCC) and ultra large crude carriers (ULCC) at lengths almost a half-kilometer long. Costing more than $100 million to construct, a single cargo can fetch as much as $50 million.[159] The largest are even too big to enter some ports, requiring at-sea, pre-dock offloading. Before being decommissioned in 2009, the world's largest oil tanker, the *Mont*, was longer than the height of the Empire State Building. Highlighting the ongoing significance to tanker traffic in the Middle East, the grounded cargo ship the *Ever Given* shut down shipping via the Suez for a week in 2021, halting 30% of global container traffic and one-tenth the daily supply of oil.

Gulf oil is also transported by pipeline, primarily the 1,200-km Petroline from Saudi Aramco's oil-processing facility at Abqaiq, south of Dammam, to the Red Sea port of Yanbu, strategically important in the case of disruption in the Strait of Hormuz, another "chokepoint" through which more than 20% of the world's daily supply passes. Six miles wide at its narrowest point, transit is periodically threatened in times of strife as during the 1987 "Tanker War" between Iran and Iraq, when the US navy was called in to escort reflagged Kuwaiti ships. Another potential chokepoint is the Bab-el-Mandeb Strait on the southwest coast of the Arabian Peninsula between Yemen and Djibouti, con-necting the Gulf of Aden to the Red Sea and linking the Indian Ocean to the Mediterranean.

With the advent of supertankers, pipeline traffic in the Middle East corres-pondingly decreased to less than 10%, although prior to 1990 the Trans-Arabian Pipeline (Tapline) also pumped 500,000 barrels per day from Abqaiq to the Mediterranean port city of Sidon, Lebanon, a distance of 1,700 km through Saudi Arabia, Jordan, Syria, and Lebanon. Laid after the end of the Red Line Agreement and British exit from Jordan, Tapline was one of the biggest engineering feats of its time, built by the American engineering and construction company Bechtel and considered "one of the great arteries of Empire," vastly reducing the 5,600-km sea route around the Arabian Peninsula and saving Aramco and its American backers millions in tanker costs and Suez tolls.

In *The Crash of Flight 3804*, American journalist Charlotte Dennett's account of her father's ill-fated, post-war spying for the CIA – then called the OSS – she noted that Tapline "completed in 1950, would change the balance of power in the whole Middle East and, for that matter, the whole world. It would help firmly establish the United States as a global superpower."[160] Given the strategic importance of Gulf oil to American interests – both for domestic consumers and the US military – maintaining the supply and distribution of oil became the primary goal of the American government in the Middle East.

Tapline also saved American oil for domestic use in the event of Cold War hostilities, as Saudi Arabia could provide Europe with petroleum instead of depleting American reserves as well as generating wealth both for the US treasury and the Saudi royal family via increased sales.[161] The newly built pipeline also cemented American priority in the region as the USA, Britain, France, and the Soviet Union – allies during the war against Germany – began to assert their influence in an increasingly fragmented and volatile post-war world as the competition heated up in a cut-throat game of "energy imperialism" and endless wars.[162]

Pipelines and politics, however, are a dangerous mix, especially in a region full of autocratic leaders, warring tribes, and ever-shifting alliances. Prior to construction, Tapline's route and terminus on the Mediterranean was the cause of increased intrigue and gamesmanship among the oil companies and between state entities, not least because of rising Jewish immigration to the region anathema to Saudi Arabia and Syria. Construction began only after a 1949, US-sponsored coup overthrew the Syrian president, who was opposed to the project, replaced by an army officer who immediately approved the route and Sidon terminus. Periodically closed during on-and-off troubles and adversely affected by increased supertanker traffic, Saudi Aramco also eventually shut down Tapline after a tariff disagreement with Syria and Jordan's support for Iraq during the 1990 Gulf War.

The same scenario may be playing itself out again in the latest Middle East civil war after Syria's president Bashar al-Assad chose a Russian-backed, Iranian-sourced pipeline proposal over a US-backed, Qatari-sourced plan to bring natural gas supplies from the South Pars-North Dome gas field in the Persian Gulf to the Mediterranean and on to Europe. Calls to renew other dormant pipeline routes from Kirkuk in northern Iraq to the Mediterranean, either to Tripoli or Haifa, have also begun despite troubles en route. Highlighting the ever-changing landscape, Iraqi oil exports significantly increased after years of strife, while Iran, which continues to produce about 5% of world oil, has seen its exports slashed during the ongoing diplomatic conflict with the United States over uranium enrichment. The Cold War may be over, but smaller wars are still raging throughout the Middle East.

Through it all, Saudi oil remains protected, first among unequals. Built near the Dammam find on the Gulf coast, the Ras Tanura terminal is tap to the world's greatest bonanza, where roughly one in six barrels of exported oil begins each day, the valve personally opened by Ibn Saud in 1939. In the desert sands oil is everywhere, yet hidden from view as Peter Maass notes in *Crude Oil*:

> In a technological sleight of hand, oil can be extracted from the deserts of Arabia, processed to eliminate water and natural gas, sent through pipelines to a terminal on the gulf, loaded onto a supertanker and shipped to a port thousands of miles away, then run through a refinery and poured into a tanker truck that delivers it to a suburban gas station, where it is pumped into an SUV – all without anyone actually glimpsing the stuff.[163]

Today, Saudi Arabia is the largest producer and exporter of oil in the world with proven reserves of 268 billion barrels (16%), second only to Venezuela's almost 300 billion barrels (18%) – although Venezuela's heavier crude is harder to refine – and ahead of Canada, Iran, and Iraq (each at about 10%). Together, the Gulf countries have proven reserves equal to almost half the world's estimated 1.6 trillion barrel supply. The prognosis of Major Frank Holmes, who stuck to his dream of finding oil under the deserts of Arabia, has more than been realized (Figure 2.9).

Persian Gulf countries also contain vast amounts of natural gas, accounting for as much as 40% of proven reserves. The Gulf countries produce about one-sixth the world's supply (almost 600 billon of a global 3.7 trillion m^3)[164] and export 13% (almost 160 billion of a global 1.2 trillion m^3),[165] led primarily by Iran and Qatar. Located under the Gulf waters between Iran and Qatar, the extraordinarily vast South Pars-North Dome Field contains more than 50 trillion cubic meters of natural gas alone, enough to cover global consumption for over a decade.

As noted by *National Geographic* in an article about the changing fortunes of a modernizing Qatar after a peaceful family coup in 1995 brought a 45-year-old

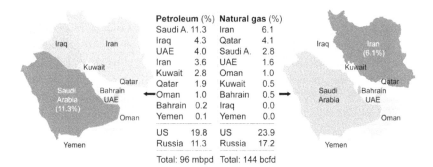

Petroleum (%)		Natural gas (%)	
Saudi A.	11.3	Iran	6.1
Iraq	4.3	Qatar	4.1
UAE	4.0	Saudi A.	2.8
Iran	3.6	UAE	1.6
Kuwait	2.8	Oman	1.0
Qatar	1.9	Kuwait	0.5
Oman	1.0	Bahrain	0.5
Bahrain	0.2	Iraq	0.0
Yemen	0.1	Yemen	0.0
US	19.8	US	23.9
Russia	11.3	Russia	17.2
Total: 96 mbpd		Total: 144 bcfd	

Figure 2.9 Middle East petroleum (*source*: US Energy Information Administration, "International: Petroleum and other liquids," 2023. www.eia.gov/international/data/world/petroleum-and-other-liquids/annual-petroleum-and-other-liquids-production) and natural gas (*source*: US Energy Information Administration, "International: Natural gas," 2023. www.eia.gov/international/data/world/natural-gas/dry-natural-gas-production) production percentages (2021).

Hamid Khalifa to the throne of one of the smallest and least-known Arab countries, the Qatari portion of the field beggars belief, holding roughly 25 trillion cubic meters of natural gas or "enough to heat all US homes for more than a century."[166] Although lagging behind the Qataris in exploiting its offshore natural gas bonanza, Iran is nonetheless central to both the Persian Gulf and Caspian Sea, which together contain about two-thirds of global petroleum reserves.

The politics of oil is about ensuring a steady supply at any cost, including war in the name of peace. But to understand today's market, one must first learn from where the oil comes, almost 80 million barrels produced and 40 million barrels exported every single day. Why is the stability of the Saudi royal family important to the West? Why are Iraq and Iran permanently in the news or at war? Why do the tiny statelets of Kuwait, Qatar, Bahrain, and Oman, with a combined population of under 13 million, exert so much influence on world affairs? Why is the Burj Khalifa, the world's tallest building since 2010, in Dubai, UAE? Why is China interested in ending the internecine conflict between Shia-led Iran and Sunni-led Saudi Arabia?

Could it be these eight countries produce and export almost one-third of all oil with trillions of dollars in annual revenues? Could it be the entire global economic system is based upon a free-flowing, protected supply of cheap, easy-to-refine oil, much of it found in the Middle East? The geopolitics of the twentieth century began the moment liquid fuel was found in the depths of a distant dessert, requiring a strong-armed enforcer to ensure its continued flow from well to home. The consumer may not be privy to the details – extraction, protection, transportation, or refining – but there is no denying the enormity of its commerce or the consequences.

2.8 Saudi Arabia, OPEC, and the Politics/Price of Oil

In the business of oil, one country stands above the others, impervious to conflict yet hastily reinventing itself in a changing energy landscape. Modern geopolitics is inextricably linked to a desert country, three times the size of Texas, floating atop the world's most coveted asset. With vast petroleum resources, one needn't look further than the Arabian Peninsula to understand a world run on black gold.

Since 1932, the Kingdom of Saudi Arabia has been run by the House of Saud, beginning with its founder Ibn Saud, followed since his death in 1953 by six of his 45 sons from 20 of his wives (he also had more than 100 daughters).[167] Little is known about the inner workings of the Saud family, a mix of brothers, half-brothers, cousins and half-cousins (perhaps as many as 15,000), ranging from flashy, nouveau-riche gamblers to ultra-conservative Wahhabi apologists, who for many years banned women from driving and outlawed cinemas.

The House of Saud has seen many financial ups and downs with the ever-changing price of oil as well as challenges from within as it seeks to balance an immense financial wealth with an austere Wahhabi ideology. In effect, the country is a family-run business without need of popular support, impervious to external stricture, although some have questioned its overly mysterious and autocratic ways. Middle East scholar Bernard Lewis noted that "The discovery of oil and the commercial exploration of oil has been a disaster for the region as a whole and it has enormously strengthened the power of the ruler, whatever kind of ruler it might be."[168] Lewis further noted "It made a world force of what otherwise would have been an extremely stringent marginal country."[169]

Much of today's geopolitics can be dated to 1960 and the start of the Organization of Petroleum Exporting Countries (OPEC), led by Saudi Arabia. Angered by continued Israeli occupation of the Sinai and Golan Heights, lost in the 1967 Six-Day War, Egypt and Syria launched a surprise attack on Israel in 1973 on the holiest of Jewish holidays, Yom Kippur. When the USA sent $2.2 billion in arms to aid Israel, Egyptian President Anwar Sadat persuaded the Saudi king Faisal – the first son to succeed his father – to use the "oil weapon" to restrict Western exports, as well as threaten a 5% monthly production cut. The price of oil increased fourfold from $2.70 to $11.00

a barrel, firmly establishing Saudi Arabia and OPEC at the center of a world dependent on oil.

The oil weapon successfully transformed international relations, giving oil-producing nations a say in global sales for the first time. Although 50–50 arrangements had already existed in a number of oil-producing states as begun in Venezuela, the oil-flush states wanted more of the pie, *their* pie as they began to clamor. As noted after OPEC's formation, "It was quite clear from the start that the price cuts might precipitate the establishment of what some delegates chose to call a cartel to confront a cartel."[170] Disruptions, however, were typically averted with deft diplomacy – during the 1957 Suez Crisis Dwight Eisenhower threatened the Saudis that the USA would look elsewhere if supplies were not secured. Even the shock of a fourfold increase in prices kept the supplies flowing.

OPEC strategies have varied over time, including production cuts to raise prices (via member quotas) or flooding the market to lower prices and increase market share to undercut the competition (via unlimited member production), for example, to derail high-cost modern producers of shale gas and oil sands and the increasingly competitive renewable technologies of wind and solar. With other oil regions producing more – Russia, the North Sea, American shale deposits, and Canadian oil sands – OPEC's influence has begun to wane of late, although the world can't help returning to the Middle East to refill its plate ever drawn to its permanently stocked, all-you-can-eat buffet. For its part, Saudi Arabia acts as the world's "swing producer" keeping up to two million barrels per day in surplus, raising or lowering production to maintain the price (or some say manipulate) to protect the market share of a precious commodity. (As of 2016, Russia agreed to act in concert with OPEC to maintain production and control prices, generally known as OPEC+.)

Oil is unlike any other product – without transparent regulation or govern-ance, one can't ensure a price based on traditional supply and demand. As in the distribution of kerosene prior to the 1911 court-ordered break-up of Standard Oil, today's oil barons seek to set their own prices. The secret for the Saudis is to get the highest price possible, yet not so high that buyers look elsewhere, what the Bedouins call good "tent management." With more viable alternative sources available to the consumer, green energies on the rise, and proper carbon pricing, oil might one day be priced according to an actual market value.

Little international scrutiny has followed the exploits of the richest, most secretive family in history. According to a former US Saudi ambassador, Chas Freeman, Saudi Arabia is a bribe-based economy full of "commissions" to senior royal family members, essentially "rakeoffs from public business."[171] In one UK deal to supply communications equipment to the Saudi national guard, the Saudi defense minister, crown prince Abdullah, was paid almost half a billion pounds, while £6 billion may have been skimmed off a secret UK–KSA arms contract for oil through US banks. Known as the Al-Yamamah project, the largest ever British financial deal has been called "the most corrupt transaction in commercial history."[172] Ironically, yamamah means "dove" in Arabic.

Now in his eighties, the current king, King Salman, is one of the last legitimate sons of Ibn Saud, and thus the first of the third generation of Saudi princes will soon become Custodian of the Two Holy Mosques. The title was given to Saladin, the first sultan of Egypt and Syria in 1137, reapplied in 1986 by Ibn Saud's fourth son Fahd to himself after trading his flamboyant ways for a more pious life – Fahd reputedly lost $6 million in a single night gambling at Monte Carlo and was known for throwing money from car windows to his fawning subjects.[173] The de-facto successor is Salman's son, crown prince Mohammed bin Salman, a.k.a. MbS, who is keen to diversify Saudi Arabia's oil-based economy. Since 2017, MbS has effectively ruled in his father's stead.

Initiated in 2016, MbS's roadmap for the future, *Vision 2030*, outlined his ambitious plans to maintain the two holy mosques, form a business hub at the crossroads of three continents – where 30% of world trade passes – and create a multi-billion-dollar sovereign wealth fund by selling up to 5% of Saudi Aramco, the world's richest company with annual profits of over $100 billion.[174] Less than a century after his grandfather's reunification of the Arabian Peninsula, followed by the first geological expeditions into the deserts of the Eastern Province by camel caravan and car convoy under the original 1933 Socal concession, MbS is tepidly modernizing the most backward yet wealthiest of countries.

The petro-states are being forced to diversify, however, because of renewable energy, global warming, and the ongoing vagaries of oil prices. When the price of crude dropped from $120 to $30 a barrel in the space of 18 months from June 2014, the losses were almost $2 billion *per day*. Long delayed by price rumblings, initial public offering (IPO) haggling, uncertainty over stability in the House of Saud, and a possible succession battle, Aramco was part-privatized in 2019, the largest stock float in history, netting the Saudis $30 billion on an initial valuation of $1.7 trillion.[175] Bahrain, Qatar, and the UAE are also diversifying into tourism, trade, and international events, financed by large, rainy-day, investment funds.

Ironically, with the eventual reduction of fossil fuels across the globe, the Middle East is set to benefit from the transition to renewables, especially in the more diversified economies of Qatar, the UAE, and Saudi Arabia. As noted by Georgetown University professor of international relations, Marwa Daoudy,

"Far from losing out in the green economy of the future, those countries are poised to reap significant gains: aggregate demand for oil is likely to increase before it falls, and they are well positioned to become major suppliers of solar energy, which will become an increasingly important resource."[176] Masdar City, the first carbon-neutral city being built just outside the UAE capital of Abu Dhabi, exemplifies a new era triumphantly run on solar power and without cars.

***

Throughout all the geopolitical gamesmanship, a flush Saudi Arabia still couldn't exist without the United States and their renegotiated deal to exchange goods and services after the 1973 oil embargo brought on by the Israeli–Arab war. The United States–Saudi Arabian Joint Commission on Economic Cooperation (JECOR) is unlike other aid programs, however, that shell out money to client states, Saudi oil profits instead used to employ American companies to build up Saudi Arabia.[177] The "building up" ensures a mutual interdependence that sees oil go to the United States, while Saudi "petrodollars" are sunk back into the American economy to "balance payments." Underwritten by US Treasury bills, it is easier to trade goods and services, including hundreds of billions of dollars of American-made arms for oil, than to instigate regime change as in the past, while the days of holding the West hostage are gone thanks to a tolerably compliant Saudi Arabia and its protected supply of exported oil.

Under the guise of global aid and fostering democracy after World War II and the subsequent Cold War, the USA was also keen to dismantle colonial management in mineral-rich countries, especially British and French colonies or protectorates. The goal was to control resources to ensure privileged American access, backed by US multinational corporations, petrodollars, and military bases. In the process, developing countries became disadvantaged by a modern, strong-arm mercantilism that sends resources and unfinished goods to the West that are then turned into higher-priced manufactured products.

Oil is the resource upon which the entire system depends, pumping up Western economies, in particular in North America, Europe, and Japan, where two-thirds of post-war growth was "due simply to an increasing use of fossil fuel."[178] The ecological footprint is also unequal as the developing world absorbs the burden of the West's extractive and polluting practices. The modern world has literally been built on the back of resource-rich developing countries. What's more, the hidden military costs to protect the supply are not included in the price at the pumps, keeping American consumers happy yet unmotivated to use alternatives while paying no heed to the environment.

One might think the 1973 oil embargo would have provided a wake-up call, Western consumers forced to question their addiction and the consequences of unchecked growth. In two televised addresses to the American public in November 1973,[179,180] US president Richard Nixon even outlined a series of measures to deal with "the most acute shortages of energy since World War II." There would be no Sunday driving ban as in Europe, but Nixon did ask citizens to reduce home heating and driving speeds, and to eliminate unnecessary lighting. He called for the closing of Sunday gasoline pumps to discourage long-distance weekend driving, adding with a wry smile that "It will mean perhaps spending a little more time at home." An expected 17% drop in supply from the crisis in the Middle East was turning "serious energy shortages . . . into a major energy crisis."

The Emergency Highway Energy Conservation Act would be passed the next year, reducing the speed limit to 50 mph for cars and 55 mph for buses, saving an estimated 200,000 barrels of gasoline a day. Noting that the average American consumed more in a week than most of the world did in a year, Nixon also proposed more efficient air travel to reduce jet fuel – better scheduling and larger passenger loads – and a curtailing of outdoor Christmas lighting. He then announced a goal of energy self-sufficiency by the end of the decade, called Project Independence, so "Americans will not have to rely on any source of energy beyond our own." Unlike his predecessors, the 37th US president instigated a national energy policy in a once bountiful country brimming with resources, stressing "the spirit of discipline, self-restraint, and unity which is the cornerstone of our great and good country."[181]

The long queues at the pumps during the 1970s eventually spurred on creative thinking about alternatives to oil, while conservation plans were encouraged by novel government initiatives. Fuel efficiency was suddenly of national importance and new standards devised to stop gas-guzzling behemoths clogging up the roads. The Corporate Average Fuel Economy (CAFE) standards increased fuel efficiency from a woeful 13 miles per gallon in 1975 to 27.5 mpg in 1986, an astonishing savings of 5 million barrels of oil per day.[182]

Amid concerns over diminishing supplies in an increasingly volatile Middle East, as well as the potential damage from burning fossil fuels, Jimmy Carter created the US Department of Energy in 1977. A former navy engineering officer with a Bachelor of Science degree from the US Naval Academy in Annapolis, Carter had considered enacting a synthetic fuels program that would convert coal and shale to gas to avert the continuing crisis, although a group of Harvard researchers noted that conservation was more effective and less polluting – reducing just 2 million barrels per day could produce a savings of $40 a barrel.[183] The math was simple as noted by fuel sciences professor Harold Schobert:

> If we had continued to conserve oil at the same rate as achieved in the 1976–1985 period, or even if we had collectively bought new cars that got 5 mpg more than they did, the United States would no longer need Persian Gulf oil. . . . Unfortunately, federal policy in the 1980s discouraged energy efficiency. . . . Even as late as the Persian Gulf War in 1991, if the first President Bush had required that the average cars get 32 mpg, that by itself would have displaced all Persian Gulf imports to the United States.[184]

Alas, US economic policy favors producers not consumers. After the ousting of the shah of Iran in 1979 in a populist, anti-American Iranian revolution that removed millions more barrels from global markets and another sharp price increase from $13 to $34 a barrel[185] – known as the Second Oil Shock after the First Oil Shock of the 1973 embargo – consumer changes in the following eight years reduced imports from the Persian Gulf by 87% and American imports by 42%, due primarily to better fuel efficiency standards. More shockingly, Ronald Reagan rolled back those same standards in 1985, doubling American oil imports from the Persian Gulf.[186] The Emergency Highway Energy Conservation Act would also be repealed in 1995 under Bill Clinton.

Despite an increased awareness of the fluctuating costs and temperamental nature of foreign supplies, oil would continue to fuel American interests, not least to ensure military readiness in a still chilly Cold War. If there was any doubt about the importance of Middle East oil for future American needs, Jimmy Carter spelled out the proclaimed American policy in his 1980 State of the Union address at the height of the Iranian hostage crisis: "An attempt by any outside force to gain control of the Persian Gulf region will be regarded as an assault on the vital interests of the United States of America and such an assault will be repelled by any means necessary including military force." The die had been cast – protection of a foreign oil source would be valued above efficiency and conservation at all costs, including war. Compliance would be bargained with military might and roughly 50,000 American troops stationed in six Gulf countries, primarily to contain Iran's influence and ensure an unfettered supply of oil to American consumers, while the USA continues to make friendly with Saudi Arabia to keep OPEC sales denominated in petrodollars.

The state-sponsored protection continues, the so-called Carter Doctrine justifying Western support for Kuwait against an invading Iraqi army in 1990 with its sights set on neighboring Riyadh as well as the 2003 invasion of Iraq to oust a recalcitrant Saddam Hussein after the World Trade Center was destroyed on September 11, 2001. The fix was in, ensuring that oil would continue to fund the world's economy. As ever, the Saudis were happy to comply, although not everyone thought that price gouging was a prudent strategy lest one anger a servile customer base. Ahmed Zaki Yamani, Saudi oil minister from 1962 to 1986, foresaw problems from excessive increases to a vibrant oil market after the 1973 embargo:

> Very high prices will have a reaction. The reaction happened. But you don't come to oil producers who can make a lot of money and tell them no don't make a lot of money. They don't consider the strategy. Strategy because when you raise the price of oil you enable the oil companies to use the extra money to explore for oil and this is what happened in the North Sea, in Mexico, and elsewhere. So the level outside OPEC took place, competing with the price of OPEC.[187]

What would become known as the "Yamani Edict" initially kept a lid on more price increases, but the reality of high-priced petroleum soon brought greater revenues to producers and distributors. Exxon doubled its revenues on the backs of hijacked consumers, while conservation and alternative energy became dirtier than oil. As Yamani predicted, however, bleeding one's customers is a failed strategy. Under safe, non-OPEC control in the UK and Norway, North Sea oil and gas fields were developed because of higher prices and increased investment. Considered "buccaneering capitalism at its most adventurous,"[188] the first major offshore wells soon began producing, eventually reducing European dependence on Middle Eastern supplies despite exploration costs ten times that of the Middle East.

France began an overhaul of national energy needs, converting almost entirely to nuclear power, while Japan looked to nuclear and natural gas that until then had been mostly flared off. West Germany also found gas, piped in from Russia. Fortuitously located on the edge of a permanently gusting North Sea, Denmark turned to wind power with added investment and government subsidies. In the United States, nuclear energy was expanded, coal reintroduced as an electricity fuel, and Colorado shale oil explored at costs up to $8 billion soon abandoned when prices fell again. Development of offshore fields, shale oil, and oil sands began to trim Saudi influence with many new players eager to lead the way.

Saudi Arabia – in effect, Saudi Aramco and the royal family – has had its way with the American consumer for over seven decades, little if any criticism offered up as public debate about undemocratic governance, support for exported terrorism, and female enslavement. But the winds are changing among the deserts of Arabia as the danger of extracting, transporting, and burning petroleum is exposed amid the devastation wrought by cheap liquid energy. Of course, old habits die hard. Quitting any addiction is easier said than done (Figure 2.10).

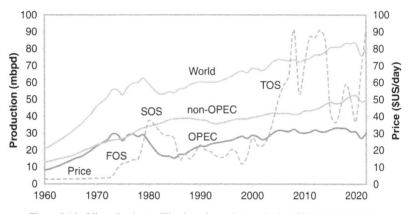

Figure 2.10 Oil production (million barrels per day) and price ($US) (1960–2022).

2.9 Oil Basics: Who, What, Where

It took five years to tap the giant Dammam field, but other "elephants" soon followed at Ghawar (1948, online in 1951) and Khurais (1963, online in 2009). Echoing the earlier analysis of Lee DeGolyer and his survey team, the US State Department described Saudi Arabia in 1945 as "a stupendous source of strategic power, and one of the greatest material prizes in world history,"[189] providing the go-ahead to subjugate the whole of the Middle East and reason enough to invade Iraq in 2003 in response to terrorist attacks on the World Trade Center and the Pentagon, ultimately pushing oil prices to record highs in what could be called the Third Oil Shock. Although not involved in the 2001 9/11 attack, Iraq had lots of oil – good quality, easily accessible, and near to the sea.

Much of the world's oil resides outside of Western influence or is located in unstable regions, including many of the world's largest fields: Saudi Arabia (Ghawar, Khurais), Iraq (West Kurna, Majnoon, Rumaila), Iran (Ahwaz), and Venezuela (Carabobo). One needn't venture far from the source to find trouble. According to the EIA, the top 10 countries with the largest proven oil reserves comprise 85% of a total estimated 1,600 billion barrels. The top five – Venezuela (18%), Saudi Arabia (16%), Canada (10%), Iran (10%), and Iraq (9%) – hold twice as much as the rest of the world combined, while OPEC controls more than three-quarters and the Gulf states almost half (Table 2.8). Now part of OPEC+, Russian influence has also become more important, not least because of its plentiful reserves of natural gas.

Table 2.8 *Petroleum (billion barrels) and natural gas (trillion cubic meters) (2019)*

No.	Country	Oil reserves (billion bbl)	Oil reserves (%)	Country	Gas reserves (trillion m³)	Gas reserves (%)
1	Venezuela	300	18.2	Russia	48.0	23.5
2	Saudi Arabia	270	16.0	Iran	34.0	16.5
3	Canada	170	10.3	Qatar	24.0	11.8
4	Iran	160	9.5	USA	15.5	7.6
5	Iraq	150	8.9	Saudi Arabia	8.5	4.2
6	Kuwait	100	6.1	Turkmenistan	7.5	3.7
7	UAE	100	5.9	UAE	6.0	3.0
8	Russia	80	4.8	Venezuela	6.0	2.8
9	Libya	50	2.9	Nigeria	5.5	2.7
10	USA	50	2.8	China	5.5	2.7
	Top 10	1,420	85.5	Top 10	160.0	78.5
	Rest of World	240	14.5	Rest of World	44.0	21.5
	OPEC	1,290	77.7	OPEC	72.0	35.2
	Gulf	800	48.2	Gulf	79.0	38.7
	Total	1,660	100.0	Total	204.0	100.0

Sources: "Oil – proved reserves (bbl) > TOP 100 – World," IndexMundi (from The CIA World Factbook). www.indexmundi.com/map/?t=100&v=97&r=xx&l=en (petroleum); "Natural gas – proved reserves (trillion cubic meters) > TOP 100 – World," IndexMundi (from The CIA World Factbook). www.indexmundi.com/map/? t=100&v=98&r=xx&l=en (natural gas).

Natural gas (~80–95% methane) is essential to Western lifestyles, especially for heating and cooking. According to the EIA, the top 10 countries with the largest proven natural gas reserves comprise 79% of an estimated 200 trillion cubic meters. Russia (24%), Iran (17%), and Qatar (12%) top the table, while OPEC controls 35% and the Gulf countries 39% (Table 2.8). Jointly controlled by Iran and Qatar, the South Pars-North Dome Field alone contains more than 50 trillion cubic meters (28% of all proven reserves!).

But what is a barrel of oil, a cubic meter of natural gas, let alone 1,600 billion barrels or 200 trillion cubic meters? How much do I use in a day, a month, a year? The world? What is the economic value? Although hard to define because consumption varies across the globe, the average American car uses about 80 gallons of gasoline per month (~$4,000/year), a smaller compact European car maybe half that, while an underground storage tank in a local gas station holds about 9,000 gallons, enough to keep customers topped up as needed. A DOT-111 railway car carries about 30,000 gallons as in each of 72

Bakken crude-filled cars that derailed in 2013 in Lac-Mégantic, Quebec, obliterating much of its downtown in the resultant explosions. The *Exxon Valdez* oil tanker was carrying 55 million gallons of piped North Slope crude when it hit a reef in 1989, one-fifth of its hold spilling out into Prince William Sound, while the *Deepwater Horizon* offshore rig poured out 200 million gallons of oil into the Gulf of Mexico after a devastating blowout preventer failure in 2010.

A typical Western home in northern Europe, northern USA, or Canada might use 20 gallons or half a barrel of heating oil per month in winter, that is, for homes not warmed by natural gas or electric heating. As for natural gas (essentially methane), a typical household might burn 200 cubic meters per year, depending on one's heating, cooking, and bathing needs. To be sure, we don't all use the same amount, but in total the world consumes almost 100 million barrels of oil and 10 billion cubic meters of natural gas per day, round numbers to help understand the massive amounts we burn, that is, about half a gallon and 1.3 m^3 each day per person (2020 numbers trended down because of the COVID-19 pandemic). If a barrel of oil costs $100, that's $10 billion a day or about $3.65 trillion a year. If natural gas costs $1/$m^3$, that's another $3.65 trillion a year.

To help gauge the output, production *percentages* give a better perspective – whether at the well, pipeline capacity, or market size. If you don't know how much something is, such as 100 million barrels of crude or 10 billion cubic meters of natural gas, working with relative amounts helps the understanding of a lot, a little, and anything in between. The world's largest reserve (Ghawar) pumps about 5 million barrels per day, that is, 5 mbpd or roughly 5% of global production, the largest offshore well (Safaniya) about 1.5 mbpd (~1.5%), while the largest-capacity pipeline (the Druzhba pipeline from Russia to Europe) typically transports more than 1 mbpd (~1%).[190]

Some of the millions of barrels of oil is en route, including "ghost tankers" stuck at sea. After Iranian oil exports were restricted when the USA withdrew from the Iranian nuclear deal in 2018, 15 million barrels or almost 15% of a global daily supply of oil languished in idle Iranian tankers, including two million barrels on a ship called *Happiness*. During the COVID-19 pandemic in April 2020, super tankers were also enlisted for storage as oil prices dropped by more than 100%. Temporarily going negative as supplies greatly exceeded demand, some ships stored up to two million barrels at a cost of $335,000 per day.[191]

***

We must also know how oil is sold on the world market. Crude oil is classified by three main attributes: geography, density, and sulfur content (the less dense and less sulfur the more desirable). The primary geographic benchmarks are West Texas Intermediate (WTI; North America), Brent (Europe, a blend of 15 North Sea crudes), and Dubai/Oman (a.k.a. Fateh, the Middle East). The delivery point for WTI is at Cushing, Oklahoma, the world's largest oil storage depot, which holds about 75 million barrels, while Brent is delivered from four different UK locations and Dubai three terminals in the Persian Gulf.

Prices vary greatly as supply, demand, and the political landscape or regulatory strategy changes, while current events such as the occasional disaster or a geopolitical crisis can significantly affect the spot price. For example, after Hurricane Harvey knocked out refineries around Houston in 2017, the Brent/WTI spread trended higher, while a drop of just 1% of global supply after the 2011 NATO-led intervention in Libya resulted in a 30% surge in Brent because of a loss of Libyan crude that was primarily sold on the European market. Oil and gas prices have been especially volatile since the start of the Ukraine war in 2022.

Crudes are also classified by density and sulfur content, indicating the ease of refining. Density is light, medium, heavy, or extra heavy, according to its API gravity (a unitless inverse ratio devised by the American Petroleum Institute although shown in degrees). Extra-heavy crudes (API gravity < 10°) sink in water, while heavy, medium, and light crudes (>10°) float.[192] Most crudes are lighter than water.

If the sulfur content is less than 0.5% by mass, a crude is labeled "sweet," otherwise "sour" – early prospectors actually tasted samples. Mostly found in the Appalachian Basin, West Texas, the Bakken, the North Sea, North Africa, Australia, and the Far East, sweet crudes are safer to extract and transport, easier to refine, and less corrosive, and thus consequently more expensive and easier to deplete.[193] Light sweet crudes (high API gravity, low sulfur content) are best for refining because more hydrocarbons are convertible to gasoline. A reduction in the amount of easy-to-refine, light sweet crudes typically raises prices, although the politics of the day often has a greater influence.

A graph of sulfur content versus API gravity for a number of commercial crudes is shown in Figure 2.11, followed by a table of data (Table 2.9) for the primary benchmark crudes WTI (42.0°, 0.45%), Brent (37.9°, 0.45%), and Dubai (31.0°, 2.04%). WTI and Brent are both sweet light crude oils (LCO), while Dubai is a medium sour crude. In early 2021, a new oil trading exchange was launched in Abu Dubai to compete with WTI, Brent, and Dubai, using the sweet light crude Murban (40.4°, 0.79%) as a benchmark, primarily produced and pumped by the Abu Dhabi National Oil Company (ADNOC).

Most importantly, if no new reserves are found and we continue to consume oil and gas as we do today, we will run out of oil in about 44 years (1.6 trillion

Table 2.9 *Benchmark crudes by API gravity, sulfur content, delivery point, and exchange*

ype	API gravity (°)	Sulfur (%)	Delivery point(s)	Exchange
WTI	42.0	0.45	Cushing, Oklahoma	New York Mercantile
Brent	37.9	0.45	Sullom Voe, Hound Point, Sture, Teesside	Intercontinental
Dubai	31.0	2.04	Dubai, Upper Zakum, Oman	Dubai Mercantile
Murban	40.4	0.79	Abu Dhabi	IFAD

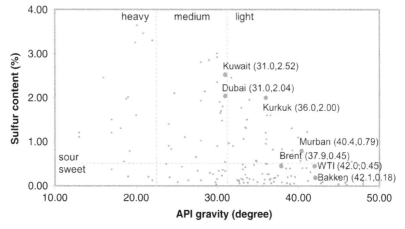

Figure 2.11 Crude oil quality by type: sulfur content (%) versus API gravity (degree). *Source:* "Crude grades," Energy Insights, McKinsey & Company. www.mckinseyenergyinsights.com/resources/refinery-reference-desk/crude-grades/.

barrels / 100 million barrels per day / 365 days per year) and natural gas in about 55 years (200 trillion cubic meters / 10 billion cubic meters per day / 365 days per year). That is, assuming consumption rates remain the same and that almost one billion people without electricity don't start plugging into a fossil-fuel-supplied grid and consuming oil and gas as in the developed world.

Underscoring the disparity between rich and poor countries, global consumption patterns vary widely. Americans consume about 7 tonnes of oil annually per capita (48 barrels), Europeans half that, while the two lowest-consuming countries are under 1 (Ethiopia and Somalia). While the nineteenth century saw an enormous increase in coal consumption, the modern petroleum era of the twentieth century has produced exponential growth in people and energy consumption, such that GDP correlates positively with energy use. As measured by the Human Development Index, quality of life also increases with energy consumption, meaning a fourfold increase in energy is needed over the next century to run a modernizing world in a style taken for granted in the West – a disaster for the atmosphere.

If you want to see the world's wealth, look at energy consumption per capita as in the "energy-intensity" graph shown in Figure 2.12. The richer the country, the higher the energy use per capita. Of 200 countries, 82 are above-average energy users per capita (20 MWh or 72 EJ): 35 in Europe, eight in the Middle East, along with the giants of Canada, the USA, South Korea, Taiwan, and Australia. Only two are in Africa (Libya and South Africa) and only one in South America (Chile), while the most energy-consuming countries per capita are Iceland, Qatar, Singapore, Bahrain, Trinidad and Tobago, and Brunei (all above 120 MWh). If the energy intensity of the countries on the bottom left – essentially Asia, Africa, and South America – rise to the top right without transitioning to clean energy, the atmosphere and the climate are doomed.

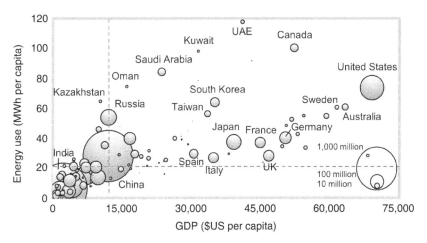

Figure 2.12 Energy use (MWh) versus GDP ($US) per capita across the globe (2021). *Source*: "Energy consumption by country 2023," World Population Review, 2023. https://worldpopulationreview.com/country-rankings/energy-con sumption-by-country.

Note the contrast between Canada (100 MWh) and the United States (78 MWh), both high energy-consuming-per-capita countries, and China (28 MWh) and India (6 MWh), near or below-average energy users per capita. China uses more *total* energy (146 EJ), however, because of its large popula- tion – almost twice the USA (88 EJ) – and is thus responsible for more overall GHG emissions, despite using much less energy and emitting much fewer emissions *per capita*. Clearly, if the poorer, more populous countries start living and emitting GHGs like the West, we're done for.

2.10 Modern Oil Extraction and a Continuing Black Gold Economy

Today, oil extraction includes more than the easy-to-get, "conventional" oil in pumped gushers. After the First Oil Shock in 1973, when petroleum prices skyrocketed and production was nationalized in many oil-producing countries, previously hard-to-get, "non-conventional" oil supplies – offshore oil and gas, shale oil and gas, and oil sands – were increasingly sought in the Gulf of Mexico, the North Sea, the Arctic, North American shale rock, and the Athabasca oil sands of northern Alberta. By the 1990s, hydraulic fracturing (fracking) of large continental shale deposits would revolutionize US produc- tion, while the breakup of the Soviet Union would facilitate new supplies of piped natural gas. Today, output from offshore, shale, and oil-sands reserves is as prominent as the tried-and-true, conventional oil and gas supplies, which we look at now and in the following two sections.

Small-scale, offshore drilling into shallow coastal waters first began in 1898 in Summerland, California, an extension of earlier onshore drilling from piers, before wooden piles were drilled directly into the water to install an offshore "rig," first employed in 1910 at Caddo Lake, Louisiana. As noted by Freudenburg and Gramling:

> Pilings were driven into the lake bottom, and platforms were constructed on top. Drilling equipment was hauled to the platforms by barge, and underwater pipelines were constructed to bring the oil ashore. This latter activity became a precursor to the extensive network of offshore and onshore pipelines that would eventually form an amazingly complex web across most of southern Louisiana.[194]

Higher, more dangerous, underground pressures sparked the need for blowout preventers – to reduce the pressure after a buildup of gas and if necessary to seal the well – and then in 1924 the use of hollow, steel-reinforced, concrete caissons instead of wooden pilings in the deeper shipworm-laden waters of

Lake Maracaibo, Venezuela, "a technology that would later evolve into the kind of platforms that became the approach of choice in the North Sea and North Atlantic."[195]

Moveable drilling barges followed in the harder-to-access, deltaic wetlands by the mouth of the Mississippi River – the *Gillaso* the first in 1932 – leading to a subsequent rise in offshore drilling in the Gulf of Mexico. "Spoil" piles littered the banks of nearby offshore sites preventing normal drainage, while tens of thousands of acres of wetlands were lost, devastating the natural barrier to the sea and causing increased damage to inland communities during hurricane season. Leakages also exacerbated coastal erosion by killing shoreline grasses.

True *offshore* drilling began in 1945 in the Gulf of Mexico, south of Morgan City, Louisiana, after the technology of submersible drilling barges was developed and the US federal government successfully claimed ownership of the sea bottom, previously considered state property. Originally settled in its favor in the Tidelands Cases (1947–50) before being overturned in favor of state control in 1953, the feds were eventually given jurisdiction over offshore waters more than 3 miles off coast (9 miles in Florida and Texas) and the right to lease as it pleased in the deeper Outer Continental Shelf (OCS).

Keen to encourage more oil development, the federal government provided generous leases for unlimited drilling in the newly defined offshore waters. Novel challenges included accessing remote locations in harsher marine environments (where higher winds and waves are common), more expensive construction and operating costs, matching larger well pressures to the drilling fluid at depth, and the complexity of rigging platforms in the open sea.

Federal leases began in 1954, changed in 1984 from tract offerings to area-wide leasing that significantly increased the amount of drilling while drastically decreasing the federal cut, "a sale situation where the buyers (major oil companies) know the potential value of a given tract, but the seller (the federal government) does not."[196] Offshore oil companies had won the oil lottery, paying both lower development costs to drill and lower royalties for commercially pumped oil than in almost any other country. In effect, American taxpayers were subsidizing offshore development costs and extraction fees, never mind a more relaxed system of regulations that turned a blind eye to safety in the harder-to-access, deep-sea waters.

On the other side of the world, the oddest of offshore sites was erected in 1947, 50 km east of Baku in the Caspian Sea. Considered the oldest oil platform by *The Guinness Book of Records*, Oily Rocks, a.k.a. Neft Daşlari, was a famous Soviet "town on stilts" containing apartments, hotels, shops, and 300 km of roads built around 600 makeshift drilling platforms and scuttled ships as depicted in a crazy helicopter saw-cutting scene in the James Bond movie *The World is Not Enough* with Pierce Brosnan and Robby Coltrane. At the height of production in the 1960s, the pride of Soviet offshore oil was home to 5,000 workers. Oily Rocks became unproductive, however, after the breakup of the USSR, the site rusting beyond repair and unable to compete with modern industry.

Elsewhere, as a result of the ongoing supply crises in the Middle East during the 1970s, fuel substitution had become an essential policy for oil-guzzling countries trying to make up the shortfall from MENA – a Pentagon term for Middle East North Africa – resulting in increased offshore exploration and production around the world. Britain was a major beneficiary of the new oil stream, its economy crippled after the First Oil Shock, although just as North Sea oil supplies were coming online the UK had to sell 20% of its resources to foreign interests to ensure IMF backing after its balance of payments collapsed in 1976.[197]

In 1965, British Petroleum's *Sea Gem*, a 5,600-ton converted steel barge, was the first to discover natural gas in the North Sea's British sector, 42 miles off the Lincolnshire coast in the West Sole field (a.k.a. Block 48/6).[198] Tragically, the fixed-bottom supported *Sea Gem* would capsize from metal fatigue 3 months later while preparing to move, killing 13 crew members. But despite the increased dangers of deeper offshore locations, drilling continued in the West Sole from a newly constructed floating platform dubbed the *Sea Quest*. In 1967, natural gas was brought ashore for the first time at the Yorkshire Easington Gas Terminal, east of Hull.

To accommodate the newfound North Sea supply, household appliances were converted from town gas to "natural" gas, which burned cleaner and had twice the heating value. Costs were passed onto consumers to offset investment in the new infrastructure. The "natural" part of what is essentially methane (CH_4) was included in the nomenclature to differentiate from dirtier town gas that had previously been derived from coal and produced in local "gasworks" as a by-product of making coke. Stored at pressure in large cylindrically scaffholded containers, urban coal-gas gasometers can still be seen today in various cities, some as converted museums and apartments.

Other North Sea countries also benefitted from the uncertainty of Middle East supplies. In 1969, the *Ocean Viking* – Norway's first North Sea rig – drilled one of the largest offshore fields ever, almost giving up after 11 tries.[199] Practically in the middle of the shared waters, on the edge of the Norwegian zone, Block 2/4, now known as Ekofisk, began exporting crude oil to England and natural gas to Germany in 1971, establishing the viability of North Sea production.

Alas, in the push to develop oil in increasingly difficult regions, safety was often compromised. In 1988, an explosion aboard Occidental Petroleum's *Piper Alpha* platform – the North Sea's most productive rig about 200 km northeast of Aberdeen – resulted in the death of 167 workers. Taking more than a month to extinguish, the blaze was so bad even the legendary Hellfighter Red Adair couldn't help. In 2010, the *Deepwater Horizon* blowout in the middle of the Gulf of Mexico saw 11 oil workers killed and 200 million gallons of oil gush out over three helpless months before eventually being sealed, setting off the worst environmental disaster in the United States since the *Exxon Valdez*.

And yet, despite the difficulty of drilling in ever deeper waters, North Sea output flourished in the final three decades of the twentieth century, helping to offset OPEC dominance, especially in energy-deficient Europe. By the 1980s, production levels were about 6 mbpd or almost 10% of world supply. Today, the Greater Ekofisk Area alone still produces about 300,000 barrels of oil equivalent per day, keeping Norway near the top of offshore petroleum extraction and underpinning the Norwegian economy, a perennial, top-10 income-per-capita performer. Norway's vast petroleum wealth also affords a political direction apart from the European Union – no need to share the fruits of the sea when one doesn't have to (Figure 2.13).

Figure 2.13 North Sea oil and gas fields (UK, Norway, the Netherlands, and Denmark).

Alaskan oil was also in demand after the 1973 Arab oil embargo that served to ramp up US development of unconventional sources. Despite the frigid climate and harsh conditions, exploration began in the 1960s on Alaska's North Slope before the Atlantic Richfield Company (ARCO) successfully brought oil to the surface in Prudhoe Bay, soon to become North America's largest oil field. The only problem was how to get the oil to market, solved by the $9 billion Trans-Alaska Pipeline System (TAPS), an 800-mile steel conduit with 12 pumping stations that took two years to build, snaking its way above and below the permafrost through the Alaska interior from the Arctic coast to the ice-free port of Valdez, east of Anchorage. Today, 1 mbpd are brought to port – down from a peak of 2 mbpd in 1987 – despite regular spills along the 48-inch-wide pipeline. The effect of the 1989 *Exxon Valdez* disaster also lingers, which poured 11 million gallons of North Slope crude into the pristine waters of Port William Sound, only 2 km from port after having just been filled.

Offshore activity is limited by the harsh conditions, but with enormous reserves the Arctic may soon become the driller's last stand – nine of the world's 25 largest gas fields are north of the Arctic Circle. The US Geological Survey estimates that the Arctic holds 13% of the world's undiscovered oil and 30% of its natural gas, while others estimate offshore Alaskan reserves at more than the rest of the United States combined. Divided equally between shallow waters, continental shelves, and deep waters over 500 m, as much as 85% of Arctic oil and gas is offshore. Exploration in the Chukchi Sea and the Beaufort Sea isn't cost-effective for now given the conditions – drilling is hard slogging even in summer in relatively shallow waters – but exploration will become relatively easier as more Arctic ice melts with increased global warming, doubly hazardous for the Earth and the environment.

Prices dictate exploration as much as the climate, while government policy and accidents affect development. Despite having invested $5 billion, Shell dumped plans to develop Arctic fields after a series of setbacks, including its *Kulluk* drilling rig running aground in a 2012 winter storm as it attempted to move outside Alaskan territorial waters to avoid state taxes.[200] In 2015, Barack Obama called the Arctic ecosystem "fragile and unique," declaring the 20 million acre Arctic National Wildlife Region (ANWR) in northeast Alaska off limits, although Donald Trump reopened the area to seismic testing in 2018 with an eye to leasing, the first step to opening up drilling that the then Interior Secretary Ryan Zinke declared was essential to establishing Trump's goal of American "energy dominance."[201]

In August 2020, the first ever drilling-lease auctions were announced for later that year, although low oil prices offered little incentive for oil and gas companies to set up operations, while banks were reticent to provide financing for increased oil investment in the wake of continuing bad news about climate change. Despite the difficult economics, in March 2023 Joe Biden approved drilling in the Willow Project area within the federal National Petroleum Reserve, a decision that prompted outcry over his previous green posturing. The decision may have had more to do with increased fuel needs for an expanding conflict in Ukraine.

Russian oil and gas exploration and mining activities in the north are also extensive, operating from a number of Arctic ports. Murmansk and Norilsk are the two largest cities above the Arctic Circle (over 200,000 inhabitants in peak season), but as the polar region continues to melt – possibly year-round ice-free by 2050[202] – Russia will drill and exploit more Arctic locations that account for 20% of its GDP. In 2020, the Russian government announced a €210 billion Arctic plan to expand drilling, improve the Northern Sea Route, and build new petrochemical facilities in the Yamal Peninsula and east Arctic.[203]

Elsewhere, offshore exploration continues pushing the engineering boundaries to deep-sea extraction as diminishing supplies and higher prices spur on drilling in more remote waters. The British government opened up exploration west of the Shetland Islands, the "final frontier" for British oil and gas that may hold 20% of remaining UK reserves, backed by a generous £3 billion field allowance to encourage development.

Sea depths can reach up to 1.5 km, where higher pressures, temperatures, and remotely operated underwater vehicles (ROVs) pose new challenges to the success and safety of offshore development. As one drilling engineer noted, "If you imagine you have got this thick hole at the bottom of the seafloor, as the waves move the vessel you've got to try and keep this thing right above the hole even when the tide is moving it. So what you need is dynamic positioning systems which link to satellites to try and keep the vessels where you want them."[204] Another offshore worker likened tapping a deep-sea well to hitting the pitcher's mound at Yankee Stadium from 30,000 feet above.

About 100 km off Norway's North Sea coast, the Ormen Lange gas field contains one of the world's largest deepwater wells, but is so difficult to access on the continental shelf that the pumping platform is mounted directly on the seabed. Well operations are remotely controlled from the mainland to transport the extracted gas by underwater pipeline, first to shore in Norway for processing and to remove liquid condensates before entering the UK via the Langeled pipeline. Originally called Britpipe, Langeled was the longest subsea pipeline upon completion in 2006 at 1,200 km. Named after a Viking longboat, Ormen Lange pumps 70 million cubic meters per day of natural gas to the UK, 20% of its supply (~1% daily global output).

Brazil broke the deepwater barrier in 1992 when the state-owned Petrobras successfully installed a rig 800 m above the ocean floor in a section of the Campos Basin, 80 km east of Rio de Janeiro. Due south, a 200-km long "super giant" holding as much as eight billion barrels was discovered in 2007 in the Santos Basin, one of the world's largest ever offshore finds. Formerly known as Tupi, the massive renamed Lula field lies 6.5 km below sea level, requiring the latest in offshore technology to traverse 2 km to the sea floor, 2 km through a salt layer (left over from when the Atlantic Ocean formed), and another 2.5 km to the underground reserve of "presalt" oil. Current production is 0.1 mbpd of oil and 5 million cubic meters per day of natural gas.

To date, the world's deepest offshore production site is the Stones field in the Gulf of Mexico, roughly 9 km down and operated by Shell. The floating production, storage, and offloading facility is moored 2.9 km above the sea-floor. But deeper doesn't necessarily mean more – the world's largest offshore reserve, Saudi Arabia's Safaniya field, sits in the shallow waters of the Persian Gulf, estimated to contain more than 50 billion barrels, 70% of which is recoverable. Saudi Aramco's offshore elephant has been pumping almost 1.5 million barrels of crude per day since 1957, while two other Saudi-owned offshore fields round out the top three, together pumping over 2.2 mbpd from under the Persian Gulf (~2% of world production).

Slow to the game but making up for lost time after the breakup of the Soviet Union, the Caspian Sea has been called the "new oil El Dorado." In 2000, the Kashagan field was discovered there at the northern tip in western Kazakhstan about 100 km south of Atyrau, the region's capital that sits at the mouth of the Ural River. Alas, the largest discovery in 40 years and world's fourth largest offshore field has been plagued from the start. Unfortunately, Kashagan oil contains 17% hydrogen sulfide (making extraction more dangerous), temperatures range from –40 to +40°C, and the shallow Caspian waters are completely covered in ice during winter, while the oil is another 5 km underground.

Located in a former Soviet republic, the politics of the region are also tricky, and the landlocked gas difficult to move to market through neighboring countries, while a pipeline leak on one of the artificial islands shut down production for 3 years. Nevertheless, after more than a decade of construction and $50 billion spent, the first phase of the Kashagan offshore field came online in 2017, pumping 200,000 barrels per day of very light crude (45° API).[205] The plan is to ramp up to 1.5 mbpd in two more phases.

Offshore sites continue to provide a significant percentage of global petroleum in both shallow and deep waters, although development costs are enormous at about $100 million for a single deep-sea exploration well. In the decade from 2000, 14,000 offshore exploratory and production wells were drilled as the global supply more than tripled to 5 mbpd.[206] At the height of production, 3,500 platforms were operating in the Gulf of Mexico, more than the rest of the world combined, including 50 deepwater rigs with over 40,000 miles of underwater pipelines.[207] At the end of 2022, there were roughly 1,800 rigs operating across the globe,[208] one-third of which were offshore. Most are in the North Sea, Gulf of Mexico, and Persian Gulf.

**

When we think of petroleum we mostly think of oil, but natural gas (methane) is giving oil a run for its money. Five countries produce more than half of all natural gas: the United States (21%), Russia (18%), Iran (6%), Qatar (5%), and Canada (4%). Although the USA recently became the world's top producer at about 2 bcmpd (~20% of world total) because of fracking (which we'll look at next), Russia has typically been the world's largest natural gas producer, much of it offshore and all of it controlled by the state-run Gazprom.

Since the breakup of the Soviet Union, the large fields of West Siberia, Sakhalin Island in the northwest Pacific, and the Kara Sea in the Arctic have been controlled by the Russian government and various state-sponsored oligarchs, who engineered a rigged takeover of the Soviet Ministry of Oil and Gas in an infamous "loans for shares" insider-trading scheme. Led by Vladimir Putin, the former KGB lieutenant-colonel turned politician, Russia grew to become a global player via Gazprom and Rosneft. Gazprom is now the world's largest natural gas company at more than 15%, including 75% of a subsidized domestic market and until recently about 40% of the European market. Rosneft is also the largest oil company, having absorbed the bankrupted Yukos, formerly run by Putin's adversary Mikhail Khodorkovsky, who ultimately lost out in a post-Soviet game of musical oil chairs, sentenced to 11 years in prison before Putin eventually pardoned him.

As in the Arctic, however, the problem with Russian oil and gas is transportation, made all the more difficult by conflict in Ukraine, the second largest European country by area, through which 80% of piped Russian gas once flowed from western Siberia to Europe.[209] The flow of Russian gas to Europe began in the 1960s, but a prize is no good if it can't be delivered as an increasingly worrisome game of pipeline politics is played out in the heart of Europe, in particular the neighboring countries of Ukraine, Belarus, Poland, and Turkey, as well as the Baltic Sea and Black Sea. Russia also controls both the supply and the pipelines, long considered illegal by the European Union.

The $10 billion Nord Stream had been the main Gazprom pipeline to Europe since opening in 2011. A 1,200-km conduit under the Baltic Sea from Vyborg north of Saint Petersburg to Greifswald in northeastern Germany, Nord Stream

(NS1) delivered about a third of Europe's natural gas supply of 0.5 bcmpd
(~5% world total). Completed in 2021, a parallel, $11 billion, Nord Stream 2
(NS2) pipeline was slated to double capacity, but remained uncertified amid
concerns over growing Russian control. Citing a precarious reliance on Russian
gas, the then Dutch prime minister Mark Rutte noted, "state-controlled outside
players like Gazprom cannot take over vital energy infrastructure on European
territory."[210] The unimpeded subsea route would have made Germany the main
European entry point for Russian gas with no transfer costs. Russia had been
paying Ukraine $2 billion a year.

In the ongoing fight to control the supply and distribution of oil and gas, NS2
was also fraught with delays and disruptions, possibly precipitating the ousting
of the democratically elected, pro-Russian, Ukrainian president Victor
Yanukovych in 2014, followed later that year by the subsequent Russian
annexation of Crimea and parts of the Donbas in eastern Ukraine. Western
sanctions immediately followed the start of construction of NS2 in 2018 after
pressure from the United States, who were lobbying for a larger slice of the fast-
growing, globalized natural-gas market, and again in 2021 over the impending
certification of the completed NS2 pipeline before the American government
eventually relented.

But just as the colder winter weather arrived in 2021, more delays followed
because of bureaucratic snags tied to renewed European and American fears
over Russian energy-dominance in Europe and concerns about Russian troops
amassing on the Ukrainian border. Claimed to be a standard military exercise
by Russia, Putin's muscle-flexing on the Ukrainian border and NATO pushback
in January 2022 may have had more to do with oil delivery routes through
Ukraine than encroaching empires in a historically belligerent part of the globe.
The pipeline politics, however, became more than just words when Russia
invaded Ukraine on February 24, 2022.[211] Prices and supplies have fluctuated
ever since in a tit-for-tat series of sanctions and retaliations.

Having pledged to ditch both coal and nuclear, Germany was especially
vulnerable to the lost pipeline traffic, yet joined in ramping up sanctions over
Russian aggression. In an obvious quid pro quo prior to the war, Russia had
offered to pipe more gas if NS2 certification was sped up, instantly lowering
prices, which could have netted Gazprom $90 billion, a more than fourfold
increase on previous years.[212] When Russia recognized the majority ethnic-
Russian Donbas territories of Luhansk and Donetsk, however, NS2 certification was immediately paused by the German chancellor Olaf Scholz (Russia
later annexed four oblasts, denounced by the West).

Although the war has made Putin an international pariah and disrupted
economies around the world, the natural gas nonetheless continued to flow

via rerouted deals amid hastily imposed sanctions. Provisions were sought elsewhere to make up the shortfall in Europe, in particular via the TurkStream pipeline and American and Qatari liquefied natural gas (LNG) that required building import facilities to handle the cryogenically shipped cargo.

It seemed unlikely that Putin would use his energy weapon for a complete shutdown, however, given potential lost revenues of over $200 million a day, but uncertainty over retaliatory Russian management measures prior to the invasion trebled heating bills in Europe, before global oil and gas prices spiked to new highs at the start of the war. Russia also halted supplies to Poland and Bulgaria for not paying in rubles, presumably to divide EU resolve, despite the gas contracts stipulating payment in euros or US dollars, further escalating prices and providing Russia with an even larger bounty. Turning on and off the tap to increase prices and profits, Europe is being held hostage by Putin's "gas diplomacy," while Ukraine is punished for its Western posturing.

The EU eventually countered by capping prices and stocking up on supplies as winter approached, followed by Russia completely closing NS1 supposedly for repairs. And then in September 2022, leaks appeared in both pipelines after a series of underwater explosions, gas bubbling up in the Baltic Sea near the Danish island of Bornholm. The unprecedented damage caused by unknown saboteurs, possibly linked to Ukraine, added to concerns over maintaining critical infrastructure and the environmental harm from the worst methane leak ever (300,000 tons). Norway is particularly worried, asked to make up the shortage to Europe via numerous subsea pipelines.

To be sure, "energy imperialism" occupies an oversized amount of foreign policy, financing, and attention, while colder winters will be a nightmare for countries beholden to Kremlin-backed oligarchs and the "endemic corruption in the region's energy sector, which is largely dominated by local oligarchs, corrupt elites, and Russian interests."[213] Relying on Putin and the Kremlin to sell oil and gas via a network of billion-dollar pipelines is as risky as in the days of Rockefeller and the Standard. Of course, if the European Union can counter Russian influence with price caps, joint buying, and importing elsewhere or transitioning faster to renewables, Putin's strategy to weaken Europe will fail. Alas, those caught up in the crossfire of modern warfare and ongoing petroleum politics are paying a heavy price. Much of eastern Ukraine has been sacrificed to another round in the Great Game for Oil.

∗∗∗

Local arms races, proxy skirmishes, and the fear of international conflict are also increasing in the former Soviet-controlled countries looking to expand their own petroleum production. Flare-ups in Azerbaijan, Georgia, and Chechnya are all related to oil extraction or pipeline routes, built through politically unstable and strife-ridden regions, while the 1,800-km Turkmenistan–China natural gas pipeline to the western Chinese province of Xinjiang – the European gateway to China – is fraught with danger over separatist intentions to establish an independent state for the majority, non-Chinese, Muslim Uighur population. As Michael T. Klare noted in *Resources Wars*, "The various pipelines routes leading out of the Caspian are likely to remain periodic sites of conflict for a long time to come," which could lead "in a worst-case situation, to the deployment of Russian and American combat troops."[214] With its ever-increasing economic might and growing energy import gap, you can add China to the pipeline politics mix and the potential for more conflict.

War continues to follow oil's sheen wherever competing interests are unresolved. After more than seven decades as the economic leader in an international order that favors American control, the USA is keen to contain Russia and China in a changing twenty-first-century geopolitical landscape. American interests are also constrained by US demands that pipeline routes be free of Russian or Iranian control. Turkmenistan's natural gas holdings has even been cited as the cause of the US occupation of Afghanistan, seen as the most profitable passage to transport Turkmenistan gas to port via Afghanistan, Pakistan, and India (a.k.a. TAPI). One hopes the ongoing wars in distant lands to maintain supplies for the home front and superiority at the trade table isn't a cynical goal to destroy yet another region. Nonetheless, increased production in the coming years will keep the Caspian Sea's five countries – Russia, Kazakhstan, Turkmenistan, Iran, and Azerbaijan – jostling for another petroleum bounty. The Caspian Sea has estimated reserves 10 times that of the North Sea.

Flowing in the other direction from the Caspian, the 1,800-km Baku–Tbilisi–Ceyhan (BTC) pipeline follows part of the Nobel brothers' original oil route west from Azerbaijan to Georgia, before turning south to Turkey and the Mediterranean Sea port of Ceyhan. Embroiled in its own regional conflicts, not least over Armenian sovereignty, the BTC pipeline serves as the main US energy corridor in Central Asia, built wholly underground to ensure safety as it "passes through some of the most volatile parts in the world."[215]

There are thousands of miles of pipelines crisscrossing Europe to protect against single-failure points that maintain the energy supply of an entire continent to keep the wheels of industry turning and millions of homes warm in winter. Built in Soviet times to supply the Communist bloc, Soviet oil first reached eastern Europe in 1962 via the 4,000-km Druzhba Pipeline from

Siberia to Czechoslovakia, while the Trans-Siberian Pipeline is now the world's longest at 4,500 km, linking the world's second-largest gas field in Urengoy just south of the Arctic Circle to Europe. Called a "big bowl of spaghetti"[216] by historian Daniel Yergin, a "bag of snakes" may be a better epithet to describe the mess of long-distance conduits, considering the slippery nature of the players. As always, to understand who's who in the oil and gas biz, one follows the money. Alas, the word "druzhba" means *friendship* in Russian, now being tested to the limits in the cruelest of games.

What's more, increased oil and gas extraction and new pipeline builds run contrary to climate concerns. Alas, with competing transnational routes and ever-changing alliances, where fragile agreements are fraught with long-term worries, pipelines and politics remains a dangerous game. Europe is in for a long ride if it can't sort out its borders or who can sell energy to whom, a problem the United States needn't worry about, thanks to its own plentiful supply of fracked petroleum as we look at now. One needn't have friends to control the flow of oil, certainly within one's own borders (for now at least).

2.11 Shale Oil Extraction: Frack, Rattle, and Roll

To understand the latest petroleum development that has radically changed the global energy picture, both economically and politically, we look to the vast shale rock formations of North America, where novel techniques have been employed to extract previously hard-to-get shale oil and shale gas. A viable source of oil since Edwin Drake's 1859 Titusville find, shale took too much work in Drake's day to recover the trapped deposits and thus "conventional" oil was extracted first before turning to the harder-to-access, "unconventional" shale oil reserves (Figure 2.14).

After the easy-to-reach, loose oil in a well is exhausted (a gusher), other methods are employed to remove the tight oil lying in the surrounding rock and in previously untapped shale formations (shale oil and gas stores). Hydraulic fracturing ("fracking") uses a pressurized fluid to remove the hard-to-get, tight oil from a shale deposit, where a mixture of water, sand, and chemicals (surfactants and friction reducers) is injected into the underground bedrock to free the trapped oil and gas. When the injection fluid pressure exceeds the formation fracture pressure, the shale is literally broken (fractured) by the high-pressure (hydraulic) fluid to shake out the decomposed dead organisms in the rock (essentially uncooked kerogen) as either oil or gas (mostly methane). The less-dense oil and gas migrates upward until trapped by an impermeable barrier before being collected in the drilling process.

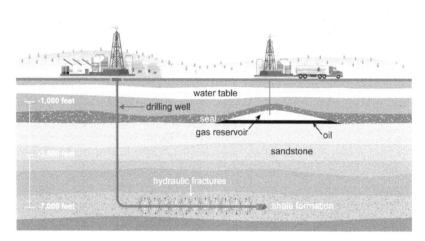

Figure 2.14 Directional drilling and hydraulic fracturing (fracking) in shale.

Controlled explosions were used in the late 1940s to fracture shale rock with limited success. In the 1970s, high-pressure water was tried, while by the 1990s directional drilling (a.k.a. horizontal or lateral drilling) steered the fluid (injected through a perforated pipe) into the shale layers to loosen the trapped oil and gas. By the late 1990s, fracking and directional drilling began showing positive results with light sand in the water mixture, known as LSF or light-sand "fraccing," which propped open the fractures to maintain sufficient rock porosity.

Thanks to a 1980 federal tax credit that provided a financial incentive to drill for unconventional natural gas as well as improved 3D seismic testing to image hard-to-find subsurface deposits, the legendary Texan oilman George Mitchell finally tapped the Barnett Shale located around Dallas and Fort Worth, success-fully "cracking the code" in 1998 after almost two decades of trying. By 2001, other "petropreneurs" began to see an increase in gas supplies coming from the Barnett Shale, prompting a buyout of Mitchell Energy by the Oklahoma firm Devon Energy, an expert in directional drilling. By 2003, as Daniel Yergin notes in *The Quest*, "Shale gas, heretofore commercially inaccessible, began to flow in significant volumes. Combining the advances in fraccing and horizontal drilling ... would unleash what became known as the unconventional gas revolution."[217]

"Fraccing," as it was then known, soon spread to other areas with known shale deposits, particularly in Louisiana, Arkansas, and Oklahoma, followed by the "mighty Marcellus" in the vast Appalachian shale region encompassing parts of New York, Pennsylvania, and West Virginia. Following the success of the Barnett Shale, the Bakken Formation – largely shared between North Dakota and Montana – was tapped. Producing most of the globally fracked oil and about 15% of all natural gas, US domestic oil and gas extraction increased for the first time since 1970, revolutionizing the global petroleum industry. Today, the largest producing shale fields are the Permian Basin in West Texas at 5.6 million barrels of oil per day (60% of US production), the Bakken at 1.3 mbpd (13%), and the Eagle Ford in South Texas at 1.3 mbpd (13%). The Marcellus Formation in Appalachia leads shale gas production with 35 million cubic feet per day (37%) although only 0.1 mbpd of oil, followed by the Permian (23%) and the Haynesville Formation on the border between Texas and Louisiana (17%).[218]

By 2021, the Permian Basin had become one of the largest producing oil fields in the world, the source of almost two-fifths of US oil and one-sixth natural gas in 15,000 active wells,[219] while the Marcellus Formation was estimated to have enough natural gas to last American consumers 50 years. In the heart of California, the Monterey Formation may hold as much as two-thirds of all shale gas in the lower 48 states, yet remains underdeveloped because of environmental concerns over pollution, leakages, and seismic instability. Up and running in 20 American states, fracking was first banned in Vermont (2012), New York (2014), and Maryland (2017). In Canada, which shares part of the Bakken Formation with the USA, fracking is underway in Saskatchewan, Alberta, and British Columbia, although moratoriums have been declared in Quebec (2011), eastern Canada (2014), and most of Yukon (2015).

Today, about two million wells have been drilled by hydraulic fracturing and directional drilling, completely transforming the oil-and-gas industry in the United States and across the globe. After the word "fracking" entered the petroleum lexicon, the puns started flying fast and furious: "Frack off," "Frack to the future," and "Meet the Frackers" just three sub-headlines appearing in *The Economist* within 3 weeks in 2013, although the most appropriate may have been a 2014 *CounterPunch* article entitled "Frack, Rattle and Roll."

Indeed, fracking is not without concern or controversy. As much as 100 million liters of high-pressure water is injected into a high-volume well during fracking, furnished either from nearby water supplies, trucked in, or piped across private and public lands. The injected water includes carcinogenic crystalline silica sand and chemical additives to improve extraction, such as kerosene and diesel fuel that contain a host of familiar toxins such as benzene, ethylbenzene, toluene, xylene, and naphthalene, as well as polycyclic aromatic hydrocarbons (PAHs), formaldehyde, hydrochloric acid, and sodium hydroxide. During and after fluid injection, the chemicals can find their way through the more porous rock into nearby water supplies.

Most shale deposits are located 6,000–10,000 feet underground, well below ground-water levels, but water tables and aquifers are still at risk from leakages with potentially devastating consequences, while the fracked natural gas floats upward through the more porous rock in enlarged vertical seams and can escape from badly concreted well casings, referred to as "fugitive emissions." Perhaps the most iconic fracking image is the "flaming tap," where water can be lit on fire as in Dimock, Pennsylvania, after methane from a nearby fracked well leaked into the local water supply, famously shown in the 2010 documentary *Gasland*. Explosions are possible at high enough concentrations.

Gas leaks are nothing new – the Delphic Oracle in ancient Greece was known for producing prophetic visions in local priestesses (the Pythia), who would sit on a tripod above a cavernous fissure at the Temple of Apollo to inhale the intoxicating rising vapors (pneuma). Those vapors are now believed to have been a mixture of light hydrocarbon gases – methane, ethane, and ethylene (a modern anesthetic) – that seeped up from a naturally occurring tectonic fault beneath the temple.[220] Alas, the fracking boom in the United States and Canada since 2000 has been shown to be responsible for an unprecedented increase in methane emissions.

A 2019 Cornell University study put the percentage of leaked, vented, or flared methane from fracking at between 2% and 6%,[221] while propane and butane have also been observed. A 2014 report entitled "Troubled Waters" by the Center for Biological Diversity noted that "10 fracking chemicals routinely used offshore in California could kill or harm a broad variety of marine species, including sea otters and fish,"[222] while a 2011 Duke University study on water-well contaminants near shale gas pads found that homes within 1 km of a fracking site are "15 or 20 times more likely to have excessive methane in their water."[223]

Methane released during drilling and pipeline transportation became EPA-regulated in 2016, yet the oil-and-gas sector was still the largest methane emitter in the USA, accounting for a third of methane emissions.[224] The goal was to cut methane emissions to 45% below 2012 levels by 2026. Although not as long-lived in the atmosphere, methane is still a worrisome GHG with 80 times the global warming potential as carbon dioxide (CH_4 absorbs more energy in the infrared part of the spectrum than CO_2).

In August 2020, however, the Trump-led EPA rescinded the prior regulations on wells drilled since 2016 to "remove the largest pipelines, storage sites and other parts of the transmission system from EPA oversight of smog and greenhouse-gas emissions," citing "improper" and "overly burdensome federal overreach."[225] The ban was overturned the following year by Joe Biden. Nonetheless, in 2022 it was reported that methane emissions across the

United States were 27% higher than the EPA had previously estimated, up to 50 times in the Permian Basin![226] Some jurisdictions restrict fracking within a certain distance of human habitation and schools, although many wonder how far is safe when it comes to health. During fracking, air is also polluted with hydrogen sulfide, a nerve toxin that can cause irreversible brain damage, as well as benzene and other carcinogenic volatile organic compounds, generating lingering problems for those who live nearby, including headaches, nose bleeds, vomiting, nausea, allergies, eczema, hives, arrhythmia, and intestinal and respiratory ailments.

After fracking began in Saskatchewan in 2004, hydrogen sulfide (H_2S) levels above 100 parts per million were recorded, enough to produce olfactory paralysis that leaves one unable to detect the tell-tale, characteristic rotten-egg smell, while continued exposure can be fatal within 2 days. At 1,000 parts per million, one can "die rapidly from respiratory paralysis, or over the course of days, from an inflammatory reaction in the lungs."[227] In 2014, high levels of airborne H_2S led to one death in Oxbow, near the North Dakota border, while others complained of chronic illnesses. Lethal H_2S levels possibly exceeding 150,000 parts per million were observed across the province, while an audit of active sour-gas facilities found that only 31% had proper emission control systems.

According to a 2018 study by the University of Colorado's School of Public Health, the risk of cancer is 8.3 times higher for those who live near an oil and gas (O&G) facility. Entitled "Ambient Nonmethane Hydrocarbon Levels Along Colorado's Northern Front Range: Acute and Chronic Health Risks," the report found that "state and federal regulatory policies may not be protective of health for populations residing near O&G facilities."[228] Worryingly, millions of Americans live near such facilities. A 2020 Yale School of Public Health study further reported that children between the ages of 2 and 7 who lived near fracking sites at birth are two to three times more likely to suffer from leukemia, primarily because of exposure to contaminated drinking water.[229]

Flaring is also a major problem, alarmingly still common, where the associated gas is burnt onsite instead of stored or piped because of inadequate storage facilities and lack of natural-gas pipelines in remote areas. Because natural gas is cheaper than oil, companies don't bother to capture the methane, under no obligation to protect the environment. Imagine a gas stove permanently left on to burn in one's kitchen, a needless waste and dangerous toxic fouling. Typically on fracking sites, the associated gas is burned rather than vented to convert methane to less-potent carbon dioxide and to remove volatile organic compounds.

Not all flares burn cleanly, however, especially at older wells, some of which have been around since the 1920s.[230] Unfortunately, a casual disregard for waste has been standard procedure in the oil industry since the start of natural-gas distribution, beginning from the Panhandle Field in Texas (1918) and the Hugoton Field in Kansas (1922), which supplied almost one-sixth of US natural gas in the twentieth century. As noted in a Federal Trade Commission report to Congress, more natural gas was wasted than consumed in the first decade of distribution during the 1920s, partly because of the lack of pipelines, but mostly because of venting into the atmosphere (4,400/3,500 billion cubic feet).[231]

Not much has changed. Today, state laws permit flaring for 10 days to help a company get its fracking house in order, yet ongoing extensions are the norm. In the Permian Basin, 3.5% of gas is flared, greater than the entire consumption of some states, while even more is lost in the Bakken. Shockingly, almost 150 billion cubic meters of natural gas is flared worldwide, led by the United States.[232] Even Russia and the Middle East are concerned about how much gas is flared by US O&G companies.[233] With no financial incentive to store unused gas and in the absence of appropriate regulations, flaring is the most cynical of oil-industry practices.

Numbered in the hundreds of thousands, wells also leak methane through the ground into the atmosphere, often overlooked when calculating the carbon budget of the fossil-fuel industry. Especially damaging are thousands of "super or ultra emitters" in the United States and Russia that spew about 10% of all methane emissions in the oil and gas industry, which itself emits about one-third of global methane emissions.[234] To be sure, leaking wells can be plugged, but only if they are found. Invisible to the naked eye, methane can be imaged using infrared sensing, including high-resolution, satellite spectral monitoring. As noted by Sharon Wilson, who lives in the Permian Basin and has been imaging methane leaks for a decade, "If you could see the emissions with your naked eye, there never would have been a fracking boom."[235]

Unfortunately, sloppy operating procedures are the norm, especially in the Permian Basin, where leakage is 60% higher than the national average.[236] A 2022 Stanford publication showed that almost 10% of methane produced in the Permian is lost, what *The Economist* called, "not just a waste. It is an assault on the climate."[237] Satellite monitoring of atmospheric methane with more precise spatial and temporal resolution is improving along with drones and ground-based observation and will help detect leakages and fugitive emissions. But without effective regulations to collect methane, standard bad practice will continue.

***

Perhaps the most controversial fracking by-product is the seismic activity triggered by injecting millions of gallons of high-pressure fluid into formerly stable rock formations, undermining the integrity of age-old geological structures. Earthquakes are related to the porosity of the rock because the greater the porosity the greater the possibility of structural change inducing a seismic response. As one might expect, the earth moves when you frack. Earthquakes have been recorded near fracked wells across the globe, including thousands in the United States and Canada where fracking is most prevalent. In Alberta, after injection-induced seismic activity was linked to fracking activity, a group of geologists studying earthquakes likened the process "to small underground explosions, shocks that travel into the rock formation and rapidly change the stress patterns within."[238]

Although headline-grabbing seismic activity can occur during high-pressure water injection, the presence of millions of liters of fracking fluid within a deposit during and after fracking also puts increased stress on the underground rock, while there is a direct correlation between an increase in earthquakes and pumping (or re-injecting) "produced" waste water into so-called "disposal" wells, especially near fault lines. The US Interior Department reported a sixfold increase in the number of earthquakes from 2000 to 2011, which was "strongly correlated to wastewater injection," and that fracking fluid will "continue to compromise the integrity of well casings, increasing the likelihood of water contamination."[239]

Believed to be caused by nearby natural-gas drilling that pierced a volcano, the Sidoarjo mudflow in 2006 killed 20 people and displaced 40,000 others in the East Java region of Indonesia, polluting the nearby Porong River with cadmium and lead. Hydrogen sulfide was also detected in the mudflow, which is still flowing after the initial blowout and is expected to continue for decades. A rise in seismic activity was observed in 2013 in an abandoned petroleum field under the Mediterranean Sea after compressed gas was injected into porous rock off the coast of Castellón, Spain, for a proposed gas storage plant. The regional government halted further injections after hundreds of tremors were recorded, including a 4.2 quake that destabilized the underground strata. No more earthquakes occurred, leaving little doubt that the gas-storage plant was to blame.

In "Frack, Rattle and Roll," Joshua Frank cited earthquakes in Pennsylvania, Oklahoma, Ohio, and Arkansas after fracking operations started. In 2011, more than 100 tremors – including a 3.9 quake on New Year's Eve – were recorded in Youngstown, Pennsylvania, which had never before experienced an earthquake since records were first kept by settlers in 1776. Later that year in Oklahoma – home to more than 10,000 active underground injection wells – a 5.6 quake hit near the town of Prague, while in 2014 a dozen earthquakes were recorded south of Lowellville, Ohio, where a disposal well was being drilled. As Frank noted, "thanks to the insatiable rush to tap every last drop of oil and gas from the depths of the Earth's crust, earthquakes are fast becoming the new norm."[240] In the past year, Earthquake Track has recorded over 1,000 earthquakes of at least 1.5 magnitude in Oklahoma and almost 30,000 in the US.[241]

In Europe, a more robust regulatory regime requires frackers to recycle water, disclose all chemicals used during fracking (to track the source of contamination), and identify all potential exposure pathways "to assess the risk of leakage or migration of drilling fluids, hydraulic fracturing fluids, naturally occurring material, hydrocarbons and gases from the well or target formation as well as of induced seismicity."[242] However, member states can easily flout EU rules through their own national legislation. Fracking permits also depend on where one lives. UK regulations require fracking to stop for 18 hours after a 0.5-Richter tremor, severely restricting drilling operations, while the limit in the US is 4.0 Richter.

The amount of injection fluid can also contribute to the size of a quake. In Lancashire, northwest England, tremors in the heart of old coal country – measured at 1.4 and 2.3 Richter – were likely caused by fracking on pre-existing fault lines. According to geophysicist Arthur McGarr, if you double the amount of injection fluid the maximum magnitude of a quake increases by 0.4 on the Richter scale.[243]

Despite the concerns, fracking was initially highly prized in England, the former UK prime minister David Cameron announcing they were "going all out for shale." Stretching from Blackpool on the west coast to Nottingham and Scarborough in the east, the half-kilometer-thick Bowland Formation is reported to have enough natural gas to supply Britain for 500 years if all of the shale can be produced. By 2019, however, only one site was actively being explored, a newly appointed government fracking tsar resigning because of low limits on tremors. Poland, Bulgaria, and France are also sitting on top of large shale reserves as are Germany, Hungary, and Ukraine, although at greater depths and thus harder to exploit. In Europe and elsewhere, there is much political pressure to reduce the reliance on Russian gas imports with some form of fracking proposed as part of the mix.

Currently, fracking is outright banned in France, the Netherlands, the Czech Republic, and Bulgaria, while Scotland banned fracking in 2017 after a public consultation that drew 65,000 responses. One commenter noted, "We have so much wind and wave power that it is retrograde in the extreme to lend any support to the fracking industry,"[244] while other general concerns cited pollution, contaminated groundwater, damaged communities, engineering validity, and dubious economic value. The ban goes with Scotland's earlier moratorium on coal-bed methane and underground coal gasification, another unconventional production method that ignites underground or offshore coal seams to capture gas on the surface.

Europe is right to be concerned. Much of the continent is more densely populated than the USA and highly susceptible to underground drilling. The largest natural gas store in Europe is the Groningen field in the northern Netherlands, providing up to 30% of European gas since conventional extraction began there in 1963. The 60-year-old site has provided billions of euros in annual revenue to the Dutch government (€428 billion adjusted for inflation[245]), but subsoil faults and earthquakes have prompted remedial action since a compromised dike, canal, or lock could be catastrophic to nearby low-lying regions. Ten times more frequent since 1992, a 3.6-Richter earthquake in 2012 finally forced Dutch authorities to shut down further extraction and the drilling zone was moved to limit structural damage to thousands of local buildings. In preparation for more drilling, houses in the village of Meedhuizen, 25 km northeast of Groningen, were examined and almost all were deemed in need of reinforcing,[246] while 10 km northeast of Groningen in the village of Overschild, 80% of homes were slated to be demolished because of subsidence from gas extraction.[247] Citizens everywhere are right to be concerned about the fragility of the underlying strata.

In the United States, however, fracking has been hailed as an industry savior. From 2007 to 2019, shale gas production shot up 20 times – 1,300 to 26,000 billion cubic feet per year[248] – providing Barack Obama with an energy boom unlike any other president before him and an almost 60% decrease in foreign oil imports.[249] As of 2015, 40 billion cubic feet per day of natural gas was being pumped, mostly from the nine largest American shale "plays." Alas, simple math undermines the fracking equation if $2 billion a month is spent to produce 1 million barrels a day, making the break-even point about $67 per barrel. At $100 a barrel the frackers can recoup their investment; at $50 they lose their shirts.

Continental Resources CEO and founder Howard Hamm, a pioneer in the shale-gas development of the Bakken Formation, put the case quite simply, stating that drillers have to "make a decent rate of return," and that any return on investment is unsustainable below a break-even point, which he pegged at $50 per barrel.[250] At the time, oil had just risen to over $50 per barrel after a prolonged period under $50, and Hamm was touting a further increase of 20% based on revised EIA estimates. In the high-cost Athabasca oil sands, break-even ranges even higher, from $60 to $65 per barrel with some projects as high as $130.

Clearly, the viability of fracking and other unconventional extraction methods requires high oil prices and free-flowing investment, but without government help or deep-pocketed financing – much of it backed by huge debt – unconventional oil is simply too expensive and prolonged negative cash-flow unsustainable. High prices make unconventional oil an attractive investment, one analysis estimating that 200 million barrels of oil sands can be commercially produced at $80/barrel, while 2 *trillion* barrels of shale is viable at $100/barrel.[251] Better technology for recoverable reserves, improved transportation infrastructure, and industry-led regulations for unconventional oil and gas development also helps lower costs, especially in remote areas such as the Bakken and even more so in the further reaches of northern Alberta, where transporting heavy oil sands to Texas via pipeline makes little sense at $50 a barrel.

Unlike conventional players who can ride out price swings in a volatile market, such as the Saudis who can extract a seemingly limitless supply of elephant oil at less than $5/barrel (or North Sea suppliers at under $10/barrel), high-end frackers play a perilous game. Even labor can become prohibitively expensive when boom and bust is the norm for unconventional plays. Such dilemmas may seem like conspiratorial OPEC maneuvering – Saudi Arabia lowering prices to pump more while earning less – or a real market at work, but the Saudis can afford to be patient when low oil prices make hard-to-reach offshore drilling, bitumen-heavy oil sands, and fracking an expensive alternative to free-flowing Arabian crude.

The Saudis must be careful, however. If supplies are restricted as in the 1970s, higher prices will encourage more development of unconventional sources and embolden renewable-energy alternatives. The same follows if prices rise because of political maneuvering, such as restricting Iranian exports or in times of war as in 2022 when the Russian invasion of Ukraine sent oil prices soaring to over $115 a barrel. A reduction in more carbon-intensive, unconventional extraction methods is good for the environment, but the ups and downs in the oil and gas market have been part of the petroleum game since pumpjacks first started sucking black gold from under the ground. Fracking is just a new name and new way to scour the Earth.

***

While fracking has been a boon to natural gas supplies, it still has numerous detractors concerned about the vast water footprint, groundwater fouling, leaked atmospheric contaminants, and associated seismic activity. At the very least, basic rules are needed to ensure the stability of rock layers, groundwater quality, limited leakages, and the health of those who live near a well pad or rely on water in nearby aquifers. One can reduce the fracking footprint by drilling multiple wells from a single pad, use seawater instead of freshwater, extract gas more slowly, monitor resulting seismic activity and leakages, ban open evaporation pits, reduce or restrict flaring (or indeed capture associated gas), limit venting, recycle fracking fluids, and list all injection chemicals (even those considered proprietary), while much more needs to be done to stop the trillions of liters of water removed from the water cycle and discarded as contaminated waste, especially in drought-prone areas. That is, if fracking shouldn't be banned altogether, albeit unlikely in a beholden political landscape.

Is fracking a new frontier, pushing the limits of the latest extraction technology yet oblivious to the damage – frack first and ask questions later? Jobs and government revenue from leases are cited as the main reasons to open up more fracking to untapped shale formations, but as with other oil jobs the work is temporary, while leases come and go, leaving the local community to pick up the pieces when the wells run dry. What happens after the oil companies are gone and the wells start to leak? Who pays for the unplugged orphaned wells?

Anthony Ingraffea, a former gas man turned anti-fracking advocate who was born and raised in Pennsylvania, summed up the worries of those who live near well pads: "We're gonna risk our property values, we're gonna risk our water, we're gonna risk our air, we're gonna risk our health, we're gonna exacerbate climate change and what do we get out of this? We lose what we have, we lose why we live here."[252] Albertan farmer Don Bester questioned the integrity of governments charged with overseeing the interests of ordinary citizens, warning that the oil and gas industry and the government are "in bed together."[253]

Natural gas is mostly made of methane (CH_4) and about 5% ethane (C_2H_6), but also includes small amounts of propane, butane, pentane, and hexane that condense at atmospheric pressure (and liquefy out at the well surface), as well as other impurities such as carbon dioxide, hydrogen, hydrogen sulfide, and nitrogen. Ethane has a higher calorific value than methane and can be left in for household burning or separated out for use in petrochemical feedstocks, while a relatively harmless sulfur-containing methanethiol (methyl mercaptan, CH_3 SH) is added to odorless methane to emit a tell-tale rotten-eggs smell for safer household use (thiol gives skunk spray its characteristic pungent smell).

But although natural gas is cleaner than coal, emitting about 50% less CO_2 per unit energy when burned, methane is still a fossil fuel with all the nasty by-products. Over 40 million natural-gas stoves, cooktops, and ovens in American homes also release methane via incomplete combustion, leaks, and at ignition.[254] Up to 1.3% is unburned methane, equivalent to the yearly carbon dioxide emissions of 500,000 cars, while over three-quarters of the methane is released even when a stove is off. Health-damaging pollutants such as nitrogen oxides (NOx) are also emitted and can trigger respiratory diseases.

An even bigger worry may be gas leaks from fugitive emissions that add to global warming, undoing any value in using cleaner natural gas as a "transition" or "bridge" fuel. Offsetting the presumed benefits, the US National Oceanic and Atmospheric Administration estimated in 2012 that 4% of the fracked natural gas was lost to the environment, while in some regions estimates were over 9%.[255] Although not as long lived in the atmosphere as CO_2, "When natural gas escapes unburned, as it often does during production and distribution, it is a big trouble maker. Its essential component, methane, is particularly pernicious – a greenhouse gas that is more than 80 times as potent as carbon dioxide over 20 years as it dissipates."[256]

Some are also advocating for natural gas as a cheap, easy-to-implement coal substitution for electricity generation that would lower carbon emissions, but as Dieter Helm acknowledges in *The Carbon Crunch*, "There is no escaping the environmental impact of both methane and shale gas production. . . . methane may be short lived in the atmosphere, but it is potent, and many gas pipelines – notably in Russia – leak a lot."[257] Alarmingly, natural gas becomes more environmentally dangerous than coal if only 3.2% of extracted methane leaks instead of being burned,[258] hardly useful as a bridge fuel. As IEA chief economist Fatih Birol notes, "A golden age for gas is not necessarily a golden age for the climate."[259]

Even Rex Tillerson, former ExxonMobil CEO and Donald Trump's first secretary of state, had his doubts about fracking after he sued a water-utility company tasked with supplying water to a fracking operation near his Texas property. The 2014 suit claimed that a partially built, 160-foot water tower was a monstrosity, affecting property values and "causing unreasonable discomfort and annoyance to persons of ordinary sensibilities."[260] Tillerson may have given the simplest reason why no one wants fracking in their neighborhood, especially run by companies that care more about profit than community health or well-being.

One must consider the effects of fracking on everyone, from a working-class Colorado family, whose children go to school amid a litany of haphazardly placed well pads, to a multi-millionaire Texan oil man cum politician who wants to safeguard his net worth. As the mechanic in the 1970s Fram oil filter commercial reminds us if we don't think about the consequences of our actions: "You can pay me now . . . or you can pay me later." The later part always costs more. With fracking, we are destroying precious water supplies, polluting the air and water, and ramping up methane consumption as we ramp down coal. Despite the lower CO_2

footprint, which helps explain lower CO_2 emissions as more gas replaces coal, methane is still a greenhouse gas with a much higher global warming factor than carbon dioxide.

Some have also likened fracking to a Ponzi scheme as new investors pay off old investors to keep the funds flowing. Amid the COVID-19 pandemic when oil prices plummeted and cheap money dried up, the bubble burst for some over-leveraged, debt-ridden companies. The pioneering Oklahoma-based Chesapeake Energy went bust with $9 billion in debt, while state budgets with high O&G exposure were especially vulnerable as taxpayers were left with the bill for asset retirement obligations (cleanups and closures) as bankruptcies increased.[261]

To be sure, new extraction methods such as offshore, fracking, and oil sands are more damaging than tapping conventional elephant oil – a staple of global oil supplies for decades, some of which are beginning to show signs of age – but all externalities must be included, especially public health and the environment. One can't just drill, steam, and pipe without paying the consequences. The whole process from deep underground to a customer's door needs to be considered: extraction, refining, transportation, and burning. Downstream is as important as upstream.

2.12 Shale Oil Blues: Production, Distribution, Pollution

Today, LNG competes with piped gas where pipelines are impractical, as in a distant and isolated Japan, the leading global importer. Known as "trains," LNG factories were built to convert gas to liquid, facilitating transport to Europe and Japan at much reduced prices. To make LNG, the gas is compressed and refrigerated at $-162°C$ (volume is proportional to temperature), becoming a colorless liquid 1/600 the size that is then stored in insulated tanks on giant cryogenic carrier ships for long-haul transport. After the merger of Exxon and Mobil in 1998, which combined Exxon's financial muscle and Mobil's natural gas know-how, Qatar's North Dome field – the world's largest, shared with Iran's South Pars underneath the Persian Gulf – was developed. By 2002, the tiny country of Qatar with a then population of 1.5 million was supplying one-third of the world's LNG.[262] Australia is also liquefying its vast deposits of coal-bed methane to produce LNG for export across Asia.

The increase in shale gas in the United States has also led to lower natural gas prices, making the USA one of the lowest-priced regions in the world. Having geared up to import LNG before the shale gas revolution, newly built import terminals are fast being turned into export terminals as supplies increase.

Primarily because of the fracking boom in large shale regions such as the Permian Basin, coupled with the 2015 lifting of the 1975 crude oil export ban, the USA has found its own oil weapon to undercut foreign players as the cryogenically fitted super tankers pass each other in the night.

At the head of the Qatari fleet, the 2009 Marshall Island-flagged *Al Rekayyat*, operated by Shell, is the size of three football fields and regularly shuttles a quarter million cubic meters of LNG from the port of Ras Laffan, north of Doha in Qatar, to India, China, and South Korea, worth up to $40 million a load at 2020 market prices of around $0.10/m^3. As natural gas prices soared 10 times that in early 2022, a single load fetched as much as $400 million. Detailed safety measures are essential because mobile phones and cameras can cause an explosive fire.[263] Shell also runs the largest floating LNG (FLNG) platform, the $14 billion, 488-m long *Prelude*, which pumps gas from under the seabed, cooled onboard for transport, supplanting the need for fixed onshore facilities.[264]

The increased importance of natural gas is seen in the Gas Exporting Countries Forum (GECF), a kind of OPEC for gas. Established in 2001 and headquartered in Doha, there are 12 GECF member and seven observer states, controlling about two-thirds of the world's natural-gas reserves, pipeline traffic, and LNG trade. The top three, Russia, Iran, and Qatar, control almost 60% of global reserves.

The Leviathan was a multi-headed biblical sea monster and the name of a 1651 Thomas Hobbes book that argued for strong central government to still dissent among rival factions. Today, Leviathan is an offshore natural gas field in the eastern Mediterranean, divided between the competing maritime nations of Lebanon, Israel, Cyprus, and Egypt. Discovered in 2011 and holding as much as 2 trillion cubic meters – 10 times British North Sea reserves or enough to satisfy Israel's energy needs for half a century – the problem as usual is getting the gas out from under the ground and transporting it to market, with the added twist that Lebanon and Israel do not have an agreed border while an ongoing territory dispute exists between Greece and Turkey, who grudgingly share the island of Cyprus.

One plan is to build an LNG plant in Cyprus and ship the gas to ports in Europe, but comes with a hefty $20 billion price tag. Another is to pipe the gas to Italy via the longest subsea pipeline ever constructed at a cost of about $7 billion. Any eventual route through contested waters will need to balance the interests of all players, some of whom are not the best of friends and/or are at war (for example, Syria/Lebanon, Syria/Israel, Palestine/Israel, Greek/Turkish Cyprus). Added to the security issues, various construction logistics await any agreed upon project as the resource competition heats up in the eastern

Mediterranean to produce and control an emerging energy hub in a region that has been a powder keg for over a century – for example, the proposed subsea pipeline route lies under 3 km of deep waters. There is no slowing development, but for any pipeline to work, long-time adversaries will have to get along. Life in the eastern Mediterranean is about to get a whole lot more complicated.

Adding to potential trouble ahead, the nearby gas fields Aphrodite (Cyprus EEZ), Tamar (Israel EEZ), and Zohr (Egypt) may total an additional 2 trillion cubic meters. The Aphrodite and Tamar fields were developed by Texas-based Noble Energy, since acquired by Chevron, while the Zohr field is operated by Italian oil and gas major Eni. Aphrodite is the goddess of love and marriage, Tamar posed as a prostitute to become pregnant with her father-in-law, and Zohr means radiant. Every skill will be called upon to maintain order and keep the players happy. Trillions of dollars are at stake in the cradle of civilization, ensuring interest for years, although another natural gas field found in 2000 about 35 km off the coast of Gaza may take more than simple diplomacy to control.

Liquefied natural gas also requires extra attention to ensure basic safety measures. As fuel sciences expert Harold Schobert notes, "Suppose a leak allowed some of the LNG to vaporize. Mixtures of 5–15% of natural gas in air are explosive. What if an LNG tanker blew up in, say, Boston harbor? The loss of life and property damage would be enormous."[265] In a hurry to join the gas game, Ghana has suffered eight LNG explosions in three years, including a tanker explosion, offloading fuel, and at a local natural-gas filling station in the capital Accra with hundreds of lost lives. Although no one was killed in a 2022 LNG explosion at a Texas export plant near Freeport, almost 20% of US LNG exports were halted.[266] No one can afford to take stored fuel for granted whatever the hydrocarbon composition.

Under the Natural Gas section of the US National Environmental Policy Act, LNG terminals must undergo an environmental review with a public commenting period to receive a permit. The process is almost always adhered to, although in the race to export fracked gas, so-called "small-scale" LNG terminals are exempt from public purview.[267] Of course, there is nothing small about LNG: "It's just making the refrigerator component itself a little bit more modular, repeatable and standardized," stated Meg Gentle, CEO of Tellurian, which claims to produce 55 million tons per annum, about 20% of LNG today.[268]

According to a 2017 PricewaterhouseCoopers report, the annual LNG market was expected to grow to about 100 million tons by 2030, primarily for marine fuel (bunkering), heavy road transport fuel, and off-grid power generation, which works out to $300 billion a year at prices of around

$3,000/ton (1 ton of natural gas is equivalent to about 1.4 billion cubic meters). The numbers are set to rise with the increased uncertainty over Russian natural gas and more newly built LNG facilities.

With reduced American regulations for LNG export terminals, the USA is gearing up to compete with Qatar and Australia in the LNG export market. Almost all of the exported natural gas will come via fracking, piped cross-country to American ports. Exporting LNG would seem unwise, however, if reserves are finite, standard policy since the 1970s as Americans worry about relying on unstable foreign supplies or covering domestic needs when prices rise. One also wonders why natural gas isn't being saved for domestic heating or to smooth out intermittent demand in "peaker" plants (power stations that switch on during high consumer demand).

Some consider natural gas the perfect transition fuel, a stop gap between yesterday's fossil fuels and tomorrow's renewables, better than oil and much better than coal, although much worse if it leaks, as we've already seen. Others believe natural gas from fracking is nothing new, just more of the same old bad thinking that maintains a failed energy infrastructure based on fossil fuels. Despite global warming concerns, the American Gas Association called natural gas "a foundational fuel for decades and decades to come."[269] Giving up coal is nothing compared to turning off the taps on the entire economic foundation of the United States.

2.13 More Dark Sides: Pipelines, Railways, Oil Sands, and Climate Catastrophe

Crude oil is dirty and hard to extract, with spills a regular occurrence. We usually only hear about the biggies – *Deepwater Horizon*, *Exxon Valdez*, or Santa Barbara, where 3 million gallons of oil were dumped into the Pacific Ocean in 1969. The Santa Barbara spill helped rally opposition against negligent industry practices, inspiring Earth Day the following year and the creation of the 1970 National Environmental Policy Act. But the damage is everywhere: in the oil-filled waterways of the Niger Delta, the toxic sludge piles of the Ecuadoran interior, and the thousands of communities near petroleum facilities the world over.

Oil finds its way into every crevice, nook, and tiny cranny of our air, water, and soil, shimmering its tell-tale rainbow swirl and pungent odor for all to see and breathe, while oil companies routinely flout regulations and regulatory control. Were *Exxon Valdez* and *Deepwater Horizon* one-offs, easily blamed on a drunken captain or a lax safety culture? More likely, the deeper reason is that safety costs money, a dangerous modus operandi. But no one can afford a tanker crash into a clearly marked reef or an offshore blowout that wipes out the coastline and the livelihoods of those who would never think to play fast and loose with the rules. In engineering, one can never assume a state of invulnerability, which ultimately leads to sloppy practice, relaxed safety measures, and a casual oversight of potential problems, a.k.a. the "atrophy of vigilance." Even with the best intentions, one can never be fully prepared.

Not all disasters occur at sea. While our waterways may be safer than in the past, pipelines are more likely to leak or spill today, especially in the United States with the world's largest pipeline network of almost 3 million miles, enough to ring the equator more than 100 times. Built in 1931 from the Texas Panhandle to Chicago, the first 1,000-mile-long natural-gas pipeline employed new techniques in electric welding, which cost less, was less laborious, and used less pipe (no need of overlapping rivets).[270] The 1,254-mile-long Big Inch – the largest pipeline at the time – began carrying oil from Longview, Texas, to New Jersey in December 1942, delivering oil to the Northeast for the war effort. Big Inch was essential to counter German U-boats destroying oil-laden American ships.

A second, almost parallel, Little Big Inch was "looped" beside Big Inch to carry gasoline, kerosene, diesel, and heating oil, beginning in January 1944. After the war, both Big Inch and Little Big Inch were converted to natural gas pipelines, supplying a strapped Northeast with a much-needed cooking-gas and winter-heating supply, both of which helped to replace coal with oil and are still operational today.[271]

Almost two incidents per day, however, occur along the 2.8 million mile national network according to the US Pipeline and Hazardous Materials Safety Administration (PHMSA), caused by corrosion, excavation damage, shoddy workmanship, and welding or equipment failure. Since 2010, there have been more than 100 deaths, 500 injuries, and $3.5 billion in damage.[272] What's more, almost a third of pipelines were built with steel segments fused in the early 1950s, using a flawed technique of low-frequency, electric-resistance welds. With the increase in fracking, natural-gas pipelines are also leaking more methane, while EcoWatch noted that over half of US pipelines are more than 45 years old and that technologies designed to detect leaks aren't reliable. In fact, federal records show that almost as many leaks are detected by a random member of the public than by advanced detection systems, such as flow-rate sensors, fiber-optic temperature monitors, and hydrocarbon-sensing cables.[273]

Pipelines must also be monitored and protected from natural and man-made interference, in particular, earthquakes, terrorism, and theft. In 2001, an Alaskan resident shot a hole in the Trans-Alaska pipeline, resulting in a 3-day shutdown and a 300,000-gallon spill, while in 2019 almost 100 people died after a botched tapping, when a gasoline pipeline exploded north of Mexico City. Mexico can regularly see up to 1,000 illegal taps per month.[274] In May 2021, the 5,000-mile-plus Colonial Pipeline system, the largest in the United States, which handles over 2 mbpd of gasoline, jet fuel, and heating oil from Texan refineries to the eastern USA, was hit by a ransomware cyberattack, disabling transmission for almost a week.

On July 25, 2010, more than 1 million gallons of diluted bitumen (a.k.a. dilbit) from the Canadian oil sands leaked into the Kalamazoo River, after a 30-inch carbon steel pipeline ruptured near Battle Creek, Michigan, the largest inland oil spill in US history. Monitored in a control room 1,500 miles away, the spill wasn't detected by the pipeline operator until members of the public notified company officials. Thinned with diluents such as benzene to facilitate flow, the diluted bitumen is the same type of oil that was slated for transport along the proposed Keystone XL pipeline, considered by some to be so acidic and abrasive that it increases corrosion. The worry is that the viscous Canadian dilbit causes ruptures from increased operational stress. Dilbit is also heavier and sinks in water, hampering cleanup efforts after a spill, while carcinogenic benzene evaporates into the air. After the Kalamazoo spill, 150 families were forced to relocate permanently, while much of the river between Marshall and Kalamazoo was closed to the public for almost 2 years.[275]

Opposition is particularly strong in the ancestral lands of the Standing Rock Sioux Tribe on the border of North and South Dakota, where protesters gathered to stop construction of the Dakota Access Pipeline (DAPL) under the Missouri River, which supplies drinking water to over 17 million people, including the indigenous Standing Rock Tribe. The 1,200-mile route that runs from the Bakken Formation in northwest North Dakota to a storage hub near Pakota, Illinois, traverses the Missouri on its southern pass from Bismarck, and now pumps 500,000 barrels a day of light crude oil from six different terminals in North Dakota, or about half of the Bakken's daily run.

Originally designed to cross the Missouri River north of Bismarck, the $3.8 billion DAPL was rerouted because of worries about potential contamination to water supplies. Skirting Standing Rock lands near Canon Ball, North Dakota, the changed route crossed under the Missouri at Lake Oahe, "the sole water supply for the Standing Rock and Cheyenne River Sioux Tribes and thousands of other people."[276] One wonders why a different route wasn't chosen, satisfying both the pipeline company and the Standing Rock Sioux, but perhaps less resistance was expected through Native lands. Caught unaware by the support in such a remote region, however, the new route sparked a standoff by the Sioux, hundreds of tribes from across the USA, and thousands of supporters, many of them local indigenous women, who set up camp for months to oppose the construction. Calling themselves "protectors" not protesters with the slogan *mní wičhóni* ("water is life"), the issue was about ensuring clean water for a population of 17 million. Understandably, many were concerned because the nearby Belle Fourche Pipeline has seen numerous incidents and a disastrous spill. More worrisome, DAPL sprung a leak in a pumping station before becoming fully operational.

Writing for CityLab, which has mapped pipeline accidents in the USA for the last 30 years, George Joseph notes, "The sheer number of incidents involving America's fossil fuel infrastructure suggests environmental concerns should go beyond Standing Rock," adding that incidents are particularly common in Texas and Louisiana, "where numerous lines carry oil and gas, extracted on- and off-shore, to serve the rest of the country."[277] After a 380,000-gallon leak on part of the Keystone Pipeline in North Dakota, the associate director of the Sierra Club, noted, "it's not a question of whether a pipeline will spill, but when."[278] Most worryingly, Keystone crosses the Yellowstone River between Billings and Glendive, towns that have already suffered river spills and lackluster industry responses. In 2021, the DAPL route was stopped pending an environmental review, upheld on appeal to the Supreme Court in 2022. For now, the water protectors have prevailed and slayed the "black snake."

Without pipelines, other options are rail and truck (or flaring on site in the case of natural gas). Expensive and time-consuming to construct, pipelines often lag production as new wells mature and delays follow a typically lengthy infrastructure approval process, forcing oil onto less safe railways and highways. According to the American Association of Railroads, "99.997 percent of all hazardous materials reach their destination without being released because of an accident."[279] That still leaves three out of every 100,000 that do.

In the early morning hours of Saturday, July 6, 2013, an unattended 74-car train carrying 2 million gallons of Bakken crude oil rolled 13 km down a hill before derailing in the town of Lac-Mégantic, Quebec, killing 47 people and obliterating much of the historic downtown in the resultant explosions. The investigation determined that the lead locomotive had been "surging and smoking," yet was inexplicably parked unattended overnight on a 1.2% incline. Idling with insufficient hand brakes applied, the pressure to the automatic air-brake system was inadvertently released after the power was cut off to the burning engine. Brian Stevens, a veteran air-brake mechanic and rail union director, blamed industry deregulation, which had drastically cut safety inspections since 1984, lamenting that "The railways write the rules."[280]

At the 2017 trial of the three Montreal Maine and Atlantic Railway employees charged with criminal negligence, court evidence revealed that rail convoys carrying crude oil through Lac-Mégantic had been almost doubling per week and that one-man crews with few safety checks had begun shortly before the crash. The cargo was also falsely identified in shipping documents as the least hazardous type of oil. Because existing pipelines have insufficient capacity and there are no refineries in North Dakota to service the Bakken, more oil is being shipped by rail, a practice that was practically nonexistent a decade previously.[281]

At the time of the Lac-Mégantic disaster, rail transport of volatile crude oil in North America had jumped almost 5,000 percent in 6 years, accounting for 1.5 million barrels per day. Montreal Fire Department division chief Gordon Routley noted that fire-fighters had to deal with "a new realm, the transportation of crude oil by rail." Known as "bomb trains," the thin-walled, high-center-of-mass DOT-111 cars were already known to have a high tank failure rate in a derailment and are no longer permitted to carry crude in Canada, although the practice still continues in the United States. Despite the ban, there have been numerous major derailments of Canadian oil trains since 2020, spilling over 8 million liters of oil, while in Saskatchewan crude loads increased sevenfold in less than 3 years.[282] Although not carrying oil, a 2023 train derailment in East Palestine, Ohio, burned toxic cargo from a number of overturned railcars for over 2 days, contaminating the soil, air, and water, and underscoring the dangers of transporting hazardous flammable material by rail.

If oil and gas is found in remote areas such as the North Slope, Gulf of Mexico, Bakken Formation, or Athabascan oil sands, the fuel must be transported to market by pipe, ship, rail, or truck. But when accidents occur at such high rates, one wonders whose interests are being served by prioritizing long-haul transportation through pristine lands and populated areas over ensuring clean water supplies, agricultural integrity, and public safety.

***

While most pipelines spills or leaks aren't as horrifying as Lac-Mégantic or any of the other major oil-related rail derailments across the USA and Canada, pipelines are still highly dangerous, posing a threat to water supplies, agricultural interests, and fuel disruptions. Safety is the major concern for those who live alongside a pipeline, surprisingly self-regulated in the pipeline industry. PHMSA only requires self-assessment every five years and an "integrity

management" plan for pipelines with known defects located in highly populated or environmentally sensitive areas.

Nowhere is the concern more worrisome than with TransCanada's Keystone
XL pipeline, the $8 billion, 1,900-mile artery designed to pump 800,000 barrels
a day of bitumen-heavy, diluted oil sands from Hardisty, Alberta, to Steele City,
Nebraska, and on through existing pipelines to coastal refineries on the Gulf of
Mexico. Simply known as KXL for its extra-large 36-inch-diameter cross-
section compared to the existing 30-inch-diameter Keystone pipeline, the
beleaguered project has rarely been out of the news because of concerns over
transporting the more volatile, corrosive, and toxic bitumen crude and
increased environmental damage to extract and refine energy-intensive oil
sands.

Many wondered why TransCanada (now called TC Energy) didn't double up
on their existing pipeline route from Hardisty east to Winnipeg and then due
south through the Dakotas, instead of transiting the farm-rich Nebraska
Sandhills and Ogallala freshwater aquifer, the largest in North America.
Perhaps, they didn't want to disturb an already profitable route or traverse
fracking country where the earth is known to move.

As the name implies, "oil sands" are a mix of oil and sand, easily seen on the
Earth's surface, although mined *in situ* belowground. To extract usable oil from
the dough-like mixture, the heavy sands are "washed" by pumping high-
pressure super-heated steam into the ground to loosen or liquefy the oil,
which lowers the hydrocarbon viscosity and increases flow. The world's largest
deposits are the Athabasca oil sands (a.k.a. tar sands) in northern Alberta about
100 miles northeast of Edmonton with an estimated 160 billion barrels, a whopping 10% of estimated global reserves and world's third largest oil patch,
lagging only the *entire* countries of Venezuela and Saudi Arabia.

A mixture of bitumen, sand, clay, and water, oil sands are especially hard to
develop because the heavy crude requires five times the energy of traditional
extraction methods to clean before being piped elsewhere for refining. Today,
most "cleaning" is done via steam-assisted gravity drainage (SAGD), while the
high-intensity energy and water production process generates almost a quarter
more GHGs than conventional oil.[283] The "dirty" oil is even too heavy for some
refineries, but despite the obvious environmental dangers, the Canadian government is keen to reap a huge bounty on the plentiful and highly sought-after
resource. As noted by an industry VP with skin in the game, the oil sands are
"going to be needed in any energy mix in any consensus, any report that's been
published."[284]

Expected to bring as much as $1 trillion to KXL's Canadian owners over
a 50-year lifetime, US Representative Henry Waxman questioned the value to

Americans, however, especially those forced to give up their land through eminent domain, stating that Keystone XL "would raise gas prices, endanger water supplies, and increase carbon emissions."[285] Traversing almost 2,000 bodies of fresh water from start to finish, the proposed pipeline primarily enhances the coffers of a foreign oil company and a few Texan refineries. Jobs are the standard mantra, but as always are only temporary in the pipeline business. Waxman spelled out the reality of favoring petroleum over newer technologies, "We're not gonna give up on oil overnight, we're not gonna give up on coal overnight, but it seems to me that we're not giving the private sector the incentives to develop technology and to use ways that will reduce these emissions and make us a more energy secure nation."[286]

It's hard to say no to the world's third-largest petroleum reserve – almost 5 years of globally consumed oil – or an industry that produces 20% of Canada's GDP, but as temperatures increase there may be no choice but to leave the dirtiest oil in the ground, or in this case on the surface. If not, one should expect to see more calamities, such as the forest fires that raged for 2 months during the spring of 2016 on the northern edge of the oil sands near Fort McMurray, forcing the entire city of 88,000 people including almost 14,000 oil workers to evacuate. In spite of an overwhelming scientific consensus, connecting global warming and climate change to carbon emissions has become a politically contentious issue about the extent of human culpability. But the connection between environmental degradation and dirty oil extraction is plain for all to see. In the heart of the Athabasca oil lands, one sees the worst.

Pollution is also significant from oil sands mining and upgraders. A 2009 study showed that at spring melt each year particulate matter and polycyclic aromatic hydrocarbons are deposited over a 50-km range, annually washing the equivalent of as much as a 13,000-barrel spill into the Athabasca River. A follow-up study in 2010 found that air pollution and watershed destruction from oil sands development also deposited heavy metals into the river, such as arsenic, thallium, and mercury at levels 30 times the permitted guidelines.[287]

Today, the likelihood is that KXL won't be built after numerous legal challenges have led to ongoing delays. Each new federal administration brings another ruling, the pipeline ping-pong game continuing after a Montana judge ruled in 2018 that another environmental review was needed after a successful lawsuit by Native and environmental groups, rescinding a prior Trump presidential permit that had overturned an earlier Obama ban. When US court approval was again given in 2019, TC Energy announced it would start preparing construction despite a few pending lawsuits, although the project remained uncertain because of continued battles, a slump in oil prices, and the lack of secured long-term refining contracts in Texas.

Perhaps signaling a final end to the controversial, cross-border project, Joe Biden rescinded Trump's previous presidential permit on the first day of his administration in 2021, citing climate change as "an existential crisis," although he didn't overturn Trump's other permits for DAPL or Enbridge's expanded Line 3 replacement pipeline from Alberta to Wisconsin. In June 2021, TC Energy finally pulled the plug, however, announcing it was cancelling the project. Albertan dilbit slated for KXL will instead be sent by existing pipelines such as Keystone and other transport means. Underscoring the constant danger, Keystone sprung a leak in December 2022, spilling 14,000 barrels into a rural creek northwest of Kansas City, contaminating farmlands and drinking water. With its latest malfunction, Keystone has now leaked more oil than any other pipeline in the United States.

Despite the potential for more disastrous spills, expansion is still the main goal of the oil companies. According to a 2022 Global Energy Monitor report, a massive build-out of new pipelines is being planned by the oil industry thanks to record profits, dwarfing all previous construction projects in the past decade. Ignoring calls to limit global warming in a time of increasing temperatures and adverse climate events, 24,000 km of new oil pipelines are in the works, led by the United States, India, China, and Russia, 40% of which are already under construction and representing assets of more than $75 billion. In a blatant disregard of worsening climate data, "some of the world's biggest consumers of fossil fuels are doubling down on oil, even as the climate crisis intensifies."[288] Upon completion, the pipelines will be associated with almost 5 billion tons per year of CO_2 for decades to come (over 10% current global annual emissions).

***

As we have seen, petroleum stands apart from other extractive industries, taking not only the subterranean wealth, but paying little heed to surrounding life – human or otherwise. Cutting corners is the way of the fossil-fuel business as industry insiders encourage deregulation. Lax regulations mean fewer restrictions on drilling, transportation, and refining, adding to profits yet increasing the environmental damage. However, safety should be everyone's concern, because there is no return after a spill, the land and water transformed beyond simple repair and the lives of local residents scarred forever. The damage is incalculable in a major accident, whether in Alaska (*Exxon Valdez*), the Gulf of Mexico (BP's *Deep Horizon*), the Niger Delta (Shell), Ecuador (Chevron), or any number of coastal spills around the world.

The Centre of Documentation, Research and Experimentation on Accidental Water Pollution (Cedre), a French non-profit organization created in 1979 after

Table 2.10 *Top 10 releases of crude oil by volume²⁸⁹ (*land based)*

#	Year	Name	Location	Cause	Quantity (10⁶ gallons)	Quantity (tonnes)
1	1991	Gulf War	Persian Gulf*	War	450	1,500,000
2	1910	Lakeview Gusher	Kern County*	Untapped	378	1,200,000
3	2010	Deepwater Horizon	Gulf of Mexico	Blowout	200	650,000
4	1979	Ixtoc I	Gulf of Mexico	Explosion	140	460,000
5	1979	Atlantic Empress	Tobago	Collision	90	290,000
6	1979	Mingbulak	Uzbekistan*	Blowout	88	280,000
7	1994	Kolva River	Russian Arctic	Rupture	84	270,000
8	1983	Nowruz oil field	Persian Gulf	War	80	260,000
9	1983	Castillo de Bellver	Cape Town	Fire	79	260,000
10	1978	Amoco Cadiz	Brittany	Weather	69	220,000
	1989	Exxon Valdez	Alaska	Grounding	11	36,000
	1969	Platform A	Santa Barbara	Leak	3	10,000

Source: Rafferty, J. P., "9 of the biggest oil spills in history," Britannica. www.britan
nica.com/list/9-of-the-biggest-oil-spills-in-history.

the *Amoco Cadiz* spill off the coast of Brittany, provides extensive data on
hundreds of maritime accidents. The top 10 marine spills by volume are listed
in Table 2.10 with the 1969 Santa Barbara and 1989 *Exxon Valdez* spills
included for comparison.

What's more, few share in the immense profits garnered by the major oil
players. In *Blowout in the Gulf*, William Freudenburg and Robert Gramling
give a particularly damning indictment: "US energy policies over the past
quarter-century have conferred most of their benefits to a handful of the world's
largest oil companies, doing so while offering little if any visible advantage for
the larger economy, and clearly creating losses for the federal treasury –
continuing to do so during decades of record federal budget deficits."[290] The
reality of the oil business is that there will be more pipeline breaks, tanker
crashes, and rail accidents, spilling toxic oil into our increasingly polluted
waters and lands, while the public foots the bill.

Furthermore, the health consequences from burnt carbon are clear. There are
many types of emissions, some quite nasty and for varying reasons as we've
already seen with gasoline and diesel – CO_2, CO, C, NO_x, VOC, S, SO_2, as well
as trace amounts of metal (especially mercury and lead). More toxic fouling
continues from fracking, bad pipelines, refining, and everyday practice. One
can't precisely measure all atmospheric pollutants, but road vehicles are
believed to be the largest contributor, responsible for 60% of carbon monoxide,
44% of hydrocarbons, and 31% of nitrogen oxides.[291] The levels are

significant, turning high-traffic cities into chemical soups on the order of 1950s London, 1960s Los Angeles, and most recently in a number of European cities, including London, Paris, and Madrid.

A 2016 World Health Organization (WHO) database of 3,000 cities noted that over 80% of people living in urban areas suffer air pollution levels exceeding WHO limits,[292] while the fast-growing cities in the developing world are worst affected.[293] Studies also showed an increase in dementia for those living near high-traffic roads, while just idling or stopping to pay a toll is damaging – after the EZPass was introduced at American toll booths, premature and low birth weights were reduced in the vicinity by over 10%.[294]

Although emissions are regulated such that new vehicles must insure minimum standards, there are still enormous health risks. According to the EPA, "Over 149 million Americans are currently experiencing unhealthy levels of air pollution which are linked with adverse health impacts such as hospital admissions, emergency room visits, and premature mortality. Motor vehicles are a particularly important source of exposure to air pollution, especially in urban areas."[295] In 2018, WHO estimated that 4.2 million people die prematurely *each year* because of outdoor air pollution. That's more than 10,000 a day, while a European Public Health Alliance report calculated that traffic pollution alone costs over €70 billion annually in Europe, "with diesel fumes responsible for three-quarters of the harm."[296]

More than 60 years ago Rachel Carson wrote in *Silent Spring*, "As man proceeds towards his announced goal of the conquest of nature, he has written a depressing record of destruction, directed not only against the earth he inhabits but against the life that shares it with him."[297] Carson's book was based on the evidence of wilderness degradation she saw growing up near Pittsburgh and on earlier books depicting the extent of our disregard for nature, such as William Vogt's *The Road to Survival* and Fairfield Osborn's *Our Plundered Planet*, both published in 1948, as well as Aldo Leopold's posthumously published 1949 *A Sand County Almanac*. Each told a similar horror story of neglect.

Vogt wrote about the Earth's natural "carrying capacity" and that we can't keep taking from the ground without ultimately paying the piper, while Osborn reminded us of the perils of blind consumption and Leopold how nature's balance is being disturbed. Leopold's diligent observations of local natural habitats in Wisconsin, where he went to work as forester and professor of wildlife management, were summed up by a new "land ethic" for all to follow:

> It is inconceivable to me that an ethical relation to land can exist without love, respect, and admiration for land, and a high regard for its value. By value, I of course

mean something far broader than mere economic value; I mean value in the philosophical sense.[298]

One can go back even further to Henry David Thoreau, the nineteenth-century New England author, philosopher, and nature convert, who wrote about serenity in Walden Pond away from the hubbub of "modern" life. That was over 150 years ago. How much longer must we endure the continual disregard of our environment? When will the devastation be enough? When do we say "no" to the senseless destruction of Mother Earth?

2.14 Carbon Capture: Industry Savior or Pie-in-the-Sky Technology?

Perhaps because we can't see the air around us, we are uncertain about its makeup. Not until the mid-eighteenth century were the two main components identified – nitrogen (78%) and oxygen (21%) – while a third, argon (0.93%), wouldn't be identified until a century later, along with two other inert "noble" gases, neon and krypton. The rest, such as carbon dioxide (0.041%), methane (0.00017%), nitrous oxide, ozone, and various other trace amounts and their exact percentages, would be identified over time and are of varying degrees of significance because of different heat-trapping properties – together they measure less than one-half of 1% along with a constantly fluctuating amount of water vapor.

Carbon dioxide (CO_2), the atmospheric gas everyone is talking about, was actually identified prior to the others in 1754 by the Scottish physician and chemist Joseph Black. Black isolated CO_2 – a colorless, odorless, and heavier-than-air gas – by heating limestone during his investigations into stomach maladies and their remedies. Limestone or chalk (calcium carbonate, $CaCO_3$) is an antacid and its effervescence is now understood as the generation of carbon dioxide. About a decade later, the English chemist Joseph Priestley collected CO_2 (which he called "fixed air") from the fermentation of beer in a Leeds brewery near to where he lived to make soda water, an eighteenth-century health fad and modern staple of the fizzy-drink industry, from which he formulated new ideas about the properties of gases. Priestley also "discovered" oxygen by isolating it in air (which he called "dephlogisticated air"). The French chemist Antoine Lavoisier soon after showed that oxygen – the name he gave to the newly isolated gas – was essential for combustion and respiration, and that the masses of the components before and after burning were the same, leading to a clear mathematical basis for a chemical reaction and the law of conservation of mass.

Initially not considered a greenhouse gas, methane (CH_4) wasn't detected in the atmosphere until 1948 and was only analyzed in the 1960s and 1970s "out of simple curiosity about geochemical cycles involving minor carbon and hydrogen compounds."[299] Found in swamp gas (decaying wetland bacteria), rice paddies (plant decomposition by microbes in anoxic soil), and belching livestock where bacteria in the stomach of a cow breaks down cellulose in grass and other feed to produce methane ("enteric" fermentation), any human contribution was originally thought to be minor.

The atmosphere also contains trace amounts of other invisible gases, whose amounts are listed in parts per million (ppm): neon 18.18, helium 5.24, krypton 1.14, and hydrogen 0.55 (by volume in dry air),[300] while carbon dioxide is at more than 420 ppm and methane above 1.9 ppm,[301] both measurably increasing since the beginning of the industrial era. Roughly 4 ppm of carbon dioxide is added to the atmosphere each year, about half staying around and half absorbed by trees, plants, soil, and the oceans. Most importantly, although the amounts are small the heat-trapping properties are not.

Today, we are aware of a "greenhouse effect" caused by water vapor and a few atmospheric gases (mostly CO_2 and CH_4) that are largely transparent to light but opaque to heat. Aptly called "one-way filters"[302] or "one-way mirrors,"[303] shorter-wave visible and UV light gets in, but longer-wave IR heat – mostly reradiated from the earth – is absorbed on the way out and can't escape. Fortunately for us, these greenhouse gases heat the earth from the ground up, otherwise the Earth would be a frigid $-18°C$ instead of the more pleasant, life-giving $15°C$ (on average), but they have also been adding to global warming since the beginning of the Industrial Revolution as CO_2 and CH_4 levels continue to rise. Worryingly, the amount of trapped heat from the "anthropogenic forcing" of greenhouse gases has been rapidly increasing of late, doubling since 2005 in 14 years.[304] Fortunately, the two main atmospheric components (molecular nitrogen and molecular oxygen) don't trap reradiated heat, although the two minor molecules that do are increasing beyond normal limits (carbon dioxide and methane).

The French mathematician Joseph Fourier, the Anglo-Irish physicist John Tyndall, and the Swedish chemist Svante Arrhenius all did early work to establish the now well-known, heat-trapping properties of water vapor, carbon dioxide, and methane. Fourier worked out that the temperature differences between night and day (and between winter and summer) were minimal because of the insulating atmospheric blanket of greenhouse gases, a term he coined, while Tyndall was more interested in explaining the mechanisms of past ice ages – evidence of which had only been recently discovered in the scarred glacial landscapes of northern Europe – concluding that atmospheric greenhouse gases could have altered the climate in a distant past. Although uninterested in analyzing the effects of atmospheric warming in his own time, Tyndall did note that the varying amounts of water vapor, carbon dioxide, and methane could indeed be responsible for past ice ages after setting up his own "artificial sky in a tube" in the basement of the London's Royal Institution.[305]

Arrhenius also worked on the theory of ice ages and anything that struck his wide scientific fancy, but it was his work on terrestrial carbon dioxide that established the first direct link between GHGs and temperature, for which he is now largely remembered. Based on his analysis, it was also known in the early 1900s that burning coal would produce more than enough atmospheric CO_2 to raise global temperatures beyond safe limits (becoming a popular meme a century later about what was known when). As noted in the 1912 *Popular Mechanics* article (entitled "Remarkable Weather of 1911" and subtitled "The Effect of the Combustion of Coal on the Climate – What Scientists Predict for the Future"), "The atmosphere contains altogether [1.5 trillion] tons of carbon dioxide. Consequently the combustion of coal at the present rate will double it in about 200 years, unless it is removed by some means in enormous quantities."[306] Alas, *Popular Mechanics* couldn't have anticipated the extraordinary growth in the fossil-fuel industry in the last century. At our current rate of annual emissions, the doubling of CO_2 would occur in just 40 years rather than 200 (1.5 trillion tons/37 billion tons per year), perhaps the most alarming evidence of the severity of the problem.

Today, seven greenhouse gases are monitored by the United Nations Framework Convention on Climate Change (UNFCCC) set up in 1992 – carbon dioxide, methane, nitrous oxide (N_2O), sulfur hexafluoride (SF_6), hydrofluorocarbons (HFCs), perfluorocarbons (PFCs), and nitrogen trifluoride (NF_3). Carbon dioxide is responsible for most of the global GHG emissions (76%) and can remain in the atmosphere for thousands of years, but the others are also significant, such as methane (16%), nitrous oxide (6%), and the ozone-depleting halocarbons (2%). The halocarbons include the dangerous, fluorine-containing refrigerants CFC-11 and CFC-12 or so-called F-gases that comprise only a relatively miniscule amount in the atmosphere but can significantly contribute to global warming.[307] Setting off alarm bells about a growing human contribution to the greenhouse effect, a.k.a. anthropogenic global warming (AGW), the atmospheric concentration of CO_2 rose above 400 ppm in 2015 for the first time since the start of the industrial era, that is, about 4 molecules per 10,000, and continues to increase by about 2 ppm per year. At the start of 2023, atmospheric CO_2 was almost 420 ppm.

The exact warming effect of each of the atmospheric gases is hard to quantify because of differing concentrations, absorption properties, and lifetimes, so a single measure was devised called the global warming potential (GWP), which measures relative contributions using CO_2 as a reference base of 1. GWP is useful for greenhouse gases that comprise smaller concentrations than CO_2, yet absorb more reradiated terrestrial heat. Methane absorbs more than carbon dioxide (primarily in the infrared), yet doesn't last as long, giving it a GWP of 84 after 20 years that drops to 28 after 100 years. Nitrous oxide has a GWP of 265 and CFC-12 8,500. Although emissions vary across the globe, total GHG emissions are now on the order of 50 billion tons per year, 37 billion or three-quarters of which is CO_2. Note, GHG concentrations are also sometimes given in terms of CO_2 equivalent emissions (CO_2e) that includes the effect of the other heat-absorbing gases, in particular CH_4 and the CFCs. Some GHGs also occur naturally in the atmosphere, but the increasing contribution from burning fossil fuels in the industrial era is altering the mix.

The four major sources that make up the 50 billion tons per year of GHG emissions are electricity and heat, agriculture, industry (especially steel and cement manufacturing), and transportation (Figure 2.15). The most significant is electricity (the grid), which if decarbonized can also be plugged into the other sectors to reduce the overall concentration. Electricity and heat (25%), agriculture (24%), and industry (21%) make up about three-quarters of all GHG emissions, although the percentages vary by location and who reports the numbers. As we have seen, China (30%), the USA (15%), and Europe (9%) are the top three regional emitters.

***

Unfortunately, long-stored carbon is increasingly being returned to the atmosphere (~2 ppm/year) and the oceans via fossil-fuel combustion and loss of forests from land change and fires, resulting in rising global temperatures that are playing havoc with terrestrial climate systems. As more carbon is oxidized to carbon dioxide via burning and organic decay, some now think capturing the emitted carbon dioxide is the answer to our GHG problems. Something has to be done to counter the rising temperatures, increasing at a rate unseen since the start of the current Holocene era at the end of the last ice age about 12,000 years ago.

"Carbon capture" has been around since the 1970s to coax more out of a dying well, such as collecting carbon dioxide in a natural gas reservoir that is then pumped into an oil well to squeeze out more production. Injecting CO_2 is the primary "enhanced" or "improved" oil recovery method (EOR/IOR) for a late-life well. Carbon dioxide can also be separated from the outflow in a power-plant smokestack via CO_2-binding amines, what some amusingly call "clean coal," although the process applies to any fossil-fuel burning or

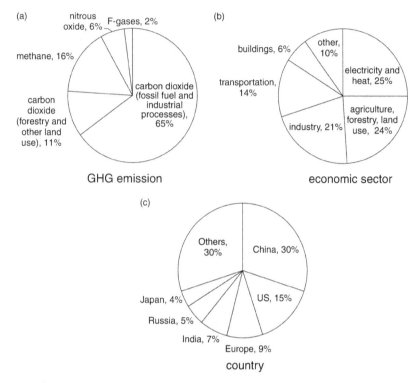

Figure 2.15 Percentage GHG emissions: (a) by gas, (b) by sector, and (c) by country (*source:* US Environmental Protection Agency, "Global Greenhouse Gas Emissions Data.")

industrial process that captures CO_2 and is easier than removing carbon from the atmosphere after the fact.[308]

Whether removed at the source (point capture) or afterwards (direct air capture), the goal is to then inject the captured CO_2 into a spent oil or gas field, deep underground rock formation, saline aquifer, or the bottom of the ocean (although the oceans are already becoming saturated). The captured carbon is typically compressed until liquid or dissolved in water, but can also be turned into inert solid carbonate or mineralized upon injection, reacting with the underlying stores of calcium, magnesium, or basalt, literally "turned to stone." Carbon capture and storage/sequester (CCS) thus reduces atmospheric CO_2 and hence global warming, although the process is staggeringly difficult and expensive. Columbia University's Jeffrey Sachs, co-founder of the Deep Decarbonization Pathway Project, noted that CCS is controversial, but if the fossil-fuel companies want to survive, "they should be investing like crazy in proving the technology."[309]

One might be surprised to learn that the "capture" part of CCS has been around for decades, but that the "storage" part hasn't been high on any list of industry retrofits. Ironically, most carbon is captured only to coax out more output from a not-yet-empty oil-and-gas store, furthering the future of fossil fuels rather than addressing the dangers of increased GHGs. The Global Carbon Capture and Storage Institute (GCCSI) calculated that about 40 million tons of CO_2 – less than 0.1% of annual emissions – is currently being captured from industrial sources, more than half of which is reused for oil recovery.[310] Not enough by any standard to make a dent. Despite the minimal uptake, the IEA has called carbon capture "a crucial part of worldwide efforts to limit global warming," estimating that CCS could provide about 20% of global GHG reductions by 2050.[311]

Exxon's Shute Creek natural gas processing facility near LaBarge, Wyoming, is the world's largest carbon-capture plant, removing 50% of out-flow CO_2 to enhance further oil recovery via its proprietary Controlled Freeze Zone (CFZ) technology, but is not employed to capture atmospheric emissions. What isn't used to squeeze out more oil from an aging well is stored in an acid gas injection unit for future use or simply vented. The first European plant to burn coal with CCS, the lignite-fueled, 30-MW Schwarze Pumpe pilot plant in southeast Germany, which started up in 2006, stopped capturing carbon in 2014 because of high costs. Fitted with state-of-the-art CCS technology, Chevron's Gorgon LNG plant in western Australia has also been touted as an industry savior, but has been plagued by technical problems, chiefly sand clogging the injection pipes. Despite the faulty capture system, CO_2 continues to be generated and vented into the atmosphere instead of buried as licensed. As of 2017, only two working power plants in North America employed CCS technology – the 110-MW Boundary Dam coal plant near Estevan, Saskatchewan, and the 240-MW Petra Nova coal plant outside Houston – removing up to 90% of CO_2. All come with heavy government backing.

Although power plants have been slow to the game, because of high costs, technical challenges, and little financial incentive (such as carbon pricing), the Norwegian state-owned gas company Equinor (formerly Statoil) has been injecting about 1 million tons of CO_2 a year since 1996 from its spent oil and gas Sleipner Field into the Utsira Formation in the North Sea. Not surprisingly, Norway was one of the first countries to introduce a carbon tax. Forming carbolic acid (C_6H_5OH) when CO_2 mixes with water, the Utsira aquifer may be large enough to store all of Europe's power-station emissions for centuries to come,[312] as well as carbon from other industries, such as the global steelmaker ArcelorMittal and the French industrial gas provider Air Liquide, although the injected CO_2 can still leak because of limited reaction with the underlying sedimentary rock. As Chris Goodall noted in *Ten Technologies to Save the Planet*, "The gas is buoyant and will try to escape upwards in an underground reservoir of any type."[313]

In 2019, an ambitious plan was announced to inject 10 million tonnes of CO_2 into two empty sandstone gas-field reservoirs two miles below the North Sea seabed, collected from ports in Rotterdam, Ghent, and Antwerp that emit a third of all GHG emissions in the Benelux. The Dutch government had issues, however, "about how the CO_2 will affect the deep subsurface."[314] A similar pilot project underway at the Drax power plant in Yorkshire – the UK's largest emitter of CO_2 – began capturing 1 ton of carbon a day that will eventually be piped into empty oil and gas reservoirs under the North Sea, literally a spit in the ocean. Nonetheless, although the amount is minimal so far, such point-capture systems are better than continuing to pump out more carbon dioxide into the atmosphere, while if carbon was properly priced or regulated such industrial sequestration would incentivize polluters to clean up their act. It is fanciful to think the oil companies will ever add costs to their core business on their own.

Highlighting the difficulty and enormous challenges of incorporating carbon-capture-and-sequester technology, there were only 26 CCS commercial facilities up and running as of 2021, 19 of which functioned only to flush out more oil in an EOR system.[315] Furthermore, as noted in a 2022 report on 13 CCS projects, a majority didn't work, its author stating that "As a solution to tackling catastrophic rising emissions in its current framework, CCS is not a climate solution."[316]

Making dirty petroleum into cleaner fuel while storing carbon is also more theory than practice, although a few companies are trying. Buoyed by almost $1 billion in subsidies from the Canadian government, a Shell-led consortium added a CCS system to an existing bitumen upgrader plant northeast of Edmonton to lower the carbon content in oil sands to reduce emissions by 35%. Fitted in 2015, the CO_2 is captured during hydrogen conversion of bitumen into synthetic crude, pressurized, piped 80 km offsite, and injected into a 2-km deep layer of limestone and salty water. According to Shell, 1 million tonnes of CO_2 is captured per year by its Quest CCS system.[317] In 2021, a similar system called Polaris was announced for another nearby Shell facility, where the CO_2 is piped 12 km to underground storage wells – operations are expected to begin by 2025 at about 70% less cost than Quest.[318] As more CCS ramps up across oil-rich Alberta, the goal is to sequester 10 million tonnes per year to clean the dirtiest of fuels and reform its bad-boy image.

CCS will be needed on a much larger scale, however, if the oil companies are serious about lowering their greenhouse gas emissions. The oil sands account for about 11% of Canada's GHG emissions, while Canada officially announced an emissions reduction up to 45% below 2005 levels by 2030 in its 2021 NDC target report to the UN.[319] Capturing 1 million tonnes today or 10 million tonnes in the future might seem a lot, but not compared to 50 *billion* tons of greenhouse gases emitted *annually*.

Oil prices are also important along with the generous government subsidies. Shell's Quest CCS is a non-profit project to make oil sands cleaner – ostensibly to meet emission-reduction targets and enhance sales – but the break-even point is around $50 per barrel, not counting government aid. The Canadian government–industry partnership for Boundary Dam cost $1.24 billion, in part to showcase the new technology, while the US federal grant for Petra Nova chipped in $190 million. Another CCS project for a mixed natural gas and coal power plant in Kemper County, Mississippi, was abandoned when estimated costs skyrocketed to $7.5 billion.[320]

Having already been priced out of the market *without* CCS, fossil-fuel power plants can't reasonably expect to compete *with* expensive carbon capture, for example, adding 30 cents/kWh to coal and 15 cents/kWh to natural gas electricity costs.[321] Even in Wyoming, in the heart of coal country where coal-power companies have invested in modern carbon capture add-ons to maintain revenues as new laws require reduced emissions in fossil-fuel plants, the technology has been deemed not "economically feasible at this time" at upwards of three times the cost to build a plant.[322] To be sure, the money would be better spent developing and installing more reliable clean energy rather than compartmentalizing unproven add-on CCS technologies to undo the flaws of burning carbon.

✻✻✻

Nonetheless, some are keen to show that the old ways can still work if we only cleaned up after ourselves, such as direct-air capture (a.k.a. "ambient scrubbing"), essentially sucking carbon dioxide out of the atmosphere to reverse planet-warming heat absorption. Direct-air capture or DAC technology uses a bank of fans to remove CO_2, typically reacted with an aqueous potassium hydroxide sorbent (a.k.a. lye or caustic potash) in a calcium caustic recovery loop, a so-called "carbon dioxide reduction reaction" (CO_2RR). Liquid alkali metal-oxide and solid amine-based sorbent materials are also being tried to remove ambient CO_2 in other so-called "negative-emission technology" (NET) systems.

Sitting atop a waste incineration facility in the hilly canton of Zurich, in Hinwil, Switzerland, the world's first commercial, carbon-dioxide removal (CDR) DAC plant filters out ambient CO_2 from the air via an alkaline-functionalized absorbent powder using low-temperature waste-heat in the desorption process. About 900 tons of CO_2 per year is captured and stored in tanks before the concentrated CO_2 is mixed with water and mineralized as carbonate in the underground basalt rock, a kind of reverse fracking or "refossilizing," where the dissolved CO_2 is stored in the open pore spaces of the rock. The cost to remove the atmospheric CO_2 via mineralization (CDR via DACS) is estimated at around $100 per ton of CO_2.[323] More reactive than sedimentary rock, basalt, magnesium, and calcium are the preferred minerals.

Located beside the 300-MW Hellisheiði geothermal plant in Iceland, the Orca CC plant also dissolves captured carbon dioxide in water, which is then injected into the bedrock, where it accumulates in the underlying basalt rock pores, accelerating the natural sediment CO_2 uptake. Four thousand tons per year can be captured, while the company, Carbfix, claims the process can store all the CO_2 generated from burning the world's remaining fossil fuels.[324]

To reach net negative emissions in keeping with the 2015 Paris Agreement, however, many *gigaton*-scale CDR plants will be needed at exceedingly prohibitive costs (1 gigaton = 1 billion tons), estimated at between 3.5 and 30 trillion dollars *per year*, equal to as much as a third of the world's annual economy![325] Nor do current DAC systems remove methane or other GHGs besides carbon dioxide, a.k.a. greenhouse gas removal (GGR). Any seques-tered CO_2 in the carbonated rock could also be released again under a future reactive process.

Highlighting the vast challenges ahead, a plant being built in the Permian Basin aims to capture 1 million tons of CO_2 per year starting in 2024, or about 1/4,000th of annual global output – another spit in the ocean. The plan is to inject the CO_2 into the ground or make synthetic fuel such as gasoline, diesel, and jet fuel, while other plants are being engineered to help grow vegetables. *Thousands* of such plants, however, will be needed just to remove current *annual* CO_2 emissions – *100 million* trailer-sized units according to physicist Klaus Lackner, director of the Center for Negative Carbon Emissions[326] – not to mention the massive amounts of accumulated carbon dioxide *already* in the atmosphere from over two centuries of unchecked emissions. As of 2021, there were just 15 DAC plants in operation worldwide.

Another CCS process in development uses ophiolite – crust and mantle exposed after subduction – to suck in CO_2 from the atmosphere to form carbon-ated minerals. On the northeast coast of Oman in the Arabian Peninsula, just south of Muscat, the Semail Ophiolite was explored by earth scientists from Columbia University, who found a "remarkable rate of such carbonate formation ... enough to sequester all the carbon dioxide produced by humans for hundreds of years."[327] The site hasn't been commercially exploited, however, as a viable carbon capture and storage method in the petroleum-rich Middle East, but new techniques are being studied to sequester carbon 10 km underground in the high-magnesium mantle. The exposed calcium- and magnesium-rich peridotite mantle rock was pushed up to the surface via plate tectonics 80 million years ago and has naturally been absorbing and petrifying about 100,000 tons of atmospheric carbon dioxide per year – one gram per cubic meter of stone.

Columbia University geologist Peter Kelemen thinks that one billion tons of carbon dioxide per cubic kilometer could be captured per year – an increase by a factor of one million – by pumping CO_2 mixed with seawater belowground to better react with the hotter subterranean rock (~100°C). The process mimics rainwater seepage to greater depths, where more mineralization (a.k.a. mineral carbonation) can occur, although with increased subterranean mass the earth will also rise.[328] With 15,000 km^3 of mantle rock, Oman has enough capacity to store three centuries of GHGs via mineralized carbonated water. In the north of Oman at Wadi Lawayni, carbon dioxide is being captured in a pilot project via aboveground DAC machines that is then mixed with water and injected underground. The founder of an Oman-based company named 44.01 – the molecular weight of CO_2 – believes the process "could one day mineralize 1.3 billion tons of CO_2 annually in Oman's mantle formation" via 5,000 injection wells.[329]

In conjunction with other sites around the world, including in Iceland where Carbfix is storing CO_2 in less-reactive basalt rock, mineralization could remove more than 10 billion tons per year, about a quarter of the annual output, or even as some have claimed as much as 3 trillion tons, twice the amount of emissions since the start of industrialization.[330] Although still in early development, a once unimaginable negative-emissions industry hopes to store enough carbon to reverse the process of burning carbon-containing fossil fuels that created the CO_2 in the first place, turning the clock back on the darkest side of the Industrial Revolution. One wonders if such "reverse fracking" will be financed with the same gusto as regular fracking.

Time will tell if such industrial CCS technologies can remove the extraordinary amount of GHGs already in the atmosphere as well as an additional output of 50 billion tonnes per year and counting, but for now small-scale CCS can certainly help both with emissions and the bottom line in various industries. Breweries are perfect for capturing site-generated CO_2 in the fermentation process to carbonate beer rather than trucking in external supplies. Many major breweries have already started their own CCS systems, bringing us full circle back to Joseph Priestley, who first noted the production of carbon dioxide in a Leeds brewery, starting him off on an illustrious career in chemistry as well as the soft-drinks industry he created by impregnating the natural effervescence of his "fixed air" in water. Small-scale CCS isn't enough, but will help keep even more carbon from entering the atmosphere.

Other novel CCS systems are also being developed as global warming worsens and more funding is available, as in the $20 million 2021 Carbon XPRIZE. One such CCS process removes carbon similar to how photosynthesis turns CO_2, water, and sunlight into sugar for plant energy. The CO_2 is converted by adding water and then electrochemically reducing the carbon dioxide to carbon monoxide, hydrocarbon fuels (ethane, methane, etc.), and feedstock to make everyday materials. The process is more efficient and faster than nature, reducing thousands of trees worth of carbon to the size of a refrigerator. More than 1,000 research teams are involved in the $100 million Carbon Removal XPRIZE set for Earth Day 2025.

It may also be easier to remove carbon dioxide from the oceans, which are saturated with atmospheric CO_2 and one of the Earth's main carbon sinks, storing up to 40% of annual anthropogenic emissions. With a concentration of dissolved CO_2 in seawater at over 100 times that of air, separating and capturing carbon dioxide from water can be much cheaper than from air, estimated in a 2023 MIT proof-of-concept proposal at $56/ton compared to $1,000/ton.[331] With lower carbon concentrations, the oceans can then reabsorb atmospheric CO_2 in a two-step removal and storage process. The system uses much less energy, avoiding the need of heat absorbers and large compressors as in a DAC system; nor does it require expensive membranes and chemicals that reduce water pH levels in other water-filter systems such as bipolar membrane electrodialysis, where molecular CO_2 is stripped out under vacuum. The MIT system uses electrochemical modulation of water pH to release the CO_2 before alkalizing the treated water, and can potentially be coupled with existing desalination plants.

"Algae farming" may also become a viable carbon-capture method (a.k.a. biofixation), which uses waste CO_2 to grow algae for biodiesel. Algae can double in size per day, potentially producing 10,000 gallons of biodiesel per acre per year (compared to 50 for soybeans, 150 for canola, or 650 for palm).[332] Conveniently located beside a steel or cement making plant, such CCS methods can mitigate carbon emissions, while helping to develop less-invasive next-generation biofuels, but may exceed system limits as in any exponential growth scheme. Another CCS technology aims to store CO_2 in concrete during the curing stage. As ever, the research is ongoing.

***

To be sure, CCS is a challenging, expensive technological fix, and certainly not as simple as some companies want us to believe in their glossy ads showing a butterfly net and the word "CO_2." The potential exists to capture atmospheric carbon, a.k.a. beyond zero or negative emissions, but removing such vast amounts is a monumental challenge, while all removal methods require energy that burns more carbon that then has to be captured.

In truth, industrial-scale CCS is more about politics than reality. The advanced carbon-intensive economies want to keep fossil-fuel plants open, while continually postponing action to reduce global warming. Decreasing the rate of increase isn't going to cut it, nor is claiming that such reduced increases will somehow magically reach zero or even "net zero" by 2050. Net zero is a fudge that concentrates on neutralizing the by-products of combustion rather than on reducing production itself, that is, just leave it in the ground. A goal of zero (or near-zero) is much better, albeit ambitious given the slow pace of change – prior to COP27 in Egypt, Climate Action Tracker reported that out of 40 systems change indicators assessed since COP26 not one was on track to meet their 2030 targets.[333]

As coal declines in the West, however, retrofitting coal plants to capture carbon is as important as ensuring that new fossil-fuel-burning plants are fitted with anti-pollution and carbon-capture measures, especially as coal-powered electricity generation rises in China, India, and elsewhere in the developing world. Although burning coal will never be clean, technology exists to make coal cleaner and is cheaper if incorporated in new builds rather than later as expensive add-ons. Of course, if governments were serious about climate change, they would start shuttering more coal plants, or at the very least enact regulations mandating source-point capture of emissions on all fossil-fuel plants.

Getting to zero emissions by 2050 and negative emissions thereafter is fanciful at best, especially since the short-term plan is business as usual before CCS kicks in on a scale needed to remove 50 billion tons of *annual* GHG emissions (three-quarters CO_2). Even if the technology was viable and up and running, changing industry practices in the time frame proposed is akin to a super tanker stopping on a dime. One can hardly expect oil companies could (or would even want to) industrially sequester almost 20 times as much carbon as all the trees, plants, grasses, soil, and oceans naturally sequester carbon. As Vox reporter David Roberts wryly noted, "Maybe we shouldn't bet the future of the human race on a gigantic mega project that we have not started and do not know if we can do. Maybe that's not a good gamble."[334] As emphatically noted in a 2022 *New York Times* op-ed, in response to CCS funding in Joe Biden's signature climate-change act, "Every dollar spent on this climate technology is a waste."[335] The main argument was that renewable-generated electricity is already far cheaper than attempting to fix the unfixable. And that government funding should be more wisely spent.

Should we also suppose that such environmental "services" could be appropriately managed as we continue to grow without restraint, unable to fit uncertain geological responses into an economy that produces a linear relation between human stress and ecological strain? A host of non-linear feedback responses are already manifesting across an increasingly fragile Earth – melting ice sheets (decreases solar reflection that increases planetary heat absorption), thawing tundra (releases methane that traps more heat), a disappearing glacial third pole (increases moisture that increases atmospheric heat), higher-temperature oceans (that absorb less CO_2) – all of which may soon put the Earth's increasingly stressed bio systems beyond repair. Someday, we may be able to capture enough carbon with great effort and money, but where on earth, literally, can we put all the stored heat coming our way now?

Mother Nature is the best at sequestering carbon in plants via photosynthesis (all that is needed is water, carbon dioxide, and sun), dissolution in water (the oceans are full of dissolved CO_2), and absorption in vegetated soils via root respiration and plant decay (harder and longer, but how we got peat and coal in the first place). The simplest CCS is a tree and the vast amounts of forests across the globe. Highlighting the effect of natural atmospheric carbon removal, thousands of lost forests that were cut down by European explorers who colonized the Americas in the fifteenth century regrew after 50 million indigenous people died from infectious disease, global temperatures eventually dropping enough to produce the Little Ice Age.

Soil also holds more carbon than the atmosphere, trees, and plants combined, while simple biochar stores may sequester carbon better than complex CSS systems, as well as improving crop yields and reducing inefficient wood burning in developing countries. "Zero-till" farming and restoring carbon-degraded pastoral grazing grounds also lowers atmospheric carbon, particularly in "brittle" drought-prone regions and overgrazed grasslands, where grass cover has lost its roots leading to desertification. Diverse pastoral grazing also increases the protein density in livestock, improving food content and reducing portion sizes.

Although the amount of CO_2 uptake in agricultural lands varies around the world, we should consider soil as a valuable – indeed essential – carbon land sink in the same way we do trees and plants, and at minimal cost compared to expensive, high-tech CCS solutions that see production only in a series of unconnected steps – some of which only undo the negative effects of the previous step – rather than an integrated, Earth-centered process.

Applying good scientific and engineering principles and practices are about addressing causes rather than symptoms as well as accounting for all effects, good and bad. Sometimes, the best solutions are decidedly low-tech, naturally derived from millennia of learned practice of working *with* the Earth. We should take care to preserve solutions we already have. And we should stop pumping 50 billion tons a year of GHGs into the air from burning fossil fuels.

2.15 Peak Oil and the Downward Side of Growth

How much oil is still in the ground is a hotly debated subject, although estimates of future supplies have been proffered since the start of the industry. One way to gauge the amount of oil in early wells was from production records, site foremen nervously calculating how much was left in a particular well as output rose and fell. Another is to count the workers.

In Pithole, Pennsylvania, an entire town grew up after oil was discovered there in 1865, reaching 15,000 inhabitants until the oil disappeared less than 2 years later, a single farm exchanging hands for $2 million at the peak before being sold for just $4.37 at the end.[336] In Spindletop, outside of Beaumont, Texas, the rise and fall was just as dramatic, yielding almost 18 million barrels a day in 1902 reduced to 10,000 a day 2 years later.[337] Spindletop gave birth to the Texas oil industry and a leg up for its most famous "wildcatter," Howard Hughes, who was born in Humble, Texas, just north of Houston. In the ever-exciting oil biz, expected output has made and lost the fortune of many an oil man as depletion inevitably follows extraction.

After the world's largest ever oil field was discovered in 1948 in eastern Saudi Arabia, counting barrels became secondary to counting dollars. Since 1951, when production began at the Ghawar field, over 80 *billion* barrels have been extracted from its six regions, by far the world's largest ever conventional field – roughly 280 km by 30 km in the En Nala anticline (an elongated dome under which oil and gas is trapped) – comprising almost half of all Saudi reserves. The Saudis have been pumping out the profits from its prized elephant for decades, with estimates of another 55 billion barrels still *in situ*. Although some believe the presence of water signals a coming depletion, current extraction is still roughly 4 million barrels per day or about 5% of global output in a single location.

The concept of "peak" oil, however, is mainly attributed to one man, the American geologist M. King Hubbert, who reported in the 1950s about a future resource crisis based on past production data. Hubbert was born in San Saba, Texas, attended the University of Chicago (which awarded him a PhD for his work on geological systems modeling), and taught at Columbia, before starting work as a geology consultant for Shell Development in Houston in 1943. At a 1956 American Petroleum Institute (API) meeting in San Antonio, Texas, Hubbert presented a paper entitled "Nuclear Energy and the Fossil Fuels" that would ultimately change how we think about oil supplies. Having observed that crude-oil production had increased at a rate of 7.9% per year from 1880 to 1930 (the output doubling every 8.7 years), Hubbert noted that "No finite resource can sustain for longer than a brief period such a rate of growth of production; therefore, although production rates tend initially to increase exponentially, physical limits present their continuing to do so."[338] He labeled his observation "peak" oil, although his conclusions were not universally accepted, in particular by those who somehow thought oil was limitless.

   Hubbert based his analysis on past production records in Ohio (where output peaked in 1896), Illinois (which had two peaks, one in 1910 using surface geology to find new reserves, and another in 1940 via more advanced seismographs), and on 40 years of records from his native Texas, which he predicted would reach peak oil "about 1965" and peak gas "about 1970," although he acknowledged that "Improved methods of secondary recovery will probably make the rate of decline less steep . . ., but are not likely seriously to postpone the date of culmination."[339] From his data, Hubbert predicted that US oil production would peak in 1970, which it did, and world oil production in 2000, a date still much debated. With the advent of offshore oil, shale oil, and oil sands, the exact end date is unknown, not to mention undiscovered conventional finds or advanced secondary recovery methods. Nonetheless, average global oil production per person peaked in 1979 at 0.6 gallons/day, further highlighting the diminishing returns of an increasing population and a decreasing resource.[340]

   Thanks to Hubbert's analysis, we rightly call oil "production" "extraction," because oil is a limited resource and not a replenishable "produced" supply. With all limited supplies, there is a beginning, a middle (the peak), and an end, the peak estimated from the changing extraction rates *prior* to the peak, all things being equal. Whether Hubbert's predicted peak date or other subsequent predictions are exact (or indeed peak coal, peak mineral, peak uranium, or peak anything) is not the point, but rather that we will eventually run out *by definition*, and that burning what's left to run inefficient internal combustion engines is not the best way to use what we still have.

   Many numbers have been quoted regarding peak oil, but only two matter: 1.6 trillion estimated barrels left in the ground and about 100 million barrels per day consumed (that is, about 35 billion barrels/year), which works out to 44 years left based on our current rate of consumption. For peak gas, the two numbers are 200 trillion cubic meters left and 10 billion cubic meters consumed per day, giving us about 55 years at our current rate of use.[341] Of course, the extraction is harder on the down side, as wells lose their natural pressure from years of pumping and thus secondary recovery methods are needed to maintain the same flow, such as increasing well pressure by pumping water, gas, or captured carbon dioxide into a declining reservoir. Conversely, if renewable energy and electric vehicles increase the use of non-petroleum-based power, lower demand will lengthen the end.

As Peter Maass succinctly puts it, "Crunch time comes long before the last drop of oil is sucked from the Arabian desert. It begins when producers are unable to increase their output."[342] What's more, although the exact date of the peak may be unknown, the future of oil is faced with a double whammy: declining output as resources diminish yet rising demand as more of the world ramps up consumption. US president Theodore Roosevelt was aware of limited supplies, stating more than 100 years ago: "it is ominously evident that these resources are in the course of rapid depletion." We might not think 100 years is "rapid," although compared to recorded history the oil extraction business hasn't been around that long. Clearly, we need to mind what we use, whatever the culmination date.

In his 1956 API paper, Hubbert puts the amount of time left into perspective with a seemingly humorous illustration (Figure 2.16), showing oil consumption on a time scale from 5,000 years in the past to 5,000 years in the future, where "the discovery, exploitation, and exhaustion of the fossil fuels will be seen to be but an ephemeral event in the span of recorded history."[343] In his figure, he included a presumed limitless supply of nuclear energy. Hubbert also predicted

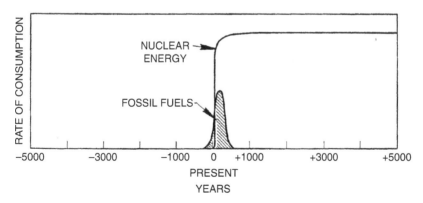

Figure 2.16 M. King Hubbert's fossil-fuels consumption rate over a 10,000-year span (*source*: Hubbert, "Nuclear energy and the fossil fuels").

coal would last 1,250 years, peaking in the year 2150 based on historic rates of extraction. Of course, many of us don't think that far ahead, a dangerous oversight. And, although the effects of burnt hydrocarbons were evident at the time of his paper, global warming was not yet on the environmental radar.

Offshore oil, shale oil, and oil sands have postponed Hubbert's predicted culmination as we've already seen, technologies that were in their infancy when he first presented his results to the API, although he did include them in his initial estimate based on the available data at the time. Found mostly in the shale fields of the United States, Hubbert put shale oil reserves at 2.5 times that of conventional oil and oil sands at almost equal. Extracted via fracking, primarily from the Bakken Formation and the Permian Basin, the recent "shale boom" (or "shale gale") has kept the USA from increasing imported oil, while Alaska, which holds as much as 10 years of global supply alone, wouldn't become a state until 3 years after his talk.

One reason for the differing estimates of the end of oil is the improved means of discovery since the 1950s. Early location techniques relied on maps that showed features of the surface and underlying strata such as seeps, faults, and anticlines, but were as much an art as a science, the land classified according to the production potential of a field. The presence of marine fossils, core drilling, magnetometers, torsion balances (measuring local gravity variations), and controlled explosions were also employed to "see" the underground geology or seismic soundings ("thumpers") that produced differentially reflected vibrations from the underground terrain to depths of as much as 30,000 feet.

Today, seismic testing searches for oil via a series of audio recorders (geophones) that measure the return of a sonic beep to map the contours of an underground deposit in 3D (not unlike sonar detection), sometimes finding oil pockets and sometimes finding air. Advanced subsurface geology and modern computing has made discovery even easier in recent years. Indeed, Hubbert didn't have computer-aided 3D imaging technology to rely on for his analysis, a.k.a. the "digital oil field."

Of course, "proven" reserves do not necessarily mean "actual" reserves, and there is still more oil to be extracted from existing wells and more to be found in untapped unconventional resources, while not all the Earth's crust has yet been explored. Predicting the end of anything is iffy, like a street prophet calling out "the end is nigh" day upon day. On the downside of extraction, "late wells" can also push out more oil with new techniques. The Kern River field north of Los Angeles was thought to be in its last days after 70 years of pumping, when in the 1970s high-pressure steam was injected into the underperforming well, pushing out almost 1 *billion* more barrels on an original estimate of only 47 million left.[344]

The amount of "proven" reserves has also fluctuated over the last half century, even in previously discovered fields, as oil companies manipulate the data to maintain share price, increase output quotas (based on reserves), and keep prices within a manageable range (for example, $25–$35/barrel), not least to discourage alternative energy development or maintain crucial leverage over any and all presumed future competitors.[345] A disputed time frame also provides ample fodder to fuel the "depletionists" versus "anti-depletionists" debate. However, as David Goodstein conjectures in *Out of Gas*, "Which comes first, Hubbert's peak or the collapse of the Saudi regime? Both would have the same effect and both seem inevitable."[346] Goodstein also cites three reasons to end our increasingly damaging addiction to oil: (1) no more dependence on unstable regions, (2) prevent irreversible climate damage, and (3) we will run out.[347]

What's more, predicting the end of oil may be no more successful than picking horses from the color of a jockey's jersey – without more information than the basics of an individual field, you might hit a winner but chances are you're left holding the bag, a very expensive bag. Peaks naturally occur in any boom-and-bust industry, but the problem with oil is we're running out of booms.

Historian Daniel Yergin cites five presumed "peaks" over 150 years of oil extraction: (1) in 1885 after the early gushers of the Pennsylvania oil region began to run out; (2) after World War I and the rise of the internal combustion engine (including "gasolineless" Sundays); (3) after World War II when the USA became a net importer for the first time; (4) during the various 1970s crises, including publication of the Club of Rome's study "The Limits to Growth," the Arab oil embargo, and the Islamic Revolution in Iran; and (5) after the rise of industrialized China with its rapid increase in oil consumption from the start of the twenty-first century.[348] All the while, the global population continues to grow, reducing the shared future supply.

Local peaks are easier to define, such as the original Pennsylvania oil patch that peaked in 1890. More recently, Gulf of Mexico output peaked in 1971,[349] Prudhoe Bay, Alaska, in 1988, and the North Sea in 1999 (Norway and the UK at about 3 mbpd each), although the extent of the tail is uncertain. The Gulf of Mexico is now in slow terminal decline – shallow fields more so than deepwater fields and gas almost completely exhausted – while Mexico's output has been steadily declining since 2004, peaking at almost 3.4 mbpd.[350] North Sea oil is in fatal decline and will end up costing billions when the end comes. Over 300 UK North Sea oil and gas installations are expected to be decommissioned in the next two decades as reserves run out with British taxpayers on the hook for up to £75 billion, an amount that will rise even higher if companies can't pay or are bankrupted.[351] Even the elephants are dying – Kuwait's Burgan field, the second largest in the world, peaked in 2010.

Deepwater deposits are also diminishing. According to Jerry Kepes, executive director for Plays and Basins research at IHS Markit, which publishes a list of top-50 energy firms annually, "The deepwater was one of the last big exploration plays on the planet. We're now looking at the second half of the global deepwater play. You can see the end of it, maybe 25 years from now."[352] That was in 2011. To further the theatrical metaphor, we're in the final act before the curtain falls, or if one prefers sports, we're fourth and long in our own end throwing a "Hail Mary" pass.

Following the decline of conventional US oil in 1970 at almost 10 million barrels per day, more or less as Hubbert predicted, renewed US output has also begun to dip after a decade of increased shale-gas extraction dampens, although lower prices may have been responsible for the most recent reduction. According to the EIA, the total amount of recoverable shale oil is only about 10% of available conventional crude and has already peaked. A long-time Amoco geologist referred to the end of shale as "more of a retirement party than a revolution,"[353] while *Oil and Gas Journal* notes that well-productivity data are a more accurate measure than output, and has compiled productivity profiles for a number of oil-producing regions that clearly show a worldwide decline across the board (Figure 2.17b).

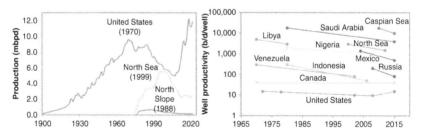

Figure 2.17 Decreasing oil extraction: (a) United States (conventional oil peak 1970 and ongoing "shale-gale"), Alaska North Slope (peak 1988), and United Kingdom and Norway North Sea (peak 1999) (*source*: "Petroleum and other liquids," US Energy Information Administration. www.eia.gov/dnav/pet/hist/ LeafHandler.ashx?n=PET&s=MCRFPUS2&f=A); and (b) well productivity (barrels per day per well) of major oil-producing regions (1970–2015) (*source*: Sandrea, R. and Goddard, D. A., "New reservoir-quality index forecasts field well-productivity worldwide," Oil and Gas Journal, December 5, 2016. www.ogj.com/ articles/print/volume-114/issue-12/drilling-production/new-reservoir-quality-index-forecasts-field-well-productivity-worldwide.html).

Estimating reserves and future extraction is always difficult, but oil company PR may be the best guide. In its 2011 "Will You Join Us" campaign, Chevron-Texaco stated "The world consumes 2 barrels of oil for every barrel of oil discovered," while BP famously rebranded itself as "Beyond Petroleum," telegraphing the end is nigh. Another indication about the approaching down-side to oil is the Gulf States collecting VAT (value-added tax a.k.a. GST or goods-and-service tax), while in late 2019 Saudi Arabia floated 2% of its state-owned cash-cow Aramco to raise $30 billion for Vision 2030. Clearly, oil-reliant governments need money. Even the Pentagon – the world's largest consumer of oil at 5% – is worried about peak oil and climate change. Today in peacetime, the US air force alone annually consumes as much jet fuel as during the whole of WWII, while up to 15% of American carbon dioxide emissions during the Cold War came from the military.[354]

Boom–bust extraction is not a new concept. The ups and downs of oil have been with us since the beginning of the modern oil industry. At the end of World War II, the United States produced 50% of the world's oil, but was soon importing more than it extracted as local wells ran dry and consumption continued to climb. Peak Oil had indeed become an existential threat to a world hooked on a free-flowing supply of petroleum, but because of the increase in shale oil since 2007 those imports were much reduced, while astonishingly the USA now exports shale gas to Latin America and Europe with plans to increase LNG exports around the world, especially to compete with Russia after the 2022 invasion of Ukraine.

Although the end days are still much debated, peak *consumption* or peak *demand* may arrive sooner as more electric vehicles and renewables come online. The oil world is preparing for the disruption, especially in the Gulf States as they begin to diversify, although as Bahrain's oil minister noted in 2018, "There is no real evidence of peak demand being achievable in the next few decades. Even if electric vehicles and renewable energy take off, we still see hydrocarbons remaining the dominant part of the fuel mix. Supply risks far outweigh demand destruction."[355] Whether demand or supply crashes first, there is little evidence the world can change its addictive ways any time soon.

Nor are oil prices so high as to make renewable energies the only investment option, but renewables have sufficiently developed to start undercutting oil (which we'll look at in Chapters 4 and 5). The uptake in electric vehicles (Chapter 6) will push the end out even further, but the free-wheeling years of petroleum are behind us. What's more, the many derived petroleum products continue to emit toxic waste and greenhouse gases, damaging the world beyond an ability to cope and prompting concerned customers to look elsewhere.

As we pass peak discovery as opposed to peak demand and begin to descend the extraction downside, oil companies will take more risks, moving to deeper waters and deeper wells in more remote locations. As the low-cost supplies tighten, exploration will expand into harder-to-tap Brazilian deepwater, the Russian Arctic, and Canadian oil-sands reserves. The main players want to keep filling their coffers for as long as they can despite the cost to the planet, social fabric of communities, or integrity of national boundaries, justifying more exploration and the expense. But the old business model won't work without guarantees of future earnings given the $100 million a pop to drill an exploratory well and a success rate between 10% and 30% as more "dry holes" than "sweet spots" are found.

One must also question the value of consuming a limited supply for something that can be got elsewhere. If we begin to reach the limits of extraction, should we save oil as a feedstock for plastics, chemicals, and asphalt? This is an important question that must be answered before it is impossible to ask. We had better be ready whenever the day of depletion comes or more realistically when the extraction rates start flat-lining and consumers feel the brunt of reduced supplies, that is, post peak. Most worryingly, we're sitting on a carbon time bomb if all proven reserves are dug up and burned – almost 3,000 billion tons of CO_2 or 60 times annual emissions will be released into the atmosphere, three times the amount estimated to limit temperatures to *only* a 2-degree increase, the so-called carbon budget.

"Leave it in the ground" is the rally cry of the twenty-first century, but not a welcome option for the multi-trillion-dollar petroleum industry. The most valuable commodity ever was pegged at over $3 trillion for a single year when oil prices reached almost $100/barrel in 2014 (WTI average $93/barrel[356]), highlighting the allure of our addiction. Oil companies will lose their collective shirts if their only product is left in the ground as "stranded assets" and reserves can't be replaced upon which their future worth is valued, estimated at up to $100 trillion. Of course, reserve estimates can be off, demand can rise from increased population or fall because of competition from green alternatives, and shifting geopolitics can shut down supply chains, all leading to a changing rate of extraction and consumption. A future *Mad Max* world of gasoline shortages and roaming tribes is unlikely, but the world will be powered by other than fossil fuels by the end of this century.

However long the oil lasts, we can't keep consuming over 1,000 barrels every *second*, burning more trapped petroleum under our feet and injecting hydrocarbon contaminants and heat-trapping greenhouse gases into the

Table 2.11 *The last battle of the fossil fuels (*anthracite, †average)*

Fossil fuel	Formula	Energy (MJ/kg)	Initial Reserves	Current Reserves	Consumption Rate	R/P (years)	CO_2 (g/MJ)
Coal	C	32.6*	3,000 Gtons	900 Gtons	8 Gtons/ year	110	112
Natural gas (methane)	CH_4	55.5	570 trillion cubic meters	200 trillion cubic meters	10 bcmd	55	49
Gasoline (octane†)	C_8H_{18}	45.8	3.8 trillion barrels	1.6 trillion barrels	100 mbpd	44	64

atmosphere as if there was no tomorrow. With 15 times the amount of oil in reserve than we can burn and still remain below the "2-degree-celsius-or-less" (3.6°F) Paris Agreement temperature limit, we can't continue with business as usual or spend vast amounts to dig up more ($50 billion/year just on oil exploration by some estimates). Nor can we return to a world where unchecked growth is built on cheap and abundant oil.

Are we seeing the beginning of the end, the twilight of our most gleaming idol? Is the next great global energy transition being forced upon us? In the words of Ahmed Zaki Yamani, Saudi oil minister from 1962 to 1986, who warned about gouging reliant consumers, "the Stone Age did not end for lack of stone and the Oil Age will end long before the world runs out of oil" (Table 2.11). Alas, short-term profits and share price is still being valued over environmental degradation and global warming.

To understand the plight of those scarred by oil, one need ask only one question: "Who owns the oil?" In *Crude World*, Peter Maass lists the options: "Is the oil owned by the [one] who works the land that sits atop the oil? The surrounding community? The state in which the community is located? The federal government in a capital hundreds or thousands of miles away? The foreign company that invested millions of dollars to find it?"[357] Maybe not a simple question, but one not enough people are asking.

In the aftermath of two Gulf wars and more than two decades of US occupation in the Middle East, one wonders if the strategic goal of those in charge is a permanently splintered world, where petroleum is easier to control and export, whatever the consequences. The United States burns about 20 million barrels *per day* or roughly 20% of global consumption, while the US military is the single largest guzzler on the planet. There can be no doubt that *Pax Americana* is designed to protect its flow. There can be no doubt that our modern Western lifestyle is the leading cause of increased environmental destruction.

Hold on for what is sure to be a white-knuckle ride as the rest of the century unfolds. The heady days of "drill, baby, drill" and ask questions later are over, but the consequences are not. To realize a better future, we must take responsibility for our actions. No time like the present to change our ways.

3

The Nuclear World: Atoms for Peace

3.1 Nuclear Fission: Turning Mass into Energy

I did my first university work term at Atomic Energy Canada Limited, calculating the radioactivity released to the atmosphere after a LOCA, a seemingly innocuous acronym for "loss of coolant accident." I was a second-year physics student at the time, in awe of the process of splitting uranium in a collision with a neutron that in turn released more neutrons to split more uranium, resulting in an ongoing chain reaction that generated lots of energy. My work entailed inputting data into computer models to simulate possible LOCA scenarios. Although a LOCA was a highly unlikely event given the independent backup safety systems in a nuclear reactor – gravity-assisted cobalt shutoff rods, cadmium injection, moderator dump – all of which would immediately stop the reactor by keeping more neutrons from splitting more uranium, the government licensing agency needed to know, just in case.

A nuclear reaction is a fairly straightforward process: a slowed-down neutron strikes a U-235 atom, which splits the uranium apart (fission) and releases energy because the resultant parts have less mass than the original atom. The energy of the mass difference is substantial as shown by Einstein's equation, $E = mc^2$. Indeed, very little mass (m) can make a lot of energy (E), because c^2 is so large. For example, 1 gram $= 9 \times 10^{13}$ joules, that is, a 9 followed by 13 zeros or 90 trillion joules! By comparison, recall that a watt is 1 joule/second and thus a standard 60-watt incandescent light bulb consumes 60 joules in one second. The hard part is to keep the process going in a controlled way that turns heat into electricity via a piped-in, heat-transfer cooling system, followed by the same electrical generation system for burning fossil fuels, where heat converts water to steam to run a turbine that creates electricity as in a conventional power plant.

Not all uranium is "burnable." The most common uranium isotope, U-238, absorbs neutrons instead of fissioning, so we have to use the much less plentiful isotope U-235 for our nuclear fuel, making the fission business that much harder. U-238 accounts for 99.3% of natural uranium as found in the ground, while the fissionable U-235 makes up only 0.7%, roughly one part in 140.[1] Separating U-235 atoms from U-238 atoms (a.k.a. enriching) isn't easy either, so neutron-absorbing uranium (U-238) is always present with fissioning uranium (U-235) in the reactor fuel.

Fortunately, a few tricks are available, such as enriching U-235 to around 3–5% or slowing down the neutrons even more to use the less-plentiful U-235 as is. Note that U-238 and U-235 have the same number of protons (92), but a different number of neutrons (146 and 143), and are known as "isotopes." The "mass number" (for example, 238 or 235) indicates the number of protons and neutrons, collectively known as nucleons.

The general nuclear fission equation is $n + X \rightarrow X^* \rightarrow X_1 + X_2 + Nn + E$, where a neutron (n) is "captured" by a fissionable atom X (for example, U-235), the X atom becomes unstable X^*, breaks into two "daughter nuclei" fission products (X_1 and X_2), ejects a number of neutrons (N = 1, 2, or 3 depending on the fission products), and releases energy (E). One possible reaction for U-235 is shown in Figure 3.1, producing the fission products barium (Ba) and krypton (Kr). Numerous other possible reactions produce different fission-product pairs, but this one has all the ingredients of a typical fission reaction.

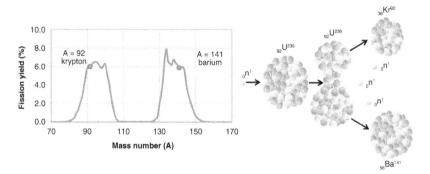

Figure 3.1 Nuclear fission: (a) U-235 fission-product yield versus mass number A (*source*: England, T. R. and Rider, B. F., "LA-UR-94–3106, ENDF-349, Evaluation and Compilation of Fission Product Yields 1993" (table 7, Set A, Mass Chain Yields, u235t), Los Alamos National Laboratory, October 1994) and (b) fission process started by a captured neutron. One possible reaction yields the fission products krypton (A = 92) and barium (A = 141).

$$_0n^1 + _{92}U^{235} \rightarrow _{92}U^{236} \rightarrow _{56}Ba^{141} + _{36}Kr^{92} + 3_0n^1 + E(170 \text{ MeV}) \qquad [\text{Eq.3.1}]$$

For an element $_ZX^A$, the atomic number (subscript Z) and mass number (superscript A) are always conserved in a nuclear reaction. So, in this example, the atomic number (number of protons), $Z = 92 = 56 + 36$, and the mass number (number of nucleons), $A = 1 + 235 = 236 = 141 + 92 + 3$, are the same before and after the reaction. Conservation means we get out what we put in.

The energy is worked out by simple atomic accounting, adding up the difference in mass of the original atom and the resultant parts. Here, the U-235 fission reaction produces 170 MeV, the mass–energy difference between a U-235 atom (plus one captured neutron) and the two fission products, in this case barium (Ba-141) and krypton (Kr-92) (plus three liberated neutrons). As a single atom is so small, the released energy is typically given in MeV (mega-electron-volts), which is equivalent to 1.6×10^{-13} joules (0.00000000000016 J), although in a reactor there are gazillions of atoms to make the energy add up.

One might think uranium would split into two equal parts of atomic number $Z = 46$ (92/2) or mass number $A = 118$ (236/2), but in fact the split is uneven (roughly 3:2). Typically, we get one large fission product near barium ($Z = 56$, $A = 141$) and another smaller fission product near krypton ($Z = 36$, $A = 92$), although there are more than 100 different possible daughter nuclei, as shown in Figure 3.1a. Most are highly unstable (that is, radioactive) because each resultant fission product is neutron heavy and must realign itself to become more stable. For example, barium ($_{56}Ba^{141}$) turns into stable praseodymium ($_{59}Pr^{141}$) after about a month and krypton ($_{36}Kr^{92}$) turns into stable zirconium ($_{40}Zr^{92}$) in about 6 hours, both via a series of "beta" decays where a neutron turns into a proton and an electron, increasing the atomic number of the decaying atom by one. Highly energetic and potentially very dangerous, alpha particles (doubly charged helium nuclei, $_2He^{4++}$) and beta particles (nuclear electrons created when a neutron turns into a proton, $_{-1}\beta°$) are both ejected during the ongoing radioactive decay of fission products, as are highly penetrating gamma rays (high-energy photons).

Nuclear fission is a statistical process with many different possible reactions, producing different radioactive fission products, a range of energies, and from one to three neutrons. Importantly, the average number of neutrons is about 2.5, and thus capable of creating a chain reaction in the uranium, while the average energy is about 200 MeV. In each case, the stored energy in the U-235 fuel is converted to the kinetic energy of the fission-product pairs, which heats the coolant to create the steam, while delayed neutrons from the decaying fission products help keep the reaction going. Astonishingly, 1 g of U-235 fuel per day in a reactor can generate almost 1 MW of power from over 3×10^{13} fissions *per second.*[2]

What's not to like? Uranium ore (UO_2) is fairly abundant in the Earth's crust, mined primarily in Kazakhstan (~40%), Canada (~20%), and Australia (~10%), while very little fuel is needed to generate lots of energy, although the uranium must first be refined, for example, milled, enriched, and packaged as rods in a fuel assembly (or bundle) in a "light-water" reactor. Fortunately, as nuclear advocates and some environmentalists are keen to note, there are no nasty carbon-containing by-products or greenhouse gases as with fossil fuels, and thus running a nuclear plant reduces carbon emissions that would otherwise be emitted in an equivalent-rated coal-, oil-, or natural-gas-burning plant.

Indeed, in those green university days, I thought nuclear power was the answer to all our energy needs, buoyed by company literature citing the impressive performance and safety record of a CANDU (CANada Deuterium Uranium), Canada's pressurized "heavy-water" reactor that uses natural uranium as fuel. As co-op students, we were sent to regular industry talks, the last slide in a presentation typically showing a picture of the nuclear power plant life cycle, the 40 years or so from green pasture to working 600-MW reactor back to green pasture again after being decommissioned, with accompanying family bike-riding pictures through the pretty, now-restored, green pasture. Nothing much was said about the waste material – what it was, how long it lasted, how much there was, where it would go – but none of us asked too many questions other than about the physics, little wondering if the talks were more PR than reality. No mention either about carbon-intensive and dangerous mining.

There are two main nuclear reactor types, depending on the concentration of U-235 and the absorbing material employed to slow down the neutrons (called a "moderator"). A typical light-water reactor uses "enriched" uranium and a light-water moderator (mostly regular H_2O water), while a heavy-water reactor uses "natural" uranium and a heavy-water moderator (mostly "heavy" D_2O water). The moderator also doubles as a coolant to take away the fission heat, the whole point of a reactor.

In the more common light-water reactor, found in the United States and most nuclear-power-producing countries, the U-235 content is increased roughly six times to between 3% and 5% to make up for the high neutron absorption of U-238 because the amount of U-235 in natural uranium is not enough to maintain a chain reaction. In a heavy-water reactor, fewer neutrons are absorbed by the U-238 atoms and instead bounce around inside the reactor, improving the likelihood of capture by U-235 atoms. The neutron "cross-section" (fissionability[3]) in heavy water is 30 times that of light water, making up for the lower percentage of U-235 in the fuel. In short, we *enrich* the U-235 content to use light water (in a more common, light-water reactor) or we *thermalize* more neutrons with heavy water to use the lower-percentage U-235 found in naturally occurring uranium (in the niche-market, heavy-water reactor).

Interestingly, water in a stream, lake, or household tap is not all regular H_2O, as not all the hydrogen is "protium" ($_1H^1$), but includes two naturally occurring hydrogen isotopes: heavy hydrogen or deuterium ($_1H^2$ or D, ~0.015%) and tritium ($_1H^3$ or T, <10^{-18}%). Still, it takes time and money to turn regular light water into heavy water, separating out the roughly 1-in-7,000 deuterium oxide molecules (D_2O) from the hydrogen oxide molecules (H_2O) by isotope separation.

There are pros and cons to both designs. The advantage of a heavy-water reactor is that we don't need to enrich the fuel, and can use the uranium more or less as dug out from the ground, separated from its impure UO_2 ore (uraninite, a.k.a. pitchblende[4]), saving money on expensive fuel-refining costs. The disadvantage is that we need to spend money making heavy water, produced by bombarding regular light water with neutrons ($H_2O + 2n = D_2O$). Although heavy water is mostly used as a neutron moderator in CANDU reactors, other applications include tracer compounds to label hydrogen in organic reactions, nuclear magnetic resonance (NMR) spectroscopy, and neutrino detectors.

Conversely, we can use regular H_2O water in a light-water reactor, but must pay to enrich the U-235, a difficult and expensive process involving gaseous UF_6 diffusion or centrifuges, the same methods employed to enrich uranium to make a nuclear bomb. Note that weapons-grade uranium contains 90% U-235, while reactor-grade uranium contains either 3–5% U-235 for light-water reactors or 0.7% U-235 for heavy-water reactors. Depleted uranium (DU) has less than 0.3% U-235, that is, almost 100% U-238, still dangerous but no good as is for power. Depleted uranium is what's left over after U-235 has been removed for enriching (Figure 3.2).

Enriching natural uranium into reactor-grade U-235 (from 0.7% to 3–5%) is done by isotope separation, based on the differences in the atomic weights of U-235/U-238 (with three more neutrons, U-238 is heavier). After milling to separate the uranium from its ore, the uranium is ground into a condensed powder known as yellowcake (mostly U_3O_8) – so-called because the separated uranium crystals are yellow – and turned into purified UO_2 by smelting, which is then fabricated into pellets for use in heavy-water reactors (thus no enriching needed) or converted into UF_6 (uranium hexafluoride or "hex") for enriching in light-water reactors. Vaporous UF_6 is better for enrichment because U-235 is more easily separated from U-238 by gaseous diffusion (hex has a low boiling

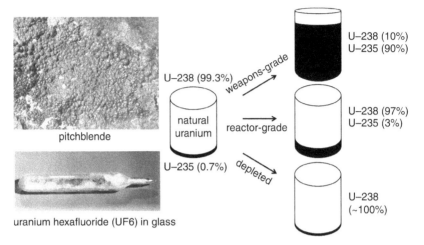

Figure 3.2 Uranium enrichment: (a) uranium ore (*source*: Geomartin CC BY-SA 3.0), (b) uranium hexafluoride (*source*: Argonne National Laboratory), and (c) natural uranium to weapons-grade uranium (90% U-235) or reactor-grade uranium (3–5% U-235).

point), where the lighter U-235 atoms move faster across a barrier.[5] Thousands of cascading units are needed because the U-235 yield at each stage is minimal.

Centrifuges require much less electricity and are more common today than gaseous diffusion, separating out the U-235 by high-speed spinning until the desired percentage is reached (the heavier U-238 atoms move to the outside), the same process for making today's weapons-grade uranium and hence an important indicator of the existence of an illegal weapons program. For use in a light-water reactor, the high-grade UF_6 is then remade into solid UO_2, before being sintered and shaped into 1-inch-long cylindrical pellets, packed into 12-feet-long, 1-inch-diameter zirconium tubes to make fuel rods (think of a Pez dispenser) and arranged in a multi-rod assembly to be lowered into a reactor for irradiation.

The fuel rods are left in the reactor core for up to 4 years as the fission heat is transferred to the circulating coolant. During burning, the U-235 fuel is replaced by neutron-absorbing, fission-product "waste" such as krypton and barium, before being removed and stored in onsite "spent fuel" pools. All fuel handling is done by machine, while the reactor is sealed in a metal pressure vessel and enclosed in a thick-walled concrete containment building. Some of the U-235 remains unburned, as much as 2% depending on the time spent in the reactor and the fueling scheme, becoming unusable along with the highly radioactive, reaction-killing, fission-product waste. The life of a CANDU fuel bundle is similar as it passes horizontally through the reactor core (the middle is the hottest), before being ejected out the other end. CANDU refueling is continuous without the need for a reactor shutdown.

Atom smashing is now big business, generating almost 5% of global electrical power in 440 nuclear power plants worldwide. One hundred are in the United States, providing almost 20% of electrical power to the US national grid. At the beginning of the Atomic Age, after the physics and engineering of harnessing U-235 atoms had been worked out, some believed nuclear power would be the answer to all our energy needs and become "too cheap to meter."

Note that "atomic" and "nuclear" physics seem interchangeable, the fundamental difference muddied by the history of nuclear power. An atom consists of a nucleus (protons and neutrons) and orbital electrons, but although "atomic" often means nuclear as in weapons and power, *atomic* physics technically refers to the electronic structure of an atom and the energy levels of the electrons that *orbit* the nucleus, while *nuclear* physics is what goes on *inside* the nucleus. Furthermore, chemical reactions involve only the orbital electrons, while nuclear reactions break apart the core of an atom and release much more energy (~ 200 MeV/eV). Alas, we are stuck with the muddied distinction as defined in the 1954 US Atomic Energy Act, which states "The term 'atomic energy' means all forms of energy released in the course of nuclear fission or nuclear transformation."

3.2 Nuclear Beginnings: Atoms for War

To mark a beginning to nuclear physics, one typically starts with Ernest Rutherford, a New Zealander who did pioneering work at the universities of Manchester, McGill, and Cambridge. He was the first to recognize during a series of experiments from 1908 to 1913 that an atom was made of a dense central "nucleus" (which he named from the Latin for "little nut"), when alpha particles fired at a gold-foil target bounced back toward the source rather than being deflected as assumed in an earlier theoretical model. In the first practical model of the material world, the hydrogen nucleus (H^+) became the building block of all elements, which Rutherford called a "proton." In the academic home of Isaac Newton, and where in 1897 J. J. Thompson discovered the first subatomic particle, the electron,[6] Rutherford built a different kind of laboratory to explore the inner workings of matter, generating new atomic stones to fling at increasingly higher velocities. Smashing stuff had become acceptable science.

Building on Rutherford's work at the Cavendish Lab in Cambridge, his research partner James Chadwick discovered the neutron in 1932, similar in mass to a proton but without an electric charge. Chadwick noted that atoms would have too much positive charge if only protons accounted for the atomic mass of an element (hence the atomic number is not equal to the atomic mass). In 1938, John Cockcroft and Ernest Walton would be the first to break apart an atom, smashing a lithium target with a stream of protons accelerated by a high voltage,[7] the original particle accelerator and "first direct quantitative check of Einstein's equation, $E = mc^2$."[8] The atomic zoo was expanding with every curious new discovery.

Elsewhere, the Joliot-Curies in Paris employed alpha particles to break apart other light elements, building on the work of Marie Curie – the other great pioneer of nuclear science who earlier isolated two highly radioactive unknown elements in uranium ore, which she named polonium and radium – before Enrico Fermi in Rome tried neutrons as his atomic bullets, differentially slowed by his marble and wooden benches (hydrogenous wood slowed more neutrons). Although they had lots of fun smashing elements with their high-velocity protons, alpha particles, and neutrons, the vast amount of energy stored in the nucleus wasn't fully understood by any of the pioneering nuclear alchemists, either physicists or chemists.

A massive array of capacitors and rectifiers accelerated the atom-smashing particles at higher speeds, while oscilloscopes, scintillation screens, and Geiger counters measured what came out after the smashing.[9] Nothing much was expected from the exploratory work other than a better understanding of the properties of matter and more insight into a previously unknown nuclear realm. As Rutherford noted, "Anyone who expects a source of power from the transformation of these atoms is talking moonshine."[10] Einstein himself was also doubtful at first, stating in 1934, "splitting the atom by bombardment is something akin to shooting birds in the dark in a place where there are only a few birds."[11]

What came next is caught up in the intricacies of World War II, ultimately pushing the United States to build the first nuclear bomb as part of the Manhattan Project, fearful that Nazi Germany was developing its own nuclear-bomb program after scientists in Berlin showed that the nuclear fission of uranium could in fact release an enormous amount of energy. Generating usable nuclear power was a by-product of the bomb.

Systematically working his way through the elements from hydrogen ($Z = 1$) on up, Fermi had created the first "transuranium" elements ($Z > 92$) in 1934 by bombarding uranium with slow neutrons, which spurred on Otto Hahn, a chemist, and Lise Meitner, a physicist, to do their own uranium smashing at the Kaiser Wilhelm Institute in Berlin. Although the German chemist Ida Noddock was the first to suggest the theory of fission – a cell biology term used by American biologist William Arnold while working with Niels Bohr in Copenhagen – no one was quite sure what was going on in the bombardment. Fermi believed he had either created neptunium (Z = 93) and plutonium (Z = 94),[12] a nuclear isomer, or various complex radioactive decay products, none of which made sense until Hahn and his young assistant Fritz Strassmann calculated that the mass number of the target uranium atoms was equal to the sum of the mass numbers of the measured reaction products.

Meitner, a Jewish-born Austrian exile who had fled Germany after the 1938 *Anschluss* to work in Stockholm, and her nephew Otto Frisch determined that the uranium atoms had been *split* apart, basing her thinking on the liquid-drop models of Bohr, George Gamow, and Carl Friedrich von Weizsäcker. As Frisch noted, "gradually the idea took shape that this was no chipping or cracking of the nucleus but rather a process to be explained by Bohr's idea that the nucleus was like a liquid drop; such a liquid drop might elongate and divide itself."[13] Imagine a water-filled balloon squeezed somewhere near the middle to form a dumbbell that then snaps in two. Indeed, when a nucleus expands after capturing a neutron, the long-range, proton–proton, repulsive forces exceed the short-range, nucleon–nucleon, attractive forces that bind the atom together, causing "fission." The resultant kinetic energy would be considerable, as much as 200 MeV, as Meitner and Frisch wrote in a *Nature* letter on January 16, 1939.

Meitner's news spread fast, prompting Fermi to continue his neutron-induced radioactivity and uranium-smashing experiments at Columbia University in New York, where he had relocated with his Jewish wife Laura by way of Sweden after receiving the 1938 Nobel Prize "for his demonstrations of the existence of new radioactive elements produced by neutron irradiation, and for his related discovery of nuclear reactions brought about by slow neutrons." Adding to the sense of urgency, the results of Hahn and Strassmann were verified by Frisch in Copenhagen, who recorded electric pulses of the fission products with an oscilloscope, and in a cloud-chamber photograph by two Berkeley physicists.[14]

The importance of the "secondary" neutrons produced in the reaction was immediately understood as a way to create a chain reaction, but whether any such reaction could be controlled was still unknown. In a now better-understood nuclear splitting, Fermi also proposed the existence of a new particle, which he called a neutrino for "little neutral one," earlier hypothesized by the Austrian physicist Wolfgang Pauli as part of the process of beta decay. The neutrino satisfied the law of conservation of energy that some had thought to abandon to explain the mysterious hidden energy of the nucleus, becoming yet another part of the growing subatomic catalogue of particles, all of which had to be carefully accounted for in any reaction.

As the atomic scientists surmised from a now credibly established theory, a "controlled" nuclear reaction via "primary" neutrons would almost certainly be able to detonate a bomb.[15] Adding to the worry of such raw power, Germany's fission research was expanding, led by Nobel physicist Werner Heisenberg, who was building an experimental reactor (*Uranmaschine*) at the Kaiser Wilhelm Institute in Berlin, fueled with uranium from Europe's only source, the Ore Mountains south of the Czech–German border.

The American quest to make the bomb would begin in earnest after the Hungarian émigré Leo Szilard – another European scientist caught in the crossfire of war, who had fled to the United States like so many others – commissioned Einstein to write a letter to President Franklin Roosevelt, warning of the likely German progress on nuclear fission. Dated August 2, 1939, the letter called for concerted government action: "In the course of the last four months it has been made probable – through the work of Joliot in France and Fermi and Szilard in America – that it may become possible to set up a nuclear chain reaction in a large mass of uranium, by which vast amounts of power and large quantities of radium-like elements would be generated." The letter goes on to list possible uranium sources in Canada, Czechoslovakia, and the Belgian Congo, and ends with an ominous declaration that Germany had stopped the sale of uranium ore from Czechoslovakian mines upon annexing the Sudetenland. One month later, the Germans would invade Poland.[16]

On September 1, 1939, prior to a ban on reporting nuclear results, Niels Bohr and his former student John Wheeler published their seminal paper "The Mechanism of Nuclear Fission" in the American journal *Physical Review*, calculating that the odd-numbered, least-abundant uranium isotope U-235 was more likely to fission than U-238, and highlighting for all the practicality of generating large amounts of energy with uranium. There was no return once the nuclear genie was loosed from its atomic bottle. That same day, the world was at war.

The $2 billion Manhattan Project was the largest research project ever, employing more than 125,000 scientists, technicians, office workers, laborers, and military personnel at its peak at various locations across the United States, including the main think tank at Los Alamos in New Mexico and two productions centers, one in Oak Ridge, Tennessee, and another in Hanford, Washington (codenamed Project Y, K-25/Y-12, and Site W). Most of the heavy thinking took place in Los Alamos (a.k.a. "the Hill"), where the top physicists of the day had been corralled by the army, such as Hans Bethe, Felix Bloch, Emilio Segrè, Edward Teller, Eugene Wigner, and a young Richard Feynman, who wowed everyone with his talent, levity, and safecracking ability, and would soon head the Theoretical Computations Group in charge of the IBM calculating machines.

After a daring escape in an open boat from occupied Denmark to England, Niels Bohr, the Danish theoretician responsible for the concept of atomic shells, made several trips to Los Alamos, having calculated that 1 kg of purified U-235 would be sufficient to make a bomb,[17] while others acted as visiting consultants from university laboratories across the USA, such as Enrico Fermi, Ernest Lawrence, and Isidor Isaac Rabi. The British national James Chadwick, who had first discovered the neutron only a decade earlier, fittingly became a part of the team after the American and British efforts were combined. Einstein, however, was considered a security risk, mostly because of his misunderstood pacifist views, and was not asked to join the project and besides wasn't a nuclear physicist, although he did do some isolated work from Princeton on isotope separation and ordinance capabilities.[18]

The thousands of dedicated workers at Los Alamos were all led by J. Robert Oppenheimer, known as Oppie, a hyper-intelligent, philosophically minded theoretical physicist from Berkeley and colleague of the pioneering American atom-smasher Ernest Lawrence, the 1939 Nobel Prize winner for the invention of the cyclotron. Having visited the Pecos Wilderness near Los Alamos as a child on family vacations, Oppie reckoned that the secluded New Mexico desert was perfect for a "secret complex of atomic-weapons laboratories."[19]

Given the uncertain outcome of working with novel materials, two bomb designs were devised to improve the chances of a successful detonation. One employed enriched uranium (U-235) and the other plutonium (Pu-239), an element found only in trace amounts in nature. Both materials were produced with great difficulty: U-235 via electromagnetic separation coupled with gaseous diffusion at Oak Ridge and Pu-239 via uranium neutron capture in the B Reactor at Hanford, based on Fermi's test pile at the University of Chicago, to where he had moved from Columbia to support the war effort. Feynman's as-yet-unfinished PhD thesis was on the separation of U-235 from U-238, although the method eventually employed to create weapons-grade U-235 was invented by Lawrence at Berkeley, while the Pu-239 was collected from inside the Hanford reactor after the transmutation of U-239.

In the "uranium-gun" design, two subcritical-mass pieces of enriched U-235 would be kept separate until triggering, while the "plutonium-implosion" bomb would be triggered by setting off plastic explosives around a subcritical spherical mass of Pu-239, condensing the core to twice its density via "shaped charges" that evenly focused the implosion like a lens. Plutonium was much easier to produce, but spontaneous fission in Pu-240 was too high for a gun design, solved by imploding the plutonium to a supercritical mass.[20] The "lens" was made of layered plastic explosives that had the consistency of taffy and could be shaped to produce a massive, evenly applied shockwave, symmetrically compressing the core. The implosion had to be perfectly uniform to avoid a dud, a.k.a. a "mangled grapefruit." The fast-to-slow detonation layers were designed by another European émigré, John von Neumann.[21]

After almost 4 years of development, the first test was prepared at Alamogordo in the New Mexico desert, about 150 miles south of Albuquerque. Codenamed Trinity, some were unsure if the atmosphere would catch fire via a nitrogen fusion chain and ignite the world.[22] Others took bets on the size of the explosion. Although the world survived, the blast heat turned the desert sand to green glass – now known as "trinitite" – in all directions for 800 yards.

The Manhattan Project (codenamed S-1) detonated three atomic bombs, the original Trinity "gadget" test, July 16, 1944, and two more on live targets less than 3 weeks later, even though Germany had already been defeated by then and were no longer involved in an atomic weapons program, negating the project's purpose. The second (Little Boy) was dropped by the B-29 bomber *Enola Gay* over the city of Hiroshima on August 6, 1945, and the third (Fat Man) by another B-29, *Bock's Car*, on Nagasaki three days later, effectively ending the war.

The Hiroshima bomb was the uranium-gun design with an explosive power of 12.5 kilotons TNT (~ 50 TJ), killing an estimated 105,000 people and destroying 54,000 buildings, while the Nagasaki bomb employed the plutonium-implosion design – as at the proof-firing Trinity test – with the explosive power of 22 kilotons TNT (~90 TJ), killing 65,000 people and destroying 14,000 buildings.[23] Almost 10,000 people per square kilometer were killed by the two roughly 1,800-feet-high "air-bursts," the explosive power of the Hiroshima bomb equal to 16,000 of the largest conventional bombs dropped from a B-17. Twenty years later, Oppenheimer would capture the magnitude of worry most of the scientists felt upon witnessing the destructive power of the world's deadliest creation that had lit up the sky like a second Sun at Alamogordo:

> We knew the world would not be the same. A few people laughed, a few people cried, most people were silent. I remembered the line from the Hindu scripture, the Bhagavad-Gita; Vishnu is trying to persuade the Prince that he should do his duty and, to impress him, takes on his multi-armed form and says, "Now I am become Death, the destroyer of worlds." I suppose we all thought that, one way or another.[24]

The ethics of dropping a nuclear bomb on live targets has been debated ever since. Some said two bombs were detonated to demonstrate the ease of deployment and that further Japanese resistance would be futile, saving countless Allied lives in a prolonged invasion of Japan (both bombs were stored together at the assembly site on Tinian Island). Some believed the Americans had to beat the Soviets into Japan to control reconstruction and post-war markets, while sending a message against communist expansion into Western Europe and Asia.

Others thought more sinister goals were at play, two billion dollars too much to spend on an untested concept. The English physicist and novelist C. P. Snow wrote in *The New Men*, "It had to be dropped in a hurry because the war will be over and there won't be another chance," while Cornell physicist Freeman Dyson noted "the whole machinery was ready." Still others thought the world should know the bomb's full potential as the United Nations was being formed. Already, the scientists and government overlords were debating the future deployment policy of the most deadly weapon ever constructed.

Niels Bohr argued that hard-won nuclear secrets should be shared with the Soviets to avoid an unnecessary and costly arms race as well as future prolifer-ation, which has indeed transpired, but the politics of the day triumphed over scientific logic. Bohr even met with Roosevelt to discuss the possibility of "atomic diplomacy" with the Soviets, nixed by British prime minister Winston Churchill, who refused to see beyond the existential challenges of the current war.[25] Bohr's "open world" policy was akin to his *complementarity* principle, where wave-particle duality was compared to stockpiling atomic bombs that made the world both safer (as a permanently unusable deterrent) yet more dangerous (in the event the unthinkable transpired).[26] The usually apolitical Fermi imagined an honest agreement with effective control measures and that "perhaps the new dangers may lead to an understanding between nations."[27] The future of warfare and life had become both freed and enslaved by science.

Expressing concern about the lack of contact between scientists and govern-ment officials, Einstein also sent a second letter to Roosevelt. Dated March 25, 1945, the letter remained unopened on the president's desk, Roosevelt dying before he could read it. Often credited as the Father of the Bomb because of his equation and first letter, despite having done nothing to build it, Einstein lamented, "Had I known the Germans would not succeed in producing an atomic bomb, I never would have lifted a finger."[28] In fact, Germany's nuclear program had been primitive, unable even to build a working reactor.[29] Einstein would spend the rest of his life campaigning for arms control, reduced mili-taries, and a "supernational" security authority. Later, he noted that signing the letter in 1939, ostensibly starting the world on the road to an unstoppable nuclear-armed future, was the greatest mistake of his life.

***

After the war was over and the physics of exploding nuclear bombs better understood, other countries built their own kiloton and then megaton weapons using the Earth as a giant test lab. The Soviet Union detonated its first atomic bomb in 1949 at the Polygon test site in northeast Kazakhstan, a plutonium fission bomb codenamed "First Lightning" (called "Joe-1" by the USA after Joseph Stalin). Soviet development was helped by German refugee and spy Klaus Fuchs, who worked with the British contingent at Los Alamos, a little-known American physicist turned spy Ted Hall, who worked on explosion experiments and was unhappy that the USA hadn't shared nuclear information with its Allies, and 300 tons of uranium dioxide found after Germany's surrender, including a small amount at the Kaiser Wilhelm Institute in Dahlem where Hahn and Strassmann had first split the uranium atom a decade earlier.

The Soviets were keen to catch up to the Americans after the fall of Berlin, igniting a race for nuclear supremacy and the Cold War, which would divide the world into rival spheres of influence and may have averted a possible American first strike on a war-weary USSR. In 1952, the British tested their first nuclear bomb in the Montebello Islands off the northwest coast of Australia. After the detonation of the first nuclear bomb, the power and pace of destruction increased more in a decade than ever before in human history.

Overseen by the tutelage of the Hungarian émigré and Manhattan Project alumnus Edward Teller, nuclear weapons today are thermonuclear bombs, where a plutonium-fission "A-bomb" is detonated to generate the million-degree temperature needed to trigger a fusion "H-bomb," fusing hydrogen nuclei to release energy rather than fissioning U-235 or Pu-239. Much more powerful than an A-bomb, the "Super" was opposed by some of the leading Los Alamos scientists because of its limitless destructive power, yet advocated by others to strengthen US military capability. The first Super, codenamed "Mike," was successfully tested by the United States on November 1, 1952, on the remote Pacific Ocean coral atoll of Eniwetok in the Marshall Islands, halfway between Hawaii and the Philippines. Redesigned to be deployed by plane, the first H-bomb that could literally wipe out an entire country was dropped on March 1, 1954, on nearby Bikini Atoll. At 15 megatons TNT, "Bravo" was more than 1,000 times as powerful as the original Hiroshima blast, creating a 250-foot deep, mile-wide crater on the ocean floor and "fallout across more than 7,000 square miles of the Pacific Ocean."[30] In the midst of a widening Cold War chasm, few questions were asked.

Between 1946 and 1958, more than 1,200 nuclear bombs were detonated, equivalent to over one Hiroshima per day, leaving a distinctive, still detectable radiation signature across the globe. More atomic muscle-flexing followed when the Soviet Union detonated a 58-megaton-TNT H-bomb in 1961 in the Russian Arctic. At almost 5,000 times the power of the original Hiroshima blast, RDS-220 or "Tsar Bomba" was the largest bomb ever exploded. The "destroyer of worlds" now contains enough destructive power to kill us all many times over.

For most of the scientists in the Manhattan Project, weapons development was morally permissible during war, but not in peacetime. As noted in the 1949 General Advisory Committee report, signed off by Oppenheimer before he washed his hands of any further development, there was "no inherent limit in the destructive power" to the H-bomb and that such a "weapon of genocide . . . should never be produced."[31] But wiser heads did not prevail, the $2 billion wartime project spiraling out of control into half a century of Cold War gamesmanship. The USA brazenly tested bombs without any military need and the Soviets marched through eastern Europe as if daring their former allies to stop them. The US price tag for testing and stockpiling their new military toys was a staggering $5.5 trillion.[32]

Presently, nine countries possess verifiable nuclear weapons – the USA, Russia, the UK, France, China, India, Pakistan, Israel, and North Korea – with an estimated 13,500 nuclear warheads either deployed on base with operational forces in a state of "launch on warning" or in reserve with some assembly required (down from a peak of 85,000 at the height of the Cold War). More than 90% are held by the USA (5,800) and Russia (6,375), totaling 20 EJ of destructive energy or almost half a million Hiroshimas.[33] Unofficially, there may be more, the wonder not that so many countries possess nuclear weapons, but that so few do or that none have been used again in anger since the end of World War II.

Furthermore, despite reducing the possibility of direct conflict between the United States and the Soviet Union cum Russia in any number of hotspots around the world, the arms race has diverted – and continues to divert – a substantial amount of spending to the military along with "a sharp increase in the influence of the military upon foreign policy."[34] It is unfathomable that trillions of dollars have and are still being spent on something that can never be used. Even if they could, how many times can one obliterate existence?

Nonetheless, thousands of unusable doomsday machines stand poised, ready to destroy, despite their essential impotence, the risk of accident ever present – operational, machine-triggered, or software-assisted. When a nuclear-armed American B-52 collided in midair with a refueling plane during a Chrome Dome regular test run in 1966, seven of 11 crew members on the two planes died and four nuclear bombs inadvertently dropped, landing near the fishing village of Palomares in southern Spain. Although the 70-kiloton-plus

H-bombs didn't detonate, plutonium was released in the resultant, non-nuclear TNT explosions of two of the bombs, spreading radiation over an 800-km radius and contaminating a 2-km^2 area with highly radioactive debris. Initially denied and then downplayed by the American and Spanish governments, one of the errant bombs was retrieved 10 weeks later from the sea, a flotilla of ships sent to find the missing "broken arrow."[35] If the nuclear bombs had exploded, a large part of southern Spain would have been destroyed and rendered completely uninhabitable. More than 50 years on, the cleanup around Palomares continues, 1.6 million tons of soil removed to the United States at a cost of $2 billion, while an estimated 50,000 m^3 of contaminated soil still remains.[36]

To be sure, atomic bookkeeping comes with its own unique concerns, too chilly to comprehend. At least two of the 32 broken arrows reported since 1957 have never been found, one accidentally jettisoned mid-flight off the coast of Georgia in 1958 after another midair collision involving a hydrogen-bomb-carrying B-47, while another rolled off the side of an aircraft carrier in 1965.[37] Both are still unaccounted for, buried somewhere under miles of murky ocean waters. Others have also likely gone missing from the secret arsenals of nuclear-armed states.

By the start of the 1960s, the "strategic" deployment of a nuclear deterrent was the *raison d'être* of the Cold War, a.k.a. "Balance of Terror," fueled by meaningless claims of a "missile gap." But the question soon changed from how a bomb is built to whether more should be built. The expense is enormous, a 1998 source estimating that the US nuclear-bomb program had cost $5 trillion to develop and maintain since 1940, while at least 5% of all commercial energy consumed in the United States and Soviet Union from 1950 to 1990 was spent on developing, stockpiling, and creating launch systems for a growing nuclear arsenal.[38] No new American bombs have been added since the 1990s, although a recently enacted US nuclear makeover is expected to cost over $1 trillion (including $56 billion for expected cost overruns!), while over $50 billion is spent each year on maintenance.

Mutually assured destruction (MAD) is the agreed outcome in a nuclear confrontation, with billions dead in minutes and billions more in the resultant radioactive fallout, followed by years of "nuclear winter," the term coined by American astrophysicist Carl Sagan and others in a 1983 paper using models previously based on the effects of volcanic eruptions. Oppenheimer likened the arms race "to two scorpions in a bottle, each capable of killing the other, but only at the risk of his own life."[39] Regardless the unthinkable outcome, there is little value in maintaining an oversized arsenal as a deterrent for future aggression, either militarily or financially. At least, for now, there are no nuclear weapons in space.

Although a horrific possible future has kept rival powers from initiating an unwinnable war, Vaclav Smil noted in *Energy and Civilization* that "the magnitude of the nuclear stockpiles amassed by the two adversaries, and hence their embedded energy cost, has gone far beyond any rationally defensible deterrent level."[40] In fact, stockpiling unusable nuclear weapons undermines spending on conventional weapons that may be needed in the event of a real war and for important industrial and social spending. And yet, undeterred by the cost and horror of total annihilation, a silent sentinel stands guard on a future no one can endure.

In the changing times that immediately followed World War II, however, the powers that be sought to implement a new policy to attempt to make amends for the devastation at Hiroshima and Nagasaki, turning to the prospect of peaceful nuclear power. Thousands of scientists and engineers retooled their abilities to tame the raw energy within the atom as power moved from the work benches of Los Alamos to the halls of government. Former bomb-making factories were converted to national research laboratories, while academic and civilian research facilities grew along with the ever-expanding military program, tasked with preventing another world war by maintaining an unthinkable, always-ready deterrent. Whether any of the modern alchemists could reorient their thinking was still unknown.

In *From Faust to Strangelove*, a book that explores the changing public perception and image of the scientist, Roslynn Haynes notes that after Hiroshima and Nagasaki physicists were no longer perceived as innocent boffins, while "it became progressively more difficult to believe in the moral superiority of scientists and even more difficult to believe in their ability to initiate a new, peaceful society."[41] Nonetheless, despite its clandestine origins, moral ambiguity ("technical arrogance" to use Freeman Dyson's characterization), and the intellectual chasm of classifying the mysterious workings of a compact "little nut" at the core of an uncertain new Atomic Age, nuclear fission would become fully vested in locomotion and everyday electrical power. How we got there is as fascinating a tale as there is in the history of science and engineering, from which the promise of endless energy rolls on.

3.3 The Origins of Nuclear Power: Atoms for War and Peace

On a brisk, wintery morning, December 2, 1942, Enrico Fermi made his way to work at the University of Chicago, where he had been enlisted in the war effort to engineer a way to make plutonium, known only as element number 94 until

earlier that March. Waiting for him at a campus doubles squash court underneath the west stands of the disused Stagg Field football stadium was a team of
scientific workers from the "Metallurgical Lab" (Met Lab), who under Fermi's
supervision had assembled a nuclear reactor as part of the top-secret Manhattan
Project. Called a "pile" for its numerous uranium-filled layers of graphite, the
goal was to prove that a geometrically assembled array of lumped uranium
could go "critical," that is, produce a self-sustaining fission chain reaction, by
which plutonium could be produced.[42]

Assembled around the clock in two 12-hour shifts over a period of a month,
the Chicago Pile One (CP-1) consisted of 45,000 highly purified graphite
blocks stacked in 57 layers in a 30-by-32-foot lattice, some of the blocks
hollowed out to hold 40 tons of hockey-puck-shaped uranium-metal discs.
Graphite (a carbon allotrope) was a readily available moderator, known from
Fermi's earlier work at Columbia, which according to his precise calculations
should sufficiently slow down enough fast-fission neutrons to kick-start the
fission process that would produce the secondary fission-product neutrons
needed for the growing pile to go critical in an ongoing chain reaction. As
each layer was added, Fermi measured the neutron count, confident of his
numbers.[43]

Containing enough graphite to provide a pencil for "each inhabitant of the
earth, man, woman, and child,"[44] as his wife Laura would later write, Fermi
called the 385 tons of graphite, 6 tons of pure uranium metal, and 34 tons of
uranium oxide, all held together in a wooden cradle standing 22 feet high, "a
crude pile of black bricks and wooden timber." To keep the pile from going
critical on its own, 14 extractable, neutron-absorbing, cadmium control rods
were horizontally inserted into the middle of the pile, while three brave scientists – called the "liquid control" or "suicide" squad – stood overhead at the
ready with buckets of cadmium salt solution in case the massive construction
caught fire.

Inch by inch and one by one, the control rods were pulled out, Fermi noting
the increasing neutron levels on a Geiger counter. The "neutron economy" was
everything as more neutrons were released by more U-235 fissions within the
pile, sufficiently slowed down by the graphite moderator to facilitate even
more. After a break for lunch, the final rod was removed and at 2:20 pm the
reaction went critical, the reproduction factor, k, greater than one (1.0006). The
neutron count was literally off the chart. In a phone call to Washington to report
on the success, the project leader Arthur Compton, who had recruited Fermi to
Chicago, declared, "The Italian navigator has just landed in the new world."[45]

The CP-1 prototype was a great success, the ever-cautious Fermi not only
happy for having built the world's first nuclear reactor, but for the ease in which

he could control the ongoing reaction, exclaiming, "To operate a pile is as easy to keep a car running on a straight road by adjusting the steering wheel when the car tends to shift right or left."[46] Although the uranium was not enriched, hence the large size for the minimal power generated (under 1 watt[47]), criticality had been proven, essential for the Manhattan Project's plutonium production and later to show that peaceful nuclear power was doable after the horrors of war.

"Atoms for Peace," as President Eisenhower styled his proposed safer world in a 1953 UN speech, would absolve "atoms for war" that saw 170,000 people die instantly or in the immediate aftermath of the Hiroshima and Nagasaki blasts, followed by hundreds of thousands more from cellular degradation and radiation-induced cancers in the ensuing years. As Fermi noted in 1952, "It was our hope during the war years that with the end of the war, the emphasis would be shifted from weapons to the development of these peaceful aims. Unfortunately, it appears that the end of the war really has not brought peace. We all hope as time goes on that it may become possible to devote more and more activity to peaceful purposes and less and less to the production of weapons."[48]

Throughout the war, nuclear scientists working on the Manhattan Project had been sidetracked in the rush to make the bomb before Germany. In the aftermath, the idea to produce peaceful power for the masses, "outside of the shadows of war," renewed their hopes for a saner future, while aiding the rehabilitation of their reputations as creators of death. The basics entailed using water or some other coolant to remove heat continuously from inside the pile instead of exploding a highly enriched core. Military research money was earmarked to remake death and destruction into energy and light, starting at the newly reformed weapons facilities cum laboratories in the USA, USSR, and UK.

CP-1 had already moved during the war away from public scrutiny and potential urban catastrophe to Argonne Woods in southwest Chicago, reassembled as CP-2 and then CP-3, the site eventually renamed the Argonne National Laboratory before moving again to nearby LaMont. Part atomic facility and part University of Chicago research lab, Fermi oversaw the research at Argonne during and after the war, building on a growing nuclear expertise there and at Los Alamos, Oak Ridge, Livermore (Lawrence's reformed Berkeley lab), and especially Hanford. The central tenets of reactor physics were worked out in the early post-war years, nuclear science transmuting from a source of curiosity about the nature of matter and the raw power within to the engineering challenge of turning mass into utility for all. No longer beyond comprehension, the atom could do what no alchemist had ever imagined:

The basic theory of a nuclear reactor was developed during the war by Fermi, Wigner, and others and tested in the military reactors. After 1945, a large amount of work was spent in completing the understanding of fission processes and developing a detailed theory of the nuclear reactor. By 1958, with the publication of Alvin Weinberg's and Eugene Wigner's authoritative *The Physical Theory of Neutron Chain Reactors*, the work of the physicists was over in this area.[49]

Capitalizing on a functionally superior, light-weight, cleaner fuel, the first operational nuclear submarine, the USS *Nautilus*, was launched in 1955, overseen by Admiral Hyman Rickover, while the USS *Enterprise* – the first nuclear-powered aircraft carrier and longest naval vessel ever built – was commissioned in 1961, launched by Eisenhower's wife Mamie. Although keen to generate nuclear power for naval propulsion, the USA still had plenty of energy from its vast oil and coal reserves, and was less inclined to instigate a national energy makeover. The navy was also eager to get in on the atomic act and not be excluded from the nuclear club by the army or air force.

Called the "Father of the Nuclear Navy," Rickover was a highly intelligent, hard-working electrical engineer who had worked his way up the ranks to admiral in 1953. During World War II, he was in charge of the electrical section of the Bureau of Ships, overseeing the change to infrared signal communication from visible light for ship-to-ship messaging "that made American ships targets for enemy bombs and torpedoes."[50] Assigned to Oak Ridge after the war, Rickover wanted to build "the ultimate fighting platform," the world's first nuclear-powered submarine that could remain undetected underwater for pro-longed periods of time.[51] Importantly, he chose water as the reactor moderator, subsequently employed in all future American commercial designs.

But despite their head start on nuclear-power technology, the Americans were more interested in making bigger bombs and designing advanced ship propulsion for a nuclear fleet unmatched in history. Built by the British near the coastal village of Seascale in northwest England, the first civilian nuclear reactor went critical on October 17, 1956. Calder Hall was a 92-MW, graphite-moderated, CO_2-gas-cooled reactor, based on wartime research at Chalk River on the Ottawa River and later at Harwell in Oxfordshire, both directed by the original atom-smasher John Cockcroft. The reactor was dual-purpose, designed to generate electrical power and make plutonium for the British bomb-making program, hence the oddest of nuclear acronyms, PIPPA (pres-surized pile producing power and plutonium).

Adapted from the propulsion reactor in Rickover's nuclear navy, the first US civilian reactor was a 60-MW pressurized light-water reactor, successfully tested – "cold-critical" without power hook-up – on December 2, 1957, the fifteenth anniversary of Fermi's original CP-1, and "grid-tied" on December 18. Constructed on the shores of the Ohio River, 25 miles northwest of Pittsburgh, the Shippingport Atomic Power Station reactor was built by Westinghouse and the

Duquesne Lighting Company under the direction of Rickover's Naval Research Group, and proclaimed the "world's first full-scale atomic electric plant devoted exclusively to peacetime uses."[52] Triumphantly fulfilling Eisenhower's earlier Atoms for Peace declaration to the UN despite the simultaneous building up of nuclear arms, Shippingport was the first commercial American nuclear power plant, following 14 previous reactors built to manufacture weapons-grade plutonium cores.[53] Clean air was also cited in contrast to a proposed coal plant on the Alleghany River. As historian Richard Rhodes noted, "no expensive precipitators for smoke control, no expensive scrubbers for sulfur-oxide control, 60 megawatts of peak-load power, and a leg up on nuclear-power technology."[54]

Rickover also made a strategic design change, employing uranium dioxide ceramic fuel clad in zirconium instead of uranium metal fuel, making the uranium more difficult to repurpose in a bomb. The United States was ready to take on the world with an unrivaled nuclear arsenal, deterrent to any and all potential belligerents, as well as a space-age atomic-power program to keep the home fires burning brightest. Shippingport operated for 20 years before being re-engineered as a light-water "breeder" reactor, where a thorium core is surrounded by a natural uranium "blanket" to create U-233 fuel (another fissile isotope of uranium), before being decommissioned in 1982.[55]

***

The pace of construction increased thereafter with two scaled-up commercial reactors in operation in the United States by the end of the 1960s, the 570-MW Connecticut Yankee "pressurized-water" reactor built by Westinghouse south of Hartford on the shores of the Connecticut River (1968) and the 640-MW Oyster Creek "boiling-water" reactor built by General Electric on the New Jersey shore (1969). In the following decade, 100 reactors were built by the two pioneering electrical companies, Westinghouse (pressurized) and GE (boiling), competing again as if electric power had come full circle to its origins. In other countries, reactor designs evolved from their own engineering and technological practices. The dominant design in the USA employed enriched U-235 fuel with light water as a moderator and coolant (pressurized or boiling), known as a light-water reactor (LWR), while graphite moderators became prevalent in the UK (air- or gas-cooled) and in the Soviet Union (light-water cooled).

Essential experience was gained in the early years of nuclear power, such as calculating the optimal fuel scheme to get the most out of every gram of U-235 and the safe operating procedures of the world's newest energy source. In the United Kingdom, nine more reactors were built in the decade after Calder Hall, supplying almost 25% of British electrical power, while in the United States 20 went online.[56] More reactors were built in the secretive and closed-off Soviet Union, starting in the town of Obninsk near Moscow with a small, dual-purpose, high-power channel reactor (*reaktor bolshoy moshchnosty kanalny* or RBMK) that went critical in 1954. By 1985, there were 25 Soviet reactors.[57]

Today, the most common reactor employs a light-water moderator/coolant, either pressurized (PWR) or boiling (BWR). In a PWR, a light-water coolant is kept under high pressure to remain liquid (2000 psi and 300°C[58]) as it passes through the core before heating a secondary coolant in a heat-exchange system to generate steam to turn the turbine. In a BWR, the light-water coolant is boiled and the steam used to turn the turbine before being condensed and returned to the reactor. A PWR is safer because the irradiated primary coolant stays within the containment structure, while the coolant in a BWR also turns the turbines and thus radioactive matter can escape more easily to the atmosphere in the event of an accident (Figure 3.3).

The other main reactor types employ either a pressurized, heavy-water moderator/coolant (PHWR) as in a CANDU, found mostly in Canada and India, a graphite moderator/gas coolant (Magnox and AGR), found exclusively in the UK, or a graphite moderator/light-water coolant (RBMK), made in Russia, while some experimental breeder reactors (a.k.a. catalytic nuclear burners) are still being tested. Before being discontinued, the Magnox reactor used natural uranium clad in magnesium oxide (hence the name), while an advanced gas-cooled reactor (AGR) uses 2%-enriched U-235 fuel. The RBMK-1000 is the same reactor as in Chernobyl, but now comes with a containment structure to reduce radioactive release in the event of a core breach as occurred in 1986 (which we'll look at later).

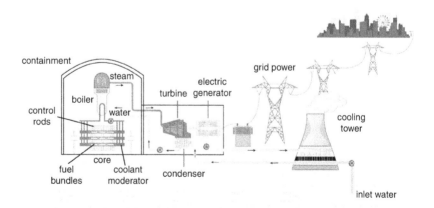

Figure 3.3 The basics of a pressurized light-water nuclear reactor (PWR).

In a "breeder" reactor, fissionable fuel is created inside the reactor core. Breeding is a two-stage process as the fuel can't fission on its own, but turns into a fuel that can fission via neutron absorption while in operation. Typically, the uranium isotope U-233 is bred from thorium and is thus called a thorium reactor ($_0n^1 + _{90}Th^{232} \rightarrow _{90}Th^{233} \rightarrow _{91}Pa^{233} + _{-1}\beta° \rightarrow _{92}U^{233} + _{-1}\beta°$). Thorium reactors are cooled with liquid sodium and need a boost to start. All reactors "breed" some nuclear fuel, providing a kick or "plutonium peak" in normal operation as U-238 transmutes into U-239 that decays into neptunium and then plutonium before fissioning, which provides about one-third the reactor power (also a way to make weapons-grade bomb material, believed to be how India made its first nuclear bomb in 1974).

Breeder reactors convert "fertile" thorium-232 or natural uranium-238 into fissionable fuel and don't need to be enriched, saving time, money, and the environmental stress of extensive refining, but have not become economically viable as originally supposed. In his 1956 address to the American Petroleum Institute, where he famously introduced the theory of Peak Oil, M. King Hubbert was overly optimistic, stating "it will be assumed that complete breeding will have become the standard practice within the comparatively near future."[59]

Whatever the reactor design, the essential neutron economy is managed by raising and lowering neutron-absorbing "control rods" to keep the reaction steady: both "prompt" *primary* neutrons emitted during fission and "delayed" *secondary* neutrons released by fission products on average about 14 seconds later that help maintain a constant criticality. As plenty of water is needed to cool the core and produce steam, nuclear power plants are constructed near rivers, lakes, or oceans, while all nuclear reactors produce a cornucopia of radioactive waste stored in onsite spent fuel pools (a.k.a. bays).

Today, there are 437 commercial nuclear plants operating worldwide in 30 countries, providing roughly 400 GW of power, 92 of which are in the USA (see Table 3.1). The American reactors are all light-water (LWR), either PWR (~67%) or BWR (~33%), while almost 85% of global nuclear reactors are LWR. The others are either PHWR (such as the CANDU, ~10%), graphite-moderated/gas-cooled (~2.5%), graphite-moderated/water-cooled (~2.5%), or fast-breeder reactors (<0.5%).

The World Nuclear Association (WNA) maintains a global nuclear power database, citing 437 working nuclear power plants in 2022 rated at a total peak capacity of 390 GW that annually requires 68,000 tonnes of uranium fuel to keep running.[60] Two countries, France (65%) and Ukraine (53%), generate more than half their electric power with nuclear (Table 3.2).

Table 3.1 *Number of nuclear power plants by reactor type, total power output, and fuel*

Reactor		#	GW	Fuel	Moderator	Coolant
Pressurized water	PWR	307	293	Enriched UO_2	Water	Water
Boiling water	BWR	60	61	Enriched UO_2	Water	Water
Pressurized heavy water	PHWR	47	24	Natural UO_2	Heavy water	Heavy water
Light water graphite	RBMK	11	7	Enriched UO_2	Graphite	Water
Gas-cooled	Magnox, AGR	8	5	Natural U metal, enriched UO_2	Graphite	CO_2
Fast neutron sodium	FBR	2	1	PuO_2 and UO_2	None	Liquid

Source: "Nuclear Power Reactors," *World Nuclear Association*, May 2023. www.world-nuclear.org/information-library/nuclear-fuel-cycle/nuclear-power-reactors/nuclear-power-reactors.aspx.

Table 3.2 *Nuclear power, 2022 (country, number, power, global and national percentage)*

Country	Number of plants	Power (GW)	Global percentage	National percentage
USA	92	95	24	18
France	56	61	16	65
China	54	52	13	5
Japan	33	32	8	6
Russia	37	28	7	19
South Korea	25	24	6	26
Canada	19	14	3	14
Ukraine	15	13	3	53
Spain	7	7	2	21
Sweden	6	7	2	32
Rest	90	58	15	
Total	437	394	100	

Source: "World Nuclear Power Reactors & Uranium Requirements," *World Nuclear Association*, August 2023. www.world-nuclear.org/information-library/facts-and-fig ures/world-nuclear-power-reactors-and-uranium-requireme.aspx.

The largest nuclear power station in the world is the 8-GW Kashiwazaki-Kariwa plant on the west coast of Japan (five BWRs, two ABWRs), while the largest in Europe is the 6-GW Zaporizhzhia plant on the Dnieper River in southeast Ukraine, comprising six water-cooled, water-moderated PWRs, a.k.a. water–water energetic reactors (VVERs). During the 2022 Russian invasion of Ukraine, the site was overrun by military forces as was the Chernobyl plant, raising alarms about possible radiation contamination in a worst-case scenario. After an anxiety-filled visit to the plant by a team of nuclear inspectors from the International Atomic Energy Agency (IAEA) after only one of the six reactors had remained online, two inspectors stayed onsite as the IAEA called for a demilitarized perimeter and nuclear safety protection zone in the embattled region, underscoring the challenge of keeping citizens safe from radiation leaks in wartime.

***

Despite many engineering firsts, dizzying technical achievements, and a seemingly simpler supply of fuel, today providing almost 5% of global electric power, atomic energy is under fire on numerous fronts, because of high construction costs, safety concerns, and the ongoing problem of radioactive waste, all a far cry from the early 1960s when everyone wanted to get into the nuclear biz. The army had their bigger and better bombs, the navy their submarines that can stay submerged for years, while the air force even wanted to build a nuclear plane, spending almost $5 billion on a prototype before admitting that flying something as bulky as a nuclear reactor was not practical nor would the damage be easy to contain in the event of a radioactive reactor falling from the sky.

Edward Teller wanted to use atomic bombs for small-scale engineering to clear land for a deepwater Alaskan port (codenamed Operation Chariot), build a Sacramento Valley canal to move water to San Francisco, and extract oil and gas from the heavy bituminous Albertan oil sands in an ill-devised 1958 nuclear fracking plan (codenamed Project Oilsands).[61] For a trained physicist, who intimately knew the secrets of the atom, one wonders how the father of the H-bomb was so ignorant about the realities of radioactive contamination. Project Plowshare was similarly designed to use "nuclear excavation technology" for large-scale landscaping, including creating a new Panama Canal (dubbed the Pan-Atomic Canal), but was abandoned for obvious safety reasons. Project Gasbuggy, however, was approved to detonate an atomic bomb to extract natural gas about a mile underground in northern New Mexico in another fracking experiment to explore peaceful nuclear uses. The site is now a no-go area. After a similar test detonation in Colorado, the gas was "too heavily contaminated with radioactive elements to be marketable."[62]

Even NASA wanted to go nuclear, hoping to power future satellites with plutonium. That is, until a plutonium-powered navigation satellite failed to

achieve orbit in a 1964 launch and fell back to Earth, disintegrating upon re-
entry (also why launching nuclear waste into the Sun or space is not such
a smart idea). A subsequent soil-sampling program found, "embarrassingly,
that radioactive debris from the satellite was present in 'all continents and at all
latitudes'."[63]

NASA also had heady plans to power a spaceship with "nuclear bomblets"
that would carry 10 times the payload of a Saturn V moon rocket, but Project
Orion literally didn't make it off the ground and was ditched in 1965 after
a decade of fruitless work.[64] Initially seeded with $1 million in funding to
a private contractor, the idea was to focus the massive propulsive power of 100
nuclear bombs detonated at half-second intervals to reach space, although no
one was quite sure how to keep the ship from blowing itself apart.[65] Project
A119 was another top-secret US plan to detonate a lunar atomic bomb as
a show of force after *Sputnik*, also discarded in favor of the PR-winning
Apollo moon landing. The lunar detonation was the brainchild of another
Manhattan Project luminary, Harold Urey, who thought the explosion would
shower the Earth with moon rocks or that the resulting debris collected by
a following missile could then be analyzed.

There was even a plan to nuke the recently discovered Van Allen radiation
belts in an unforgettable fireworks display to show the world the full power of
the American atomic arsenal, in particular the Soviets. Unfortunately, this
demented scheme – dubbed the Rainbow Bomb – went ahead and on July 8,
1962, a 1.4-megaton H-bomb was exploded 250 miles above a Pacific Ocean
launch site on Johnston Island. In his catalogue of bizarre experiments conducted in the name of science, Alex Boese noted that "Huge amounts of high-
energy particles flew in all directions. On Hawaii, people first saw a brilliant
white flash that burned through the clouds. There was no sound, just the light.
Then as the particles descended into the atmosphere, glowing streaks of green
and red appeared."[66]

The man-made atomic light show lasted 7 minutes, but also knocked out
seven of 21 satellites in orbit at the time and temporarily damaged the Earth's
magnetic field. Launched the previous day, AT&T's *Telstar I* also became
a casualty of the EM pulse months later, while radiation levels around the
Van Allen belts took 2 years to settle. On the plus side, one of the most
irresponsible displays of military hubris brought an end to atmospheric, underwater, and space testing of atomic bombs as the United States and the Soviet
Union signed the Partial Test Ban Treaty the next year. Apparently, filling the
skies with nuclear radiation was a step too bold even for science.

In the early heyday of reactor building, many thought nuclear power was as
simple as igniting an atomic-powered Batmobile. Fortunately, calmer heads

prevailed and we aren't now cleaning up a permanent radioactive mess created by atomic planes, nuclear blasters, or plutonium-powered satellites and rockets. Nonetheless, we are still dealing with the radiation fallout from decades of weapons testing, a grim reminder of the madness of unchecked technological warfare. There are also restrictions about where to build a nuclear power plant, such as not on seismically active ground, above water tables, or near large populations. Reactor containment buildings must also be able to withstand a jet plane crash or a large-scale earthquake. Alas, not all of the world's 400 or so reactors fit the bill.

Decommissioning is an especially sticky issue, where after 60 years of operation, radiation-degraded materials are no longer fit for service. The Nuclear Regulatory Commission (the US atomic licensing agency) grants a 40-year operating period that can be renewed up to a further 20 years. The first commercial US nuclear power plant – Oyster Creek Nuclear Generating Station near the Jersey shore – was shut down in 2018 after almost 50 years, while the Connecticut Yankee Nuclear Generating Station built in 1968 was taken offline in 1996 and decommissioned 8 years later. In the absence of a permanent long-term nuclear waste-disposal plan, the spent fuel rods remain onsite, stored in a so-called Independent Spent Fuel Storage Installation (ISFSI).

The USS *Enterprise* was deactivated in 2012, although its reactors are still intact with no decision yet on full decommissioning, while the USS *Nautilus* was retired in 1973 after 15 years of service and is now a museum in New London, Connecticut. The first vessel to reach the North Pole – in 1958 on its third attempt – *Nautilus* logged over 500,000 miles, more than seven times the 20,000 leagues of its Jules Verne-inspired namesake.

Having ignored the realities of nuclear aging by continuously kicking the decommissioning can down the road, today's fleet of 40- to 60-year-old reactors are showing their age. The WNA estimates that at least 100 plants will be shut down by 2040,[67] although in early 2021 the US Nuclear Regulatory Commission called for American reactor lifetimes to be extended to 100 years ("Life Beyond Eighty"), increasing the threat to public safety and the likeli-hood of an accident. Despite ongoing concerns about cost, safety, waste disposal, and what to do when a reactor is no longer fit for service, however, 60 new reactors are under construction in 16 countries (half in China and India), roughly 100 are on order or planned (with a total capacity over 100 GW), and more than 300 are in the works.[68]

Large capital costs are also a deterrent for would-be operators, particularly when amortized over decades or subjected to expensive cost overruns from delayed site licensing that can take years depending on local opposition.

Nuclear NIMBY ("not in my backyard") is a powerful force against building a nuclear plant, especially after the headline-grabbing accidents of Three Mile Island (1979), Chernobyl (1986), and Fukushima (2011). No new nuclear power plants have been built in the USA since 1979, although two under construction in Georgia are expected online soon according to the latest revised estimates (Vogtle 3 and Vogtle 4).

To make up for the shortfall as old units go offline and no new replacement reactors are built, more energy is being squeezed out of existing reactors at a rating higher than the original design (called "uprating"), of which 149 of 150 requests in the USA have been approved.[69] Uprating and less downtime have kept the overall percentage of nuclear power in the United States at around 20% over the last 40 years despite an ever-growing population and no new reactors being built on American soil in four decades.

***

Although the physics is basically the same, new reactor designs have been employed to improve safety. Gen-III or "advanced" nuclear power reactors are fitted with "passive" safety features (for example, natural water convection) and designed to work without electricity in the event of a failure, retroactively added to the older Gen-I and Gen-II reactors to avoid a Fukushima-type loss of power and coolant. EDF's EPR, Hitachi's ABWR, Toshiba's AP1000, and KEPCO's APR1400 are essentially Gen-III upgrades of older Gen-II reactors.

Experimental Gen-IV reactors are also in various stages of development, such as the very-high-temperature reactor (VHTR) and molten salt reactor (MSR) designs. A VHTR is a helium-gas-cooled, graphite-moderated reactor with an exceedingly high coolant temperature over 1,000°C that uses a tri-isotropic (TRISO)-coated particle fuel comprised of a uranium, carbon, and oxygen kernel, which resembles a billiard-ball-sized pebble, hence the name "pebble-bed reactor" (PBR). The reaction heat can also be used to generate high temperatures required in the chemical, oil, and iron industries or to produce hydrogen. MSRs employ liquid-sodium cooling that circulates at a lower temperature and pressure to run in a simpler, less-expensive reactor design. Trials are starting up in China and the USA. Advanced fuel types such as thorium, plutonium, and mixed oxide (MOx) have also been proposed.

Small-scale nuclear reactors are also being designed to limit construction costs that can run to at least $10 billion per gigawatt, not including overruns. The so-called SMRs – small modular reactors – have lower power ratings at up to 350 MW, less upfront financing, are built in sections in a factory and shipped on site for easy assembly (nuclear IKEA!), and require fewer operators and less maintenance. Those cooled with sodium don't need water and thus aren't restricted to coastal or riverside sites. Just add more to expand, although the average cost is still high without any obvious economy of scale.

One proposed "micro-reactor" was the Toshiba 4S (Super-Safe, Small, and Simple), a 10-MW cigar-shaped, liquid-sodium-cooled design, run on highly enriched fuel (19.9% U-235) and a fast-neutron graphite reflector to control the reaction (fission stops upon opening the reflector). The coolant circulates via electromagnetic pumps, while more fuel is burnt than in a conventional nuclear reactor, producing less waste and less plutonium. The underground setup also increases safety in the event of an accident, and is meant to run for 30 years with minimal supervision, although weapons-grade uranium can more easily be made from the higher enriched uranium. Especially suited to remote areas, a 4S mini-reactor was planned for Galena, Alaska, a 500-strong town on the Yukon River with only 4 hours of daylight in the depths of winter, where customers pay well above the going price for diesel. Alas, the Galena project was cancelled and no commercial mini-nukes have yet been built as the search continues to make nuclear power safe and affordable.

Funded by Microsoft co-founder Bill Gates, TerraPower is investing in small-scale nuclear, such as the traveling wave reactor (TWR), a prototype, liquid-sodium-cooled, fast-breeder "Natrium" reactor that uses fertile depleted uranium for fuel. Fertile material is not fissionable on its own, but can be converted into fissionable fuel by absorbing neutrons within the reactor; for example, U-238 first absorbs a neutron to become U-239 that beta-decays to Np-239 and then fissionable Pu-239. A small amount of concentrated fissile U-235 "initiator" fuel is needed to start a breeder reactor as in a traditional nuclear reactor before the fertile U-238 atoms absorb enough neutrons to begin producing Pu-239. Other fertile/fissile fuel combinations are possible besides the U-238/Pu-239 cycle, such as thorium-232 and uranium-233 in a Th-232/ U-233 cycle. A steady "breed-burn" wave is maintained by moving the fuel within the reactor to ensure a constant neutron flux.

Such low-pressure, non-light-water reactors (NLWRs) have never been proven, however, ironically noted by one economist as "PowerPoint reactors – it looks nice on the slide but they're far from an operating pilot plant. We are more than a decade away from anything on the ground."[70] Safety and maintenance is also a concern. Hyman Rickover avoided sodium-cooled reactors for the US navy because of high volatility, leaks, radiation exposure, and excess repair time, reporting to Congress in 1957 that "Sodium becomes 30,000 times as radioactive as water. Furthermore, sodium has a half-life of 14.7 hours, while water has a half-life of about 8 seconds."[71]

Nonetheless, in 2021, a Chinese research group in Wuwei, Gansu, completed construction on the first liquid-sodium-cooled reactor since Oak Ridge

National Laboratory's 7-MW, U-233-fueled, molten-salt test device that was shut down in 1969 after 5 years.[72] Expected to produce only 2 MW and fueled for the first time with thorium, the Chinese MSR could take a decade to commercialize to a working 300-MW reactor. TerraPower is also hoping to build a $4 billion, 345-MW, depleted-uranium demonstration reactor in the coal-mining town of Kemmerer, Wyoming (population 2,656), which could be up and running by 2030 subject to the testing process.[73] Half of the seed money is coming from the US Department of Energy, angering critics of the unproven technology, while no long-term, waste-disposal solutions were included in the plan.

Today, the only operational SMR is a floating reactor that brings nuclear power to the consumer. In 2018, Russia's state nuclear power company, Rosatom, started up the *Akademik Lomonosov*, a $480 million, 70-MW, barge-mounted, pressurized LWR, held in place by tether to a wharf behind a storm- and tsunami-safe breakwater in Kola Bay near the northern city of Murmansk. Factory-assembled and mass-produced in a shipyard, costs were reduced by about one-third and construction time by more than half. Used to provide district heating to the town of Pevek via steam-outlet heat transfer, one resident replied when asked about a possible radiation leak or accident, "We try not think about it, honestly."[74]

Resembling more a container ship than nuclear plant, a floating reactor can be towed where needed to provide short-term power. Although an emergency cooling system is not required at sea, the design has sparked obvious concerns – tsunamis, typhoons, collisions, and pirates offer new challenges to the usual worries about nuclear power. Despite the added operational responsibility, China plans to power artificial islands in the South China Sea with 20 floating reactors rated up to 200 MW each. There isn't much scope for new designs in the high-priced nuclear market, but after a number of small-scale demonstra-tions of the novel, low-temperature, "pool type" Yanlong reactor, China also plans to build a 400-MW unit to heat 200,000 homes via district heating with water as a moderator, coolant, and radiation shield. Underwater and seabed reactors have also been proposed.

Despite the increased interest in small-scale nuclear reactors, SMRs aren't cheap, however, primarily because of the uncertain, first-of-a-kind technology. A 2017 report predicted that SMRs have a "low likelihood of eventual take-up, and will have a minimal impact when they do arrive," while a 2018 report noted that extra costs are always expected with new plant designs.[75] Just as Admiral Rickover noted in 1957, when a congressional committee questioned the efficiency of his first civilian reactor in Shippingport: "Any plant you haven't built is always more efficient than the one you have built."[76] Getting new technology right is never easy, doubly so with nuclear technology. Neither has the nuclear industry delivered on its promise of cheap power, instead saddling the world with an expensive, uncertain design, fraught with peril from the proliferation of weapons to ongoing safety concerns in a radioactive release or accident.

The attraction to nuclear science has always been connected to a belief that we are masters of our material world and that the power of the atom is the solution to all our energy needs. In his 1914 novel *The World Set Free*, H. G. Wells championed the brave new world of artificial radioactivity, signaling the coming era of radium as an elixir for cancer treatment and nuclear power as a way to break free of our earthly bonds and embrace a more modern future, while at the outset of civilian nuclear power in 1956 M. King Hubbert confidently stated, "we may have at least found an energy supply adequate for our needs for at least the next few centuries of the 'foreseeable future'."[77] It would take a new understanding of the dangers of what we can't see to slow that dream.

3.4 Radiation: What You Don't See Is What You Get

Uranium ($Z = 92$) is the last element of the periodic table that naturally occurs in large-scale amounts. *Transuranic* elements ($Z > 92$) need to be artificially produced, for example, in a cyclotron where the elements from $Z = 93$ to 106 (neptunium to seaborgium) were first created. They are also highly unstable, as are their "daughter nuclei." In fact, all elements above lead are unstable to varying degrees, different isotopes radioactively decaying in a chain of cascading transmutations, emitting alpha (α) and/or beta (β) particles in an ongoing quest for nuclear stability. The end goal is the stable nucleus of lead ($Z = 82$).

The reason is simple geometry and the ongoing competition between protons and neutrons in such a small space. As C. P. Snow noted, "If an atom were expanded to the size of the dome of St. Paul's Cathedral, virtually all its mass would lie within a central nucleus no larger than an orange."[78] Inside that tiny core, nuclear packing is at a premium as positively charged protons repel each other, mitigated by neutrally charged neutrons to maintain stability. It is a deadly three-dimensional balance to keep the long-range, repulsive, proton–proton Coulomb force ($1/r^2$) in check with the short-range, attractive, strong nuclear force ($< 10^{-15}$ m). To keep the balance in higher-Z atoms, more neutrons are needed. Indeed, for $Z = 2$ (helium, $_2\text{He}^4$) there is an equal number of protons and neutrons – a trend that continues for low-Z atoms – while by $Z = 92$ (uranium, $_{92}\text{U}^{238}$), 146 neutrons are needed to balance just 92 protons. If three is a crowd in a relationship, 238 is way too much for an atomic party.

Radioactive decay redresses the ongoing imbalance by ejecting nuclear matter from the core in an atomic tantrum: either an alpha or beta particle as well as the illusive neutrino that Pauli and Fermi first proposed. For example, U-238 eventually decays to stable lead after a series of alpha- and beta-particle emissions, accompanied by high-energy, electromagnetic radiation called gamma radiation (γ), essentially high-energy x-rays.

The time for half of any material to radioactively decay is called a "half-life" ($t_{1/2}$), and is an exponentially decreasing function, that is, after each $t_{1/2}$ period of time only half of the initial material still remains: 1, 1/2, 1/4, 1/8, 1/16, 1/32, 1/64, 1/128 (less than 1% after seven half-lives), ..., etc., before nothing is left.[79] We can plot the course by simple nucleon subtraction (α-particle emission) or proton addition (β-particle emission). For example, U-238 decays to Th-234 via alpha-particle emission ($_{92}U^{238} \rightarrow {}_{90}Th^{234} + {}_2He^4$), taking about 4.5 billion years for half of the uranium to turn into thorium. The by-products also have corresponding half-lives, thorium-234 turning into protactinium-234 via beta decay after 24.1 days ($_{90}Th^{234} \rightarrow {}_{91}Pa^{234} + {}_{-1}\beta°$). More alpha and beta emissions follow in a continuing series of decays before reaching the end of the line at stable lead (Pb-206), 14 stages in all (Figure 3.4).

Radioactivity is a random physical process originating in all unstable nuclei and occurs everywhere in the natural background of our material world: in the rocks, soil, and atmosphere (enhanced since the 1950s by nuclear-test fallout). Defined by Marie Curie as "spontaneous emission of radiation" as opposed to neutron-induced emission of radiation following nuclear fission, natural "low-level" radioactivity goes unnoticed in our daily lives and is relatively harmless because our cells can constantly repair themselves. Enriching or refining radioactive material and neutron-induced radioactivity emitted by fission products in nuclear fuel and bomb detonation (including a "dirty" bomb that

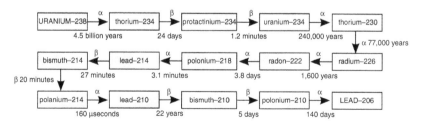

Figure 3.4 The 14 decay stages of uranium-238 to stable lead-206.

disperses radioactive material in a conventional explosion) are not at all harmless, our bodies' cell repair mechanisms unable to overcome the damage as DNA is destroyed.

Capable of breaking apart atoms in genes by high-frequency "jiggling," the radiation type – often referred to as "ionizing" for its chemical effect on absorbing material[80] – as well as the half-life are important for health and safety, for example, "An α-emitter with a very long half-life (for example, the common isotope of uranium, $_{92}U^{238}$) is relatively safe. A γ-emitter with a short half-life is very dangerous."[81] The potential biological damage depends on the absorbing material: skin, tissue (especially lymphoid), bone (and bone marrow), blood, and cells. However, if alpha particles penetrate human tissue or an alpha emitter is ingested they are 10 times as dangerous as gamma radiation. Note that a short half-life indicates higher activity, measured in decays per second (becquerels), although the amount of radioactive material is important (one atom of Pu-239 is not as dangerous as 1 kg of U-235). Curies are also used: 1 curie (Ci) = 3.7×10^{10} becquerels (Bq), equal to the activity of 1 g of radium.

Over 100 different fission products are made in a nuclear reactor, all producing α, β, and γ emission (Figure 3.5): alpha particles (highly ionizing though weakly penetrating), beta particles (less ionizing but more penetrating), and gamma rays (weakly ionizing but highly penetrating). The radiation is contained within the reactor building by thick shielding, but after being removed from the core the spent fuel rods are still highly radioactive – as seen onsite in lead-lined, concrete fuel-storage pools where the water turns an eerie deep blue from the beta-ray excited molecules emitting higher-than-normal-energy photons (called Cherenkov radiation) – and must be permanently stored or reprocessed. All fission products are either highly radioactive or very long-lived, not to mention the deadly release of radioactive debris to the environment in the event of a reactor breech or the fallout from a nuclear bomb blast.

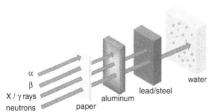

Radiation	Symbol	Description
alpha particle	$_2He^4$	helium nucleus: 2 protons, 2 neutrons
beta particle	$_{-1}\beta^0$	electron
gamma ray	γ	high-energy EM radiation
neutron	$_0n^1$	no charge, mass slightly less than a proton

Figure 3.5 Penetrating power of ionizing radiation (α, β, x/γ, n).

Somatic effects (direct exposure) and hereditary effects (gamete damage passed onto a descendant through sperm and/or eggs) are both possible and can be acute (large single dose) or chronic (small repeated doses). The acute effects are reasonably well understood; long-term, low-level exposure less so. Radiation is always nicking away at our DNA and proteins, even sunlight, but the amount of damage depends on the rate of exposure versus the rate of our cells' repair mechanism (DNA) and/or replacement mechanism (proteins/lipids). If the damage exceeds the ability to repair, we have a problem. Radiation sickness is at one end of the spectrum, while keratosis (a.k.a. age spots) is closer to the other. Furthermore, most cancers tend to come later in life because our repair mechanisms aren't as efficient as we age, as well as adding over time to any accumulated damage that never healed. But just because one doesn't get "sick" after radiation exposure or even develop cancer within a few years doesn't mean no damage was done. One must always be wary, including about childhood summer sunburns, a good reason to always use sufficient sunscreen.

Radiation dangers were learned on the job by early experimentalists and other industry workers. Henri Becquerel was burned by a sealed vial of radium salt he carried in his pocket. Marie Curie, who purified radium in the pitchblende residues brought from mines in St. Joachimsthal, Bohemia, and her daughter, Irène Joliot-Curie, both died of leukemia, a cancer of the white blood cells produced in bone marrow that limits the body's ability to fight infection and may have resulted from the "long exposure to radiation that both of them experienced during their scientific careers,"[82] while Pierre Curie may have been dying of radiation poisoning when he was killed in a horrific carriage accident crossing a street in Paris. He noted in his 1905 Nobel lecture that after a 2-week exposure to even a small amount of radium, "redness will appear on the epidermis, and then a sore which will be very difficult to heal. A more prolonged action could lead to paralysis and death."[83]

Four of the five workers who started the study of radium under the Curies died of radium poisoning,[84] while luminous dial painters who shaped the tips of their radium-laced brushes with their tongues to paint thin lines and the smallest of dots onto fancy, glow-in-the-dark watch faces (called "lip pointing") died of cancer, anemia, and necrosis of the jaw, before the dangers were understood and the practice curtailed. Early uranium miners suffered increased levels of lung cancer, likely from ingesting radon gas (a uranium and radium decay-chain nuclide), whose alpha-emitting "radon daughters" radioactively decayed inside their lungs, or from directly inhaling alpha particles attached to dust in poorly ventilated areas. For miners in the Elliot Lake region between Sudbury and Sault Ste. Marie in northern

Ontario, where uranium has been mined since the 1950s, the incidence of lung cancer was found to be twice as high as in the general population. Elsewhere, a young uranium refinery worker exposed to alpha-particle dust died within 2 years from reticulum cell sarcoma (bone lymphoma), as did many Japanese bomb victims.[85]

Soon after World War II ended, two scientists at Los Alamos died of acute radiation poisoning in separate incidents when the same plutonium test pile accidentally went supercritical – Louis Slotin died after 9 days and Harry Daghlian within a month. A number of Manhattan Project scientists died of cancer, including Oppenheimer, Fermi, and Herbert Anderson, who developed beryllium disease from inhaling radioactive dust particles created during the sawing, grinding, and cutting of beryllium (an early reactor moderator and neutron trigger). Over 40% of the 220 cast and crew of the 1956 epic *The Conqueror* produced by Howard Hughes and starring John Wayne contracted cancer, 47 of whom died. The Utah film location was downwind of a nuclear test site in Yucca Falls, Nevada, while 60 tons of radioactive-laden dust was transported to Hollywood to finish the picture on a studio lot.[86] After a reactor coolant leak aboard the first-generation Soviet nuclear submarine K-19, which was patrolling the north Atlantic on its maiden voyage in 1961 in the midst of heightened Cold War tensions, eight engineering crew members died within a few days and another 20 within a few years. (Perhaps the first ever LOCA, the event was depicted in the film *K-19: The Widowmaker* starring Harrison Ford and Liam Neeson.)

Fortunately, alpha particles don't easily penetrate the skin, but can still be inhaled or ingested, becoming especially dangerous, for example, the poisoning in a London hotel of the Russian émigré Alexander Litvinenko, who was given tea laced with polonium-210, a high-intensity alpha emitter. As Penny Sanger notes in *Blind Faith: The nuclear industry in one small town* (p. 46), "Alpha particles carry an electrical charge and cling to dust particles that may be swallowed, inhaled, or even ... work their way into the body through the pores of the skin. They in particular are the source of the lung dose that may cause the lung cancer associated with radon gas." Whatever the type of radiation (α, β, or γ), one must be ever cautious with nuclear material.

3.5 Radiation: Long-Term Effects

As for the long term, the lingering effects are well known from the hundreds of thousands of surviving *hibakusha* (bomb-affected people) in Hiroshima and Nagasaki in the years and decades after the first atomic bomb detonations, at the time called "radiation sickness." The total number of deaths from the blasts and immediate aftermath are estimated to be between 110,000 and 210,000,[87] but only 5% of the radiation came from the initial blast energy, 10% from residual radiation, while the majority was from fallout, brought to the earth by "black rain" and responsible for the most acute radiation symptoms, including outside the bomb area.[88] Despite a post-war, US senate commission report by General Leslie Groves, the ever-pragmatic military head of the Manhattan Project, stating that high-dose radiation was "without undue suffering" and "a very pleasant way to die,"[89] numerous *hibakusha* died of painful leukemia and other cancers after the fact, while many others were severely disfigured and suffered for decades, afraid to show themselves in public or marry and have children because of a fear of offspring birth defects.

Today, two of the most worrisome radioisotopes are strontium-90 (Sr-90) and iodine-131 (I-131), both biologically active and relatively long-lived. Sr-90 has a half-life of 28.8 years (decaying to Zr-90) and can replace calcium in bones causing leukemia (group II elements, strontium and calcium are chemically similar).[90] I-131 has a half-life of 8 days (decaying to Xe-131), but can accumulate in the thyroid gland, which regulates metabolism, especially dangerous for children with active thyroids. I-131 also increases the risk of thyroid cancer.

Prior to the 1963 Limited Test Ban Treaty banning aboveground nuclear detonations, radiation fallout from nuclear weapons testing resulted in well-documented increases in cancers around the world, particularly leukemia and thyroid cancer, while local cancers have been traced to contaminated milk wherever cows ate strontium-90-laden feed. After a 1957 reactor fire at a UK plutonium-making plant, 200 square miles of countryside was contaminated, while half a century later local leukemia rates were roughly 10 times the UK average.[91] In the event of a nuclear accident, taking calcium supplements and iodine tablets are recommended to saturate the body, protecting against radioactive Sr-90 and I-131 retention.

In 2002, after growing public concerns over the Sellafield nuclear fuel reprocessing plant located at the converted site of the former Calder Hall power-producing reactor and Windscale plutonium bomb-making reactor on the northwest coast of England, every household in Ireland was sent a box of potassium iodate (KI) pills, delivered by regular post to their doors (six 85-g tablets containing 50 g each of stable iodine). I still have the small orange box that eerily reads "FOR USE ONLY IN THE EVENT OF A NUCLEAR ACCIDENT" with instructions that "these tablets work by 'topping up' the thyroid gland with stable iodine in order to prevent it from accumulating any radioactive iodine that may have been released into the environment." Pregnant women, breast-feeding mothers, newborns, and children up to 16 years are considered the most vulnerable, while only those allergic to iodine are advised not to take them (or if the tablets turn brown). The effects are statistical, any absorbed/ingested radioactive strontium/iodine competing for binding sites with the ingested calcium/iodine, which doesn't always work. Fortunately, I haven't had to open the contents. More worrisome, iodine tablets were distributed in the summer of 2022 to people living near the Zaporizhzhia nuclear power plant in southern Ukraine after possible threats of radiation breaches from ongoing shelling during the Ukraine war.

Cesium-134 and cesium-137 are also common radioactive fission products released to the atmosphere in a nuclear accident, soaking into the ground after mixing with rain. Both gamma-ray emitters causing burns, acute radiation sickness, and cancer, Cs-134 is more active ($t_{1/2}$ = 2 years), while Cs-137 is of greater long-term concern ($t_{1/2}$ = 30 years; both have further high-energy decay series). Almost half of the radioactive Cs-137 released in the 1986 Chernobyl reactor meltdown in Ukraine remains in the environment, found "to some extent" in every country in the northern hemisphere, including in the "raw milk from a Minnesota dairy."[92] Government restrictions on the production, transportation, and consumption of some food remained in effect for decades afterwards in many countries, including on UK sheep and Swedish and Finnish reindeer, as well as mushrooms, berries, and fish throughout much of Europe. An insoluble, orally administered "ion-exchange" drug decorporates cesium from the body.

Radon gas is a leading cause of lung cancer, especially pernicious because gas is more easily inhaled than other radioisotopes. Radon is odorless, tasteless, invisible, chemically inert, and exists in highly variable concentrations, making accurate detection difficult. An intermediate radioisotope in the decay chain of uranium and thorium, radon gas is the first daughter nuclei of radium ($_{88}Rd^{226} \rightarrow {}_{86}Ra^{222} + {}_2He^4$), while its own decay products are also radioactive – the eerie sounding radon daughters polonium-218 and lead-214 – present wherever radon gas lingers, continuing to emit alpha, beta, and gamma radiation after inhalation or ingestion. The most stable (that is, least unstable) isotope, radon-222, has a half-life of 3.8 days. Radon gas is found in pitchblende mines, a by-product of nuclear fuel enriching and refining, and in basements and other poorly ventilated areas, having seeped up through the cracks in the underlying rock, soil, and concrete from deep within the earth.

Plutonium-239 ($t_{1/2}$ = 24,000 years) accumulates in the lungs, liver, and bones as well as the gonads, causing damage to reproductive organs. Tritium or hydrogen-3 ($t_{1/2}$ = 12.5 years) affects the whole body and is dangerous if ingested (for example, pure tritiated water, T_2O, or diluted HTO). Chelating agents are used to eliminate internal contamination with transuranic metals, while drinking lots of ordinary (uncontaminated) water will flush out radioactive tritium. If inhaled, ingested, or absorbed through the skin, americium-241 ($t_{1/2}$ = 432 years) – minute amounts of which are used in everyday smoke detectors – accumulates in the lungs, liver, and bones, while iridium-192 ($t_{1/2}$ = 74 days) accumulates in the spleen.

When used in controlled ways, nuclear radiation has practical applications similar to x-rays. Cobalt-60 is a common cancer treatment radioisotope, where a highly focused external beam of gamma rays from Co-60 decay ($t_{1/2}$ = 5.3 years) is directed at a cancerous tumor to destroy malignant cells, called a "gamma knife," although some nearby healthy cells are killed in the process, often causing radiation sickness (nausea, fatigue, hair loss). Co-60 gamma radiation is also used to sterilize medical equipment (gowns, masks, bandages, syringes) and to irradiate food (a controversial way to kill viruses, bacteria, and germs such as the microorganisms *Escherichia coli*, *Listeria*, and *Salmonella* as well as insects and parasites), while Co-60 pellets can be inserted directly into the body to destroy cancerous tumors. Ironically, ionizing radiation can indiscriminately cause cancer, but when used locally kills cancerous cells.

Marie Curie's daughter Irène won the 1935 Nobel Prize in Chemistry for creating short-lived artificial radioactive isotopes (a.k.a. radioisotopes) to treat cancer, employed in implants or as external radiation sources, which were later incorporated in medical imaging techniques for radiotherapy. Short-lived sources are most effective, employed before they completely decay, yet not sufficiently long-lasting to cause ongoing damage. Today, there are 20 million nuclear procedures per year in the United States, Tc-99m the most common radioactive tracer employed in about two-thirds of all nuclear procedures involving one-third of all hospital patients.[93] A radioactive tracer that "tags" molecules to image their path through the body, Tc-99m is very short-lived – the m is for metastable, while its γ-emission half-life is 6 hours, essentially gone within two days. Carbon-14 and tritium are also employed, while phosphorous-32 marks cancer cells that absorb more phosphates than healthy cells.

The most biologically worrisome radioisotopes (a.k.a. radionuclides), their likely source, radiation type, half-life, and main health concerns are summarized in Table 3.3 as reported by the US Centers for Disease Control and Prevention. Radium and tritium are also included. Further radioactive decay follows from each daughter nuclei, emitting more ionizing radiation (treatment countermeasures are available from the US Department of Health).

There is much debate over the advantages and disadvantages of low-level radiation. Radium (Z = 88) was once thought to be a cure-all for various ailments such as arthritis and cancer, a presumed wonder mineral with

Table 3.3 *Important radioactive isotopes (radioisotopes)*

Radioisotope	$_ZX^A$	Source	Type	Half-life	Main health concerns
Hydrogen-3 (tritium)	$_1H^3$ (T)	Reactor	β	12.5 years	Cancer
Cobalt-60	$_{27}Co^{60}$	Natural, reactor	$\beta\gamma$	5.3 years	Burns, acute radiation sickness, cancer
Strontium-90	$_{38}Sr^{90}$	Fallout, reactor	β	29 years	Burns, cancer (bone and bone marrow)
Iodine-131	$_{53}I^{131}$	Reactor	$\beta\gamma$	8 days	Burns, thyroid cancer
Cesium-137	$_{55}Cs^{137}$	Fallout, reactor	$\beta\gamma$	30 years	Burns, acute radiation sickness, cancer
Iridium-192	$_{77}Ir^{192}$	Reactor	$\beta\gamma$	74 days	Burns, acute radiation sickness
Radium-226	$_{88}Ra^{226}$	Decay	α	1,600 years	Cancer
Uranium-235	$_{92}U^{235}$	Reactor	α	7.0×10^6 years	Bone and lung cancer
Uranium-238	$_{92}U^{238}$	Reactor	α	4.5×10^9 years	Bone and lung cancer
Plutonium-239	$_{94}Pu^{239}$	Reactor	α	24,000 years	Lung disease, cancer
Americium-241	$_{95}Am^{241}$	Fallout	$\alpha\gamma$	432 years	Cancer

Source: "Emergency Preparedness and Response: Radioactive Isotopes," *Centers for Disease Control and Prevention*, Atlanta, Georgia. https://emergency.cdc.gov/radi ation/isotopes/index.asp.

miraculous medicinal properties that gently massaged under the skin in small doses, understood now to be low-level radiation. Marie Curie called the glow from the radium she painstakingly extracted from her uranium ores "faint, fairy light." Unfortunately, she was slowly poisoned by the more sinister effects when concentrated. Her and her husband's Paris office is still radioactive today, showing up localized "hot spots" on a Geiger counter almost 100 years after her pioneering experiments on radioactive substances.

Hot spas have also been known to soothe various aches and pains, including rheumatism, gout, and neuralgia. Such geothermal spas are warmed by trapped heat within the earth and vary in radioactivity and temperature depending on location (some are also heated by volcanic sources). Others, now known as radium spas, are heated by radon gas from deep within the Earth's interior. One of the most active radium spas is in Bad Gastein in Austria ($\sim$40 kBq/m^3), while the ancient spas on the Greek island of Ikaria are famous for their therapeutic value, having regularly been used by Byzantium royals, "superheated" by temperatures up to 60°C and accompanied by weak to strong radioactivity ($\sim$8 kBq/m^3). Named after Icarus, who flew too close to the Sun and plunged into the nearby Aegean Sea when his wax wings melted, the island locals are said to live longer on average than the rest of Europe, although diet, location, and lifestyle likely explains the increased longevity.

Alas, one must be careful with what one can't see. The adverse effects are clear, especially if alpha emitters are inhaled or ingested or one is exposed to high levels of gamma or x-ray radiation. Radiologists have a 5.2-year lower life span on average than other doctors as well as a 70% higher incidence of cardiovascular disease and some cancers compared to the general population, while their rate of leukemia is 8 times that of other medical doctors.[94] In fact, 2% of all cancers are caused by medical x-rays, primarily CT scans (3D x-ray images produced via computed tomography). Like radioactivity itself, radiation disease is statistical and hard to determine an exact cause, while body types and immune systems differ greatly in the general population. An acceptable dose is difficult to quantify.

Two analogies help explain the confusing dilemma between beneficial radiation (for example, directed medical isotopes) and harmful radiation (for example, from uranium refining, leaky spent fuel, accidental release, and nuclear fallout): water and hammer taps. One drop of water on the skin is inconsequential as are a series of drops, while a steady flow soothes as in a hot shower. A non-stop series of high-frequency drops, however, will ultimately damage the skin. Replace the water with a gentle tap from a hammer and the body can still recover if subjected to a small number, but tap every second for an entire day – even very lightly – and the skin will blister and bleed before eventually dying. If the taps continue, the hand will even fall off.

A further analogy is the Sun, especially for gamma radiation. Too much exposure to high-energy photons damages the upper layers of the skin such that without sufficient time for the epidermis and dermis to heal, the skin will not only burn but die, increasing the risk of cancer (especially melanoma). Like Icarus, we must all be wary of overexposure to the Sun.

Although there is no agreed safe level – clinical or otherwise – the risk of cancer increases with the level, length, and location of exposure. Furthermore, not all low-level exposure leads to cancer, but can produce cardiovascular and circulatory problems, often excluded when analyzing clustered data. Counting only cancers is not sufficient to measure all health effects of radiation, as "radiation can change the whole chemistry of the body, making it more susceptible to other disease."[95] Although hard to measure, hereditary effects are also possible.

Dr. Rosalie Bertell, a noted cancer research scientist who specialized in the health effects of x-rays, disagreed with the basic assumption that low-level nuclear radiation is harmless. She noted that while normal cellular repair can occur after radiation exposure stops, some cellular damage is irreparable and that on top of the observable clinical effects, radiation produces "a generalized, systemic 'aging' effect on the body."[96]

The experiments of Herman Muller, the pioneering geneticist and outspoken 1946 Nobel laureate, showed that genetic mutations were produced in fruit flies after x-ray exposure, suggesting a no-threshold dose and increased damage with more exposure. Although contrary evidence suggested the possibility of a low-dose threshold, Muller's results popularized the idea of a linear no-threshold (LNT) relationship between radiation exposure and harmful effects, precipitating the introduction of shielding, at least to protect the reproductive organs.[97] In the 1954 president's address to the American Association for the Advancement of Science, Caltech professor and genetics pioneer Alfred Sturtevant reiterated the general understanding that "any level whatever" is "at least genetically harmful to human beings" and that low-level, high-energy radiation is a "biological hazard."[98]

***

To understand better the potentially lethal health effects and requisite safety guidelines, we must understand the units, which unfortunately are a bit confusing. So far, we have seen the becquerel (Bq) and the curie (Ci), named after Henri Becquerel and Marie Curie, which relate to the *physics* of radiation, that is, the statistical disintegration of a nucleus as measured in decays per second heard in the clicking of a Geiger counter (1 Bq or 2.7×10^{-110} Ci). For example, 1 g of radium-226 decays 37 billion times per second, defined as 1 curie (37 billion Bq = 1 Ci), while 1 g of plutonium-239 decays 2.3 billion times per second (2.3 billion Bq or 0.063 Ci).[99]

Other units measure the *biological* effect of radiation, that is, the harm to a cell rather than a click on a Geiger counter. Rads and rems are non-standard but are still sometimes used, where a rad (radiation absorbed dose) measures the physical exposure per mass and a rem (radiation equivalent man) the biological effect. Grays and sieverts (Sv) are the SI equivalent units, where 1 Sv equals 100 rems. A rem or sievert does roughly the same damage whatever the type of radiation. To keep things simple, we will stick to sieverts or microsieverts (μSv) wherever we can.

Another measure, the Q-factor (Q) also helps clarify the differing effects of the main ionizing culprits: alpha radiation is the worst (Q = 20) because it travels the shortest distance and thus does the most localized internal damage if ingested, while beta, gamma, and x-rays are comparatively less damaging (Q = 1). The damage also depends on the organ and its cell division rate, the most susceptible being the gonads, thyroid, skin, eyes, lungs, and bone marrow.

The recommended annual maximum dose for the general public is 5,000 μS, while the average background radiation is typically around 1,000–3,500 μSv per year: 500 from cosmic rays, 500 from soil, and 2,500 from one's surroundings depending on where one lives. For example, Denver is near to local uranium-laden granite, close to the mountains, and at high elevation (the Mile High City), and thus receives more background radiation than average. In our daily lives, most of us don't come close to exceeding the limit, even with a regular dental x-ray (5 μSv), chest x-ray (100 μSv), mammogram (400 μSv), or occasional 2-hour plane trip (5 μSv) where cosmic rays are less shielded at higher altitudes, although a CT scan is especially high (~7,000 μSv) and thus the number of scans a patient receives is closely monitored. Wherever possible, ultrasound or magnetic resonance imaging is preferred to a high-dose PET-CT scan.

The recommended maximum dose for a US nuclear worker is 50,000 μSv/ year (10 times the general public limit), while a cardiologist or radiologist who performs 100 operations in a year might receive 10,000 μSv. An astronaut on the International Space Station gets about 160,000 μSv/year (more than 30 times the general public limit) as do cigarette smokers from trace amounts of radioactive lead and polonium in tobacco (another good reason to quit). Few of us will ever reach 1 Sv (1 million μSv) in our lifetime from normal activity (200 years of the maximum public dose), although at 32 times the average annual dose smoking clearly increases exposure and the risk of cancer (other cigarette toxins also add to the risk). Importantly, 1 Sv indicates a 5.5% increased chance of cancer, although receiving a 1-Sv dose all at once is extremely dangerous.

The US Nuclear Regulatory Commission (NRC) provides a "personal annual radiation dose" calculator to work out one's exposure depending on location and lifestyle (for example, amount of plane travel, plutonium-powered pacemaker, x-rays).[100] Mine is 3,400 μSv (340 mrems), well below the recommended annual maximum public dose of 5,000 μSv (500 mrems), more than half coming from atmospheric radon. According to the NRC, Americans receive an average annual radiation dose of about 6,200 μSv (620 mrems).

Some of the most radioactive places on Earth are listed in Table 3.4, the data collected via a hand-held Geiger counter. The basement of the Pripyat hospital – where the fire fighters retreated after the 1986 Chernobyl reactor accident – produced the highest surveyed level, still delivering 1,000 μSv/h after 30 years, 1,800 times the annual recommended maximum dose, which would equal about 1 Sv after 6 weeks of continuous exposure. That's the equivalent of 60 years of smoking in one year or 100 packs a day for a year!

Table 3.4 *A survey of some of the most radioactive places on Earth*

Radioactive source	μSv/h	× annual dose
Hiroshima peace dome	0.3	0.53
Marie Curie office (door knob)	1.5	2.6
Jáchymov uranium mine	1.7	3.0
Alamogordo Trinity bomb site	0.8	1.4
Alamogordo trinitite mineral	2.1	3.7
Airplane flight (33,000 feet)	>2	>3.5
Chernobyl reactor #4 exterior	5	8.8
Fukushima exclusion zone	10	18
Astronaut	18	32
Smoker's lung (polonium)	18	32
Chernobyl hospital basement	1,000	1,800

Source: "The Most Radioactive Places on Earth" [documentary], presented by Derek Muller, *Veritasium*, December 17, 2014. www.youtube.com/watch?v=TRL7o2kPqw0.

Not included in the survey, the Rongelap Atoll in the Marshall Islands may be the most contaminated place on Earth, adjacent to the site of 67 nuclear weapons tests from 1946 to 1958, including the 15-megaton-TNT Bravo thermonuclear blast to the west at Bikini Atoll. The picturesque coral atoll is still uninhabitable more than a half century on, the evacuated residents promised a return that never comes. Some believe the US government used the tests to measure the effects of radiation on humans (Project 4.1 Biomedical Studies). The Soviets also exposed 1.5 million people during almost 500 atomic tests in northeast Kazakhstan, where locals continue to suffer increased rates of miscarriage, skin disease, breast, throat, and lung cancer, aplastic and hemolytic anemia, physical deformities, nervous disorders, schizophrenia, mental retardation, hereditary disease, and oncological disorders at twice the rate of Hiroshima and Nagasaki. There are no plans to clean up either sites, while no one has cleaned up Marie Curie's suburban Paris lab more than 100 years after she painstakingly purified pitchblende to extract radium.

Radioactivity is not easily understood, inside or outside of the nuclear industry. Built up from more than a century of gold mining, waste heaps on the outskirts of Johannesburg, South Africa, contain almost 600,000 tons of uranium (50 g per ton of waste tailings), 30 times the natural background level of 0.10 μSv/h, comparable to the exclusion zone in Chernobyl. The open sites also include heavy metals at over 300 times the level allowed for arsenic and 80 times lead in what has been called "one of the biggest environmental disasters in the world."[101]

Although the effects of nuclear radiation are still debated, basic occupational precautions are essential: masks and adequate ventilation for uranium miners and refinery workers, full-body lead aprons for medical workers, and restricted exposure in high-risk jobs (limiting the number of surgeries, high-altitude work, or power-plant exposure). Dosimeter badges must be worn at all times, but don't necessarily measure full exposure. For example, surgical aprons don't cover the head, hands, or legs (although the gonads are the most susceptible).

While most of us needn't worry about annual radiation doses, accident exclusion zones, or abandoned test sites, there are still everyday concerns for those who live near a uranium refinery, fuel reprocessing facility, or nuclear power plant that regularly release radioactive material into the air, soil, and water during normal operations. Groundwater is also contaminated by radio-activity from corroded pipes and spills from old storage tanks, while careless waste-management practice continues to cause ongoing environmental damage and increased health problems. In some locales, the lack of action is beyond comprehension.

3.6 Nuclear Waste Products: Impossible to Ignore

A 2010 report on the state of American reactors, entitled *Leak First, Fix Later*, listed 102 radioactive leaks since 1963, mostly tritium, cobalt-60, and stron-tium-90. The report noted that a reactor facility contains up to 20 miles of buried pipes of varying durability that carry radioactive water under buildings, foundations, and parking lots, and can fail from corrosion, erosion, or seismic activity. The radioactivity is found only after the damage is done.

After discovering radioactive ditchwater beside the Exelon Braidwood Generating Station, about 50 km southwest of Chicago, further studies counted 22 spills from 1996 to 2000 over a "four and a half mile-long pipe running from the nuclear station to a dilution discharge point on the Kankakee River."[102] Two of the spills dumped six million gallons of radioactive reactor water that contaminated ponds, wells, and groundwater with tritium and cobalt-60. In 2009, an Exelon reactor at Oyster Creek, New Jersey, spilled tritium-laced water at 50 times the state standard, contaminating local drinking supplies.[103] Similar to the petroleum industry, the nuclear industry is not obliged to report leaks and is subject only to voluntary self-reporting.

Radioactive waste is categorized either as high-level waste (such as spent fuel rods and bomb material), intermediate-level decommissioning material (for example, reactor core parts), or low-level waste (such as mill tailings, nuclear plant water, and reactor building parts as well as clothes and hospital waste). The United States produces 15 million tons per year of mill tailings, the "largest source of any form of radioactive waste."[104] A typical GW-reactor produces about 3 m^3 of radioactive waste per year, which puts the USA at about 300 m^3 for its 100 reactors or more than 2,000 tons a year. A 2009 *Scientific American* article pegged the accumulated American spent fuel waste at 64,000 tons, enough to cover a football field 7 yards deep.[105] By 2023, the amount was over 90,000 tons, adding almost another 3 yards to the problem.

To be sure, the waste isn't going to disappear anytime soon, nor are disposal solutions getting any easier. Indeed, one of nuclear power's biggest problems is where to put the accumulating waste, some with radioactive half-lives in the millions of years. The issue can't be swept under the rug any longer despite refineries still storing low-level waste in plain view under black tarpaulin covers in temporary sites, easily able to leach into the soil or groundwater or even blowing off with the wind.

The story of one small town on the north shore of Lake Ontario is a cautionary tale that beggars belief. Located 100 km east of Toronto, Port Hope became part of the nuclear family in 1932 after a local mining company, Eldorado Gold Mines, began producing radium for the burgeoning international medical market. Led by a co-worker of Marie Curie, who had been lured from France to manage the lucrative new radium trade, pitchblende was dug out of inhospitable mines in Great Bear Lake in the Northwest Territories and laboriously turned into pure radium, refined in the much more accessible and hospitable town of Port Hope, then a lakeshore paradise of about 5,000 inhabitants. One gram of radium, which comprises less than a one-millionth part of pitchblende – the shiny, black, tar-like uranium- and radium-containing ore – was produced from 6.5 tons of ore, requiring 7 tons of chemicals to dissolve and filter the caustic slurry.[106] The prize at the end was a single gram of radium, fetching as much as $125,000 and undercutting the world's only other radium refinery in Czechoslovakia, where Marie Curie first collected uranium ore to produce polonium and radium and which Hitler had commandeered in 1939.[107] At the time, the unwanted uranium was discarded in the process.

Tasked by the Canadian government to supply bomb material for the Manhattan Project in World War II, Eldorado retooled its radium production to provide almost 700 tons of uranium oxide to the war effort, becoming one of the world's largest uranium refineries by the end of the hostilities. Reformed as a crown corporation, the company then began producing uranium and UF$_6$ feedstock for civilian reactors – natural-occurring uranium fuel bundles for Canada's home-grown, heavy-water reactors and converted uranium for export to manufacture enriched U-235 fuel rods in American light-water reactors. Strategically important as an international uranium supplier and the town's then largest employer, Eldorado Nuclear Limited was part privatized in 1988 and renamed the Canadian Mining and Energy Company (a.k.a. Cameco), before being fully privatized in 2002 to become the world's largest publicly traded uranium company, producing almost one-third of Western uranium supplies.[108] Having lost its medical luster and of questionable therapeutic value, radium production was phased out in 1954.

During decades of operation at the mouth of the Ganaraska River that cuts the picturesque town roughly in half, Eldorado refined uranium to make fuel for the nuclear industry, often disregarding sensible safety procedures to dispose of mill tailings produced in the process. Heavy-metal wastes such as arsenic were casually discarded in unfenced, poorly signed, and leaky dumps around the town and in other nearby locales. Lax or non-existent regulations even permitted landfill to be carted off by local builders, ultimately contaminating various parts of the community with toxic and radioactive waste, some that had decayed into carcinogenic radon gas and accumulated in poorly ventilated areas.

By 1975, elevated levels of radioactivity were discovered throughout the town, including in an elementary school built with Eldorado landfill, while ravines, beaches, and the harbor all became contaminated. Numerous hot spots showed high radiation and toxicity levels – radon 100 times, arsenic 200 times, and gamma radiation 300 times normal levels – forcing the school's closure for almost 2 years, while elevated levels of radon gas were detected in more than 500 homes, alarming citizens about the health effects of long-term radiation exposure, especially for children who are more susceptible than adults.

Little was done to alleviate the worries of those concerned about the health effects of low-level radiation, setting off a decades-long battle between pro-nuclear advocates concerned about jobs and anti-nuclear activists wanting the waste cleaned up and the government to commission a study into the damage and potential future effects to their health. Originally dubbed "The town that radiates friendliness" in a slick 1930s radium marketing slogan, one can now take a "toxic tour" around town that starts the Geiger counter ominously clicking.

On-site construction regulations were also found to be substandard, Eldorado given preferential treatment in 1981 to build a new UF6 plant without any of the usual planning permission. Gobbling up lake-front real estate in one of the most postcard-perfect southern Ontario towns, a community beach, picnic area, and baseball diamond were all appropriated. Cementing the power of the state over communal lands, the once-popular West Beach became inaccessible to the public as Eldorado expanded without community input or regulatory oversight.

National nuclear security (or its converted civilian nuclear power form) is uniquely advantaged to create an energy strategy outside the public forum or agreed public good. In Canada, a crown corporation isn't required to submit an environmental assessment plan, and thus remains beyond the reach of standard scrutiny, stifling discussion about operations, safety, and strategic goals, while denying open debate. Typically, the public can't even participate in regular safety reviews. Elsewhere, for example, in the UK, the regulator doesn't even have to consult the public on any issues about nuclear safety.[109]

By 1995, the then renamed Cameco also wanted to store one million tonnes of nuclear waste in a proposed $200 million, 85-foot-deep, cavern complex at the previously appropriated West Beach site to house mill tailings from its operations as well as other radioactive waste from communities across Canada. Perhaps banking on a similar acquiescence from a seemingly uninformed public, pro-industry advocates glossed over the dangers of low-level radio-activity for almost a year, assuming approval was a formality.

However, Port Hopers had other ideas, tired of Eldorado's overreach. One resident summed up the growing outrage in a Letter to the Editor of the town newspaper about building a waste disposal site beside the town water pumping station, 500 m from an intake water pipe: "Sign me crazy but not stupid." Another letter from a former Eldorado mining engineer cum anti-dumping activist aptly captured the concern within the community: "The one certainty is that once you get underground is that nothing is certain. In X number of years, God only knows what might happen."[110] A simple calculation illustrated the magnitude of the proposed operation: "200,000 truck loads, or one every ten minutes, ten hours a day for ten years!"[111] Long-time resident and award-winning author Farley Mowat branded the proposal "Canada's Nuclear Sinkhole," while the geneticist and science broadcaster David Suzuki, who came to speak at a community meeting at the height of a town-wide discussion on the issue, clarified the reality of burying nuclear waste: "If it's so safe, why is the nuclear industry protected from legal responsibility in the event of an accident?"[112]

Local opposition mounted, until the ill-advised plan to build an underground nuclear waste site on the shores of Lake Ontario, half a kilometer from the center of a picturesque town that boasts one of the best-preserved Victorian streetscapes in Canada, was nixed, scuttling for a few more years the federal government's plan to construct a permanent long-term national nuclear waste site. A paltry $11 million in proposed compensation was clearly not enough to convince anyone that the government had their best interests in mind.

In 2004, elevated levels of radiation were again discovered at another Port Hope school, where gamma radiation and radon gas was detected at 125 times accepted limits (over 500 picocuries). One long-time local activist wondered what the authorities were hiding, stating "They knew this place was dangerous and still let kids in."[113] Countering the government dogma about safe levels of low-level radiation, an epidemiologist also found "higher than normal rates of leukemia and childhood deaths, as well as an elevated incidence of brain, lung and colon cancers for certain demographics and time periods."[114] Three years later, uranium was found in the urine samples of a group of former nuclear workers concerned about unexplained illnesses. Armed with more data and further study, the fight began anew.

Two recent studies further verified the dangers for uranium workers and those living near fuel-conversion facilities. A 2010 study of almost 18,000 Eldorado workers reported a linearly increasing rate of lung cancer incidence and cancer deaths with radon decay product (RDP) exposure, although a correlation between overall mortality and RDP exposure wasn't found, attributed to the healthy worker effect.[115] The workers were employed by Eldorado from 1932 to 1980, mostly at three main sites: Port Radium (Northwest Territories), Beaverlodge (northern Saskatchewan), and Port Hope's fuel conversion facility (southern Ontario). Another study in 2013 showed a correlation between women living in the Port Hope area and lung cancer, but this may have been related to the socioeconomic characteristics of the region.[116]

There is little doubt that the radioactive waste strewn across the town came from the conversion plant, caused by careless Eldorado/Cameco management over a prolonged period. Alas, few agreed on how to account for past actions, pitting property values, tourism, and growth versus public health. The small town, which had at one time served as the capital of a rapidly growing Upper Canada prior to Confederation, was becoming a popular commuter town and weekend tourist destination for the bulging modern capital of Toronto despite its nuclear problems. The issue, as usual, was about money, although when a government is the landlord in a tenant dispute the public is oddly in opposition to itself.

It's difficult to say no to industry in a small town, doubly so when that industry is the main employer and in this case backed by the federal govern-ment, one that is less than forthcoming with the details of its "strategically important" activities. Add in an ever-present element of secrecy to nuclear power and a stifling amount of jargon, and we have a recipe for fear and suspicion. What exactly is going on and in whose interests does a strategically important crown corporation or converted for-profit publicly traded company act? Why does responsibility stop at a company fence when the company's activities clearly extend beyond the physical boundary of the plant and include invisible though detectable radiation? Why is the government protecting a private company?

***

Of course, we have to deal with the waste *somehow*, a seemingly endless game of political hot potato. In 2002, another small town was proposed as the site of Canada's officially proclaimed national deep geologic repository (DGR) after Port Hope and Deep River – located upriver from the site of Canada's first nuclear reactor at Chalk River north of Ottawa – declined. No stranger either to the nuclear industry, Kincardine, Ontario, is a short drive from Bruce Power Station, one of Ontario's three nuclear power-generating facilities and world's largest fully operational nuclear plant. On the eastern shore of Lake Huron south of the Bruce Peninsula about 200 km northwest of Toronto, Kincardine had been spared as an earlier choice of "host community" for Canada's permanent nuclear waste site until the others both refused. That is, until the crown corporation in charge of the project, Ontario Power Generation (OPG), unveiled a plan to store 200,000 m^3 of low and intermediate nuclear waste from Ontario's 18 CANDU reactors in a proposed $1 billion storage facility, 680 m belowground in Kincardine. As in Port Hope, the proposal was met with fierce local opposition.

Precipitating numerous Canadian and American objections, the Kincardine DGR debate has raged for more than two decades. Proponents of the plan champion the stability of the underlying low-permeability limestone rock, while opponents counter that building an underground nuclear waste site 1 km from the shores of Lake Huron is crazy, not least because the Great Lakes is the world's largest body of fresh water, providing clean drinking water for almost 40 million Canadians and Americans and where a radioactive leak would have devastating consequences. Even more worrisome, the Great Lakes are only 10,000 years old, where seismically active "pop-ups" still regularly appear on the lake bottom from ongoing external plate stress, while the Kincardine DGR is expected to operate as a safe repository for more than one million years. Someone hasn't done the most basic math.

Common sense dictates that water supplies and radioactive waste don't mix. Dr. Gordon Edwards, co-founder of the Canadian Coalition for Nuclear Responsibility, called the plan "absurd," noting that "water is the biggest single threat to the safe long-term storage of nuclear waste. Water floods underground mines, corrodes containers, promotes chemical reactions, generates gas pressure, and carries radioactive poisons back into the food chain. Of all the places to dump nuclear wastes, the Great Lakes drainage basin would seem to be one of the very worst."[117] Dr. Frank Greening, a former nuclear scientist and one-time OPG employee, called the plan "idiotic" and "dangerous," noting that the official estimates of expected radiation are "1,000 times lower." Resolutions opposing the plan were passed in over 150 communities on both sides of the border, including Chicago and Toronto.

It seems Port Hope, Kincardine, and Deep River were chosen because they already "host" nuclear facilities and would presumably be more accepting. But the site of an existing uranium refinery or nuclear power plant – chosen because of a *proximity* to water – is a major no-no for nuclear *storage*. What's more, some communities may be used to their own nuclear waste, albeit begrudgingly, but are not keen to welcome other people's radioactive junk, trucked in from afar. In early 2020, the local Saugeen Ojibway Nation, whose land comprises the area of the proposed Bruce Peninsula DGR, rejected OPG's $150 million proposal to build on their territory.[118] A policy of "rolling stewardship" will instead store the retrievable radioactive waste from Ontario's reactors aboveground on site until an acceptable storage solution is found.

In 2022, the Canadian government announced two more possibilities for a national storage site, one in South Bruce east of Kincardine and the other in Ignace near the Manitoba border in western Ontario. The plan is to store 5.5 million spent fuel rods from reactors in four provinces in a $23 billion, forever nuclear vault. If approved, the site will take four decades to fill, starting sometime after 2040, requiring 30,000 shipments by road and/or rail, that is, two shipments a day, every day, for 40 years.[119] Expect the opposition to be ready, especially in South Bruce, the more populous of the two sites and still within the Great Lakes basin.

There are no guarantees either that a DGR will work. In 2014, a "waste isolation pilot plant" (WIPP) in Carlsbad, New Mexico, sprung a leak after a storage container burst, having been mistakenly packed with organic matter (kitty litter!), costing almost $2 billion to clean up when trace amounts of plutonium and americium were found about a half-mile away at aboveground air-quality stations.[120] The nearby Carlsbad Caverns should have been an obvious showstopper, famous for their underground chambers formed by the dissolution of limestone by groundwater, another dubious feature for the long-term storage of nuclear waste. In fact, there have only been three DGRs constructed so far to store nuclear waste, all of which have failed, two in Germany and the Carlsbad WIPP.

Hoping to buck the odds, the Finnish government is building a DGR at the Olkiluoto Nuclear Power Plant, located on an island off the west coast of Finland and 200 km northwest of Helsinki. Called Onkalo (Finnish for "cavity"), the ambitious plan is to store plutonium-rich spent fuel rods at a depth of

500 m, employing spiral tunnels and a "multi-barrier" system that includes corrosion-resistant copper canisters, a bed of water-absorbing bentonite clay, and continuous infill. At the end of the proposed 200-year lifetime, the entire structure will be sealed for all time ("Nuclear Eternity") or until the next ice age cracks it open. Alas, one Finnish opponent to the €3 billion nuclear tomb stated that the site was not chosen for its geology, but because "Eurajoki was the first municipality to say 'ok, we can take it', and there wasn't an active nuclear opposition in this area."[121] Who is to say what will cause the *next* accident and what damage will result? Unlike, coal, oil, and gas, we can't recover as easily from a nuclear mistake. Onkalo is expected to receive its first waste in 2025.

In the UK, the magnesium-oxide (Magnox) and advanced gas-cooled (AGR) reactors are particularly bad at waste. In 1990, the chairman of UK National Power noted that "British reactors had produced more waste than all the rest of the nuclear industries combined,"[122] yet without a permanent waste storage facility the spent fuel from 15 British nuclear reactors has been stored at Sellafield. Situated beside the world's first commercial nuclear power plant, Sellafield has been called the riskiest nuclear waste site in the world, while a former UK minister of energy revealed they are only starting to "get to grips with the legacy after decades of inaction."[123]

Located on the Cumbrian coast just west of the Lake District National Park, Sellafield contains a 100-m fuel-waste storage pond from the original UK bomb-making program as well as hundreds of tonnes of other highly radioactive material, while a 21-m high silo of fuel cladding has been full since 1964. As noted in *New Scientist*, "The decaying structures are cracking, leaking waste into the soil, and are at risk of explosions from gases created by corrosion."[124] Nuclear cleaning isn't cheap either, the Sellafield job earmarked at a whopping £80 billion, while the problems persist over finding a permanent storage site. The cost of the UK's most recently proposed long-term DGR has ballooned to over £50 billion, half to be paid by taxpayers, even though Cumbria doesn't want the stuff.[125] Next to never letting nuclear waste in, ensuring no more arrives is as good as it gets. As the saying goes, it's not enough to cook uranium and serve it to the public, one also has to do the dishes.

In northwest Russia, $200 million has already been spent to decommission old Soviet nuclear storage sites. Following years of neglect after the breakup of the USSR, spent fuel from more than 22,000 Soviet-era nuclear submarine canisters stored in the Arctic town of Andreeva Bay, located about 50 km east of Murmansk, are being reprocessed in Mayak, site of the Soviet Union's original bomb-making program and world's first major nuclear accident. The partly damaged fuel elements at the world's largest spent fuel store were previously kept in leaky dry storage units after leaking from on-site pools and now constitute enough radioactive material to fill 400 40-ton containers, equivalent to 100 reactor cores. A former naval officer who has monitored the site for years commented on the lax policy, "I've been all over the world to pretty much every country that uses nuclear power and I've never seen anything so awful before. With nuclear material, everything should be done very carefully, and here they just took the material and threw it into an even more dangerous situation."[126] The fuel will now be transported by ship from Andreeva Bay to nearby Murmansk and then by train to Mayak across almost half of Russia.

As bad as nuclear waste storage sites are in the former Soviet Union and as hard as it is to agree on the means or even a location to store nuclear waste in Canada or the UK, the United States hasn't solved its problem either of where to put long-term waste. The nuclear NIMBY battle wages on in the heartland of atomic power, a hot potato hotter than the waste. As noted by the World Nuclear Association:

> The question of how to store and eventually dispose of high-level nuclear waste has been the subject of policy debate in the USA for several decades and is still unresolved. As well as civil high-level wastes (essentially all US used fuel plus research reactor used fuel of US origin) there is a significant amount of military high-level radioactive waste which Congress intends to share the same geological repository. Naval used fuel is stored at the Idaho National Laboratory.
>   Since the beginning of the commercial use of nuclear power in the USA, used fuel assemblies have been stored under water in pools (and later in dry casks as well) at reactor sites, and remained the responsibility of the plant owners. The prohibition of used fuel reprocessing in 1977, combined with the continued accumulation, brought the question of permanent underground disposal to the forefront.[127]

Originally proposed as a centralized, high-level radioactive waste site in 1987, Nevada's Yucca Mountain was meant to put the US nuclear waste disposal issue out of sight for good if not out of mind, but the issue has flip-flopped for decades. Harry Reid, the former Democratic Senate leader, was against building an underground nuclear vault in his home state, calling the plan the "Screw Nevada Bill," while 80% of Nevadans, including local Native American tribes, oppose the idea. Nevadans were especially annoyed that Nevada doesn't even have a nuclear power plant and had previously been used for weapons testing without consultation.

What's more, only 12 of the 99 reactors in the USA are west of the Mississippi, making Nevada a poor logistical choice on distance alone, never mind that Nevada is the fourth most geologically active state. Without sufficient political push (that is, a large compensation/bribe), nobody wants the waste, estimated at over 90,000 tons and growing by 2,200 tons a year (a 10-yard deep football field).

Of course, if the point of building a storage site in the desert is that fewer people will be affected in the event of an accident, it doesn't hurt that there are fewer people to object.

Concerned about the stability of the seismically and volcanically active region as well as the oxidizing environment, Allison Macfarlane, a former NRC chairman (2012–2014), stated that Yucca is unsuitable.[128] Another former NRC chairman (2009–2012), Gregory Jaczko, believes the nuclear industry should clean up its own mess rather than forcing taxpayers to foot the bill. "No other industry is able to complain so loudly that someone else has failed to take care of its waste," he wrote in *Confessions of a Rogue Nuclear Regulator*.[129] Not surprisingly, the Yucca Mountain proposal was nixed by Barack Obama in 2011, before being restarted by Donald Trump in 2017. One expects the flip-flopping to continue as the world's #1 nuclear waster continues to dodge an issue it has been avoiding since 1945.

Another highly contentious issue is transportation, often glossed over by top-down system builders who think that waste disposal can be solved with a wish and a pen, where highly radioactive material is hauled by train or truck across thousands of miles through unprotected towns and countryside. Known as "glow" trains and "glow" trucks, some opponents call the transport routes "Fukushima Freeway."

More than 60 years since the first civilian reactors came online, high-level radioactive waste (HLRW) from spent fuel and decommissioned bomb material represents the most significant problem for the nuclear industry, never mind thousands of tons of low-level radioactive waste (LLRW) from mill tailings (including acids, nitrates, fluorides, arsenic, and other chemical poisons) that must constantly be monitored for leaching, leaks, elemental disruption, and even burrowing animals. So, what to do with the waste, most of which lasts billions of years?

Common sense dictates a dump should not be built near a high water table, public waterway (for example, the Great Lakes!), unstable or eroding ground, earthquake zone, permeable soil, or agricultural land, and should be protected from the elements and outside disruption. Long-term tunnel stability (or other means of access) must be assured, while future needs (and costs) must be included in any design; 24/7 monitoring is imperative, which given the extraordinarily long half-lives of much of the nuclear waste is essentially permanent. For now, the best solution is to leave the waste on site until a better solution comes along – not much consolation, but better than doing something really stupid.

Without solving the waste issue, one wonders how nuclear power can ever be a viable energy solution. As noted in the Flowers Report, a 1976 UK commission

on environmental pollution: "There should be no commitment to a large programme of nuclear fission power until it has been demonstrated beyond reasonable doubt that a method exists to ensure the safe containment of long-lived,
highly radioactive waste for the indefinite future."[130] Clearly, we aren't learning
from our mistakes. In the rush to install peaceful nuclear power and make amends
for the horrors of Hiroshima and Nagasaki, nuclear waste is still an afterthought,
saddling future generations with even more worry.

At least we don't dump radioactive waste into the ocean as in the early days of
nuclear power, while some HLRW can be reprocessed by reburning in a reactor.
In the aftermath of the Cold War, 250 tons of diluted Soviet bomb-grade material
(about 10,000 warheads) were sold to the USA for recycling, which helped
power "the lights of Boston."[131] Under a 1993 non-proliferation agreement
between the US and Russia, the highly enriched uranium (HEU) from as many
as 20,000 warheads was reprocessed, providing half of all uranium fuel used in
American reactors until 2013. Although the novel "Megatons for Megawatts"
program has since been discontinued, "downblended" HEU weapons and other
spent fuel can still be reprocessed to reduce the ongoing accumulation of high-
level nuclear waste. American weapons were also reprocessed, providing 5% of
US nuclear power. Without other solutions, it makes sense at least to burn the
most dangerous uranium material in a reactor.

Of course, by continuing to kick the can down the road, the problems only
worsen as more reactors are decommissioned with no place to go. What's more,
there is little political will to demand that the nuclear industry pay for long-term
storage, and thus the public must broker the cost and responsibility of an endless
ongoing cleanup, the state providing cover for an industry that hasn't worked out
the basics of its own business. In the United States, the 1982 Nuclear Waste
Policy Act (NWPA) assumes government responsibility for radioactive waste,
essentially a free license for the nuclear industry to continue polluting at will,
while the 1954 Price-Anderson Act also limits liability to $700 million, cement-
ing a cozy, decades-long, government backing for ineptitude.

In Port Hope, after decades of living with the unseen dangers of radioactive
waste, the Canadian government at least decided to remove the mess created by
the former crown corporation Eldorado Nuclear Limited cum publicly traded
Cameco. Set up in 2001, the Port Hope Area Initiative (PHAI) oversees the
removal of low-level radioactive waste to a long-term, aboveground, storage
site, 10 km north of the town (already in operation for Cameco waste). There is
no guarantee of future safety, but first removals began in 2018. The
$1.28 billion project aims to remove 2.5 million cubic meters of contaminated
material such as uranium, thorium, radium, and arsenic, while 5,000 homes will
be tested. That works out to $75,000 per person, a high price to clean up what

was previously deemed safe by government officials, although no one is quite sure the damage can ever be undone. Some doubters wryly recommended giving the money to each resident instead, concerned about suspect government sampling methods and dodgy analyses.

After the last truckload, possibly within the next decade after a series of administrative delays in part because of an absence of local testing facilities, the town hopes to move on from almost a century of "radiating friendliness." At least for Port Hopers, the problem will lie under someone else's rug. Hopefully, they do a better job. Nonetheless, with one of the world's largest uranium refineries still dominating its lakefront and a nearby waste storage site still a concern, Port Hope will continue to endure its long nuclear legacy. The Canadian government may have finally seen the error of letting an unbridled nuclear industry have its way, but one doesn't have to try hard to come up with a snazzy new PR slogan to gloss over the past – "Scenic Port Hope: still glowing after all these years."

Nuclear power incites passionate followers for and against. The standard debate is "jobs versus the environment" as with other industries, although nuclear power is more worrisome because of the uncertain long-term effects of unseen and unknown amounts of radiation, the public unable to gauge their own safety given the secrecy and jargon. One must also wonder how a government can act contrary to the interests of its own citizens by denying basic safety information. Nuclear secrecy should never apply to health. At the very least, poorly regulated, for-profit companies should be excluded from deciding the public interest.

There is only one question about radioactive waste – would you want it in your backyard? Few of us would say yes, although in the absence of an honest public broker one must become better informed about the risks and learn as much as possible about the potential harm. The Earth is not infinite nor a dump to compensate our negligence. Nor is nuclear power "too cheap to meter" as originally advertised. It's well past time to re-evaluate an industry PR that for over a century has held the public captive to the mysteries of the atom.

If not for the waste, nuclear power could well be the answer to our energy needs, but the ongoing waste problems haven't been solved *75 years on*. Even more worrisome, nuclear risk is always skewed – there may not be as many accidents as with oil and gas, but they are always horrific (as we look at now).

3.7 Nuclear Safety Systems: Can We Ever Be 100% Sure?

More worrisome than radioactive waste is the potential for an accident, which for nuclear power is more dangerous than any other power-generating technology. Since the start of the Atomic Age, when Fermi's CP-1 pile first went critical on December 2, 1942, three dates have become stamped in the annals of nuclear catastrophe: March 28, 1979 (Three Mile Island partial meltdown), April 26, 1986 (Chernobyl explosion), and March 11, 2011 (Fukushima multiple explosions and triple meltdown). Anyone who lived near Three Mile Island (TMI) in southern Pennsylvania, Chernobyl in northern Ukraine, or Fukushima Prefecture in the Tōhoku region of Honshu about 200 km north of Tokyo can tell you what they were doing the moment they learned of the news and were forced to evacuate.

To standardize the reporting of a so-called nuclear "incident" or "event" the International Atomic Energy Agency created the International Nuclear Event Scale (INES) in 1990, which records all radioactive releases ranked from 0 ("no safety significance") to 7 ("major accident"). In the first 70 years of civilian nuclear power, 20 facilities have reported at least a "serious incident" (INES-3) with another 15 "incidents" (INES-2) or "anomalies" (INES-1). To date, the two most serious events are both labeled 7s – "Major Release: Widespread health and environmental effects," the first at Chernobyl, Ukraine, in 1986 (Reactor No. 4 fuel meltdown and fire) and second at Fukushima, Japan, in 2011 (Daiichi 1, 2, 3 fuel damage and radiation release), both precipitating mass evacuations of the local population.

The world's first "serious nuclear accident" (INES-6) occurred in 1957 in Mayak, Russia, because of unexpected criticality in a reprocessing plant ("Significant Release: Full implementation of local emergency plans"). There have also been four 5s ("Limited Release: Partial implementation of local emergency plans or severe damage to reactor core or to radiological barriers"): at Chalk River, Canada, in 1952 (core damage); Windscale, UK, in 1957 (fire); Three Mile Island, USA, in 1979 (fuel melting); and Goiânia, Brazil, in 1987 (hospital theft). Given the nuclear industry's track record one should expect more.

It's not easy to grade every event on a simple scale. Richard Wakeford, editor of the *Journal of Radiological Protection*, notes that Mayak should be a 7 and Windscale a 6 if the latest estimates of polonium-210 release are included.[132] Nonetheless, INES helps put the damage into perspective. The causes of the most serious events include fuel meltdown and fire, fuel damage, radiation release and evacuation, reprocessing plant criticality (onsite nuclear fission), graphite overheating, criticality in fuel plant for an experimental reactor, fuel pond overheating, cooling interruption, turbine fire, and severe corrosion.[133] Table 3.5 lists the location and date of an event labeled at least a serious incident at a nuclear facility, not counting unreported incidents in Russia and elsewhere (note some suffered multiple events).

Table 3.5 *International Nuclear Event Scale (INES) of at least a "serious incident"*

Level	Event category	Location (year)
7	Major accident	Chernobyl (1986), Fukushima (2011)
6	Serious accident	Mayak (1957)
5	Accident with wider consequences	Chalk River (1952), Windscale (1957), Three Mile Island (1979), Goiânia (1987)
4	Accident with local consequences	Sellafield (5 from 1955 to 1979), Idaho Falls (1961), Saint-Laurent (1969, 1980), Lucens (1969), Jaslovské Bohunice (1977), Andreev Bay (1982), Buenos Aires (1983), Tōkai (1999), Mayapuri (2010)
3	Serious incident	Vandellòs (1989), Oak Harbor (2002), Paks (2003), Sellafield (2005)

We aren't always privy either to all the facts regarding radioactive release, alarmingly so after a nuclear accident. At the Mayak nuclear complex in the Russian province of Chelyabinsk, in the closed town of Ozersk, scene of the world's first significant accident, a spent fuel reprocessing plant went critical, producing a large release of radioactivity and full implementation of local emergency plans, yet remained unknown for decades in the West because of Cold War secrecy. We know now that the cooling failed in a storage tank at the joint military–civilian nuclear installation where Soviet weapons-grade plutonium was produced, resulting in a chemical explosion and a 74×10^{15} Bq (74 PBq) release of radionuclides over a 300 km × 50 km area of the Southern Urals. Now called the East Urals Radioactive Trace (EURT), access is restricted.

Known as the "Kyshtym Accident" because there was no official town in the area and labeled "Chelyabinsk-65" for administrative reasons, 5,000 workers received more than 1 Sv in the first few hours, more than 250,000 area residents received doses above the recommended annual limit, and almost 11,000 people were evacuated[134] (recall that 1 Sv equals a 5.5% increased chance of cancer). Although suspicions filtered into the West of a nuclear accident behind the Iron Curtain, details weren't officially reported until 32 years later in 1989.

After the accident, radioactive material poured into the Urals, the Techa River, and Lake Karachay. Considered the most radioactive body of water on Earth, Lake Karachay had been used as a nuclear waste dump since the start of the Soviet-era weapons program, but was eventually filled in with concrete blocks and completely covered in rock and soil in 2015, becoming Russia's de-facto permanent nuclear waste storage facility. No need for decades of debate on aboveground or belowground waste sites in autocratic Russia or which unlucky town gets to host the stuff. Although the Russian state nuclear company Rosatom announced that radiation levels had decreased around Karachay, a 1,000-times spike in ruthenium-106 levels was also detected in 2017 in central Europe, believed to have come from the Mayak nuclear complex. Russia claimed the increased levels were from the batteries of a burnt satellite.

After the Mayak accident (as we now know), another significant accident occurred only a few weeks later on October 10, 1957, when a graphite-moderated reactor core caught fire at the Windscale plant on the Cumbrian coast in the UK, 100 km northwest of Liverpool. The now-graded INES-5 event was denied at first by the British government, who were keen to sign a nuclear weapons technology sharing agreement with the USA.

Operational since 1952, the two-pile plant was built to produce plutonium and tritium for the UK's post-war, nuclear bomb program and was partly designed by John Cockcroft, Britain's original atom smasher. No. 1 Pile was graphite-moderated and air-cooled, the air expelled up a tall chimney and filtered by "Cockcroft's Folly," retroactively added to stop the release of radioactive debris in the event of a fire. Unfortunately, in a rush to join the USA in the exclusive H-bomb club before a presumed worldwide nuclear weapons test ban, production of plutonium and tritium at Windscale was increased beyond safe operating levels. As a result, an isotope cartridge caught fire in one of the reactor channels, which spread through the graphite-moderated reactor after the blaze was fanned by *increasing* the air coolant to the core. Radioactivity poured out over the Cumbrian countryside and the Irish Sea for three days, contaminating 200 square miles of farmland, the fire eventually extinguished by shutting off the air "coolant." Thanks to Cockcroft's ad-hoc safety fix, Cumbria is not now a no-go nuclear exclusion area like Fukushima, Chernobyl, and EURT, although the Irish Sea is the world's most radioactivity-filled waterway after Fukushima.

Designed to produce utility electricity and transition to peaceful nuclear power, the world's first civilian nuclear power plant, Calder Hall, was built beside Windscale, opened in 1956 by Queen Elizabeth, and where nuclear weapons material continues to be manufactured. Renamed Sellafield in 1981, the Windscale and Calder Hall site still produces plutonium for the UK nuclear weapons program, but also reprocesses spent fuel from 15 British nuclear reactors as well as from other countries. In 2005, more than 26 kg of plutonium went missing (enough for seven nuclear bombs), the UK Atomic Energy Authority claiming an auditing error in the accounts. As we saw earlier, no permanent storage site has yet been agreed to handle Britain's long-term waste that includes over 50 kg of plutonium, the cleanup costs alone estimated at around £80 billion and counting while the NIMBY haggling continues. The next most significant nuclear accident was at TMI in Pennsylvania, located on the Susquehanna River about 10 miles southeast of the state capital Harrisburg. TMI was the first accident to focus the American public on the dangers of nuclear power, caused by a combination of design flaws and human errors. At 4:37 a.m. on March 28, 1979, the cooling water system pump at the 900-MW, TMI-2 PWR failed, and because the water supply for the three backup pumps had been shut off during a routine maintenance operation 2 weeks earlier and not turned back on the reactor heat couldn't be removed. The temperature and pressure in Reactor 2 immediately began to rise, which triggered an automatic shutdown where all the control rods are inserted into the reactor, called "scram" for "safety control rod actuator mechanism," or as some have conjectured, "get the hell out."

Without any core coolant to remove the heat, the pressure exceeded 2,250 psi, opening a pressure-relief valve that then failed to close, lowering the pressure below its design load of 1,600 psi, which automatically triggered another backup measure, the emergency core coolant system or ECCS (a high-pressured spray injection). Thinking that the original backup coolant was working, however, as shown on an incorrect water-level gauge, the TMI-2 operators turned off the ECCS to avoid flooding. Nothing was cooling the reactor and so the core overheated, the zirconium-clad fuel rods ruptured, and the uranium pellets melted, indicating a temperature of at least 5,100°F. Melting almost half the core, a complete core meltdown was averted "only by a last-minute rush of cooling water."[135] Alas, during the cooling rush the overheated zirconium cracked and reacted with water and steam, creating a dangerous "hydrogen bubble" that fortunately didn't explode, although radioactive water flooded everywhere – 400,000 gallons through the stuck relief valve – and radioactive gas vented into the air.

Essentially, four design flaws and two human errors (one from mistaken information) accounted for the worst nuclear accident in American history, completely destroying the almost-new TMI-2 reactor and causing a minor panic as roughly 150,000 people voluntarily evacuated after the forced evacuation of pregnant women and children in the area. Although the radiation release was considered minimal, the damage could have been much worse.

The analysis of the TMI event showed that some safety systems worked as designed while others did not, the operators unfortunately acting on wrong information to tackle the emergency. Almost four decades after the Italian navigator had landed in the new world, running a nuclear reactor was not as easy as turning a car right or left as Fermi first boasted. In release just 2 weeks before the TMI partial meltdown, the controversial movie *The China Syndrome* (starring Jack Lemmon as a nervous nuclear operator and Jane Fonda as the dogged TV news reporter in search of a story) heightened public worry about a potentially disastrous nuclear accident. Although "China Syndrome" incorrectly refers to the direction and distance melted reactor fuel will flow – China is not geographically opposite the USA, while melted core material will eventually solidify about 6 feet below ground – liquefied uranium in a reactor spells serious trouble.

***

Underplaying the severity of previous events and the chances of more, a well-financed nuclear industry wants the public to think everything is safe, green, and economical. But nuclear-generated electricity is not as foolproof as the PR claims, as we know now from the two most dangerous nuclear events yet, each labeled a 7 on INES. Both Chernobyl and Fukushima continue to spread radioactivity across the world and remain no-go exclusion zones to hundreds of thousands of former residents in ghost towns permanently cut-off from human habitation. Call it Nuclear Murphy: when things go wrong, they go very wrong.

   Chernobyl is still the world's worst nuclear accident to date, occurring in the former Soviet-controlled republic of Ukraine, located on the Pripyat River about 100 km north of the capital Kyiv and 5 km south of the Belarus border.[136] Just after midnight on April 26, 1986, during what was intended to be a simple safety test, Chernobyl Reactor No. 4 – an 800-MW, graphite-moderated, light-water-cooled RBMK reactor – was destroyed by a steam explosion that blew the lid off the inner biological shield and then another more powerful hydrogen explosion a few seconds later that breeched the outer containment. The hydrogen blast, which to those in the control room felt like an earthquake, threw up parts of the fuel and blocks of graphite moderator across the whole of the plant and onto the roof of the neighboring Reactor 3. The resulting graphite blaze killed 31 firefighters in the first few days, followed by thousands more in the radioactive aftermath.

   Prior to the breakup of the Soviet Union, Cold War industrial competition was fierce and nuclear methods not readily shared, including test data and LOCA simulations. Today, we know that the world's worst-ever nuclear accident was caused by a series of horrendous errors during an ill-advised turbine test to determine the lowest power level at which the reactor could restart after a shutdown, ostensibly to measure the safety systems in the event of a *real* sudden loss of power! Such a test – to see if a spinning down turbine could provide electricity to the emergency backup systems (for example, the ECCS) in the event of a sudden loss of external power – would have been deemed far too dangerous to try in the West. Step after step at Chernobyl was either a stupid mistake or a violation of legal operating procedures, even in the USSR where pushy party bosses routinely stressed production over safety. One has to wonder what was going on and who was in charge.

After the test started going haywire, all the control rods were inexplicably removed from the core, causing a massive upsurge in power at almost 500 times the design rating (384 GW/0.8 GW), instantly vaporizing the twice-normal volume of coolant water (called flashing) and disintegrating the zirconium-clad fuel rods, which then created a hydrogen bubble from oxidized water (as at TMI when water reacts with overheated zirconium). That, in addition to the pure carbon graphite reacting with steam to make more hydrogen and carbon dioxide gas, literally blew the 2,200-ton top off the pressure vessel, venting the reactor contents to the atmosphere and ejecting flaming reactor parts all over the reactor complex.[137] The hydrogen explosion shot radioactive material 2 km high as clouds of radioactive gases and debris expanded outwards up to 9 km away, while 100% of the Xe-133, 65% of the I-131, and 40% of the Cs-137 inventory within the core were released into the atmosphere.[138,139]

Fueled by 1,700 tons of burning graphite, the resulting chemical fire lasted 8 days, releasing more volatile radionuclides across much of Europe and the northern hemisphere. Worries about the state of the reactor core continued after the initial explosions as workers in hovering helicopters dropped 30 tons of boron- and lead-laden sand and clay over the broken reactor building, valiantly trying to squelch the still fissioning uranium fuel. Some wondered whether another blast would "release radioactive clouds large enough to make a good part of Europe uninhabitable."[140] If the overheating core had melted through the reactor's foundations, another real possibility was the radioactive poisoning of the local water table, the nearby Dnieper River (Ukraine's major waterway), and even the world's oceans. Upon entering the water supply, the melted fuel would have also created radioactive steam that would have been impossible to contain and been released to the atmosphere.

Three days after the disaster, elevated radiation levels were detected over 1,000 km away at the Forsmark nuclear power plant outside Stockholm, triggering worries about a local radiation release before attention turned to the Soviet Union from where the winds were blowing. High levels of radioactive fallout were eventually measured in the USA, North Africa, and even in Hiroshima over 8,000 km away. Although Ukraine and the neighboring countries of Belarus and Russia were most affected, more than half of Chernobyl's volatile inventory landed elsewhere. Adding to the severity of the radioactive release, the RBMK design had no outer containment building, but such a structure would likely have been breached anyway given the severity of the explosion, although the overall release would have been greatly reduced and perhaps more easily staunched.

In the immediate aftermath, iodine-131 was the main concern ($t_{1/2}$ = 8 days) – readily absorbed by leafy vegetables and grass eaten by farm animals – followed by cesium-134 ($t_{1/2}$ = 2 years) and cesium-137 ($t_{1/2}$ = 30 years). Two-fifths of the surface area of Europe was contaminated above 4,000 Bq/m^2 and 2.3% above 40,000 Bq/m^2, almost 10 times the level in Chernobyl (the EU limit of cesium-137 in dairy foods is 600 Bq/kg).[141] According to the World Health Organization, the total radioactivity emitted was an astounding 200 times more than the atomic bomb blasts on Hiroshima and Nagasaki combined.[142] Recent numbers put the release at 50 million curies "the equivalent of 500 Hiroshima bombs."[143]

After the initial accident, it took 36 hours to start evacuations from the nearby city of Pripyat, just northwest of the reactor complex (one of nine "atomgrads" or atomic cities built across the Soviet Union to house nuclear workers and their families). All 50,000 residents were eventually evacuated within 2 days in a series of grim bus runs, while every building in nearby Kyiv – the Ukrainian capital 100 km to the south – was hosed down for at least a month after the accident. Roughly 2,500 km^2 around the reactor remains contaminated (an area the size of Luxembourg). The former closed nuclear workers' city of Pripyat is now a ghost town as is the city of Chernobyl 15 km southeast of the reactor complex, both lying completely within the 30-km exclusion zone, officially called the Chernobyl Nuclear Power Plant Zone of Alienation from where 350,000 people in both Ukraine and Belarus have been displaced, never to return.[144]

Thirty years later, in an attempt to stem the dispersal of radiation from the still highly radioactive broken core, a massive, $1.8 billion, steel arch "sarcophagus" – the largest movable metal structure in the world – was inched into place atop the wrecked reactor and hastily assembled and crumbling, original, steel-and-concrete sarcophagus. In *Chernobyl: History of a Tragedy*, the Ukrainian-born Harvard University history professor Serhii Plokhy called the high-tech shelter, "a warning to societies that put military or economic object-ives above environmental and health concerns."[145] Despite reducing the still active radiation by 90%, the surrounding area will be uninhabitable for at least another 20,000 years. Initially estimated at $10 billion, financial losses for the accident have now reached more than $200 billion. More than two decades after the accident, the neighboring country of Belarus directly to the north was still spending $1 million per day to deal with the consequences.

The number of deaths related to Chernobyl is widely disputed as one might expect, anywhere from 4,000 to 90,000 in the general literature, while a 40% increase of cancer was observed in Belarus up to 20 years later.[146] Russian biologist Alexey Yablokov estimates that one million people may have died, stating there is no reason to neglect "the consequences of radioactive contamination in other countries, which received more than 50% of the Chernobyl radionuclides."[147] According to numbers released by the Ukraine government after the breakup of the Soviet Union, 3.3 million Ukrainians were categorized as "sufferers," while 38,000 km^2 or 5% of the country was contaminated; in Belarus, more than 44,000 km^2 was "severely contaminated, accounting for 23% of the republic's territory and 19% of its population."[148]

On the thirtieth anniversary of the accident, Kim Hjelmgaard documented the plight of those who still suffer from the ongoing effects in a series of *USA Today* articles. He recounts the stories of exiles who worried about the effects on their children and grandchildren, the difficult times for those who live in the economically depressed areas of neighboring Belarus (often overlooked in the disaster), and the condition of the 800,000 so-called "liquidators" and "biorobots" between the ages of 18 and 22 – brave Russians who worked to contain the initial damage – 160,000 of whom may have died for their bravery before reaching their fortieth birthdays. A grim reality of life in the aftermath of nuclear disaster was still very much evident 30 years on, including almost half a million children born after the accident, who have "respiratory, digestive, musculoskeletal, eye diseases, blood diseases, cancer, congenital malformations, genetic abnormalities, trauma." Chillingly, as Hjelmgaard noted, "There are 2,397,863 people registered with Ukraine's health ministry to receive ongoing Chernobyl-related health care. Of these, 453,391 are children."[149]

Today, feral animals roam freely in the exclusion zone surrounding Chernobyl, while a few brave souls have unofficially returned to their abandoned homes to live out the end of their lives, known as "Samosely" (self-settlers). Officially, no one can be let back in. A "red forest" grows unimpeded by human activity throughout "the Zone," the hottest on earth. Continuing radiation damage is not the only worry, which now includes the threat of radioactive forest fires spewing out more radiation across the open skies and the spread of genetically mutated migrating animal species across Europe. During the 2022 invasion of Ukraine, Russian soldiers were also exposed to unknown amounts of radiation after digging trenches in the evacuation zone without any protective gear.

Although the Chernobyl reactor site is a complete write off and nearby Pripyat a permanent ghost town, one can take a guided tour within the Zone arranged as a day trip from Kyiv to Chernobyl and back. On the site of the world's worst nuclear disaster, you can poke your nose around various deserted sites – such as the Reactor No. 4 control room, the Pripyat hospital, and the iconic Ferris wheel in the eerily abandoned local amusement park – learn about the science of the accident, and see the still unchanged Soviet-era administrative setup frozen in time. Access to some areas is restricted, especially near the reactor, while one must wear a gamma-ray dosimeter to monitor radiation levels inside the exclusion zone. Eating and drinking in the open air, touching things, gathering plants, sitting on the ground, or setting personal items down are all strictly forbidden in this uniquely modern theme park.

Serhii Plokhy described his experiences on one such tour that prompted him to write *Chernobyl: History of a Tragedy*, where he compares Chernobyl to a modern-day Pompeii, noting, "It was not the heat or magma of a volcano that claimed and stopped life there, but invisible particles of radiation, which drove out the inhabitants but spared most of the vegetation."[150] He also cites the April 26, 1986, accident at Chernobyl as the beginning of the end of the Soviet Union, a "technological disaster that brought down not only the Soviet nuclear industry but the Soviet system as a whole."[151] *Glasnost* (openness) was a direct result of a concerned public wanting to know more, followed by *perestroika* (restructuring) and the fall of the Berlin Wall three and a half years later. One shouldn't have to suffer a nuclear disaster to see the end of authoritarianism. Nor should anyone be permanently excluded from their home (Figure 3.6).

3.8 Fukushima: The Last Nuclear Disaster or Portent of More?

Although the chilling effects of Chernobyl are still with us more than three decades on and much more is now known about what happened, the horrifying aftermath of a triple meltdown on the east coast of Japan in early 2011 is still unknown and potentially more dangerous. Located in the Fukushima Prefecture about 200 km north of Tokyo and 100 km south of Sendai, the Daiichi Nuclear Power Plant comprised six General Electric Mark I boiling-water reactors with a total electrical generating power of 4.7 GW. Units 1, 2, and 3 were all in operation on March 11, 2011, while units 4, 5, and 6 had been shut down for maintenance when a 9.0-Richter earthquake struck in the Pacific Ocean, centered 80 km east of Sendai, the largest in Japanese history.

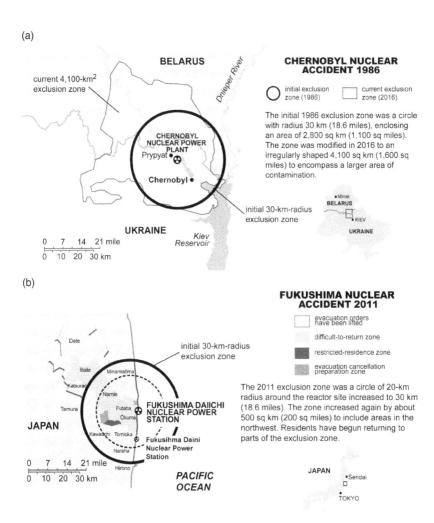

Figure 3.6 Nuclear accident exclusion zones: (a) Chernobyl and (b) Fukushima.

Reactors 1, 2, and 3 immediately "scrammed" as designed after the earthquake, the back-up diesel generators kicking in to continue powering the
cooling pumps to keep the cores from overheating. Back-up power also maintained the cooling to the spent fuel pool in Unit 4 (located above the reactor),
where the entire core inventory of 548 fuel rods had been temporarily stored.
The Daiichi plant was completely swamped, however, when a devastating 10–
15-m high tsunami hit Japan's east coast within the hour. Seawater flooded

everywhere, knocking out all of the back-up diesel generators that powered the essential cooling to the three active cores and four spent fuel pools in Units 1–4. As at TMI and Chernobyl, when cooling is lost the fuel starts melting, generating hydrogen gas as the zirconium-clad fuel rods overheat and react with steam, risking an explosion and a radioactive release. Twenty-four hours after the tsunami, Reactor 1 blew as hundreds of workers valiantly fought to pour water over the core to avoid a full meltdown. Two days later, Reactor 3 blew, followed another day later by Reactor 2. Reactor 2 was especially dangerous as it had just started burning plutonium fuel.

All three cores in Reactors 1, 2, and 3 eventually completely melted, producing lava-like fuel-containing material (a.k.a. corium), a hot, radioactive soup of melted fuel rods and cladding. Without any cooling to remove the still simmering heat, the spent fuel in Unit 4 also caught fire, releasing 0.4 Sv/h into the atmosphere. Units 5 and 6, which had been built further inland on higher ground, both survived mostly unscathed.

The fuel meltdowns in Daiichi 1, 2, and 3 and the spent fuel pool fire in Daiichi 4 is now considered one of the worst nuclear accidents ever, collateral damage of a Pacific Ocean tsunami that pounded the east coast of Japan for over 10 minutes and saw almost 20,000 people die amid massive flooding. Although the three reactors successfully shut down as designed after an earthquake, without any back-up power in the aftermath of the resulting tsunami, the fuel overheated and all three reactor cores melted, sending plumes of radioactive material into the surrounding countryside and radioactivity-laced water into the Pacific Ocean.

The overall amount of fission-product inventory released to the atmosphere was estimated at about 10% that of Chernobyl, which had no outer containment building, releasing almost 8 tons (5%) of its core inventory primarily during the initial explosions. According to the official IAEA report, the upper estimate of radioactive releases from Fukushima were 25% of the iodine-131 and 45% of the cesium-137 compared to Chernobyl, with minimal release of the medium volatility isotopes Sr^{90}, Ru^{103}, and Ba^{140}, although radioactive noble gases were about twice as high.[152] Proximity to the ocean, however, clearly made the ongoing dispersal of radioactive material into the groundwater and marine environment much worse, especially because the location of the corium was unknown, a nightmare scenario that could devastatingly contaminate both local and international waters.

Within 2 weeks, 160,000 residents were forced to evacuate from a 20-km radius exclusion zone, soon after expanded to 30 km, while as many as 300,000 were evacuated from Fukushima Prefecture, 200,000 of whom were not able to return to their homes more than 5 years later. A decade after the accident, two-thirds of evacuees had no plan to come back. In nearby Namie, where 80% of the town is still restricted, only 1,500 of 21,000 former residents have returned.[153] At the time of the accident, the evacuation of Tokyo with a population of 37 million people had been considered a real possibility.

If Fukushima tells us anything, it is that we cannot design against a loss-of-coolant accident, even for a reactor built on the shores of an ocean (although saltwater is too corrosive to use in regular operation). Even worse, the reactor buildings were inexplicably built belowground to protect against an earthquake that coupled with an insufficiently high sea barrier made for a grossly inadequate tsunami defense. What's more, without an independent back-up power supply located on higher ground, as in international best practice, the Daiichi plant was helpless to cool the enormous reaction heat in its reactors and spent fuel pools after the onsite power failed. Unfortunately, the reality of nuclear power is that we can't design against all eventualities – some only obvious after the fact – yet essential given the potential for a major radioactive release and long-term evacuation of those unlucky enough to live "up-weather" within an indeterminate range of blowing winds and air currents.

Mistakes happen with any technology, such as inserting a control rod upside down, forgetting a welding rig inside a pressure vessel, or from simple everyday corrosion, rust, and leaks, never mind building a nuclear power plant below sea level on the coast in an active tectonic region known to be susceptible to tsunamis. Giant waves are an integral part of Japanese culture, best exemplified by Katsushika Hokusai's famous woodblock print "The Great Wave off Kanagawa" that gracefully shows the power of the sea engulfing three boats full of fishermen and the country's tallest peak Mount Fuji. One must wonder about basic planning principles that overlooked the inevitable. The resultant clean up in and around Fukushima could cost more than $500 billion and take over four decades to complete.

As Tokyo Electric Power Company (TEPCO), owner/operator of the Fukushima Daiichi plant, began to assess the damage to the reactors and melted fuel, Arnie Gundersen, a former nuclear engineer turned whistleblower, listed the most serious problems: (1) the below sea-level reactor basements, where leaking radioactive material continues to mix with groundwater flowing directly into the ocean; (2) leakage from an aboveground onsite "tank farm," where radioactive water was being stored during the ongoing clean up; and (3) the potential for much wider devastation in the event of another earthquake. The radioactive water storage tank farm would eventually exceed 1.3 million tons in 1,000 tanks, while up to 400 tons per day of contaminated groundwater had been leaking into the ocean before an "ice wall" built by TEPCO reduced the flow by half. Gundersen also recommended a zeolyte wall to keep the groundwater from entering the plant and to absorb radioactive cesium.

A decade after the accident, TEPCO announced plans to empty the overflowing store of radioactive water into the Pacific Ocean starting in 2023, because of lack of onsite space, instead of building more tanks outside the plant perimeter, angering both local fishermen and neighboring countries. TEPCO claimed radioactive isotopes such as Sr-90 and Cs-137 were removed from all "treated" water, despite a 2018 analysis showing a 70% contamination level that cast doubt on the company's honesty.[154] The treatment process also leaves behind a radioactive slurry particularly high in strontium that quickly degrades the containment liners and must be regularly replaced. As of 2022, TEPCO had yet to announce any "acceptable plans for dealing with the necessary transfer of slurry from weakening, almost deteriorated containers, into fresh, new containers."[155] Highly carcinogenic tritium also remains, worrying fisherman and farmers alike, who continue to suffer from loss of catch and produce. With a half-life of 12.5 years, radioactive tritium will last another century and "is a carcinogen (causes cancer), a mutagen (causes genetic mutation), and a teratogen (causes malformation of an embryo)."[156] Carbon-14 and cobalt-60 also pose risks.

Allowing the water to evaporate was also proposed, but dilution via dumping was deemed the cheapest option. One wonders why TEPCO doesn't just add more reinforced tanks, either onsite or nearby, instead of risking more damage by dumping radioactive water into the ocean or at least wait until levels decrease. What's the hurry to undo what essentially can't be undone? After much criticism from Pacific Rim countries, Japan finally announced in early 2023 that it would delay dumping radioactive water until more support was garnered. Later in the year, the IAEA released a report minimizing the dangers over neighboring countries' objections and dumping finally began.

To this day, TEPCO still doesn't know the condition of the fuel in the three damaged reactor buildings or exactly where the fuel ended up after melting into the foundations, far too hot to attempt to find even with volunteers wearing radiation suits. In 2017, a robot equipped with a camera and radiation gauges – called Scorpion because of its folding tail of instruments – was unable to venture very far into Unit 2, considered the least radioactive of the destroyed reactors. Scorpion registered 250 Sv/h before dying, likely because of a fried circuit board from the high levels of radiation, while an earlier probe fitted with a remote-controlled camera recorded levels up to 650 Sv/h at about the same spot, "enough to kill a human within a minute."[157]

In early 2022, a ROV-A robot was finally able to send back pictures from inside Unit 1, showing what looked like melted nuclear fuel at the bottom of the reactor submerged in cooling water. Roughly 900 tons of melted fuel remains inside Units 1, 2, and 3, and is expected to take at least 30 years to remove, possibly much more.[158] Containing radioactive damage from one reactor in Chernobyl was difficult enough; three is almost unimaginable.

The full biological impact remains unknown and is much debated. While thyroid cancers begin within 3 years, an increase in solid cancers is expected after a decade as the radiation accumulates over time in the body's organs. Blood cancers have already started being detected in children and in the clean-up crews, while 4 years afterwards "off the coast of Oregon and California every Bluefin tuna caught in the last year has tested positive for radioactive Cesium 137."[159] As much as 80% of the overall radiation may have been deposited (and is still being deposited) into the Pacific Ocean, which will severely affect marine life and the food chain for years to come as the radiation concentrates in algae, crustaceans, and on up to smaller and then larger fish.

If Chernobyl is a guide, the numbers will rise and be in dispute by those in favor of nuclear power versus those concerned about ongoing health problems. Alas, life is anything but normal for those still barred from their homes, while the sale and consumption of foods containing more than 100 Bq/kg of cesium-134 and cesium-137 is still restricted throughout Japan. To be sure, despite the high levels of radiation detected outside the Daiichi plant in numerous locations, the number of deaths is difficult to count. But you know levels are serious when highways around Fukushima are posted with LED signs showing radiation readings in μSv/h.

Berkeley physics professor Richard Muller noted that the number of initial "radiation deaths" (<100) was small compared to the "tsunami deaths" (>15,000) and is keen to point out the human consequences of forced evacuation,[160] although Ian Fairlie, the scientific secretary to the former UK Committee Examining Radiation Risks of Internal Emitters (CERRIE), believes the numbers will continue to rise. Using a 10% per sievert fatal risk of cancer, he estimates that 5,000 people in Japan will die from future cancers (72% in Fukushima Prefecture, the most contaminated part of Japan, and 28% in the rest of the country), in keeping with reports by both UNSCEAR (the UN Scientific Committee on the Effects of Atomic Radiation) and the IPPNW (International Physicians for the Prevention of Nuclear War).[161]

Pieter Franken, co-founder of the Tokyo-based NGO Safecast, mistrusts official numbers, literally taking a Geiger counter for a ride after the meltdown to measure street-by-street levels unrecorded by government sources: "On my first drive, the readings I was getting were significantly higher than those being reported on TV."[162] The numbers showed higher levels in some towns further from the plant, indicating that atmospheric carry is more complicated than simply devising concentric radiation zones outwards from the source. Franken also spearheaded a citizen-scientist movement that uses homemade Geiger counters to measure radioactivity for oneself, the street-level analysis deterring some locals from returning home, thinking that radiation levels were still too high.

One dairy farmer from the village of Iitate, about 30 km northwest of the Daiichi plant, began returning every month to take his own readings since he and almost 6,000 residents were forced to evacuate, finding the levels "consistently higher than those from the government monitoring posts, and are not falling anywhere near quick enough, despite the decontamination efforts." The mayor of Iitate, also a dairy farmer, believed the danger was exaggerated, but admitted they will be unable to grow food for "years to come," and hoped returning farmers will "switch to flowers and other crops not for human consumption."[163]

Although the government cleared away topsoil for long-term storage, collecting 3 million bags in a few months and 22 million cubic meters over a decade, only one-fifth of those displaced from the picturesque village of Iitate care to return. Consumer confidence in food harvested within Fukushima Prefecture also remains low, making a return to farm life extremely difficult for displaced farmers. A decade after the accident, 15 of 54 countries that had banned Japanese food imports maintained those restrictions.

Decommissioning the Daiichi reactors will also take decades, the main challenges being the removal of melted fuel and debris from inside each of the three damaged reactors as well as the spent fuel pools. By 2015, the still hot fuel from Unit 4 had been successfully emptied, removed to the nearby Daini nuclear plant, south of Daiichi, which suffered much less damage in the "3/11" tsunami. Retrieving all the melted fuel from Units 1, 2, and 3 was expected to last until 2024, but original industry estimates were hugely optimistic. A revised time frame for removal is now set to take at least until 2050.

A Greenpeace nuclear specialist in Japan described the challenges as "unprecedented and almost beyond comprehension," noting that TEPCO's decommissioning schedule was "never realistic or credible."[164] Dr. Helen Caldicott, the Australian pediatrician and author of the 1982 Academy Award winning documentary *If You Love this Planet* about the dangers of nuclear power, noted that because of the accumulation of long-lived radioactive isotopes in the food chain and the body, a nuclear accident never ends, stating that Fukushima was "by orders of magnitude many times worse than Chernobyl."[165]

The damage was not restricted to just Japanese citizens. In 2017, the United States appeals court ruled that 318 American sailors could sue the Japanese government and TEPCO for illnesses caused by radioactive exposure. Mostly stationed aboard the USS *Ronald Reagan* aircraft carrier, dispatched to Japan's northeast coast to help with the relief efforts at Daiichi, the sailors complained of numerous ailments, "ranging from leukaemia to ulcers, brain cancer, brain tumours, testicular cancer, thyroid illnesses and stomach complaints,"[166] while some of their children were born with birth defects. Former North Carolina senator John Edwards represented the sailors against TEPCO and GE, the supplier of the four boiling-water reactors that recommended the low-lying coastal location rather than a safer, more inland site on higher ground. Although the case was rejected in 2020, expect more litigation to continue.

Rebecca Johnson, former senior advisor to the Weapons of Mass Destruction Commission chaired by International Atomic Energy Agency head Hans Blix, stated that Fukushima "was both avoidable and inevitable," noting that "the natural disaster of an unusually large tsunami was turned into a nuclear catastrophe by the systemic failings of the nuclear plant's operators, the Tokyo Electric Power Company (TEPCO), inadequate technical and emergency preparations and back-ups, bureaucratic complacency and weak regulations."[167] A Japanese parliamentary report into the failings at Fukushima found that a culture of complacency led to the "man-made disaster," citing an overly familiar relationship between Japan's nuclear industry and the government regulatory bodies.[168]

In the aftermath of the accident, some questioned the wisdom of TEPCO running the clean-up operation, preferring an international body given the obvious self-interest, while others claimed information was withheld because of preparations for the 2020 Tokyo Summer Olympics, egged on by the then president Shinzō Abe, and that resource-poor Japan wanted to restart its nuclear plants as soon as possible. Openness and transparency doesn't come easy in the nuclear world. Certainly, a better firewall is needed between industry and government.

In 2012, The Carnegie Endowment for International Peace published a report entitled "Why Fukushima was Preventable," condemning the lack of independence of the Japanese nuclear regulator NISA and the nuclear industry as well as the governing body that "deterred NISA from asserting its authority to make rules, order safety improvements, and enforce its decisions." In Japan, the practice of *amakudari* ("descent from heaven") appoints utility executives as regulators, while *amaagari* ("ascent to heaven") employs industry experts in government agencies, a cozy arrangement that promotes a lax safety culture free from open criticism and curtails lateral thinking, a.k.a. "the revolving door" in the West.

The Carnegie report added that if TEPCO and NISA had followed international best practice with regard to safety "they might have realized that the tsunami threat to Fukushima Daiichi had been underestimated."[169] Placing back-up generators below a reactor and storing spent fuel above a reactor were both major design flaws, making it difficult to supply water in the event of a LOCA. Furthermore, the approach to safety in Japan was undervalued, unlike in Europe where safety is "enhanced" by ordering necessary modifications as needed, while in the USA safety is "maintained" through a practice of "backfitting," where costs and benefits are weighed against safety margins, a system that nonetheless "discourages safety upgrades requiring expensive engineering changes."[170] As hard as it is to anticipate a disaster, with nuclear power one must always stay ahead of the curve. Alas, one can be certain the insufficiently conservative Daiichi "design-basis" was improperly vetted given the frequency of tsunamis in the region.

Some attempts at restitution have at least begun over a decade since the tragic events of March 11, 2011. In 2022, four TEPCO executives including the then president were ordered by a Japanese court to pay 13 trillion yen to 48 TEPCO shareholders, the first case to find company liability for damages from the accident. Not surprisingly, the court found that TEPCO's countermeasures against a tsunami "fundamentally lacked safety awareness and a sense of responsibility," likely because of damning evidence that TEPCO had predicted the plant was susceptible to tsunami waves up to 15.7 m after a major offshore earthquake.[171] It is unlikely, however, that the former executives will be able to pay the full $94 billion in damages awarded. They were also acquitted in 2023 in a separate trial of professional negligence resulting in death.

Richard Broinowski, author of *Fallout from Fukushima*, is concerned that the "industry has been allowed to develop as far as it has without as many checks and balances as it should have,"[172] citing administrative secrecy from the top, as in a mafia or nuclear priesthood. In the USA, the independence of the Nuclear Regulatory Commission has been routinely called into question, while in Canada the Canadian Nuclear Safety Commission is primarily funded by the nuclear industry, which can easily create conflicts of interest with regard to licensing and plant renewals. Foxes guarding the henhouse, lunatics running the asylum, voluntary self-reporting in the world's most dangerous industry – what could go wrong?

***

After a nuclear accident there is always a backlash and rethinking of energy policy. Since Three Mile Island in 1979, no new nuclear power plants have been built in the USA. What's more, 100 orders were cancelled after the TMI partial core meltdown, while a completely finished $6 billion Long Island plant was mothballed and sold for $1.[173] After the Fukushima disaster, plans for a hoped-for US nuclear renaissance were dashed, although two new plants are under construction in Georgia despite numerous delays and massive cost overruns (as we'll see next).

In the immediate aftermath of Fukushima, Japan turned off almost all of its 54 reactors, representing 30% of Japanese grid power, relying instead on natural gas to make up the shortfall.[174] Some reactors have since restarted, although ambitious expansion plans have so far been shelved. The Japanese have lived with the unseen hand of nuclear fallout far longer than anyone and must now worry even more.

Germany began exiting the nuclear business immediately after Fukushima, unwilling to risk a catastrophic accident on German soil or, as some pundits claimed, to bolster the CDU party of Chancellor Angela Merkel, which needed Green votes to rescue her ailing coalition. Within days of the three-core meltdown, Merkel – a physical chemist by training and former minister for nuclear safety – ordered all 17 German nuclear plants to be closed, eight reactors immediately and nine more in phases by 2022.

The move increased Germany's reliance on coal in the short term – easily imported from neighboring Poland – and Russian natural gas, while accelerating their pledge to ramp up an already ambitious renewables sector. Siemens, the manufacturer of Germany's 17 reactors, announced it would exit the nuclear business altogether. After the decision, the Swedish power provider Vattenfall sued the German government for €4.6 billion for "expropriation," eventually losing the dispute not because of Fukushima angst, but from unresolved plant problems, ultimately receiving €340 million in compensation. Having also pledged to end coal by 2038, Germany is the first modern industrial nation to quit both nuclear and coal to stake its energy future on renewables.

In late 2021, Belgium stated it too would exit the nuclear biz, announcing the shutdown of all seven of its reactors by 2025, a decision postponed almost immediately for 10 years after Russia invaded Ukraine in February 2022. In the wake of the ongoing war, however, Germany temporarily restarted a shuttered coal plant, underscoring the danger of relying on imported energy and the need for more home-grown renewables. The nuclear exit, a.k.a. *Atomaussteig*, nonetheless remained intact, the coalition-party Greens refusing to change its long-standing, anti-nuclear policy. Although two of the last three reactors slated for closure at the end of 2022 (Isar 2 and Neckarwestheim 2 in the south) were kept in reserve to be connected if needed over the winter, all three (including Emsland in the north) closed on April 15, 2023, ending Germany's 60-year nuclear industry, the loss of power covered by other sources.

Some advocates also claim that closing reactors after an accident can cause more harm than the initial damage, for example, by increasing coal emissions, but in the absence of ironclad safety assurances governments must first act to guard against more accidents. Many of the world's nuclear reactors are built near major cities. The oldest French nuclear plant is built near a seismic fault line on the west bank of the Rhine on the French-German border and was ordered shut after Fukushima, but was still operating a decade later. One wonders how long we can keep playing nuclear roulette.

Despite promising to cut back its nuclear-generated electrical power by about a third, from around 75% to 50% by 2025, the French have instead decided to explore more nuclear power via a proposed shift to small modular reactors. As the world leader in home-grown nuclear power generation, France continues to stake its energy future on nuclear technology in a quest for continued energy independence. In the wake of the OPEC-fueled oil crisis in the 1970s, France became the world's leading nuclear nation by percentage, building most of its 56 reactors in the following decade under the 1974 Messmer plan, determined to hedge against rising oil prices rather than worry about a potential accident. Today, France still routinely exports nuclear-generated electricity to its neighbors, although plants were forced offline in the summer of 2022 as inlet water temperatures rose to unacceptably high levels during the increasingly hotter weather.

Armenia's Soviet-era Metsamor reactor may be the world's most dangerous, a 440-MW VVER with no containment building. Located on the Armenian–Turkish border, the town of Metsamor was created to service the 1976-commissioned VVER that together with a second 1980 unit supply 40% of the country's electrical power. Built in an active seismic area, the reactors were temporarily shut down after the 1988 Armenian earthquake that killed more than 25,000 people. Apparently, some inhabitants would rather live with the consequences of a potential accident than be without power, having suffered miserably without electricity through the "bone-chilling cold and dark days when the plant was closed down for several years."[175]

Calling the plant "a danger to the entire region," the European Union funded a number of fixes, while six similar plants in Bulgaria and Slovakia were closed as a condition of joining the EU. But despite the worry and an offer of €200 million to close the plant, the Armenian government has decided to squeeze out a few more years from Metsamor, the only VVER 440 in operation outside Russia, until a newer VVER 1000 is built as a replacement. One hopes the next version has sufficient earthquake defenses and a containment building.

We don't have to travel to the dilapidated former Soviet Union, however, to uncover potential danger. In early 2016, the aging 2-GW, two-unit, LWR Indian Point Energy Center, located 35 miles from downtown Manhattan on the east bank of the Hudson River, was temporarily shut down after tritium-laced water entered the river and was detected in the groundwater at 650 times normal levels. Later in the year, hundreds of faulty bolts were discovered while a weld leak prompted another unexpected shutdown, forcing then Governor Andrew Cuomo to announce "the aging and wearing away of important components at the facility are having a direct and unacceptable impact on safety, and is further proof that the plant is not a reliable generation resource."

A meltdown close to a large city would be unfathomable, perhaps an 8 on the INES scale. And yet, operations at Indian Point continued despite the problems and *expiration* of the original 40-year licenses for both units. Since 2000, the four nearest counties to Indian Point showed a 60% increase in thyroid cancer compared to the US average and underactive thyroid glands twice the national rate for babies.[176] Finally, after numerous leaks and spills, recertification for 20 more years was denied and the 58-year-old plant closed in 2021.

More worrisome may be the dry-cask, spent-fuel storage facility at the San Onofre Nuclear Generation Station, 50 miles north of San Diego, which sits about 100 feet from the Pacific Ocean. Closed in 2013 and in the process of decommissioning, 73, 20-foot long, half-inch-thick, stainless-steel canisters is all that keeps the waste intact. Each canister contains an amount of cesium-137 equal to the *entire* Chernobyl release, where a breach would be beyond disastrous, whether caused by an earthquake (the site is next to a fault line), high-tide flooding, or something unexpected. As one nuclear energy consultant noted, "The most toxic substance on Earth is separated from exposure to society by one-half inch of steel encased in a canister."[177] One must have complete trust and unrelenting faith in the nuclear business.

While Hiroshima, Nagasaki, and the 1957 plutonium bomb-making reactor releases in Mayak and Windscale/Sellafield belong to a different era than the civilian nuclear power plants of Chernobyl and Fukushima, accidents still happen with potentially catastrophic consequences. The permanent no-go areas surrounding these devastated sites, which have upended the lives of so many forced to flee in panic, should serve as more than ghostly reminders of past errors, but a recognition of the damage awaiting anyone unlucky enough to live near the next accident.

Nuclear waste facilities and power stations are even more vulnerable in times of strife as seen during the war in Ukraine after a fire broke out in 2022 from Russian shelling at the Zaporizhzhia plant. A sudden power failure would be devastating, leading to a loss of coolant to the reactor core and emergency shutdown systems, while a direct hit to any of the six reactors would be beyond catastrophic. At least after a mining accident or oil spill, most of us can return home. After a nuclear accident, there is no home to return to.

3.9   The State of the Art: Is There a Nuclear Future?

So what is the state of today's nuclear industry? In the West, not so great, while in China and India more nuclear reactors than ever are being built. Hoping to kick-start a dormant American industry after a 25-year hiatus, the US Congress even extended the 1957 Nuclear Industries Indemnity Act in 2005, limiting company liability in the event of a "nuclear incident." The reworked deal provided better incentives for new builds, $18 billion in loan guarantees, $2 billion indemnification against overruns, and $1 billion in tax breaks, while the Nuclear Regulatory Commission relaxed its lengthy licensing procedures to help with the construction hurdle. Later, the Obama administration added more regulatory relief, guaranteed federal loans, and increased tax incentives for new builds. But even with all the generous government backing, nuclear is not easy to get up and running.

If they ever go online, two plants under construction near Augusta, Georgia, will be the first new nuclear reactors to start up in the USA since the 1979 partial meltdown at Three Mile Island. Georgia Power is the majority owner of two 1.1-GW advanced pressurized reactors (AP1000) being built by Bechtel and based on a Westinghouse PWR design that includes so-called Gen-III+ passive safety features and modular construction. Given a license in 2012 after years of wrangling, the final cost is expected to run to over $27 billion or double the initial estimate.[178] In 2017, two similar AP1000 plants being built near Columbia, South Carolina, were axed after almost a decade of construction with $9 billion of a $25 billion estimated price tag *already* spent.

As a result of the mounting losses from numerous delays and high cost overruns in Georgia and South Carolina, which included a number of safety overhauls for the newer Gen-III designs, Westinghouse Electric Company filed for Chapter 11 bankruptcy in 2017, citing $10 billion in debts. Spun off from Westinghouse's Pittsburgh-based electric powerhouse in 1998 with a majority stake bought by Toshiba in 2006, the demise of one of the original flagship nuclear power generating companies underscores the problem of large-scale construction in today's nuclear industry, particularly in an era of increased public scrutiny and safety concerns. As for the other great American electric pioneer, GE's nuclear division merged with Hitachi in 2007, further highlighting the difficulty of adding power to the domestic US grid via nuclear energy. Compounding the ignominy, GE was removed from the Dow Jones Index in 2018, the last nineteenth-century company to be delisted.

Although the original players are no longer calling the shots in a modern, Asia-centric, nuclear world, the Gen-III Toshiba/Westinghouse advanced power reactor (APR) and the Hitachi/GE advanced boiling water reactor

(ABWR) still reflect the initial pressurized- and boiling-water designs made by the two pioneering rivals of the 1950s. GE is also trumpeting a streamlined BWR model – boldly called a "steam dryer" or "economic simplified boiling-water reactor" (ESBWR) – although the project has been beset by design issues, while Hitachi/GE was fined for lying to the regulator.

As problems mounted for the latest advanced Gen-III designs, the two 1.1-GW AP1000 reactors in Georgia slipped further behind schedule. Named Vogtle 3 and 4 after a former power-company president – adding to two earlier 1.2-GW, Westinghouse PWR reactors on the same site near Augusta – the start dates were pushed back by at least 4 years to 2021 (Vogtle 3) and 2022 (Vogtle 4), doubling the proposed construction time. After more problems surfaced over poor building practices, the start times were again delayed to 2022 and then 2023. One must have all one's ducks in a row to add nuclear power to the grid these days – iron-clad financing, advanced safety systems, public backing, reliable construction, and a competitive price. Alas, natural gas and renewables have blown up the business plan despite nuclear's rosy "carbon-free" and "cheap" electricity mantra.

In a bizarre financing twist to help cover the costs of the not-yet-built reactors, Georgians were even required to pay for their "premade" electricity, on the hook for almost a 10% increase in utility prices, despite not having received one iota of power. As noted in *The Ecologist*, "The utilities have been paying for individual elements of the two new plants as they are completed. The long delays mean that the interest costs are higher than expected and the regulator has already granted rate increases to compensate the eventual owners."[179] Even more bizarre, customers in South Carolina were tapped for rate increases from their two *cancelled* reactors at the Virgil C. Summer power plant near Columbus. Named for another former power-company chairman, the public is now permanently beholden to the largest pair of nuclear white elephants.

Not only has the time and cost to build a reactor become prohibitive – taking more than a decade from breaking ground to first grid-tied watt – nuclear power is more expensive at every turn as operators look for increased government bailouts to "compete" with cheaper natural gas and renewables, costing taxpayers billions more. In 2016, Exelon was given $2.3 billion to keep afloat two aging reactors in Illinois and asked for another $7.6 billion a year later to keep three more loss-making behemoths running in New York.[180] In the absence of a state bailout in Ohio, a FirstEnergy reactor near Toledo, however, will close before its license expires, likely bankrupting its owner because of mounting losses.

Saddled by poor decision-making, Pacific Gas and Electric (PG&E) has also had its own financial problems related to ongoing wildfires in California, declaring bankruptcy in 2019 before restructuring in 2020. Although not initially bailed out, PG&E was permitted to keep its Diablo Canyon reactors running until 2025, the only two nuclear power stations in California, both 1.1-GW PWRs that sit precariously within 500 m of the San Andreas Fault (the largest in North America, where the North American plate meets the Pacific plate). Although scheduled for shutdown in 2025 after years of private–public debate over ridding the state of nuclear power, PG&E was given the option to keep the two 40-year-old reactors open until 2029 and 2030, ostensibly to counter a sharp increase in energy demand. The de-facto bailout will be paid for with a generous $1.4 billion California state loan covered by federal funding through the Civil Nuclear Credit Program of the Biden administration's 2022 Bipartisan Infrastructure Law, a classic "robbing Peter to pay Paul" accounting strategy.

With renewables championed as safer, cleaner, and cheaper than nuclear power, getting a new project off the ground is almost impossible, typically a deal breaker at upwards of $10 billion per gigawatt. Ironically, because of Donald Trump's position on climate change, V. C. Summer in South Carolina and Vogtle in Georgia were denied claim benefits via Barack Obama's 2015 Clean Energy Plan to help soften the bill. When it comes to energy, changing policies over an extended construction period makes building new plants even harder as the loan interest for long builds exceed the principal. Nonetheless, the USA is still the world's premier producer of nuclear power (95 GWe in 92 reactors), followed by France (61 GWe, 56 plants), China (52 GWe, 54 plants), Japan, Russia, South Korea, Canada, Ukraine, Spain, and Sweden.[181] Currently, 32 countries generate nuclear power for a total of almost 400 GW from 437 plants.

In part to help offset an expected 60% reduction in energy generation by 2030 from decommissioning old nuclear plants and a continued phase-out of coal (the very long goodbye), Hinkley Point C (HPC) – now under construction on the south side of the Bristol Channel in Somerset, southwest England – will be the first nuclear reactor built in the UK since Sizewell B, a 1.25-GW PWR constructed in 1995 on the Suffolk coast in southeast England. After passing the final EU regulatory hurdle over a proposed large government subsidy, Hinkley's two European pressurized reactors (EPR) will provide 3.2 GW of power or about 7% of UK needs at an initial estimated cost of £24 billion. To help get Hinkley going, the UK government backtracked on a promise of no new nuclear subsidies, guaranteeing a wholesale price of £92.50 per MWh for 35 years indexed to inflation, more than double the going rate, as well as

£17 billion in financing. At the same time, the strike price for onshore wind/ solar was £80/MWh and offshore wind £56/MWh.

If the past is any measure, cost overruns will be enormous at final grid tie-in, which will eat into governmental budgets and curtail green spending. By 2021, Hinkley was already £3 billion over budget, while a year later *The Economist* reported that HPC was 2 years behind schedule and £10 billion over budget.[182] The director of an independent energy price comparison website thinks the project could result in British ratepayers collectively paying "an additional £5.2bn per year on top of their current electricity costs."[183] Nonetheless, nuclear advocates keep pushing for a dubious "nuclear renaissance." Ironically, when Hinkley is completed, solar and wind will be far cheaper.

After four decades of wrangling, Hinkley C is being built by the state-owned French utility Électricité de France (EDF), keen to maintain its precarious dominance in Europe. Initially part-financed by the Chinese state-run General Nuclear Power Group (CGN), who were eager to branch out from their home-grown nuclear backyard after years of domestic construction, many are doubtful about a technology that may be obsolete by the time of completion, while some are not sure the design will even work. The so-called "next-generation" EPR is unproven, beset by technical problems and cost overruns, and has yet to produce a *single* watt other than from two Chinese reactors north of Hong Kong, the first of which suffered a fuel-rod radiation leak in June 2021.

The details were sketchy, but a rise in noble gas concentration in the primary circuit was detected because of problems with the fuel-rod casings, prompting a hurried memo to the US Department of Energy that warned of an "imminent radiological threat."[184] For safety purposes, CGN shut down the reactor to investigate and replace the damaged fuel. What's more, China was removed from the Hinkley management group in 2021 by the UK government, who were worried about rising Chinese economic clout and possible national security issues. The buyout assumed CGN's 20% stake and put an end to further Chinese involvement in two more proposed nuclear plants at Sizewell and Bradwell east of London.

The Guardian has flat out predicted Hinkley will be the "world's most expensive power plant"[185] when finished, although others are gunning for the dubious distinction. Overlooking the English Channel in Normandy on the northwest coast of France, construction began in 2007 on the 1.6-GW Flamanville 3 EPR, variously delayed by bad welds, low steel strength, cooling-system valve failures, slow component delivery, and poor quality control, the €3.3 billion price tag ballooning to over €10 billion. After more problems appeared in 2018 in almost half of the inspected welds, more delays were announced, again pushing up costs, while major metallurgical flaws were found in the massive stainless steel reactor vessel and dome, preventing the scheduled completion.[186] The reactor was slated to be grid-connected in 2019, delayed until 2020 and then 2021. As of 2023, Flamanville 3 was still not operational, with an estimated final bill approaching €20 billion.

France was hoping the EPR design would provide "a safer, more powerful and long-lasting nuclear reactor that would replace its ageing fleet and boost French nuclear exports," but has instead "become a byword for its failings and an embarrassment for the government, which owns 84% of EDF."[187] The fault may lie in the design as another EPR – the 1.6-GW Olkiluoto 3 (OLK3) in the municipality of Eurajoki in western Finland – began construction in 2005 and was expected to be up and running in 5 years, but was finally grid-tied in 2023, 18 years after first breaking ground. Problems at Olkiluoto included watery concrete and monitor and control systems issues. Both plants are well over budget with a planning-to-operation time of over two decades.

China, however, continues to ramp up its nuclear presence, away from the West's strict regulatory control, capital-raising issues, and endless construction delays. Beginning with two reactors in 1994, 54 nuclear plants are now online, pushing 52 GW of power on to the Chinese grid.[188] Providing much cheaper electricity than the West, building costs are about $2,000/kWh or one-fifth that of the USA, hopefully without compromised safety.[189] While Toshiba/ Westinghouse, Hitachi/GE, and EDF have seen their business models shredded in the West, China leads the nuclear charge into the twenty-first century, expecting to add another 77 GW from 69 more reactors either under construction or in planning, little impeded by cost or public opposition. Another 156 units have been proposed at almost 200 GW.[190]

India is also enjoying a surge in new builds to modernize its grid and provide reliable baseload power in a developing economy. Having built its first CANDU reactor at Rajasthan in 1973 – one of nine eventual plants that some claim were used to make a nuclear bomb – India has added eight reactors in the last two decades with another eight planned by 2026, hoping to double its share of nuclear power from 2.6% to 12 GW.[191]

Over a third nuclear, South Korea has been aggressively selling its state-run KEPCO expertise, including a possible new build at Bradwell, east of London. KEPCO is also involved in a joint venture with Emirates Nuclear Energy Corporation to build the first reactors on the Arabian Peninsula as part of a $30 billion, 5.6-GW project at Barakah, 300 km west of Abu Dhabi, where the site is "dominated by a series of enormous domes like an industrialised version of a mosque."[192] The four APR1400 reactors are expected to provide 25% of electrical power to the UAE in a region once awash in oil. As proudly noted in the UAE's English-language newspaper *The National*, one uranium pellet is about half the size of a dirham coin, the local UAE currency, but is equal to 471 liters of oil, 481 m^3 of natural gas, or 1 tonne of coal. Not mentioned is the over 100 billion dirham coins needed to build the Barakah reactors.

Richard Muller is a cheerleader for both natural gas and nuclear, arguing that gas is better than coal because of lower carbon emissions (about half) and that nuclear is best of all despite sketchy decommissioning plans, long-term waste problems, and the possibility of a catastrophic accident. He does cite higher construction costs, however – twice that of a coal plant and four times a natural gas plant – but champions the abundance of minable uranium and reduced nuclear fuel requirements (1 kg of U-235 = 12,000 barrels of oil).[193]

M. King Hubbert was excited about the potential for unlimited nuclear power, primarily because of reduced fuel needs, calculating that 358,000 metric tons of uranium was equal to all the fossil-fuel reserves in the USA, of which only 553 metric tons (0.15%) were consumed in 1956. According to the World Nuclear Association, the initial fuel load is just 3% of overall equipment costs in a reactor, while fuel operating costs are only 14% (half from refining) compared to 78% for coal and 87% for natural gas.[194] Transportation is also easier and cheaper for the more concentrated uranium fuel, yet potentially much more dangerous.

***

So, what is the real cost of nuclear power, that is, when all the accounting is done? To compare nuclear power with other power-generating technologies, we must account for every expenditure, including overruns (more expensive for a long, amortized build), lifetime maintenance and fuel (40–60 years), waste management (indeterminate and endless), and a still unknown decommissioning bill.

Germany has budgeted €38 billion to decommission its 17 nuclear reactors, phased out after Angela Merkel's Fukushima moratorium, while France's state-run EDF will likely go bankrupt if it has to stump up more than the €100 billion earmarked for its 58 aging reactors.[195] Almost all US reactors have exceeded their original 40-year lease period, and will soon need to undergo expensive shutdowns, amounting to hundreds of billions of dollars not currently on the books. To be sure, there are enormous hidden costs to nuclear power.

Shockingly, President Eisenhower knew nuclear power was more expensive than conventional power during his solemn 1953 "Atoms for Peace" speech to the UN that proposed to repurpose fissionable uranium supplies for atomic weapons to nuclear power. An internal classified report stated, "Nuclear power plants may cost twice as much to operate and as much as 50 percent more to build and equip than conventional thermal plants."[196] The first dozen US reactors collectively lost almost $1 billion, Westinghouse and GE hoping such "loss leaders" would eventually become cost-effective.[197]

Nonetheless, the nuclear industry pressed on, subsidized to the hilt, which continues today despite cheaper alternatives. Considered "the most success-ful nuclear scale up experience in an industrialized country," even the vaunted French reactor program incurred "a substantial escalation of real-term con-struction costs," resulting in an *increase* rather than decrease in costs over time, that is, "negative learning."[198] Not only has nuclear power not been too cheap to meter, but the customers are stuck with an ever-increasing *l'addi-tion*. Even without the extensive peripheral costs, such as waste storage, clean ups, decommissioning, and subsidies, nuclear has never been as cheap as advertised.

Channeling the upside-down absurdity of Alice's Wonderland in *Through the Looking Glass*, a former head of the UK Atomic Energy Authority noted that nuclear power "has been, and continues to be, a case of 'jam tomorrow, but never today.'"[199] Shockingly, half of all US nuclear reactors are losing money,[200] while a 2019 study – based on 674 nuclear power plants built worldwide since 1951 – calculated that a 1-GW plant loses almost €5 billion on average. Furthermore, not one was built with "private capital under com-petitive conditions," but were all heavily subsidized and tied to military objectives, which continue to be unprofitable.[201] Even the US Federal Energy Regulatory Commission chairmen from 2009 to 2013, Jon Wellinghoff, called nuclear "too expensive," adding "We may not need any, ever."[202]

We must also include the cost of subsidies not on offer to other technologies. According to a 2011 report on US government subsidies, nuclear power received $3.5 billion annually since 1947 ($175 billion over the first 50 years) compared to $4.86 billion per year for oil and gas (almost $500 billion since 1918).[203] Furthermore, the percentage of subsidies in the first 15 years when financial support is most needed – measured as a percentage of the federal budget – is through the roof for nuclear power, with taxpayers also on the hook for future long-term waste disposal, essentially another subsidy whose total costs are still unknown. Conversely, renewable-energy technologies received only $0.37 billion per year since 1994, under $6 billion in total, hardly a level playing field, although biofuels received $1.08 billion annually since 1980.

Subsidies are essential to develop new technology, but one must question why the nuclear industry along with the oil and gas industry still receive billions of dollars in annual government handouts, many times more than renewables. Permanent funding only helps to extend old-world thinking, restrict innovation, and prop up an inefficiently mispriced economy. Alarmingly, as we fall further behind in installing renewable energy, bailouts continue for loss-making nuclear plants, while fossil-fuel subsidies continue to rise unabated. As already noted, annual government subsidies for fossil fuels across the globe doubled from the previous year in 2021 to almost $700 billion.

The imbalance isn't only in the USA. In Canada, Atomic Energy Canada Limited annually received subsidies of $170 million until 1997,[204] while in the UK an 11% nuclear levy on electricity bills from 1990 to 1998 intended to cover future decommissioning and waste costs was instead used to build Sizewell B, an estimated £9.1 billion subsidy.[205] Worldwide, the nuclear industry has received 60% of subsidies compared to 25% for coal, oil, and gas, and only 12% for renewables, while generous tax breaks are also available.

A full accounting must also include clean-up costs. Who pays for the $1 billion clean up at Three Mile Island, the $1.3 billion clean up in Port Hope, the $2 billion mess in and around Palomares, the $2 billion WIPP leaks in Carlsbad, New Mexico? What about the $2.4 billion compensation for American deaths and illnesses (especially from thyroid cancer) for those living downwind of the Trinity and other atomic weapons test sites in Nevada?[206] The $7 billion Deep River, Ontario, clean up, the devastation to Mayak and Windscale, the $27 billion Rokkasho plutonium reprocessing facility in Japan to "recycle" spent fuel still not functional after 20 years, the £80 billion Sellafield shambles that still costs more than £3 billion a year to secure? The £132 billion to decommission 14 former UK nuclear sites?

Even the proposals for future clean ups are exorbitant – the Yucca Mountain waste storage site in Nevada has already cost over $15 billion without having stored a single ounce of nuclear waste and may never. The test sites in the Pacific, Kazakhstan, and Nevada are riddled with radioactivity and may be the largest nuclear waste sites in the world. Who pays to make them safe for displaced citizens to return, if ever?

The UN pegged the Chernobyl damage at more than $200 billion, not including the $1.8 billion giant steel arch installed 30 years later, while in 2021 new evidence suggested the broken core may be fissioning again, requiring even more expensive containment work. The total cost to clean up Fukushima may ultimately exceed half a *trillion* dollars. Are these costs included in the nuclear power bill? What about decommissioning almost half of all current nuclear reactors over the next 25 years? Will that be included in a too-cheap-to-meter bill?

Who pays for the estimated $660 billion clean up of liquid wastes stored in 177 underground tanks at the Hanford Manhattan Project plutonium production reactors, located roughly 2 miles from the Columbia River?[207] Six of the nine reactors have already been encased in steel and cement for long-term storage, while the last two reactors – K-East and K-West that closed in 1970 and 1971 – will similarly be "cocooned" over the next 75 years before being dismantled. Nuclear power has become a never-ending money pit, an endless drain on public finances, restricting needed investments on more profitable clean-energy projects.

Nor is nuclear as carbon free as some claim, requiring huge amounts of concrete and water as well as carbon-intensive uranium mining, resulting "in up to 25 times more carbon emissions than wind energy, when reactor construction and uranium refining and transport are considered."[208] As Stanford University engineering professor Mark Jacobson notes, "There is no such thing as a zero- or close-to-zero emission nuclear power plant. Even existing plants emit due to the continuous mining and refining of uranium needed for the plant."[209]

Neither can nuclear be labeled "green," in particular because of the "environmental impact of nuclear waste."[210] A legacy of wartime competition and ongoing post-war diplomatic failure, the nuclear lobby continues to be well funded, however, as seen when the European Commission included nuclear and natural gas in the EU "taxonomy of environmentally sustainable economic activities." Calls of greenwashing rang out across Europe as the pro-nuclear government of France squared off against the now anti-nuclear government of Germany. Even if nuclear power was as green as claimed, building renewable energy is faster and cheaper watt for watt, exactly what is needed now as global temperatures increase. A nuclear power plant costs 10 times as much, takes more than a decade to build, and produces much more emissions per kWh than a wind- or solar-power installation.

Energy analyst and writer Amory Lovins, who has advocated for more conservation to reduce our overreliance on energy (coining the term "negawatts"), states that so-called "low-carbon" nuclear power actually makes global warming worse, hampering climate protection precisely because of the high costs and the either–or nature of investment, where "Costly options save less carbon per dollar than cheaper options."[211] As Lovins notes, "It is essential to look at nuclear power's climate performance compared to its or its competitors' cost and speed. That comparison is at the core of answering the question about whether to include nuclear power in climate mitigation."[212] The environment is on short notice because of increasing carbon emissions, but reducing upstream dirt for more downstream damage is no solution, just more of the same upside-down thinking as we continue to pay through the nose for a broken dream.

Nuclear power is pretty good at staying on, however, with a capacity factor of about 90% thanks to the uprating of older reactors. Not quite plug-and-play, but fewer headaches to maintain baseload grid levels. Although a nuclear reactor can't be used to stabilize the grid, some investors claim that small modular reactors will be able to ramp up and down to augment increased variability from intermittent sources. Still, nuclear power takes the longest to start from an off position, while costs are high to start or stop a fissioning reactor. By comparison, solar and wind output will always be intermittent, but a flatter output can be maintained with improved battery technology and interactive grid sharing in the same way electricity is drawn from numerous power stations to maintain a constant instantaneous supply (which we'll look at in Part II). Most importantly, solar- and wind-power installations take very little time to construct compared to nuclear.

So what is the real cost compared to other energy technologies when everything is included: planning, financing, materials, equipment, building, operations, fuel, waste management, decommissioning, salvage, abandonment (never mind the overlooked cleanup costs)? The levelized cost of energy (LCOE) is a single metric in dollars per unit of power ($/MWh) that compares the overall costs of different power-generating technologies. LCOE is a life-cycle analysis that includes all present and future costs, essentially building and operating costs divided by lifetime power output.

The LCOE analysis from 2009 to 2021 by the American investment bank Lazard (Figure 3.7) shows that nuclear power has been more expensive than oil and gas for decades – an *inverse* learning curve – while photovoltaic (PV) solar

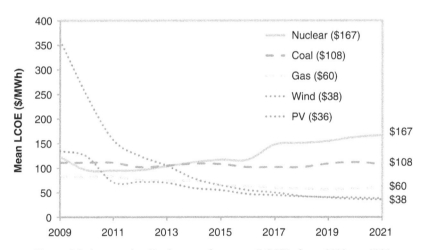

Figure 3.7 Average levelized cost of energy (LCOE) from 2009 to 2021 ($/MWh) (*source*: "Lazard's Levelized Cost of Energy Analysis 2021," *Lazard*, Version 16.0, April 2023. https://www.lazard.com/media/typdgxmm/lazards-lcoeplus -april-2023.pdf).

power has been cheaper than nuclear since 2013, already having reached "parity" and now a quarter of the cost. Throw in all the extras and the comparison isn't even close. What's more, the levelized cost of nuclear has increased by a third since 2009 while solar has dropped 90%.

Some think nuclear power has remained expensive because a single design has not been agreed, undermining the standard learning curve of a new technology that continually reduces costs with improved engineering efficiency, a.k.a. the "virtuous circle." More likely, the increased costs of regulatory oversight, safety refits, and delayed construction has pushed nuclear-generated power beyond practical limits, while safeguarding sensitive atomic technology precludes sharing. Maybe it is time to give up on a power-generating technology that gets more expensive over time and can never deliver on its promises.

Science writer Fred Pearce wonders if nuclear is in fatal decline, citing the ongoing financial problems of Westinghouse, Toshiba, and EDF: "While gas and renewables get cheaper, the price of nuclear power only rises. This is in large part to meet safety concerns linked to past reactor disasters like Chernobyl and Fukushima and to post-9/11 security worries, and also a result of utilities factoring in the costs of decommissioning their aging reactors."[213] Cheaper alternatives, unresolved waste issues, and fear of an accident laying waste to more of the earth continues to keep nuclear power from ever becoming a viable future energy source.

The public has clearly had enough of our ongoing nuclear folly as the era of top-down energy gives way to a call for a cleaner, safer future. According to a 2011 poll in 24 countries, a majority of the public opposes nuclear power: 81% in Italy, 79% in Germany, and 51% in the UK.[214] In the 24 countries surveyed, only 38% were in favor of nuclear power (solar 97%, wind 93%, hydro 91%, natural gas 80%, coal 48%, and nuclear 38%), while only 31% support building more reactors, 44% in the USA and 43% in the UK. And yet nuclear power is still controlled and nurtured by the state, given preferential treatment without being tasked to maintain public safety beyond the lifetime of a reactor. That the state can close off a nuclear site for all time as in Mayak, Chernobyl, and Fukushima, and make taxpayers pay for the privilege of being excluded, shows the real power of nuclear energy. How many permanent exclusion zones must be created before we stop throwing good money after bad to make right past mistakes?

Occupying a ruggedly beautiful stretch of coastline near the wilderness of Big Sur, where Jack Kerouac praised the rhythmic sound of the "ocean motor," Diablo Canyon may be the only viable solution to an aging fleet of nuclear reactors – closure. Completed in 1985, the two Westinghouse 1.1-GW PWRs provide 10% of state power, but were slated to close by 2025 at the end of their original licenses. Alas, the last nuclear power plants operating in California were extended (a.k.a. "second license renewal") to ease the latest so-called "energy crunch" despite the potential devastation from an earthquake that could release radiation across California, endangering 40 million inhabitants. In the absence of new builds, dangerous extensions are all that is left for the nuclear industry. One day, we may finally see the end, at least in California.

The development of nuclear power was an extraordinary achievement, involving thousands of scientists and engineers, decades of research, and trillions of dollars in construction costs, in part to assuage the guilt of having destroyed two cities, but has led to a world we are neither ready for nor can endure. It is fitting to give the last words to Albert Einstein, who so ably championed and then cautioned the world about the raw power of the atom: "Nuclear power is a hell of a way to boil water."

3.10 Fusion: The Power of the Future, Coming Soon at Long Last?

As a first-year university physics student in 1979, I was told that a working nuclear fusion reactor was only a matter of time and would eventually solve all our energy problems. Fifty years was given as a possible time frame. Turns out the same estimate was given to my professor decades earlier when he was an undergraduate, while today's students are given a similar story. As the old joke goes: fusion, the energy of the future . . . and always will be.

A working fusion reactor may happen one day, even within the next 50 years, but there are still numerous challenges to overcome, not least how to confine a 150 million degree nuclear reaction in a container that doesn't melt. Forget about boiling water to run a steam turbine; with fusion we can't even hold the fire. It's even harder than trying to put the Sun in a box. How to keep the fusion process going for a sufficient length of time and dissipate the reaction heat are also major obstacles to producing abundant, low-waste power, as is getting more out than we put in – making fusion work in a box on earth is one thing, breaking even is a whole other dream.

By contrast, a fission reaction burns at less than 1,000 degrees, easily contained as long as the heat is continuously removed by a coolant (for example, water, carbon dioxide, air, liquid sodium). Even a major fire is peanuts by comparison, such as burning zirconium fuel rods and uranium fuel melting the reactor vessel and concrete containment structure, assumed to fall unimpeded through the Earth. Popularly known since Three Mile Island as the China Syndrome, the seemingly unstoppable overheated liquid blob will still be cooled by the ground about 6 feet under. Alas, at temperatures 10 times that of the Sun needed for fusion, no material box could ever do the job. We need something else to contain the energy – a magnetic field (as in a tokamak or stellarator) or a way to remove the heat almost instantaneously (as in laser confinement). A temperature over 5,000 degrees would melt any container.

As we saw earlier, fission breaks heavy elements apart (for example, U → Kr + Ba). In a fusion reaction, lighter elements join together to make a heavier element, the simplest a joining of four hydrogen atoms into a helium atom (for example, H + H + H + H → He), where at 150 million degrees the hydrogen atoms overcome their large repulsive forces to fuse. Most importantly, because the combined masses of four hydrogen atoms (4_1H^1) is greater than the mass of one helium atom ($_2He^4$), the difference is made up in energy, lots of it according to Einstein's famous $E = mc^2$ equation ($E \approx (4m_H - m_{He})c^2$). If you do the math, a helium atom has about 1% less mass than its component parts (a.k.a. the "mass defect").[215] We get even more if we fuse deuterium ($_1H^2$) and tritium ($_1H^3$), the two heavy hydrogen isotopes D and T, where $D^2 + T^3 \rightarrow He^4$ + n and $E \approx (m_D + m_T - m_{He} - m_n)c^2$). A DT fusion reaction generates more energy at a lower temperature than 4H fusion because of the lower Coulomb repulsion (neutrons have no charge).

Throughout the universe, higher-Z atoms are continuously being fused in the stars – nature's "element factory" – to form elements as high as iron (Z = 26). No longer able to create elements beyond iron, a star eventually collapses from its own gravity, imploding and then exploding. Fusion is the origin of most of our material world, while elements higher than iron are created by supernovae and neutron star collisions. One obvious fusion source is our own local star, the Sun, which continuously burns hydrogen (or its heavy hydrogen isotope D) to make helium and energy, a.k.a. stellar nucleosynthesis. In the 1920s, the British astronomer Arthur Eddington suggested that solar mass is converted to energy by fusing hydrogen into helium, giving us our always-on sunlight and over-turning the notion that the Sun is powered in any conventional sense like a fossil fuel that would soon burn itself out.

The German-American physicist Hans Bethe would win the 1967 Nobel Prize in Physics for "contributions to the theory of nuclear reactions, especially his discoveries concerning the energy production in stars," in particular the proton–proton (PP) reaction chain of fainter stars such as our own relatively average, G-type, main-sequence Sun or the carbon–nitrogen fusion cycle of brighter stars. Surmising that carbon must be created in stars because all life is carbon, British astronomer Fred Hoyle worked out the intermediate helium–carbon fusion cycle or "triple-alpha process" ($He^{2+} + He^{2+} \rightarrow Be$, $Be + He^{2+} \rightarrow C$).

Given the massive size of our Sun, the energy is enormous. In fact, because of solar fusion the Sun becomes 4 million tons lighter every *second*, although there is no need to worry about the hydrogen fuel running out any time soon. As Bob Berman writes in *Zapped*, "given that the Sun has a total mass of two nonillion – that's the number 2 followed by 27 zeros – tons, its ongoing loss of mass is not noticeable. It'll be billions of years before any serious consequences ensue."[216]

Note that at such high temperatures, atoms exist only in a plasma state, the so-called fourth state of matter (solid → liquid → gas → plasma), where bound electrons have been stripped from their atoms to form a soup of comingling, oppositely charged, ionized nuclei and electrons, that is, a highly electrically conductive charged gas or plasma. To get an idea how hot the universe is, 99.9% of all matter is plasma, more than 90% hydrogen (90% of all stars one sees in the night sky burn hydrogen). Containing such a hot plasma at temperatures greater than the Sun and extracting useful energy is the fusion holy grail.

The nuclear "binding curve" shows us which elements fuse, which fission, and the energy released in the process. Because of how nucleons vie for stable atomic space, elements below iron can fuse (Fe, Z = 26) if the material temperature and density is high enough, while elements above iron can fission when split by a neutron (as we saw earlier). Iron has the highest nuclear binding energy in the periodic table and is thus the most stable element. The idea is that we get fusion going left to right in the periodic table for elements *lighter* than iron because the mass of the fused sum is less than the mass of the initial parts, while we get fission going right to left for elements *heavier* than iron because the mass of the fission products is less than the mass of the original atom. The reaction energy is generated from the missing mass in both cases. In nuclear physics, the sum of the parts does not equal the whole without accounting for the stored nuclear energy (Figure 3.8).

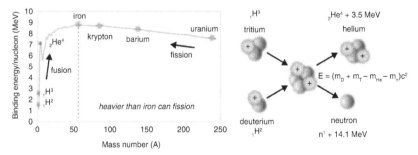

Figure 3.8 Nuclear fusion. (a) The nuclear binding curve and (b) deuterium–tritium (DT) fusion reaction.

One can see how enticing fusion is from the binding curve, because we get about 200 MeV when uranium turns into krypton and barium via fission (n + U→ Kr + Ba + 3 n + 170 MeV), roughly 1 MeV per nucleon, but much more when we fuse deuterium and tritium into helium (D + T → He + n + 17.6 MeV), or about 4 MeV per nucleon. The steeper slope for fusion indicates more released energy.

Unfortunately, fusion is not easy here on Earth. Without the Sun's gravity to help break through the repulsive atomic barrier of the reaction protons – an enormous force at a distance of 10^{-15} m (0.000000000000001 m!) – we have to heat a DT plasma to at least 100 million degrees, that is, six times hotter than in the solar core, to speed up the nuclear projectiles and help them fuse. That we can do with an electric current (~resistive heating), radiowaves (~microwave heating), or a particle beam. But without the Sun's massive gravity to confine the reactions, we also have to devise a way to hold the plasma without burning down the house, for example, via a large magnetic field. A magnetic field also compresses the plasma, increasing the density and probability of fusion.

A particle accelerator works by shaping the path of two atomic particles using a series of bending magnets and smashing them into each other to see what comes out, such as the illusive Higgs boson (the so-called God particle). In the famous 2012 experiment at CERN's Large Hadron Collider (LHC), the world's largest particle accelerator, two beams of protons were accelerated to just under the speed of light by a massive array of superconducting electro-magnets along a 27-km long circular tunnel and then smashed together. In one of the most fascinating (and expensive!) experiments ever, the magnetic field of the LHC bent the whizzing protons into two almost-circular beams to smash them head on.

But keeping stable a whole reactor full of hot protons and electrons that want to go this way and that, while hovering inside a large and fluctuating magnetic field, is a whole other game. Heated ions naturally want to recombine and cool down and fall to the ground, so the hydrogen plasma must remain in circular motion, shaped in mid-air to keep from hitting the walls. At temperatures approaching 100 million degrees, one might as well try to hold a falling star without touching it.

***

In the 1950s, a hot hydrogen plasma was generated in the Zero Energy Thermonuclear Assembly (ZETA) at the Atomic Energy Research Establishment in Didcot in the UK, the first major attempt to produce large-scale hydrogen fusion, overseen again by the original atom smasher John Cockcroft. To maintain the circulating plasma, two parallel currents were "pinched" via electromagnetic attraction (known as a "z-pinch"), but led to large instabilities. Nonetheless, a large hydrogen plasma – heated by an applied electric current – had been produced for the first time, albeit lasting only for about 1 millisecond. Initially declared to be the world's first "artificial Sun" calculations soon showed, however, that the 5 million degree plasma was not nearly hot enough to generate fusion.

Other designs soon pushed the fusion frontiers, such as a stellarator, the Model A first built in Princeton in 1953 by American astrophysicist Lyman Spitzer, the father of US fusion research. A stellarator twists the plasma in a figure-of-eight as it circulates to even out the magnetic field and keep the hydrogen nuclei and electrons stable while in motion, but Spitzer's design was discarded as too small, the plasma drifting too quickly to the walls. Despite the lack of success, however, the z-pinch and stellarator both helped establish the fundamental principles of terrestrial fusion, and that a bigger box would be needed to maintain plasma stability whirling around in an earthly cage.

Around the same time, another fusion device was built in Russia by physicists Igor Tamm and Andrei Sakharov, employing 2 magnetic fields to confine the plasma instead of a single field as in a twisty stellarator. Called a tokamak – a Russian acronym for "toroidal chamber magnetic coils" – the details were initially unknown to the West because of Cold War secrecy, but the results eventually showed a workable design (Sakharov was the Soviet bomb developer turned human-rights and peace activist who won the 1975 Nobel Peace Prize).

In a tokamak, the plasma moves in a helical path that wraps around itself in a donut-shaped chamber. An infinitely long straight tunnel is theoretically the best design, but is obviously impossible, so the cylindrical container is bent into

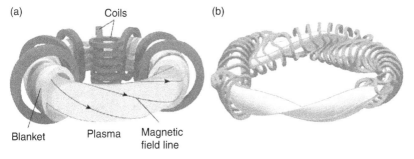

Figure 3.9 Magnetic confined fusion reactor schematics: (a) tokamak and (b) stellarator (*source*: Xu, Y., "A general comparison between tokamak and stellarator plasmas," Matter and Radiation at Extremes 1: 192, 2016. https://doi .org/10.1016/j.mre.2016.07.001. CC BY-NC-ND 4.0).

a torus through which the corkscrewing plasma can continuously flow, essentially a tube without ends (in mathematical parlance a donut is a torus). The two magnetic fields in a tokamak act in perpendicular directions, called toroidal (B_ϕ) and poloidal (B_θ), where the toroidal plasma current creates the compressive poloidal field. The aspect ratio (R/a) is typically around 4, where R is the donut's radial size (major radius) and a its thickness (minor radius). The first tokamak, T1, built in 1958 at Moscow's famed Kurchatov Institute, had a major radius R of 0.67 meters and a minor radius a of 0.17 meters, producing a toroidal field B_ϕ of 1.5 tesla. In 1968, a larger tokamak, T3 ($R = 2$ m, $a = 0.4$ m), heated a millisecond plasma to 10 million degrees, jump-starting international research on the Russian tokamak design.

But despite the success of the early tokamaks, keeping the plasma flow stable was still a deal breaker as the charged particles drifted to the walls since the bunched-up magnetic field lines in a torus-shaped device are stronger at the core (the donut hole) and weaker at the edge. Bigger plasma "bottles" were needed to maintain stability – the bigger the better to minimize bending and drift, which of course adds to the expense. What's more, since confinement is short-lived because of fluid instabilities, a fusion reactor is pulsed; that is, we evacuate the air (no heat transfer in a vacuum), turn on the magnets, introduce hydrogen into the chamber (typically a 50–50 DT mix) that forms a plasma after the applied current heats the fuel, which then ignites the fusion reaction when temperatures exceed 100 million K. When the plasma becomes unstable, we do it all over again, while if anything goes wrong the system immediately shuts itself down.

Located just south of Oxford at an old air force base in Culham, the Joint European Torus (JET) has been running since 1984, and was the first fusion device to sustain a plasma for almost 2 seconds before loss of confinement. In 1997, JET also recorded the most powerful plasma ever, generating 16 MW of fusion power from an initial heating of 24 MW, alas a net loss because the out/in ratio is less than one (Q = P_{out}/P_{in} = 16 MW / 24 MW = 0.67). Highlighting the difficulty of making fusion work, no heat is extracted from JET – we're still figuring out how to keep things stable and make a plasma last.

I was fortunate to see how JET works as part of an Oxford plasma program. On a tour of the facility, our group was told that before starting a run, the operators call the local utility company as a courtesy, knowing that the initial power could overwhelm the grid. A JET experiment can hog more than 1/6 of UK electrical power at the start, and although a run typically lasts only a few seconds the spike is too much to handle. The almost instantaneous 24-MW draw needed to start the process shows how hard it is to mimic the Sun and get the action going. Occasionally, as we were told, the operators were asked to wait until *Coronation Street* was over or until everyone had plugged out their kettles at the end of the long-running British soap. By comparison, the Manhattan Project took 20% of the US grid to run the Oak Ridge enrichment plant, an extraordinary technological and logistics challenge.

As with most large-scale science research, progress is incremental, where more is learned with each new design and iteration. The latest and greatest fusion device is the International Tokamak Experimental Reactor (ITER), located in the Cadarache research center in the south of France, 100 kilometers due east of the city of Arles, where Vincent van Gogh painted some of his most memorable works as he captured the vibrancy in his pastoral landscapes on canvas in his own whirling dynamics.

ITER (pronounced "eater") hopes to generate "first plasma" by 2025 after almost 20 years of planning and construction, although nothing is ever easy with fusion or goes to plan – the project is more than 10 years behind schedule and three times over budget. At roughly the size of a football field including all adjunct buildings, the biggest fusion device ever will produce the highest yield yet, while the reaction chamber itself will contain a plasma volume of 840 cubic meters – ten times larger than JET – with an outer chamber wall diameter of 16.4 meters and an inner donut wall diameter of 4 meters ($R = 6.2$ m, $a = 2.0$ m).

As one might expect, development costs are too much for one research group, and thus international cooperation is needed. The seven ITER Agreement members (a.k.a. domestic agencies) are Europe, China, India, Japan, Korea, Russia, and the United States, who share costs and intellectual property, bringing together scientists and engineers from around the world. At roughly $20 billion and counting, some have called ITER the most expensive experiment ever, more than the $13 billion spent by CERN to build and run the LHC that successfully detected the Higgs boson by colliding protons at super-high speeds.

As in the name, ITER is a tokamak design (previously the T was for "thermonuclear" but was changed to "tokamak" to sound safer), while "experimental" means we're still working out the kinks, that is, a proof of concept to show how to contain a fusion reaction for long enough to get more energy out than we put in, known as the "net energy" or Q factor, where the total power generated (P_{out}) is greater than the thermal power injected into the chamber to heat the DT fuel to a plasma (P_{in}) and initiate fusion. ITER also means "journey" or "route" in Latin, highlighting the many steps to reach the end. Considered half-completed at the end of 2018, the bill could be a whopping $40 billion by the time of first plasma.

The goal of JET was to show that a "burning plasma" could be stably contained for longer periods, while the goal of ITER is to produce a Q of 10, producing 500 MW of fusion power from 50 MW of initial heating power. In a burning plasma, alpha particle reaction products become the main source of heating, thus producing further fusion reactions (D,T: α, n). ITER is not designed to generate electricity, but to demonstrate that net energy can be achieved, thus preparing the way for the next design that will extract GW-scale electrical output in a working fusion reactor. No one should ever under-estimate the challenges of bottling the Sun.

In 2035, 10 years after the projected first plasma, ITER is expected to demonstrate how to sustain a DT reaction by internal heating (thus greatly reducing the input energy) as well as breed tritium, because tritium is expensive, short-lived ($t_{1/2}$ = 12 years), and available only in limited supply (now delayed until the 2040s). Costing $30,000 per gram, the only source is collected as a by-product in a CANDU reactor, regularly detritiated from the heavy-water moderator. If all goes well, the next iteration after ITER will be a working fusion reactor, already in the early planning stages called DEMO, where high-speed neutron reaction products collide with a blanket-lined reactor wall, transferring kinetic energy to a circulating fluid to generate heat (as in a conventional power station). Originally planned to generate 500 MW, DEMO has been scaled back to 200 MW.

ITER is by far the biggest and most complex experiment ever devised and has seen many setbacks since breaking ground in 2008. Just to create the magnets for the tokamak chamber, 100,000 km of niobium-tin superconducting strands of metal were manufactured by nine suppliers over seven years (the agency agreement requires duplicate member production).[217] Built on top of a cryogenic base that is liquid-helium cooled to a temperature just above absolute zero (-269°C or 4.15 K), the largest electromagnets ever made will create the magnetic fields needed to keep the super-heated plasma moving on its merry curved helical way. The stronger the magnetic field the better – by doubling the magnetic field, the plasma volume needed to produce an equivalent power is reduced by a factor of 16.

In the meantime, about 30 fusion projects are up and running around the world, all of various shapes and sizes (such as a spherical tokamak that has less surface area and doesn't need as large a magnetic field), while the stellarator has made a comeback, including the Large Helical Device (LHD) in Japan and the Wendelstein 7-X (a.k.a. "star in a jar") in Greifswald, Germany, each hoping to find the right geometry and dimensions to imitate the Sun, or at least add to our understanding of controlled terrestrial fusion.

W7-X is the largest ever constructed stellarator, designed with the aid of advanced computer simulations to work out the precise shape of its 2 million component parts to create a mostly toroidal helical coil in an extraordinarily futuristic-looking twisty construction. As in all magnetically confined plasmas, the goal is to minimize drift by keeping the plasma suspended inside the walls, ever hovering without touching. W7-X's first hydrogen plasma was achieved in 2016, measuring 80 million degrees and lasting a quarter of a second. Flipping the switch, the then German chancellor and PhD scientist Angela Merkel noted "When we look at nuclear fusion, we realize how much time and effort is needed in basic research. In addition to knowledge, a good deal of stamina, creativity, and audacity is required."[218] A 16-month upgrade completed in 2017 produced a world-record plasma discharge lasting over 100 seconds, with further plans to generate continuous power in a 30-minute plasma.

In December 2021, the Chinese HT-7U Experimental Advanced Super conducting Tokamak (EAST) produced the longest hot plasma yet, a record 70 million degrees for 1,056 seconds (17 minutes, 36 seconds), adding to its record plasma temperature milestone earlier the same year of 120 million degrees for 100 seconds. EAST had eclipsed the previous record of 100 million degrees for 30 seconds by the Korea Superconducting Advanced Research (KSTAR) device. In December 2021, JET also generated a record 59 MJ of heat in a 5-second fusion burst, the maximum possible using copper magnets and more than twice their 1997 best.[219] The first fusion tests in over 2 decades at JET – now billed as "little ITER" to better simulate the ITER setup – significantly lengthened the plasma lifetime of a few milliseconds by refitting the hydrogen-absorbing graphite materials in the reactor chamber with tungsten and beryllium. The redesign at JET is helping to confirm the more advanced goals of ITER.

***

Not all fusion projects employ magnetic confinement fusion (MCF). Others generate a controlled thermonuclear reaction (CTR) via confinement on a small scale by laser ignition, where opposing laser beams compress a DT fuel pellet to such a small size that the crunching breaks through the repulsive barrier of the hydrogen atoms to trigger fusion, known as "inertial confinement fusion" (ICF). Today, most major fusion research is either by magnetic containment (tokamak or stellarator) or inertial confinement (for example, laser ignition).

In 1960, only days after Ted Maiman demonstrated the world's first laser – the "ruby" laser emitting red light at 694.3 nm – a physicist at Lawrence Livermore National Laboratory (LLNL) imagined how a laser system could be designed to compress hydrogen atoms and generate the same high temperature and density needed for fusion ($\sim 10^8$ K, $\sim 1,000$ times solid). LLNL's John Nuckolls published his ideas in a 1972 *Nature* article entitled "Laser Compression of Matter to Super-High Densities: Thermonuclear (CTR)

Applications," noting that "Hydrogen may be compressed to more than 10,000 times liquid density by an implosion system energized by a high energy laser. This scheme makes possible efficient thermonuclear burn of small pellets of heavy hydrogen isotopes, and makes feasible fusion power reactors using practical lasers."[220] The article had been delayed for 12 years because of national security restrictions.

In ICF, symmetrical implosion of a spherical DT fuel pellet is created by splitting a laser pulse into a number of opposing beams, a.k.a. "laser pistons," the many-sided impact – called a "shot" – applied uniformly to the target, both spatially and temporally. For example, 6 beams are sufficient to produce target compression – 2 along each of the 3 Cartesian axes acting in opposite directions – although the more the merrier. X-ray implosion occurs as the opposite reaction to the pellet surface vaporizing outward, igniting the fusion reaction.

The eventual goal is to exploit the process in a working fusion reactor by continuously dropping DT pellets via gravity, each irradiated by the high-energy opposing beams. Most of the energy will come from hot reaction neutrons absorbed in a lithium blanket, where cooling pipes transfer heat as in a conventional power plant or are converted directly to electricity in a magnetic field. The expensive tritium can also be recovered in a lithium blanket or cheaper DD fuel used with a more energetic laser pulse, although a higher ignition temperature is needed to overcome the higher Coulomb repulsion force.

On the same site as Nuckolls's Livermore lab, the $3.5 billion National Ignition Facility (NIF) was completed in 2009 to harness inertial-confined fusion energy. Boasting the world's largest and most powerful laser ever built, NIF can deliver a 1.9-MJ shot (at 351 nm in the UV) from 192 beams for a total deliverable power of 500 TW, 60 times more than global grid capacity although only for a fraction of a second. To get your head around the enormity of a laser-ignition facility, the setup occupies roughly 16,000 m^3 (3 football fields) and employs 38,000 optical units (mirrors, beam splitters, and lenses) that make and shape the 192 beams to produce the shot. Although the shot lasts only a few nanoseconds (10^{-9} s or 1 billionth of a second), NIF is 500 times more powerful than any other laser and delivers more power than "all the sunlight falling on the Earth."[221]

To produce an even compression, the 192 beam lines are directed onto a cylindrical gold container about the size of a pencil eraser known as a hohlraum (German for "cavity"), generating x-rays that create uniform shock waves at more than 10 million atmospheres, imploding the peppercorn-sized pellet that fuses the deuterium and tritium fuel inside. The goal is to compress the DT target, initiate fusion, and extract the energy before

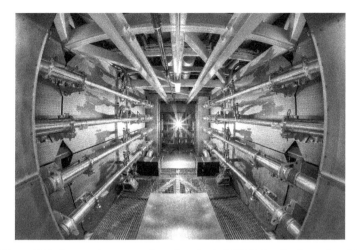

Figure 3.10 Symmetric compression of a spherical DT fuel pellet for an inertial confined fusion experiment at NIF (*source:* National Ignition Facility).

everything is obliterated, achieving a self-sustaining reaction or "ignition" in fusion parlance. As noted by Mike Dunne, director of the UK Central Laser Facility and former fusion director at NIF, "Getting it right requires a lot of effort; for example, the target chamber is under vacuum to allow the lasers to be focused down to spots just 1 mm in diameter, and the fuel pellet itself has to be extremely round and smooth, as any imperfection is exponentially amplified in the course of the implosion."[222]

Incomplete implosion is a fusion no-no, where the slightest surface imperfection in the target stops uniform compression as in the mangled grapefruits at Los Alamos. An asymmetric implosion reduces the percentage of kinetic energy converted to heat – at low laser intensities, we get hydrodynamic instability because of surface roughness, while at high intensities, plasma oscillations retard laser absorption. As such, the conditions for successful ignition were not initially achieved (only 1/3 necessary compression) and NIF was mostly used instead for material studies, astrophysics analysis, and stockpile security for advanced weapons testing (required since the 1996 Comprehensive Nuclear Test Ban Treaty), reducing overall "beam time" for fusion research from 60% to under 30%.

After 3,000 shots, the target and laser pulse shape were redesigned in 2020 and the imaging diagnostics improved to view the compression better, allowing researchers to close in on a burning plasma, the interim self-heating stage essential for ignition and eventual "runaway energy gain."[223] While 100 kJ is needed to achieve a burning plasma, the initial yields were below 60 kJ, although on August 8, 2021, a self-sustaining output energy of 1.35 MJ was achieved (70% of the 1.9-MJ input laser energy), 8 times more than earlier results and producing NIF's first artificial star albeit only for 1 trillionth of a second.

Different hohlraum shapes were then investigated to improve the laser focus, including a double-walled capsule to trap and transfer x-ray energy more efficiently and a foam-soaked fuel pellet to produce a higher-temperature central hot spot. Alas, the breakthrough ICF results could not repeated in subsequent experiments because of difficulties at the point of ignition, casting doubts about the design and the "inability to understand, engineer and predict experiments at these energies with precision."[224]

Some questioned the inherent limitations of the ICF design because of target inconsistencies and wanted to see a complete overhaul of the setup to concentrate on the next-generation laser on the "road to ignition." Others thought the knowledge learned at NIF was essential to achieving eventual ignition, requiring only more time and money. Despite the concerns, NIF finally achieved "break-even" in December 2022 with a Q of 1.5 (3 MJ out from the 1.9-MJ laser shot), although the shot still took 300 MJ to generate, repeated with greater gain six months later.

Even with successful ignition and a fudged Q, the current laser repetition rate is insufficient – 10 times per second is needed for GW-power station output – NIF firing only about once an hour. Each hohlraum also costs over one million dollars and so a redesign is in order to make a working system. As is typical in large research projects, funding is the rate-determining step, especially when the goal is to build and contain a star. As such, ICF is being explored in a number of other novel projects, including HiPER (High Power laser Energy Research) to study "fast ignition" with smaller and thus much cheaper lasers and ELI (Extreme Light Infrastructure), both working to improve implosion and increase the repetition rates needed for power-plant scale output.

Chirped pulse amplification (CPA) – a laser confinement system for which Gérard Mourou and Donna Strickland won the 2018 Nobel Prize in Physics – is being tried on a hydrogen-boron mixture (HB11 or p-B11) in a cylindrical target that fuses via directed motion in an axial "burn wave." The p-B11 fuel ignites at much higher temperatures, on the order of 1 billion degrees or 10 times DT fusion, where the magnetic bottle is created by the spinning plasma. A prototype power plant could be built in a decade for under $100 million.[225]

An Oxford University spinoff, First Light Fusion, is also investigating smaller-scale "projectile" ICF, in which a 200,000-volt, 14 million amp, 500-nanosecond electromagnetic discharge pulse shoots a high-velocity slab at a gas-filled DT target. Rather than a perfectly symmetrical target implosion, which is difficult to achieve, First Light's Machine 3 device generates an intense shock wave that induces asymmetrical collapse of the fusion-fuel target. At roughly £4 million, the pulsed projectile device costs a fraction of the $3.5 billion NIF laser, while producing a shot equivalent to 500 simultaneous lightning strikes.[226]

Researchers at MIT and the spinoff company Commonwealth Fusion Systems (CFS) in Cambridge, Massachusetts, are also developing improved superconducting magnets to dramatically decrease a tokamak chamber size to less than 2% ITER as in their SPARC device, hoping to generate 200 MW in 10-second pulses at a Q ratio of 10 or more. To keep a lid on material costs, cheaper liquid nitrogen (77 K) is used instead of liquid helium (4.15 K), the high-temperature superconductors producing larger magnetic fields in a smaller chamber volume.[227] Smaller, less-expensive designs can help work out the mysteries of terrestrial solar power without getting bogged down by the overwhelming costs of a large-scale device. Comparing nascent fusion development to the proliferation of independent space-flight research after NASA ended the shuttle program, Bob Mumgaard, chief executive of CFS noted, "Companies are starting to build things at the level of what governments can build."[228] SPARC could become the first tokamak device to produce net energy, while Tokamak Energy has also produced gain in a small spherical tokamak (ST 25).

Slow to finance a national fusion program beyond basic research (such as at NIF), the United States announced a plan in 2021 to fund a prototype fusion power plant on a much smaller scale than ITER, hopefully starting within 2 decades "to move forward with fusion on a time scale that can impact climate change."[229] As always, budget constraints present a major obstacle. The US Department of Energy's Fusion Energy Sciences annual budget is a miniscule $671 million, more than one-third of which already goes to ITER.

Fusion is certainly not cheap, but another novel design is reducing costs by merging magnetic and inertial confinement to produce magnetic target fusion (MTF). A mix between magnetic containment and inertial compression, the fusion chamber is heated *and* compressed, the hot reaction neutrons absorbed by liquid metal to produce power. The confined plasma is compressed and allowed to expand, a process continuously repeated every second, generating fusion heat during the compression stage.

MTF was initially proposed at the US Naval Research Laboratory in the 1970s as a "compromise between the energy-intensive high magnetic fields needed to confine a tokamak plasma, and the energy-intensive shock waves, lasers or other methods used to rapidly compress plasma in inertial-confinement designs." A British company, GF, based in Culham, Oxfordshire, near the original JET research site, hopes to have a demonstration plant running by 2025 that will "power homes, businesses and industry with clean, reliable and affordable fusion energy by the early 2030s."[230] One should never overestimate the claims of private companies keen to satisfy short-term investment goals, nor the perpetually bold predictions of limitless fusion in our lifetime, but it is still exciting to imagine the possibilities.

∗∗∗

Fusion also sees a fair share of outside-the-box thinking, creativity as important as science when it comes to inventing. Even simple ideas can help, such as a sand pile informing researchers about criticality in a fusion containment chamber. For example, by continuously dropping sand grains on a flat surface, one sees the pile rise with each drop, building up in regular fashion despite occasional grain clusters sliding down the sides, until all of a sudden the whole pile fails. Learning more about how a sand pile fails can help us understand more about how criticality is lost inside a fusion reactor that was stable up to the last moment.

Another "table-top" fusion device called a "fusor" has been employed in a number of university labs to create a working neutron source. One such device was built by Taylor Wilson at the University of Nevada, Reno, costing $100,000, orders of magnitude less expensive than the $3.5 billion NIF, $13 billion JET, or $20 billion ITER (so far). A fusor "shoots" positively charged deuterium ions at each other from the inside wall of a spherical chamber that are attracted at high speeds to a central, negatively charged, golf-ball-size tungsten/tantalum grid, igniting a fusion reaction when the ions collide (some collide, some miss, some come back for another pass), seen in the tell-tale, blue-white color of the generated plasma. All of this is done under a vacuum comparable to interstellar space, that is, 10^{25} times less dense than on Earth, and requires a highly pure, ultra-thin meshed grid.

Although not designed to generate electricity, the fusor advances our understanding of practical terrestrial fusion and can be used to make affordable radioisotopes if the resultant neutron output is sufficiently focused. Short-lived artificial radioisotopes for nuclear medicine are currently made off site and transported to hospitals, a time-constrained delivery process that would benefit from a cheaper, readily available, onsite device. What's more, the fusion process can be analyzed at a fraction of the cost of Big Science research. A fusor is almost child's play, especially as Taylor Wilson was only 14 years old when he built one at the University of Nevada, begging the question about why more money isn't spent on Small Science.

In *The Boy Who Played with Fusion*, the extraordinary story about Taylor's nuclear obsession, which includes building his first fusion reactor in his parents' Arkansas garage before graduating high school, Tom Clynes notes that "While the amateurs' experiments aren't nearly as advanced as those done in multibillion-dollar facilities, it's conceivable that developers of these homebrew reactors could play a vital role in moving fusion forward, as citizen scientists have in other realms."[231] Taylor would go on to wow the science community at the Intel International Science and Engineering Fair, further explaining in a few TED talks how fusion can help radiotherapy and bomb detection.

We should also mention cold fusion, although one shouldn't be misled by the hype. "Cold fusion" continues to be studied in various research labs, despite being ignominiously slammed after electrochemists Stanley Pons and Martin Fleischmann claimed to have solved the seemingly unsolvable in 1989. Three decades after their famously ill-advised press conference at the University of Utah, which was hurried to get the news out before another competing group could steal their presumed glory, cold fusion is still spoken of in whispered terms. Of course, cold fusion is not a fusion process, but rather a chemically assisted or low-energy nuclear reaction (CANR or LENR), doomed by its misleading moniker.

Some form of reaction clearly occurs when an electric current is passed through a jug of heavy water (D_2O) containing a palladium and a platinum electrode in a typical CANR/LENR experiment – liberating more heat than chemically expected – but what exactly is going on is uncertain. The results have been hard to replicate consistently, the cornerstone of science, thus hampering a full understanding and more funding.

Some believe the status of "cold-fusion" technology is similar to that of semiconductors in the 1950s,[232] while others note that the research is valuable but mislabeled. Hal Fox, the editor of *New Energy News*, was adamant about Pons and Fleischmann's original discovery:

[T]he discovery of cold fusion, although vigorously attacked (especially by hot fusion lobbyists), marked the beginning of a series of discoveries of low-energy nuclear reactions. Pons and Fleischmann deserve a Nobel Prize. The nuclear reactions are complex and not, as yet, fully explained.[233]

Nonetheless, if they can get the science worked out and the results replicated (and explained), palladium is very limited and expensive, and thus a working LENR device would be hard to scale up to a usable MW-size device, although one possibility is to use a non-platinum group metal catalyst. As always, the future awaits with fusion or fusion-like research.

There are many purported "new" energies on the horizon that never quite live up to their billing, touted by those who have big ideas but can't find a way to share them. Most are of the snake-oil variety or variations on a perpetual-energy device with or without the hidden wires. One company claims to have tapped into a sub-ground-state energy of the hydrogen atom, offering the power of "2,000 Suns in a coffee cup." Another didn't know why their supposed over-unity device "worked," but stated "It absolutely does." If it looks like a duck, walks like a duck, and quacks like a duck, ...

There are also fusion detractors, who think neutron-induced radioactivity and tritium storage are potential problems, presenting more challenges to ensure the safe operation of a working fusion reactor. Radioactive reactor components, lithium blankets, and plutonium waste will all need to be safe-guarded and disposed of, although their collection and storage won't create anywhere near the problems of long-lasting radioactive fission products. A fusion reactor is also expensive to build, can be repurposed to collect weapons material (Pu-239), and still requires long-distance grid transmission as in any large central thermal power plant model.

Why all the effort for something that may never work and requires a large fortune to build? Despite the extraordinarily high development costs and uncertain results, fusion-generated electricity could run as low as 0.001 cent per kWh, 10,000 times less expensive compared to the roughly 10 cents per kWh in today's power plants. If we can get the mega machines to work, the price tag will be worth every penny to run a power station on water.

Indeed, fusion fuel costs are minor compared to fission. Deuterium is fairly easily obtained – D_2O exists naturally in about 1 in 7,000 parts H_2O – and the more expensive tritium in a heavy-water reactor or as is hoped on the go in a neutron-multiplying reactor blanket of a next-generation fusion device. A working 1-GW fusion reactor, however, will need about 50 kg per year of tritium at about $1.5 billion per year if tritium breeding proves difficult, another obstacle as only about 500 grams per year is currently available.[234] Machine costs are naturally expensive for a developing technology, but replacement costs should be reasonable. Reactor walls will need to be replaced after becoming radioactive and radiation-degraded by prolonged neutron bombardment.

No one said bottling a star would be easy, but fusion may well be the answer to all our energy needs and become too cheap to meter . . . one day – all with few contaminants and water as the only fuel. But many challenges still remain, chiefly that more energy is needed to fuse hydrogen atoms than we get out, a losing venture any way you slice it. Containing the Sun on Earth is still not possible beyond the hugely complex and very expensive.

There are plenty of reasons to keep trying, such as cheap, clean, and abundant energy, but there is still much to do to contain the heat of a man-made Sun. It seems nothing is as good as the real deal (as we look at now).

Part I Coal, Oil, Nuclear Milestones

Table I.1 *Milestones in the age of coal and steam*

1776	James Watt invents the first general-purpose steam engine with separate condenser at the University of Glasgow
1798	Benjamin Thompson publishes "An Experimental Enquiry Concerning the Source of the Heat which is Excited by Friction"
1807	Outdoor street lighting in London begins with coal gas
1811	Robert Fulton runs a steamboat *up* the Mississippi River
1830	The first official passenger train service runs from Liverpool to Manchester
1838	SS *Sirius* makes the first transatlantic crossing faster than sail
1869	The Union Pacific and Central Pacific railways meet at Promontory Summit, Utah
1879	Thomas Edison produces a long-lasting incandescent electric light bulb at his lab in Menlo Park, New Jersey
1882	The first commercial electric power plant starts operating in Pearl Street, Lower Manhattan, powered by Edison's coal-fired Jumbo dynamos
1884	The first dynamo turbine is designed by Charles Parson at Holborn Street, London

Table I.2 *Milestones in the age of oil*

1859	"Colonel" Edwin Drake finds oil at 69.5 feet in Titusville, Pennsylvania, after drilling for months
1864	Nikolaus Otto designs a 4-stroke internal combustion engine run on piped-in coal gas
1882	John D. Rockefeller forms Standard Oil Trust, a centrally organized "corporation of corporations" to circumvent state laws
1900	The 35-hp Mercedes racing car is introduced by Wilhelm Maybach
1909	Model T Fords roll off the assembly line in Dearborn, Michigan
1932	A former WWI British army quartermaster, Major Frank Holmes, finds oil off the Saudi Arabian coast in Bahrain
1944	Returning from the Yalta conference, Franklin Roosevelt meets the Saudi Arabian king Ibn Saud aboard the USS *Quincy* in the Suez Canal
1956	M. King Hubbert presents his "peak oil" paper at an American Petroleum Institute meeting in San Antonio, Texas
1960	OPEC is created following discussions between Juan Pablo Pérez and Abdullah Tariki, representatives of Venezuela and Saudi Arabia
1973 / 1979	Oil prices rocket from the First Oil Shock after an Arab oil embargo and the Second Oil Shock after the toppling of the last shah of Iran
1989 / 2010	*Exxon Valdez* spills 11 million gallons of crude into Prince William Sound after hitting a reef. *Deepwater Horizon* spills 200 million gallons of crude into the Gulf of Mexico after a blowout preventer failure

Table I.3 *Milestones in the nuclear age*

1903	Marie Curie shares the Nobel Prize in Physics with Pierre Curie and Henri Becquerel for the discovery and research on spontaneous radiation
1909	Ernest Rutherford postulates a dense, positively charged, central nucleus after alpha particles fired at gold leaf were deflected back at the source
1932	James Chadwick discovers the neutron at the Cavendish Lab in Cambridge, roughly equal in mass to a proton but without any charge
1938	John Cockcroft and Ernest Walton build the first atom smasher, breaking apart lithium atoms with protons
1939	Lise Meitner calculates the considerable energy released from fission of a U-235 uranium atom bombarded by neutrons (200 MeV)
1939	Albert Einstein signs a letter to President Roosevelt warning of a possible German nuclear bomb-making program
1942	Enrico Fermi builds the first nuclear reactor (CP-1) in a squash court at the University of Chicago
1945	The first atomic bombs are dropped on August 6 on Hiroshima (uranium-gun design) and on August 9 on Nagasaki (plutonium-implosion design)
1953	President Eisenhower gives his "Atoms for Peace" speech to the UN General Assembly
1955	The first nuclear-powered submarine, the USS *Nautilus*, is launched
1956	The first civilian nuclear reactor, Calder Hall, goes online, a 92-MW, graphite-moderated, gas-cooled reactor in northwest England
1957	The first major nuclear accident at the Mayak Soviet nuclear weapons complex
1979	Three Mile Island partial meltdown (March 28)
1986	Chernobyl hydrogen explosion (April 26)
2011	Fukushima multiple explosions and triple meltdown (March 11)

Table I.4 *A few fusion milestones*

1905	$E = mc^2$ equation appears in Einstein's "Does the Inertia of a Body Depend Upon Its Energy Content?" *Annalen der Physik* paper
1920	Arthur Eddington suggests the Sun is a fusion reactor
1950s	ZETA z-pinch tested at Didcot, UK
1958	First Russian tokamak T1 constructed at the Kurchatov Institute
1960	Ted Maiman demonstrates the world's first laser at Hughes Aircraft
1967	Hans Bethe wins a Noble Prize for explaining how stars burn
1972	John Nuckolls publishes ICF fundamentals in *Nature*
1984	Joint European Torus (JET) in Culham, Oxfordshire
1991	Almost 2-second plasma at JET
2008	ITER breaks ground in Cadarache, France
2009	First NIF shot breaks laser power record
2016	First W-7X hydrogen plasma at world's largest stellarator
2022	NIF "break-even" with Q of 1.5
2025	ITER's planned first plasma
2035	ITER's planned DT internal heating
20??	DEMO up and running, a working fusion reactor

PART II

In with the New

Any sufficiently advanced technology is indistinguishable from magic.

Arthur C. Clarke

If the facts don't fit the theory, change the facts.

Albert Einstein

Science is organized knowledge. Wisdom is organized life.

Immanuel Kant

4

Old to New: The Sun and All Its Glory

4.1 Our Massive Solar Source

Solar has arrived. In fact, solar has been arriving for 4.5 billion years, about 170 petajoules every second from a massive star at the center of our solar system, enough to power the world's energy needs many times over. Our Sun is the perfect source, continuously fusing hydrogen into helium in a nuclear reaction at its core, where the temperature is almost 16 million degrees. In the process, mass is converted into energy ($E = mc^2$), emitted primarily as electromagnetic radiation from the Sun's surface, along with various charged and uncharged particles. Currently consisting of 71% hydrogen and 27% helium by mass,[1] by the end of its life the Sun will have burned off most of its hydrogen fuel before expanding as a red giant to encompass the inner planets and the Earth. Not to worry, classified as a "G2 main-sequence yellow dwarf," our Sun is happily middle-aged and will continue to shine and emit useful energy for at least another billion years, until its rising luminosity turns the Earth into an uninhabitable, Venus-like planet.

What we see here on Earth is a distribution of photons of varying energies (wavelengths), comprising the visible spectrum (the "viz") from lower-energy red to higher-energy violet, appearing as a mélange of white light. Infrared (IR) and ultraviolet (UV) radiation is also emitted beyond the viz, the complete electromagnetic spectrum characteristic of the temperature of the Sun's surface (photosphere) at 5,800 K. The visible range – roughly 380–780 nanometers (nm) in wavelength – accounts for 43% of the radiation, the longer-wave IR (>780 nm) 53%, and the shorter-wave UV (<380 nm) 4% (as shown in Figure 4.1).[2]

After being ejected from the photosphere, each photon takes a little more than 8 minutes to travel the vacuum of space to Earth, easily calculated because the Sun is 150 million kilometers away (defined as 1 astronomical unit or 1 AU)

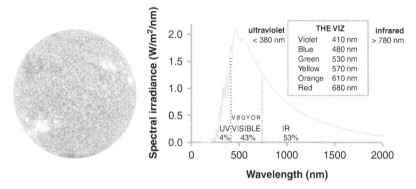

Figure 4.1 The Sun and its electromagnetic spectrum: ultraviolet (UV), visible
(violet, blue, green, yellow, orange, red), and infrared (IR). (*Note*: each color has
a discernible range, but only one wavelength is listed, for example, red (680 nm)
covers about 620–780 nm.)

and all electromagnetic radiation travels at the speed of light, c = 3.0 × 10^8 m/s
(about 1 billion km/h). As such, the time for a solar photon to reach us here on
Earth is the distance to the Sun divided by the speed of light, that is, 150 ×
10^9 m/3.0 × 10^8 m/s = 500 seconds or about 8.3 minutes.[3]

Only a miniscule percentage of the Sun's emitted photons, however, make
the 8-plus-minute journey to Earth, because solar energy is radiated outward in
all directions and our tiny planet is but a pin prick in the vastness of space (the
Earth's radius, R_E, is 6,400 km while 1 AU is 150 million km). We can work out
the percentage from the ratio of the Earth's exposed area to the surface area of
a propagating radiation sphere out from the Sun at a radius of 1 AU, that is,
$\pi R_E^2 / 4\pi AU^2 = 0.46 \times 10^{-9}$ or roughly one 2-billionth of the Sun's continuously
emitted energy – very little, relatively speaking. Fortunately for us, there are
gazillions of photons traveling our way every second from our massive central
star.

One problem with working out numbers in the solar system is the relative
scale. We can't easily represent the Sun and Earth together, because the Sun is
a distance of about 12,000 times the Earth's diameter but is more than
one million times larger by volume, that is, 1.3 million Earths can fit into one
Sun, but only 12,000 Earths need to be laid end to end to reach the Sun. There
are various analogies with different-sized fruits, vegetables, and other objects
to show the relative sizes together. Mine is a ten-pin bowling ball and a pinhead,
who are also playing tennis. If the Sun is a bowling ball ($r \sim 110$ mm) serving
a stream of photons from the baseline of a tennis court, then the Earth is a tiny
pinhead ($r \sim 1$ mm) receiving the continuous electromagnetic energy just inside
the opposite baseline (shown in Figure 4.2).

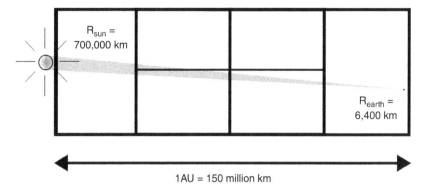

Figure 4.2 Solar radiation reaching a distant and tiny Earth (one 2-billionth of the energy radiated from the Sun's 5,800-K surface). (The Earth is shown 100 times larger at this scale.)

Despite the miniscule percentage reaching us here on Earth, the Sun's energy is more than sufficient to sustain life, directly as sunlight and through various chemical and physical processes: (1) solar energy trapped by photosynthesis is stored in plants and animals, forming fossil fuels over long periods of time, a.k.a. "fossilized sunshine" in peat, coal, oil, and natural gas (as well as plant biomass); (2) differentially heating the Earth's atmosphere and oceans, giving rise to wind, currents, and weather; (3) thermally heating various storage media (for example, rock, sand, water); and (4) converting photons to electrons in a modern photo-voltaic solar cell. The Sun's energy also increases and decreases during the day as the Earth rotates – strongest when the Sun is highest in the sky (local noon) and weakest at sunrise and sunset – as well as throughout the year depending on the day of the year and one's latitude.

Energy, wavelength, frequency, and color are all interchangeable in the physics of light.[4] Interestingly, unlike our red- and blue-colored water taps, blue (480 nm) is hotter than red (680 nm), as the English potter Josiah Wedgwood first noted when he saw that temperature was directly related to the color of the fired clay in his heated kilns that reached up to 1400°C, running through the spectrum from red to blue as the temperature increased.

We can see the same correspondence in a fireplace or campfire when wood or coals turn from red to blue to white as the fire burns hotter. The Sun's energy can also be seen in the color of the sky: blue during the day as shorter-wavelength photons are preferentially scattered in the thinner, overhead atmos-phere and red at sunrise or sunset as more longer-wavelength photons are scattered in the thicker, oblique-angled atmosphere (more than 10 times the

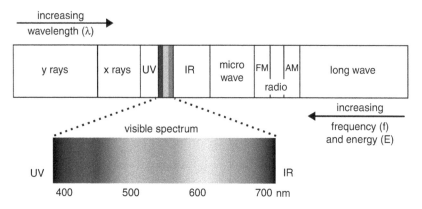

Figure 4.3 The electromagnetic spectrum in wavelength (nm) and frequency (hertz). Energy depends on frequency (E ∝ f) and inversely on wavelength (E ∝ 1/λ).

distance), while clouds are white because larger water molecules scatter all incoming light.[5]

Electromagnetic (EM) radiation exists across a range of wavelengths (shown in Figure 4.3), visible light just a small part. The visible region is from violet to red (380–780 nm), demonstrated by Isaac Newton in 1665 in his famous prism experiment where he separated white light into component colors in the study of his Woolsthorpe Manor home about 100 miles north of London. Generally, we think of six spectral colors, although the number is somewhat arbitrary. Newton had asked a friend to divide his newly understood spectrum and when the friend came up with only six colors, Newton added a seventh, indigo, in harmony with his religious superstitions.[6] Christiaan Huygens, the Dutch lens grinder, astronomer, and early theorist on the wave properties of light,[7] wanted to know what caused the refraction of white light into separate colors – chromatic aberration is a significant problem with telescope lenses – but the ever-pragmatic Newton was unconcerned, writing to Huygens that separating and recombining colors of the rainbow was "a tedious and difficult task . . . . But to examine how Colors may be explain'd *hypothetically*, is besides my purpose."[8]

The German–English composer turned telescope maker and astronomer William Herschel and his astronomer sister Caroline were the first to posit invisible energy *outside* of the visible range, after splitting sunlight with a prism and observing that "light" beyond the red end of the spectrum heated their detector. Originally called "calorific" rays by Herschel because of the obvious heating effect, the newly discovered invisible energy band became known as infrared (IR), "infra" meaning "beyond." More than half of all solar radiation is infrared. At the other end of the visible spectrum, another band of invisible energy was discovered by the German chemist Johan Ritter, called "ultra"-violet (UV), that is, higher in energy than violet, and accounts for about 4% of the Sun's irradiance. Today, we commonly speak of IR (unseen thermal waves) and potentially dangerous UV reaching us on a beach, particularly wary of the higher-energy (shorter-wavelength) UV radiation damaging our skin cells.

Adapted to function in the Sun's radiant energy range over millions of years, we see only a small part of the overall EM spectrum, evolution having made our eyes sensitive to the dominant part of the Sun's energy, that is, the "visible" range from violet to red. A 2020 Princeton-led study reported that humming-birds perceive non-spectral colors and thus UV light, thanks to four light-sensitive color cone types in their eyes (red, green, blue, and a fourth, purple) as do other birds, fish, and reptiles (as did dinosaurs), important for signaling and foraging.[9] Humans with only three types are stuck seeing from violet to red.[10]

Interestingly, peak solar energy is in the blue part of the spectrum (~475 nm), but selective absorption in the atmosphere transmits more green (~530 nm) to the Earth' surface. Plants, however, don't use green light and only appear green because they absorb other ranges and not green (chlorophyll absorbs primarily in the blue and red and thus plants reflect back in the green range). Fortunately, ozone (O_3) in the Earth's upper atmosphere absorbs most of the Sun's higher-energy UV light (~380–200 nm), which also explains why you can't get a tan through a car window because glass is primarily made of silicon dioxide which absorbs UV. Higher-energy x-rays (<10 nm) are also absorbed by air, similarly to how calcium preferentially absorbs x-rays to image bones in the body, showing that a terrestrial x-ray laser weapon is the stuff of science fiction because x-rays are absorbed by air.

As seen in a high-tech military operation or police raid on TV, night-vision goggles detect IR radiation, which we feel as heat, presenting a false-color map to image objects in the dark (red is shown hotter than blue!). Natural gas pipeline monitors also operate in a similar fashion to detect methane leaks radiating out as heat (IR). Although the eye cannot image light beyond about 780 nm on the red end of the spectrum, if you put a television remote in the camera frame of a mobile/cell phone, you can "see" the infrared signal winking like an eye when you press a button. The camera's CCD detector is made of silicon, which can detect IR photons to about 1,100 nm (a typical TV remote sends IR signals at 940 nm).

Perhaps on another planet with another atmosphere that revolves around another sun with a different surface temperature than our own 5,800-K Sun, an extraterrestrial eye will have adapted to see other parts of the EM spectrum, say, 1,100 nm (IR) from a 2,600-K surface-temperature star, such as one of the seven potentially habitable "extrasolar" planets orbiting the recently discovered TRAPPIST-1 red dwarf, or 320 nm (UV) from a 9,000-K star.[11] The 30 billion or so other K- and M-class red dwarf stars at between 2,500 and 4,000 K in our galaxy could also be ripe for non-human, infrared eyes. Here on Earth, however, the eye has adapted to the 380–780 nm band of visible solar radiation because of our Sun's 5,800-K photosphere and selective absorption by component gases in our planet's atmosphere.

Auroras and localized discharges of lightning also show us nature's color range, indicating the composition of the Earth's atmosphere – mostly nitrogen (78%), oxygen (21%), and argon (0.9%), with trace amounts of CO_2, CH_4, NOx, O_3, and water vapor. Electrons and ions from the solar wind are funneled toward the Earth's poles where the magnetic field lines are strongest, producing pulsating wisps of colorful aurora emanating near both poles (borealis at the North Pole and australis at the South Pole). As the Sun's extended atmosphere interacts with the Earth's atmosphere, the "excited" electrons return to their "ground" states to emit gorgeously eerie greens and reds (characteristic of atmospheric oxygen) and blues (characteristic of atmospheric nitrogen). Green is the most common color, produced by atmospheric oxygen between 90 and 200 km up. Lightning causes the same electron transitions to light the sky in wondrous atomic color.

But how do we know how much power the Sun emits or how much energy reaches the Earth as useful EM radiation? Here, we consider the Sun as a continuously radiating blob, emitting power relative to its size and surface temperature, called a *black-body* radiator in physics (think of a heated black bowling ball).[12] Power, P, is proportional only to the surface area ($A = 4\pi R^2$) and the fourth power of the temperature (T^4) of a radiating black body, according to the Stefan–Boltzmann law, that is, $P = 4\pi R^2 \sigma T^4$ (σ is the Stefan–Boltzmann constant in units of $W/m^2/K^4$). Assuming that one-third of the Sun's radiation is absorbed in the Earth's atmosphere, the Slovene physicist Josef Stefan worked backwards from this relation to calculate that the temperature of the Sun's surface was 5,700 K, remarkably close to today's accepted value of 5,778 K, while his student, the enigmatic Austrian physicist Ludwig Boltzmann, derived the theory.

So, for a black-body radiator the size of our massive Sun ($R_{Sun} = 695,700$ km!) with a surface temperature of 5,778 K, we get 384×10^{24} watts of solar radiation, about one 2-billionth of which is incident on Earth (as we saw above), producing a total solar power at the earth's upper atmosphere of 173 × 10^{15} watts (173 petawatts). The roughly 170 petawatts of continuous sunshine is more than 20,000 times the current electrical capacity here on Earth. By comparison, a bowling ball ($r \sim 110$ mm) heated to the same temperature as the Sun's photosphere would emit about 10 MW (assuming it didn't melt in the process), while the human body at rest radiates about 100 W (assuming normal body and room temperatures).[13]

To convert the incident solar power into a more useful number, we divide by the Earth's exposed area (πR_E^2) to get an average solar flux of 1,370 watts per square meter (W/m^2), known as the "solar constant," that is, the amount of solar energy crossing the Earth per square meter every second before entering the atmosphere (Figure 4.4).[14] By comparison, Mercury receives almost 9,000 W/m^2 while distant Pluto receives only 1 W/m^2, the two extremes of our vast solar system.

Not everywhere on Earth, however, is the same, *radiatively* speaking. The time of year, time of day, and latitude increases or decreases the Sun's energy at a given place on a spherical Earth as does the landscape, weather, and the Earth's roughly 100-km thick atmosphere. The solar constant also changes from year to year (about ±0.5 W/m^2 per year over an 11-year sunspot cycle[15]), but 1,370 W/m^2 is a useful reference for the solar flux, increased or decreased by atmospheric reflection/absorption, latitudinal variation, and seasonal differences. If we could convert all of the Sun's incident power on 10 m^2 of rooftop solar panels into electricity, we could generate 1,370 W/m^2 × 10 m^2 or 13.7 kW, enough for daily consumption in most homes. With more than a billion roofs around the world that's a lot of converted sunshine.

About 20% of the Sun's energy, however, is reflected back into space from the upper atmosphere, while cloud cover can absorb up to 30% more, reducing the net energy at the Earth's surface by about half. Neither are solar panels 100% efficient (which we'll look at later), but 10 m^2 isn't that big an area. A typical Sun-facing rooftop might be 100 m^2, converting plenty of solar

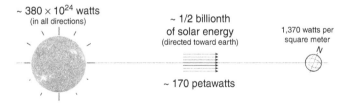

Figure 4.4 Solar radiation incident on a distant Earth: 170 × 10^{15} W transmitted through the vastness of space to become 1,370 W/m^2 (the solar constant).

energy into useful electrical power. If only 2% of the Sahara Desert (0.02 × 9.4 million km^2) receiving half of the average solar flux of 1,370 W/m^2 was covered with low-cost solar panels at about 20% efficiency (that is, converting one-fifth of the incident sunlight into electricity) – and the generated electric power could be effectively distributed – we could provide enough electricity to satisfy the world's needs. Double the efficiency and the power doubles; doable, but more expensive.

4.2 The Solar Source on Earth

A 2008 *Scientific American* article entitled "A Solar Grand Plan" showed how solar power could end American dependence on foreign oil and significantly reduce greenhouse gas emissions, stating that "Solar potential's energy is off the chart" and "The energy in sunlight striking the earth in 40 minutes is equivalent to global energy consumption for a year."[16] Tokuji Hayakawa the founder of the Japanese electronics company Sharp, which produced the world's first solar-powered calculator in 1976, stated as early as 1970, "I believe the biggest issue for the future is the accumulation of solar heat and light. While all living things enjoy the blessings of the Sun, we have to rely on electricity from power stations. With magnificent heat and light streaming down on us, we must think of ways of using those blessings. This is where solar cells come in."[17]

To calculate the average solar energy in any location from the incident solar radiation (or "insolation"), the solar constant (1,370 W/m^2) is prorated by a factor representing the latitude and ground conditions, thus accounting for the varying sun angles on a revolving and orbiting Earth. To be sure, sun in the Sahara (latitude ~ 25 °N) is not the same as sun in Munich, Germany (48.1 °N), or Anchorage, Alaska (61.2 °N), but the same factor roughly applies at similar latitudes in both hemispheres, so Los Angeles (34.0 °N) receives about the same average insolation as Sydney (33.9 °S). The length of day also affects the available solar energy, stretching and shrinking between the equinoxes (typically March 21 and September 21), with as little as 9 hours of sun in winter and as much as 15 hours in summer for mid-latitude locations. And, of course, there is no sun at night.

A world solar energy map highlights the high-insolation regions in and around the tropics, where the Sun's rays are most vertical during the year. Only about 35% of the world's population live in the tropics, however, so connecting available solar energy to existing populations is essential to implement large-scale distribution of solar power – 3% live south of the Tropic of

Capricorn (<23.4 °S), while a whopping 62% live north of the Tropic of Cancer (>23.4 °N), not surprisingly given the greater land mass. As can be seen, Saudi Arabia is blessed with an abundance of sun at the edge of the tropics to go with its plentiful petroleum reserves. The southwest USA, Mexico, and northern Africa are also equally sun-kissed, as are parts of South America, South Africa, and Australia (Figure 4.5).

Note that temperatures are not hotter in summer because the Earth is closer to the Sun as one might casually think, but because the Earth is tilted on an axis, which orients each hemisphere more towards the Sun during summer and less during winter (Figure 4.6). In fact, the Earth is closest to the Sun on or about January 3 (perihelion = 147 million km) and furthest from the Sun on or about July 4 (aphelion = 152 million km), and thus the Earth receives more overall

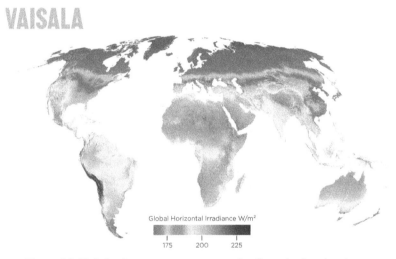

Figure 4.5 Global solar energy map: average irradiance by location (*source*: Vaisala).

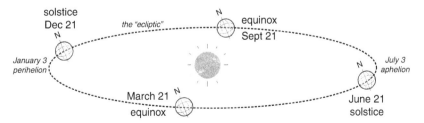

Figure 4.6 The Earth's annual journey around the Sun (angled at 23.4°).

solar energy in January than in July, the solar constant varying from perihelion to aphelion by almost 7% (1,412–1,321 W/m²). The perihelion also travels from year to year as the tilt, axial precession, and shape of the orbit (eccentricity) change. When the three orbital cycles line up to produce the coldest northern summer, we get an ice age.[18]

Originating about 4.5 billion years ago from the "big splat" – an asteroid knocked off a chunk of the Earth to form the moon and adjusted the Earth's polar axis in the process – the Earth's 23.4-degree tilt gives us our seasons, the tropics (23.4 °S to 23.4 °N), and an ever-changing overhead sun angle.[19] The tilt also varies between 22.1° and 24.5° over a 41,000-year period, changing season dates and terrestrial markers. For example, the Arctic Circle is currently moving north by about 15 meters a year as the tilt decreases towards its minimum.

As one can attest to from everyday experience, the angle of the Sun is a major factor in solar efficiency, lower in winter (more horizontal) and higher in summer (more vertical), changing by 7.8 degrees each month (2 × 23.4°/6) and ranging from a minimum at the start of winter to a maximum at the start of summer (roughly December 21 and June 21). The angular difference between the two solstices is twice the Earth's tilt, while the lowest is 0° (only possible in the polar regions, thus producing a 24-hour night in winter) and highest directly overhead at right angles to the ground (only in the tropics).[20] Where my wife and I live in northern Spain (43.5 °N), the Sun varies from about 23° to 70° producing 2.4 times maximum power at summer peak to winter bleak (Figure 4.7).[21]

Calculating sun angles on a rotating and orbiting Earth is confusing at best, reminiscent of a tortuous geometry or trigonometry class, but if one looks due south in the northern hemisphere or due north in the southern hemisphere, you will see how high the Sun is above the horizon or below the zenith at any time of the year, best worked out at local noon when the Sun is at its highest.[22]

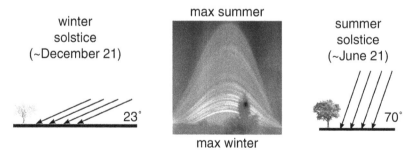

Figure 4.7 Changing solar angle during the year (*source*: Elekes Andor CC BY-SA 4.0).

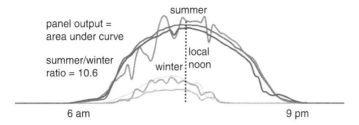

panel output =
area under curve

summer/winter
ratio = 10.6

summer

local
noon

winter

6 am   9 pm

Figure 4.8 Solar output during the day from a fixed rooftop solar panel in summer
and winter.

Today, numerous websites list daily solar data at any latitude, although you can
easily construct a simple instrument to work out the changing sun angles during
the year by hanging a bobbed string from the middle of an upside-down
protractor's straight edge and aligning the edge with your eye and the Sun
(note, one should never look directly at the Sun). The same instrument can be
used to calculate the altitude of a star, essentially a simple quadrant that helped
early explorers navigate the seas.

Highlighting the enormous effects of seasonal variation and cloud cover,
Figure 4.8 shows the measured output from a fixed rooftop solar array over
three consecutive days in summer and winter. Here, one sees the difference
between the summer and winter peaks and increased output in summer
because of more sun and a longer day – the peak is roughly 2.5 times in
summer, while total power is more than 10 times (areas under the curves).[23]
Note the one-hour time shift of Daylight Savings on peak power (local
noon) and the ragged effect of intermittent clouding through the day.
Figure 4.9 shows the insolation at three different cities in the USA, indicating the changing solar output from south to north and through the year. Both
figures elegantly explain the effect on the ground of a constantly changing sun.

There is plenty to visualize about changing sun angles on a spherical surface,
but to maximize incident light on a solar panel or solar collector one clearly
orients a panel perpendicular to the Sun. Turning a solar panel from side to side
during the day and up and down over the year to intersect the Sun at the
optimum right angle also increases the conversion efficiency of incident solar
energy to electrical power, standard practice in a solar farm. Proper orientation
is essential to maximize the output as is regularly hosing down panels with
water to remove dust and debris, especially on flat-roof surfaces (and clearing
snow in winter climes).

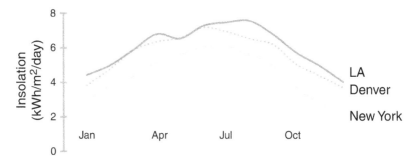

Figure 4.9 Solar output (kWh/m²/day) during the year from a fixed rooftop solar panel in Los Angeles, Denver, and New York (*source*: "PVWatts Calculator (My Location)," V8, US Department of Energy National Renewable Energy Laboratory (NREL). https://pvwatts.nrel.gov/pvwatts.php).

To be sure, we can't use solar power 24 hours a day as with other electric power-generating technologies, because a solar cell doesn't operate in the dark, but any electrical energy produced during "sun-up" operation can be stored in a chemical battery or other storage media, much improving technologies as we'll see in Chapters 5 and 6. The theoretical maximum power of a solar panel is not always achievable either, down-rated because of time of day (low sun in the morning and evening), time of year (reduced winter power), cloud cover, maintenance, dust, etc., which reduces the overall performance from the rated peak value (W_p) to its operating power (W). Using a 21.7% "capacity factor" to compare different energy technologies, Berkeley physicist Richard Muller noted that "Even in a cloudless desert, the average solar power is only 25% of the peak."[24] Nonetheless, solar is particularly important in the afternoon when more air conditioners and factories operate, important to maximize output.[25]

Selective absorption by component gases in the lower atmosphere also reduces the amount of solar radiation reaching the Earth's surface by up to 50%, depending on the local temperature, pressure, and composition of the absorbing gases, mostly found in the warm, humid, oxygen-rich troposphere within 1 km of the Earth (the atmosphere diminishes with height and is about 100 km thick, that is, the edge of space[26]). Complicating things a bit, the absorption also depends on wavelength: for example, molecular nitrogen and oxygen absorb all energy below 290 nm – fortunately for life on Earth – while ionized gases in the upper atmosphere reflect incoming long-wave radiation above 20 m. Within these two opaque ranges, most atmospheric absorption is either in the UV (200–350 nm) or the IR (780–2,500 nm).

To simplify, we can compare the reduced solar flux to washing lettuce in the sink – some water splashes back at all angles (20%), some attaches to the lettuce (30%), while the rest (50%) gets through, representing back reflection/ scattering, absorption, and the transmission of light. Furthermore, it might appear that all sunlight is white light, but we receive a wide range of wavelengths of differing strengths as Newton, the Herschels, and Ritter first demonstrated, some of which is good for us (indeed essential) and some bad (≤UVC). Note that you can still tan on a cloudy day, because much of the incident energy passes through, including the more dangerous shortwave UVA (95%) and UVB (5%) rays, although thankfully UVC is completely blocked.

What's more, some of the sunlight is reradiated by the Earth (mostly IR) and absorbed in the atmosphere, providing us with life-giving heat, a.k.a. the greenhouse effect. About 10% of the heat is stopped within 10 feet of the Earth's surface, creating an aqueous vapor blanket that the Anglo-Irish physicist John Tyndall noted was "more necessary to the vegetable life of England than clothing is to man."[27] Tyndall also showed in his early experiments on the properties of different atmospheric gases that water vapor, carbon dioxide, and methane absorb EM radiation, although the absorption is now being enhanced by anthropogenic carbon dioxide and methane, the most famous of the greenhouse gases, raising global temperatures faster than in the past. The Earth can adapt to a naturally changing solar irradiance that alternates between warmer interglaciation periods and cooler ice ages, but add the effects of burning fossil fuels and the changes are coming too fast.[28]

Analyzing the Sun's composition wasn't easy. The French philosopher Auguste Comte famously claimed in 1835 that we would never know its makeup or that of the distant stars. But thanks to spectral analysis – an atomic fingerprint showing the amount of energy emitted at each wavelength, called a spectral "line" – we now know what the Sun and other stars are made of. Prior to Comte's famous retort, the German scientist Joseph von Fraunhofer had observed numerous lines in the Sun's spectrum with a prism and was the first to measure their respective wavelengths, although he had no idea the solar composition. Working in Heidelberg from the 1850s, two more German scientists, Robert Bunsen and Gustav Kirchhoff, developed the science of spectral analysis, conjecturing that the spectrum of a heated substance could be matched to the existence of elements within the substance. They saw the same spectral lines in the Sun that Fraunhofer had, but were able to identify sodium among other elements in the thousands of lines.[29] They also discovered two new elements with their more advanced spectroscopic methods (cesium in 1860 and rubidium in 1861).[30]

A dark spectral line was then observed during an 1868 solar eclipse in the yellow part of the Sun's spectrum, but couldn't be matched to any known element. Ridiculed at first, some scientists speculated about the existence of a new element, previously unknown on Earth. The theory of an as-yet-undiscovered element was eventually accepted and helium joined the pantheon of elements in the periodic table, named after the Greek word for sun, "helios" (atomic number 2). In the 1920s, the English-born Harvard astrophysicist Cecilia Helena Payne would seal the deal, positing that the Sun was largely composed of hydrogen (~ 90%) and helium atoms, based on the strengths of their spectral lines and excited ions. The amount of hydrogen and helium was also initially underestimated in the highly ionized solar plasma because of their weak spectral lines (there are no spectral lines without orbital electrons[31]).

Helium was eventually isolated on Earth about 30 years later, leading to the discovery of a whole new row of elements called the noble gases. When thermonuclear fusion was finally understood as the main mechanism in solar luminosity, thanks in part to Payne's earlier spectral analysis, the Sun was no longer thought of as a mysterious, fuel-burning, terrestrial knock-off as initially assumed, but a monstrous, hydrogen-fusing, photon-spitting fire ball, continuously fusing hydrogen to helium and bathing us all in glorious wonder (as we saw in the last chapter). Happily, we don't have to worry about running out any time soon, but if we could harness just a small fraction of that radiant energy, our everyday power needs would be solved. Alas, as we have already seen, trying to reproduce solar fusion in a terrestrial tokamak or stellarator is the closest we've come to reproducing the Sun on Earth, and then only for a few seconds.

The alternative is to use what we can for now: 170 petajoules of radiated solar energy arriving every second in a continuous stream of heat and light, reduced roughly by half before reaching us on the ground,[32] passively heating our homes, generating wind and ocean currents, producing motive steam from heated water in a concentrated solar power plant, or by directly converting photons into electrons in a photovoltaic solar panel or power plant (a.k.a. solar farm).

4.3 The First Solar Revolution: Passive Sun Power

As youngsters, many of us played around with a magnifying glass, trying to focus sunlight on a piece of paper or a dried leaf to make it burn or at least smoke. I'm not sure how many of us thought the Sun's rays could one day power the world – at least not from the direct heat of the Sun's energy shining down on us – but certainly we were aware of a welcome respite from the cold, particularly those of us who have ever lived through the depths of a bitterly cold winter with first-hand experience of months of freezing temperatures.

If you pass through the Turkish countryside, however, you will see the simplest of solar heating devices dotting rooftops everywhere: a thermal solar water heater, which can be as rudimentary as a metal tank – water filled by hand pumping and directly heated by the Sun – or as fancy as a closed-loop system with a horizontal tank above an angled flat-plate solar collector (heat-absorbing metal fin tubes). The heated water rises via "thermosiphon" for smaller systems (under 3,000 liters per day) or forced flow for larger systems where solar-heated water is circulated through a water storage tank to transfer the heat to the household water supply. Today, solar water heaters are found in many sun-drenched Mediterranean locales, perfect for daily dish washing, laundry, and bathing (Figure 4.10).

Widespread since the early 2000s in Turkey, Cyprus, Greece, Israel, Austria, and Japan, solar water heaters can be installed on a rooftop for as little as $2,000 with a reduction of up to 80% on hot-water costs.[33] Huge savings are available, helping to reduce electrical power consumption from hot-water heating, which accounts for 25% of energy use in households and almost 12% in commercial buildings.[34] Global installed capacity is roughly 500 GW from over 100 million operational units, with China (73%) leading the way followed by Turkey (4%) and the USA (4%), two-thirds pumped and one-third thermosiphon.[35]

As noted by a California research group, "in the late 1800s, before oil and gas became available in the West, more than one third of Pasadena residents had solar hot-water systems," while prior to the 1920s solar water heaters were common in California until natural gas lowered fuel prices.[36] As late as the

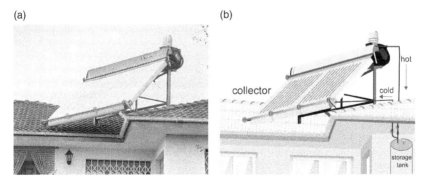

Figure 4.10 Thermal solar: (a) rooftop thermosiphon solar water heater and (b) schematic.

1950s, 80% of Florida homes also had solar water heaters before declining in the free-wheeling, electricity-friendly 1960s.[37]

Solar water heating is less popular the closer one lives to the poles – antifreeze and a heat-exchange water system are needed in below-zero climes. Indeed, there is no mystery as to why the Egyptians worshipped a sun god, Ra, while the Scandinavians preferred warrior deities (Cairo latitude 30°, Oslo 60°). Alas, *electric* water heating is the norm, even in places with plentiful sun, an entrenched modern convenience in a utility-powered world.

The Sun has long been revered by the ancients to keep time. The apparent sun path in the sky is caused by the Earth's annual orbit and daily rotation, precisely followed to indicate optimal planting and harvesting as well as everyday timekeeping as measured by a sundial from a simple shadow-casting gnomon, where the time of the shortest shadow indicates local noon, halfway between sunrise and sunset. The annual return from the cold of a long winter is marked by elaborate stone constructions in such diverse places as Newgrange on the banks of the River Boyne in County Meath, Ireland, Pueblo Bonito in New Mexico, and Stonehenge in southern England.

A Neolithic passage tomb built around 2500 BCE, and thus older than the pyramids, Newgrange is aligned with the rising Sun on and about the winter solstice, such that a narrow beam of light passes through a roof box to fill a small inner chamber at the exact moment of sunrise, intended to take whoever is buried inside to the heavens,[38] while the D-shaped Pueblo Bonito great house in Chaco Canyon, New Mexico, marks the summer solstice as light enters the center of a spiral petroglyph. On Salisbury Plain, Stonehenge also functions as a celestial marker, as do other dwellings around the world that told the ancients when and where the Sun would rise as if a giant alarm clock (Figure 4.11a). For example, a shadow "snake" appears on the Mayan Kukulcán Pyramid at Chichen Itza in the early afternoon on each of the two equinoxes.

Around 240 BCE, the Greek philosopher Eratosthenes calculated the circumference of an assumed spherical Earth from the angle of the Sun in two different locations along the River Nile (Figure 4.11b). Knowing that the Sun shone directly overhead at local noon on the summer solstice in Syene – casting no shadow and completely filling a well – he measured the Sun's angle at local noon in Alexandria where he was head of the famous library, a distance of 5,000 *stade* and practically on the same longitude as Syene. In Alexandria, the Sun's noon shadow cast an angle of 7.5° and so by extrapolating the distance between Syene and Alexandria to a sphere, Eratosthenes knew that the Earth's circumference was 48 times longer, that is, 240,000 stade (5,000 stade × 360°/7.5°) or 44,400 km in today's units, remarkably close to the actual distance of 40,200 km (the ancient Greek unit of distance, the stade, was about 185 m).[39]

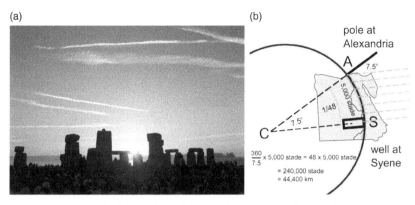

Figure 4.11 Celestial markers: (a) Stonehenge at sunrise on the summer solstice (*source*: Andrew Dunn CC BY-2.0) and (b) Eratosthenes calculates the circumference of a spherical earth at Alexandria (not to scale).

Note that Syene (modern Aswan) is at ~23.5° latitude and Alexandria ~31° latitude, thus one-forty-eighth the circumference of the Earth apart (31° – 23.5° = 7.5° and 360/7.5 = 48).[40]

Taking advantage of the Sun's obvious directional advantages was essential to align an ancient home for warmth. Add sun-facing glass or mica windows to let light in during the day and trap the reradiated heat inside at night and we have the makings of a rudimentary "direct gain" passive solar house. During a discussion with his pupil Aristippus of Cyrene about the nature of beauty, Socrates explained how to make a house both pleasant and warm based on the Sun's direction:

> Now in houses with a south aspect, the Sun's rays penetrate into the porticos in winter, but in the summer, the path of the Sun is right over our heads and above the roof, so that there is shade. If then this is the best arrangement, we should build the south side loftier to get the winter Sun and the north side lower to keep out the winter winds. To put it shortly, the house in which the owner can find a pleasant retreat at all seasons and can store his belongings safely is presumably at once the pleasantest and the most beautiful.[41]

In 213 BCE, the Greek mathematician Archimedes famously focused the Sun to repel the Roman navy at Syracuse in southeast Sicily where he was born, or so the story goes. But after recreating the setup in 2004, the American TV show *Mythbusters* initially declared the claim "busted." Consisting of 61 angled mirrors on a large circular wooden frame focused to a point 60 feet away, their made-for-television contraption managed only a tepid 200°F, hardly a useable "death ray" to repel an invading navy. Co-host Adam Savage noted that even if the reflected heat system could work, "You could only use it at a certain time of day when it is most effective, it's difficult to aim, what are you gonna do ask your enemy to please show up around noon."[42]

In 2005, a group of MIT students redid the experiment with 129 1-foot-square, flat, hand-held mirrors, producing "wisps of smoke" on a bright but cloud-filled day, although they later managed a sustained flame within 10 minutes of clear sky, using mounted mirrors positioned in a two-tiered, semi-circular arrangement.[43] The temperature of the volatiles igniting from the wood was estimated at 1,100°F. This ancient solar heat-making contraption may not have destroyed the Roman navy, but was certainly capable of generating large amounts of heat.

Indeed, the Sun is recognized everywhere for its warmth, ability to store heat (for example, in water or sand), and timekeeping, as well as a celestial marker of our place in the heavens. Native Americans included the Sun in their plans when they built south-facing mountain pueblos as early as 2,000 years ago, their adobe (or "mudbrick") homes known for their high thermal mass (ability to store energy), efficiently absorbing sunlight during the day (energy in) and releasing heat at night (energy out). Access to the Sun was enshrined in Rome's sixth-century Codex Justinianus, rights still not encoded in modern European or American law.

But we must fast forward to the mid-eighteenth century and the Swiss physicist Horace de Saussure to see direct solar energy in action as a solar cooker and other heat-storing devices. Covering an insulated rectangular box with layered glass, de Saussure created a simple solar device to heat water to over 200°F. Today, solar cookers employ parabolic or funnel mirrors to produce temperatures as high as 750°F (400°C), essential gear for an eco-camping trip or hip backyard BBQ. As an avid alpinist, de Saussure also measured the humidity of the Earth's atmosphere at various heights, noting how the Sun equally heated water whatever the altitude.

Clarence Kemp of Baltimore would receive the first patent for a solar water heater. As Kemp noted in his 1890 patent description, the mechanism was straightforward: "the invention consists, primarily, in exposing tanks contain-ing water within a glass-covered box to the heat-rays of the Sun and in providing the said tanks with suitable pipes whereby water is made to pass from one tank to another and be finally discharged in a heated condition."[44] By 1900, Kemp had sold 1,600 "Climax" solar heaters in southern California alone.[45] The wonder is what took so long – it is child's play to warm water by concentrating the Sun's rays, easily seen by leaving a water bottle outside on a hot day and even more so if the bottle is black.

The former Carnegie Steel engineer William Bailey introduced narrower pipes to heat the water faster and added insulation to Kemp's tank to store hot water for nighttime operation, selling thousands of units of his "Day and Night" solar heaters in the 1920s.[46] An early advert egged-on buyers with the enticing slogan "What so Rare as a Cloudy Day in Arizona." Clearly, one can save money via solar-generated heat, as was simply stated in another 1935 *Popular Mechanics* ad: "reduces Gas Bills." Until cheap oil came along, the solar water heater was well-established in many countries, and is still popular today in less-developed countries where the cost to install capital-intensive electric infrastructure is more onerous.

Others found novel ways to "run the sun" by heating water in a steam engine. Inspired by de Saussure, the French mathematics teacher Augustin Mouchot built a sun-fueled, steam-powered plant that employed a "truncated solar dish" to better focus the Sun's rays on a tank of water to increase the collected solar power (Figure 4.12a). Granted the first solar engine patent in 1861, funded in part by Napoleon III, the reflective solar collector was described as "an inverted lamp shade coated on the inside with very thin silver leaf" and the water boiler as "an enormous thimble made of blackened copper and covered with a glass bell," while a tracking mechanism followed both the Sun's altitude and azimuth in the sky.[47] A scaled-up version was exhibited in Tours in 1872 that produced half a horsepower to pump water, while a larger device was installed in Constantine, Algeria, with a multi-tube boiler to increase steam pressure and performance, much like the boiler improvements made on early steam trains.

Representing Algeria, for whom he made both solar-irrigation and distillation pumps in the plentiful sunshine of the Algerian desert, Mouchot wowed visitors at the 1878 Paris World Exposition by running an ice-making machine

(a)   (b)

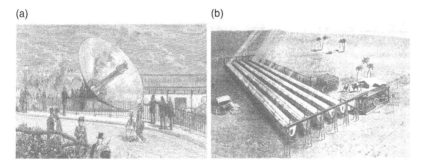

Figure 4.12 Concentrated solar power: (a) Augustin Mouchot's solar-steam-powered ice-making machine at the Paris World Exposition (1878) and (b) Frank Shuman's parabolic trough water-pumping plant on the Nile (1913).

from a solar-powered steam engine, employing a 4-m diameter reflecting mirror with an 80-liter boiler, for which he won the Gold Medal. It seemed crazy, but the Sun's rays could run an engine to make ice!

In 1913, Frank Shuman, an American engineer and the inventor of automobile safety glass, constructed the first large-scale, concentrated solar power plant, which employed a solar-steam-powered engine to pump 6,000 gallons of water per minute for irrigation (Figure 4.12b). After trying various early models in his hometown of Philadelphia – cheaply set up in conditions that were too humid to be efficient – he built his novel "Sun Engine" generator to irrigate fields along the River Nile just south of Cairo. The design was well-suited for dry equatorial regions – steam is reduced by humidity and atmospheric impurities – significantly improving on expensive coal and oil as well as hand pumping in remote areas such as along the Nile or in southern American states that average 90% sunlight. Although initially costing more to set up the absorbers, Shuman noted in a 1911 *Scientific American* article that "the great economy occurs in the item of fuel"[48] – the inexhaustible Sun. Worried about the efficiency of various parts in his early thermal engine, Shuman nevertheless saw that solar power was unlimited and could one day "displace all other forms of mechanical power over at least 10 per cent of the earth's land surface."[49] Alas, World War I and the rise of cheap petroleum derailed his plans, while his unique solar-steam-powered plant on the edge of the Nile was dismantled for war parts.

As petroleum became the primary means of producing power – plentiful, cheap, and relatively easy to extract – further development of a new, sun-powered, energy paradigm stalled, although passive solar technology in building construction slowly emerged after the hostilities ended. Shutters, blinds, drapes, adjustable awnings, plants, even the color of interior materials to absorb or reflect light during constantly changing sun conditions were all employed to improve the thermal properties in buildings (black absorbs ~95% of incident radiation). Passive solar was limited only by the imagination.

The playwright George Bernard Shaw designed a rotating writing hut in the garden of his rural Hertfordshire home, maximizing heat and ventilation throughout the year in a simple passive solar design. As anyone with a poorly designed solarium can attest, sloped roof windows allow more heat in during summer, but lose more heat at night and overheat in warm weather, essentially acting as a greenhouse by letting sunlight in yet trapping the reradiated heat. Simplest ideas are often best. Deciduous trees strategically planted outside let more sunlight in during winter (no leaves) and provide needed shade during summer (full foliage), while coniferous trees keep out winds in known directions the whole year round.

Based on the ideas of Buckminster Fuller, solar domes became popular with the Back-to-the-Land Movement of the 1960s. Concrete slabs, dirt, water drums, mirrored curtains, and Trombe walls (high-heat-capacity interior walls adjacent to a window) all provided efficient "indirect gain" to heat and to ventilate a space completely off grid. Multi-layered windows provided better insulation to trap reradiated heat as do double and triple glazing, now standard in most window designs with a high resistance (R) to conductive heat flow and low emissivity (E).

By the late twentieth century, large office buildings were being designed to reuse the radiant heat emitted from bodies, photocopiers, and overhead lighting. The excess thermal energy was stored in a rooftop or interior pool and recirculated through the building as needed. Today, large atria, waterfalls, and "solar chimneys" all redistribute heat and cool air in interior spaces to moderate temperatures. "Net zero" buildings are all the rage and begin with high-R insulation, producing as much energy as they consume. Note that "net zero" is not *zero* emissions, but a good start to getting there.

Unfortunately, passive solar is still dismissed as a poor man's solution in affluent countries or an esoteric architectural design beyond the reach of most. We may understand the importance of *feng shui* in the home – objects arranged in harmony with the wind (*feng*) and the water (*shui*) as in Chinese tradition – but we don't use the Sun to full advantage or even as Socrates first explained to his student Aristippus. Electricity is how most of us heat and cool our interior spaces, permanently plugged into a utility supply that manages power as part of an external network. Preferring to heat water or dry clothes with electricity on a blistering hot summer's day in the southern USA (and paying to do so!) or in southern Spain, the hottest region in Europe, is a systematic failure in how we design and build our homes. Excluding the Sun in our daily lives is strikingly at odds with simple solutions from ancient Greece, Rome, and the Sun-facing pueblos of the Americas, which have been with us since humans started building homes.

Employing solar power in a spectacularly new way, however, would soon begin after World War II – direct *electrical* conversion of sunlight via engineered "solid-state" materials. The advent of *photo-voltaic* (PV) solar energy would need no proxy fuel to make steam to generate electricity; only photons in a truly transformative technology to harness the radiant energy of the Sun. After a new crop of post-war scientists figured out how to convert photons into electrons – not dissimilar to how plants convert sunlight into energy in photosynthesis – a blossoming electronic age would morph into today's modern photonic world, seen in how we calculate, how we communicate, and – as is becoming more common – how we create energy.

Making the transition to photovoltaic solar power, however, would require an understanding of the quantum properties of matter that followed the extensive radar research into crystals during World War II. The technology that emerged from this new material understanding would fundamentally change a heat-centered, mechanical world, where power is generated by burning the stored energy of hydrocarbon fuel, into an electric wonder, where photons are used to manipulate electrons. The time was ripe to "harvest" electricity from the Sun.

4.4 What Is Photo-Voltaic? The Science of "Harvesting the Sun"

The primary component of a solar cell is the same as in sand, which is mostly made of silica, a.k.a. silicon dioxide (SiO_2). Turning ocean dirt into an efficient silicon solar cell (an electric-current-generating photovoltaic device), however, would require a completely new understanding of matter to control conduction electrons in a solid. After being separated from its oxide, the silicon would also be purified to exacting standards and preferentially "doped" to create an electric field to generate a current from the liberated orbital electrons.

Silicon dioxide has been used to make glass as far back as the third millennium BCE in the time of the Egyptian pharaohs, melted at high temperatures into a liquid and then shaped into small beads. Improved glass-blowing techniques and grinding produced spectacles from the mid-fourteenth century, followed by lenses in the fifteenth century to focus light and the first practical demonstration of the telescope from atop a tower of the Binnenhof in The Hague in 1608. Two years later in Padua, Galileo confirmed Copernicus's paradigm-changing heliocentric solar system with his own two-lens telescope, starting with the discovery of four of Jupiter's moons that clearly didn't orbit the Earth. Glass windows and mirrors, however, remained a rarity for anyone other than the rich until well into the 1800s and the introduction of plate glass, especially in the immense Crystal Palace of 1851, before the invention of the float-glass process in 1952 ushered in the modern era of glass-clad skyscrapers.

Elemental silicon, however, was only first prepared in 1833, by Jöns Berzelius, the Swedish chemist responsible for assigning our modern chemical nomenclature from the initials of an element's name, for example, H for hydrogen and O for oxygen, or, in the event of repeats, an initial plus a second letter, for example, Au for gold (*aurum*) and Ag for silver (*argentum*). An early advocate of the English chemist John Dalton's atomic theory,

Berzelius proposed the dualistic nature of compounds held together by opposing positive and negative electrical charges (for example, $SiO_2 = Si^{+2} + O_2^{-2}$), determining the elemental proportions in such compounds and thus giving chemistry a much-needed mathematical and stoichiometric foundation (for example, $SiO_2 = 1Si + 2O$). A meticulous experimenter, Berzelius also calculated the atomic weights of almost all of the then accepted 49 elements and discovered three new ones as well (cerium, selenium, and thorium).[50] His assistants added six more, including lithium (the lightest metal) and vanadium (the same metal Henry Ford would use in a novel steel alloy to double the strength and dramatically lower the weight of a car). No one at the time, however, could have imagined the future importance of elemental silicon, gleaned from everyday sand.

After oxygen (47%), silicon is the second most abundant element in the Earth's crust (28%). In pure crystalline form, silicon is gray in color with a metallic luster, and is now found in materials as diverse as clay, glass, cement, breast implants (gel form), and as an inert rubber-like household sealant (polymer silicone form). In today's multi-trillion-dollar semiconductor industry, silicon is also essential to manufacture circuit patterns (primarily transistors) etched into meter-wide wafers in a giant, dust-free, fabrication or "fab" plant, cut up into integrated circuit (IC) microchips to run our computers, mobile phones, and USB flash drives.[51] But before we see how silicon has become the main ingredient in today's solar cell and computer chip, we first look at some early light-sensitive materials that converted light into electricity, although with only limited success.

In 1839, Edmond Becquerel discovered the photovoltaic effect (technically, a photochemical or photogalvanic effect) – the basis for converting photons into electrons – at the age of 19 in his father's Paris laboratory. While experimenting with solid electrodes in electrolytic solutions, he generated a voltage by illuminating a silver-chloride-coated platinum electrode immersed in aqueous nitric acid. The son of a pioneering scientist in the field of luminescence and electricity, Edmond would go on to discover light-sensitive silver halides needed to darken film to form a photographic image and was the father of Henri, co-recipient of the 1903 Nobel Prize in Physics with Pierre and Marie Curie for his discovery of radioactivity in a sample of uranium salt.

In 1873, the English electrical engineer Willoughby Smith, chief electrician of the Gutta Percha Company, which manufactured the first undersea transatlantic cable in 1858, saw that light increased the conduction of electrons in thin selenium bars hermetically sealed in a glass tube with platinum contact wires, after observing a doubling of the conductivity from night to day when testing selenium as a possible material to improve electrical transmission in underwater sea cables.[52] About his discovery, Smith noted, "When the bars were fixed in a box with a sliding cover, so as to exclude all light, their resistance was at its highest, and remained very constant, fulfilling all the conditions necessary to my requirements; but immediately the cover of the box was removed, the conductivity increased from 15 to 100 per cent; according to the intensity of the light falling on the bar."[53] Ultimately of little use for long-distance, electric-signal transmission, selenium would become perfectly suited to calculate the appropriate exposure value for photography in a light-measuring device and after 1923 to record sound on film, leading to the first "talkies" (the generated current is proportional to the light).

The interesting light-sensitive electrical properties of selenium were also used to explore possible power sources, although with minimal results. In 1881, the American electrician Charles Fritts built the first working "solar cell," made of thin selenium wafers covered with semi-transparent gold wires, which he fashioned into panels. In 1884, he installed the world's first solar array on a New York City rooftop just 2 years after Edison had started up his Pearl Street power station, noting that the output was "continuous, constant and of considerable force not only by exposure to sunlight but also to dim, diffused daylight, and even to lamplight."[54] The device contained the basic functioning of a current-generating solar cell, and although German inventor and industrialist Werner von Siemens called the invention "scientifically of the most far-reaching importance,"[55] it had a woeful 1% efficiency.

Canadian inventor George Cove also built a rudimentary solar power device in 1905, producing 240 watts from four 4.5-m^2 panels to charge an array of lead–acid batteries. Cove's solar "thermoelectric generator" unwittingly used a "peculiar composition" of metallic plugs, one end exposed to light and the other cooled and sheltered, capturing both light and heat, which he demonstrated atop the Metropole Building in Halifax and then like Fritts on a New York rooftop.[56] Despite an increased efficiency, Cove was also unsure how his device worked.

Regardless the lack of understanding, light could be turned into electricity under certain conditions in some elements despite their curious properties and baffling limitations. The theoretical basis of the "photo-electric" effect wouldn't be fully understood until after quantum theory was established via Max Planck's discrete statistical solution to the black-body spectrum in 1901, Albert Einstein's explanation of quantized light (later to be called "photons") in the first of his famous four 1905 *Annus Mirabilis* papers ("On a Heuristic Viewpoint Concerning the Production and Transformation of Light"), and then in the 1930s with the theory of electronic band structure and band gaps.

Published in the *Annalen der Physik* journal when he was a lowly 25-year-old third-grade Swiss patent clerk, Einstein would eventually win the 1921 Nobel Prize in Physics for the photoelectric effect, showing that the energy of light is proportional to its frequency ($E = hf$, where h is Planck's constant), and that Planck's mathematical fudge factor was in fact a fundamental constant (soon to be called h). Einstein's ideas stood classical physics on its head – light was somehow both a wave *and* a particle – but most importantly he explained how the energy of a photon could liberate electrons in a solid to produce a current.

The photoelectric effect is straightforward: if incident light on a metal conductor is of a sufficient "threshold" energy – that is, when a photon has more energy than the metal's "work function" (a kind of internal atomic resistance) – the metal gives up its bound orbital electrons (Figure 4.13). The energy is enough to knock electrons from their outer atomic shell to produce a current. A typical first-year university physics experiment demonstrates the effect by shining light of three different energies indicated by their colors – usually yellow, green, and blue – on a photocathode detector, by which one can calculate the kinetic energy of the liberated "photoelectrons" from the applied electric force needed to counteract the current (as well as determine the value of Planck's constant).[57]

The photoelectric effect wouldn't immediately be put to use, however, despite a growing commercial need to control the flow of electrons in an electronic circuit, either by blocking them or turning them on and off to manipulate radio waves. With the advent of the telephone and radio in the early twentieth century, vacuum tubes instead evolved to both "rectify" a current, for example, to convert alternating current (AC) to direct current (DC) to drive a speaker, and to "amplify" weak or "feeble" signals that dissipate over long distances to relay long-distance communications.

***

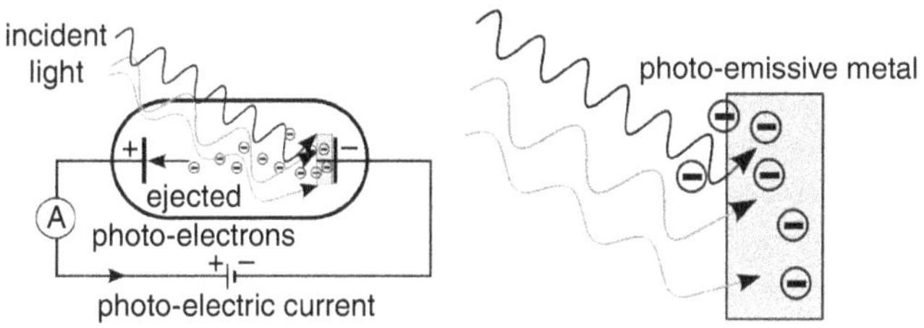

Figure 4.13 The photoelectric effect creates an electric current via liberated photoelectrons when light of sufficient energy is incident on a metal substance.

Prior to the success of solid-state electronics, radio waves were rectified with a vacuum tube "diode," also known as a "valve" for its ability to turn a current on or off depending on the direction of flow. Employing thermionic emission in an evacuated glass bulb via two electrodes (cathode and anode), which had evolved from the cathode ray tubes used to investigate the properties of the recently discovered electromagnetic waves, a diode converts AC to DC (rectification). Invented in 1904 by the English electrical engineer John Ambrose Fleming, rectification allows electrons to flow in one direction ("forward" from cathode to anode) but not the other ("reverse" from anode to cathode), which Fleming patented as an "oscillation valve."[58] The effect had been initially discovered by Edison, based on his work on heating filaments for his electric light bulb, but as usual he was too busy to follow up on any commercial value.

In 1906, the American inventor Lee De Forest, a dedicated experimenter like Edison but with a PhD from Yale, added a third electrode (grid) between the hot cathode filament and the anode plate to create a vacuum tube "triode" (a.k.a. "audion") that regulated the flow of electrons to amplify a weak input signal, essential for relaying broadcast signals over longer distances (Figure 4.14). Three metal prongs protruded from the tube, evacuated to 1 millionth of an atmosphere (in other words, a vacuum) to improve conduction, such that the cathode–anode current (between prongs 1 and 3) can be manipulated by applying a voltage to the grid (prong 2), producing an amplified signal. The lower the grid potential, the more current flows from cathode to anode (a.k.a.

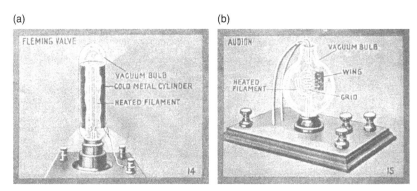

Figure 4.14 (a) The diode or Fleming valve (for forward and reverse current) and (b) the triode or audion (grid moderates cathode–anode current from heated filament to wing) (*source*: Hemour, S. and Wu, K., "Radio-frequency rectifier for electromagnetic energy harvesting: Development path and future outlook," Proceedings of the IEEE, 102(11): 1667–1691, November 2014.)

heated filament to wing). Think of a bouncer at a club opening and closing a door to let more or fewer people in – the flow of people is much greater than the movement of his hand, yet, importantly, the flow is proportional and thus an analog amplification of his moving hand.

Research into the conduction of electrons in solid-state materials, however, would ultimately be stifled by an already commercially viable vacuum-tube technology, despite German electrical engineer Karl Ferdinand Braun having shown that antenna reception could be improved with a crude crystal receiver made of a thin metal wire emanating from the lead ore galena (known as a "cat whisker"). Expanded long-distance radio communication would eventually lead to increased curiosity about solid-state electron conduction, but not before the vacuum tube was fully exploited in the burgeoning field of radio communications. The race was on to improve both the transmission and reception of radio signals (and soon after higher-frequency television signals and early computing), employing the existing vacuum-tube technology, while solid-state rectifiers and amplifiers remained a finicky laboratory curiosity. The modern entertainment industry was born on the back of the vacuum tube, becoming an industry stalwart such that by 1910 "De Forest was broadcasting the voice of fabled tenor Enrico Caruso from New York's Metropolitan Opera."[59]

Long-distance telephone service was also possible because of vacuum-tube "repeaters," spaced miles apart to amplify attenuated signals in a growing national telephone system that kept calls from being drowned in static. The first-ever transcontinental phone call was placed in July 1914, recreated 6 months later as part of a publicity stunt to celebrate Bell Telephone's introduction of its new, long-distance, vacuum-tube signal amplification system. Located in New York City, the company's founder Alexander Graham Bell repeated his famous cry for help to his former assistant Thomas Watson, who was at the other end of the line in San Francisco. The signals were transmitted and their voices heard thanks to vacuum-tube repeaters in Pittsburgh, Omaha, and Salt Lake City.[60] Proving their worth many times over, vacuum tubes made Ma Bell a fortune.

Solid-state signal detection, however, was still little more than black magic to most researchers, and depended on the orientation of the crystal, despite contact-based, metal-semiconductor rectifiers being employed during World War I to tune radio signals, in particular Braun's cat whisker crystal detector. The unknown variability of "hot spots" in the crystalline materials led vacuum-tube experts "to regard the art of crystal rectification as being close to disreputable."[61] Even with increased research into radar in World War II, development was slow as engineers preferred to work with their trusted vacuum tubes, a much more mature and better understood technology.

Alas, vacuum tubes were bulky and prone to breakdowns, and better alternatives were sought to rectify and amplify signals, especially in the growing field of solid-state physics (now called "condensed matter" physics to include solids and liquids). In 1939, working at Bell Telephone's radio lab in Holmdel, New Jersey, the electrochemist Russell Ohl discovered a large voltage in a piece of commercial-grade silicon during his radar research on signal detectors after a current flowed under flashlight illumination. Although the signal was varied, Walter Brattain, working at Bell's main labs in nearby Murray Hill, noted, "this was the first time that anybody had ever found a photovoltaic effect in elementary material."[62]

Assuming the undesirable signal variability resulted from his poor-quality, low-grade commercial sample, Ohl purified the silicon by baking it in inert helium, cooling it, and cutting it into ingots, obtaining 99.8% pure silicon rods.[63] Fortuitously, Ohl's silicon had differing levels and types of impurities embedded in the purified section and an adjacent commercial-grade section, which incredibly produced an electric field across the interface that allowed current to flow in one direction, exhibiting a "phenomenal photoelectromotive force."[64]

An expert in atomic and crystal structure, who had been working for over a decade on the odd properties of "column-IV" elements, Ohl had fashioned the first-ever working semiconductor diode by chance, later dubbed a "p-n" junction for the positive and negative charged impurity regions, considered the "most important circuit element in our present-day semiconductor electronics."[65] Following his work on semiconductor diodes, Ohl received the first patent in 1941 for a "light-sensitive electric device."[66] The forerunner of the modern solar cell, Ohl's curious new solid-state-constructed material generated electricity when exposed to light, exhibiting 10 times the voltage as other photocells such as selenium and copper oxide (an earlier, primitive metal semiconductor rectifier).[67] Newly developed fabrication techniques would eventually improve the conversion efficiency of light to electricity, paving the way to today's integrated computer circuits, solar cells, and light-emitting diodes, all thanks to the novel p-n junction.

Alas, solar's time had not yet arrived, the manufacturing process too imma-ture to develop commercially, while another world war would again redirect the motivation of scientists and engineers, in particular radar that required all hands on deck and the expertise of many of the early solid-state researchers at Bell Labs, Western Electric, and MIT. Vacuum tubes would continue to switch and amplify electronic and radio signals for the duration of World War II, essential for radar and field communications, pushing back the Sun's appearance in electronics by more than a decade. The stakes were too high to commit more research on a not-yet-ready-for-prime-time technology, even one with the potential to remake an entire economy.

4.5 Solid State Fundamentals: Redirecting Electrons

To understand how an everyday solid-state device is made – whether a semiconductor diode that rectifies a current in a circuit, a light-emitting diode (LED) that converts electricity to light, a solar cell that converts incident light into electricity (basically an LED in reverse), or a semiconductor triode (transistor) that amplifies weak signals – we turn to the basic structure of solids, especially in crystalline form. The science has to do with the availability of electrons coaxed from their outer atomic shells into usable current.

An amorphous solid such as glass has no regular atomic structure, while others are more ordered and thus exhibit a higher electrical conductivity in all directions. In copper (Cu, $Z = 29$: 2 8 18 **1**), the single outer-shell electron in each atom can easily move in the presence of an electric field to produce a current, while other crystalline materials have varying conductive properties. For example, in quartz, a crystalline mineral used to make glass and ceramics, the electrons are much more tightly bound (the resistance of quartz is 10^{24} times that of copper[68]). Insulators such as glass and rubber have fewer free electrons to conduct a current.

In between an insulator and a conductor, silicon (Si, $Z = 14$: 2 8 **4**) acts as a conduction "facilitator" because of its half-full, half-empty outer shell of four electrons, easily donating electrons *to* or accepting electrons *from* other elements. Neither a metallic conductor (free outer-shell electrons as in copper) or an insulator (tightly bound electrons as in quartz), the column-IV periodic table element silicon is an "in between" element or *semi*-conductor. Another column-IV element, germanium (Ge, $Z = 32$: 2 8 18 **4**), also exhibits the same donor–acceptor properties. Importantly, with a half-full outer shell of four electrons, semiconductors like silicon and germanium provide little conductivity at room temperature, but if the temperature is increased or an impurity (say 1 part in a million) is introduced into its crystal lattice the outer electrons can move more freely and the resultant conductivity is much higher.

To insert a desired impurity into a semiconductor's crystal lattice, called "doping," we use a neighboring column-V element (for example, N, P, As, or Sb, all with valence +5) for "*n*-type" doping or a column-III element (for example, B, Al, Ga, or In, all with valence +3) for "*p*-type" doping. A column-V element has one *excess* outer-shell (or valence) electron per atom (for example, phosphorous, $Z = 15$: 2 8 **5**), thus producing an excess of *negative* charge carriers and *more* conduction electrons to the conduction band, hence the name "*n*-type" doping. Note, these are the same electrons that form chemical bonds when shared with other atoms. Conversely, for "*p*-type" doping of a semiconductor, an impurity is introduced from a column-III element (for example, boron, Z = 5: 2 **3**), providing an excess of *positive* outer-shell charge carriers (called "holes," where the absence of an electron behaves like a positive charge). If a single crystal of silicon or germanium is then differentially doped with boron on one half and phosphorous on the other half, we get a "*p-n*" junction as Ohl discovered, exhibiting characteristics similar to the vacuum-tube diode, where current flows easily in one direction but not in the other.

Ohl and his metallurgist colleague Jack Scaff coined the terms "*n*-type" and "*p*-type"[69] (initially called excess and deficit conductivity) to describe their novel, two-region device, created with differing types and levels of impurities. Although unknown at the time, later research showed that Ohl's first silicon semiconductor contained phosphorous in the *n*-type region and boron and aluminum in the *p*-type region.

In effect, a diode is a semiconductor sandwich, combining two differently doped regions of semiconductor material, one doped with a heavier, higher-valence element (one column to the right of silicon and germanium) to provide excess *electrons* and the other doped with a lighter, lower-valence element (one column to the left of silicon and germanium) to provide excess *holes*, which produces an electric field across the junction, current flowing easily in one direction (forward) but not in the other (reverse) as seen in Figures 4.15–4.17. Typically, a column-IV semiconductor is doped in what is known as a "diffusion furnace" (for *n*-type) or during crystallization (for *p*-type) to replace the substrate atoms with the required impurities, although a technique known as "ion implantation" can also be applied directly to the substrate.

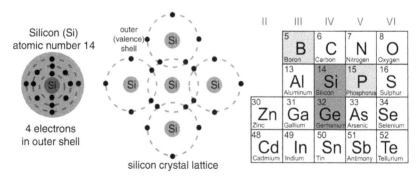

Figure 4.15 (a) Electron orbital occupancy of a column-IV element such as silicon and (b) neighboring elements in columns II, III, V, and VI of the periodic table. Silicon has four outer shell electrons, phosphorus five (one extra electron), and boron three (one less electron).

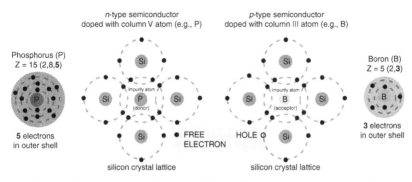

Figure 4.16 (a) Examples of *n*-type doping with a column-V element such as phosphorous and (b) *p*-type doping with a column-III element such as boron.

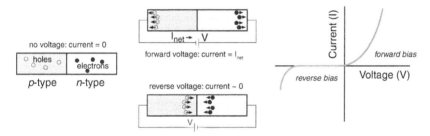

Figure 4.17 A semiconductor diode "switch" (a) with forward voltage (current on) and reverse voltage (current off until breakdown) and (b) characteristic "I–V" curve for a diode.

Almost from the day World War II ended, engineers and scientists at Bell Labs – Bell Telephone's research wing located at various New Jersey sites – began in earnest to investigate the intriguing characteristics of their novel solid-state materials. As if to commemorate the inventive tradition of local hero Thomas Edison, Bell had already moved its main research operations from 463 West Street along the Hudson River in Manhattan to a modern, state-of-the-art industrial research facility in Murray Hill, a suburban New Jersey site only 10 miles north of Edison's historic Menlo Park, where "New devices were waiting to be discovered through the combination of basic research and the intellectual force of physicists and engineers working together."[70]

Braun's primitive World War I cat whisker detector, which had irritatingly baffled early electrical workers, had been remade as Ohl's less mysterious *p-n* device, substantially improved and upgraded with silicon purified to 99.999%, this time with the desired level of donor and acceptor impurities purposely included. Although the age of vacuum-tube technology would ably last for a few decades more, the new physics was fast winning out in a modern electronics game. Solid-state current rectification was no longer considered black magic, even if the process had started out with a bit of old-fashioned luck in the baking. Alas, the solar-conversion capability of the new device would not be sufficiently exploited until after solid-state amplification was worked out, an essential part of Bell's core business.

***

Building on the improvements in solid-state conduction, especially Russell Ohl's earlier work on microwave crystal detectors, the "point contact" transistor was next invented at Murray Hill by Walter Brattain and John Bardeen, two Bell Labs physicists working under William "Bill" Shockley. Successfully demonstrated for the first time on December 16, 1947, and demoed to the press 6 months later, the crude, solid-state transistor (a portmanteau of "trans resistor") amplified the current as in a vacuum-tube triode, but more importantly was made of a solid piece of doped germanium less than 1/16-inch thick. The first test device looked like a metallic-lined arrowhead precariously wedged into a rectangular slab of germanium, which worked by injecting positive-charged, minority-carrier holes into the semiconductor surface at the point contact (actually two gold contacts 2 mm apart).

From that moment on, electronics was never the same. A truly solid-state amplifier would ultimately assign the vacuum tube to the dustbin of history, the three Bell Labs physicists Shockley, Brattain, and Bardeen sharing the 1956 Nobel Prize "For their researches on semiconductors and their discovery of the transistor effect."[71] In keeping with famous words uttered at the moment of invention, such as Archimedes' "Eureka" and Bell's "Mr Watson, come here, I want to see you," Brattain dryly exclaimed during the first successful test of the point-contact transistor: "This thing's got gain."

However, although much more robust, smaller, and ultimately cheaper to make than a vacuum tube, the germanium point-contact transistor was too delicate to mass produce easily and would soon be supplanted by the "junction" transistor, which overcame the fragility of Brattain and Bardeen's point-contact construction via a three-region, n-type, p-type, n-type ("npn") doped crystal to increase conductivity across the semiconductor material.[72] The theory of the more practical junction transistor was worked out over the Christmas holidays that year in a Chicago hotel room by Shockley, miffed he hadn't been more involved in Brattain and Bardeen's earlier research, although a working model couldn't be fashioned until after the process of artificial crystal growth from seed had been perfected three years later.[73] Essentially, the unreliable point contacts were replaced by much better performance *p-n* junctions, made via a precisely controlled, single-crystal growing technique and the resulting higher-purity germanium. When the palm-sized device was first tested, the output was almost 20 times the input, required one-millionth the power of a vacuum tube, and generated much less waste heat, a critical concern as circuit sizes grew.

Alas, germanium was too rare a substance, unreliable at high temperatures, and unduly finicky for large-scale transistor fabrication. The first *silicon* transistor would be manufactured in January 1954, although it too was initially difficult to mass produce until a Bell Labs chemist, Morris Tanenbaum, doped the silicon crystal by diffusion, a process invented by another Bell Labs chemist Calvin Fuller, where high-concentration dopants were added to the molten silicon during crystal growth (called activation), thus manipulating "the concentrations of impurities in silicon with remarkable precision."[74] At 800°C, the high-temperature diffusion process would require an ultraclean fabrication facility to keep out unwanted impurities, especially copper. Tanenbaum was shocked that the device worked so much better than a germanium transistor, writing in his notebook, "This looks like the transistor we've been waiting for. It should be a cinch to make."[75] Indeed it was, and the rest as they say is history – the date was March 17, 1955.

Beyond amplifying telephone signals, the possible applications of Bell's "very manufacturable" silicon transistor were not fully appreciated at first, such as digital computing and information theory, although telecommunications would immediately benefit as networks evolved from copper wires and electric currents to fiber optics and light waves. The solar cell too would have its day, almost an afterthought to the pioneering breakthroughs in transistors, developed primarily through the efforts of three more Bell Labs scientists, physicists Daryl Chapin and Gerald Pearson and the physical chemist Fuller.

Chapin had been trying to develop a power source for telephone systems in isolated humid environments where dry-cell batteries rapidly degrade and had hooked up with Pearson and Fuller, both of whom were working on improving gallium-doped semiconductors. Within the Solid State Physics group at Murray Hill, Pearson had worked with Shockley's Semiconductor subgroup with the practical experimentalist Brattain and the prodigious theorist Bardeen and might even have shared in their Nobel Prize if the rules allowed for more than three recipients, while Fuller had developed the essential diffusion doping process. After numerous attempts failed to reach a functional level of efficiency, deemed to be at least 6%, Chapin recalled Einstein's theory of light quanta, and engineered the *p-n* junction closer to the surface, where "more powerful photons belonging to light of shorter wavelengths could effectively move electrons to where they could be harvested as electricity."[76] Chapin also coated the silicon surface with a dull transparent plastic to reduce reflection.

Having manipulated the electric-conduction properties of differentially doped semiconductor materials to improve the rectification and amplification of electronic signals for commercial purposes in solid-state diodes and triodes – the primary interest of telephony – Bell Labs had invented the modern solar cell, based on Ohl's original *p-n* junction. Although the ground-breaking research on transistors would receive most of the accolades and development money – considered by some to be the most important invention of the last half of the twentieth century – quietly and without much commercial consideration, Bell Labs had produced a thin, silicon-doped wafer that converted light into electricity at 6% efficiency.

The three scientists linked together several of the 0.6-V solid-state devices – doped with boron (III) and arsenic (V) – to create a "solar battery" (Figure 4.18a). With surprisingly little fanfare, the world's first practical solar cell was demonstrated at Bell Labs on April 25, 1954, "to power a small toy Ferris wheel and a solar powered radio transmitter," the inventors noting that the 6% conversion efficiency "compares favorably with the efficiency of steam and gasoline engines, in contrast with other photoelectric devices which have never been rated higher than 1%."[77]

(a)   (b)

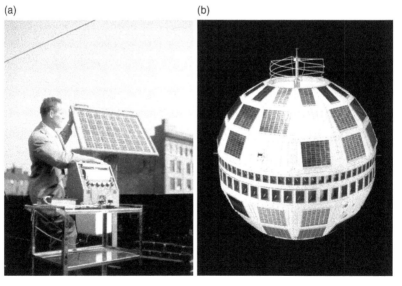

Figure 4.18 Early solar cells (*source*: Nokia Bell Labs): (a) Bell's 1954 6%-efficient, silicon-doped "solar battery" and (b) AT&T's 1962 *Telstar* producing 14 watts from 3,600 modules.

Solar modules were soon available at $300 per watt, alas too expensive even for Bell to use as intended. But despite being well beyond most budgets, the modern solar age had begun. As noted by *The New York Times* shortly after the initial demonstration:

> The new device is a simple-looking apparatus made of strips of silicon, a principal ingredient of common sand. It may mark the beginning of a new era, leading eventually to the realization of mankind's most cherished dreams – the harnessing of the sun for the uses of civilization. The sun pours out daily more than a quadrillion (1,000,000,000,000,000) kilowatt hours of energy, greater than the energy content of all the reserves of coal, oil, natural gas and uranium in the earth's crust.[78]

Alas, there was no time to rest on any scientific laurels, despite having invented a functional new form of energy with a fuel efficiency (sunlight to electricity) comparable to steam and gasoline: *Sputnik* changed everything. Launched in 1957 with a ballistic-missile carrier rocket that horrified the unprepared West, the tiny *sputnik* – meaning "satellite" or "fellow traveler" in Russian – created an urgency to make on-board computers smaller and lighter in a newly declared space race with the Soviet Union, rather than churning out sun-powered toys or batteries. The practicality of ringing the Earth with solar-powered communication satellites – the brainchild of Bell Labs engineer John Pierce and based on an earlier idea by the science-fiction writer Arthur C. Clarke[79] – would also have to wait. Neither satellites nor the rockets to launch them had yet been built by the startled Americans.

Under the guidance of the physical chemist and former Bell Labs PhD scientist Gordon Teal, Texas Instruments (TI) was the first to exploit the mass fabrication of silicon transistors, which performed better than germanium at the high temperatures needed for military weapons. Teal applied a modified crystal-pulling growth technique to the more abundant silicon, which was also better for switching and not as leaky. Having previously relied on military contracts, TI branched out into commercial applications at the end of the Korean War, launching the Regency TR1 pocket transistor radio in time for Christmas 1954, considered by some to be the birth of popular culture if not our modern high-tech commercial world.[80] Transistor sales increased from $27 million in 1953 to $233 million by 1960.[81]

The next big thing would be to miniaturize the cumbersome circuit boards, arduously made by cutting out individual transistors, attaching electrodes (the prongs[82]), and reconnecting each part, before Jack Kilby of Texas Instruments and Bob Noyce of Fairchild Semiconductor each thought to "wire" a whole circuit on a single piece of semiconductor material. The first "integrated circuit" (IC) consisted of one transistor, one resistor, and one capacitor in a crude germanium prototype, before silicon became the semiconductor material of choice.

Intel (INTegrated ELectronics) – the Silicon Valley company Noyce co-founded after leaving Fairchild – was then commandeered to provide ICs for the Gemini and Apollo space programs that led to their game-changing 1103 memory chip. Small became even smaller as mini morphed into micro and the Space Age became the Information Age, financially aided by President Kennedy's aim to put a man on the moon within a decade and a leg up at last on the Soviets. Component sizes continued to shrink as increasingly finer circuit patterns of conductors, insulators, semiconductors, and their metal connectors were geometrically etched onto a single silicon wafer via layered diffusion and an advanced photolithographic process. Technique turned into a mass-produced automated assembly as art and technology merged. In 1971, Intel produced the first "processor on a chip," the essence of the modern personal computer as PCs rather than PVs became the dominant product of a new age.

With the basics of IC manufacturing solved, the number of components per chip doubled thereafter every 18 months, that is, a 67% increase in density per year, producing more chips, faster switching speeds, and decreased consumer costs, now known as Moore's Law from a 1964 prediction by Intel co-founder Gordon Moore.[83] Moore, a PhD physical chemist from Caltech, and Noyce, a PhD physicist from MIT, were part of the "traitorous eight" who had left Shockley's fledgling, Palo Alto, high-technology company, Shockley Semiconductor Laboratory to start up Fairchild that would eventually spinoff into Intel, AMD, and other chipmakers (the so-called "Fairchildren").[84] Starting in 1968, the number of transistors on a single chip increased from 1,000 to more than 4 billion over the next four decades (now over 8 billion with the latest 2-nm EUVL manufactured chips), powering our modern way of life and the technology of the future, unthinkable without the ubiquitous silicon chip.

The impact of photonics on a previously fragile thermionic electronics world was nothing short of spectacular, as noted by American author Tom Wolfe in an essay on Noyce, his choice as the spiritual father of Silicon Valley:

The vacuum tube was based on the lightbulb, but the vacuum tube opened up fields the lightbulb did not even suggest: long-distance radio and telephone communication. . . . The integrated circuit was based on the transistor, but the integrated circuit opened up fields the transistor did not even suggest. The integrated circuit made it possible to create miniature computers, to put all the functions of the mighty ENIAC on a panel the size of a playing card.[85]

Today, one could add that the semiconductor diode was based on the vacuum-tube valve to rectify currents, opening up fields neither the valve nor the diode could suggest – modern lighting and photovoltaics via the LED and complimentary solar cell, two game changers in a revamped, twenty-first-century energy market. At the time, any hoped-for PV world was slow to materialize, however, stuck in university and industry research labs, space-flight miniaturization, and satellite battery-power units, where government funding was readily available to explore the potential of preferentially modifying the solid state. Commercial potential was not as forthcoming, although simple consumer applications would slowly appear.

Initially considered "potentially viable," PV now has "plentiful value" in the vibrant, modern fields of lighting (as we look at now) and power generation (as we look at next). Sometimes invention falls in clearly defined steps; sometimes we have to clear a path to see the next big thing.

4.6 From Chips and Solar Cells to LEDs and Light

Ohl's first diode was made of doped silicon and Bardeen and Brattain's first transistor doped germanium, both exploiting the addition of controlled or "functional" impurities of less than one part in a million. Today, semiconductors are still made using these two main column-IV elements or one of the many compound substrates, for example, a column III–V GaAs *heterojunction*. Importantly, the average charge number of electrons is still four (a half-full outer shell), providing a range of "band-gap" energies, the difference in energy between a material's valence band and conduction band.[86] Gallium arsenic (GaAs) is a standard III–V *binary* compound substrate, while *tertiary* compounds such as $Al_{1-x}Ga_xAs$ and *quaternary* compounds such as $Ga_xIn_{1-x}As_yP_{1-y}$ are also common. As before, column-V/III dopants produce the *n*- and *p*-type regions.

All semiconductors are characterized by their band gap (E_g), either an *indirect* band gap for elemental semiconductors such as silicon and germanium (requiring *phonon*-assisted conversion) or *direct* band gap for compound semiconductors such as GaAs and InSb, where photons convert valence electrons to conduction electrons. In an LED, the band gap corresponds to the emitted photon energy in electron volts (eV) or the wavelength in nanometers, while in a solar cell the band gap is the absorbed photon energy or wavelength, showing again how energy and wavelength (or color) are intimately related, and that an LED and a solar cell are complimentary devices. In an LED, electricity is converted to atomic transitions that then emit photons of a given energy, while in a solar cell photons are absorbed that liberate electrons to create electricity. To quantify the range of band-gap energies, we can use wavelength or, to keep life simple, color. The famous GaAsP diode has a wavelength of 626 nm (1.98 eV), emitting the familiar red seen on many electronic panels and displays.[87] Early "light" emitting diodes were all in the infrared range (InSb 7,290 nm, GaAs 867 nm, and CdSe 713 nm), but visible LEDs would soon become commercially viable, popular with electronic kit makers and in a range of new instrumentation. To help understand how an LED or a reverse LED solar cell works, the band-gap energy, E_g, equivalent wavelength, and color of a number of common semiconductors are shown in Table 4.1.

The first LED to emit in the visible range was the ubiquitous red LED at 626 nm, developed in 1962 from a mixture of gallium arsenide and gallium phosphide on a GaAs substrate (called GaAsP). Created by General Electric engineer Nick Holonyak – John Bardeen's first PhD student after leaving Bell Labs in 1951 for the University of Illinois, Urbana–Champaign – the red LED was nicknamed "the magic one" and started a revolution in modern lighting. Holonyak also invented the red laser-diode in DVD players and checkout counters. Following in 1972, the first yellow LED was made at Monsanto by one of Holonyak's students and soon after a green LED (GaInN). By adjusting

Table 4.1 *Band gaps (eV), equivalent wavelengths (nm), and EM range (infrared or color) in various common semiconductors*

Material	Energy gap at 300 K (eV)	Wavelength (nm) (and EM range)
Ge (IV)	0.66	1,880 (IR)
Si (IV)	1.11	1,120 (IR)
InSb (III–V)	0.17	7,290 (IR)
InAs (III–V)	0.36	3,440 (IR)
GaSb (III–V)	0.68	1,820 (IR)
InP (III–V)	1.27	976 (IR)
GaAs (III–V)	1.43	867 (IR)
CdTe (II–VI)	1.44	861 (IR)
CdSe (II–VI)	1.74	713 (IR)
GaAsP (III–V)	1.98	626 (red)
GaP (III–V)	2.25	551 (green)
ZnO (II–VI)	3.20	387 (blue)
GaN (III–V)	3.40	364 (blue)
ZnS (II–VI)	3.60	344 (blue)

Source: Kittel, C., Introduction to Solid State Physics, 6th ed., p. 185, John Wiley, New York, NY, 1986. Hyperphysics: http://hyperphysics.phy-astr.gsu.edu/hbase/Tables/Semgap.html.

the arsenic-to-phosphorus ratio, a GaAsP LED can also be customized to emit from 1.4 to 2.3 eV.

In 1993, the first bright blue LED was invented by Shuji Nakamura at the Japanese engineering company Nichia, now the world's largest supplier of LEDs. Nakamura used gallium nitride (GaN), enabling the production of white light by passing blue LED light through a yellow phosphor. Today, red, green, and blue LEDs are assembled in concert to provide a full spectrum of colors for lamps as well as computer, television, and smart-phone screens. Although Holonyak was oddly overlooked, Nakamura shared the 2014 Nobel Prize in Physics with Isamu Akasaki and Hiroshi Amano "for the invention of efficient blue light-emitting diodes which has enabled bright and energy-saving white light sources."

When you look at an LED display – for example, on a digital clock or microwave timer – think of how the electric current is converted to light, the reverse of how a solar cell converts light into electricity (technically, in a solar cell a photon creates an electron–hole pair, which both move to create the electric current). Today, visible-range LEDs are commonly found in instrument panels, traffic lights, car brake lights, and everyday digital displays, and are much more efficient and rugged than incandescent tungsten-filament bulbs or fluorescent lighting (Figure 4.19). For example, LEDs "last 5 to 10 times longer than traditional incandescent bulbs, and use only 20% of the energy for the same light output."[88] It is also estimated that 20% of global electrical use comes from lighting, so LEDs can reduce our overall electrical consumption to 4%, a massive energy and GHG savings for a global population of more than eight billion.[89]

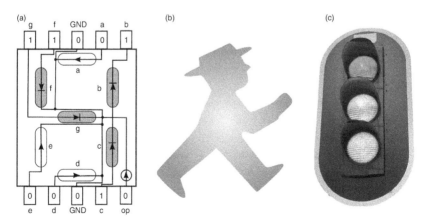

Figure 4.19 LEDs: seven-segment display (lighting up a "4"), *Ampelmännchen*, and traffic light.

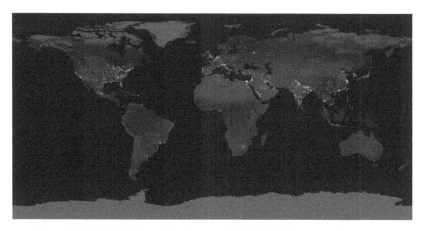

Figure 4.20 The lit Earth at night (*source*: NASA Earth Observatory).

One measure of progress in lighting technology is the increased luminous efficiency. A lumen is the SI unit of flux or brightness, that is, the amount of visible light reaching the eye. Measured by source power converted into light as expressed in lumens per watt (lm/W), the luminous efficiency of various kinds of lighting is 0.3 for candles, 1–2 for gas, 10–15 for incandescent (<5 in 1900), 100 for compact fluorescent, 100–150 for LED, and 200 for low-pressure sodium in yellow street lights, while the cost of a lumen of light has decreased by almost 600 times in the last century and 10,000 times since the Middle Ages.[90]

Huge savings are available with LEDs, especially seen in the night sky. As shown in the lit Earth at night from NASA's Earth Observatory (Figure 4.20), thousands of twinkling cities display a web-like interconnectedness across the once vast darkness of rural life. A hypnotic order appears from the seemingly anarchic expanse of human settlement, displaying the hidden contours of development and the progress of modern civilization. To some, lighting the night is the pinnacle of achievement, guiding us to our destinations and protecting us from our fears; to others another form of man-made pollution that further isolates us from a visible yet cruelly unknowable universe.

The US Department of Energy (DOE) believes that converting to LED lighting can save as much as $40 billion annually by 2030 from reduced primary energy consumption of 4.5 quads or "more than two times the amount of electricity the Energy Information Administration predicts will be produced by wind power or 20 times what will be generated by solar power in 2030,"[91] greatly reducing our carbon footprint and helping to become more energy secure. A quad is equal to 1.055 *quintillion* joules, that is, 10^{18} J or 1 exajoule (1 EJ)! According to the DOE:

> Greater LED efficacy enables more light to be produced with less electrical power at lower operating temperatures. As a result, lighting manufacturers and designers can choose to reduce the light source size, decrease the number of LEDs, increase light output for a given source size, reduce electrical power input, reduce the amount of generated heat, or many combinations of the above.[92]

Chase Manhattan's lighting bill was cut in half by retrofitting almost 5,000 branches with over 1.4 million energy-efficient LEDs. Each bulb comes with an estimated 5.7-year lifetime (50,000 hours), while the world's largest single-order retrofit is expected to save 184,000 kilowatt-hours (equivalent to over 100,000 acres of preserved forest, 25,000 fewer vehicles, or 45,000 fewer tons of landfill waste).[93] To get an idea of the high cost of lighting, even Bell Labs had to cut back during lean times in 2000 after being forced to split from AT&T, turning off every other light in its vast Murray Hill, New Jersey, complex.[94] Unfortunately, the preeminent American innovation factory hadn't discovered the LED among its many impressive inventions.

Less expensive because electricity is converted into light without heating a filament as in an incandescent bulb (only 10% of the electricity is converted to light while 90% becomes heat), or a fluorescent gas discharge, LEDs are becoming standard in our modern lighted world. One already sees street lighting being replaced with programmable smart LEDs that can be tailored to local conditions to remake night-time streetscapes with huge savings, including many impressive Christmas light installations, as cash-strapped cities around the world turn to cheaper LEDs.

Reducing lighting costs is especially important during peak electricity demand in the early evening, known as the "magic hour" as lights turn on all at once at twilight, even more so in winter when heating systems are also running. If everyone plugs in at the same time and the generating system can't respond to the rapid drain, a sudden loss of power can even bring down the grid. At 5:16 p.m. on the evening of November 9, 1965, a relay tripped at the SAB hydroelectric power plant near Niagara Falls, knocking out electricity to 30 million people on the east coast of North America, a much more unlikely event today with more efficient LEDs. Behold the beginning of another lighting revolution, just as significant as Edison's incandescent light bulb of the 1880s or Standard Oil's kerosene-fueled lamps before that.

The Internet is also run on LEDs (and lasers), coded signals switched on and off at unimaginably high frequencies, the theory of which was developed by Claude Shannon, another paradigm-changing Bell Labs engineer-inventor.

Infrared Leds (up to 1,500 nm) transmit the digital 1s and 0s in kilobyte packets sent from node to node along almost 1 million km of optical fiber. The CRAY-3 was the first supercomputer to use integrated circuits made entirely of GaAs logic circuitry, GaAs Leds providing the fastest signal speeds.[95] We've come a long way from Edison's inefficient incandescent bulb – 90% of the energy lost as heat – to today's highly efficient diffuse Leds, switched on to light the world or switched on and off to inform us of the latest news in an ever-shrinking global village. LED technology is also revolutionizing farming with cheap 24/7 lighting that can substantially increase growth and output in indoor climate-controlled operations.

There seems to be no end to new semiconductor uses (ICs, detectors, lasers, Leds, CCDs), although the simple solar cell's journey has been more arduous. After half a century of being underappreciated, however, the solar cell is finally getting its due, no longer sidetracked in the race to make smaller components in the commercial development of telephones, radio, radar, computers, digital networks, camera imaging, and modern lighting. As we have seen, the science is straightforward: differential impurity doping of a semiconductor crystal to create a *p-n* junction that generates electricity when a photon is incident. As we look at now, all one needs is a photovoltaic cell, some sun-facing open space, and the Sun.

4.7 Enter the Solar Panel: The Sun for Hire

A solar cell is essentially a backwards LED – electron–hole pairs form across a *p-n* junction when light is incident, producing an electric current that can power an external circuit (the load). A solar cell is a light-*absorbing* diode (or photodiode), while an LED is a light-*emitting* diode. Like an LED, a solar cell has no mechanical moving parts, allowing for less maintenance and longer lifetime; only the liberation of electrons by photons at the atomic scale, turning incident sunlight into electricity. Multiple solar cells connected in series can generate enough power to run a home.

Early photovoltaic solar cells powered remote generators (off-grid), satellites (way off-grid!), calculators (low power), watches (small and low power), radios, streetlamp sensors, as well as various children's novelty toys and hand-waving figurines. Now a thriving cottage industry, online catalogues include the original solar-waving Queen Elizabeth, a Canadian Mountie, Napoleon, the Pope, and, fittingly it would seem, Albert Einstein.

Preferring dependable, long-life, solar batteries to heavy chemical batteries that lasted only a few weeks, James Van Allen, the principle investigator of the cosmic ray detector on the first American satellite, *Explorer 1*, estimated that solar batteries could reduce a satellite's total weight by 20 pounds, a critical mission concern. *Explorer 1* was launched by NASA in January 1958, four months after *Sputnik*, followed in March of the same year by the navy's *Vanguard 1*.[96] Since then, almost all satellites and spacecraft have employed solar cells to power onboard equipment. Although a few problems had to be solved before solar batteries were put in space, such as overcoming voltage loss as the temperature rose to 80°C, the potential was enormous. As noted in *The NASA Historical Series*, "The wisdom of providing for solar power in future American satellites seemed self-evident, despite the additional cost consequent upon the longer period of time during which radio tracking stations and data reduction centers would have to operate."[97] Tracking stations would be set up across the globe, ready to relay data as the orbiting satellites passed by, over and over again.

In a hastily ramped-up space race, discovery was ripe for the picking. James Van Allen discovered the radiation belts within the magnetosphere surrounding the Earth, which now bear his name thanks to a non-functioning Geiger counter aboard *Sputnik*. As the story goes, when the CIA learned that *Sputnik* contained an onboard Geiger counter, NASA decided to include one of its own on *Explorer 1*, but because *Sputnik*'s Geiger counter didn't work while *Explorer*'s did, Van Allen got the scientific billing. The data relayed from *Vanguard 1* also showed that the Earth was not perfectly round, but slightly pear-shaped with an equatorial bulge, verifying why a pendulum clock runs 2.5 minutes a day slower at the poles than the equator and confirming Newton's 1687 calculation of a 230:231 length ratio of polar to equatorial axes, which he determined from the different pendulum arcs due to gravity.[98] In 1960, the x-ray solar detectors aboard *Solrad 1* showed the connection between disruptions to terrestrial radio communications and the Sun, while much later in 2014 one of the 26 Galileo global navigation system satellites was accidently launched into an elliptical orbit, the onboard atomic clock elegantly confirming Einstein's theory of relativity as it ticked faster and slower in ever-changing orbital speeds.[99]

Solar-cell development, however, progressed slowly despite the apparent urgency for more space-based spying on Soviet military readiness. But with each new launch the cells got bigger and better, as did satellite technology. *Vanguard I* (1958) produced less than 1 watt from 108 PV cells in eight exterior panels, while *Explorer VI* (1959) had 9,600 cells (1 cm × 2 cm each).

In 1962, AT&T's *Telstar* – the first telecommunications satellite with near-instantaneous, point-to-point, transatlantic communication – produced an initial output of 14 watts from 3,600 12-cell modules (Figure 4.18b). About as big as an oversized beach ball at 17-inches diameter and weighing 170 pounds, *Telstar* was launched into medium-earth orbit (MEO) between 1,000 and 6,000 km. A bejeweled wonder with 15,000 parts, *Telstar* was the culmination of decades of innovation at Bell Labs, covering patents on semiconductor fabrication, transistors, wave guides, maser amplification, and solar cells. Shrinking the world with live images almost immediately upon launch, TV signals were broadcast for the first time between the USA and Europe, including the news such as President Kennedy announcing *Telstar*'s launch, an event *The New York Times* regarded "as rivaling in significance the first telegraphed transmission by Samuel F. B. Morse more than a century ago."[100] Although still in orbit around the Earth, *Telstar* no longer transmits data, electrically dead since a nuclear test destroyed its electronics. *Vanguard I* is also still in orbit but electrically dead since 1964.

In 1966, the first orbiting astronomical observatory (OAO-1) was powered by a 1-kW PV array to measure UV and x-ray radiation above the Earth's absorbing atmosphere, while by the 1970s solar cells were powering devices beyond the limits of affordable grid connection, such as "navigation warning lights and horns on many offshore gas and oil rigs, lighthouses, railroad crossings."[101] Today, the International Space Station (ISS) sports four double-pronged folding-blanket solar wings that rotate on gimbals to face the Sun (Figure 4.21). Each wing is 115 feet long by 38 feet wide, weighs more than

Figure 4.21 The International Space Station with four rotating 115-feet by 38-feet double-pronged folding-blanket solar array wings (*source:* NASA).

2,400 pounds, and comprises 32,800 solar cells that can generate up to 30 kW (120 kW in total).[102] In permanent low-earth orbit (LEO) at about 400 km above the Earth, the ISS power system is entirely self-contained.

Of course, discovery is not cheap. Device-grade silicon cost five times as much as gold in 1958,[103] while NASA spent $50 million on photovoltaic development during the first decade of the race to the moon.[104] The *Telstar* launch alone cost $3 million.[105] But thanks to the many technological advances worked out on each successive mission, yesterday's high-cost, high-spec solar cells are now available to everyday consumers and not just an elite engineering few.

***

Most solar cells in a rooftop panel are made of crystalline silicon (c-Si), fabricated either as "monocrystalline" or "polycrystalline." Less pure than the silicon substrate used to make micro-sized transistors etched on an integrated circuit in a modern computer, "mono-Si" is cut from a round-grown ingot baked at 2,500°F (slowly pulled out as a stick from a liquid broth that cools to a solid) and appears black or blue, while "poly-Si" is cast-molded, baked at 1,800°F, and looks like metal-flaked patterned chipboard.[106] Fabrication includes a number of steps: *p*-type dopants are added during crystallization, ingots are cut into 15 cm × 15 cm wafers up to 500 μm thick, *n*-type dopants are diffused on one side to make the *p-n* junctions, anti-reflective coatings are deposited, metal conductors screen-printed, and finally the cells are fashioned into working commercial modules.[107]

The final assembly is protected from the elements by tempered glass and water-resistant foil, and sealed in an aluminum frame with external electrical connections (Figure 4.22). Mono-Si is generally more efficient and expensive

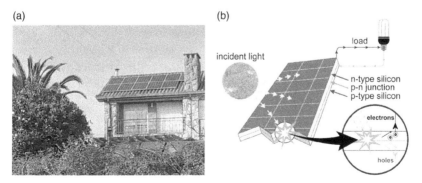

Figure 4.22 Harvesting the Sun: (a) 11 rooftop solar panels and (b) electron current flow in a solar cell, converting incident sunlight into electricity.

(and more environmentally friendly), thus taking less space in a panel, and also handles heat better and functions better in shady conditions or on hazy days. The market share is about 50–50, generally divided by efficiency and price. Over 90% of demand today is from China, which dominates modern solar-cell manufacturing.

Early solar cells had short lifetimes and low efficiencies, measured by the percentage power degradation per year and the percentage of incident energy converted to electricity (for example, Bell Labs' original 6% solar battery). In the 1980s, Stanford engineering professor Richard Swanson – founder of SunPower and considered the father of US solar – shrunk the metal contacts that blocked incident light and placed them at the back, increasing a cell's working surface and efficiency, "the biggest changes since the beginning of the solar cell business."[108]

More innovations followed with more funding and increased research and development. In 1992, the University of South Florida developed a "thin-film" cell made with a cadmium–telluride II–VI heterojunction (CdTe, a.k.a. "cad-tell") that had a higher light-absorption coefficient than crystalline silicon, thus requiring much less material to fabricate. About 1/30th as thick as silicon, the 3–4-μm CdTe cell was the first to crack the 15% efficiency barrier. Applied via chemical vapor deposition or "sputtering" between a transparent conducting oxide (TCO) surface and a metallic contact, thin-film cells capture more sunlight because of their long, skinny absorbing area, saving on material costs and fab time, and can operate at higher temperatures. Although thin-film cells are generally less efficient because long-wave IR photons pass through, they are much cheaper, using less semiconductor material (~1–2%).

Traditional crystalline silicon cells (either mono or poly) are still preferred for smaller rooftop sites and ground-based arrays – c-Si solar panels account for about 90% of the total PV market – while the newer thin-film technologies such as CdTe, amorphous silicon (a-Si), and copper indium gallium diselenide (CIGS) are more suited for larger installations (Figure 4.23). C-Si is about twice as efficient as a-Si, thus taking up less space, but a-Si is better in overcast, shady, and high-temperature conditions and is cheaper. Mounted on flexible substrates, a-Si and CIGS cells can even be bent, making them suitable for novel applications such as building-integrated PV (BIPV), solar shingles or slaters, and wearable solar.

A single-axis solar tracker changes a panel's tilt angle up and down throughout the year to optimize seasonal sunlight – a minimum at the winter solstice to a maximum at the summer solstice – while a dual-axis solar tracker also rotates the panel from side to side to match the changing sun angle from dawn to dusk, keeping the panel always at right angles to the incident light throughout the day.

(a)   (b)

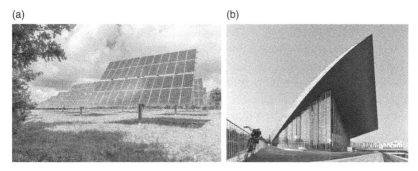

Figure 4.23 A few PV installations: (a) ground array (*source*: Sebastian Ganso) and (b) building-integrated (*source*: Ramoul CC BY-SA 3.0).

Sun-tracker technology that tilts and rotates a flat-plate panel perpendicular to the Sun to maximize solar incidence can increase a cell's output by as much as 55% in summer and 20% in winter.[109] Rotating side to side during the day and up or down throughout the year will maximize performance, but if a panel can't be angled in real time as in a dual-axis, ground-based panel or PV farm, the best fixed direction is sun-facing at the equinox angle to maximize total annual insolation at 90°– L (local latitude). Adjustable two-state racking will also change the angle to optimize summer and winter sunlight like putting on or taking off the winter storms.

Various adjunct technologies add to overall performance, such as anti-reflective (AR) texturing and coatings applied during module fabrication, as 30% of incident photons can be reflected at a panel's surface, while bypass diodes stop current loss during partial shading (similar to overriding a broken bulb in a series string of Christmas lights), although retail buyers needn't worry about the details. Panels should always be installed in a shade-free location and periodically cleaned to remove any obstructions such as fallen debris and dust – an occasional cold-water surface rinse is perfect (self-cleaning glass is also an option).

Passivated emitter and rear cell (PERC) technology increases efficiency via a reflective layer on the back of a panel, providing unabsorbed light a second pass. Invented by Martin Green, director of the Australian Centre for Advanced Photovoltaics at the University of New South Wales, the world's largest university PV research group, PERC modules now account for about 25% of silicon-cell manufacturing at over $10 billion/year, expected to rise to $1 trillion by 2040.[110] The first to reach 20% energy conversion in 1989 and 40% in 2014, Green shared the 2018 Global Energy Prize, having "revolution-ized the efficiency and costs of solar photovoltaics, making this now the lowest cost option for bulk electricity supply."

Bifacial modules go a step further, capturing photons in both directions (direct sunlight and ambient reflected light). Suitable for ground-based structures, such as parking lots or gazebos, a bifacial installation costs more upfront, but realizes significant fixed-cost savings and increases output by as much as 30%. Bifacial panels especially benefit from increased reflective surfaces such as snow. The current mono/bifacial mix is roughly 50–50.

A concentrator photovoltaic (CPV) cell increases the conversion efficiency by focusing more sunlight onto the cell surface, employing add-on reflective or refractive optics, classified into three magnifying ranges: low, medium, and high (<10×, 10–100×, >100×). The idea is simple, although as ever the devil is in the detail. Less efficient because of non-uniform illumination and lack of a precise sun-facing orientation, low-concentration PV (LCPV) systems use single-junction cells and are well-suited for geographically tuned BIPV. Operating at higher temperatures, high-concentration PV (HCPV) systems require dual-axis, sun-tracking technology to precisely align the direct light, as well as thermal cooling for ventilation, and typically work best in more arid regions of the tropics with high annual direct irradiation. Most CPV farms are HCPV at over 300 Suns (300–1,000×) with dual-axis tracking, offering a smaller footprint and thus lower costs, while output doesn't decline as much at high temperatures compared to c-Si modules.

CPV is generally cheaper than regular PV because less semiconductor material is employed and is more common in larger utility-scale solar installations, although CPV cells also work on the small kW-scale. Although not as viable for rooftop solar, CPV has recorded some of the highest PV efficiencies with multi-junction cells, such as the Fraunhofer Institute's 43.4% single full-glass lens and wafer-bonded, four-junction concentrator cell.[111]

A single-junction PV cell employs only one semiconductor, limiting the response to photons of energy equal or greater to the band gap, acting as a "radiation-controlled" switch, but a solar panel can perform better if different spectral-range cells are stacked on top of each other in descending band-gap order (for example, InGaP, InGaAs, and Ge to cover the UV, visible, and IR ranges). In a multi-junction (MJ) cell, higher-energy photons are captured by the top UV cell, while lower-energy photons pass through in turn to the visible- and IR-absorbing cells underneath. Such "cascade" or "tandem" cells convert more of the solar spectrum into electricity and are thus more efficient (~40%), but are also more expensive, obviously using more material in a layered device.[112]

Note that photons are absorbed at the band-gap energy and fractionally above, although the maximum conversion efficiency depends on the light-absorbing material (shown in Figure 4.24c for common cell types). For example, crystalline

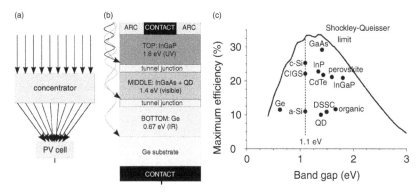

Figure 4.24 High-efficiency cells: (a) concentrator, (b) multi-junction tandem cell, and (c) solar cell conversion efficiencies versus semiconductor band-gap energy (eV).

silicon absorbs at or above 1.1 eV (that is, less than ~1,100 nm) as the photoelectrons are excited across the band gap between the valence and conduction bands, good for the near IR and visible range to a theoretical maximum efficiency of 33% under standard test conditions (STC).[113] Above the band-gap energy, the spectral response is linearly reduced as more higher-energy photons are absorbed as heat. What isn't converted to electricity is either reflected at the surface, absorbed (as heat), or passes through (photons less than the band-gap energy).

The US National Renewable Energy Laboratory (NREL) publishes a comprehensive graph of conversion efficiency over time, showing the evolution of different research cells from 1976 for the main PV technologies such as crystalline silicon, single-junction GaAs, thin-film, multi-junction, and emerging third-generation PV (organic, dye-sensitized, quantum dots, perovskite), some of which we'll look at after we've covered the basics. Although there is plenty of information to parse from almost 50 years of research by the world's leading developers and manufacturers – RCA, IBM, Boeing-Spectrolab, Westinghouse, ARCO, UNSW, NREL, Sanyo, Panasonic, Sharp, Fraunhofer, etc. – one sees how conversion efficiencies continue to improve over a wide range of competing technologies. One can also trace the progress of a particular research group or company over the years (now available online in interactive form[114]).

Behind each timeline is the usual story of long hours, money problems, and the dream of designing the next big thing. Today, conversion efficiencies range from roughly 25% (lower-cost, single-junction cells) to 47.6% (higher-cost, multi-junction, concentrator "champion" cells), while efficiencies continue to increase with corresponding cost reductions, making solar cells more viable. The future of solar power is being written by the players on NREL's cell-efficiency map, while the winner (or winners) of the solar sweepstakes will be well-positioned to harvest the Sun and determine the shape of energy production for years to come. The money spent here on R&D could translate into massive energy savings down the road.

4.8 The PV Market: At Home and in the Lab

Average PV costs have generally followed Swanson's Law, halving every decade (also stated as module prices drop 20% as shipped volume doubles). In 1975, a PV cell cost more than $100/watt, while by 2020 the best-made Chinese cells were down to $0.25/watt. Note that installed costs are higher than wholesale costs (roughly twice) because of various soft costs; nor is electrical output the same as rated output because of atmospheric absorption, diurnal fluctuations, and system inefficiencies. Average power (W) is about one-eighth the peak power (W_p), typically reduced one-half from intermittent clouds and one-quarter from a varying daily sun.[115]

Correspondingly, total installed capacity has increased each year with decreased PV costs. Almost as much PV capacity was installed in just 2 years (2012–2013) than in the 60 years following the invention of the solar cell at Bell Labs.[116] By 2018, the global installed PV capacity was almost 500 GW, exceeding the nameplate generating capacity of the world's 400 nuclear reactors. By 2022, global capacity had doubled again after an annual increase of 40% in 2021. As in any fast-growing market, such rapid expansion cannot continue indefinitely; however, growth has been exponential for almost two decades. A 2018 Google Maps survey showed almost 1.5 million US homes with solar panels, rising to 2 million less than a year later, almost 1 million of which are in California, while the numbers in other states are now starting to rise even faster.

Upfront costs are still high, however, and the installation process daunting to many. As a former Berkeley City Hall chief of staff noted when he decided to install solar panels on his home to kick-start his own Kyoto-inspired reduction in GHG emissions, "I got two or three bids, looked at what it would take, and had the same sort of realization that everybody else had. Which was, that's a really big check you have to write."[117] More than two decades on, the process is still not simple, although costs continue to drop as the technology improves year on year, a.k.a. the "virtuous circle," while the return on investment (ROI) is now as low as 4 years in some areas. Savings also vary greatly

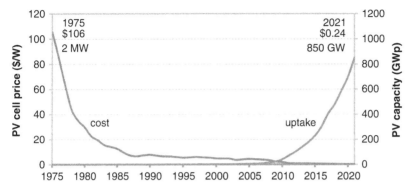

Figure 4.25 Solar panel price per watt ($/W) versus global installation (GW) (1975–2021).

with location because of differing utility prices, government incentives, and tariffs. As for any maturing technology, the market is still being worked out. Figure 4.25 shows the increased PV uptake and corresponding reduction in prices from 1975 ($106/W, 2 MW) to 2021 ($0.24/W, 850 GW).

There are conflicting ideas about how best to encourage PV uptake, especially in the USA. In the mid-2000s, the city of Berkeley decided to finance the high upfront costs of solar-panel installation with public and then private loans to be repaid by property taxes (for example, $1,000 a year over 20 years). As noted by science and technology writer Bob Johnstone in *Switching to Solar*, "If the home were sold, any remaining taxes would simply be paid by the new owners."[118] The novel plan spread throughout California, but was nixed because of concerns by the federal mortgage loan companies Fannie Mae and Freddie Mac.

Palm Desert tried another tack, hoping to allay worries about excessive long-term debt – a "feed-in tariff" (FIT) that cut the payback time in half. A feed-in tariff allows consumers to sell power to a utility company at a prescribed rate, encouraging homeowners to generate electricity beyond their own needs, which lowers the payback time on all upfront costs yet more importantly encourages rooftop installations wherever suited – warehouses, malls, parking lots, arenas, anywhere with an open sunny area.

Palm Desert's then mayor Jim Ferguson noted that a feed-in tariff "reduces your capital cost return from about fourteen or fifteen years down to six or seven years. That makes it more financeable for your average homeowner."[119] Unfortunately, Palm Desert couldn't get its monopoly utility to play ball, ending up instead with "net metering" – a credit system that reduces electricity bills to at most zero, but with no guaranteed buyback of any excess generated power. As a result, without the incentive of an early return on investment the final PV uptake was much less than planned. Indeed, there are numerous obstacles, including wading through the mounds of paperwork and regulations to allow a home or business to operate essentially as a micro-utility.

Although there are many ways to incentivize solar – FITs, net metering, rebates, tax credits, peak-time buybacks, group discounts, loans, leasing, pay as you go (PAYG), and power purchase agreements (PPAs) – installed PV is greatest where guaranteed FITs match utility rates. Fittingly, the first American FIT was introduced in the Sunshine State, modeled after a successful German system.[120] In 2009, the city of Gainesville, Florida, chose a rate of return of 5%, guaranteeing utility purchase of excess electricity for 20 years at $0.32/kWh, estimated to add 70 cents to a homeowner's monthly bill. Within days, the program had orders for 40 MW, 20 times Florida's existing installed PV.[121]

Alas, the American government doesn't make life easy to avail of home-grown technology and the USA has fallen behind markets in China, Japan, and Germany. As Bob Johnstone noted, "Streamlined administration procedures were thus the main reason why installing PV in Germany was so much cheaper. In the US, as one bemused German observed, it was easier to buy a gun than to install a solar system on your roof."[122] A culture of centralized control prevails in the well-established American, utility-run system, top-down metering preferred to bottom-up, do-it-yourself creativity, surprisingly odd in a country that prides itself on rugged individualism and laissez-faire economics.

Regulated monopsony utilities that sell on power are in fact monopoly structures that hold court in the US energy market, at great disadvantage to the consumer, a "profoundly suboptimal and harmful"[123] arrangement that provides inefficient power to beholden consumers. In the heady times of post-war suburban expansion during the 1950s and 1960s, General Electric even threatened not to connect new subdivisions to the grid if alternative electric power was included in the construction plans, saddling homes with expensive electric heating "without any technological necessity for it."[124]

The century-old, established paradigm of producing centralized remote power is difficult to nudge – change is rarely welcome to those whose bottom line is vulnerable. Utility providers think in terms of single GW plants rather than multiple MW warehouse- or kW home-powered systems, a mentality that leaves millions of customers disadvantaged by an outdated management model. Utility-scale solar farms – both PV and concentrated solar power plants – are better than burning fossil fuels, but still follow the central-plant transmission model. Old habits die hard, but one has to look beyond how energy was created in the past to see how small-scale PV is becoming more viable, both residential and commercial, and that distributed energy resources (DERs) are cropping up everywhere.

***

Despite the mind-boggling complexity to connect household PV in some areas, ready-to-go panels can be purchased and installed fairly easily on one's own, offering efficiencies of around 20% at less than $1.00/watt (a simple lighting device is about 30 cents/watt). A 5-kW home installation will cost around $5,000 plus fixed costs, a.k.a. balance of system (BOS) or balance of system and services (BOSS). The installation requires inverters to convert low-voltage DC to high-voltage, grid-tied AC (for example, 12 V DC to 110 V AC), cabling, connections, breakers, optional batteries with charge controllers (to prevent overcharging/discharging), as well as monitors and maintenance that all contribute to the final cost. Add labor if you're not up to it on your own, the final bill depending on roof quality and access.

You will also need enough space – configurations vary, but a 5-kW installation (for example, 25 200-W panels) might need 500 square feet of roof (~50 m^2). If space is limited, more-efficient cells will be needed, which will cost more, or higher-output panels to reduce the system footprint (for example, ten 500-W panels). A general rule of thumb used to be 100 square feet per kilowatt (10 m^2/kW), but continues to improve year on year. A ground-mounted system solves any nagging roof problems if enough open space is available. Any soft costs will also increase the final bill, such as permitting, regulatory fees, financing, and overhead.

It's not immediately obvious the number of panels needed to power a home. To work out how many will cover your electrical needs, one must first calculate how much electricity you use, an often difficult task given today's hard-to-understand electric bills. It's a little easier to work backwards from the monthly statement, as most of us know what we pay in dollars and cents. For example, if I pay $100 per month for electricity, costing 10 cents per kilowatt-hour, I use 1,000 kWh every month or roughly 33 kWh every day ($100 per $0.10/kWh per 30 days). You can also work out how much energy you use by adding up the individual power ratings (in watts) and total operating time per day (in hours) of all the appliances/devices you run, which should come out to around the same 33 kWh per day (something we do in Chapter 7 to find ways to help save money).

Next, we need to know how much the Sun shines where you live. The Internet is full of solar calculators showing the available sunlight in any region throughout the year based on latitude and cloud cover. Where we live (43.5° N), we get about 5 hours of sunlight on average, so we need 6.6 kW worth of panels (33 kWh/5 h). It's a good idea to add 20% for various system inefficiencies (equipment, dirt, shading, non-peak exposure), bringing the total to almost 8 kilowatts. Of course, you get more juice in summer, which can be stored with a battery backup system or sold back to the grid where available, while in winter you will need the grid or batteries to make up for the low-sun shortfall. You can also oversize the array to make up for winter loss. In northern climes, summer to winter output can be 10:1 (see Figure 4.8).

One can ask a local solar company to do the calculations and installation, but for an off-the-shelf panel rated at 200 W that works out to 40 panels to produce 8,000 W (or 20 400-W panels). You might want to start off with a few to get your feet wet and build the system as you go – PV is highly scalable. For example, a starter system could power just the lighting. The ballpark numbers to remember are 33 kWh per day, 5 hours average sunlight and 8 kW of panels, good for a typical mid-latitude home, assuming normal Western power-consumption habits. In less-developed countries, average power consumption will be substantially less: for example, the average Chinese home uses 20% as much power as in the USA, while Colombia is half that again.[125] Consumption also varies according to location within any country (see Figure 4.9).

There is much to get your head round, even in a simple home installation: type of panel, number of panels, local sun hours, and the various BOS add-ons, but the understanding becomes easier with time and practice. A complete PV system also manages the changing irradiance through the day to regulate current and voltage in real-time, from variable sun angles, clouds, shading, and temperatures. The angle of the Sun changes 7.8° each month, while even partial shading can reduce panel output by more than half. An installation should be mounted at the optimum angle to maximize direct sunlight during the year, although mounts can be tailored to one's needs – a fully upright wall-mounted panel maximizes the morning or evening sun and a flat-roof panel the midday sun. Tweaking the setup is easier with experience, whether a stand-alone (off-grid), grid-tie, grid-tie with battery backup (for example, Powerwall), or grid fallback (grid when needed) system. You should get an electrician to configure the final installation, legally required in some jurisdictions.

Many consumers wonder in the midst of a fast-changing market whether they should buy now or wait until next year's model. Although costs keep decreasing as efficiencies rise, one can still find plenty of affordable and reliable solar panels today, financed with a 25-year or longer warranty. Annual performance degradation is less than 0.7% or 27% reduction in 25 years. One must be cautious, however, as some solar companies won't be around in 25 years, while inverters and batteries don't come with the same

Table 4.2 *Solar calculations for three different sites (1.25 efficiency factor)*

Location	Monthly bill ($)	Cost (¢/kWh)	Monthly use (kWh)	Daily use (kWh)	Peak sun (h)	Panel power (kW)
Example	100	10.00	1,000	33.3	5.0	8.4
New York	100	16.57	873	24.4	3.2	15.3
Los Angeles	100	11.60	1,351	42.2	4.2	12.2

guarantees as panels (typically lasting 10 years). The best thing is to shop around and ask lots of questions.

You will also need a smart meter to measure the electricity coming into the home (as in standard practice) as well as the electricity going out from your new micro-power generator, either for net metering (meters already go forwards and backwards anyway) or with a FIT paid by the utility at a set rate for any excess generated power, turning yesterday's consumer into a modern *prosumer*. As PV becomes more prevalent, smart meters will manage the real-time injection of large amounts of excess home-made power into the grid to avoid spikes and maintain safety as millions of prosumers simultaneously produce electricity (which we'll look at later). Fortunately, in high-PV regions, peak sun typically corresponds to peak air conditioning. NREL publishes solar data for maximizing PV installations (for example, see *PVWatts*). Sample installation data plus our simple example are shown in Table 4.2.

4.9 The Solar Field: Bigger and Better Year on Year

The race is definitely on to become the next big thing as manufacturers, suppliers, installers, and service providers all vie for market share, none yet having set themselves apart from the crowd. Founded in 1999 by Harold McMaster and based in Tempe, Arizona, First Solar was the first to break the $1/watt barrier, employing thin-film technology (CdTe), better economy of scale, and large reductions in manufacturing costs, originally deriving 94% of sales in the more amenable European market (74% in Germany), away from the overly bureaucratic, no feed-in-tariff USA.[126] First Solar was also the first company to implement a global, end-of-life, PV-module recycling program, not least because elemental cadmium is toxic.

Rebuilding its economy after decades of stagnation during the Cultural Revolution, China is now the world's leading panel manufacturer with about

70% of all global sales, boasting the world's top companies and first solar billionaire, Shi Zhengrong. Four Chinese companies – JinkoSolar, JA Solar, Trina Solar, and LONGi GET – have jockeyed for top spot for years and now offer 400-W range c-Si panels at over 20% efficiency for both residential and commercial use.

By 2015, the top 10 c-Si manufacturers had installed more than 150 GW, about half of all global installed PV, while 2018 saw the first 100-GW installation year. In 2020, 130 GW of PV solar panels were shipped, over 70% supplied by the top 10 manufacturers. Most belong to the Silicon Module Super League (SMSL), a group of leading c-Si companies, along with Yingli Green and GCL (the world's largest poly-Si manufacturer). Almost all are based in China, reflecting the country's dominance in the field. To gauge the extraordinarily changing manufacturing landscape, in 2016 Yingli started making more PV in a *single* day than it had previously made in four years, increasing output from 3 MW to 6 GW.[127]

Hardly household names to those outside the industry, the top c-Si panel suppliers by shipped production are listed in Table 4.3 for 2015 and 2019–2022. Output continues to rise year on year, increasing from 27 GW in 2015 to 248 GW by 2022. The two largest American suppliers – First Solar (1999, Tempe, Arizona) and SunPower Corp (1985, San Jose, California) – fell out of

Table 4.3 *Top 10 global solar cell suppliers (GW) in 2015 and 2019 to 2022*

#	Company	Year	HQ	2015*	2019*	2020*	2021[†]	2022[‡]
1	LONGi	2000	Xi'an, China	–	11.0	14.7	39	45
2	Trina Solar	1997	Jiangsu, China	3.6	6.0	9.0	25	43
3	JinkoSolar	2006	Shanghai, China	2.4	9.7	8.7	22	43
4	JA Solar	2006	Shanghai, China	3.6	7.5	10.8	24	41
5	Canadian Solar	2001	Guelph, Canada	2.7	8.6	8.3	15	21
6	Risen Energy	1986	Ninghai, China	–	–	–	8	16
7	AstroEnergy	2006	Hangzhou, China	–	–	7.4	6	14
8	Tongwei Solar	2009	Hefei, China	–	12.8	12.1	–	9
9	DAS Solar	2018	Quzhou, China	–	–	–	–	9
10	SunTech	2001	Wuxi, China	–	–	6.3	7	7
	Total top 10			27	78	93	>146	248

Sources: * Feldman, D., Wu, K., and Margo, R., "H1 2021, Solar Industry Update," p. 43, NREL, June 22, 2021. www.nrel.gov/docs/fy21osti/80427.pdf; [†]Lee, M., "Jinko Solar jumps to top spot in Q1 2022 module shipment ranking," *Solarbe Global*, May 6, 2022. www.solarbeglobal.com/jinko-solar-jumps-to-top-spot-in-q1-2022-module-ship ment-ranking/; [‡]Lee, M., "Module shipment ranking 2022: Top 10 manufacturers shipped 240 GW globally," *Solarbe Global*, January 13, 2023. www.solarbeglobal.com/ module-shipment-ranking-2022-top-10-manufacturers-shipped-240-gw-globally/.

the top 10 in 2016, although First Solar later reappeared. Others have also fallen in and out of the top 10 suppliers' list, including three Chinese companies, Aiko Solar, GCLSI, and Shunfeng, and the Korean manufacturer Hanwha Q-Cells.

On the installation side, SolarCity started out by offering popular group discounts and leasing options in the San Francisco area, increasing its brand recognition in 2016 after being bought out by Tesla, and is one of the largest American installers of solar panels (now called Tesla Energy). Streamlining the residential installation market with a new line of solar-powered shingles, Tesla built on Subhendu Guha's 1998 creation of flexible amorphous-silicon (a-Si). Solar-clad roofs needn't be an ugly afterthought and can even be a thing of beauty with a frameless design, as sleek as they are functional. As Tesla CEO Elon Musk succinctly noted, "It's not a thing on the roof. It *is* the roof."[128]

The challenge is to make the panels both structurally and electrically functional, but also physically appealing by laminating shingle components in a solar "cheese melt" sandwich. Manufacturing the latest in solar style, tile models include textured, slate, Tuscan, and smooth glass. As a newly integrated solar cell and battery company, Tesla hopes to provide elegant solar-powered roofing for the estimated 5 million American homes reroofed each year. But despite the improved look, production numbers remain low and marketing costs high. As the head of operations noted, "We have a product, we have the customers, we are just ramping it up to a point where [the business] is sustainable."[129]

Not everyone likes how a solar panel looks, the aesthetics as important as the output. Film director James Cameron installed a 260-kWh array of "sunflowers" at his wife's K-12 school in Malibu, California, stating "The idea was to unify form and function with this life-affirming image that anyone looking at it would instantly get."[130] Standing atop a 33-foot tall metal stem, each flower is composed of 14 solar petals surrounding a central face that tracks the Sun through the day like a sunflower following the Sun from dawn to dusk. Dotted around the grounds, Cameron's sculptured solar flowers provide between 75 and 100% of the school's electrical needs depending on the season.

An Australian company used a similar concept to create a portable "smart flower," an attractive ground-based installation with collapsible petals, internal battery storage, and GPS tracking that provides plenty of off-grid household juice. In Dubai, solar "palm trees" have been installed in over 100 locations along its beaches, providing solar-powered battery charging, Wi-Fi, and information screens. More and more, solar-power devices are functioning as part of the background, integrated into a variety of landscapes, built environments, and everyday architecture. In a stroke of a pen in 2022, the French government decreed that all parking lots of at least 80 spaces must be covered with solar panels by 2028, expected to provide power savings equivalent to the output of 10 nuclear plants.[131]

The MIT start-up Ubiquitous Energy has developed a solar-powered window, a transparent luminescent solar concentrator that passes visible light as per normal, yet filters the UV and IR rays via organic salts guided to a PV cell. Selective absorption of incident sunlight is naturally less efficient because the visible part of the spectrum is not converted into electricity, but solar cells that function both as a window and a power source can provide a wealth of untapped energy. Imagine a building powered – in part at least – by its own windows.

In 2021, Michigan State University installed a 100-square-foot transparent test array above the entrance to one of its campus buildings. The conversion efficiency is only 10%, but the array generates enough power to light the atrium. As noted by Ubiquitous Energy's co-founder and MSU professor Richard Lunt, "Transparent solar glass expands the options of solar power tremendously and changes the way we think about generating power. There is no longer a tradeoff between aesthetics and renewable energy. You could turn nearly every surface of a building or landscape into a solar array and generate power right where you use it without even knowing that it's there."[132] In 2018, MSU also covered 5,000 parking spaces with solar panels, generating 5% of campus power and providing seasonal shade (for example, as seen in Figure 4.26a).[133]

Traditional solar cells and partial-spectrum window cells have huge potential to create power in everyday architecture, such as skyscrapers, domes, skylights, and archways. The arched roof of the Stillwell Avenue Terminal Train Shed in Coney Island, New York, is covered with 2,730 thin-film PV panels, generating almost 15% of the terminal's annual power. Incorporated into the sweeping overhead structure, the panels are 5-foot square, triple-laminated, and 5% transparent, providing more than 200 kW, although maintenance-related budget problems have taken them offline.[134] The low-level voltage in the first-of-its-kind structure even helped to keep the pigeons away!

(a)   (b)

Figure 4.26 More PV installations: (a) parking-lot canopy (*source*: Hanjin CC BY-SA 3.0) and (b) canal canopy (*source*: Solar AquaGrid).

Copenhagen's International School is now powered by 12,000 colored solar panels, the world's largest solar façade, providing 300 MWh a year or half of the building's electrical power. Apple's circular Cupertino HQ is clad with 16 MW of rooftop capacity, while GE's Boston HQ has a sail-like "solar veil."[135] Vertically clad solar is not optimally angled to the Sun, but the plentiful surface area makes up for the lower efficiency.

Thin-film cells can now be affixed to almost anything, fashioned onto the roofs of cars, buses, and trains. Charged during daylight hours, the power is consumed as needed by the on-board electrical system. In 2021, a double-occupant camper van affixed with 8.75 m^2 of solar cells (doubled at full spread) was driven 2,500 km across Europe by a team of Eindhoven University students. The Stella Vita "house on wheels" managed 730 km at a go without charging. Although surface solar cells may be better suited to buses, trains, and ships with larger areas to affix the panels, according to Toyota a simple solar roof increases efficiency by up to 10%, seamlessly adding extra range to a standard electric-powered vehicle.[136]

Rooftop solar is also becoming more innovative, such as solar-and-heat systems that increase efficiency by circulating waste heat behind a solar panel, known as building-integrated photovoltaic/thermal (BIPVT). As noted by the head of an Australian BIPVT project in Sydney, the "innovative thermal duct system warms and cools air to supplement air conditioning in the homes."[137] Solar panels have also been installed over irrigation canals to increase pumping power, decrease evaporation, and double on space as in one novel program in Gujarat, India. The efficiency of such "agrivoltaic" greenhouse farming is improved by plant respiration that cools the panel's underside at the same time, while aiding plant growth by shading and maintaining soil moisture, especially useful in arid regions. Making panels more efficient and durable will only increase adoption in the built environment.

In 2022, the Turlock Irrigation District (TID) announced Project Nexus, a pilot program in California's Central Valley to build solar canopies over part of the TID's 4,000-mile canal system (Figure 4.26b). The first of its kind in the USA, the $20 million public–private–academic partnership deals with reduced water, canal maintenance, and renewable power generation, which if expanded to the entire canal system could generate 13 GW and save 63 billion gallons of water per year. As noted in the initial study, "mounting solar panels over open canals can result in significant water, energy, and cost savings when compared to ground-mounted solar systems, including added efficiency resulting from an exponential shading/cooling effect."[138]

***

While "first-generation" c-Si cells have become commercially viable in rooftop installations – already at grid parity with coal, oil, and natural gas in some areas – and "second-generation" thin-film cells have found their own low-cost niche markets, new "third-generation" cells are also starting to vie for market share, such as organic (OPV[139]), dye-sensitized (DSSC[140]), quantum dot (QD[141]), and perovskite solar cells, which may one day "dominate the collection of light energy for generating electricity."[142] Most are still in the development phase.

One particular third-generation cell, however, is fast becoming a PV darling. Named after the Russian mineralogist Lev Perovski, the perovskite solar cell (PSC) is relatively simple and cheap to make compared to crystalline solar cells, harvesting light in an active layer made of a hybrid, organic–inorganic, metal-halide-based material (lead, tin, or metal-like iodine). Just half a micrometer thick, a perovskite cell has a broad absorption spectrum across most of the visible range and absorbs light 400 times better than silicon, but still isn't sufficiently stable or durable.[143] Manufactured in a lightweight, roll-to-roll process and deposited via inkjet or spin coating on a flexible substrate, perovskite cells are literally ejected like sheets of electric paper from a printing press or mixed into a solution and sprayed onto a surface. If the material limitations can be worked out, perovskite solar cells are especially suited for BIPV and could revolutionize the industry.

After the University of New South Wales achieved a record 12.1% efficiency on a 16-cm^2 perovskite cell, Hiroshi Segawa of the University of Tokyo declared perovskites "the front-runner of low-cost solar cell technologies."[144] In another novel, multi-junction construction similar to an organic/quantum dot tandem cell and known as a perovskite-on-silicon tandem, the perovskite cell covers the blue end of the spectrum while a traditional silicon cell covers the red end, absorbing a wider spectral range to exceed 25% conversion efficiency. The theoretical limit is about 40%.

Organic polymers are also improving and being fitted to various structures, such as the first large-scale printed solar array installed on a covered roof in Sydney, Australia. The 75-μm thick liquid organic polymer cells are lightweight and easy to apply via simple adhesive, although overall efficiency (2%) and durability (2 years) are still low.[145] Another test device made of organic polymer film and quantum dot construction converts wasted, higher-energy, blue and green photons to lower-energy IR photons via "singlet exciton fission," which are absorbed in silicon to increase efficiency by about a half to 35%.[146]

Other non-silicon-based solar-cell projects include ferroelectric crystals that do not require p-n doping and are thus easier to manufacture. Researchers at

Martin Luther University in Halle, Germany, recorded a thousand-fold increase in the photovoltaic effect of a thin-film "superlattice" structure that alternates vaporized titanate (TiO_3) layers of strontium, barium, and calcium. One 200-nm thick test device consisted of 500 layers, producing a much higher electric permittivity and generated photocurrent.[147] Bismuth-based nanocrystals engineered via cation disorder have also been integrated into an "ultrathin" solar cell that increases the absorption coefficient up to 10 times across a wider spectral range. A joint research project at UCL, ICFO, and Imperial College achieved a conversion efficiency of 9% for a device only 100 nm thick, much less than current "thin-film" cells.[148]

Other potential uses are being exploited, such as matching a cell's absorption spectrum to household lighting, where indoor surfaces are laminated with electricity-generating coatings. One simple idea is to punch a lattice of small holes into a panel to let light through, not unlike a semi-transparent, bus-window advertisement. All sorts of emerging solar cells are being dreamed up – carbon nanotubes, 3D architectures, PV liquids, bi-triggering, hot-carrier cells – although low efficiencies have yet to make them competitive with the established crystalline silicon PV market. Each year sees more innovation and increased growth with advances in crystal growth techniques, shrinking material widths, improved light-trapping, and better manufacturing.

As consumer solar continues to expand, there are concerns about material supply. Made from sand, silicon is available in abundance, keeping c-Si at the forefront of a fast-growing market for years to come, while thin-film technology may suffer from limited availability. Tellurium and indium are both rare, respectively obtained during copper and zinc refining, while gallium and arsenic are considered precious metals, likely curtailing continued growth of second-gen CdTe and CIGS cells. Much less material is needed to make higher light-absorbing thin-film materials, but manufacturing could become prohibitively expensive as stocks dwindle. Cadmium on its own is also toxic, although fine in compound form as CdTe, almost all of which can be recycled.

To the victor go the spoils, which are becoming more plentiful in today's PV market. Time will tell how much the evolving third-generation market can carve out from the established mono-Si, poly-Si, and thin-film leaders (Figure 4.27). But we can expect to see much more solar-integrated structures, employing traditional c-Si panels, novel BIPV, and other new constructions.

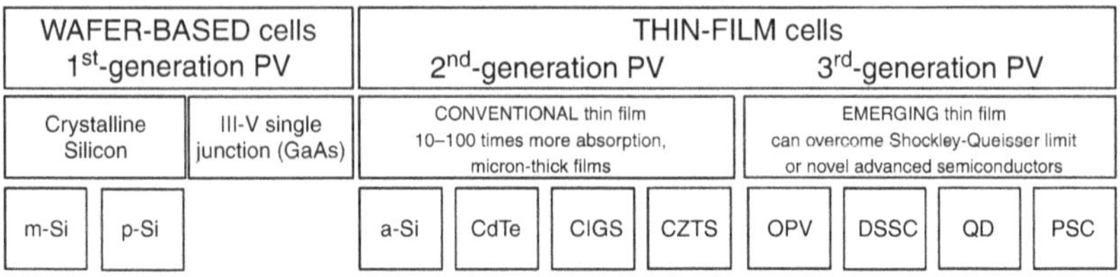

| WAFER-BASED cells | THIN-FILM cells | |
| 1st-generation PV | 2nd-generation PV | 3rd-generation PV |

| Crystalline Silicon | III-V single junction (Gaas) | CONVENTIONAL thin film 10–100 times more absorption, micron-thick films | EMERGING thin film can overcome Shockley-Queisser limit or novel advanced semiconductors |

| m-Si | p-Si | | a-Si | CdTe | CIGS | CZTS | OPV | DSSC | QD | PSC |

Figure 4.27 Photovoltaic technology: traditional wafer-based and thin-film solar cells.

4.10 Utility-Scale Solar: The New and the Old Revamped

Photovoltaics is ideal for grid-independent power applications, such as kilowatt household or warehouse rooftop systems, but as the price of solar continues to drop more grid-tied PV farms are cropping up on a megawatt and even gigawatt scale. In many regions, solar has already passed grid parity with coal (that is, less than $0.10/kWh), while more growth and savings are expected. PV technology is easy to scale as a solar "farm" adds panels in phases to increase output, while manufacturing, automation, and innovation continues to improve, spurring on even more growth as in the nascent boom times of computers, mobile phones, and digital cameras. Installing new systems is like Lego, Tom Buttgenbach of 8 Minute Energy calling PV "really boring – panels, panels, panels."[149]

Racing against a government-funded project that hoped to build the first ever large-scale solar power plant in the early 1980s, the world's first megawatt-capacity PV installation was built in 1982 northeast of Los Angeles at Hesperia in the Mojave Desert by the Atlantic Richfield Company (ARCO), generating 1 MW for the first time in a single location. ARCO's Lugo plant was made of panels earmarked for rooftop use and set on a single-axis (S/A) sun-tracking system to maximize daily insolation: "The whole thing – from design to construction to getting the plant into operation and online – was done in a matter of months."[150] As noted by Bob Johnstone in Switching to Solar (p. 71):

> The resultant plant, seen in the late afternoon sun with the low sierra in the background, looked eerily beautiful. Lined up facing west in rows, like giant playing cards tilted at an angle of about 80 degrees, the 108 arrays bore a slight but unmistakable resemblance to the stone heads on Easter Island. Capable of supplying enough energy to power up four hundred homes, Hesperia Plains was three times larger than any other PV system in the world at the time.

ARCO's next solar installation was a 5.6-MW array north of Santa Barbara at Carrizo Plain. The output from each panel was optimized with large, dual-axis

(D/A) computer-controlled trackers designed for an earlier solar–thermal power plant to control the direction of sun-reflecting mirrors. The electricity was sold at a loss-making 3–4 cents per kilowatt-hour,[151] but more design wrinkles were worked out in the following decades to increase the generated electric output and provide a reliable, competitively priced power source.

Building on early designs, a solar farm installation is now arranged to match optimal sun conditions with sufficient row spacing to minimize inter-row shading, while dual tracking maximizes global horizontal irradiance (GHI), both direct normal irradiance (DNI) and diffuse horizontal irradiance (DHI). A tracker system typically generates 30% more power than a fixed system and up to 40% in the middle of the day when utilities are pushed to the limits and electricity costs are highest.[152] Storage add-ons also smooth out fluctuations from reduced sunlight and dark time (as we'll see in the next two chapters).

Solar farms can also help to flatten the load by "oversizing" or "overbuilding," that is, exceeding the needed capacity but holding back output from the inverter at peak sun, useful for reduced output on cloudy days. Oversizing will become more commonplace as manufacturing costs continue to fall. As Chris Goodall notes in *The Switch*, "Solar cells will make electricity eventually so cheap that it won't matter if we over-install generating capacity in fields around the world in order to ensure that we always have enough during daytime hours."[153]

Each year brings a record number of new installations. In 2015, more than 3,000 PV solar farms were in operation worldwide, generating roughly 50 GW with another 75 GW in the works from 1,500 more sites, while by the end of 2017 the number had jumped to over 5,000 with a total installed capacity over 140 GW.[154] In 2022, 268 GW of panels were installed globally, bringing total PV capacity to over 1 TW.

The world's first 1-GW plant – the Yanchi Solar PV Station in Qinghai, China – was constructed in 2016 by Huawei, built in the Tibetan Plateau to avail of better sun conditions in the high-altitude and thin-air of "the rooftop of the world." The massive 4 million panel solar park is arranged over a 27-km^2 grid, resembling an army of modern Terracotta warriors. As at other large solar farms, panels are robot-cleaned to remove debris and dust. Competitive PV power plants are no longer a dream at 1 GW, signifying a new era of tapping the Sun as each year outperforms the last.

The evolution of four decades of record-breaking solar installations is shown in Table 4.4, from ARCO's 1-MW Lugo plant (built in 1982 and now decommissioned) to the first 1-GW plant in 2016 (Yanchi Solar in China) and first 2-GW plant in 2019 (Pavagada in India). Note the rapid growth in large, utility-scale farms since 2014 with the 550-MW Topaz Solar and 579-MW Solar Star

Table 4.4 *Record photovoltaic (PV) solar farms (¹decommisioned, ²fixed tilt)*

Year	Plant	Capacity (MW)	Location	Operator
1982	Lugo[1]	1.0	Hesperia, California	ARCO
1983	Carrizo Plain[1]	5.6	San Luis Obispo County, California	ARCO
2005	Bavaria Solarpark (Mühlhausen)	6.3	Bavaria	SunPower
2006	Erlasee Solar Park	11.4	Bavaria	S.A.G. Solarstrom
2008	Olmedilla Photovoltaic Park	60	Castile-La Mancha, Spain	Nobesol
2010	Sarnia Photovoltaic Power Plant	97	Sarnia, Ontario	First Solar
2011	Golmud Solar Park	200	Qinghai, China	Huanghe Hydropower
2012	Agua Caliente Solar Project[2]	290	Yuma County, Arizona	First Solar
2014	Topaz Solar[2]	550	San Luis Obispo County, California	First Solar
2015	Solar Star	579	Rosamond, California	SunPower
2016	Kamuthi Solar Power Project	648	Tamil Nadu, India	Adani
2016	Yanchi Solar PV Station	1,000	Qinghai, China	Greencells
2017	Tengger Desert Solar Park	1,547	Ningxia, China	CNG/Zhongwei
2019	Pavagada Solar Park	2,050	Karnataka, India	KSPDCL
2020	Bhadla Solar Park	2,240	Rajasthan, India	RSPDCL

plants built to serve Los Angeles. In 2021, a massive 4-GW solar farm was announced for the Philippines that will add 3.5 GW to an existing 500-MW installation on the island of Luzon. No one can doubt anymore a renewables-integrated future (Figure 4.28).

China is especially keen to exploit technologies discovered elsewhere. The first-ever floating PV farm – a 400-kW$_p$ array on a 3,000 m^2 pond – was installed in 2008 in the Far Niente winery in Napa Valley, where land is limited and expensive. Multiplying that a hundred-fold a decade later, the Chinese inverter company Sungrow installed a 40-MW floating PV plant in a flooded former coal mine in Huainan, Anhui Province, the largest of its kind until a year later when the Three Gorges New Energy Company built a 150-MW floating

(a)   (b)

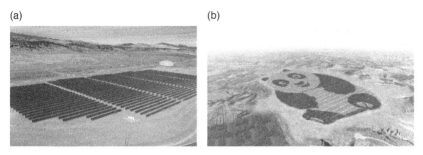

Figure 4.28 (a) PV solar farm (*source*: US DOE) and (b) Panda Green Energy's 100-MW Datong Panda plant in Shanxi Province, China (*source*: UNDP).

PV plant atop a lake formed by the collapse of another old mine in nearby Huainan City.[155]

Floating PV (a.k.a. "floatovoltaics") has the added benefit of cooling panels and reducing water evaporation, essential for agriculture. Dam reservoirs are also potential floating PV sites that could triple hydropower output and more easily grid tied to existing transmission infrastructure, while "high-wave" offshore solar is expanding as PV farms are installed at sea. Less subject to land restrictions and NIMBY concerns, some can act in conjunction with existing or proposed wind farms, a.k.a. the "Blue Economy."

In 2022, a first-of-its-kind open-sea floating commercial system was expanded to 50 kW in the challenging conditions of the North Sea as part of a Dutch pilot project 15 km offshore from The Hague. Lighter, flexible panels that bend with the waves use less material and are thus much cheaper, but as noted by the project manager, "It is technically very challenging to install large floating systems … and keep them operational for decades."[156] The system passed its first major test after weathering two major storms in a month.

***

Today's utility-scale plants also use solar energy to make electricity the old-fashioned way by heating water to make steam to turn a turbine that generates electricity (Figure 4.29). There are two main "concentrated solar power" (CSP) thermal–solar systems: parabolic trough and power tower. In a parabolic-trough system, large adjustable curved mirrors focus the incoming sunlight onto a liquid-filled, stainless-steel pipe, where temperatures reach over 500°C to create the steam via a heat exchanger. In a power-tower system, a large circular field of flat-mirrors called "heliostats" reflect the Sun onto a tower-mounted receiver, built sufficiently high to capture sunlight without obstruction, the temperatures reaching as high as 1,000°C. The heliostat mounts are

(a)   (b)

Figure 4.29 Concentrated solar power: (a) parabolic trough (SEGS, California) and (b) power tower (PS10 and PS20, Seville) (*source*: Koza1983 CC BY 3.0).

computer-controlled to direct the Sun at the optimum angle for maximum concentration up to 10,000 Suns, although on very sunny days the solar flux is curtailed to avoid melting the tower. The central receiver is filled with liquid (for example, liquid sodium), gas, or solid particles, with the steam generated again via heat transfer.

A parabolic-trough system is built on a north–south line to track the Sun from sunrise to sunset, a modern version of Frank Shuman's solar-steam Nile irrigation water pump, while a power-tower system tracks the Sun from side to side and up and down to concentrate the most amount of thermal energy onto the receiving tower, a modern version of Archimedes' famous though likely apocryphal mirror weapon. Other novel "high-heat" power-tower systems can also produce temperatures above 1,000°C by precise AI-guided focusing, generating sufficient heat to manufacture steel, cement, and glass, processes that require high temperatures and create large amounts of carbon dioxide via current practices – industrial steelmaking and cement manufacturing each account for about 7% of GHG emissions.[157]

A CSP plant can also store generated heat in a liquid or solid material for future use. Water is the preferred storage medium because of its high specific heat capacity, low cost, and non-toxicity: "In addition, the energy can be transported by the storage water itself, without the need for additional heat exchangers."[158] Other solid storage materials include packed beds of stone and phase-change materials that absorb and release latent heat, such as molten salts for high temperatures. Heat storage is perfect for down-sun conditions, allowing a CSP plant to function 24/7.

The first modern, grid-connected, parabolic-trough power plant was built in 1984 as part of the Solar Energy Generating Systems (SEGS) facility in the Mojave Desert, California, now comprising nine separate plants in Daggett,

Kramer Junction, and Harper Lake, all still running at an output of 359 MW, while the first power-tower plant, Solar One, was part of a pilot project in Daggett operating from 1982 to 1986. In 1995, Solar One was converted to Solar Two to incorporate molten-salt heat storage (a low-loss radiator) for down-sun power.

Based on the early test concepts of Solar One and Solar Two, the world's first commercial power-tower system, Planta Solar 10 (PS10), was built in 2007 about 20 km west of the Andalusian capital of Seville in southern Spain, which receives over 3,000 hours of sunlight a year, and produces 11 MW. The adjacent 20-MW Gemasolar plant was the first to run 24/7 with molten-salt heat storage, able to operate for 15 hours in winter and 24 hours in summer, perfect for baseline power. The nearby Solnova parabolic-trough solar–thermal plant produces a further 150 MW.

Although not as popular as PV and harder to set up and operate, there are 114 operational CSP plants worldwide with a total capacity of 6.3 GW, of which 61% are parabolic troughs and 22% power towers – 51 in Spain, 14 in China, and 10 in the USA – with another 10 under construction.[159] Less-common CSP types are linear Fresnel refractors (8%), Sterling dish engines, and solar chimneys. Power-tower systems that heat water via direct steam generation operate at higher temperatures and are thus more efficient than parabolic-trough systems that utilize an intermediate heat transfer fluid such as thermal oil, while the conversion efficiency almost doubles to over 60% as the receiving tower temperature increases from 800 K to 3,000 K.[160]

As one can readily imagine, concentrated solar power works best in very hot places with more direct sunlight, such as the 110-MW Crescent Dunes power-tower plant built in Nevada in 2014 at the northern edge of the Mojave Desert and the 160-MW Noor 1 parabolic-trough plant in Ouarzazate, Morocco, built in 2015 at "the door of the desert" on the western edge of the Sahara Desert. Desert-based CSP is an attractive renewable-energy option for countries in North Africa, the Middle East, China (the Gobi Desert), and Australia. Plenty of water is needed – although once procured it can be recycled over the lifetime of the plant – as are nearby power lines, not always available in desert locations.

Until recently, the world's two largest CSP plants were the 392-MW Ivanpah Solar Power Facility in San Bernardino County, California (power tower), and the still-working 359-MW SEGS facility (parabolic trough). Upon completion of Noor 2 and 3, the Ouarzazate plant in Morocco will become the largest at 580 MW, providing almost one-fifth of electrical power generation in Morocco. In a nod to its source, *noor* means "light" in Arabic. An interesting but now discarded idea was to power Europe from a string of CSP plants in North Africa via subsea transmission lines under the Mediterranean, but the "Desertec" plan

was deemed unworkable, primarily because of the vast infrastructure required –
up to 30 high-voltage transmission lines at $2 billion each[161] – never mind the
large number of CSP plants.

Solar–thermal is especially suited for peak demand in summer and in the
middle of the day when air conditioning is typically highest (as is PV), while
net conversion efficiency continues to increase with material advances, such as
heliostats made of lightweight polymers instead of glass and receiving pipes
coated with improved heat-absorbing material, such as black ceramic. But
although CSP has become more feasible and capable of 24/7 operation with
heat storage, PV plants are much easier to build and operate. The first two
ARCO PV plants were run with no human operators, impossible even with
a small CSP plant.

PV plants also have lower upfront costs and are not as susceptible to
intermittent sunlight which can completely shut down a CSP plant without
heat storage. Reductions in PV module fabrication costs are also continuing to
lower electricity prices for the consumer. Tom Buttgenbach of 8 Minute Energy
bluntly stated the reality of large-scale solar: "CSP is getting killed by the
chip."

To help visualize the footprint of a modern solar farm, utility-scale power
plants can easily be seen in satellite pictures such as Google Maps. Just west of
Seville in Sanlúcar de Mayor, the tell-tale semicircle mirror arrays of the 10-
MW PS10 and 20-MW PS20 power-tower CSP plants can clearly be seen, as
can the 150-MW Solnova parabolic-trough CSP plant with its obvious lined
configuration. From the sky, it's hard to tell a parabolic-trough setup from
a vineyard because of the similarly spaced parallel piping. Europe's first CSP
parabolic-trough plant – the almost square, 150-MW Andasol power station
near Guadix that employs molten-salt heat storage – can also be seen north of
the Sierra Nevada mountain range, roughly 300 km due east of Granada. The
symmetric majesty is a wonder to behold.

***

Ideally, one builds a power plant near where people live, but optimal solar sites
with predictable weather are typically found in sparsely populated and desert
locations. Growing at more than 40% per year for over a decade, the majority of
large-scale solar installations are still located in the desert followed by farms.
Many infrastructure challenges exist without local distribution, but the goal is
to find a suitable open area with low urban density close to existing power lines.
Transmission technology has improved, but high losses – roughly 3% loss per
1,000 km – still make sending power over long distances difficult.

Although the Desertec plan was nixed because of the high costs of sending power from North Africa to Europe, the increasing number of solar installations requires more efficient, long-distance transmission lines to keep the generated power from being stranded. Such ambitious mega projects can introduce mega problems, however, because of the uncertainty of building in politically fragile regions. Although Desertec was technically feasible, securing the political will and acquiring investment is as problematic as in any installation.

A not dissimilar idea was proposed for the USA with large wind and solar farms in Texas and other southwestern states sending power to the northeast along high-voltage transmission lines, but came with an unworkable $160 billion price tag. The value of supporting large, centralized, unidirectional power plants over local, distributed, peer-to-peer power systems was also questioned.[162] Planning to export PV-generated power along 4,200 km of HVDC subsea cables from Australia's Northern Territory to Singapore, another project hoped to become the first large-scale solar farm to send power via long-distance transmission lines, providing 15% of the electrical needs for almost 6 million people. Unfortunately, Sun Cable's $30 billion, 3.2-GW AAPowerLink went into voluntary receivership in early 2023.[163] It seems the large upfront costs of long-distance power transmission will keep the more ambitious projects in the design phase for now.

In the early 1980s, NREL's director Roland Hulstrom calculated that American electrical needs could be powered by a 100 mile × 100 mile-square PV farm, publishing his idea in a 1999 *Science* article to illustrate the theoretical simplicity.[164] To be sure, 10,000 square miles is a small percentage of available land in the USA (~0.3%), less than the amount of paved roads (~1%) and miniscule compared to the amount of land needed for coal, oil, and gas. As stated by Hulstrom, the size of the square would also shrink if supplies of wind, geothermal, and hydroelectric were added to the mix. With today's improved PV technology, an area about half the size of Hulstrom's square could cover the American market, for example, in a 70 mile × 70 mile Texan PV farm. Similar squares apply to the electrical needs of Europe, Asia, Africa, South America, and Australia.

Clearly there is value in providing *centralized*, megawatt (or GW!) utility-scale solar-generated electricity where feasible, but distributed, *home-made*, kilowatt solar power is also viable given the enormous number of rooftop sites. As noted in the seemingly futuristic 2008 *Scientific American* article "A Solar Grand Plan," "The greatest obstacle to implementing a renewable US energy system is not technology or money … It is the lack of public awareness that solar power is a practical alternative – and one that can fuel transportation as well. Forward-looking thinkers should try to inspire US citizens, and their political and scientific leaders, about solar power's incredible potential."[165] Simplifying the strategy in typical *Terminator* style, the then California governor Arnold Schwarzenegger emphatically stated the same idea on a *60 Minutes* program later that year: "green technology is where it's at."[166]

Despite the slow uptake in the past 40 years, PV is starting to make a dent in overall power generation, whether in residential, non-residential (commercial and community), or utility (fixed-tilt and tracking) sites. Today, utility installations generate 70% of all PV output, followed by "behind-the-meter" residential (20%) and commercial (10%), although the number of residential and commercial sites could be much higher because off-grid PV power is unmetered (Table 4.5). Many hands make light work or open sites and roofs (120 million in the USA alone).

As of 2023, PV installations in the USA reached more than 155 GW, enough to power 27 million homes, while the amount of added annual solar capacity now accounts for more than 40% of electrical power, exceeding wind (32%) and natural gas (17%). As noted by Wood Mackenzie and the Solar Energy Industries Association, passage of the historic 2022 IRA was also "a massive growth catalyst . . . for every segment of the solar industry," expected to boost further deployment by 40%.[167] What's more, because of the provisions in the Solar Energy Manufacturing for America Act (SEMA), domestic solar will be able to avail of production and investment tax credits for the first time, prompting companies to build within the USA and secure valuable local supply chains.

Are the fossil fuel companies running scared because of guaranteed and stable PV prices at grid parity? SUNY journalism professor Karl Grossman stated "There are those who seek to profit from expensive electricity generated by oil, gas, coal and nuclear power – and they would try to suppress the

Table 4.5 *Utility, commercial, residential PV solar (peak power, 2022 costs, percent)*

Installation	Peak power	Cost ($/$W_{DC}$)	Percent
Utility	1 MW–2 GW	1.07	~70
Commercial	~500 kW	1.71	~10
Residential	~5 kW	3.25	~20

Source: "US Solar Market Insight Executive Summary," Q3 2022, Wood Mackenzie/SEIA, September 2022. www.woodmac.com/industry/power-and-renewables/us-solar-market-insight/.

renewable energy revolution now underway. They must be stopped, and the windfall of safe, green, inexpensive electricity be allowed to flow."[168] *Washington Post* journalist Joby Warrick concurred, adding that the "grave new threat to operators of America's electric grid [is] not superstorms or cyberattacks, but rooftop solar panels."[169]

As with any new technology, there have been commercial failures along the way. Solyndra used thin-film CIGS cells applied to a hollow glass cylinder and went bankrupt in 2011 with half a billion dollars in outstanding government loans, souring more than a few dreams and angering free-market pundits wary of misguided government help. But no matter how bitter the debate over using public versus private money, the writing is on the wall. Since 1975, the cost of photovoltaic cells has dropped over a hundred-fold, while installation capacity has correspondingly risen over a thousand-fold. The 2022 *International Technology Roadmap for Photovoltaic* predicts there could be as much as 22 TW_p of installed PV by 2050 generating 38 PWh and dwarfing today's current global grid capacity of about 8 TW.[170]

While it's true that the Age of Petroleum might have come crashing down on its own in 50 years or so, having passed the point of no return on the downside of Peak Oil (as we saw in Chapter 2), renewable energy is giving oil a push. Although we need petroleum for plastics, chemicals, pharmaceuticals, and other household products, and for the foreseeable future during the transition from fossil fuels to renewables, at least to power the manufacturing of high-energy solar fab plants and industrial-scale wind turbines, the end of oil is in sight, beginning with local and utility-scale energy farms and the banning of gasmobiles from cities across the world. We are on the upside of change, not yet aware of the dizzying heights ahead. As with all revolutions, the future eventually becomes unstoppable.

4.11 Change: The End and the Beginning of the Revolutions

You can't hold back the Sun. Today, one can install a 1-kW rooftop solar-panel system for $2,000 guaranteed for 25 years that pays for itself in under 5 years, expanding at will to suit your needs, even going completely off-grid if you can manage. The average American home consumes about 30 kWh of electricity per day, which if fitted with off-the-shelf solar panels would completely remake the grid, upending our petroleum-run world and vastly reducing GHG emissions and pollution.

Starting in California and soon after in Colorado, Nevada, and Texas, new homes built by Lenna, the second-largest US home builder, now include

SunStreet solar panels with a guaranteed lower electricity rate – for example, 20% reduction for 20 years – although one can choose instead to buy the panels outright. Local community solar farms are also becoming popular, where homeowners rent or buy shared panel output to offset household electrical use, providing a reasonable return of investment and cheaper electricity. As part of a local experiment to turn homes and neighborhoods into virtual power plants, Green Mountain Power is pioneering community energy in Vermont by installing solar panels, batteries, insulation, and heat pumps for customers.[171] Financially sound, environmentally friendly, and drawing less power from the grid, small is good where "self-generation" is the key.

Solar is cropping up everywhere, in vehicle roofs, windows, and awnings (power and shade), while bike paths, sidewalks, and roadside sound barriers are all undergoing a sun-kissed makeover in rich and poor countries alike. Ambient light can be absorbed in millimeter-thick surface cells to power any number of electric gadgets from headphones and speakers to bike lighting and body wear. Developing regions are even bypassing the fossil-fuel-based utility-power generation model altogether. It is fitting that the word *revolution* comes from Nicolaus Copernicus, who first published the theory of a heliocentric solar system to explain the seemingly curious motion of the planets relative to the Earth. Published after his death in 1543 for fear that his ideas would upset the ruling authorities, *The Revolution of Heavenly Bodies* changed the world.

Although the battle between carbon and silicon is heating up, one wonders how long Big Oil can dominate the market as more renewable energy is added year on year. In *More with Less*, two Dutch energy consultants believe "the moment of equivalence" between fossil fuels and renewable energy could happen as early as 2040, dubbed "50–50–40," primarily because of better energy efficiency, increased renewable energy, and a slower-growth population trend, culminating in peak energy consumption and peak population between 9 and 10 billion.[172] We have only scratched the surface, but as solar- and wind-generated power expands to meet the electrical needs of more consumers, new-energy technologies are beginning to follow the same exponential adoption paths as personal computers, cameras, and telecommunications.

Sweden hopes to be 100% renewable by 2040 and was the first country to reach 50% annual renewable power in one year, primarily via onshore wind,[173] while Denmark and Portugal have already achieved 100%-wind-powered days (as we will see in the next chapter). In Queensland, Australia, almost 25% of homes generate electricity via solar panels, while in South Australia 260% grid power was potentially available via renewables in 2023. Two countries (Suriname and Bhutan) reached net zero before the 2021 Glasgow COP, while 16 others made pledges for 2050.[174]

Amidst the ongoing energy makeover, one small island in the sunny South Pacific was the first to go 100% solar. Long dependent on American diesel to supply its energy needs, the easternmost Samoan island of T'au is ideally suited to tap the Sun with as much as 8 hours of sunshine per day, and is now powered by 5,328 SolarCity solar panels and 60 Tesla Powerpacks. The 1.4-MW microgrid provides 6 MW of storage, enough to supply the electrical needs of its 600 residents, charged in 7 hours and good for 3 days of down-sun.[175] As Auckland University engineering professor Ashton Partridge noted, "The cost of setup for solar is high and there has been a push-back against that. But it is ideal if governments absorb that cost, especially for these remote communities that would otherwise be totally reliant on non-renewable energy sources."[176]

In 2017, Tesla also built a large solar plus battery storage farm on 50 acres of the Hawaiian island of Kauai – 54,978 panels, 13 MW @ 13.9 cents/kWh, 52 MWh from 272 batteries – annually offsetting 1.6 million gallons of diesel fuel, and helping Hawaii reach its goal of 100% renewably sourced electricity by 2045 as now required by law. "If this can be done in Hawaii, it can be replicated anywhere," noted Martha Symko-Davies, an NREL Energy Systems Integration Facility program manager.[177]

In 2021, the state of Hawaii also announced that all new PV solar installations must include a 4-hour battery energy storage system (BESS), although new wind projects still have the option of no storage or either a 2- or 4-hour backup system. Renewable plus storage systems must charge directly from the grid and are eligible for an investment tax credit that lowers capital costs by a third. As noted by *Energy News*, "The state is targeting 100% clean energy by 2045 and given that all fossil fuel used on its islands is imported at great cost, renewables and storage can largely deliver power more cheaply."[178]

In a bold move to establish a foothold in the burgeoning charge-storage market, Tesla also built the world's largest battery system in South Australia after a rash of power failures that saw thousands of homes regularly lose power. Backed by an ironclad guarantee from CEO Elon Musk, who stated that "Tesla will get the system installed and working 100 days from contract signature or it is free," the $50 million, 129-MWh Hornsdale Power Reserve (HPR) Li-ion battery array went online on December 1, 2017, in the middle of the hot Australian summer. Referred to as "one of this century's first great engineering marvels and a potential solution to the country's energy woes,"[179] the grid-scale HPR battery located about 200 km north of Adelaide came online as advertised with time to spare. The system immediately proved its worth during a 2019 hurricane that knocked out the main interconnector between South Australia and Victoria.[180]

The HPR battery now powers 30,000 homes on its own for up to an hour with instant backup for the occasionally uncertain and intermittent network, stabilizing the South Australian grid at peak times and in emergency. The argument about unreliability and downtime loses its shine when industrial-size batteries can ably fill the gap. Dubbed the "Tesla Big Battery," capacity was expanded in 2020 by 50% to 185 MWh, while other locations across Australia are adding more battery storage to support an increasingly greener grid. In 2022, HPR's grid inertia capability was upped again, providing increased emergency stability at a capacity of 150 MW/193.5 MWh (1.29 hours).

Today, the entire country of Costa Rica is almost completely powered by renewable energy, using a mix of hydro, geothermal, wind, and solar to generate electricity, thanks to a plentiful supply of rain for hydropower and over 100 volcanoes, hotspots, cinder cones, and geothermal hot springs. Costa Rica hoped to be carbon neutral in time to celebrate its 200th birthday in 2021, balancing carbon emissions with sequestered offsets, but petroleum is still needed for transportation.[181] In the Caribbean, more countries are also investing in localized solar power after suffering regular power outages during the frequent hurricane seasons of late that can isolate whole regions from an increasingly unreliable grid.

In 2017, India began installing rooftop solar at thousands of railway stations, announcing a plan to power almost all of its vast network of over 7,000 stations nationwide with solar. Ideal in countries like India, with favorable sunlight and where 300 million people live without electricity, many of whom still use kerosene for lighting, rural areas can now bypass the grid entirely. Instead of building expensive fossil-fuel plants that require dirty extraction and laborious transportation and whose costs can't be guaranteed because of a constantly fluctuating oil market, the developing world is increasingly being lit by local wind and solar. As the EPI's Lester Brown noted:

> Switching from kerosene lamps to solar cells is particularly helpful in fighting climate change. Although the estimated 1.5 billion kerosene lamps used worldwide provide less than 1 percent of all residential lighting, they account for 29 percent of the lighting sector's carbon dioxide emissions. Kerosene lamps burn the equivalent of 1.3 million barrels of oil per day, equal to roughly half the daily oil production of Kuwait. With the price of kerosene rising and the cost of solar cells declining, the decision to make the switch becomes progressively easier.[182]

Chile passed its goal of 20% renewables before 2025, helped by a 110-MW CSP plant with storage high in the Atacama Desert. Latin America's first solar thermal plant with 10,600 heliostats, the Cerro Dominador plant enjoys an average solar irradiation roughly double lower-altitude countries such as

Spain.[183] Solar plants in the north of the country, however, are still partly idle because of limited transmission capacity to higher-population regions. The island province of Tasmania, which escaped the ravages of increased global warming as seen in a rash of "black summer" bush fires across continental Australia, now provides all of its electricity from renewables, exporting excess power to the mainland via a subsea cable.

Even the oil-rich Saudis are getting in on the act, planning to produce 30% of power from renewables by 2030. The first phase called for 400 MW of wind- and 300 MW of solar-generated power, increasing to 10 GW by 2023 at a cost of $1 million per MW. The Saudi energy minister Khalid Al-Falih called the 10-GW target, "only the beginning."[184] In Dubai, a PV project by the Dubai Electricity and Water Authority (DEWA) can provide electrical power at 7.63 cents/kWh for CSP and 2.7 cents/kWh for PV,[185] lower than natural gas in a country where gas was once king. When finished, Dubai's world's largest CSP plant will boast the tallest solar power tower at roughly 260 m and the largest thermal storage capacity. Another solar project in neighboring Abu Dhabi is even more competitive with a 2.42 cents/kWh PPA.[186] For fossil fuels to compete, oil would have to sell at under $5 per barrel.

The sky is the limit, whether PV on rooftops, the ground, or at sea. Indeed, in the not too distant future, energy could even be beamed from a space-based network of solar panels, launched into geostationary orbit, the photoelectric current converted into high-frequency radio waves and transmitted to Earth as if straight out of a Bond movie. A 20-year time frame was proposed, although no green technology yet exists to launch the components.[187] Not all ideas will pan out, but the future is limited only by the imagination.

New developments are breathtaking in their scope, not unlike the creation of the US grid at the turn of the twentieth century that powered the dreams of an emerging superpower or the building of the interstate highway system in the 1950s that saw more than 40,000 miles of newly paved road carved into the earth and soul of a nation. Few could imagine how the grid would completely transform everyday existence or how the open road would become an integral part of life, including weekend getaways in the family sedan, cross-country summer vacations, and a suburban lifestyle many of us now take for granted.

Of course, far from ensuring a level playing field for competing interests, many governments are still beholden to Big Oil. Having funded the infrastructure of the petroleum industry on the backs of citizens – railways, highways, power plants – the status quo remains firmly rooted. Lazy, old-world, centralized thinking rules, with its rigorously controlled distribution system. In the USA, teams of well-organized lobbyists still call the shots, where three out of four O&G lobbyists are former congress members, "who served on the committees that oversee and regulate the industry, or worked for various federal agencies responsible for regulating the energy industry."[188] What's more, the same lobbyists actively work against climate-change legislation, while promoting a misinformation campaign designed to convince the public that global warming is a myth. Despite the opposition, the operational energy landscape continues to change, especially at the subnational level that is less restricted than national governments, adding new solutions where viable.

Elsewhere, utility companies are fighting back, citing low natural gas stocks to raise electricity prices, even warning customers not to buy electric vehicles in a volatile market, for example, in Spain after rates tripled there in 2021 because of uncertainty in the Russian gas market. Shockingly, the federal Australian government attacked South Australia's progressive energy strategy, choosing instead to support an outdated fossil-fuel-first policy at odds with the obvious geographic reality and need for more renewables. In *The Great Transition: Shifting from fossil fuels to solar and wind energy*, Lester Brown noted, however, that "Utilities trying to stifle solar power may soon realize that the effort is futile."[189] Having underestimated the threat from distributed solar power and battery storage, Barclays Bank downgraded the entire US electricity sector in 2014. As one analyst wrote, "whatever roadblocks utilities try to toss up ... it's already too late."[190] There is no hiding the IEA's estimate that $4 trillion will be spent funding renewables in the next decade.

No one should underestimate the challenge. Local power generation and distributed energy is an existential threat to the petroleum industry. More renewables and electrification means less oil and fewer profits in the pockets of its masters. The corporate, investor-backed, world is not planning to let go without a fight.

4.12 More Change: New Energy Technology

Albert Einstein wasn't understood by many, if anyone, when he published his theories on special and general relativity in 1904 and 1915, respectively; few of Isaac Newton's contemporaries understood his 1687 *Principia*, which perfected Francis Bacon's ideas of modern reason by quantitative experiment codifying man's mastery of previously unknown natural phenomena in universal laws – in effect placing science at the core of Western civilization – and Nicolaus Copernicus was so afraid of his heliocentric repositioning of the heavens that his 1543 work wasn't published until after his death. Alas, renewable energy is not complicated – the basic science of wind power has been around for over a millennium (which we will look at next) – although solar cells and real-time cooperative energy trading are relatively new to the game. Reaching an agreed understanding of the direction of new technology, however, takes time, no matter how easy or hard to assimilate the central idea.

Nonetheless, the renewable energy revolution marches on, ramping higher as more green infrastructure comes online by the day. The percentage of renewables in the global primary energy supply has increased 10 times to 2% in the last half century, while solar and wind-generated electrical power has increased almost 100 times to generate 10% of total grid output in the past two decades. Clearly, the old will continue to coexist with the new as green and brown exist in tandem as one increases and the other declines, just as wood, coal, and oil shared the limelight over the last century.

Even in the USA, a beacon of innovation and change, the old takes time to give way. Wood energy was dominant until 1900 before being replaced by coal that was then replaced 65 years later by oil. The pages of history turn slowly as competing sources battle for supremacy. In *Energy and Civilization*, Vaclav Smil noted that "Draft animals, water power, and steam engines coexisted in industrialized Europe and North America for more than a century," while "Only by 1963, when America's tractor power was nearly 12 times the record draft animal capacity of 1920, did the US Department of Agriculture stop counting draft animals."[191]

Figure 4.30 shows the percentage of the world's major energy sources over the past two centuries, highlighting the period of transition and the date at which each was replaced by the next, for example, wood to coal (~1900) and coal to oil (~1965), followed by the appearance of oil and natural gas (and uranium in countries such as the USA and France). This is the history of the world prior to the twenty-first century.

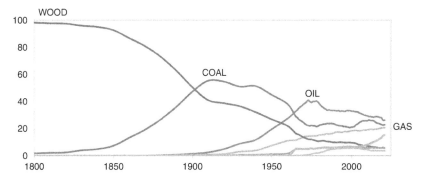

Figure 4.30 Global energy percentage from 1800 to present showing the arrival (and in some cases departure) of the main energy sources: biomass (wood), coal, oil, and natural gas (*source*: Smil, Energy and Civilization, p. 396).

Germany's energy transformation has been remarkable, moving from coal and gas to wind and solar in a few decades, such that 35% of Germany's energy now comes from renewables (including 15% wind and 7.5% solar). In many areas, renewable energy includes better and smarter infrastructure such as high-voltage, low-loss, DC transmission lines, smart grids/cities, and connected devices with modern sensors, cameras, adaptive learning, and AI, all helping to make the grid more efficient. Most importantly, economic growth is being decoupled from energy use, stimulating new kinds of investment.

But we are not in the midst of universal transition for all. Coal is still the dominant energy source across much of the globe because of low cost and easy access, continuing to grow each year. Even wood (a.k.a. modern biomass) is being burned at greater amounts, hardly what one would call a transition. In fact, energy *transition* is not a wholly appropriate term, but rather energy *addition* as noted by French historians Christophe Bonneuil and Jean-Baptiste Fressoz in *The Shock of the Anthropocene*.[192] Perhaps in the richer, developed world, where alternatives are easier to implement, we can applaud efforts to replace the dirty with the clean and speak of a transition from old to new and from brown to green.

What's more, consumption must be addressed if we are truly to transition from the destructive burning of carbon-based fuels. We can all use less and still be comfortable in our modernity. Nor is nature an externality that can be ignored in the calculation of economic value, largely codified by GDP and quantity of goods as if quality of life doesn't matter. We can no longer exclude greenhouse gas emissions, pollution, life expectancy, mental health, and other non-monetized services from the energy equation.

Some are calling for increased government efforts as in a modern Manhattan Project or Marshall Plan. Named after the Kennedy-era Moon Shot, the 2011 US DOE's "Sun Shot Initiative" declared a goal of $0.06/kWh utility-scale solar power within a decade, which was realized three years earlier than planned and revised to $0.03/kWh solar by 2030. A proposed Green New Deal aims to decarbonize the US economy by converting to 100% renewable energy by 2035, zero net emissions by 2050, and 10 million new energy jobs. Although it is difficult to muster the troops in today's fractured political climate, money can be made available if the desire and interest is there, especially with global warming and pollution putting more people's lives at risk. Sufficient financing will drive the transition, helping to build essential infrastructure, while increased subsidies can help establish renewable energy as a viable alternative to fossil fuels and reset the market to at least a level playing field.

In the face of a perceived existential crisis during World War II, the US government spent $2 billion to build an atomic bomb over a period of 6 years, while NASA's expenditure was more than $100 billion in the decade it took to put an American astronaut on the moon and reclaim its leadership among post-war superpowers after the lingering shock of *Sputnik*.[193] At the height of the Cold War, a single shuttle launch cost 1.5 billion dollars ($192 billion for 131 flights[194]). As *Apollo 16* astronaut Charley Duke, the youngest person to walk on the moon, succinctly noted about the realities of achieving the most challenging of goals, "400,000 people and an unlimited budget – you can do a lot."

Recognizing the urgency to tackle global warming after years of higher temperatures, a significant clean-energy funding deal was finally signed by the US government in August 2022. Passed by the slimmest of margins in a divided Congress, Joe Biden's pared-down Build Back Better legislation had originally set aside $1.5 trillion in funding, but after stiff opposition was rejigged as the Inflation Reduction Act (IRA), providing $369 billion over 10 years for renewable energy, including credits for clean electricity, solar, and wind. Global warming activist and former vice president Al Gore called the act a "true game changer" that "will create jobs, lower costs, increase US competitiveness, reduce air pollution,"[195] while economist Paul Krugman noted that "The growth of renewable energy leads to the creation of relatively high paying jobs, which are more often than not located in areas that stand to lose from a decline in fossil fuel extraction jobs."[196] Not considered enough by some, the funding has nonetheless kick-started a previously stagnant federal response to the increasingly dangerous climate crisis causing dozens of disasters per year in the USA alone.

If the market can remake an entrenched petroleum-based economy on its own without an overly bureaucratic picking of winners and losers, the renewables revolution will allow for a more organic path. Bottom-up transition is preferred, where winners are chosen by consumers, but all innovation needs a financial nudge and a steady regulatory hand to ensure success and safety for all. The standard argument is about who pays for the infrastructure – invested private interests or an overtaxed public – although private interests typically claim the rewards in the end. Vested financial interests, special-interest groups, and industry lobbyists work to maintain the status quo, while curtailing innovation that would undermine corporate wealth and long-established supply chains. That is, one supposes, until the financiers can manage the change on their terms and reap the benefits for themselves.

Bureaucracy is often cited as a deterrent, but unfettered solar power keeps private the details of individual energy use – how can anyone know what you do in your own home with your own home-grown, off-grid energy? In fact, energy statistics may become obsolete, perhaps part of the backlash for those afraid of government overreach. More individual use means less control and fewer taxes, while monopoly is a recipe for more top-heavy management. Ironically, it is not *world* government that worries some naysayers but *no* government. As the amount of distributed energy resources increase, circumventing long-established supply chains and management structures, even traditional governance is at risk.

PV technology continues to improve, becoming cheaper and more efficient each year. Innovations since 1975 have made PV viable and attractive to both users and investors, such that solar power has now reached grid parity with retail natural gas (~$0.20/kWh) and coal (~$0.10/kWh). Emanuel Sachs, MIT professor of mechanical engineering, predicted in 2009 that 7% of the global electricity supply would come from PV by 2020, adding that once the storage technology is solved to compensate for intermittency, "PV will become the largest manufacturing industry in history."[197] We're not quite there yet, although it looks like he was right on the money in some advanced countries. When all factors are included, solar energy is cheaper, safer, and more sustainable. A green economy is becoming the economic driver of the future, where fuel is manufactured and not burned.

Oddly and perhaps alarmingly for those in the West who want to see more investment in clean energy, the developing world is leading the way, in some cases building from scratch. India and China continue to flip-flop as world-record holders for largest solar plant. In 2016, India topped the list (648-MW Kamuthi PV plant in Tamil Nadu), followed two years later by China (1,000-MW Yanchi plant in Qinghai) and then two record-shattering 2-GW-plus solar plants in Karnataka, southwest India, one later the same year (2,050-MW Pavagada Solar Park) and the other in 2020 in Rajasthan, northwest India (2,245-MW Bhadla Solar Park). The pace of change is limited only by the will to build and a concerted goal to succeed where others have failed.

Part of the challenge is to adapt our well-established ways to the specifications of the different technologies, such as using more power in the middle of the day patterned on the motion of the Sun. Indeed, with more new energy our everyday habits will change, in the same way other watershed technologies made old ways obsolete. Instead of planning around low-cost/low-demand electricity times at night, we will use more power in the middle of the day, doing or in some cases scheduling the laundry, cleaning, and household chores at peak sun. Whereas lower overnight rates once helped offset daytime demand – for example, for thermal storage and water heating – we will use more energy during the day.

Increased chemical battery storage will also help balance the grid, essential with a fundamentally intermittent supply. Conversely, dormant electric-vehicle batteries can be charged at night when wind power is more plentiful, helping to "flatten the load." A twenty-first-century interactive grid is emerging from a thoroughly modern renewable-energy supply (as we will see in the next two chapters). There will be no holding back the Sun when solar panels become so cheap even non-sun-facing installations are feasible.

We've come a long way since *Vanguard I* and *Telstar*, when only small and expensive solar cells were available to a select few. We stand at the edge of a new era, able to tap the vast amounts of energy promised in the glimpse of Bell's first 6%-efficient "solar battery." Today, the energy from just one hour of sunlight is equal to the total global energy consumed in a single year. All that is missing is the will to harvest what we need.

What was once considered alternative energy to go with an alternative lifestyle and touted in the pages of *Mother Jones* and *The Whole Earth Catalog*, tapping into a free and plentiful Sun is fast becoming a way of life. We are at the dawning of the Age of Renewables, the challenge now to live within the limits of our needs and not expand our lifestyles to meet the new abundance. As always, we must live and use energy responsibly.

5

The Old Becomes New Again: More Sustainable Energy

5.1 The Winds of Change

Water and wind have been powering mills for centuries, grinding wheat or corn to produce flour or cornmeal (especially to make bread), seeds to make spice, and plants to create colored pigments – for example, beets (red), turmeric (yellow), and cinnamon (brown) – as well as for pumping water and sawing wood. Commissioned in 1086 by William the Conqueror to survey his new English lands, *The Domesday Book* counted 5,624 watermills, each servicing an average of 50 households, more so in agricultural areas with good water flow, while by the nineteenth century as many as 100,000 windmills dotted the European landscape, providing substantial revenues to millers.[1] Don Quixote famously called the windmills of La Mancha "wild giants" against whom he intended to do battle, his trusted sidekick breathlessly trying to convince him otherwise: "What seems to be arms are just their sails, which go around in the wind and turn the millstone," pleaded Sancho Panza.[2]

As the first inanimate "prime movers," watermills started to impact Europe and Asia by 200 CE and windmills by 900 CE.[3] Believed to have originated in Persia, wind power was adapted from early sailing boats, before being brought by the Moors through Spain to northern Europe in the 1200s. Converting moving air into rotating sails, the wind-generated power was initially employed to grind grain into flour and seeds into powder between two millstones – a stationary bedstone and rotating runner stone – before the Dutch improved on the design to pump water via an Archimedes screw in the 1400s to keep their water-logged marshlands dry. In the late 1500s, the windmill owner Cornelis Corneliszoon devised a crankshaft to convert the horizontal turning of the internal rotor to the vertical up-and-down motion of a saw that could cut logs into planks at a rate 30 times faster than man.

The next major innovation was the turret that divided a windmill into a sturdy fixed bottom and a freely turning top to avail of fast-changing conditions. Not unlike a boomed triangular sail and sternpost rudder that zig-zagged a ship in any direction by tacking rather than a one-dimensional square rigging that could only travel with the wind – without which Columbus could not have crossed the Atlantic[4] – the unconstrained turret was more efficient, while bigger sails created more lift to generate more rotor power whichever way the wind blew. Possibly inspired by his experiments on human flight, the turret mill was invented in 1500 by Leonardo da Vinci and improved upon by the Flemish mathematician and engineer Simon Stevin,[5] who added transmission gears to control the speed and calculated the sail size needed to raise water a specified height.[6] Around 1600, the Dutch also added "canted leading-edge boards to previously flat blades" to produce more lift and less drag.[7] The amount of sailcloth was increased or decreased depending on the strength of the wind.

Designs have changed over the years to optimize the rotational speed, swept area (sail circle), and stability, including reducing the number of sails (a.k.a. vanes or blades) from four to three and shaping the blades. Early windmills had four sails because it was easiest to fasten two crossed pieces of wood together (the mill rods) on a common turning axis, while today most industrial windmills employ three blades, fashioned with composite, epoxy-impregnated, glass fibers, rounded and pitched to produce optimal lift in changing wind conditions.

One blade is theoretically the most aerodynamically efficient but is unstable, two blades create too much wobble, while three blades limit the stress yet still produce sufficient rotational speed. A fantail attached at the back winds a gear to automatically turn the sails into the wind (a simple negative feedback mechanism[8]), although large, old-style windmills must still be turned manually to face the wind. Pre-industrial designs topped out at about 30% efficiency, while the maximum theoretically extractable power is 59.3% of the moving air's kinetic energy (the Betz limit). As University of Manitoba environmental science professor Vaclav Smil aptly noted, "no wind machine can extract all of the available wind power: this would require a complete stopping of the airstream!"[9]

The Netherlands is perhaps best known for its windmills that continuously pump water lest the plentiful "lowlands" be swallowed by the sea. With minimal height difference over a mostly flat terrain to generate sufficient power from a watermill, wind power was essential prior to steam, diesel, and electricity as roughly half the Dutch countryside is below sea level. A series of polders, pumping stations, and canals keep the water at bay in a stepped drainage system (a polder is a stretch of fertile, below-sea-level, reclaimed land where water is continually drained[10]). Flat and exposed to the prominent North Sea winds, the Netherlands is perfectly suited for exploiting the constantly blowing conditions.

Until giving way to steam and then petroleum in the early twentieth century, there were 9,000 Dutch windmills that either pumped water (polder mills), ground seeds and grains (oil, spice, and flour mills), or cut wood (sawmills), but only about 1,000 traditional old-style windmills still remain in various states of operation, some maintained as museums and even homes. Old-style windmills in the Netherlands turn counterclockwise by custom, the sails kept slightly before vertical when not in operation (to signal joy) or slightly after vertical (to signal sadness), a.k.a. "coming" and "going." Now a UNESCO World Heritage Site in the Albasserwaard region southeast of Rotterdam, the Kinderdijk comprises 19 windmills and three pumping stations to drain water from a lower polder to a reservoir to a higher polder via scoop wheels and then on to the River Lek at low tide (Figure 5.1).

In the USA, windmills aided westward expansion, pumping well water on farms and at railway stations for locomotive storage in areas with a scarcity of streams and low rainfall. The opening scene of Sergio Leone's *Once Upon a Time in the West* shows a fan-tailed windmill servicing a railway water tower, a necessity before the advent of diesel and electrification. The iconic creaking Halladay model pumped water across much of an expanding rural America, capacity doubling from 320 MW in 1849 in the early days of the railway to 625 MW by 1919.[11] Rural electrification would limit wind power to mostly

(a)   (b)

Figure 5.1 Wind power: (a) pumping water at Kinderdijk, South Holland; (b) modern electricity-generating wind farm in southern California (*source*: Erik Wilde CC BY-SA 2.0).

pumping water, although as late as 1930 only 10% of farms were hooked up to the grid.[12]

The Scottish natural philosophy professor James Blyth devised the first electricity-producing windmill in 1887, a vertical-axis device for his home in Marykirk in northeast Scotland, while a year later the American engineer Charles F. Brush hooked up the first auto-operated wind turbine to power his Cleveland mansion. The 50-m diameter, 144 rotor blade, horizontal-axis device generated 12 kW that also charged 408 batteries in the basement, operating for over a decade.

Today, windmills range from small, self-regulating, multi-vane farm units for pumping well water to an offshore behemoth that can generate over 10 MW with 80-m long blades mounted at height to capture more wind. Think of a kite – the higher the better, where the wind at 80 m is twice as strong as at 10 m.

Today's 3-MW industrial turbine typically stands 80 m high (the hub height) – the usual height for a standard commercial unit – with three 50-m, fiber-reinforced polymer blades spanning over 100 m, pitched to catch the wind at the optimum angle. A longer blade span provides more swept area, capturing more wind and producing more torque on the rotor, thus more electrical work. Power (P) is proportional to the swept area (A) and the wind velocity (v) cubed ($P \propto Av^3$).[13] The challenge is to get the blades turning fast enough, but not too fast to compromise safety, that is, between the "cut-in" and "cut-out" (a.k.a. "survival") speeds. The optimum wind speed is about 12 m/s (~50 km/h), while a turbine isn't designed to operate at wind speeds over 25 m/s (~90 km/h). A pitch angle from about 5 to 15 degrees is best, changing with the wind speed to maximize power output (Figure 5.2).

Wind is technically a solar resource because the Sun differentially heats the air around the globe, for example, from the hot tropics to the cooler poles resulting in local breezes, aided in some places by the jagged terrain, while the mean wind speeds are much higher north and south of 30° latitude than at the equator.[14] Some hills are perfect to capture the wind as compressed air on the windy side expands when it reaches the peak. In theory, there is enough wind to supply the world's daily supply of electricity many times over (global grid capacity is about 8 TW), but the wind doesn't blow the same everywhere and conditions can be drastically different from one location to the next. For example, in Galicia in the northwest of Spain the wind blows mightily off the Atlantic Ocean, yet not as much in the southern Andalusian heartland of Seville and Granada.

Identifying a suitable location with good wind is paramount, but all sites must have appropriate soil strata to provide a stable foundation, limited

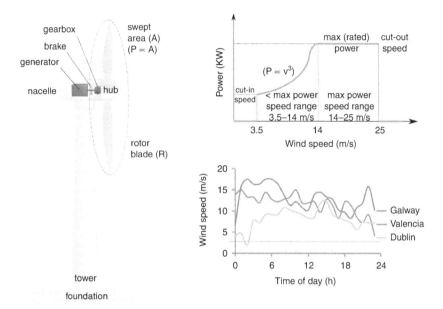

Figure 5.2 (a) Wind turbine schematic, (b) operational wind speed range, and (c) sample wind speeds at three different locations around Ireland.

landscape obstruction, and minimal environmental impact that respect land rights, local biodiversity, and natural wildlife habitat, especially avoiding natural bird migration patterns. Transporting large components to remote areas can be tricky, but a wind turbine (WT) can be installed in parts, comprising a steel-reinforced concrete foundation, sectional steel tower, turbine assembly, and composite blades. The turning rotor generates electric current in the stator, typically stepped up to 20 kV for transmission along a grid-tied power line.

The tower is made in sections that narrow with height, while the turbine assembly contains a *nacelle* that houses the drivetrain (gearbox, generator, and yaw controls) and a *hub* that provides blade pitch control, all of which are accessible by internal ladder for maintenance. Designed to last for 25 years and millions of revolutions, the blades are manufactured by joining two epoxy half-shell pieces, shaped in a mold and carefully layered with reinforcing fiberglass strips for strength. General wear and tear from wind particles on the leading edges – mostly dust, rain, and salt water – delaminates the blades over time, reducing performance, but is minimized by adding protective tape and paint. The longer the blades the more important the strength-to-weight ratio, an important issue as installation sizes increase.

External sensors measure the wind direction and speed to turn the blades into the wind (active yaw) and change the blade angle (pitch control) to optimize rotational velocity in varying conditions, including to an "off" position during excessively high winds. The nacelle, hub, and blades rotate together around the tower, covering 360 degrees to face the wind whatever the direction, while real-time data are remotely monitored.

At low incoming wind-field speeds, there is insufficient torque to turn the blades, but as the wind picks up the blades start moving, turning the rotor to produce power in the generator windings (the cut-in speed is around 3–4 m/s). As the wind speed increases, the generated power increases rapidly ($P \propto v^3$) to a maximum for safe operation at the generator factory rating. Above the maximum operational limit wind speed (~14 m/s), blade pitch controls maintain a constant rotor speed to produce electricity at the rated output, beyond which damage can occur. Above the cut-out speed (~25 m/s), internal brakes are applied to disable the rotor and avoid any system damage. Standard maintenance is required, although modern gearless turbines need much less attention.

Longer blades mean more power, capturing more of the wind because power depends on the swept area ($P \propto A$). A 50-m blade turbine generates 3 MW and an 80-m blade turbine 8 MW, roughly 2.5 times the power [$(\sim 80/50)^2$]. Rated at 9.5 MW, the Vestas V164 spans 164 m, sweeping out an area roughly equal to 20,000 m^2, providing enough power *in one rotation* to run a home for a day! By comparison, the spoked 67.5-m radius London Eye sweeps out an area of 15,000 m^2, which if designed to generate electricity would produce about 8 MW or 3/4 the Vestas V164, albeit at a speed too high for tourists. Built for pleasure, the London Eye turns at a more pedestrian rate of one revolution per half hour (1/30 rpm) compared to about 20 rpm for a working WT.[15]

One often sees a series of grouped turbines these days when traveling in the countryside, some whose blades are idle or seemingly underperforming. Clearly, the wind field can change quickly and when no wind or little wind is available a turbine can't generate any power. The wind may also be too strong for safe operation, for example, in a storm, while "curtailment" occurs when too much grid power is available, such as at night when excess wind power can't be sold on, no storage is available, or the price dips below a contracted minimum to baseload stations paid to operate 24/7. If the wind is constant or averaged over a large number of sites, wind power helps to stabilize the grid, while real-time fluctuations are typically managed with complimentary oil and gas generators in "spinning reserve" that go online as needed.

On average, a 2-MW industrial WT provides bulk energy for 2,000 people at a total cost of $2 million with a lifetime of around 25 years, although the metrics of an individual installation depend on location and site-specific operating conditions. In a breakdown of costs, the turbine accounts for about 75%, the foundation 10%, grid connections 10%, and land 5%. The five-turbine, 11.5-MW El Hierro wind farm in the Canary Islands generates almost all of the electrical power for an island community of 10,000 people, and provides a good rule of thumb: 1 MW costs $1 million for 1,000 people (~250–330 homes). Although the wind is better offshore, most installations today are still onshore (95%), where start-up costs and transmission infrastructure are cheaper and easier (foundation costs can triple at sea). Despite the increased logistics and planning, however, offshore sites are growing with bigger turbines at reduced costs, while larger blades can more easily be transported at sea.

Since the 1990s, the nominal or nameplate capacity has been steadily increasing, especially for offshore WTs. In 2019, GE introduced its Haliade-X prototype to the Port of Rotterdam, the world's most powerful WT at the time, generating 12 MW via a 220-m rotor. Each 107-m blade is longer than a football field! In late 2021, Siemens Gamesa unveiled its 14-MW, SG 14–222 DD prototype with 108-m blades and 222-m rotor off the coast of northwest Denmark. The company announced that 30 such WTs could generate enough electricity to power the entire city of Bilbao, Siemens Gamesa's headquarters at the time prior to being bought out by Siemens.[16] In 2023, China's Haizhuang Wind Power announced a prototype 18-MW WT with 260-m rotor diameter, enough to supply the annual electricity needs of 40,000 households.[17]

A continental wind map (Figure 5.3a) indicates the wind-field strengths around the world, where many local wind pockets exist on coastlines and in continental interiors, including Greenland, Iceland, Ireland, the UK, Norway, Denmark, northern Russia, western Alaska, the Great Lakes, eastern Canada, along the Andes in Chile and southern Argentina, Africa (Mauritania, Western Sahara, Chad, the Horn, coastal South Africa), the Kashmir, western China, southern New Zealand, and Tasmania. A US installation map (Figure 5.3b) highlights the plentiful concentration of wind farms in the American Midwest thanks to the large, wide-open continental plains with its famed Tornado Alley and Wizard of Oz-type weather. Texas, Oklahoma, and California generate the most wind-powered electricity in the USA. More wind farm locations are also dotted around the North Sea with the recent growth in offshore sites, split between the UK, Germany, Denmark, Norway, Belgium, and the Netherlands, while some of the best wind conditions are located in remote areas where waters are excessively deep and conditions harsh. Matching a new installation to an optimal wind-field region is essential to maximize output.

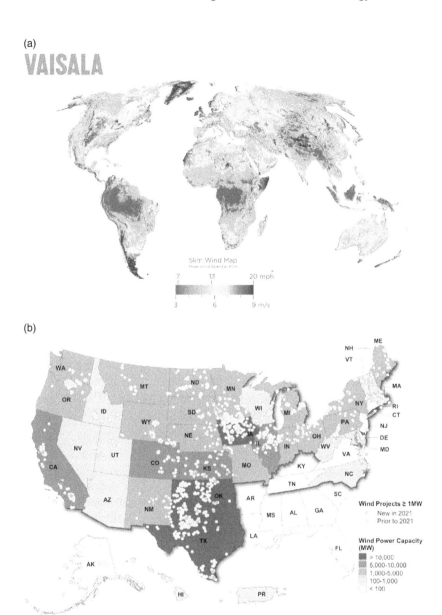

Figure 5.3 The wind field: (a) global continental wind map (*source*: Vaisala) and (b) US onshore wind installation map (*source*: Lawrence Livermore National Laboratory; Wiser, R., et al., Land-based Wind Market Report: 2022 Edition, figure 6, p. 23, US Department of Energy, August 2022).

5.2 Onshore, Offshore, NIMBY, PIMBY

As of 2022, 837 GW of wind power has been installed across the globe (roughly 10% of total installed electrical power). Onshore sites account for 93% of installed wind power (780 GW), led by China (40%), followed by the USA (17%), Germany (7%), India (5%), and Spain (4%) as seen in Figure 5.4, while almost half of offshore installations (57 GW) are in China (48%), followed by the UK (22%), Germany 13%, the Netherlands (5%), and Denmark (4%).[18] Located on the windy North Sea coast, Denmark holds the percentage annual wind-power record at 47% in 2019, including regular 100% days.[19] Availing of similarly advantageous northern winds, four German states can generate over half their electrical power from wind, while Portugal, Ireland, Scotland, and Spain each generate about 25%. In 2013, Spain became the first country where the main source of electricity was generated by wind power over an entire year, producing 20.9% from wind compared to 20.8% from nuclear, thanks primarily to the ideal wind fields blowing in from the Atlantic Ocean across the Galician coast.

In the USA, some of the best wind is in West Texas, where in the land of the fabled nodding donkey, wind power is already on par with oil and gas. Barack Obama noted at a 2016 DNC reception in Dallas: "Right now, here in Texas, wind power is already cheaper than dirty fossil fuels," a statement PolitiFact rated as "True," noting that wind-generated power ranged from 3.6 to 5.1 cents/kWh, while the national average was between 6.5 and 15 cents/kWh for coal-fired electricity and 5.2–21.8 cents/kWh for natural gas depending on plant

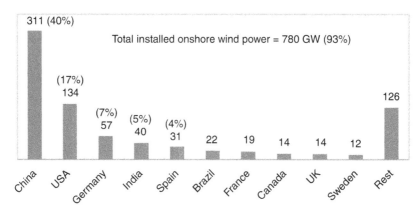

Figure 5.4 Installed onshore wind capacity (GW) by country, 2021 (*source*: Lee, J. and Zhao, F., "Global Wind Report 2022," Global Wind Energy Council, Brussels, April 4, 2022).

type.[20] PolitiFact also rated the claim that the USA was number one in the world as "Mostly True," as the USA generates more wind power than other countries despite China having more installed capacity. China's lagging transmission line and grid infrastructure has kept many rural Chinese wind farms offline or underperforming.

By 2016, 12 American states had achieved at least 10% wind penetration, led by Iowa (31%) and South Dakota (25%), while Texas generated almost 18 GW, more than all but the five highest *countries* and enough to power 6 million homes.[21] Since 2019, Iowa and Kansas have both produced more electricity from wind than any other power source, the first American states to do so – Iowa generated over 10 GW or 40% of the state grid supply, providing more than 9,000 jobs (second behind Texas).[22]

Lester Brown of the Earth Policy Institute estimated that, in the USA, "three wind-rich states – North Dakota, Kansas, and Texas – have enough harnessable wind energy to collectively satisfy national electricity needs."[23] Brown also noted that "Wind energy per acre is off the charts," calculating that a farmer in northern Iowa can earn 300 times more from wind power than raising corn for fuel-grade ethanol, receiving anywhere from $3,000 to $10,000 per turbine per year.[24] With the relatively small footprint of a turbine, farmers can also "double-crop," allowing animals to graze on the same site, especially helpful to lower-income rural farming communities. A continuous income from selling power to the grid also provides needed insurance during the regular ups and downs of managing a farm.

Billionaire investor Philip Anschutz, who started working in his father's oil-drilling company, is building a 1,000-turbine wind farm – the largest in the USA – in Carbon County, Wyoming, with plans to sell the electricity to an energy-hungry California that hopes to meet its goal of 50% renewable energy by 2030. At 3 GW, the Chokecherry and Sierra Madre Wind Energy Project includes a $3 billion power line to transmit the generated power 730 miles from Wyoming through northern Colorado, Utah, and southern Nevada to the Eldorado Substation near Las Vegas. There, the power can readily slot into the existing grid to reach California and its 40 million strong population, although the HVDC line has been held up by regulatory permitting that includes NIMBY concerns, habitat protection, and Native land encroachment. Long-distance transmission or battery storage is essential to keep power from being stranded because the excess bounty isn't needed in Wyoming, the least populous American state at 580,000. Unconcerned about a petroleum magnate turning his hand to renewable energy, the head of Anschutz Corporation's oil and gas business stated, "It's just a natural resource."[25] Carbon County may have to change its name.

Despite the growing pains of building new and upgraded infrastructure across a vast country, two-fifths of all installed wind power is located in China (all at least 1 GW capacity), including the world's five largest installations, topped by the 6.8-GW Gansu Wind Farm near the city of Jiuquan, Gansu Province, in north-central China (a.k.a. Jiuquan Wind Power Base). In 2020, more than half of all new installations were in China, underscoring the Chinese commitment to renewable energy that continues to tap more wind power year on year, as much as one new turbine an hour.

Beginning with a test phase in the 1980s, China now boasts more wind farms than any other country thanks to its ambitious Wind Base program, with plans to reach 400 GW by 2030 and 1,000 GW (1 TW!) by 2050, two-thirds of its existing grid! The best wind is in the Gobi Desert in central China (especially western Inner Mongolia, Gansu, and Xinjiang), although some rural wind farms remain underutilized because of low local demand and lack of high-voltage transmission lines to transport the generated power to urban consumers, leaving plants dormant and the power stranded. For example, the Jiuquan Wind Power Base on the edge of the Gobi Desert generates almost 8 GW (of a planned 20 GW from 7,000+ turbines), but typically operates at only 40% capacity, although more high-voltage power lines are being built to service the higher urban demand in the east of the country.[26]

The manufacturing market is ever changing, although Vestas (Denmark, 15%) was the largest supplier of wind turbines in 2021 (on and offshore), followed by Goldwind (China, 12%), Siemens Gamesa (Spain, 9%), Envision Energy (China, 9%), and GE (USA, 8%). Vestas was also the global leader for onshore installations, followed by GE, Envision, Windey, and Siemens Gamesa. The top five turbine makers accounted for roughly half of the record 2021 installation market of 99.2 GW, almost double 5 years previously (Table 5.1).

In similar fashion to NREL director Roland Hulstrom's calculation that a 100-mile-by-100-mile-square PV solar farm could power the US grid, Colorado School of Mines professor John Fanchi calculated that 12.7 million turbines rated at 4 MW could supply the entire global energy demand, assuming a 200,000 MJ/year consumption per person. The total coverage would roughly equal the size of Alaska, each site comprising a 10-square-mile footprint installed in approximately 50,000, 1-GW farms.[27] Double the turbine rating, more easily doable today, and the number of farms and land would be halved. Simplifying the numbers via percentages, a 2005 Stanford University study reported that harnessing just 20% of available wind power could generate enough power to satisfy our global demand for energy, with only 3% needed to cover electrical demand.[28]

Table 5.1 *Top 10 global wind turbine manufacturers (2016 and 2021)*

#	Company	Year	HQ	2016 (GW)	2021 (GW)
1	Vestas	1898	Aarhus, Denmark	9.0	15.2
2	Goldwind	1998	Beijing, China	6.6	12.0
3	Siemens Gamesa	1976	Bilbao, Spain	7.6	8.6
4	Envision Energy	2007	Shanghai, China	2.0	8.5
5	General Electric	1892	Boston, Massachusetts	6.9	8.3
6	Windey	2001	Zhejiang, China	–	7.7
7	Ming Yang	2006	Zhingshan, China	2.0	7.5
8	Nordex Group	1985	Hamburg, Germany	2.7	6.8
9	Shanghai Electric	2006	Shanghai, China	–	5.3
10	Dongfang Electric	1984	Sichuan, China	–	3.4
	Total installed			57.8	99.2

Sources: 2016, "Vestas reclaims top spot in annual ranking of wind turbine makers," Bloomberg New Energy Finance, February 22, 2017. https://about.bnef.com/blog/ves tas-reclaims-top-spot-annual-ranking-wind-turbine-makers/.; 2021, "Vestas leaves competitors trailing as wind industry posts another record year of almost 100 giga-watts," Bloomberg New Energy Finance, March 23, 2022. https://about.bnef.com/blog/vestas-leaves-competitors-trailing-as-wind-industry-posts-another-record-year-of-almost-100-gigawatts/.

There are still concerns, however, for those who live near a wind farm or a proposed wind farm. The blades are noisy and can cause EM interference, although improved designs are shaped to reduce noise. Most installations are built away from populated areas, typically on higher ground to capture optimal wind conditions, but the constant whoosh and steady low-resonating hum (around 50 Hz) can annoy the most resolute of neighbors.[29] Inaudible infra-sound frequencies (below 20 Hz) can disturb the inner ear, while "shadow flicker" can also unnerve some nearby residents. Some also dislike seeing rows of towering white behemoths scarring an otherwise tranquil landscape, although others see only giants of sublime majesty. Indeed, some are more adverse to wind farms, while others are happy to embrace the inherent beauty of green energy amid the towering aesthetic as they ponder the importance of a changing energy paradigm.

There have been occasional failures, including the collapse of an 80-m span Enercon turbine in 2016 in Cape Breton, Nova Scotia (the first of its kind in Canada, which now has over 6,000 turbines), a 120-m span Vestas turbine in 2015 at the Harvest 1 project in Elkton, Michigan (the first US Vestas fail), and a 100-m span Nordex turbine in 2015 at the Screggagh wind farm in County Tyrone, Northern Ireland, which buckled after it spun out of control.[30] No one was injured in any of the crashes, and wind turbines are extremely safe, having suffered only a rare failure in over half a million now operating worldwide. In the early kilowatt-models of the 1990s, only the occasional improperly secured blade fell off.

Some are also concerned that wind farms (and solar plants) are dangerous to avian life, yet a small percentage of birds are killed by windmills compared to cats, building strikes, or poison. An estimated 330,000 birds are killed each year by wind farms in the USA, less than 0.01% of a total 5 billion estimated annual bird deaths, while as many as 3.7 billion are killed by cats, 1 billion from crashing into buildings, 174 million from power lines, 6.8 million from flying into communications towers, and up to 1 million from oil and gas fluid waste pits.[31] By comparison, the world's largest concentrated solar plant reports about 500 avian fatalities a year, less than half related to concentrated solar flux where temperatures can reach over 800°C.

Nonetheless, there are ways to limit bird and bat deaths from turning blades (or singeing by focused solar), such as choosing sites away from known migratory paths, ultrasonic noise alerts (via "acoustic lighthouses"), radar shutdown as birds approach, less attractive roosting blades, and even colorful painting.[32] Bird deaths were also a concern to animal safety advocates in the early days of electric street lighting, alarmed at the gruesome electrocutions whenever a bird attempted to perch on a newly installed high-voltage power line, although some felled fowl were sold by police to local restaurants to supplement their income.[33]

***

While the location of a new onshore wind farm is much debated, an offshore site solves the primary complaint of a scarred landscape, hardly a blight on any land-lover's fancy at sea, as well as the noise. Offshore installations are unheard by all but the hardiest sailors. The wind is also stronger and more consistent at sea, providing a higher return on investment, although an offshore wind farm is more expensive to build than its onshore counterpart, requiring the sturdiest monopole foundations, a transformer substation, and long-distance underwater cabling. Other offshore concerns include waves, currents, ice, water depth, and marine life.

Denmark has long been a world leader in wind power, connecting the first turbine to the grid in 1919, northwest of Copenhagen, and typically generates 40% of its overall electricity needs from wind, with a one-day world record of 140% on July 9, 2015. Excess power is sold on to neighboring Germany, Sweden, and Norway. The Danes also built the world's first offshore wind farm, 2 km off the coast of Vindeby in southern Denmark, erected in 1991 and decommissioned in 2017 after a successful 25-year run. The 0.45-MW, 11-turbine installation

helped spur on Danish investment in wind-generated power, currently estimated at more than 5 GW (almost 50%). Roughly 20% more efficient than
onshore power,[34] offshore turbines are today's biggest, topped by the 164-m
diameter Vestas V164 with a "nameplate" capacity of 10 MW. Optimal
spacing of roughly 25 times the rotor diameter minimizes the effect of wake
turbulence.

The cooperatively owned Middelgrunden Wind Farm is a 40-MW, 20-
turbine offshore installation in the shallow coastal waters of Copenhagen, the
world's largest upon construction in 2000. Along with 10 other onshore
turbines located in the neighboring seaport, the 30 turbines operate at over
90% efficiency and provide about 6% of Copenhagen's electrical power.
Nearby across the Öresund Bridge that connects Denmark and Sweden is the
Lillgrund Wind Farm, Sweden's largest offshore installation and the world's
third largest when commissioned in 2008. The 48 2.3-MW Siemens turbines
provide an overall capacity of 110 MW and 300 GWh of annual output.

For regions with shallow coastal waters, sufficient wind speeds, and nearby
populations, offshore wind power is fast becoming the most economical way to
generate utility-scale electricity. The North Sea, the east coast of the USA, the
Gulf of Mexico, and parts of the east coast of China are all well suited for
offshore wind. According to the US Department of Energy, "the shallow waters
off the East Coast [of the USA] are capable of hosting 530,000 megawatts of
wind generating capacity, enough to satisfy 40 percent of the country's electricity needs."[35]

Clearly, we are only scratching the surface of what's available to service our
everyday energy requirements. In 2015, global offshore wind capacity was only
12 GW, the London Array in the outer Thames estuary – the then world's largest
offshore wind farm – accounting for 630 MW from 175 3.6-MW turbines,
enough to power one-quarter of all London homes at peak output.[36] Since
then, global offshore power has steadily grown, increasing to more than 20
GW by 2019 and 35 GW by 2020, mostly from Europe's nearly 5,000 grid-
tied turbines. There is still much greater onshore capacity, but more offshore sites
with much larger-capacity turbines are being built, including further out to sea.

Much of the current UK offshore wind capacity comes from the East of
England Energy Zone (EEEZ), where conditions are especially good because
of shallow waters, an optimum seabed, and strong winds, while more than two
dozen sites beyond the London Array now generate almost 40% of UK
electricity. The UK and North Sea coast totaled 25 GW, a quarter owned by
the British monarch, whose Crown Estate leases the seabed to 14 km from the
coast (measured at mean low water). In October 2020, the British government
announced an ambitious plan to power every UK home via offshore wind by

2030 (a 40-GW target) as part of a "green industrial revolution" that aims to reach net zero emissions by 2050.[37]

Based on the success of offshore installations and decreasing costs, three power companies from the Netherlands, Denmark, and Germany are planning to build a 6 km^2 artificial "power link" island in the middle of the North Sea, located in the shallow waters off Dogger Bank. Well-situated between the partner countries, Dogger Bank connected mainland Europe to Britain before becoming submerged about 10,000 years ago at the end of the last ice age. The North Sea Wind Power Hub plans to generate up to 100 GW via the coordinated electrical output from thousands of installed turbines, literally an "energy island." A modular "hub and spoke" system will distribute power as needed to each member country, converting offshore wind to onshore electricity, and could also generate hydrogen gas during high-wind, low-demand periods for storage (a.k.a. power to gas or P2G). The site is expected to provide electrical power to 80 million people in the largest wind farm ever built. There is still much to do, but it would be fitting to see wind replace oil in the North Sea, where oil replaced coal, more than two centuries on from the start of the Industrial Revolution.

The American offshore market has been slow to develop, however, despite continuing growth across Europe, which now generates power from more than 100 offshore wind farms in 11 countries.[38] The first commercial onshore American wind turbine was built in 1941 near Rutland, Vermont, the world's first megawatt turbine at 1.25 MW that was still one of the largest until the 1970s. But it took until 2016 for the USA to erect its first commercial offshore facility, the five-turbine, 30-MW Block Island Wind Farm off the south coast of Rhode Island about 10 miles east of Long Island, New York. Spaced about half a mile apart in water 90 feet deep, the turbines replaced sputtering diesel generators that once provided power to the roughly 1,500 year-round island residents, an annual savings of 1 million gallons of diesel.[39]

Built by Rhode Island-based Deepwater Wind, Block Island initially sold power at 24.4 cents/kWh and has plans for a 90-MW extension in the same leased waters. Acquired by Danish energy giant Ørsted A/S, the first-of-its-kind US installation kick-started an important new supply chain, expanding the offshore workforce, increasing economies of scale, and sparking the construction of improved port facilities along the east coast. Underscoring the problems in any new venture, however, the high-voltage underwater transmission cables weren't initially buried deep enough and required a $30 million fix. Nonetheless, after a slow start, the USA is poised to avail of three decades of European advances.

The call now is to install more wind farms in the relatively shallow Atlantic coast – dubbed "the Saudi Arabia of offshore wind" – including Cape Wind, a 470-MW array of 130 3.6-MW turbines first proposed in 2000, but still awaiting approval more than two decades on. Located south of Cape Cod between Martha's Vineyard and Nantucket Island in the shallow waters of Horseshoe Shoal (near the middle of Nantucket Sound), Cape Wind has been dogged by numerous legal challenges, led by wealthy residents who live or lived nearby, such as the Kennedy family, Walter Cronkite, John Kerry, and oil billionaire Bill Koch, who helped fund the Alliance to Protect Nantucket Sound.[40] Despite over half of voters supporting the project, Cronkite dryly noted in typical NIMBY style, "There must be other locations to make it possible."[41] The US senator from Tennessee, Lamar Alexander, who built a vacation home on Nantucket Island also complained about the proposed size, flashing red lights, and spoiled vistas, declaring that "the windmills we are talking about today are not our grandmother's windmills."[42]

About 15 miles south of Nantucket, another long-suffering project will now be the first large offshore US wind farm, more than 25 times the size of Block Island and selling power at 6.5 cents/kWh to over 400,000 Massachusetts homes. After construction began in 2019, the 800-MW Vineyard Wind site is powered by 62 13.6-MW GE Haliade-X turbines and operated by the Bilbao-based Spanish energy company Iberdrola. In the summer of 2022, the project was given the final go-ahead after residents of the seaside village of Centerville in the town of Barnstable, Cape Cod, agreed to host the onshore substation and transmission infrastructure, connected to the site via a 35-mile long 220-kV subsea cable.

Comparable in size to a typical European wind farm, the US offshore market will finally realize some of its vast potential, thanks to lower development costs and larger turbines anchored further out to sea away from objecting eyes. Ocean wind is more consistent too, without terrestrial obstacles such as mountains, trees, and buildings, while offering "peak coincidence" – output matching high mid-afternoon demand.

In early 2021, the Biden administration announced further support for more large-scale offshore wind farms, including the Ocean Wind project in New Jersey waters (1.1 GW) and two proposed California sites (4.6 GW) – the first commercial wind farms on the Pacific coast – netting the federal government $775 million in auctioned leases a year later. As rates continue to decline, wind power could generate more than 20% of US energy needs by 2030, a future too good to ignore.

Not to be left out, in late 2021 the US government started looking for interested companies to build wind farms in a 30 million acre area of the

Gulf of Mexico as part of its plan to provide 30 GW of offshore wind by 2030.
The Gulf is more challenging than either the Atlantic or Pacific because of
lower wind speeds and a regular hurricane season, but both Louisiana and
Texas hope to capitalize on decades of oil-and-gas experience. With the largest
US onshore market, Texan companies will have a leg up when the leases are
auctioned. One renewable energy executive noted that "experience in offshore
operations and maintenance is what will make offshore wind a success."[43]
Aging oil and gas infrastructure can also be repurposed to accommodate wind
power, while a vast supply chain already exists for platform construction in
numerous Gulf ports.

***

The debate continues, mostly of the NIMBY kind, as opponents typically cite
scarred landscapes, avian deaths, and noise. To be sure, large wind turbines
should only be built on appropriate sites after full evaluation and diligent
planning – away from homes, picturesque habitats, and avian flyways. Better
PR wouldn't hurt either, where companies encourage discussion, involve communities in development, and allow residents to visit the inside of a turbine.

The problem of intermittent power is also cited, but while there will be lulls
in any location, a.k.a. the "dark doldrums," occasional shutdowns do not negate
the effectiveness of a wind farm. As more installations are built in areas with
varying wind patterns, the overall output can be flattened and grid stability
improved in the same way multiple grid-tied power stations maintain
a continuous supply. Wind power applies to a region much larger than a local
weather system, the up-and-down intermittency easily covered. As former US
energy secretary and Nobel prize-winning physicist Steven Chu aptly noted,
"the larger the area you collect energy, the more you can even things out."[44]

Despite lulls occurring from week to week, the minute-to-minute and year-to-
year performance is well known, while changing conditions can be forecast and
backup power arranged during temporary downtimes via peakers, interconnectors, and storage batteries. So-called "day-ahead" forecasting is also highly
predictable, while wind-generated power as a percentage of overall demand is
increased by integrating more output from wind farms, solar farms, and interconnectors. In a world first, Ireland's grid percentage from variable energy
sources rose from 50% to 75% in a decade via better system management,
with a goal of 95% by 2030.[45]

However, wind is more vulnerable to downtime than solar, where prolonged
calm conditions can stop a turbine for extended periods, although unlike solar
panels wind power can operate 24/7. The "capacity factor" averages between
30% and 50% (output measured as percentage of a theoretical maximum),

while global weather patterns can influence performance such as climate change or an annually varying El Niño/La Niña. All power plants experience temporary shutdowns, reducing the maximum nameplate capacity (for example, a nuclear power plant typically operates at between 67% and 90% of the rated capacity). To be sure, high-capacity coal, oil, gas, and nuclear stations provide important grid stability, but so does instantaneous pumped storage hydro, backup batteries, and interconnectors that are more efficient, cheaper, and cleaner (which we'll look at in the next chapter).

The westernmost Canary Island, El Hierro, is now almost 100% renewable, using a world-first mix of wind power and pumped storage hydro. Once considered the end of the world before Cristóbal Colón made his famous 1492 voyage to the Bahamas, El Hierro is too remote to be hooked up to an external power grid, having relied instead on shipped-in diesel for decades. But thanks to five 2.3-MW wind turbines installed in 2011 in a novel 11.5-MW hybrid wind and pumped-hydropower plant, electricity generation is now almost all-green on an island of 10,000 inhabitants. In the middle of the Atlantic, El Hierro has its fill of windy days, but when the wind field is too weak, stored water is released from a natural volcanic crater 700 m above sea level to a sea-level lake to keep the power going. In times of excess wind, the storage crater is refilled by pumping the water back up the almost 1 km height. Having guaranteed energy security for the future, El Hierro wants to go 100% green by transitioning to only electric cars.[46]

As with solar, wind is not a one-size-fits-all answer to renewable energy. One must also count all costs, including decommissioning, cleanups, and waste. A hydroelectric plant may use natural geologic formations to provide clean, safe power, but is no good if the surrounding areas and fish habitats are destroyed. Coal plant emissions can be scrubbed, but pollutants and GHG emissions are still released. Oil and gas is cheap, but fuel handling, spills, leaks, and pollution from burning toxic hydrocarbons cause significant damage, while nuclear power comes with long-lasting waste and possible disaster.

Although some pollution and GHG emissions are created during manufacturing (primarily making the steel towers), wind and solar are as green as it gets. Wind farms (composite blades) and solar farms (silicon and various metals) will, however, need to be decommissioned at the end of a 25–30-year lifetime. Most of the materials can be recycled, except for the composite blades that comprise about 10%. Some blades can also be reused where possible, while methods to strip out the carbon and glass fibers are improving. Newer, recyclable, thermoplastic composite blades are more easily manufactured, more efficient, and last longer than traditional epoxy blades. Replacing older turbines with newer, more powerful models at existing sites, a.k.a. "repowering" or "returbining," has already begun in Europe after 20 years of operation (for example, 1 MW to 5 MW).

Floating turbines are the latest innovation, providing easier access to deeper waters. The world's first floating wind farm was built in the midst of Scotland's North Sea oil patch, 25 km from Peterhead. Developed by Equinor (formerly Statoil), the 30-MW Hywind Scotland comprises five 6-MW turbines, each anchored via three 1,200-ton chains to the seabed 80 m down.[47] The world's first floating wind farm became operational in October 2017, providing power to almost 20,000 homes. Costs are significantly reduced without expensive deep-water foundations and fixed-bottom monopoles, either tethered to the seabed or supported on a three-column surface frame where the buoyancy balances the ballast. Floating turbines can also be partially erected via a self-lifting, compressed gas, telescoping tower that doesn't require as much heavy lifting or adjunct vessels as in a world-first installation off the coast of Gran Canaria.

Building on the success of floating wind-turbine technology, the Portuguese company WindPlus is building the world's second floating wind farm, installing the biggest semi-submersible offshore wind turbine 20 km off the coast of Viana do Castelo in northern Portugal. The Windfloat Atlantic project will supply 25 MW of power via three 8.4-MW Vestas turbines, adding to Portugal's already impressive wind-power capacity. Giles Dickson, CEO of WindEurope, noted that "Floating offshore wind now stands at the cusp of large-scale commercialization. With the right policy measures, it could really take off over the next five to ten years."[48] According to Dickson, "bottom-fixed" floating offshore costs could dramatically drop from €0.18–0.20/kWh to €0.04–0.06/kWh by 2030.

Higher construction and installation costs make offshore wind power uncompetitive without subsidies, but costs are falling with increased uptake and scaling. The first unsubsidized offshore wind farm – the Hollandse Kust Zuid 1 & 2 off the Dutch coast near The Hague – has a capacity of 750 MW and is slated to provide 2% of Dutch electrical needs.

***

As with solar, not all wind power is utility scale. Small, independent wind turbines up to 100 kW (a.k.a. microwind turbines) can power remote locations or small urban sites from a roof or backyard, employing a fixed-pitch rotor and a tail boom to align the device towards the wind (passive yaw). Coupled with reduced costs and affordable battery backup to cover low-wind-field conditions, off-grid installations are becoming more popular, while unused open spaces on building roofs can also tap previously unrealized energy. As anyone who has ever made a child's pinwheel knows, all one needs is a good wind.

For smaller installations, a vertical-axis wind turbine is easier to manage, with a turning axis perpendicular to the ground like a revolving door (think of a scaled-up anemometer that measures wind speed by catching air in horizontal cups). Derived from a ship's sail, a vertical-axis wind turbine (VAWT) is less efficient than a horizontal-axis wind turbine (HAWT), but is virtually silent, takes us less space, and doesn't need to face the wind to rotate – any direction will do because one of the sails/cups is always facing the wind.

Invented in Finland in 1922, the Savonius drag-type VAWT is a simple, inexpensive way to generate power, for example, made from the welded halves of cut oil drums, but doesn't scale well for large devices. Invented in 1926 by a French aeronautical engineer, the Darrieus VAWT looks like a giant egg beater and uses lift to rotate two airfoils around a torque tube. A Savonius VAWT is better for pumping water, while a Darrieus VAWT can turn faster than the wind and is better at generating electricity. A variation Darrieus VAWT employs three blades in a helical foil, while other VAWTs include an H-type Darrieus, a straight-bladed gyromill, and a cycloturbine. A combi Savonius–Darrieus device avails of high- and low-wind strengths (Figure 5.5).

Off-grid wind turbines now power lights, televisions, and mobile phones in rural areas, while small systems have been installed in various buildings, including a 50-kW HAWT at one English primary school that covers 100% of its electrical demands. In 2015, two 3.2-kW helical-foil VAWTs were installed on the Eiffel Tower, sufficient to power the famed tourist attraction's first-floor commercial area. Perhaps more PR than power, the stylish helical devices match the design elegance of the iconic steel tower. In 2021, VAWTs were installed on an Istanbul highway to capture the rushing air from passing vehicles. Named Enlil for the Mesopotamian god of the atmosphere, a single device topped with a solar panel generates 1 kWh, enough to power two homes.

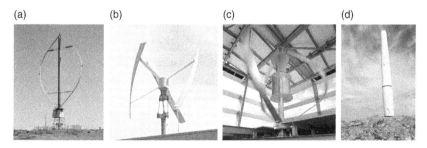

Figure 5.5 Vertical-axis wind turbines: (a) Darrieus, (b) helix-shape, (c) combination Savonius–Darrieus (*source*: BrabantBrandBox), and (d) vibration (*source*: Vortex Bladeless).

The roadside devices also double as smart sensors to track the local temperature, humidity, and carbon.

In 2022, a combination VAWT system with Venturi funnels was installed on the top floor of a 19-story apartment building in the former Philips Strijp-S industrial area of Eindhoven. The four-turbine array along with almost 300 complimentary bifacial solar panels supply a minimum of 140 MWh/year to power the building's communal lighting and elevators, selling any excess back to the grid. A "vertical forest" also adorns the exterior of a nearby apartment building that promotes biodiversity as part of the area's urban regeneration.

Ship propulsion is especially suited to integrated, wind-assisted VAWT technology. The acclaimed oceanographer and filmmaker Jacques Cousteau employed his own turbo sail – a modified airplane wing with flaps that open and close – to capture the wind at sea and reduce the need to burn expensive ship fuel. A modern version was redesigned by Dutch-based Econowind with fixed and foldable "ventifoils" to produce a large, additional propulsion force for cargo vehicles.

To overcome the rising costs and increasing weight requirements that employ larger blades at ever higher heights, VAWTs are also being developed for offshore power generation. After testing a 30-kW offshore VAWT (S1) off the coast of Lysekil in western Sweden, the Swedish company SeaTwirl plans to build a 1-MW commercial unit (S2x) off the coast of Bokn, Norway, scaled up to 6- or 10-MW by 2025.[49] Vertical-axis blades use less space than a horizontal-axis device and don't block as much incoming wind, allowing for closer spacing and a smaller footprint.

Nicknamed the "Skybrator," a recent micro-wind innovation even comes *without* blades. Created by the Spanish start-up Vortex Bladeless, a freestanding, flexible rod inside an outer tube oscillates in the wind, acting like a building dampener that turns incident vibrations into electricity via a modified alternator. Vortex's co-founder was influenced by the 1940 Tacoma Narrows Bridge collapse in Washington State that set off the structure's normal modes of oscillation, destroying the supported span because of its symmetric design. Using a similar principle, the Skybrator instead turns the vibrations into controlled motion. Individual ground- or roof-mounted versions are inexpensive (about half the cost of a standard three-blade turbine), easy to maintain (no contact between moving parts), and relatively quiet, although a bladeless turbine converts less kinetic energy (70%) into electricity than a conventional wind turbine (80%–90%). As with all new technologies, the goal is to scale up the prototype to increase power and test the limits of the design.

Piezoelectric materials can also turn the wind into usable energy. Common in microphones and speakers to convert pressure waves (sound) into electricity, piezoelectric "carpets" can be attached to the outside of a building to capture the blowing wind (*piezein* means "press" in Greek). One company, Spinetic, has built a fence-like wind "panel," perfect for small, easy-to-install applications. The appearance impacts much less on the surroundings.

One novel device employs an extendable tethered kite, flying in a figure of eight that turns a ground-based or offshore pylon spool to generate electricity, before retracting to repeat the cycle. Called a possible "magical solution" by Bill Gates, UK Kite Power Systems tested a 500-kW device at a former Scottish RAF base with plans to build a 3-MW system of 10 devices (one generates power as another retracts). The California-based Makani also developed a tethered kite with Shell, using wing-mounted rotors to transmit electricity along the tether to a floating ground station. In 2019, a 26-m span demo device transmitted 600 kW to an offshore location in 220-m deep Norwegian waters too deep to install fixed-seabed monopoles. Unfortunately, Makani lost out to the cheaper economies of scale of traditional WTs that offered power at $0.05/kWh and folded soon after.[50] Not all new technologies will be profitable, but despite the setback the sky is literally the limit for new ideas.

Wind power has advanced significantly since Leonardo da Vinci's four-sail turret mill of the 1500s, using modern technology to power 10% of today's global electrical needs (25–50% in parts of Europe). Although total penetration is uneven across the globe, the Earth Policy Institute's Lester Brown thinks that previously entrenched NIMBY thinking is beginning to give way to a more enlightened PIMBY ideal – *put* in my backyard. While some countries and regions have been slow to develop wind-power potential, either from lack of investment or planning problems, others now generate more than half their total electrical power from the wind.

The outlook for increased wind-generated power continues to improve. Since the advent of modern turbine design in the 1980s, wind-powered electricity costs have decreased more than 10-fold in under two decades, primarily because of a vastly improved economy of scale. Onshore wind is now cheaper than natural gas while offshore wind is cheaper than nuclear. Perhaps the simplest endorsement comes from another type of power – US currency. Louis Brooks, an enterprising Texan who tried his hand at ranching, wheat farming, and selling horse insurance on his 18,000-acre property in the plentiful West Texas wind-energy region of Sweetwater, operates 78 turbines and thinks wind farming is a godsend, likening the turning blades to counting money: "That noise they make – it's kind of like a cash register."[51] Who can argue with such simple green success?

5.3 The Hot Earth Rises: Geothermal Comes of Age

In Iceland, almost no one pays for home heating, while electricity prices are among the cheapest in the world. The heat and electrical power comes courtesy of a geyser, Icelandic for gusher. Three geysers to be precise, located deep under the Icelandic ground. And thanks to a large underground geothermal network, Icelanders have been enjoying clean energy for decades, while many of us still think of geothermal power as a strange mystery. Even in the dead of winter, Icelanders like to live it up in an outdoor spa, the relaxing waters warmed by nature. South of the capital Reykjavik, the Blue Lagoon is an especially popular thermal bathing spot for locals and tourists alike, while other naturally heated city pools are full of regulars out for a daily dip.

Iceland is uniquely located on a fissure zone between the North American and Eurasian Plates, called the Mid-Atlantic Ridge, which runs right through the country and bubbles up hot mineral water from under the ground as the plates are pulled apart by about 2 cm per year. Thermal spas have long inspired wonder in those looking to lead a healthier life, while some believe life itself may have begun in a hot spring or thermal vent that contains all the essential ingredients – organic molecules, water, and heat – the initial spark of life created in a flow of early protons.

To see where all the *geo*-heat comes from, we have to know a bit about the Earth. Formed around 4.5 billion years ago, the Earth is mostly comprised of oxygen (47%), silicon (27%), aluminum (8%), and iron (5%), with other various metals and minerals mixed in. There are three main layers: a 3,500-km core (further divided into a spinning solid-iron inner core and molten-iron /nickel outer core that produces the Earth's magnetic field), a 2,900-km middle mantle made mostly of rock, and a thin floating outer crust between 10 and 70 km (if the Earth were the size of a basketball, the crust would be as thick as a sheet of paper).

The top layer has dramatically changed over eons of tectonic movements, erosion, and climate change (for example, the Himalayas were created after the Indian subcontinent and Asia collided). Most importantly, the Earth contains radioactively decaying uranium, trapped core heat from the time of formation, and sinking core fragments that warm whatever is near. The thermal energy radiates upward, heating rock layers, liquid surface rock (magma), and underground pockets of water, in some cases hot enough to create steam.[52]

The temperature of the Earth increases with depth as any miner rightly knows, roughly 1°C for every 40 m to about 1,000°C at the edge of the crust with a peak of about 5,000°C in the core. Basic thermodynamics rearranges the energy, that is, trapped Earth heat is transferred to colder parts, building up in constricted areas that occasionally blow at a weak point as in a steaming covered stove pot. A simple heat pump validates the principle that energy is related to a change in temperature, and that heat flows from the hotter depths to the colder surface, a.k.a. the second law of thermodynamics. Some of the eruptions are more violent than others, such as along the famous oceanic "Ring of Fire," home to three-quarters of the world's volcanoes. Underground hot spots can also be dissipated along subterranean waterways, tapped to distribute heat or generate electrical power via a traditional steam turbine. Geothermal energy is everywhere underneath us – all we have to do is harness the hot spots or make use of the temperature differences to exploit the energy in a controlled way.

In his 1864 epic novel *Journey to the Centre of the Earth*, Jules Verne reckoned that the Earth had a radius of 9,000 km and was full of bizarre prehistoric plants and animals. Technically an oblate spheroid, the Earth bulges a bit around the equator, but is essentially a sphere around 6,400-km thick. Thankfully, we don't have to go as far as the center to get to the heat – most of the energy can be tapped within a few kilometers of the surface. Up to about 500 m, "shallow" geothermal energy (SGE) is relatively easy to extract, whereas "deep" geothermal energy (DGE) fields require more intensive drilling.

The closest most of us get to earth heat is the nightly news after an occasional volcanic eruption is reported on from any number of global hot spots, such as Kilauea blowing its top on Hawaii's Big Island in May 2018 or the Canary Island of La Palma in September 2021, highlighting the enormous energy below us and the power of stored heat as flowing hot magma engulfs anything and everything in its path. Lasting 85 days, the Cumbre Vieja eruption on La Palma forced the evacuation of 7,000 people and destroyed thousands of homes in an almost daily display of apocalyptic horror. The Old Faithful geyser in Yellowstone is another source of visible geothermal energy, performing like clockwork for lucky tourists, spouting hot water on average every 74 minutes from a massive 80 million gallon magma-heated underground reservoir, centered about 120 feet below the surface.

Other regions are famed for their hot springs, such as the Blue Lagoon in Iceland, the picturesque western Bohemian town of Karlovy Vary (Carlsbad) in the Czech Republic, and the Banff Upper Hot Springs in the scenic foothills of the Rocky Mountains. In New Zealand, the Māori even cook in their bubbling hot springs, as do the locals in the breathtakingly desolate Timanfaya National Park on Lanzarote, another of the volcanic Canary Islands. When it comes to geothermal energy, location is everything. In this case, tectonically active regions are a feature.

A fairly simple geothermal or ground source heat pump (GSHP) can operate from as little as 3 m down in some locations, extracting radiant ground heat to warm a building in winter and acting in reverse to return heat in summer, especially important as tens of millions of new air conditioners are added to the grid each year. Only the temperature difference matters as heat flows from hot to cold.[53] All that is needed to exploit the difference is a stable underground heat source for the "geo-exchange" – either rock, water, or steam – and a closed-loop piping system through the borefield, which can be set up for as little as $1,000.

An award-winning 1,600-square-foot house in Ontario's Hockley Valley avails of the relative differences in temperature to heat and cool a single-family home by raising energy from the ground in winter and releasing energy back into the ground in summer. Partly built below ground on the north and east sides to minimize heat loss and maximize solar heating, the more than two-decades-old house also includes solar panels, a small 100-watt wind turbine, R8 heat-mirror windows, passive heating floor tiles, and a Scandinavian wood-burning masonry heater, and has never been connected to the grid. In early 2022, a neighborhood Montreal co-op began heating seven homes retrofitted with geothermal heat pumps. To reduce costs and comply with local bylaws, eight 150-m deep geothermal wells were dug in a private backyard rather than the public alleyway and shared among each of the homes with a goal to connect 50 residences. Although heat is generally the greater concern during the frigid cold of a Canadian winter, air conditioning is also available when needed.

Depending on the underground characteristics, various shallow geothermal energy systems are possible in urban settings, some of which are easier to implement. As shown in an illustrative display outside the European parliament in Brussels, where one of 14 city test sites is being explored to learn more about the latent energy below our feet, numerous types of earth heat are possible: borehole heat exchangers, groundwater well heat exchangers, district heating and cooling, and flooded mines. Since 2013, geothermal heating and cooling has been pumped into the nearby House of European History, while more buildings around Brussels are being added to the mix.

Hot water or steam further down can also be tapped by drilling deeper into the underlying strata, a hot geothermal "fluid" piped to the surface and distributed as heat either directly via a heat exchanger or to run a turbine in an electrical power station as in Iceland. Think of a set of pipes winding underground and back into a building to provide natural heating. The whole setup is conceptually similar to a household radiator system.

Famous in ancient times for its therapeutic hot springs and used as baths by the Romans, the world's first geothermal power generator was built in the still geologically active region of Larderello in Tuscany about 50 km southwest of Florence. In the world's first practical demonstration to turn vented underground steam into electrical power, five light bulbs were lit in 1904 followed in 1911 by the first geothermal power plant just a few miles south in the famed Valle del Diavolo, where hot granite rocks near the surface heated pumped water to steam at 220°C (428°F).

Not much has changed in a modern geothermal plant: the vented steam drives a turbine in a conventional plant or the geothermal fluid (water or steam) vaporizes a secondary "working" fluid with a boiling point lower than water to drive a turbine in a "binary" plant. Both loops are closed in a binary plant, the geothermal fluid reinjected into the heated reservoir, while the secondary loop employs an organic fluid (Rankine cycle) or non-organic fluid (Kalina cycle). Water can also be pumped through the hot rock and returned for everyday heating. Importantly, geothermal is 24/7, ideal for always-on "firm" baseload grid power.

Maintaining a presence in the birthplace of geothermal power, the Italian energy giant Enel today operates 34 geothermal plants in the Tuscan provinces of Pisa, Grosseto, and Siena, producing nearly 6 billion kWh of electricity for more than 2 million households as well as heating almost 10,000 homes, six businesses, 30 hectares of greenhouses, and two dairies, "fuelling an important agricultural, gastronomic and touristic supply chain."[54] Enel Green Power also runs the Stillwater Solar Geothermal Hybrid Plant, east of Reno, Nevada, the world's first hybrid plant to integrate three renewable energy technologies in the same place: a 33-MW binary geothermal power station (2008), a 26-MW PV solar farm (2012), and a 2-MW parabolic-trough CSP plant (2014). The geothermal part pumps underground brine to heat a secondary liquid with much lower boiling point than water to turn the turbine (for example, isobutene flashed into super-heated vapor). The PV section works whenever the geothermal output drops, producing a complimentary output, while the CSP section raises the brine inlet temperature. Enel's Cornia 2 plant in Tuscany also burns biomass to up the initial geothermal steam temperature from 150 to 380°C. As Enel vice president William Price noted, "Combining technologies is a major step forward in the future of renewable energy technologies."[55] A constant output is the goal in all next-generation, assisted-technology hybrid designs.

In some regions, the underground fluid is ripe with minerals. For example, in the Salton Sea geothermal power stations of southern California, the concentrated brine contains about 30% of dissolved solids after the heat and steam have been removed. The leftover geothermal brine is full of extractable minerals, such as manganese, zinc, and a potentially viable supply of lithium for electric charge-storage batteries up to 10 times current US demand[56] (which we'll look at in the next chapter).

Of course, Iceland is everyone's favorite geothermal wonder, ever exciting the imagination about a curious subterranean power below our feet. The tiny country is a volcanic archipelago in a string of north Atlantic islands, where underground geothermal reservoirs provide heat to 90% of homes for warmth, bathing, laundry, and cooking. The main swimming pool in the capital Reykjavik is often full of happy swimmers as are over 100 other pools on the island, all heated by pumped geothermal energy. The geothermal energy heats homes, fish farms, swimming pools, greenhouses, and even melts sidewalk snow (Figure 5.6).

Geothermal power plants such as the 74-MW Svartsengi Power Station beside the Blue Lagoon produce 25% of Icelandic electrical needs, saving billions of dollars in energy costs since the oil crisis of the 1970s and trans-forming Iceland "from one of the poorest countries in the [European Economic Area] to one of the most productive in the world in terms of GDP per capita and quality of life rankings."[57] Meaning "Black Meadow" in Icelandic, Svartsengi was the world's first combined geothermal power-generating and district-heating plant, beginning operations in 1976 before reaching its current capacity in 2008. A total of six plants are serviced by 13 boreholes leading to five shallow steam wells and eight steam/brine mixture wells.

Building on decades of experience, engineers now want to drill beyond the 2.5-km depth of the three main Icelandic geothermal fields, hoping to realize a 10-fold increase in power at 5 km, where hydrothermal fluids can reach temperatures up to 600°C, the hottest in the world (a high-temperature field is

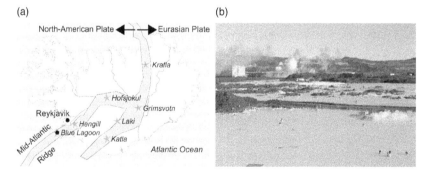

Figure 5.6 (a) Iceland's main fissure zone, showing a few of the island's many volcanoes and (b) the 74-MW Svartsengi geothermal power plant near the Blue Lagoon thermal spa southwest of Reykjavik (*source*: Maarten Visser CC BY SA-2.0).

defined as greater than 180°C). The Iceland Deep Drilling Project (IDDP) will use supercritical hydrothermal fluids to exploit even more energy in a true voyage of discovery to exploit the vast supply of trapped earth heat.

Elsewhere, earth heat is fit to service. In the Netherlands, geothermal energy has become an important part of the Dutch farming industry (world number 2 food exporter), where LED-lit climate-controlled greenhouses ensure year-long crop production, heated by underground aquifers located under almost half the country. Tapping into one aquifer, a farm near Delft generates its entire power needs from geothermal energy, winning "Best Tomato Grower of the World" in 2015. Warm water is brought to the surface, the heat transferred to a secondary loop that circulates through the greenhouses, which heats the farm and keeps the crops safe from frost in cold weather. The water is then pumped back underground and warmed again for reuse. Throughout the Netherlands, geothermal heat is expected to play a major part in the transition to sustainable energy, projected to increase from 3.5 to 50 PJ/year in 2030 and 200 PJ/year in 2050, providing as much as 25% of the country's total heat energy.[58]

Copenhagen also uses geothermal energy, along with incinerator waste heat and CHP plants, to run its citywide district-heating system, providing all the heating to the Danish capital. Rather than released into the sea as in the past, the surplus heat is diverted into over 1,000 km of insulated underground pipes, annually reducing household heating bills by €1,400 per person and saving over 2 million barrels of oil per year since its inception in 1984.[59] In an architectural first, one of the incinerators doubles as a 90-m high artificial turf urban ski slope with a 450-m run, called CopenHill, presenting a cleaner side to waste conversion that emphasizes industrial heat and power generation as fun and positive, labeled "hedonistic sustainability," while a neighboring power plant now burns wood pellets instead of coal. Copenhagen's lord mayor noted, "The central location of our combined heat and power plants is import-ant because it minimises the length of transportation of district heating and thus the heat loss."[60] Depending on your perspective, CopenHill is a novel district-heating idea or a gimmick to hide an incinerator.

In Odense, Denmark's third-largest city with 200,000 inhabitants, waste heat from thousands of servers in Facebook's 50,000 m^2 hyperscale data center is also used to heat water for the local district-heating system. The generated heat is transferred to copper coils in 176 cooling units, where the temperature is raised from 27 to 70°C (the data center is also 100% run on wind-generated power).[61] Edison's original 1882 Pearl Street power station also recycled waste heat as district heating by circulating hot vapor to radiators in nearby homes and offices before being replaced by home furnaces and electric heating.

District heating certainly makes more sense than individual centrally heated oil boilers if the hot air is readily available via geothermal heat or from the waste output of an industrial plant. For a first-of-its-kind district-heating system in the town of Mieres in a former mining region of Asturias in northern Spain, a disused mine now provides input heating water at 23°C that is then increased to 85°C. Developed and operated by energy provider and mining company Grupo Hunosa, the 6-MW plant heats the town hospital, a university building, and various apartment buildings in the largest district-heating system in Spain.

In 2019, a geothermal pilot project in Alberta began extracting heat at a former drilling site, installing a 2.5-km closed loop between existing wells. The $10 million, first-of-its-kind system is on a much larger scale than a standard home unit, but doesn't require any new thinking to distribute heat or generate electricity. An MIT plasma and fusion engineer has also been investigating how to convert abandoned coal and gas power plants to geother-mal plants by drilling *in situ* and reusing dormant turbine equipment and existing transmission lines. The plan is to use a gyrotron – a microwave cutting device previously employed in fusion experiments to heat and vaporize mater-ials – to drill deeper into the ground, thus expanding the limited possible sources beyond the more easily accessible shallow hot spots. As noted by Matt Houde, co-founder of the company charged with commercializing the technology, "We believe, if we can drill down to 20 kilometers, we can access these super-hot temperatures in greater than 90 percent of locations across the globe."[62]

In 2022, the US Department of Energy announced $20 million in funding to develop faster drilling technologies to increase penetration rates at potential geothermal sites, often more than half the installation cost. The goal is to encourage fossil-fuel companies to convert already existing infrastructure into clean renewable energy sites. As Secretary of Energy Jennifer Granholm noted, "There is incredible, untapped potential to use the heat beneath our feet to meet our energy demands with a renewable source that can be found in all pockets of this country. Not only is the use of geothermal energy a significant asset for reaching a carbon-free grid by 2035, it can drive the creation of good paying jobs in energy communities as the country transitions to cleaner, more reliable energy sources."[63]

Begun in 1960, the largest geothermal power plant in the world is the Geysers complex in the Mayacamas Mountains just west of Sacramento, California – 22 plants totaling 1.5 GW – followed by Enel's Tuscany Larderello complex with 34 plants generating 770 MW, while China uses the most direct geothermal heat at over 20 TWh, followed by Turkey, Iceland, and

Japan, accounting for more than half of all direct geothermal heating. The USA produces the most geothermal power with over 30% or 3.5 GW of a 13.2-GW global capacity, generated mostly via heat pumps, followed by the Philippines (15%), Indonesia (13%), Mexico (8%), New Zealand (8%), Italy (7%), Iceland (6%), Turkey (5%), Kenya (5%), and Japan (4%), the top 10 countries of a geologically select group.[64]

One must be careful, however, when tapping into a heated reservoir or creating boreholes to pump water through a hot spot (a.k.a. dry-rock geothermal), because ground stability can be compromised if water or steam is removed near an existing fault line or if the boreholes create excessive fracturing. Monitoring seismic activity in the California Geysers Geothermal Field since 1975, the US Geological Society (USGS) has recorded almost 4,000 quakes per year greater than 1.0 and others as high as 4.5 directly linked to geothermal-power production. An earthquake threshold of 2.0 ensures that fractures are small and that water flow rates are kept steady. As USGS seismologist David Oppenheimer notes, "Unfortunately, areas that are less tectonically active also have less subterranean heat sources."[65]

Geothermal energy is more plentiful than many of us might think at 50,000 times the amount of O&G resources within 10 km underground. Not everyone can avail of an industrial-scale geothermal reservoir to run a power plant or heat a farm or town, but many of us can tap into the energy under our feet to heat our homes. In some cases, it's as easy as sticking a pipe in the ground.

5.4 Hydroelectric Power: Earth and Water Make Electricity

After studying as an engineer in Graz and Prague, working as a technician for an electric lighting company installing an Edison-licensed telephone exchange in Budapest and fixing DC motors to run Strasbourg's railway station lighting system for Edison's Paris-based Continental Company, a young, Croatian-born, Serbian engineer boarded the SS *City of Richmond* bound for New York in the summer of 1884, armed only with his wits and a letter of introduction from his boss to Thomas Edison. The letter read: "I know two great men and you are one of them; the other is this man." The 28-year-old Nikola Tesla immediately impressed and would work for Edison for 6 months, repairing and redesigning DC dynamos in New York as well as developing a high-voltage, arc-light system for outdoor street illumination (alas never used), ultimately leaving Edison's employ after not being appropriately remunerated for his perceived contribution and prodigious talent.

In the meantime, the young Tesla tinkered on his own as he sought money to make real a visionary idea he had had 3 years earlier walking at sunset in Budapest's City Park, and which was the bane of all who tried to build a long-lasting, spark-free, electric motor. His inspiration lead to an AC induction motor using a rotating magnetic field without physical commutators (mounted contacts on a movable rotor). With the financial backing of Charles Peck and Alfred Brown, two New York businessmen keen to see his ideas put into action, Tesla and his new partners licensed his patents to the Westinghouse Electrical Manufacturing Company, which had been looking for an AC motor amid the fast-expanding electric-lighting and dynamo market.

Moving to Westinghouse's headquarters in Pittsburgh, Tesla worked as a consultant at $2,000 a month for the boss himself, George Westinghouse, and his "Westinghouse boys," building as many as 1,000 split-phase AC motors in 1889, netting him and his partners $2.50 per horsepower as stipulated in their license agreement. The deal would soon be voided, however, after Westinghouse suspended work on the motor the following year because of financial troubles.[66] Nonetheless, Tesla's dream of turning electricity into motion for the benefit of humankind had been realized beyond the test motors he had built for his patent preparations.

Unlike other failed electric motor designs, Tesla's AC induction motor employed an outer stationary coiled-magnet "stator" that generated a force on a moveable magnetized "rotor" via multiple alternating currents ("poly-phase"), which Mark Twain would later call "the most valuable patent since the telephone." Most importantly, the same principle could be used in reverse to generate electricity, where a moving rotor induces a current in the outer stator.

Fast on the heels of the trailblazing Edison, the world had its next genius-inventor, but despite being billed as life-long enemies in the press, Edison and Tesla were always more interested in turning their revolutionary ideas into practical machines than building up a business, much to the detriment of their success and annoyance of impatient financial backers. Forever true to their curious natures, Edison and Tesla both delighted in promising the world.

Having brought to life two major inventions before the age of 35 – the AC induction motor (four-wire multiphase and the more practical two-wire split-phase) as well as wireless high-frequency AC lighting, which he believed could be used to generate power[67] – both preceded by popular public lectures and eerie "mad scientist" demonstrations, Tesla next turned his attention to large-scale power generation. Hoping to use his AC induction motor in reverse, Tesla wanted to harness the moving waters of Niagara Falls and fulfill his childhood dream of bringing electrical power to the masses. Born according to family legend at midnight in an electrical storm – his mother stating he was a "child of light" – Tesla never stopped dreaming of freeing the labors of man with the wonders of electrical power.

Aside from the electric part, the concept wasn't entirely new. Water wheels have operated since at least Roman times, falling water turning a wheel for irrigation or household use or to raise and lower a block for crushing, pounding, and cutting. The third-century CE imperial grain mill and aqueduct at Barbegal in southern France was the largest mechanical-power complex in the ancient world, while various kinds of dams have been around for millennia.[68] In the 1870s, the American inventor Lester Pelton designed a split, off-center double-cup to capture falling water more efficiently than a single, centered cup, now known as a Pelton wheel, while around the same time the British–American engineer James Francis revolutionized hydropower by changing the wheel cups into spats and turning the water wheel on its side, similar to what Vitruvius had done in Roman times, while employing gears to control speeds in the simplest of designs.

With a large turning rotor and Tesla's AC generator, however, a massive volume of diverted falling water could now generate hydro-*electric* power. Availing of a continuous and controlled water flow, the power is constant, mechanically reliable, and efficient, while almost all the kinetic energy of the moving water is converted to electrical energy via induction – 90% compared to about 30% of the chemical energy from burning coal and 1% in Newcomen's original coal-fired water pump. The blueprint for modernity starts with the electrical genius of Tesla and a waterwheel turned on its side (Figure 5.7).

The scale of the Niagara project was the envy of many, including the Manhattan bankers who put together the $6 million in funding, led by J. P. Morgan who came on board only after his hand-picked man Edward Dean Adams was put in charge, a one-time major Edison stockholder and descendant of two US presidents.[69] Those at the helm of the Cataract

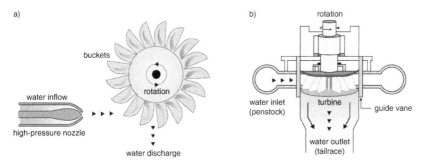

Figure 5.7 Impulse water turbines: (a) Pelton wheel and (b) Francis turbine.

Construction Company of Niagara Falls were undecided at first about which type of current should be generated and whether Edison or Westinghouse was best for the job, but they eventually settled on Westinghouse and Tesla's two-phase AC (two currents, 90 degrees out of phase) to facilitate the long-distance transmission of electricity proposed for the nearby city of Buffalo 22 miles away. The eminent Sir William Thompson, knighted for his work on laying the first trans-Atlantic cable and soon to be ennobled as Lord Kelvin in part for his work on thermodynamics, had initially believed direct current was the only viable option.

The eventual choice was in large part due to Tesla's tireless promotion of his visionary ideas, but also the success of his large-scale, polyphase, AC designs already built by Westinghouse: a Colorado gold mine transmitted 3-kV electricity from a 320-foot waterfall 3 miles away, while a 10-kV system in southern California reached over 30 miles. AC was fast becoming the leading electrical system for generators, transmission, and industrial applications following developments in high-voltage transformers and polyphase induction motors, including power systems in Europe such as Sebastian de Ferranti's 1889 Deptford Power Station on the Thames River near Greenwich that lit the City of London less than 2 miles away. The 1893 Chicago World's Fair – a.k.a. the World's Columbian Exposition to celebrate the 400th anniversary of Columbus's arrival in the New World – sealed the deal, where 180,000 lamps were lit in the illuminated marvel of White City, a contract Westinghouse had also won over Edison.

Starting in 1893, three 5,000-hp Westinghouse/Tesla generators, each 13-feet tall and weighing 85 tons, were installed as part of a planned 20 (100,000 hp in total). The generators were five times more powerful than the enormous 1,000-hp generators already in operation at the World's Fair that were wowing millions of visitors who flocked there to see the modern marvel of electric power up close. In *Empires of Light*, the historian Jill Jonnes noted the magnitude of the cataract's power station design:

> ... Niagara's mighty green waters were to be diverted into the powerhouse, funneled into eight-foot-wide penstocks (giant pipes), fall 140 feet straight down, rush around a crooked "elbow," and then roar into the double wheels of gigantic twenty-nine-ton turbines, the world's largest. These perpetually whirling turbines in the deepest basements of central stations would turn attached vertical steel shafts that would turn the electrical generators on the main floor. Having powered the turbines, Niagara's waters would then begin their three-minute journey back to the river, *whooshing* along at twenty miles an hour through the 6,800-foot sloping tailrace tunnel. The new plan was extraordinary in its simplicity.[70]

Construction on the tailrace tunnel and power station continued for almost 3 years as 3% of the Niagara River between Lake Erie and Lake Ontario was diverted beyond the Falls. Hydroelectric power is proportional to the reservoir head height and the volume of falling water, although the effective "head" height is lower because of friction and turbulence in the penstock. The Falls itself is 54 m high, but the turbine hall was situated above the lower gorge, providing roughly 40 m of effective head.[71] The turbines were finally turned on for local use on August 26, 1895, before the switch was flipped precisely at 12:01 on November 16, 1895, to begin the historic first electric journey to Buffalo.[72] If we can set a date at which the modern world began, mark November 16, 1895, when Tesla's genius converted running water to running electrons transmitted afar at utility scale.

Valiantly ushering in the twentieth century, Niagara Power Station No. 1 would become the largest power plant in the world, producing four times more electricity than any other power station at the time and as much as all US power plants combined (the last unit came online in 1904). "Upon the broad foundations laid by the pioneers at Niagara Falls has been erected the great electrical structure we see today,"[73] wrote Adams, after whom the power station would later be named, while Tesla's nine patents for motors, generators, transformers, and transmission lines are still on proud display in the decommissioned powerhouse (shut in 1961). Ironically, the long-distance, AC power lines were built by Edison's General Electric with 10.7-kV Westinghouse transformers, although so much power was used locally the new industries could have run on DC (as we saw in Chapter 1).

First described in 1683 by Europeans exploring a vast continental interior and known to the local Iroquois as Onguiaahra ("the Strait"), Niagara Falls is the world's second most voluminous cataract after Victoria Falls, characteristically horseshoe-shaped on the Canadian side with a breadth three times its height. Most importantly, the precipitous location was ideal to produce electrical power for those who lived both near and far, fashioning from nature something more than a postcard-perfect site for daredevil barrel runners, death-defying tightrope walkers, and museum tourist oddities.

Harnessed via diverted water in an engineering feat unlike any before, the waters of Niagara provided a store of potential energy to generate electricity for a host of new industries, first in the adjacent town of Niagara Falls, New York, and then in Buffalo and beyond. At a flip of the switch, cheap electricity at a distance was created, illuminating homes and businesses and powering the manufacturing of various electrochemical and electrometallurgical products. Aluminum, silicon carbide abrasives, and graphite could all be mass-produced in industrial quantities, while electric furnaces could run at higher temperatures to fuse materials (for example, clay and carbon to make silicon carbide a.k.a. carborundum[74]) and electrolytic processes to separate materials (for example, caustic soda and chlorine from salt).

Soon, Niagara Falls hosted the world's largest electrochemical complex, industrial development increasing as an "unexpected by-product."[75] Charles Martin Hall's Pittsburgh Reducing Company started manufacturing much cheaper industrial aluminum, the Earth's most abundant metallic element that requires vast amounts of electricity to separate from clay via electrolysis. Becoming a giant of modern American industry and renamed the Aluminum Company of America (Alcoa) a decade later, 1,500 hp was initially contracted from the Niagara Falls Power Company[76] (a major electricity hog, aluminum now accounts for 4% of global electrical consumption). Most importantly, large-scale utility power could be sent for the first time *to* people rather than people having to come to the power. With long-distance transmission, electricity would soon power everything, from lights and appliances to the elevators of an increasing number of skyscrapers in a fast-changing new century.

The area around the Falls flourished because of cheap electrical power, attracting 11 major companies by 1900 and 14 more a decade later to provide consumers with new industrial-scale products, such as "acetylene, alkalis, sodium, bleaches, caustic soda, chlorine – a devil's brew of chemicals produced by electrolysis or electrical processes."[77] The location also spawned the development of high-voltage electrical power transmission for those who lived further afield. In 1895, one-fifth of the American population lived within 400 miles of the Falls, fortuitously located about half-way between New York City and Chicago.[78]

Not everyone was so easily amused, however, by the wonder of the famed cataract. The Irish playwright Oscar Wilde famously quipped while visiting the Canadian side during an 1882 American lecture tour, "it is the first great disappointment in the married life of many Americans, who spend their honeymoons there."[79] But no one could deny the veritable wonder of such majestically harnessed power. The race was on to harness more power and to transmit electricity to all.

***

Early electrical systems were anything but standard, offering different currents (DC, single, dual, and multiphase AC), voltages (100–2,000 V), and AC frequencies (25–125 Hz). Because of the overwhelming technical success of the massive "dynamo-electrical" power plant at Niagara Falls, most utilities soon changed to Tesla's polyphase AC, now standard around the world. The most common is three-phase AC, where three sinusoidal voltages are out of phase with each other by 120 degrees. Two-terminal household power would be provided at 240 V in North America, 220 V in Europe and Asia, and 200 V in Japan.

North American electrical frequency was standardized at 60 cycles per second (60 hertz or Hz) for lighting, although the rest of the world adopted 50 Hz, perhaps more metric-friendly. Anything below 40 Hz flickers, while motors hum at a higher frequency. Although 25 Hz was generated at Niagara Falls, 30 Hz is standard for motors, a compromise between the competing proposals of 16 2/3 and 33 1/3.[80] Voltage regulators, fuses, and three-wire current distribution soon followed.[81] To charge for every watt-hour consumed, the electric meter became essential for distributed electrical power, developed in 1888 by who else but Edison.

Having previously relied on small local power stations, trams immediately benefitted from remotely generated power, after starting operations in Cleveland (1884), San Francisco (1885), Sarajevo (1885), and other cities around the world. The *Buffalo Gazette* noted that the transition was seamless after long-distance electrical power was transmitted from Niagara Falls on that fated first day in 1895: "The cars of the Buffalo Railway Company began yesterday to move by water-power. ... There was no hitch. There were no delays. The experiment was an experiment no longer. It was a complete success. Not many of the thousands of passengers, who rode through Main Street, knew the cars which they rode were propelled by power generated 26 miles away, and borrowed from the world's mightiest cataract."[82]

After Niagara Falls, remote power plant construction expanded rapidly, allaying concerns about the need to build numerous local plants and spelling the end of Edison's DC dream. As Tesla himself explained to Adams, "the Niagara Falls enterprise was the real starting impulse in the great movement inaugurated for the transmission and transformation of energy on a huge scale,"[83] all powered by his polyphase AC that he claimed was 97% efficient, sent over great distances via wire and transformer even though the electricity was available only at first to urban businesses and homes. The derived SI unit of magnetic induction that measures the field strength of a magnet, the tesla (T), would be named in his honor in 1960.

The key to the success of alternating current was the "transformer" that converted electrical power to higher voltages (that is, transforming), which produced less transmission line loss, an essential element because transmission accounts for about one-third of overall costs.[84] Increasing the voltage by decreasing the current minimized wasteful Joule heating in a conducting wire, allowing for much longer transmission distances. Developed by William Stanley, a Brooklyn-born Yale physicist and early pioneer of electrical installations, the transformer ultimately coaxes more effective transmission distance by upping the voltage that is then decreased at the other end, limiting line loss from transmission. Overseen by Stanley, the first AC lighting system and central plant to employ a transformer was built in 1885 in Great Barrington, Massachusetts, powering 13 stores, two hotels, two doctor's offices, a barber shop, the telephone office, and the post office.[85]

By 1897, transformer ratings reached 60 kV before doubling again to more than 120 kV to "step up" and "step down" the voltage, while every year brought more innovation and more installed power (early insulation was insufficient for voltages above 20 kV).[86] As a young engineer working for Westinghouse, Stanley noted that the transformer was "the heart of the alternating-current system. The reason ... lies in its capabilities for simple transformation of voltage over almost any required range, from hundreds of thousands of volts to almost nothing."[87] Converting low-voltage electrical power to high-voltage transmitted power and back again – which saw such spectacular success at Niagara – still resides at the core of our modern electrical power grid. Today, high-voltage AC (HVAC) power is transmitted for distances under about 700 km, while high-voltage DC (HVDC) power is more economical at longer distances and for underwater cables.[88]

On the other side of the Niagara River, the Ontario government sought to create its own hydroelectric plant to harness the energy of the Falls, ending southern Ontario's dependence on American coal for steam power and winter heating. Spearheaded by Adam Beck, a former London, Ontario, mayor and founding chairman of Ontario Hydro, the world's first publicly owned utility provided "people's power" to the emerging Canadian market, creating a government-run system in contrast to the free-wheeling setup south of the border. An autocratic yet tireless community-minded chairman, Beck organ-ized the construction of the first major hydroelectric station on Niagara's Canadian side, commissioned in 1911 to service communities as far away as Kitchener and Toronto and fulfilling the government's declaration that "the water power at Niagara should be as free as the air."[89]

Beck was also instrumental in constructing the 450-MW Queenston–Chippawa hydroelectric plant that used diverted water from a 20-km canal on the Welland River, the largest power station in the world when opened in 1922. Although the $84 million cost for 550,000 hp vastly exceeded the originally estimated $10 million and 100,000 hp,[90] the power of the Falls had finally been tapped on both sides of the Niagara River, exceeding all expectations. Despite the overruns and technological challenges, the business and construction blue-print for more large-scale hydroelectric projects was drawn, soon to be realized by the Tennessee Valley Authority (1924), Hoover Dam (1935), Grand Coulee Dam (1942), and other running-water sources across the globe.

Renamed in 1950 after the champion of "people's power," the 1922 Ontario Hydro plant is now known as Beck 1 and a second 1954 add-on Beck 2, a.k.a. SAB1 and SAB2 for Sir Adam Beck. Two more diversion tunnels were built in 1954, updated in 1958 to implement a storage reservoir, while a more recent upgrade delivers water from a 10.2-km long tunnel. The flagship public utility power station now generates 2 GW, producing electricity throughout the whole of southern Ontario. SAB's pumped storage pools also hold enough diverted water to allow a sufficiently flowing daily supply for tourist viewing at the Falls throughout the year without unduly diminishing the majesty of one of the world's most famous cataracts.

Understanding how power and nature can be in competition, Canada and the USA signed a treaty in 1907 to keep the natural beauty of Niagara Falls intact. Amended in 1950, the international pact limits the amount of diverted water for power production, stipulating that at least 100,000 cubic feet of water per second must flow over the Falls from 8 a.m. to 10 p.m. each day during the tourist season (April 1 to September 15) otherwise the flow can be reduced to half that.[91]

After a disaster on the American side on June 7, 1956, which saw the 1922 Schoellkopf Power Station partially crumble into the Niagara River, possibly weakened by an earthquake a decade earlier, the USA built the Robert Moses Niagara Hydroelectric Power Station further downriver in Lewiston, directly across from SAB1 and SAB2. The plant was named for the career government official and imperial head of the New York State Public Authority, who made no apologies for spending public money despite accusations of "creeping socialism" by rival private interests.

The publicly funded Moses plant was signed into law by President Eisenhower as part of the 1957 Niagara Redevelopment Act at a cost of $750 million, which also included a parkway and public park to beautify the area, ravaged by decades of unregulated industry. Despite remaking what manufacturing had spoiled over more than half a century, the riverside had become "an industrial horror story" as described by a local newspaper covering the legacy of unearthed toxins in the 1970s, in particular south of the Falls at Love Canal.[92] Scattered along the riverside, other trouble spots continue to leach chemicals from the past, but fortunately the legacy of Niagara is more than industrial waste.

At the time, the Moses plant was the largest construction project in the Western world, delivering more electric power than anywhere else to an ever-increasing customer base. Today, the plant is still the number one producer of

electricity in New York State and the fifth largest in the USA, while Niagara
Falls in total generates almost 5 GW for American and Canadian customers
combined.

5.5 Hydroelectric Power: A Twentieth-Century Explosion

Following modest beginnings at Edison's 600-kW Pearl Street coal-powered
station in 1882 – the world's first commercial electricity-generating power
plant that lit 100 businesses in Lower Manhattan – Edison started up his next
power plant later that same year in Appleton, Wisconsin. Home of the first
American hydroelectric power plant and rated at just 12.5 kW, Appleton lacked
even voltmeters, ammeters, or fuses, while the lights would dim and brighten
with the changing amount of flow from low summer drought to high spring
melt.[93] But with the advent of long-distance transmission as perfected at
Niagara, "the grid" expanded, delivering electricity to the doorstep of millions
for the first time.

As more large-scale, connected, power-generating stations were constructed,
grid electricity reinvented the home, ultimately defining modern life. By 1899,
there were 500 power plants rated for 150 MW of electricity, running lamps,
motors, and more than 600 miles of tram tracks. By 1906 San Francisco was
powered by a 10-MW station 147 miles away and Oakland an 11-MW station
142 miles away.[94] AC electric power soon became the new norm, transmitted
over ever greater distances. As noted by historian Maury Klein in *The Power
Makers*:

> Between 1890 and 1905 the output of electric power in the United States increased
> a hundredfold. By revolutionizing production and manufacturing, electricity made
> possible the rise of the consumer economy that was to dominate the twentieth
> century and transform every corner of American life.[95]

Soon, other cities built their own power plants in all shapes and sizes, while
more large-scale hydroelectric power stations were built near sufficiently large
heads of water. Starting in 1902 with 1 GW of power – mostly generated at
Niagara Falls – American hydropower doubled from 5 GW in 1922 to 10 GW in
1932, 20 GW in 1952, 40 GW in 1963, and 80 GW in 1985.[96]

From 1931 to 1935, the 2-GW Hoover Dam was built with public money and
cheap Depression-era labor, the unpredictable Colorado River dammed at
a newly created Lake Mead reservoir, named for the then commissioner of
the US Bureau of Reclamation. Built in arch-dam style to spread the weight of
the stored water to the river banks, concrete was poured around the clock for 3

years. Often called Boulder Dam amid a back-and-forth congressional sparring, the vertiginous Hoover Dam would be renamed for the 31st American president and was the largest of its kind when commissioned, producing a 220-m head of water – twice that of any previous dam – to run 17 turbines in two powerhouses. Electric power, flood control, a regular domestic water supply, and irrigation had come to the desert.

Thirty miles to the west, a small town of 3,000 people that had started as a refilling stop for early steam trains rose up from the desert, beginning the transformation of Las Vegas into today's flashy gambling Mecca of over 3 million residents and countless more tourists throughout the year. Thanks to the water-powered electricity bonanza, more communities grew across the nearby Mojave Desert and further afoot. Without Lake Mead – still the largest American reservoir at about 250 square miles – and Hoover Dam there would be no modern, power-gobbling Las Vegas, Phoenix, or southern California. To commemorate the 112 men who lost their lives during construction, many by falling rocks, explosions, and dehydration in the searing heat, a plaque was erected at the bottom of the majestic concrete curtain: "They died to make the desert bloom."

Following Hoover, the 6.8-GW Grand Coulee Dam was built between 1933 and 1942, part of a 29-dam system on the Columbia River, servicing 1 million homes and generating more electricity than any other US hydroelectric facility. Rising in the Rocky Mountains in Canada and emptying into the Pacific Ocean near Portland, Oregon – defining much of the Washington-Oregon border along the way – the mighty Columbia drops 2 feet per mile and contains one-third of all American hydropower potential. Built in part during the Depression, the benefits were enormous, both to the region and the wider US economy. As noted in *National Geographic*, the Grand Coulee Dam "put 7,000 people to work, created a reservoir for the biggest irrigation project the country has ever seen, provided flood control, and produced electricity that would power America's war effort. At the time, it was the largest concrete structure ever built."[97]

The Columbia River dams also provided irrigation for local farmers, whose potato crops now produce 40% of America's frozen potatoes and French fries, and facilitated easy barging of supplies to Portland, from where 40% of American wheat is exported. At the same time, however, the oversized dams and higher temperatures disrupted age-old salmon runs that once provided 10,000 jobs to the region, salmon numbers vastly diminished including 60-lb ocean-migrating chinooks. The project is also heavily subsidized, selling electricity at below-average prices, while huge debts have been racked up. All power systems are economically and environmentally disruptive, but flooding an entire countryside to create a sufficiently large water store to power distant cities alters the lives of thousands if not millions of settled inhabitants, given no choice but to relocate.

Begun in 1960, the Aswan High Dam helped to create modern Egypt after millennia of uncertainty, taming at long last the temperamental Nile River where 95% of Egyptians live, its fertile banks elegantly outlined as seen in night-time satellite images. The Aswan High Dam transformed Egypt, providing not only a stable 2-GW electrical supply that refashioned one of the oldest civilizations, but protection from an annually flooding river and mitigation of regular droughts, despite enormous consequences for those forced to move from their ancient homelands and the loss of naturally fertilizing silt from upstream Ethiopia. After the loss of silt from the building of Aswan, Egyptian farmers had to use artificial fertilizers for the first time ever.[98] In *The Winter Vault*, a story about the hundreds of thousands of people dispossessed from their ancient lands and relocation of the two giant 1250 BCE Abu Simbel temples to higher ground, which would have become submerged upon filling the artificial Lake Nasser reservoir during the building of Aswan, author and poet Anne Michaels asks whether "we belong to the place where we are born, or to the place where we are buried,"[99] seeing a magnitude of futures unequally shared in the promise of the new.

Highlighting the importance of friendly neighbors, the Nile's once-illusive source originates primarily from the Blue Nile (~85%) near Lake Tana in the Ethiopian highlands at 500 m above sea level and the White Nile (~15%) in the Lake Victoria Basin, shared by Tanzania, Uganda, and Kenya. Known as the Black Nile, the Atbara River also originates near Lake Tana and joins the Nile north of Khartoum, adding to the downstream flow during heavy summer rains.

In 2017, after 7 years of construction, Ethiopia started tapping the falling waters of the Blue Nile in the upper regions at Lake Tana, sparking regional tensions over control of the river's shared bounty, alarming governments in downstream Khartoum and Cairo. Upon completion in 2022, the $5 billion, 6-GW Grand Ethiopian Renaissance Dam (GERD) began generating electricity, yet won't become fully operational until the reservoir fills completely and the accompanying electrical infrastructure is finished. Located over 1,000 km upstream from Aswan and 500 km northwest of the Ethiopian capital of Addis Ababa, GERD is by far the largest hydroelectric power project in Africa and will triple the installed capacity of a country in which only 25% of the population previously had access to electricity. Without GERD, most Ethiopians would still be burning wood, dung, and biomass, increasing pollution, deforestation, and soil erosion.[100]

There are concerns, however, that the over-engineered dam will harm down-stream river life, primarily because of loss of silt and increased seawater encroachment in the Nile River Delta as occurred after construction of the Aswan High Dam. Euphemistically known as Ethiopia's greatest export, almost half of the 140 million tons of water required to fill GERD was initially held back, reducing the nutrient-rich silt essential for agriculture in neighboring Sudan and Egypt as well as downstream power generation. The most signifi-cant drawback to large-scale dam construction is a permanent loss of land, affecting those who live in the vicinity and from where water is diverted, but also causes problems downstream where waters that flowed for centuries are no longer the same. No legally binding agreement has yet been agreed to regulate seasonal flow or manage drought, further precipitating mistrust among the Nile countries and in the wider Horn of Africa.

No one should underestimate the importance of water to downstream liveli-hoods. The Colorado River has over 20 dams over its 2,000-km run, limiting access for those who live and work along its banks, while much of the water is diverted to farmlands and cities such that most years the river no longer reaches its mouth in the Gulf of California. The goal of any energy project is to produce power, although the trade-off as ever is between energy and nature. Moreover, who owns the water and land and is electricity a commodity or essential infrastructure?

***

Today, hydroelectric power plants generate 1,360 GW worldwide, including diverted waterfalls as at Niagara, gravity and arch dams as at Hoover (that increase the head and volume of stored water[101]), and runs-of-the-river (with varying high- and low-water marks) as at the 3.6-GW Santo Antônio dam on the Rio Madeira in western Brazil that started generating power in 2012. China (27%) leads the way, followed by Brazil (8%), the USA (7.5%), Canada (6%), and India (3.7%).[102] Roughly 10,000 m^3 of water is stored to service the world's dams, five times the amount of fresh water in all of the Earth's rivers. Located mostly in the northern hemisphere, the repositioned weight has even changed the Earth's moment of inertia, slowing the daily rotation and making each day a smidgeon longer.

With plentiful lakes and rivers atop a vast rock-lined Canadian Shield, Canada generates almost 60% of its electricity from water. One of the world's largest hydroelectric facilities is in Quebec, where an extraordinary 95% of the province's electricity is generated by hydropower. Started in 1971 and built over 25 years, the James Bay Project in northwestern Quebec diverted three main rivers into La Grande River, creating a system of dams, reservoirs, and watersheds across an area the size of New York State. Today, the entire La Grande Complex contains nine hydroelectric power stations, totaling 17 GW, with the potential to add another 10 GW.

Run by the largest hydroelectric company in the world, Hydro-Québec – a public crown corporation created in the 1920s to counter foreign investors gouging residential consumers and which by 1965 had ultimately nationalized all power generation in the province – the James Bay Project was called *la projet du siècle*, and is considered a French-Canadian engineering triumph and nationalist symbol. Overseen by Robert Bourassa, a future premier who dreamed of breaking the "vicious circle" of unemployment in his province and after whom the largest of the power stations, La Grande-2, is now named, Phase I was completed in 1984, while Phase II was finished in 1996, making the entire La Grande Complex the second largest hydroelectric facility in the world.

The James Bay Project has not come without controversy, however, resulting in a vastly changed water flow, large animal displacements, contaminated fish habitats, and lost wetlands. As a result of the diversions, the total catchment area and mean discharge of La Grande River doubled, uprooting many of the local Cree and Inuit population and destroying livelihoods, while 10,000 migrating caribou were drowned during the filling of the Caniapiscau Reservoir between 1981 and 1984. Ultimately, compensation was reached with the local indigenous First Nations in exchange for giving up any future land rights in the region.

Sending hydroelectric power to where consumers live also requires a network of transmission lines that typically criss-cross pristine rural landscapes. The further the distance the greater the challenge, as electric wires produce heat loss along the way and thus reduced power, although line loss via Joule heating is limited by stepping up the voltage before transmission, stepped down at the business end for safe use (as we saw earlier). Slung across 50-m high steel towers, the high-voltage power is transmitted through steel-reinforced aluminum wires.

The electrical power is stepped up to hundreds of kilovolts in a substation for long-distance transmission and stepped down to a usable, two-terminal 240 volts in the home, first at a local substation and then neighborhood pole-top transformer as seen on the street and indeed heard in a distinctive low-frequency hum. In the James Bay Project, hundreds of kilometers of high-voltage transmission lines feed the large populations of Montreal, Quebec City, and Gatineau in southern Quebec, some damaged or destroyed during a 1998 ice storm when excessive freezing rain snapped the wires.

The Rio Madeira HVDC system sends electrical power from two newly built run-of-the-river dams, designed to lessen the environmental impact on the Madeira River, the largest tributary of the Amazon, although the area has suffered biodiversity loss, flooding, crop damage, deforestation, and loss of livelihood for locals.[103] Completed in 2014, Brazil's 2,400-km Rio Madeira HVDC system is now the world's longest and highest-capacity power line, transmitting 7.1-GW of hydropower at ±600 kV from Porto Velho in western Brazil to the state of São Paulo in the more populous east, surpassing the 2,000-km long Xiangjiaba–Shanghai HVDC power line built in 2010 from Xiangjiaba in central China to Shanghai that transmits 6.4 GW at ±800 kV. Hydropower supplies two-thirds of the electricity to Brazil's national grid.

More applicable to lighting and AC power, the original 22-kV, long-distance power line from Niagara Falls to Buffalo transmitted AC to better accommodate Tesla's alternating-current induction motors and ensure the exclusive use of patented Westinghouse equipment. Although Tesla knew HVDC could be used, he was more interested in a constant AC supply to better match the current in his own motors. The global electrical grid now consists of 2 million kilometers of interconnected, high-voltage power lines (of at least 144 kV), enough to wrap 50 times around the equator.[104]

Completed in 2012 after almost 20 years of construction, the largest hydroelectric plant in the world is the 22.5-GW Three Gorges Dam in Hubei on the Yangtze River (Table 5.2). Called "the grandest project the Chinese people have undertaken in thousands of years,"[105] the project cost a whopping $33 billion and employed 40,000 workers. During construction, 1.3 million

Table 5.2 *Top 10 hydroelectric power plants by capacity (GW)*

#	Power Plant	Year	Location	Height (m)	Power (GW)
1	Three Gorges	2012	Yangtze River, China	175	22.5
2	Itaipú	1982	Parana River, Brazil/Uruguay	196	14.0
3	Xiluoda	2014	Jinsha River, China	286	13.9
4	Simón Bolívar (Guri)	1986	Caroni River, Venezuela	162	10.2
5	Tucuruí	1984	Tocantins River, Brazil	78	8.4
6	Grand Coulee	1941	Columbia River, United States	168	6.8
7	Sayano-Shushenskaya	1978	Yenisei River, Russia	242	6.4
8	Longtan	2009	Hongshui River, China	216	6.3
9	Krasnoyarsk	1972	Yenisei River, Russia	124	6.0
10	Robert-Bourassa	1981	La Grande River, Canada	175	5.5

people were displaced, while 13 cities, 140 towns, and more than 1,600 villages were submerged. A modern engineering wonder, the 2.4-km long, 175-m head gravity dam is the largest concrete structure in the world, powering 32 700-MW turbo generators, the world's biggest Francis turbines. By far the largest power plant ever built, annually producing about 90 TWh, the Three Gorges Dam generates more electric power than all but 33 countries.

The scale is mind boggling, radically changing both the energy and physical landscape of eastern China to provide 22.5 GW of power (10 times Hoover Dam!), a five-tier shipping canal to navigate the Yangtze River, and drought relief during the dry season, especially in downstream rice-growing areas (Figure 5.8). Designed to generate power and protect against floods that have regularly impacted the area for millennia, downstream water flow can be manipulated during heavy rain or drought by increasing or decreasing the reservoir volume with a controlled opening and closing of internal sluice gates. Nonetheless, little can be done during extreme conditions as in the summer of 2020 when the Yangtze's lower reaches still flooded despite the upstream countermeasures.

Not everyone is happy to dam so many rivers, citing the outsized scale, environmental disruption, and unsightly transmission lines, although power lines can be buried underground (for example, under existing railway tracks). Large, centralized power systems also foster reliance on an external supply and tend to increase consumption. Decoupling demand from supply is difficult, especially when utility companies are happy to sell as much power as possible, including beyond regional and national borders. Without remote power stations and high-voltage transmission lines, our cities today would be much smaller, tied to closer sources of power, whether coal, oil, wind, or water.

(a)   (b)

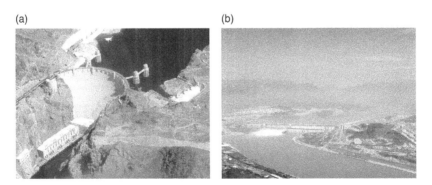

Figure 5.8 Historic dams (a) Hoover Dam, Arizona–Nevada (1935) and (b) Three Gorges Dam, Hubei (2012) (*source*: Xiaoyang Liu/Corbis Documentary/Getty Images).

The green credentials are also suspect when so much land is appropriated, while the rights of locals and wildlife are often disregarded. Natural habitats are damaged or destroyed beyond repair, blocking the migration of fish and other aquatic life in rivers and wetlands, although fish ladders, stepped pools, and opening barriers during migration periods can lessen the impact in some areas. Unmaintained dams are also dangerous as seen in Derna, Libya, in September 2023, when two dams collapsed in the aftermath of Storm Daniel, leading to the deaths of over 15,000 people. Dams also require a large amount of concrete – the Three Gorges Dam contains 16 million cubic meters of concrete, five times that of Hoover Dam, while the amount of rock in the Aswan High Dam is equivalent to 17 Great Pyramids.

Operationally, hydropower is a dream for the utility companies because the turbines can be adjusted to accommodate a changing load as in a "wicket-gate" governor system or via "free-spinning" turbines. Hydropower can also operate at 95% efficiency, essentially 24/7 – fossil fuels are about 30% efficient – and provide immediate access during peak time compared to 12 hours to start up a coal- or oil-fired plant or 24 hours for a nuclear plant.

Providing reliable and fast-acting foundation energy synchronized to grid frequency, hydroelectric power is especially useful for managing a constantly fluctuating electric load and can easily "self-start." Increasing or decreasing reservoir heights and pumped storage hydro also helps manage the constantly changing consumer demand. La Muela II near Valencia, Spain – Europe's largest hydroelectric facility at 2 GW – is a reversible power plant, where water in an upper reservoir falls to a lower reservoir during peak demand and is pumped back up during low demand, available for future on-demand use and to manage the inherent intermittency of renewables.

Hydroelectric power is a mature technology, providing 2.6% of global energy needs (~15% of electricity). Relatively easy to convert the gravitational potential energy of stored water at a height to the kinetic energy of falling water to turn a turbine, many large rivers around the world have now been dammed and fewer untapped sites remain. To increase output and efficiency, old dams are now being "returbined" with new technology. In some places, dams are also being removed, returning a river to its original state.

Hydropower is the progenitor of our modern, switch-flipping lifestyle, first made real below the roaring waters of Onguiaahra. Despite the large upfront costs, requiring public spending to underwrite a decade-long construction and the ecological impact on natural surroundings, we could not revel in the simple modernity we do today without Tesla's dream and which we take for granted is always there. Our modern thinking is implanted with the idea of energy on demand, created elsewhere without thought to the damage to local and down-stream inhabitants or the environment.

5.6 Marine Energy: Not Just Falling Water

Smaller, integrated, hydroelectric systems impact the environment less and are
more easily incorporated as "in-stream" energy rather than drastically reshaping nature via a large dam. Today, water is being employed in ever more exotic
schemes as old ideas are refashioned to produce electricity. Engineers have
long wanted to tame the seas, hoping to generate electric power from the
Earth's massive tidal and wave resources thanks to our always turning Earth
and revolving moon. As early as 1924, the US Federal Power Commission
explored ways to add to its success at Niagara Falls, poetically explaining how
tidal power "is limited only by the vastness of the seven seas and by the eternal
journey of the moon around the earth!"[106] Maybe not eternal, but certainly
sufficient for everyday needs.

A vastly untapped energy source, marine energy could provide 300 GW
globally by 2050 (100 GW in Europe) with the possibility of almost 10 TW.[107]
Writing in *Popular Science*, Erik Sofge noted that "If engineers can harness its
energy, water holds great potential: about 1,420 terawatt-hours per year, or
roughly a third of US annual electricity usage."[108] Lying between the northern
tip of mainland Scotland and the southern Orkney Islands, the immense tidal
resources of the Pentland Firth have been described as the "the Saudi Arabia of
marine power." With 20 GW of potentially convertible power flowing past at
peak tide, various designs are being developed to exploit the regular patterns of
the sea. As long as the moon is in the sky, water will have huge potential to
create power from a constantly flowing tidal current and the perpetual motion
of the waves.

Tidal energy conversion (TEC) and wave energy conversion (WEC) devices
both generate power from an oscillating water column (OWC), using either the
regular tidal "range" – the vertical difference between a high and low tide – or
the up-and-down strength of moving waves, where the amplitude of the peaks
and troughs largely depend on the weather. Ocean thermal energy conversion
(OTEC) and salinity gradient (SG) systems also exploit the differences in
temperature between warm surface water (25°C) and colder undersea water
at depths to about 1 km (5°C) or the chemical pressure difference between two
bodies of water, for example, fresh and salt water (a.k.a. osmotic power), but
currently have few applications.

Underscoring the diversity of engineering inventiveness, there are thousands
of patent proposals for various types of tidal- and wave-energy converters, such
as turbines, floats, buoyant paddle wheels, pitching devices, and cylindrical air
shafts.[109] Although limited by location to where tidal and wave power is strong
enough to be commercially exploited, water-based turbines can operate 24/7/

365, whether in a dam-like river barrage or in a floating, semi-submerged, or seabed-mounted device. The UK (especially northern Scotland), Canada (the Bay of Fundy), Chile, Portugal, South Africa, and Australia are particularly suited to tap the constant power of an ever-fluctuating ocean.

The world's first hydroelectric tidal-power station has been generating electricity since 1966 at the mouth of the Rance River in northern Brittany after replacing a series of medieval tidal mills.[110] As part of a 750-m wide bridgewater barrage between St-Malo and Dinard, 24 10-MW bulb turbines take advantage of an 8-m average height difference between the twice-daily high and low tides of the English Channel. Sluice gates hold back the river at high tide, essentially acting as a temporary dam, while the greater the range between high and low tide the more stored potential energy is available to turn the turbines to create more electricity. After over 50 years of operation, the Rance Tidal Power Station still generates 240 MW at peak times, enough to satisfy local needs.

A barrage can work both ways, but generally only the exiting tide is exploited. Spring tides provide maximum output (greater height) at roughly double the power of the minimal neap tides. A twice-monthly spring tide occurs when the Sun and moon are aligned at full and new moon, while a neap tide occurs in between when the Sun and moon are perpendicular at the first and third phase of each monthly lunar orbit. Although tidal power is naturally intermittent, periodic patterns (and hence power output) are easily forecast from local tidal tables.

On the north coast of Spain, the town of Mutriku in the Basque Country installed Europe's first commercial wave-power plant in 2011. The Mutriku Wave Energy Plant has a capacity of 300 kW, using incoming waves of 1–6 m in a 440-m long breakwater to compress air that turns a turbine. Because of its unique location, "waves roll in pretty much all year-round – the same reason surfers from all over the world visit the coastline from Bilbao to Biarritz."[111] Originally built to protect the town from incoming storms, the converted breakwater is also used now as a test site for marine-energy prototypes.

Other tidal and wave devices can be installed directly into a river or sea, either attached to the sea floor, submerged in-stream, or floating. In essence, many in-stream devices act like underwater wind turbines, but can theoretically generate more power than flowing air in a wind farm despite lower tidal speeds, since water is about 1,000 times heavier than air.

Wave energy has huge potential and is more evenly distributed than localized tides that are limited to geographically select regions. The best waves are typically hundreds of kilometers out to sea, but some coastal areas are especially good, such as along the south coast of Chile, the west coast of Australia, and in Portugal and South Africa.[112] Waves are more violent than tides, so a floating wave generator must be designed to withstand the harshest of conditions when being thrashed about because a sudden surge or storm can destroy a water-borne device in seconds.[113] Theoretically, about 30 MW per km of coast is possible.[114]

The Pelamis wave-energy converter was the first to generate grid-connected electricity from an offshore floating device, based on a 1970s' prototype called "Salter's Duck," created by University of Edinburgh engineering design professor Stephen Salter. Auto-aligned perpendicular to the waves, a series of semi-submerged cylinders move relative to each other as the waves move up and down, generating electricity along a connecting wire.

Also called a "sea snake" because of its shape, a 120-m long, 3.5-m diameter prototype was first connected to the UK grid in 2004 and tested until 2007 at the European Marine Energy Centre (EMEC) in Orkney, followed in 2010 by a 180-m long device rated at 750 kW (Figure 5.9a). Installed 5 km out to sea along the Portuguese coast, another Pelamis test device operated for 2 months,

Figure 5.9 A few wave- and tidal-energy devices: (a) floating wave 750-kW Pelamis P2 "sea-snake" (*source*: Scottish Government CC BY 2.0), (b) floating tidal 2-MW Orbital O2 (*source*: S. Clark CC BY-SA 4.0), (c) in-stream tidal 1-MW Cape Sharp "open center" turbine (*source*: British High Commission CC BY-SA 4.0), and (d) seabed tidal 1-MW Sabella D10-1000 turbine (*source*: G. Mannaerts CC BY-SA 4.0).

alas at well below design specifications. The company eventually went bankrupt, highlighting the challenges of succeeding in an ever-evolving and unforgiving market. Innovation is no guarantee of success without sufficient financial backing or government support.

Harkening back to the concept of Salter's Duck, a wave-energy system created in 2000 by two Danish sailors converted the regular rise and fall of waves into electricity using floats that move up and down on the fluctuating sea. Attached to a platform anchored to the seabed, the first Wavestar device employed two 5-m diameter floats before producing 1 MW from a 20-float commercial installation. An Italian start-up company, 40South Energy, successfully installed a wave-energy converter that automatically finds the optimal underwater depth to generate power in four helical turbines whatever the current direction. The first 150-kW commercial unit was deployed near Tuscany in 2013 with plans for a 2-MW version. Such floating wave-energy devices can also be used in conjunction with offshore wind.

Survival is always an issue in rugged ocean environments, although variable-geometry designs can change shape to accommodate the fast-changing water flow and protect a wave device against possible destruction. One oscillating-surge wave-energy converter was designed with windows to help control the destructive hydrodynamics and reduce the full brunt of the ocean force, increasing the chance of survival as well as lowering material costs. Various companies are developing wave-energy systems, including AWS Ocean, Mocean Energy, OceanEnergy, Marine Power Systems, and Carnegie. All are indebted to the lessons learned from the original Salter's Duck and Pelamis seasnake design.

Generating electricity via a modified in-stream wind turbine that uses localized tides is also possible and is more advanced than wave energy, having incorporated the basics of wind-power technology into the design. Availing of regular tidal patterns in select locations rather than relying on the inherent variability of ocean waves, tidal-energy systems also have a greater chance of commercial success. Companies developing tidal-energy systems include Orbital Marine Power, Magallanes, Nova Innovation, and SIMEC Atlantis.

Operating out of Orkney, Orbital Marine Power (formerly Scotrenewables Tidal Power) has found success taming the sea via tidal-stream energy, launching the first large-scale floating-tidal turbine in 2011, the SR250, a two-blade, 250-kW prototype that successfully connected to the grid after testing at EMEC from 2011 to 2013.[115] Less expensive than a seabed-mounted turbine, the 2-MW SR2000 was launched in 2016, a utility-scale device tested at EMEC from 2016 to 2018. During a measured week of generation, the SR2000 provided almost 8% of the total electricity demand of the Orkney Islands. In

2021, Orbital's next iteration, the 72-m, 680-tonne, 2-MW Orbital O2, was launched in 35 m of water in EMEC's test site in the Fall of Warness, Orkney, transmitting regular output via subsea cable to power 3,000 homes and an onshore green hydrogen electrolyzer (Figure 5.9b).

Further north in the coastal waters of the Shetland Islands, a community-owned project became the world's first offshore tidal array, beginning operations in 2016.[116] The Shetland Tidal Array in Bluemull Sound consists of four seabed-mounted, two-blade, 9-m rotor turbines held in place by gravity, which can generate 100 kW rated for 2 m/s tides. A variable-speed rotor converts the tidal flow in both directions without having to yaw the turbine or pitch the blades, two potential failure modes in a wind turbine. As more devices are spread along the ocean floor, a seabed-tidal farm occupies less space than a comparable wind farm, generating more energy than a wind turbine, although access is naturally more difficult.

Laid out over 3.5 km of waters in the Pentland Forth between the Orkney Islands and mainland Scotland, the first phase of a proposed, 400-MW seabed tidal plant started up in 2017. Deriving its name from a nearby castle in Caithness, the MeyGen Tidal Stream Project is now the world's largest. Designed by Simec Atlantis Energy, four 1.5-MW, 16-m rotor seabed-mounted turbines are affixed to 250–350 tonne foundations with six ballast blocks weighing 1,200 tonnes to provide horizontal stability. The low-voltage electrical output is transmitted via subsea cable to the shore at Ness of Quoys, converted to 33 kV for export via the local power distribution network. The plan is to increase capacity in two more phases, "repurposing jobs from the oil and gas sector and placing Scotland at the forefront of an estimated 25GW global export market for decades to come, as well as significantly reducing LCOE [levelized cost of energy]."[117]

At over 15 m from low to high tide (about 2 m/h), the largest tides in the world are in the Bay of Fundy, a 100-km wide opening between the maritime Canadian provinces of New Brunswick and Nova Scotia. On the east side of the bay, the land-based Annapolis Royal Generating Station has been capturing tidal water since 1984 through a system of dams and ponds that turn a ground-based turbine, although in 2016 a novel in-stream system was installed directly on the seabed to harness some of the 160 billion tonnes of water passing in and out of the bay every day, equal to almost 2 years of falling water at Niagara Falls (Figure 5.9c). The two 1,000-tonne, 10-blade, 2-MW Cape Sharp Tidal turbines were the first of their kind in North America – part of a two-turbine test system costing $15 million – with ambitious plans to expand to a 7-GW network of underwater turbines.[118]

Unfortunately, Cape Sharp Tidal also went bankrupt from lack of financing, although the Canadian government continues to lease coastal waters for new projects, keen to tap into the vast supply of marine energy in its abundant waters. As Elisa Obermann, the executive director of Marine Renewables Canada, noted, "The tidal energy resources alone is estimated to be 40,000 MW with over 200 sites across the country. Adding wave and river current, the potential soars to over 340 GW."[119]

As with wind and solar, marine-energy installations needn't be large scale. Small hydropower plants (250 W to 2 MW) with fish-friendly turbines and protection for fish migration such as simple fish ladders can be employed to tap local rivers with a sufficiently large natural head, especially where watermills once operated (typically at natural weirs or on fast-flowing rivers). Small-scale projects reduce the downstream impact, helping to diversify local energy needs, replace dirty diesel power generators, and lower carbon emissions. Power can be delivered using an Archimedes screw, Pelton wheel, cross flow, or diverted water scheme. A basic run-of-river setup doesn't require any head, employing natural river flow via a small instream turbine affixed to the riverbank, where a fast-running river at 2 m/s can generate as much as 250 W. Local fish and marine life need not be disturbed.

However the electricity is generated, whether in a large hydroelectric power plant such as Niagara, the tens of thousands of power stations that followed – be they hydro, coal, oil, solar, wind, geothermal – or in any of the myriad novel schemes that now generate power from moving water, the demand for more continues. We all consume electricity to run our lives, while the advent of electric vehicles (as we will see in the next chapter) and the batteries and grid power needed to charge them further increases demand for readily available "dispatchable" power (as we look at now).

5.7 Yesterday's Electrical Power System: Top-Down Management

It is hard to imagine life today without plug-and-play electricity, spearheaded by three pioneers of electrical power technology: the practical hands-on inventor Thomas A. Edison, who brought us the light bulb and the DC power station, the single-minded organizer George Westinghouse, who helped shepherd in a new world of AC lighting and long-distance power, and the Serbian visionary Nikola Tesla, who made real his dream of polyphase induction motors and electric generators in the world's first large-scale power station at Niagara Falls.

We must also remember the Anglo-Irish engineer Charles Parson, whose more efficient industrial turbine didn't find its way over the pond into a working power station until 1900. First used in 1884 at the Holborn Viaduct power station in central London, Westinghouse had one installed in a 2-MW commercial dynamo in Hartford, Connecticut, producing more power at lower cost and at 20% the size.[120] Further turbine improvements by GE in 1903 would see coal use drop by 75% at their Chicago power plant, the most important advance in steam engine efficiency since James Watt's separate condenser was added to Newcomen's reciprocating atmospheric pump.[121] The turbine completed the last piece in the puzzle toward rapid implementation of electrical power transmission in the twentieth century.

Many more contributed, but perhaps Tesla stands above the others after making real the visionary thoughts of his youth to explain the mysterious workings in the interplay of electricity, magnetism, and motion, the greatest step in a long ladder to modernity. As historian and Tesla biographer W. Bernard Carlson noted:

> In a nutshell, Tesla's AC inventions were essential to make electricity a service that could be mass-produced and mass-distributed; his inventions set the stage for the ways in which we produce and consume electricity today. For all these reasons modern versions of AC motors can be found running households appliances, powering industrial machinery, and even keeping the hard disks of laptop computers spinning.[122]

The global electrical supply effectively created our modern world, testament to the hard work of those who first turned falling water into electric power, transmitted to more and more customers via an interconnected, long-distance grid. Never before had the world been so utterly transformed by a technology that was so little understood. Steam power is reasonably comprehensible – steam lifts a weight, which can be engineered to move barges, ships, and trains. But the power of the electron, initially theorized in 1897, is not so readily assimilated in the mind's eye. And yet, electrical power at a distance appeared as if by magic, starting with Edison's crude, kilowatt coal-fired power plant covering no more than a few blocks in the business district of Lower Manhattan, rapidly expanding thereafter to more than 2 million kilometers in today's interconnected system of national and transnational grids.

All four were instrumental in creating the grid-tied power plant, giving us our standard "gospel of consumption" that defines modernity. Without easy access to electric power that keeps the machines of industry turning and frees us from the drudgery of household labor, the twentieth century would not have produced our current world of abundance.

***

Electricity is found throughout nature: in a simple touch on a dry day, coursing through an animal's nervous system, or released in the terror of a lightning storm. Many of us are frightened by its raw power, afraid to look too closely for fear of some unknown Promethean structure to its invisible makeup. But thanks to countless scientists, inventors, and engineers who helped tame its mystery over the last two centuries, electricity – that is, an electric current or "flow" of electrons and charged ions – is a part of our everyday lives, surging around us in overhead lines, underground cables, and in the walls of our homes and buildings. All we have to do is plug in.

The Savoy Theatre in London was the first commercial building lit entirely by electric light, powered by an in-house generator for a December 28, 1881, performance of Gilbert and Sullivan's comic opera *Patience*, while the first series of buildings lit by an external central power station was constructed the following year along Holborn Viaduct, north of London Bridge, by the Edison Electric Company. Inaugurated in 1882, the first commercial electrical power network was built in the financial district of Lower Manhattan, designed to provide lighting to about 100 buildings around Wall Street, the electricity supplied by Edison's coal-fired Pearl Street power station (as we saw in Chapter 1). As centralized power became more available, local electricity-generating stations were integrated into a larger connected network – the grid.

Edison's original DC system came with circuit breakers to ensure that any sudden power surge wouldn't blow out the delicate filament lamps, apparently overlooked by British Airways whose entire computer system was brought down by a power surge in 2017. Edison also created a unique feeder-and-mains system to conduct electricity over longer distances, significantly reducing the use of expensive copper wire. Every single foot counts because the energy is more than halved before the electricity reaches the wall plug, lost because of the low thermal conversion of the fuel (especially coal) and transmission-line (Joule) heating. William Stanley, the Westinghouse engineer who developed the transformer for a growing US grid, illustrated the problem of using DC over long distances by reciting a popular adage that the conducting wire to light Fifth Avenue from 14th to 59th Street would need to be "as large as a man's leg."[123]

As seen in a simple parallel configuration in Figure 5.10, the voltage (and thus illumination of each light) diminishes with distance from the power source (top), whereas each light (or building) in Edison's feeder-and-mains configuration maintains the same voltage because the power loss occurs only in the feeder line, made of thinner copper wire to save money (bottom). As Jill Jonnes notes in *Empires of Light*:

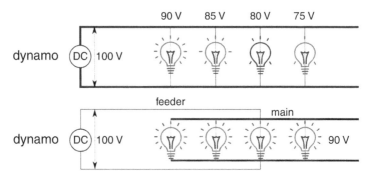

Figure 5.10 Early electric-power distribution systems: (a) a simple parallel configuration and (b) Edison's more efficient feeder-and-mains configuration.

Essentially, instead of one or two very thick (and costly) copper trunks carrying electricity forth and then branching off to each individual building, Edison proposed a network of much thinner multiple "feeder" copper wires coming from the central station DC dynamo and intersecting with many small mains that lit large clusters of lights, thereby eliminating the bulk of the copper.[124]

But despite Edison's ingenuity, his DC system still lacked sufficient range for anything other than very densely populated areas, subject to insurmountable power losses in the copper feeder wire that limited the effective distance to about a mile.[125] Edison's next installation used a three-wire, 330-V distribution system, tripling the deliverable area of his two-wire system and cutting the cost of the copper wiring from $25 to under $1.50 per lamp,[126] but his DC delivery system still couldn't compete with AC. Built 5 years later at Niagara Falls, Westinghouse's pioneering *alternating*-current power station slashed copper wire costs by 99% as more coal, oil, and hydropower stations began to transmit cheaper, high-voltage AC power along a rapidly expanding grid to satisfy an ever-increasing consumer demand.

The proliferation of early home appliances, such as the electric washing machine (1907), vacuum cleaner (1908), and home refrigerator (1912), added to the demand for more, as did the advent of radio in the 1920s and television in the 1940s, while the need for aluminum planes in World War II further increased demand. In the twentieth century, global electrical supply would increase *500-fold*: 11% every year from 1900 to 1935 and 9% annually from 1936 to the early 1970s, before declining 3.5% a year until 2000.[127]

Much of the early engineering work on dynamos and balancing external loads from different power stations was done by the seat of the pants and born of necessity, the newly understood theory lagging practical solutions. Early electric motors also went through an intense period of evolution from sparkless operation to adapted gear transmission, but AC transmission ultimately won the day, helped by GE and Westinghouse pooling their patents in an historic 1896 pact that ended the "patent wars" hampering both companies with suits and countersuits. Even electric railways and manufacturing that ran on DC motors would become powered by AC mains, internally converted to DC as needed.

Today, the grid is a vast, intricately interconnected network of power stations, transmission lines, and substations, providing electricity to meet a variety of needs, instantaneously tailoring supply to match a constantly changing demand, especially the daily and seasonal ups and downs of regular consumption, for example, extra morning/evening use or more heating and lighting in winter. Spread out everywhere to the four corners of the planet, albeit unevenly, global grid capacity is on the order of 8,000 GW (8 TW). In the USA, the grid comprises about 1,200 GW (1.2 TW) of installed capacity, divided into three regions: eastern, western, and Texas, while the smaller UK grid is almost 80 GW. The US grid consists of roughly 8,000 central power stations of at least 1 MW, 5,000 substations to step-up/step-down transmission voltage, and 6,500 high-voltage transmission lines (from 115-kV intercity to 500-kV interstate lines) criss-crossing the country to bring electricity to the doorstep of each and every consumer. In the UK, 2,000 power plants, 500 substations, and 8,000 km of transmission lines ensure the lights always come on as needed.

Whatever the time, day or night, production balances consumption, turning burnt carbon-containing fuels, falling water, fissioning uranium, blowing winds, and absorbed photons into electric current that travels from source to sink at near the speed of light. Be it a bedside lamp, television, computer, or a few minutes of a boiling kettle to make the morning coffee, the response is essentially instantaneous. Typically, electrical demand peaks at around 8 a.m. as people wake up, turn on lights, use hot water, and make breakfast, and then again between 6 and 8 p.m. when they return home from work and settle in for the night. Demand also changes during the year, depending on local climate as household heating or air conditioners are switched on.

There are three types of "load" – base, intermediate, and peaking. "Base" load is 24/7 and never dips below a minimum level (for example, 20 GW for the UK grid), while "intermediate" load comes on in the morning and evening during regular high usage. A "peaking" load occurs during times of surging demand, such as a particularly hot or cold day when more air conditioners or heaters are switched on or as we trample to the fridge and microwave at breaks during a popular sporting event (a.k.a. "TV pickup"), and is often unpredictable. Based on standard everyday use, historical data, and known weather

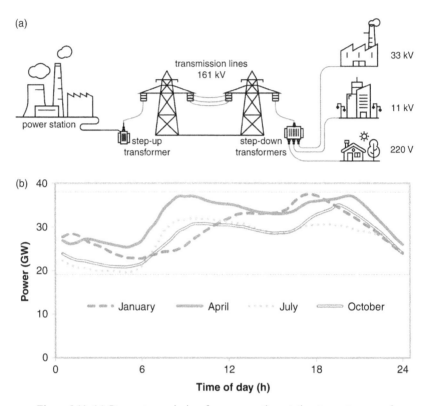

Figure 5.11 (a) Power transmission from generating station to customer and (b) sample electric power use on four different days throughout the year (*source*: "DemandData_2011," *Data explorer*, National Grid. https://demand forecast.nationalgrid.com/efs_demand_forecast/faces/DataExplorer).

conditions, regularly changing loads can be reliably forecast, allowing utility companies to schedule enough power for any particular time or day.

The power transmission setup – from generating station to customer – and the real-time grid output on four different days during the year for the UK is shown in Figure 5.11, quantifying the changing yet regular power demand (note the morning and evening peaks and varying seasonal consumption[128]). A similar consumption pattern exists throughout most of the Western world.

To cover the supplemental demand during periods of excess load (for example, 35 GW), a power provider fires up a local "peaker" plant, although electricity can also be brought in from afar. Different jurisdictions transfer power by agreement to accommodate temporary increases; for example, Denmark sells excess wind-generated power to its northern European neighbors, Quebec sells excess hydro-generated power to New York, while utilities all over the world routinely share power with each other as needed via interconnectors.

Mixing and matching power production to an ever-changing load across vast interconnected systems is a carefully balanced operation between local and regional plants. Operating as the PNJ in 1927 to pool electric supplies from Philadelphia to New Jersey, the world's formerly largest centrally controlled electricity market today delivers electricity across more than 80,000 miles of transmission lines, enough to circle the globe almost four times. The renamed PJM Interconnection now services 65 million people along the US east coast. In Europe, a 50-Hz synchronous grid (formerly called the UCTE grid) seamlessly delivers electricity to 400 million people in 24 countries no matter where the power was generated.

On occasion, a part of the grid fails after an unexpected mismatch of supply and demand, causing power surges over a widening region and potential blackouts. In contrast, a "brownout" is an intentional voltage reduction to limit power over a short period to prevent a blackout, usually seen as a sudden dimming of lights. The Great Northeast Blackout of 1965 knocked out power to roughly 30 million people for 12 hours in Ontario, New York, and much of New England after a transmission line near Niagara Falls tripped, overloading other nearby lines and isolating almost 2 GW of power. New Yorkers were hit hardest as thousands of helpless commuters were stuck overnight in subways, atop skyscrapers, or in elevators, punished for their urban mojo.

The failure started just as night fell (the "magic hour") on a cold November day after local demand exceeded 375 MW at SAB1, before spreading to SAB2 and then the Robert Moses plant on the American side of the river. Power losses were pushed out further, triggering shutdowns in other connected stations on the international CANUSE system, while the increasing, unresolved demand exceeded failsafe limits, knocking out plant after plant "like boxcars piling up after an engine jumps the tracks."[129] Removing even one power station in a highly interconnected network can produce rolling failures as more plants disconnect from the grid to avoid equipment damage during the rapid increase in demand along fewer transmission lines, initiating a cascading series of shutdowns.

Lessons were learned about how to incorporate sudden spikes, but the system is still vulnerable if too many failures occur at the same time. The largest-ever US power outage occurred in August 2003, when a sagging high-voltage power line in northern Ohio shorted to ground after touching an overgrown tree branch (literally "to ground"). In fact, three overgrown trees grounded the grid voltage at different locations in quick succession, producing a number of cascading line failures and a high-speed wobble from too much current on too few transmission lines, while a software bug stopped the auto refresh on local utility control-room consoles from showing the fault in time.[130] Power to over 50 million people was knocked out, disabling airports, transit systems, and even Cleveland's water supply. During the blackout, which cost $6 billion in lost revenues over 2 days, nine nuclear reactors were taken offline, requiring a whole day to restart and reconnect to the grid.

After the fact, it is easier to spot the errors. In 2003, the lax attitude to fixing common problems such as regular tree trimming contributed to the local utility's hands-off policy, as did basic system faults (trees are still the main cause of blackouts, especially falling branches during storms).[131] The US Federal Energy Regulatory Commission also separated power generation and distribution in 2000 to integrate new sources into the grid and support energy trading. Alas, a lack of distribution – a.k.a. "wheeling" imported power when a sudden surge exceeds supply – can be dangerous, as witnessed during a particularly disastrous recent Texas winter.

After a below-zero cold spell in February 2021, parts of Texas lost power for a week, knocking out electricity to more than 4 million homes and businesses. At roughly 90 GW capacity, most of which is coal, oil, and natural gas (wind is about 6 GW or 7%), the Texas grid is dialed-down about a third during the usually mild, southern US winter as natural-gas plants are taken offline for maintenance (full capacity is needed for air conditioning in the hot summers). Brought on by an increasingly wavy, climate-change-induced polar vortex, the reduced grid was not prepared, however, for the rapid upsurge in demand as people turned up their heat all at once – demand rising 50% higher than usual – while water pipes froze in poorly insulated homes, adding to the emergency.

The main culprits were the non-functioning natural-gas power plants, frozen wellheads and pipelines, and a woefully unprepared state-wide grid system that limits outside wheeling, although in the midst of the blackout anyone and everyone was called to task. The outages were blamed on bad planning (seasonally offline plants remained unavailable despite the cold forecast), deregulation that incentivized cheap power without safeguarding essential infrastructure (electric power obviously works in cold-weather regions when the mercury drops), and a state government that favors private markets over public safety, which had ignored earlier winterizing recommendations after a similar freeze in 2011. Even the former president and one-time governor of Texas, George Bush, was blamed for implementing a deregulated Enron-led system that created the Lone Star state's free-market energy pricing, lampooned by some as a "freeze" market. At the height of the crisis, tempers flared when Texas senator Ted Cruz flew to Cancun for a family holiday to escape the "FREEZING" temperatures, while the governor went on television to blame underperforming solar panels, wind turbines, and even a recently proposed yet unrealized Green New Deal.

In fact, the Texas island grid was mostly to blame, purposely designed to limit wheeling electricity from outside to avoid federal taxes, although some juice can cross over the border from Mexico when demand spikes. Highlighting the lack of overall preparedness, connected parts of the Panhandle and El Paso that had updated infrastructure after the 2011 freeze were left unscathed. Ever wary of federal oversight and government regulations, many oil-rich Texans prefer to go their own way, even to the detriment of safety and reliability. One bemused University of Texas at Austin researcher noted, "There's a lot of excess power in this country, but we just don't have the extension cords to bring it here."[132]

As if to hammer home the need for backup power at the ready (and the dangers of MacGyvering heat in an emergency), an all-electric Tesla includes "Camp Mode" to provide clean electric heating in a pinch as opposed to the grim possibility of carbon monoxide poisoning in a gasoline-powered car. During the sub-zero conditions at night, some Texans slept safely in their own battery-powered, four-wheeled emergency shelters.

But despite the headline-grabbing blackouts that send us scurrying for the candles or other more dangerous options during a prolonged outage, the grid is normally quite stable, boringly so, fortunately for billions of consumers who think nothing about the source of the running electrons in their wall plugs. Since the beginning of transmitted power, the grid has operated reliably as a one-way, centrally run system, consisting of a large number of users and smaller number of interconnected stations and regional systems – 60 to 1 in Edison's 1882 Lower Manhattan network or roughly 60,000 to 1 in today's US grid. What can't be generated locally is sent across greater distances, made in low-population regions to hungry customers elsewhere, such as Amazon hydropower to São Paulo, Jiuquan wind to Shanghai, and James Bay hydro-power to Montreal, as we've seen.

Costs, politics, and community opposition, however, are often the limiting factors to ensure access to more power – in particular eminent domain, rural despoilment, and environmental damage – as in Desertec or the mammoth $160 billion transmission line proposed to connect the northeastern USA to Texas. Even relatively modest plans such as the failed $4 billion privately funded Plains & Eastern Line from Oklahoma to Tennessee that would have sent 4 GW of wind-generated electricity along 720 miles of HVDC power lines are difficult to build as various regulators debate land rights, cross-border legality, and the integrity of outside providers (not that oil and gas pipelines, interstate highways, or railways are similarly scrutinized). A particularly divisive 430-mile DC power line between North Dakota and Minnesota even spawned a rebellion in the 1970s among local farmers.

Today, however, the old top-down, utility-run model with its hodgepodge of regional and long-distance transmission lines is being radically transformed. Designed to run on a steady, predictable supply of stock resources – burnable fuels such as coal, oil, natural gas, and uranium – new sources of power and a new type of producer-consumer (a.k.a. "prosumer") is beginning to generate home-grown electricity that can be slotted back onto the grid for sale as part of a feed-in tariff or net-metering scheme. With the addition of more renewable energy, the old-style one-way grid is slowly being upgraded to a modern, transactional, "bi-directional" grid. Dubbed the "energy superhighway" or "smart grid," the two-way model is evolving with multi-user input and sophisticated, real-time software. Subject to the standard growing pains of general inertia, system limitations, and money, one hopes the millions of miles of wire are up to the task.

5.8 Tomorrow's Electrical Power System: The Smart Grid from Macro to Micro

Controlled from above to manage the complex network of power-generating facilities, transmission lines, and delocalized users, the old grid demanded uniformity, a condition being stood on its head by the increase of renewables and low-level prosumers. Instead of a rigid and hierarchic monopolistic structure, a new real-time control system must now manage an increasing number of diverse, small-scale, generators – you, me, other residential and commercial users – seamlessly integrating the lower layers of the distribution network back onto the grid or into localized mini and macro grids. On top of the usual safety features, the "smart" grid must also be robust and resilient, able to guard against failures, unwanted intrusions, and targeted attacks.

In ways similar to how mobile phones were integrated into a wider digital information network, a new means of electrical power distribution is being managed on the fly. As smart digital networks dynamically integrate the demands of thousands of users in real time, vast amounts of instantaneous data are being analyzed with sophisticated, self-learning predictive algorithms instead of the standard 12 static readings a year in a monthly bill. Electricity prices have also become a tradable commodity, in part to accommodate the growing number of diverse inputs, such as renewables, natural gas, and other smaller distributed energy resources, ostensibly tied to the rules of supply and demand rather than rigidly regulated from above.

Supported by batteries to flatten out intermittent supply (especially solar and wind) and a constantly changing load from independent prosumers, utilities are being forced to change their monolithic, centralized ways to incorporate smaller, flexible inputs. Simon Hackett of Redflow Advanced Energy Storage thinks that the challenge will not come from excessive competition, but rather keeping up with demand, calling the coming grid a "bi-directional energy backbone," with "distributed command and control."[133] Although not yet disruptive, the changes are starting to move the needle.

Balancing consumption over a widely varying production base also requires utilities to play a dangerous game of arbitrage as wind and solar begin to undercut a more-than-century-old method of electric power generation coupled to long-distance transmission, even reselling power at a loss when too much supply becomes available or paying customers to disconnect at peak times. On one excessively windy day in West Texas in September 2015, the spot price dropped to negative 64 cents per MWh.[134] Prolonged shortages also hurt fixed-priced sellers who have to buy energy at a higher floating rate, drastically in some instances, as in the autumn of 2021 when a number of British energy companies went broke after dwindling summer natural-gas reserves couldn't be restocked. A single, long-term price or at best a two-tier peak/non-peak price is no longer the norm, upsetting the traditional market between producers and consumers.

Caught in an economic "death spiral," utilities have to raise prices to compete with cheaper renewables, some even made by their own customers, that then encourages more renewables, raising prices even more. Power plants are facing a "triple whammy" from lower demand, lower wholesale prices, and increased renewable energy sources. No longer able to deal with a flatter, less-profitable intraday price curve or negative prices, Malcolm Keay of the Oxford Institute for Energy Studies believes "the utility business model is broken, and markets are, too,"[135] noting that "Many conventional plants, even efficient, clean new plants, are having to close because they are losing money; in most cases they are not covering their fixed costs, and in many cases they are not even covering operating costs."[136]

We must be wary, however, about the scope and style of revolutionary change that manifests itself in haphazard ways, especially as capital competes with capacity. Historian Maury Klein noted, "The railroad mania literally created the modern American capital market," while the de-facto divvying up of the electric market between the two powerhouses GE and Westinghouse removed burdensome restrictions to facilitate a more orderly rolling out of universal electrical infrastructure, creating "not a lack of orders but a rush of business that overwhelmed their ability to meet it without constant expansion of facilities."[137]

One wonders if the usual growing pains that accompany new invention will follow after the major players have had their say, but as the smart grid expands what lessons will be forgotten in the quest to make distributed power more accessible? The scale of change will swamp anything seen in the last five decades as smart cities and "island" grids begin to generate their own electricity without having to rely on piped-in power from remote power stations, a.k.a. the "virtual power plant" (VPP). In the next few decades, innovative distribution models – in particular localized microgrids – will be born of necessity to tackle the expanding smart grid, renewable-energy market, and charge-storage paradigm. As Audrey Zebelman, chairwoman of the New York State Public Service Commission, noted, "It takes a central procurer – in this case, historically, the utility – out of the mix."[138]

Although solar is more easily scalable than other power technologies and can more easily slot into a modernizing grid than the heady days of early grid building and central power plant construction at the turn of the twentieth century, there are many similarities and past lessons to heed. Today's clamoring over lithium and cobalt supplies for electric storage batteries, for example, mirrors attempts to corner the copper market and eventual stockpiling of up to 160 million pounds of copper per year.[139] Wealthy prosumers – the term coined by futurist Alvin Toffler in his 1980 book *The Third Wave* about user participation and cooperative exchange – will also be able to isolate completely from the vagaries of the energy market and price gouging by utilities focused only on the bottom line, while others are left behind.

In Germany, the smart-energy company Sonnen began administering an 8,000-strong energy-trading network, while neighbors in Brooklyn are already exchanging energy across a community network of solar-powered homes, buying and selling electricity to each other via a virtual peer-to-peer (P2P) trading platform. Harkening back to Edison's first community network that connected 100 buildings in Lower Manhattan, the Siemens and LO3 Energy Brooklyn Microgrid started with 50 customers.

Leapfrogging an increasingly limited centralized distribution system, microgrids also make power more accessible to both developing and remote regions, without needing to rely on expensive utility-scale solutions. As noted in 2022 in *The Economist*, more than half the population of sub-Saharan Africa has no access to electricity, where "In rural areas, stand-alone, 'mini grids' linked to small generators such as a solar park are often the cheapest way for villages to get connected. Solar home-systems are booming."[140] More than one billion people are still unconnected across the globe, mostly in developing countries, but even in Canada 200,000 people still have no access to the North American grid, having instead to burn dirty diesel.[141] Ideally suited for smaller-scale distribution, microgrids are easy to install when designed to power a thousand rather than a million homes – not unlike Edison's one-off, single-user, private plants. Nor are they as readily brought down by blackouts or system breakdowns.

Apple, Google, Facebook, and other big-data companies, whose existence depends on always-on power, have already built their own microgrids, ensuring resiliency and reliability in the event of a temporary shutdown or sudden loss of service. Partly in response to Hurricane Sandy in 2012 and to improve cyber security, NJ Transit installed a large microgrid to power trains, stations, and offices, becoming the first traction company to generate its own power since the beginning of the grid.[142] During the week-long, Sandy-fueled power outages across New York City, SUNY's Stony Brook campus stayed on via its own microgrid (albeit by burning natural gas) as did Southern Oaks Hospital on Long Island (cogeneration) that needed to maintain essential healthcare infrastructure.[143] With easily slotable renewables, more microgrids will begin to appear, adding to and in some cases subtracting from the grid.

The US military is investing heavily in microgrids to keep mission-critical systems on in all conditions, whether at home or in the field. The goal is to increase efficiency, maintain stealth (silent solar over noisy, smoke-signaling diesel), and reduce dangerous fuel supply chains (a staggering 70% of army fuel is consumed just transporting the fuel).[144] On the research side, the University of California at San Diego received a $39 million grant to build a novel testbed to integrate a campus microgrid of 2,500 distributed energy resources (DERs), including solar panels, wind turbines, smart buildings, EV batteries, and fuel cells.

More than just off-grid environmentalists, ecovillagers and futurist "preppers" are switching on by switching off, building their own "grid edge" power systems to provide everyday juice and resilient energy islands that can cover a blackout or prolonged power outage. The old "macrogrid" is slowly being supplanted by microgrids (typically under 50 MW) that can plug in or plug out as needed to provide their own power and safeguard against failure, while pushing more power onto low-voltage, local transmission lines. With the separation of power generation and transmission enacted in a liberalized energy market, DERs are now being managed more like information, digitally switched on and off as needed, subject to the vagaries of real-time pricing.

Hoping to become the world's largest energy company and underscoring the importance of smart energy in a radically shifting decarbonizing market, Shell purchased Sonnen in early 2019, announcing that it could outperform industry returns by optimizing and trading on intermittent supply for home battery storage and electric-vehicle charging. One of Shell's directors noted that "Electrification is the biggest trend in energy in the coming 10 to 15 years because it's by far the easiest way to decarbonise energy usage. . . . It will grow faster than any of the other energy markets and it is easy to grow *in* growing markets."[145] Small is beautiful again, almost by default to accommodate a changed energy paradigm, 50 years after the British economist E. F. Schumacher sang the praises of a bottom-up, human-directed economy that must stay within nature's "tolerance margins."

***

For those used to a conventional, central-station macrogrid model the new technology is daunting, while much of the new-energy infrastructure is underpinned by Internet transactions between anonymous independent users. Blockchain is the platform of choice, particularly scary for those wary of settling payments across a remote and murky Internet. First imagined in a 2008 whitepaper "Bitcoin: A Peer to Peer Electronic Cash System" by a still unknown Satoshi Nakamoto, Blockchain is a distributed database that creates, manages, and verifies encrypted digital transactions in real time, typically at remote sites across the whole of the Internet, including banking, credit, retail, customs, supply chains, real estate, licensing, identity management (for example, non-fungible tokens or NFTs), hotel accommodation, medical histories, and insurance, as well as digital currencies, finance, and energy. Once posted, a transaction can't be changed without leaving a trace, making fraud almost impossible (although every use eventually finds its own abuse).

Referred to by some as Cloud 2.0, Blockchain is made up of hardware sites across the Internet (the blocks), many leased and linked together (the chain) to run decentralized applications (dapps). Considered an almost frictionless society, decentralized computing minimizes the need for local staff in favor of "miners," who lease equipment to run the applications, the ultimate e-trading house, only without the house and without the middlemen. As applied to energy, Blockchain allows consumers to buy energy (or prosumers to buy *and* sell energy) in real time, recording each transaction for immediate settlement. Digital currencies (such as bitcoin) have gotten most of the press because of the novelty of a virtual non-fiat currency and roller-coaster volatility, but energy investment and energy trading are both ripe for distributed recording.

By 2018, more than 10% of financial institutions were using Blockchain, jumpstarting a $500-billion business. Blockchain can also optimize the best time to buy or sell energy based on a changing spot price, whether an essentially static two-tiered day/night rate that already applies to many consumers or a dynamic, instantaneous rate, based on the real-time power generation of solar, wind, natural gas, and other changing inputs. Predicting the weather could have huge financial implications, creating opportunities for companies selling energy contracts at a set price so others can ensure a fixed rate in an uncertain up-and-down market, not unlike the foreign-exchange market. Blockchain brings market speed, advanced analytics, and efficiency to energy trading based on real-time demand and seasonal patterns (think of an instantaneous billing system that analyzes purchase patterns to restock shelves).

Although Blockchain may be more suited to energy trading and real-time analysis of continuously metered output, the platform also serves as an interface for investors and utility companies to finance distributed energy through connected technology. For example, rented solar panels can be managed from remote locations, anonymous absentee investors receiving payment for the generated output in bitcoin or a local currency. As noted by Abraham Cambridge, founder and CEO of SunExchange, "With Blockchain, we are able to increase the accessibility and inclusivity of solar ownership. We are making solar mainstream."[146]

As more new-energy capacity comes on line, the grid will keep getting smarter, especially with the introduction of new protocols and standards, such as the Energy Web Foundation (EWF) Blockchain that identifies and matches consumers and prosumers. There is still much to implement, upgrade, and integrate, however, including convincing the public of the value of buying and selling real-time energy and the safety in doing so.

Many utilities already provide "time-of-use" (TOU) rates or real-time pricing with smart meters. In the Netherlands, home battery systems stabilize the grid by storing energy at times of high generation that is then sold back at times of high demand. After a 2010 test run, the Dutch utility Eneco installed 300,000 smart-thermostat energy monitors in homes, cutting the energy costs of some users by a third as customers controlled their home heating settings via a smartphone app that also itemized their "electricity and natural gas consumption in detail, along with other information like weather forecasts."[147] The devices also facilitate easy servicing of repairs as needed.

In California, almost all homes have a smart meter that sends household energy data to a utility company, allowing continuous grid management and incentivized TOU pricing (peak, off-peak, and mid-peak). In the UK, customers can choose to pay nothing for electricity on weekends during low-demand times in exchange for their user data. Ontario Hydro charges twice as much peak to off-peak, while Hawaii's most populous island Oahu proposed a 3× peak/off-peak rate to push electrical use into the middle of the day at high solar hours. Significantly, a smart network is greener as less idle reserve power is needed to cover unexpected spikes at peak demand times (a.k.a. peak shaving), which means burning less, easily dispatchable, dirty coal and natural gas that must be kept on permanent standby.[148] Smart systems also help companies to sell energy services rather than just energy, a declining revenue source as more homes become fully fledged micro-energy providers.

With real savings on offer, consumers' habits will change as users become accustomed to variable pricing, made easier with automated connectivity and the Internet of things (IoT), enhanced by ultrafast 5G networks (~1 Gb/s) and eventually even faster, smarter, and more efficient 6G networks (~100 Gb/s). Some customers, however, are rightly concerned about "surge pricing" in a deregulated free-for-all, as well as the personal overhead required to watch for spikes in a constantly changing rate. During the February 2021 Texas power failure, energy prices skyrocketed, saddling non-fixed-rate customers with astronomical bills, some in excess of $10,000 for a single week!

As smart pricing becomes near instantaneous, however, home storage will ultimately make compliance a no-brainer as well as help flatten the load, turning the famous power-versus-time "duck curve" into a ripple. As Chris Goodall notes in *The Switch*, "There may be some resistance to this change, but within ten years some of us will load our dishwasher and let the grid operator choose when to start it. . . . We can expect that all non-time-sensitive electricity use will be eventually pushed into periods of peak renewables output. The pleasant consequence will be that electricity bills go down, not up."[149]

Further flexible charging will not only help manage changing loads, but will also reduce spikes as customers employ smart charging and smart meters to integrate vehicle batteries to avail of optimum charge times, flatten the load, and sell power back to the grid (which we'll look at in the next chapter). A potentially disruptive innovation, vehicle-to-grid technology (V2G) and vehicle–grid integration (VGI) will completely change how power is bought and sold by utilizing idle electric-vehicle batteries or indeed any idle energy-storage device. After all, at its core an electric vehicle is essentially a battery on wheels.

We must be careful, however, about implementing a system that puts more technology between the output (electricity) and users (us humans) or serves only to trade a commodity rather than making the commodity more efficient and affordable. We must also remember that homes aren't machines and that some people may not want to cede control of their appliances, whether a dishwasher, dehumidifier, or robot vacuum booting up in the middle of the night, despite the environmental benefits. Although replacing analog meters with smart digital meters improves overall grid efficiency and reduces electric bills as well as helping utility companies to repair outages and identify faults more quickly, leading to less down time, people are naturally wary about their household patterns being monitored, including by the government or some unknown, algorithmic, decision-making authority. In the wrong hands, distributed energy can be controlled or manipulated just as easily as information. What's next – a creepy HAL 9000 glitch or system takeover? Imagine being unable to open a door or window because the "smart" system or a cyberjacker won't allow it.

Smart interfaces also come with an inevitable loss of privacy on top of a loss of meaning as in notable data breaches at a number of Internet companies, including Facebook, Yahoo, and Ashley Madison. As lawyer Henry Drummond noted in the 1960 film *Inherit the Wind* (played by Spencer Tracy), "All right, you can have a telephone but you lose privacy and the charm of distance." Or Mae Holland, the naïve customer experience manager in *The Circle* (played by Emma Watson), who discovered that imbedding chips in bones may protect children from kidnapping but comes at an extraordinary surveillance cost, as do remotely monitoring patients versus seeing intimate behavior, capturing fugitives versus stalking, or providing automated services that make millions unemployed. The changing ethics from increased connectivity and data sharing are only just being examined in a growing surveillance-capitalism technopoly.

Many of today's changes are indeed scary. Witness an older generation distrusting or confused by online banking, unable to reserve a seat on a train or a plane, or even order a meal at a restaurant without Internet or smart-phone access. Human-less service is fine if everything works, but can be disastrous when the system fails, as in a broken turnstile in an empty subway or train station that locks passengers in or out without assistance at hand.

No matter how beneficial, we must be cautious about any technology that promotes a widening digital divide or digital exclusion, compounded by costly wireless Internet and smart-phone transactions and soulless service, ultimately hijacking the customer experience in the absence of any regulated interface. As more connected systems begin to control our lives, technology must work to our satisfaction (the happiness factor) and for as many as possible (narrowing the digital divide), while also ensuring that Big Brother and Uncle Sam don't leave us reeling in a Kafkaesque nightmare none of us want or can afford.

For now, the extent of the smart grid and number of app-controlled micro-grids is small. About 200 systems in the USA provide roughly 0.2% of grid power, limited mostly to test programs. But the numbers are expected to increase over the next few years, initiating a wave of new infrastructure unseen since the urban renewal schemes of the 1950s, when city planners prioritized cars over people by extending multi-lane highways into city cores and building Brutalist prefab corridor commuter apartments for all. At least with concrete, bricks, and mortar, one can see the changes and complain or demonstrate against a flawed design.

Many established tech companies are at the forefront of the broadening smart grid, such as Apple, Microsoft, Google, and Facebook, all of whom rely on connectivity to survive and are remaking their workplaces and data centers with smart systems. A changed *Metropolis* world is here, our twentieth-century murky brown past morphing into a twenty-first-century, green, tech-driven future as a range of streamlined companies welcome us to a new machine. Bottom-up, renewable-energy power and advanced, smart-phone connectivity is already remaking an aging, top-down, monopolistic grid – for now, humans are in control.

Macro connections are also expanding to connect an increasing number of regional and international renewable-energy installations, a.k.a. the "super grid," whether solar, wind, or hydro. In 2021, the UK began importing Norwegian hydropower along a 724-km subsea interconnector cable, the pumped reservoir filled by British wind and solar installations, thus increasing the range of localized wind and solar. Further interconnectors are being built to add more readily dispatchable stored power to help smooth disparate outputs across a wider, variable weather range. As renewable energy becomes cheaper, the issue is no longer how to generate power, but how to get the power to the customer without interruption, exactly the same as in the time of a growing grid after Niagara Falls.

To advance the transition, however, we must find a way to store what hasn't been stored before in any significant capacity, accessible at a moment's notice. Although first held in small, temporary storage jars by early "electricians" such as Pieter van Musschenbroek in the Netherlands and Benjamin Franklin in the USA, industrial-sized, long-duration charge jars are needed to help manage the load. The solution may depend on one simple ingredient – salt, which some are already calling "white petroleum" or "white gold." Without the ability to store energy easily and safely, we will never tame tomorrow's grid.

5.9 The Future of Electricity: More Efficient Charge Storage

A battery exploits the chemical reaction between two different metals to separate charge carriers (positively charged ions and negatively charged electrons), storing an excess of ions in a positive electrode (cathode) that then cross an electrolyte separator to a negative electrode (anode), while at the same time the much smaller electrons circulate in a closed circuit to run an external load such as an electronic device until the battery is discharged. Recharging with an external voltage (for example, a wall socket) pushes the ions back to the cathode for use again. The process continues until the battery degrades when the internal ion flow becomes too gunged up. Corrosion also occurs over time, especially in humid environments – best to remove a battery from a device if left unused for a long period, such as over winter at a cottage.

The Italian scientist Alessandro Volta made the first chemical battery in 1799 after two different metals – brass and iron – accidentally pressed across a frog's spinal cord and generated a current that caused involuntary muscle contractions. In part to disprove the theory of "animal electricity" proposed by his rival Luigi Galvani, Volta produced a continuous electric current and in the process a way to store electric charge. Today, the battery is the holy grail of modern energy: saving charge for future use.

In Volta's original "wet" battery, 49 alternating pairs of zinc and copper discs were separated by pieces of cloth dampened with salt water, the zinc (Zn) and copper (Cu) atoms losing electrons to become Zn^{2+} and Cu^{2+} ions (importantly, zinc loses more electrons). When the ends of the two electrodes were connected to a circuit, a potential difference of roughly 1 volt was generated as the Zn atoms lost electrons and the Cu^{2+} ions regained electrons. Chemically, the multiple cells form a single Zn–Cu battery as Cu^{2+} ions are reduced and Zn atoms are oxidized in a so-called "redox" reaction, the essence of all electric charge-storing batteries, where positive ions move through the intermediate electrolyte (a conductive substance that dissociates into positive and negative charge carriers when dissolved in a liquid) and the negative electrons supply the external current between the two oppositely charged electrodes.

Although much improved, today's modern battery is essentially the same as Volta's: two different metal electrodes (the electrochemical couple) are sandwiched around a liquid or solid electrolyte, causing the positive and negative charges to separate and form a potential difference (a.k.a. voltage) between the two metal electrodes (anode and cathode). Attached to a circuit (or load), an electric current is produced that continues until the oxidizing agent runs out, subtracting from one electrode and adding to the other.

Today, manganese (Mn) is used instead of copper to make the ubiquitous 1.5-volt, Zn–Mn alkaline "dry-cell" found in flashlights, toys, remote controls, and everyday portable devices. Pick up a store-bought battery and you'll see the + and – ends clearly marked (indicating the manganese dioxide cathode and the zinc anode electrodes) as well as an "alkaline" label (indicating the electrolyte paste between them, typically made of potassium hydroxide). The positive end (+) also protrudes from the casing, making it easier to recognize and slot into place. Standard sizes are AAA, AA, C, and D (from small to large size and low to high drain).

The rectangular, snap-connected, 9-volt battery is essentially six 1.5-volt Zn–Mn batteries connected in series and was popular for transistor radios, but is more commonly found today in smoke detectors. The shiny, round, 3-volt, lithium-ion (Li-ion) "button" battery is made of metallic lithium oxide (cathode) and porous graphite (anode) with a lithium salt electrolyte (for example, $LiPF_6$) for use in portable devices (watches, gate openers, laser pointers) as well as high-current electrical equipment (for example, a camera flash), but is now changing the world as the power behind today's computers, mobile/cell phones, and electric vehicles.

The first commercial Li-ion battery in 1991 powered Sony's CCD-TR1 Camcorder, an instant hit with the home video crowd, which spilled over to computers, phones, and other digital devices, before being chained together to power the drivetrain in an electric-engine automobile. Lead acid, zinc carbide, and other chemical cathode–electrolyte–anode batteries have all been tried in industry-scale propulsion, but are too bulky to provide sufficient energy density, typically measured in watt-hours per kilogram (Wh/kg).

Fortunately, home battery costs are minimal. A primary (that is, nonrechargeable) 1.5-V AAA is about $1, a 3-V button $3, and a 9-V $10, hardly a dent in today's household budget that might use two dozen batteries a year for a total cost of $40 (mostly AAAs for remotes), even less if bought in bulk. Indeed, we take batteries for granted, unconcerned about composition, voltage, or energy density, not noticing until the charge is gone and we need to buy more. At annual sales of around $8 billion, however, today's household battery sellers such as Duracell, Energizer, Rayovac, and Panasonic are making massive profits.

Secondary (that is, rechargeable) batteries that reverse the redox process when plugged in are about twice as expensive as primary batteries and naturally last longer until the oxidizing metal becomes too corroded or the electrolyte deteriorates after repeated use (~5 years). Early rechargeable batteries were primarily nickel–cadmium (NiCd) or nickel–metal hydride (NiMH), using fewer resources and cheaper in the long run.[150]

Invented in the 1950s, the first implantable heart pacemaker ran on a NiCd battery, as did early Earth orbiters. Both are now powered by rechargeable Li-ion batteries, as was the original Mars rover.[151] Pioneered in the 1970s at Oxford University by solid-state physicist and future Nobel Prize-winner in Chemistry John B. Goodenough, Sony started developing Li-ion batteries in the 1990s for portable phones and computers (30 minutes on 10 hours charge), while tooth brushes, razors, and video cameras still use the cheaper and more durable NiMH batteries (Figure 5.12).

Today, secondary batteries are found in many portable devices, especially mobile phones, cameras, and laptops, most of which are made of lithium-ion (Li-ion) or lithium–polymer (Li-Po) chemistries, recharged *in situ* and not as a separate battery unit. Although you can swap in a battery pack on a camera (3.6 V) or a laptop (10.8 V) as needed, most mobile phones (3.7 or 4.2 V) are hard-wired and not designed to swap the batteries in and out.

The rechargeable battery market is huge, especially in the ultra-competitive mobile-phone business, where small size, light weight, fast charging, and long life are make-or-break features. Keen to compete with Apple's top-of-the-line

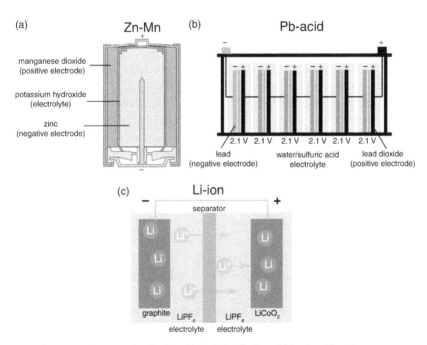

Figure 5.12 Battery chemistries: (a) Zn–Mn alkaline, (b) lead–acid, and (c) lithium-ion.

model (~$700), Samsung lost about $1 billion and perhaps $14 billion in market share[152] when it recalled 2.5 million just-launched Galaxy Note 7 smart phones in 2016[153] after the lithium-ion battery was found to catch fire during recharging, the positive and negative electrodes merging because of an excessive fast-charging voltage on the thinner, larger battery. Continuous Internet access for Pokemon Go may be attractive to some users, but exploding batteries are never good for business. After being banned on various airlines, Samsung eventually halted production.

In the past, rechargeable batteries were designed to be fully recharged after each complete discharge (that is, 0% to 100%), but Li-ion batteries have minimal "memory effect" and perform better in a state of charge (SOC) between 20% and 80%, where the chemical reaction is most effective.[154] Charging is faster from 20% to 80% (for example, about 40 minutes) and slower otherwise (for example, 40 minutes for the last 20%), a function of the difficulty of pushing more electric charge into a tighter space as the charge fills.[155] To optimize capacity and improve longevity, a full discharge or charge should be avoided.

A good analogy is putting a sweater on – the first bit is easy as one slips the sweater over one's head (lots of wiggle space), while the final bits are harder as one struggles to stuff one's arms in (much tighter space). Damage occurs if you push too much, while degradation comes from high temperatures (>30°C) and normal aging after about 500 charge/discharge cycles, a few years' worth if you recharge a phone every other day, although one might question having to upgrade a 2-year-old phone just because the battery dies.

The quantity of electrons or total energy is measured in amp-hours (Ah) for storage batteries and electric vehicles, but milliamp-hours (mAh) is better for phone, tablet, and laptop batteries.[156] Kilowatt-hours (kWh) is also used for electric vehicles, although a more practical measure is the distance driven per amount of charging time, for example, 50 km/h (or 200 km in 15 minutes from high-voltage fast charging).

Many of us are similarly unconcerned about car batteries, other than needing a boost when one goes dead. Created in 1859 by French physicist Gaston Planté, the lead–acid (PbA) "wet-cell" battery was the first rechargeable storage device. The chemistry is not unlike a household battery, but uses lead and lead oxide as the two electrodes with a sulfuric acid–water electrolyte. A standard 12-V PbA car battery – six linked 2-V cells – is typically employed to start an internal combustion engine rather than hand cranking, requiring about 450 amps (cold-crank) and costs about $100.

After Planté's early PbA battery, the smaller "carbon–zinc" battery became popular with the advent of household electrical devices, particularly the transistor radio, before being supplanted by today's Zn–Mn alkaline battery (with carbon–manganese paste electrolyte). Today, however, the greatest challenge is to power electric vehicles, homes, and back-up electrical-storage generators that require much more bang for the buck. Li-ion batteries have a much higher energy density than PbA batteries (~250 Wh/kg versus ~30–40 Wh/kg) and can thus store more charge in less material, essential for large portable units and heavier vehicles. Off-the-shelf lead–acid batteries with sufficient capacity, however, are still found in smaller electric vehicles, such as wheelchairs, granny buggies, and golf carts (as well as off-grid battery packs), while larger electric vehicles require the much more powerful lithium-ion chemistries (for example, Li-ion/cobalt oxide in a lithium-salt organic solution).

Lightweight with very high energy density, lithium-ion batteries require monitoring, balancing, and cooling, and can overheat during charging, possibly igniting the flammable electrolyte, as famously witnessed by a few unfortunate Samsung customers and a problem for high-capacity automotive batteries if not properly cooled. After a couple of high-profile early fires, Tesla devised a liquid glycol cooling system that was enmeshed in its vehicle battery packs and added an underside aluminum/titanium shield to ensure safety of the batteries in the event of an accident.[157]

With the advent of new cell chemistries, high-capacity batteries are rapidly improving and becoming more affordable to both industry and commercial users for mobile and stationary use. Although the energy-storage market is dominated almost entirely by Li-ion batteries, other technologies are also being deployed and continue to be developed, such as lead–acid, sodium, flow-vanadium, flow-zinc, and aqueous batteries. Cutting-edge R&D for both mobile and stationary batteries includes innovations in lithium-ion, aluminum-ion, solid state, lithium–sulfur, and metal–air. Recalling Volta's original classification of metals by electronegativity, by which others could design their own batteries, the US Department of Energy created a materials database of over 16,000 molecules to help match the properties of different electrode pairs and electrolytes in the quest for improved battery chemistries.

Today, high-capacity, lightweight rechargeable batteries are all the rage, spurred on by the transition to hybrid and all-electric vehicles, stand-alone home storage, and variable grid back-up systems, increasingly so as more intermittent renewable energy is added to the mix. Prices dropped six-fold from $1,200/kWh in 2009 to $200/kWh in 2016 and are expected to fall further to under $50/kWh in the coming years,[158] while the storage market increased seven-fold from 300 MW to 2.1 GW between 2016 and 2021[159] and is expected to reach $250 billion by 2040.[160]

Presciently writing in *The Solar Economy* in 2002, Herman Scheer noted that "If manageable and cost-effective power storage technologies are available, then the revolutionary transition from the fossil fuel to the solar economy will be unstoppable."[161] Already, stand-alone batteries are remaking the grid, while electric cars have become feasible because of improved battery technology, and are making a dent in overall car sales with new models on offer from most of the main players, such as the early market leaders Tesla (Models S/X/3/Y), Nissan (Leaf), and Chevy (Bolt).

The main concerns for battery-powered electric vehicles are cost (currently more than gasmobiles although fuel costs are much less), range (more than 200 km for use as other than a daily runabout), recharging time (6 hours is okay for overnight or during work hours, but cumbersome for long journeys without fast charging or battery swapping), and small vehicle size, although larger SUVs and trucks are now appearing. Range is reduced in colder temperatures because of the slower battery chemistry and the need to run the heating system, a deal-breaker for all but the most ardent cold-climate users, but battery pack wrapping and heating wires are available. At below freezing, the normally fluid electrolyte hardens to a gel, slowing the redox processes at the electrodes, resulting in a lower voltage, for example, as much as a 50% decrease at –20°C.

Mass production continues to lower prices as electric vehicles become more successful – passing 10% of all vehicles in 2022 – and as the home/industrial battery market grows. Zero emissions is the goal or reduced emissions in a hybrid (oddly called a "partial zero emission" vehicle to reflect its dual personality), although emissions won't be reduced if electric vehicles and storage batteries are charged with electricity generated elsewhere in coal-, oil-, or gas-fueled power plants.

Panasonic's "18650" cylindrical battery was the initial industry standard – 18 mm round × 65 mm long with an energy density of almost 250 Wh/kg – adapted for home and industrial use by linking multiple cells together. Building on the success of portable electronic devices, the Asian giants have led the way: Panasonic, CATL, LG Chem, BYD, and Samsung at over 100 GWh/year, while Tesla has also championed the "21700" with 50% greater capacity in its Panasonic-partnered Nevada Gigafactory, where it plans to output over 250 GWh/year worth of Li-ion battery cells and packs.[162]

In the evolving, highly competitive industrial market, the essence of high-capacity stationary batteries is to store grid energy, bought at night when cost is cheap and sold during the day when utility rates are higher. The electric load on a shared grid is generally higher during the day and lower at night, which is why utilities cut night-time prices to incentivize consumption at non-peak times, even offering *negative* pricing at high-supply, low-demand times (as we've already seen). Many of us take advantage of the preferential rates (energy arbitrage), running a dishwasher or dehumidifier at night or doing the laundry on weekends when utility costs are cheaper. More advanced systems can program the energy schedule, everything from the morning coffee pot to heating and lighting, while smart systems can even "download" electricity at cheap night-time rates and "upload" stored electricity during the day or at other advantageous times, all of which is becoming more viable because of the improved chemistries of new charge-storage devices that store the most energy in the least volume.

Solar and wind power also fluctuate independent of user demand, and can be challenging to integrate into the grid. Electrical storage is essential to manage solar-generated power for night-time use and on cloudy days as well as wind-generated power in calm periods. On top of providing a 24/7 baseload supply, utility companies regularly store backup power to accommodate the constantly varying load, while homeowners can store excess power for their own use or sell on to others as needed. So-called "peaker" plants or "load-following" generating stations are expensive, and other means of energy storage are essential (as we'll see in the next chapter).

We are at a crossroads. How we made energy defined the twentieth century; how we make and store renewable energy within a flexible interactive grid will define the twenty-first. How we store energy has never been more important, whether in a portable household device, a backup battery, or in the one product that continues to shape and define our modern world (as we look at now) and is currently undergoing a revolutionary makeover, none of which would be possible without cheap and efficient batteries.

6

Driving the Revolution Revolution: From Volta to Tesla and Back

6.1 Driving the Future: Electric Charge Powers the Way

Coal-fired steam power was the progenitor of the Industrial Revolution, bringing to life railways, steamships, and mechanical power stations for on-site manufacturing. Long-distance, electric-power transmission lines from remote generating plants also brought electrification to the masses, not least to power the incandescent light bulb that had replaced the kerosene lamp, such that by the 1950s electricity had completely modernized our world, spurring on an increase in on-demand power, public transport, and the rapid growth of cities. Improved mobility became available via electric trams and gasoline-fueled automobiles that supplanted the horse-drawn carriages of our increasingly cramped inner cities, helping to create the suburbs as lines spiraled out in all directions from a central hub.

Powered by lead–acid batteries, the first electric streetcar was Radcliffe Ward's "tramcar," demonstrated in east London in 1851, the same year as the Crystal Palace Exhibition in Hyde Park.[1] The first electrified train was tested by Werner von Siemens at the 1879 Berlin Industrial Exposition, carrying 1,000 people a day around a 300-m circuit, using the rails as conductors.[2] Edison improved upon the design the following year at Menlo Park, but didn't think there was a future for electrified public transport, missing an opportunity to convert steam trains to electricity. As he later admitted, "I had too many other things to attend to, especially in connection with electric lighting."[3] Siemens would also be the first to run a commercial electrified train service in 1881 along a 2.5-km track in Berlin, followed 3 years later in the USA in Cleveland, both systems powered by poles connected to overhead wires.[4]

Frank Sprague, a navy engineer and former Edison employee at Menlo Park, who had experience making motors for factories and mills, would go on to solve many of the early problems of the electric tram – overhead conductors,

456

commutator wear, arduous gearing, adapting sufficient power for steep grades –
beginning the first commercial electrified streetcar system in the USA in 1888
in Richmond, Virginia, a subsequent model for future transit systems.[5] As
historian Maury Klein observed, "Sprague had set in motion a revolution in
the electric industry, and all the major players scrambled to climb aboard."[6] By
1889, there were 200 electric streetcar systems up and running around the
world, while by the start of the twentieth century electricity had completely
transformed inner-city public transportation.

In the excitement that followed the rise of petroleum, the internal combustion engine, and electric trams and trains, we forget that prior to the success of
gasoline, engineers had already built a battery-operated electric automobile.
Thomas Parker, an electrical engineer from Shropshire, England, built the first
self-propelled electric car in 1884, powered by Planté's lead–acid battery,
thereafter working to install early electrified public transport systems, including the London Underground and the tramways of Port Rush, Birmingham, and
Liverpool. By the start of the twentieth century in New York, Chicago, and
Boston, electric cars were even competing with gasmobiles that suffered from
a lack of gas stations.[7] Then as now, however, the problem was battery
performance. Cleaner than running a gasoline-powered engine and already
well-established since the 1893 Columbian Exposition, charge storage was
still the rate-determining step to keep an electric engine going.

Even though he didn't capitalize on electric trams, the electric battery was
a natural progression for Edison, who had fashioned a working electric light
bulb in 1879 and then the first commercial electricity-generating power station
in 1882 to run his revolutionary invention. Batteries were soon "his new, all-
consuming preoccupation" as he sought to make electric cars "the poor-man's
vehicle."[8] In 1903, after having moved from Menlo Park to a new laboratory in
nearby West Orange and almost 4 years of research, Edison developed
a functional storage battery, applying his legendary trial-and-error skills to
determine the right chemistry to power more than just a string of lights. He
claimed his battery-powered electric car could outrun a gasoline equivalent.

Joining forces with Henry Ford, who had started out as an engineer at the
Edison Illuminating Co. in Detroit before turning his attention to mass-
producing gasoline-powered cars – encouraged in part by the Wizard himself
after their first meeting at an 1896 company convention dinner – the two
engineering giants became good friends, and in 1912 Ford enlisted Edison to
design an electrical system for his famous Model T, which until then was hand-
cranked, "at best inconvenient to use, and when it kicked back, dangerous."[9]
Ford would finance the world's most famous inventor, seemingly always
strapped for cash, although as Edison biographer Randall Stross noted,

"Edison did not abandon his previous ambitions to make a success of an electric car; he simply made Henry Ford his new partner."[10]

Besides the electric starter motor, plans were also hatched to manufacture a line of electric Fords, powered by Edison's lightweight battery. Although nothing remains of their work from this period, a 1913 photograph shows a prototype of the Detroit Electric in front of the Highland Park Ford plant, packed with three large nickel-iron batteries under the seat, while another photo from 1914 shows an electrified Model T with a range of up to 100 miles for a proposed production line selling between $500 and $750.[11]

Alas, neither the battery nor their business partnership achieved the success one might have hoped for or even expected from the collaboration of two industry titans. Edison's alkaline storage battery was deemed "insufficient to power the car lights, let alone a starter motor."[12] After dumping the Edison Storage Battery Company for a competitor, Ford would later call Edison the "world's worst businessman." More engineer than businessman, Edison was always busy with other projects while working on his battery, preoccupied by movie projectors, concrete homes, phonograph upgrades (including changing the recording format from cylinders to discs to compete with the new market-leading Victrola), or whatever struck his fancy.

After Ford severed their business ties, Edison continued to improve his battery despite setbacks from leakage and poor recharging, eventually developing a novel, multi-use, nickel–iron–alkaline battery that would power *half* of all American delivery trucks prior to the rise of the gasmobile.[13] His uncanny ability to get results, even from initially failing ventures, would result in another great technological achievement. Edison's most profitable invention was not the light bulb, phonograph, or film projector, but the electric storage battery, which would power the electric systems of ships and trains as well as lighting buoys and miner's safety lanterns, becoming the dominant battery for the next 60 years.

Having ably competed at the start of the car industry, stand-alone electric propulsion would nonetheless become practically nonexistent by the 1920s, displaced by the economics and simplicity of oil, primarily because of limited range and insufficient power of the battery.[14] Despite one-third of automobiles being electric – the Pope Waverly had a top speed of 20 mph and a range of 20 miles, while the Baker Electric (the first production electric car) could travel 80 miles on a lead–acid battery – batteries would become more suited for telegraphs, telephones, and medical and experimental use. By the time petroleum took over, the electric car was doomed, even as a local delivery runabout.

Indeed, early electric-car batteries were expensive, heavy, and fragile, and by the advent of World War I gasoline was destined to be the fuel of the future, derailing further research and development on electric cars. Inner-city trams would instead become the dominant electrified street vehicle, powered by an overhead conductive electrical system to run converted steam-powered trams. Interurban commuter trains also appeared – the first in 1892 covering a distance of 16 miles between Portland and Oregon City – followed by the newly elevated ("L" or "el") transit systems in the fast-growing metropolises of Chicago (1892), New York (1903), and Philadelphia (1907), all electrified by a third rail.[15]

The world was not yet ready for a clean, green, stand-alone, all-electric, battery-powered performance car. Unable to compete with the power and convenience of gasoline, electric propulsion would become limited to golf carts, granny carts, shuttle buggies, and wheelchairs, all run on heavy, low-range, lead–acid batteries. Ironically, the lead–acid starter motor helped end the practicality of electric vehicles by eliminating the need to hand-crank a combustion engine, a dangerous and time-consuming job. For the next century, the success of the electric car or lack thereof depended on only one thing – the battery.

Many of us have seen an electric golf cart in action, typically powered by a 6-V, 250-Ah (20 h) lead–acid (PbA) battery – perfect for a round of golf, depending on the difficulty of the course and one's golfing prowess tee to green. Having fooled around with electric motors during gasoline rationing in World War II, the electric golf cart was invented in 1951 by Merle Williams and launched in Long Beach, California. As noted in a 1954 *People Today* article, his first model immediately solved the problem of the lack of caddies and busy golfers, "headed by Pres. Eisenhower, who frequently uses a Golfmobile to cut playing time in half."[16] Powered by four heavy-duty PbA batteries, generating 24 V, the Golfmobile had a top speed of 14 mph, while the charge lasted about 36 holes. Perry Como, Bing Crosby, and Bob Hope were just a few of the happy golfers willing to shell out $990 for the luxury of playing in presidential carting style. Further marketed as an everyday runabout, Como called his electric cart, "the greatest thing since sliced bread"[17] (another novelty, introduced to the American public in 1928).

***

Over time, various small vehicles were powered by electric batteries, including airport transporters, inner-city runabouts, and local delivery vans, harkening back to earlier cleaner times before the gasmobiles took over. To help reduce inner-city pollution, taxis in many urban areas also became gasoline–electric "hybrids," perfect for low-mileage clean driving without having to worry about any niggling "range anxiety." Ferdinand Porsche had built the first gasoline–electric hybrid in 1898 while working for Mercedes-Benz, consisting of a petrol engine that charged an electric wheel hub motor on the fly. Originally dubbed *Semper Vivus* ("Always Alive"), Porsche's front-wheel-drive hybrid-electric Lohner-Porsche Mixte debuted at the 1901 Paris Motor Show and won the Exelberg Hill Climb near Vienna that same year. But even hybrids were no match for the power and convenience of petroleum-fueled transport, while Porsche would go on to design and build the iconic Volkswagen Beetle, a cheap, mass-produced, rear-engine, gasoline-powered car.

Today, the electric half of a hybrid vehicle runs on a nickel–metal hydride (NiMH) battery. Launched in 1997 in Japan and in 2000 in the USA, the Toyota Prius took the car industry by storm, fast becoming the world's top hybrid, annually selling more than 100,000 units. Originally sold at a loss, the game-changing Prius came with high-tech software, sleek aerodynamics, and "regen-erative" braking that converts kinetic energy otherwise lost as heat into electricity to help recharge the battery, extending the range by up to 30%.[18] At the heart of the hybrid is two motors – an electric motor for shorter city trips and a gasoline motor for longer highway journeys – all but eliminating range anxiety for extended travel.

Each motor compliments the other, available during low charge/low gas or the gasoline engine kicking in to give extra oomph to the electric motor when accelerating to pass. The NiMH battery also stores power during driving to improve overall "mileage" – when one releases the accelerator, energy is transferred from the wheels to the battery (for example, downhill or cruising) instead of from the battery to the wheels (for example, uphill or under normal driving conditions), similar to storing/using energy when going up and down a hill on a bicycle.

With no emissions in electric mode, the Prius was an early green champion despite the hefty price tag, scoring more than a few high-profile celebrity advocates such as Leonardo DiCaprio, George Clooney, and Gwyneth Paltrow. Although not fully electric, the Prius did much of the heavy lifting to help transition to an all-electric car as battery technology began to improve and costs dropped. The International Energy Agency estimated that today's hybrid vehicle adds $3,000 to the sticker price, ultimately recouped by lower fuel costs.[19]

One early complaint was a lack of noise, a problem to unsuspecting pas-sersby who couldn't hear a Prius coming – typically at intersections, in parking lots, or at low speeds – solved by a built-in noisemaker.[20] The irony of an electric car is that it was too quiet, hardly an important safety issue during the rise of the gasmobile. What's more, engine power was no longer measured only in horsepower, but also in torque – the force causing an object to rotate about an axis – instantly applied using one continuous gear.

An all-electric performance car, however, was another story. Earlier in 1996, General Motors had introduced the world's first modern all-electric car, but was doomed from the start, limited by its low 60-mile range. Automotive journalist Dan Neil called the now-famous Chevy EV1 a "marvel of engineering, absolutely the best electric vehicle anyone had ever seen," yet still added the EV1 to his *Time Magazine* list of "The 50 Worst Cars of All Time," stating that the "battery technology at the time was nowhere near ready to replace the piston-powered engine." Although battery weight was significantly reduced, the range, durability, and cost were still deal-breaking bugaboos:

> The early car's lead–acid bats, and even the later nickel–metal hydride batteries, couldn't supply the range or durability required by the mass market. The car itself was a tiny, super-light two-seater, not exactly what American consumers were looking for. And the EV1 was horrifically expensive to build, which was why GM's execs terminated the program . . .[21]

Available only to lease as part of an evaluation program, the Chevy EV1 was recalled in 2003 and much of the 1,117 production-run crushed, GM execs citing excessive manufacturing costs. Despite its demise, an early owner, actor Danny DeVito, called the EV1 "the coolest car I ever had," saying he would still be driving one today if it hadn't been recalled: "I was taking care of the planet. I wasn't gunking up the air. It was a fantastic ride. It was fast."[22] But without reliable battery power, eco-saving cool was not enough to save the world.

6.2 Democratizing the Car Industry from the Top Down

The lithium-ion battery changed everything, sparking a renewed development in reliable, long-lasting charge storage for electric vehicles as well as stationary back-up grid power. Originally manufactured in 1990 by Sony for laptops and other portable devices, the small-format lithium-ion (Li-ion) batteries can be chained together to power an electric vehicle. Having already been field-tested in the computer industry, Li-ion batteries were reliable, long-lasting, lightweight (lithium is the lightest metal with atomic number 3), available in large supply, and most importantly affordable with reasonably short charge times (as we saw in Chapter 5). The advantages of Li-ion batteries include being lighter and smaller than PbA batteries with a high specific energy as much as four times (160 versus 40 Wh/kg) and low self-discharge, although they require circuit protection, degrade at high temperatures, and can't rapidly be charged at or below freezing.[23]

One company was instrumental in leading the electric way – Tesla Motors (now Tesla, Inc.). Tesla's first all-electric vehicle was powered by 6,381 small-format, Panasonic 18650 cylindrical Li-ion batteries (18 mm diameter × 65 mm length), connected in a single battery pack originally located in the back, but later moved to the chassis. Named for the AC induction motor's visionary engineer-inventor, Nikola Tesla, the Silicon Valley start-up was led by its own visionary inventor and high-tech entrepreneur, Elon Musk, who made his fortune as one of the brains behind Zip2 and PayPal,[24] becoming a millionaire before the age of 30, a billionaire by 40, and the world's richest person by 50.

Musk has been called the Thomas Edison of his era and is into space flight, satellites, hyperloops, neural interfaces, and renewable energy, as well as electric cars, and has made no secret of his ambition to change the world. Tesla's two-seater, luxury sports car, the 2008 Roadster, quickly became an automotive star, the first to use Li-ion batteries and travel over 200 miles (320 km) on a single charge. Criticized for its $100,000-plus price tag – despite Musk's stated goal to transition from a high-cost, luxury car to a low-cost, mass-market car in his 2006 *Master Plan* – 2,400 Roadsters were manufactured before being discontinued. One now orbits the Sun after Musk fashioned his own red Roadster as a unique dummy payload in a 2018 SpaceX Falcon Heavy test launch, driven by a mannequin called Starman.

Based on the ground-hugging Lotus Elise body, the proof-of-concept Roadster solved many early engineering problems, successfully employing for the first time a lightweight, longer-lasting battery pack for intra-city driving that also included cell-level fusing to avoid overheating, in situ cooling in a thermal management system, and eventually a titanium safety shield between the flat-chassis battery pack and the vehicle cab. Most importantly, the AC Propulsion *tzero* "powertrain" was all-electric, in effect powered by Nikola Tesla's AC induction motor.

Founded in 2003, Tesla Motors would single-handedly reinvent the electric-car market with its high-tech, start-up ideas and ambitious green thinking. Historically, a company tries to increase market share by touting new designs and state-of-the-art technology, but Tesla appeared to have another goal, even making company patents publicly available to the competition to help develop clean technology, a move that admittedly also freed itself from paying for other future innovative technology. Having spent $50 million prior to the 2008 rollout of the first Roadster, Musk provocatively asked one simple question: "Are we making a difference in the world?"

Tesla's green ethos was the envy of the automotive world, mimicking GM's entry into automobile manufacturing a century earlier when it sliced into Ford's seemingly unassailable market share, not by doing what its competitors were doing but by doing what they weren't – no longer offering incrementally different annual models but an environmentally friendly *wow* car designed to solve the problems of getting from A to B *and* the economics of oil.

Combining an Edison-like diversity of interests with Ford's steely eye for product development, Musk was seen as the quintessential twenty-first-century entrepreneur: aspiring and conscientious. Undermining the idea of a "ladder of consumption" by financing from the top down, Tesla created an entirely new market under Musk's guidance, waiting until the high end was won before concentrating on the "democratization of the automobile," knowing that an all-green car would beat *any* car despite the price tag, while an *affordable* all-green car could redefine the world.

The road to invention is never easy, however, and nor is changing the world. After spending $100 million in 2 years and building only 100 Roadsters, Tesla Motors almost went bankrupt because of ongoing battery problems, a faulty two-speed transmission design, and excessive production costs, surviving only by raising the price on undelivered pre-sold cars and then doubling down on its next model, pre-selling thousands more. While some believed the flashy upstart was overextended, Tesla's second electric vehicle, the Model S, saved the day, allowing Tesla to go public in 2010 with a quarter-billion dollar IPO and a hefty new cash flow. A $50 million infusion by German carmaker Daimler also helped after it bought 10% of the fledgling enterprise in May 2009 in a deal to electrify the first Smart car.[25]

Following on from the ground-breaking 2008 Roadster and with the aid of a $465 million US Department of Energy loan used to acquire an abandoned, 500,000 m^2 GM-Toyota assembly plant in Fremont, California, Tesla began delivering the Model S in 2012, the world's first all-electric sedan based on the Mercedes CLS – the chassis designer called the Model S "just a CLS with a Roadster motor and batteries stuffed wherever they could put them."[26] The Model S was rated "Car of the Year" by *Motor Trend* for 2013, while *Car and Driver* awarded it "Car of the Century" in 2015. That same year, the Model X was launched, an SUV with falcon-wing back doors for easy third-row seat access. In the hyper-competitive car market, the SUV is essential, typically double the sales of a sedan.

Building on its enormous success (30% of all electric vehicles), Tesla eventually moved on from the niche luxury electric-car market it had so impressively won, launching the Model 3 in late 2017 at a more financially friendly $44,000 with an initial production target of half a million per year by

2018, and promising a basic, $35,000, lower-range version in 2019, although cars purchased outside the USA were much more expensive due to the higher freight costs to distant foreign markets. Called "the biggest consumer product launch ever," the Model 3 racked up an estimated $10 billion in sales in the first 2 days, ultimately becoming the world's best-selling electric vehicle.

Extraordinarily, in early 2017, Tesla became the highest-valued American carmaker at $50 billion, despite having built only 76,000 cars the previous year, compared to 7.5 million by GM and 6.4 million by Ford, and not having turned an annual profit. GM would soon reclaim the crown after introducing more late-to-the-game electric models, but Tesla would retake the top spot in 2020, besting VW and Toyota, despite selling one-tenth as many cars. In December 2020, Tesla Inc.'s stock (TSLA) was added to the S&P 500 index, the highest-valued company ever added.

There is no denying the sleek design of a Tesla. Versions come in "performance" (P) mode (faster with shorter-range battery) and "long-range" mode (slower with longer-range battery), although one can have both if you pay, while dual-motor (D), all-wheel drive is optional. Available battery capacities run from 60 to 100 kWh as indicated on the back right trunk door on some cars, good for about 300 miles (30 kWh ~ 100 miles).

The Model 3 uses larger 21700 Li-ion batteries that increase storage capacity by more than 25%, although the S and X models continue to use the original 18650 batteries. At the start of 2022, the first Panasonic 4680 batteries were delivered to Tesla for the model Y and pilot production of Tesla's first truck. Ever the prankster, Musk initially wanted to call the Model 3 a Model E to spell out SEX (and eventually SEXY) in his all-electric car line, but the Model E was already taken by Ford. In homage to the 1984 cult film *This Is Spinal Tap*, the volume-control systems on all Teslas go to 11, underlying Musk's humor along with his inventiveness.

***

Having brazenly taken on the major players and pushing for total dominance despite the huge challenges of starting up a new car company, the 2008 Roadster and subsequent models prompted other carmakers to develop their own electric cars, the underperforming Nissan even staking its future on the electric-vehicle market. Carlos Ghosn, former CEO of Nissan, summed up his company's revamped green thinking at the 2010 global launch of its first all-electric car, the Leaf: "The solution is mass-marketed, zero-emission mobility on a global scale."[27] Indeed, one doesn't change the world by thinking small. With plans to make 150,000 electric vehicles and 200,000 batteries per year,

Nissan built a production plant and battery factory in Smyrna, Tennessee, and now happily calls its flagship Leaf "the first mass-produced all-electric car." Despite having built the first modern electric car – the doomed EV1 – GM struggled to keep up and was eventually bailed out by the US government to the tune of $30 billion after the global credit crunch of 2008. Finally launching its first plug-in hybrid, called the Volt, in 2010 and a longer-range, mass-market, all-electric Bolt in 2017, GM oddly discontinued the Volt at the end of 2018 – Chevy's top-seller in the Bay Area[28] – exiting the hybrid business altogether, ostensibly to concentrate on all-electric autonomous vehicles, such as "robot-axis" and ride sharing (GM would also abandon the Bolt in 2023).

Despite taking the early lead in hybrid and electric-engine technology, GM was scooped by Toyota, Hyundai, Nissan, and other smaller companies, losing valuable market share by being too slow and unwilling to prioritize fuel efficiency in the early 2000s. One wonders if their thinking was more about the politics of oil, a former assistant US Department of Energy secretary calling GM's strategy "one of the major blunders in automotive history."[29] As a leading opponent of fuel-economy standards over the years and having spent millions lobbying Congress on behalf of gasoline-powered cars, some wondered if GM was in the electric-vehicle biz just for the green cred or to meet average fleet-emission regulations so they could keep on selling their high-volume, gas-guzzling pickup trucks and SUVs. *The Detroit News* even reported that GM expected to lose up to $9,000 on every Bolt sold.[30] Raising questions about its real intentions, GM also supported the Trump administration's attempts to overturn California's clean-air standards.

The stakes are enormous with more than one billion cars worldwide and another billion expected over the next two decades. More disruptive year on year with 5% of new car sales in 2020 and almost double that in 2022, going electric is changing the economics of transport as well as the rules of the road: low center of gravity (safer), shaped to maximize range (improved aerodynamics and lightweight construction), full-power instant throttle (better and safer passing), less maintenance (reduced tune-ups, oil changes, and gunk), and most importantly no exhaust or toxic fluids. We are at the start of a revolution revolution: zero emission, zero noise, zero dirt. As the Wizard of Fremont noted in typically brash fashion, "Driving a gasoline car is going to feel like yesterday."[31] Maserati China's managing director stated the reality less poetically, "EV cars are actually more exciting. The torque is higher and the speed."[32] Two early electric vehicles (the Chevy EV1 and Tesla Roadster) are shown in Figure 6.1 followed by lithium-ion battery-pack sizes for a Nissan Leaf, Chevy Volt, and Tesla Model S in Figure 6.2. Table 6.1 lists the

Table 6.1 *A few paradigm-changing electric vehicles through the years (company, model, battery, range, top speed, and cost)*

Year	Company	Model	Battery	Range (miles)	Top speed (mph)	Cost ($)
1901	Porsche	Mixte	Lead–acid	124	22	1,400
1954	Autoette	Golfmaster	Lead–acid	10	14	990
1996	GM	EV1	Lead–acid	70–90	80	34,000
1997	Toyota	Prius	NiMH	11	112	20,000
2008	Tesla	Roadster	Li-ion	200+	125	100,000+
2010	Nissan	Leaf	Li-ion	75	92	29,000
2018	Tesla	Model 3	Li-ion	250	130	44,000

(a)   (b)

Figure 6.1 Two early EV stalwarts: (a) 1996 GM lead–acid powered Chevy EV1 (*source*: Rick Rowan, CC BY-SA 2.0) and (b) 2008 lithium-ion-powered Tesla Roadster.

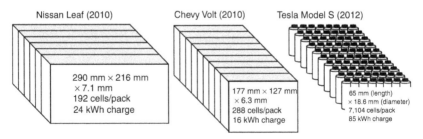

Figure 6.2 Lithium-ion battery pack of first-gen Nissan Leaf, Chevy Volt, and Tesla Model S.

specifications for major new models on the market since the lead–acid Porsche Mixte in 1901.

In *Revenge of the Electric Car*, a 2011 sequel to the 2006 documentary *Who Killed the Electric Car?*, journalist Dan Neil summed up the new thinking about automotive technology: "I spent my childhood driving fast cars, working on them, writing about it. I love gasoline horsepower, but I have come to the conclusion that I will never buy another gasoline-powered car as long as I live."[33] Importantly, electric vehicles are cheaper to drive, while connectivity and autonomous driving will only add to the dust heap of the past.

Stanford engineer and author Tony Seba believes the gasoline engine has run its course and will soon be unable to compete with electric vehicles. He stated that "gasoline cars will be the 21st century equivalent of the horse carriages by 2030,"[34] referring to the next generation of cars as "computers on wheels" (even phones on wheels!) and believes the tipping point will be a 200-mile-range electric vehicle priced at under $20,000: "We went from film cameras to digital photos in a couple of years. What the resource-based industries don't get is that technology adoption happens in an exponential manner. It never happens in linear fashion."[35] Of course, a car comes with a heftier price tag than a mobile phone, but for many is still an essential must-have possession.

Seba also predicted that all modes of land transport – cars, buses, tractors, and vans – will be electrified by 2025, creating chaos for both the oil and automotive industries as oil prices are halved to $25 a barrel and traditional manufacturing is transformed. A complete overhaul might not come as fast as some have hoped or predicted, but the electric dominoes are falling. Electric vehicles may even have been the tipping point at a 2019 United Auto Workers (UAW) strike at GM and subsequent 2023 strike at GM, Ford, and Stellantis, the disagreement more about saving jobs than money because electric vehicles are much easier to assemble – a Chevy Bolt has 125 fewer parts than a VW Golf[36] – while a Tesla can be rolled off the factory line in 10 hours. As electric vehicles start outselling gasmobiles by 2030, the UAW expects a loss of 35,000 jobs.

First demonstrated in 1831 by Michael Faraday at the Royal Society of London to generate electricity from motion, before being practically engineered and patented in 1896 by Nikola Tesla to produce motion from electricity, the principle of an electric motor is not rocket science: a current produces a magnetic field in an outer coiled stationary "stator" forcing an inner magnetized "rotor" to turn, essentially the reverse of an electric generator (Figure 6.3).[37] No need for hundreds of greased moving parts, spark plugs, exhaust systems, mufflers, toxic lubricating fluids, endless fixes, or dirty fuel. And no need for hundreds of thousands of patents.

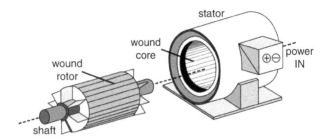

Figure 6.3 Induction motor schematic with coiled copper windings in the stator and rotor.

The oomph is instantaneous, so you don't even need to change gears to reach top speed – just put your foot down on the "Go" pedal. An electrical engine also weighs about half that of a gasoline engine, while the power is about double, making electric propulsion as exciting as it is clean. Although the batteries add weight, they can be evenly balanced throughout the car as ballast to improve handling.

Perhaps most importantly, the efficiency of an electric motor is higher than a gasoline/petrol motor, where two-thirds of the energy of a gasmobile is lost just burning the fuel. As noted by Martin Eberhard, one of five official co-founders and first CEO of Tesla, "If you took the energy in a gallon of gas and used it to spin a turbine, you'd get enough electricity to drive an electric car 100 miles."[38] No wonder the oil companies are twitchy – a 100-mpg (equivalent) mass-market electric vehicle will destroy the more-than-century-old market for gasmobiles.

The electric engine is not only more efficient; electricity is also much cheaper than gasoline. To compare like with like, we convert the fuel costs to miles per dollar: an American gasmobile averages roughly 25 mpg or about 7 miles per dollar (at $3.50/gallon), while the average electric vehicle gets about 4 miles per kWh or 40 miles per dollar (at $0.10/kWh), about 6 times better for the consumer. According to a 2018 University of Michigan study, the average electric filling costs per year are less than half that of a gasmobile ($485 EV/$1,117 gas), while the required fuel economy for a gasmobile to be cheaper than driving a battery-powered car in the USA is 57.6 mpg.[39] Indeed, many drivers have reported reduced fuel costs up to 90%, even more if low-rate overnight home charging or free charging is available. Outside the USA, savings are even higher because minimal gasoline taxes keeps prices lower for American consumers than in other high-tax regions.

A new industry argot is forming as we learn the basics and jargon of induction motors and charge storage: with gasmobiles we had engine perform-ance (hp), cylinder speed (rpms), fuel efficiency (mpg), while now we talk in torque (Nm), battery capacity (Wh/kg or kWh), charging times (for example, trickle or fast charging), and range (km/kWh). We can still be in love with our car as in Queen's 1975 anthem to revved-up muscle machines, but there is no longer the need to buy a new carburetor to go cruising in overdrive. The clean machines have arrived, better for the pocket and the environment.

6.3 A New EV World: The Future Is Already Here

In the fast-changing, electric-vehicle (EV) field, TLAs (three-letter acronyms) abound and are now part of the everyday propulsion lexicon. An old gasoline/diesel car battery is an SLI (starting, lighting, ignition), HEV a hybrid electric vehicle, BEV a battery electric vehicle, AFV an alternate fuel vehicle, NEV a new energy vehicle, and LIB a lithium-ion battery. The AFLA (another four-letter acronym) is also popular: PHEV is a plug-in hybrid electric vehicle, BOEV a battery-only electric vehicle, while FCEV is a fuel-cell electric vehicle.

All major carmakers now offer an EV in their standard line, the earliest off the mark Tesla S/X/3, Nissan Leaf, Chevrolet Volt/Bolt/Spark, Mitsubishi Outlander/i-MiEV, Renault Twizy/Zoe, BMW i3/8/Active-E, Mercedes B, Volkswagen E-Golf/E-Up, Ford C-Max/Focus, Audi A3 e-Tron, Kia Soul, Fiat 500e, Hyundai Ioniq, Honda Clarity FSX, Opel Ampera-e, and Mahindra e20.

By 2016, the world's top-selling EV was the 5–7-seater Tesla Model S with impressive specs: 0 to 60 mph in 6 seconds (even faster in "Ludicrous" mode), a range of around 300 miles (475 km) – almost enough to get you from San Francisco to LA on a single charge – hip recessed pop-out door handles, and all for $71,500. Next was the more pedestrian, five-seater Nissan Leaf: 0 to 60 in 10.2 seconds, an 84-mile range (135 km) suitable for everyday inner-city driving, and a much more affordable $29,000 sticker price, while the best-selling PHEVs included the Chevy Volt at $33,000, the Mitsubishi Outlander at $24,000, and the granddaddy Toyota Prius at $24,000 (almost 4 million units sold in its first decade).

Today, one can buy an EV for as little as $12,500 – the Renault Twizy, a 17-hp, fully electric two-seater with a 50 mph (80 km/h) top speed and 50 mile (80 km) range, ideal for an urban runabout if you don't mind the micro size (launched in 2012) – or as much as $135,700 for a top-tier 228-hp, four-seater PHEV BMW i8 (launched in 2014). *EVObsession* regularly rates the EV market and provides monthly sales figures. In 2016, 600,000 EV units were sold in the three main markets: China (280,000), Europe (210,000), and the USA (110,000).[40] In 2020, the number of units sold passed 2 million (~2%) for the first time, while 2 years later more than 8 million all-electric and hybrid vehicles were sold. Today, more than 300 EV or hybrid models are on offer.

Tesla still gets most of the press with thousands of pre-orders per day that are impossible to fill. At the Model 3 prelaunch, Tesla's ever-ebullient CEO sang the praises of its more affordable, easier-to-make, "mass-market" EV: 0 to 60 mph (0 to 100 km/h) in under 6 seconds, a range of 215 miles (346 km), supercharging capability, and comfortable seating for five adults with room for a 7-foot surfboard.[41] Not quite for everyone at a still hefty $44,000, but at least availability was more assured with increased production, thanks in part to an improved cash flow from $14 billion in sales the first week. Teething problems continue, however, as the carmaker with a conscience tries to tame the intricacies of profits and people on an industrial assembly line and the challenges of increased workforce automation.

Tesla initially aimed to produce 500,000 cars annually by 2018 (10,000/week) to compete with the established high-volume manufacturers, but scaling up from thousands to millions isn't easy. Despite the difficulty for a new carmaker in a crowded automotive world, Bloomberg noted that "Tesla created something no other automaker could claim: an electric car that hundreds of thousands of people lined up to buy. The only problem, at least in the beginning, was that Tesla couldn't produce enough of them."[42] By the end of 2018, Tesla had made only 125,000 Model 3s, well below consumer demand and company expectations. After a price cut in March 2019 from $44,000 to $35,000 for a 220-mile-range Model 3, in part to counteract the phase-out of a federal EV subsidy, Tesla again stated that sales could reach 500,000 a year.

To speed delivery and save money, new Tesla 3s were available only online and without a test drive, although a car could be returned for free after 7 days or 1,000 miles, underscoring how a century-old industry is being transformed into a high-tech business. In a fast-paced, Internet-driven marketplace, cutting out the middle man is just smart business. By the end of 2020, Tesla eventually reached 499,550 Model 3s, tantalizing close to its milestone half-million goal. By 2022, annual production was over 1 million. Time to start selling shares in Midas, Mr. Transmission, Mr. Lube, and other long-established service companies – Tesla repairs are mostly fixing faulty battery cells.

As the upstart hi-tech usurper takes on the titans of transport within the car industry, Tesla's share price is instructive. Highlighting the rapid rise of a quintessentially twenty-first-century technology company posing as a carmaker, Tesla stock (TSLA) is shown in Figure 6.4 since its 2010 initial public

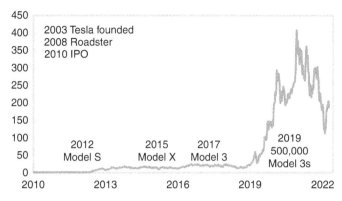

Figure 6.4 Tesla (TSLA) share price from 2010 to 2023 ($US).

offering, the first car company to be listed in 50 years. A 100-dollar investment at Tesla's inception in 2010 would fetch almost $3,000 a decade later, while in October 2021 the company was valued at over $1 trillion, buoyed by an order of 100,000 vehicles from the car-rental giant Hertz and equaling as much as the other top-10 car companies combined. In 2022, however, supply-chain problems, missed production targets, and increased competition (for example, VW's ID.7) saw the stock price drop by 50%. Despite the losses, no one was predicting Tesla's ultimate demise any longer.

***

To increase EV adoption, governments are doing their bit to grease the electric wheels. Various incentives encourage sales, such as rebates and tax credits that significantly reduce the showroom sticker price or, as has become more common, the online click price. In the USA, a federal rebate offers up to $7,500 for a full EV, while California kicks in another $2,500, which together cover the cost of the battery – a 100-kWh battery at $100/kWh costs $10,000 (good for about 300 miles of range).[43] Ontario initially offered rebates up to $14,000, although the policy was repealed after a 2018 change in government, underscoring the politics of resistance. In the UK, company-purchased EVs are tax-exempt for a year, while in the Netherlands most EVs are company-leased thanks to attractive government subsidies.

Roadside charging is still a problem because of limited filling infrastructure, although enterprising businesses and green-minded governments are providing more charging points to allay nagging range anxiety, while phone apps and onboard computers help locate the nearest open charger. For those who can't recharge at home or work, the dreaded range anxiety is, in fact, charger anxiety if the distribution of charge points is insufficient. Who owns the charge points as well as the etiquette of sharing roadside stalls is still being worked out: to be "ICEd" is when a gasmobile parks in an EV-only spot, the height of modern roadside rudeness.

Charging incentives are also enticing new EV buyers. Spanish drivers can charge for free at a number of charge points (maximum 3 hours), park for free on city streets, and skip various highway tolls. In Oslo, where EV adoption has already passed 50%, free tolls, free parking, and no VAT are available. EVs can also access the fast lane during rush hour in California and other states, while free charging, special parking privileges, and preferred lane access are also available in other jurisdictions.

In 2013, Tesla began installing a network of high-voltage "Supercharger" stations across much of the USA and parts of Europe, free to all Tesla drivers and a major incentive after almost going bust again when its envious pre-ordered sales failed to close.[44] The stations were intended to come with "solar canopies" that would generate more energy than the car could use, but rolling out the electric part of the infrastructure was challenging enough. A standard 250-kWh V3 supercharger can now fill a 60-kWh EV in 15 minutes or add 50 miles of range in under 5 minutes – not yet as convenient as liquid fuel, but no longer a long-distance travel deterrent.

Even without the solar-powered charging stations, Tesla had installed enough of a national charging network by 2014 to attempt a transcontinental trip across the USA. Driving from Los Angeles to New York, a company rally crew completed the all-electric, 3,427-mile journey in 76 hours and 5 minutes, saving an impressive $435 in gas.[45] The next year, another crew in a Tesla Model S P85D established a Guinness World Record for "shortest charging time to cross the United States in an electric vehicle" (12 h, 48 m, 19 s). By comparison, the first gasoline-powered, US cross-country journey in 1903 took 63 days, suffering from limited fueling infrastructure and bad roads, harkening back to Bertha Benz's long-distance jaunt in 1886 from Mannheim to Pforzheim when she had to stop at various local pharmacies to fill up with Ligroin fuel. Further highlighting how infrastructure lags invention, Lt. Colonel Dwight Eisenhower's 1919 "Through Darkest America with Truck and Tank" convoy from Washington, DC, to San Francisco took two arduous months to complete.

Six years after the first Supercharger was demonstrated in Hawthorne, California, Tesla had increased its charging infrastructure to 12,000 stalls over a global network of 1,400 sites (automatically adapted to host-country voltage), turning range anxiety into range envy. Some are now solar-powered, saving millions of gallons of gasoline and tens of millions of tons in burnt fossil-fuel emissions. The competition is slowly catching up, while in 2021 Tesla announced it would open its charging stations to other car brands, ending the so-called "walled garden" and providing another envious revenue stream. As of 2024, both GM and Ford could access 12,000 of Tesla's 19,000 charge points, using its smaller and more powerful North American Charging Standard (NACS) instead of the Combined Charge System (CCS).

In 2021, the Biden administration announced a federal plan to install 500,000 charge points across the country by 2030 at a cost of $7.5 billion, more than tripling current numbers and egging on the private sector to follow suit. Mired in the politics of the day, the budget was eventually approved as part of the wider 2022 Bipartisan Infrastructure Law. Ideally, the network of national charging stations will also incorporate more public space into the driving experience, making long-distance travel even greener. Once fast-charging capability matches the simplicity of liquid refueling, cross-country driving will be no harder than it is today.

***

Most EV charging stations use a standard 240-V system with a three-prong plug similar to a clothes dryer, internally converting AC to DC, and can top up a battery in 6 hours via "trickle" charge, ideal for commuting workers or overnight. Some outlets provide 480-V "fast chargers" that can charge up to 80% in 30 minutes, providing 300 miles of range per hour (RPH). Fast charging is becoming more common, but increases battery degradation, especially in hot climates.

Fast charging protocol (FCP) and slow charging protocol (SCP) is now standard for EVs, although exact charging times depend on the company, type of station, and available grid power for rapid charging, while range depends on driving speed, conditions, and terrain.[46] Tesla supercharging adds 50 miles of range in under 5 minutes and can top up in 15 minutes as do other proprietary fast-charging systems. Charging more than 80% on the road is inadvisable, however, because the extra charge doesn't justify the extra time – the last 1% can take 30 minutes and incur an overstay fee.

Home charging is difficult for those without garages or driveways, but EV stations are becoming mandatory in many new buildings, while hotels and motels provide the obvious location for charging stations. To encourage sales, overnight charging is free in some areas, likely only for a limited time and which governments will eventually charge for or tax, probably via a distance-based formula rather than a percentage of the fuel. A typical 100-mile "fill-up" might cost about $3, for example, 30 kWh × $0.10/kWh.[47]

Because of the limited availability and slower filling times, most EV charging is still done at home, estimated at about 98%, but more electronic charging station (ECS) options are now available as EV adoption increases, both for slow (5–22 kW) and fast (50–150 kW) charging. Other than a dedicated, slow-charging, single-unit, home charger with free "granny" cable that comes with the vehicle, one can charge at a shared wall- or pillar-mounted multi-unit, a marked roadside street charger, electrified street furniture such as a retrofitted lamppost (especially important if residential off-street parking is not available), a branded "destination" charger such as at a local shopping mall (typically free to customers), or a highway electric "pump" at a converted roadside service station or quick top-up parking bay (more expensive with overstay fees), as well as at a private company charging station such as a Tesla supercharger (V3 is 250 kW).[48]

Despite being used in the 1900s in the first electric taxis, swappable battery packs have been less successful, bankrupting at least one company hoping to incorporate a quick-change option into the evolving EV mix. Tesla discarded the idea of battery swapping because of a lack of service infrastructure, initially planning to offer a quick-fix charging solution in 90 seconds for a fee. The high-tech Silicon Valley industry model is good at systems, but isn't wired for service, a standard money-maker for branded garage stations over the last century. Service with a smile in the vast network of yesterday's "full-service" stations has been replaced by remote "over-the-air" software upgrades.

Of course, cautious consumers wonder what's down the rosy renewable road if power companies increase electricity rates without a filling alternative as with expensive residential electric heating. The goal is to compete with gasoline by reducing the overall cost of driving and let the market decide, but there are still concerns about tying oneself to the grid and giving more power to utility companies. Fortune 500 financing and venture capital may provide seed money to kick-start a nascent battery industry and charging infrastructure, but once the goods are in place investors will seek to maximize profits. Understandably, consumers are wary of having to pay through the nose when the dust settles or if governments look to collect more tax as they usually do.

EVs also remain a luxury for most consumers as do fast home-charging systems. Half of a government fund to promote EV sales in the Netherlands went to Tesla and Jaguar drivers, netting 12,500 of 25,000 vehicles sold, prompting one politician to state that "prosecco-drinking Tesla drivers" were profiting at the "expense of the ordinary man in the street."[49] Not surprisingly, Tesla sales rose 260% thanks to the Dutch government's environmental largesse. A 2022 Irish transport study found that EV grants and charging locations also favored high-income people in urban areas, essentially "luxury goods."[50]

EVs have a long way to go to be affordable, especially with gasoline-powered vehicle bans coming as early as 2025 in some cities. Pollution and emissions costs are high, but can the regular consumer afford the alternative to the internal combustion engine of a tried-and-true gasmobile?

The ultimate high-tech Uberman and gadget whiz James Bond now drives a $250,000 all-electric Aston Martin Rapid E, capable of 300 miles of range in an hour, 950 Nm of torque, and a sub-4-second, 0-to-60-mph start, alas, available only to a select few. Amusingly called "The Spy who Plugged Me" by an English tabloid, gasoline/petrol cars were also expensive at the start, but without inexpensive alternatives for everyday consumers, the market will continue as a rich man's game. Increased sales follow improved engineering – first we prove the technology and then we make the product better, but to increase adoption one obviously needs to offer cheaper options and improve infrastructure. The destination is no good if one can't afford the journey (spyware optional).

***

Clearly the EV market is being built on the go, but the genius of Tesla's strategy was the subsequent foray into charge storage for residential and industrial use, another potentially massive industry. In 2020, the International Energy Agency called for 10 TW of batteries and other types of energy storage by 2040 to meet sustainable energy and climate goals, 50 times current market size.[51] Not only an essential part of the electric-car business, charge storage is also reinventing the grid (as we saw in the last chapter).

Derived from its Model S Li-ion battery, Tesla released the $3,500 Powerwall in 2015, making large-scale, on-demand household power readily available to consumers, although prices were increased after demand greatly exceeded supply. Roughly the size of a refrigerator door, the 3 foot × 4 foot × 7 inch wall-mounted battery pack weighs 214 pounds (97 kg), has a 10-year lifetime, and stores 6.4 kWh of energy, sufficient for most household daily use.[52] Packs can also be chained together to store more charge as needed. By comparison, EV batteries store about 30 kWh/100 miles, in essence its own Powerwall, providing another lucrative opportunity to sell charge to others (as we will see).

Tesla then acquired SolarCity in 2016, the one-time darling of Wall Street and largest American manufacturer of solar panels at the time, announcing plans to achieve 33% more power with low-cost Silevo panels.[53] Coupled with a Powerwall, the storage part of the clean-energy puzzle can easily be pieced together as "augmented" solar-power for green-minded consumers. Helped by Tesla's high-tech expertise and experience, SolarCity's business model changed from expensive marketing, advertising, and door-to-door sales to a more slimmed-down, web-based Internet operation.

To kick-start the mass production of vehicle and residential-scale energy-storage systems, Tesla then built a $5-billion battery "Gigafactory" in the desert just east of Reno, Nevada, $1.6 billion in investment coming from long-time battery partner Panasonic. Having again one-upped the competition, Tesla now boasts an impressive line-up of electric vehicles and solar panels with enough battery capacity to drive an expanding fleet and store terawatt-hours of on-demand energy.

Fittingly, the world's first Gigafactory (GF1), a.k.a. Giga Nevada, has the world's largest building footprint at 10 million square feet, slated to be 100% site-powered by a 70-MW rooftop and ground solar array. Production began in 2017 with 150 GWh of batteries, good for 1.2 million cars and 10,000 jobs.[54] Putting everything under one roof cut battery costs by about one-third and assembly costs by 10%, thanks to an improved economy of scale (roughly one-fifth of a car's sticker price is the battery). Annually producing more Li-ion batteries than were made in 2013, half of Giga Nevada's manufactured battery output is designated for energy-storage packs, hastening Tesla's move from sustainable transport to sustainable power generation and its emergence as a multi-billion-dollar *energy* company.

In 2017, Tesla opened GF2 near SolarCity's headquarters in Buffalo, where it makes a line of solar roof tiles, and GF3 in 2019 south of Shanghai, a.k.a. Giga Shanghai, where it began rolling out EVs for the Asian market. GF4 was announced in 2020 for Grünheide, Germany, a village southeast of Berlin, which began operations a year later in conjunction with Tesla's first full-assembly European factory. Giga Berlin was notably not built near the major German carmakers – BMW in Bavaria, Volkswagen in Lower Saxony, or Daimler in Baden-Württemberg – all of whom are gearing up to challenge Tesla's dominance. Tesla's Berlin factory even cut back on costs, avoiding pandemic supply-chain problems to assemble an EV in a third of the time compared to a Volkswagen, while eliminating Germany's codetermination system where workers are represented on the board.[55] Located in Austin, Giga Texas (GF5) started up in 2022, followed by GF6 in Quebec near to a dormant lithium mine 550 km northwest of Montreal. In 2022, Tesla also announced another Shanghai manufacturing plant near to its current factory that will more than double EV production in China to 2 million cars annually, supplying the Asian, European, and Australian markets.

None of Tesla's moves are surprising, certainly not to those who have followed the company's progress since its inception. CEO Musk made clear his intentions to transition to solar and then battery storage via electric vehicles after achieving essential technological advances and desired economies of scale. His original 2006 *Master Plan* stated "1. Create a low volume car, which would necessarily be expensive; 2. Use that money to develop a medium volume car at a lower price; 3. Use that money to create an affordable, high volume car; 4 And … Provide solar power," while his 2016 *Master Plan Part Deux* proposed to expand the blossoming EV product line to "Create stunning solar roofs with seamlessly integrated battery storage."[56] Build a better mousetrap and the world will beat a path to your door. Build a better solar roof and battery system and you will have enough on-demand power at your fingertips to change the world.

Thanks to the emergence of large-scale battery storage, self-powered homes and vehicle–grid integration are no longer science fiction, never mind just competing with gasmobiles or replacing the internal combustion engine. As noted by Steve McBee, CEO of NRG Home, one of SolarCity's rivals, "If your goal is to build a meaningful solar business that is durable over time, you have to assume that the solar business is going to morph into a solar-plus storage solution. That will be mandatory at some point."[57] Indeed, the road to the future passes directly from electric vehicles and solar panels to in-house charge storage.

Despite starting from zero and lacking any large-scale production capacity, Tesla continues to maintain a competitive advantage in the EV world, primarily by manufacturing with an EV-only platform rather than jimmying existing combustion architecture. An industry report noted that the Tesla Model 3 was "years ahead of its peers" with "next-generation military-grade tech."[58] In keeping with the original plans, Musk also reiterated his planet-saving ethos at the 2019 Model Y launch, a $39,000, 480-km range compact SUV aimed at the so-called "crossover" market – is it a sports utility vehicle (SUV) or a compact utility vehicle (CUV)? The revolution is rolling on when the debate is about the pros and cons of luggage space, headroom, and aerodynamic efficiency (that is, range) in the latest EV model rather than whether the gasmobile is still king.

The real challenge, however, is to provide a car for more than just the rich. Henry Ford was able to cut Model T prices by over a half with improved high-volume production, creating affordable private transportation for middle- and working-class families ($850 in 1908 slashed to $360 in 1916[59]). With other established carmakers getting in on the act, the low-end EV market is there for the taking, but luxury can only become utilitarian when the tools become affordable, not unlike original gasmobiles as playthings for the rich before turning into essentials for all. Alas, a low-cost Tesla has yet to materialize, beset by supply-chain issues (especially chips), core profitability, and lack of manufacturing.

Nonetheless, the electric landscape continues to grow each year, despite the challenges of ramping up from zero to competitive in a notoriously difficult market. Once a major obstacle to EV adoption, batteries are now transforming the industry. As electricity prices drop below $0.10/kWh, batteries will change both the EV and charge-storage market to bring more on-demand tailored power to consumers, while cumbersome recharging and range anxiety is no longer a rate-determining step to electrify the road.

With ongoing improvements in the energy density of Li-ion battery cells (which we'll look at later), battery pack costs continue to drop, called "the end of the beginning" for EV adoption. In the decade since 2010, energy density has almost tripled (100–300 Wh/kg) while costs have dropped almost 90% (1,183–156 $/kWh).[60] With the same 500-kg battery pack, more energy can be extracted, generating more range, while lighter packs provide the same stored energy and range (for example, 100 kWh for over 300 miles), electrifying anything and everything that moves.

The transition to all-electric transport is just beginning. At the same time, our roads are becoming cleaner, while the potential exists to reduce GHG emissions with more green-powered electricity and on-demand charge storage. Edison and Ford would be proud to see their ideas in action more than a century after they first dreamed of building an all-electric car.

6.4 The Revolution Revolution: Exhaust Optional

For now, Tesla, Nissan, and Toyota are leading the EV way, pioneering engineering advances and expanding sales, while futuristic innovations are being incorporated into everyday driving, such as auto steer, adaptive cruise control, voice navigation, auto parking for those who dislike reversing into tight spaces, and autonomous driving, all of which make a car safer than under human control.

As computer and smart-phone technology become more integrated into standard automobile design, however, another company wanted to bring even more change to the industry. With a market capitalization over $1 trillion, the world's highest-valued company prior to Saudi Aramco's 2019 listing, Apple brought "killer-app" software to both the portable music player and phone industries. Intending to bring advanced software and decades of high-tech experience to the automotive industry, Apple also thought about putting a car on the road by 2020 in a hush-hush venture called Project Titan. Not such a far-fetched idea as former

Apple VP Tony Fadell noted: "A car has batteries; it has a computer; it has a motor; and it has mechanical structure. If you look at an iPhone it has all the same things."[61] Rather than starting from scratch as in other past ventures, Apple hoped to piggyback off an existing company, initiating talks with BMW, Daimler, and McLaren, although no new Apple car has yet to make an appearance. As one analyst stated about the challenge of translating past technological successes into automotive success, "it wasn't a huge jump to go from the Mac to the iPod to the iPhone. But you can't take an iPhone person and say, 'go make a car.'"[62]

Nonetheless, high-tech innovations are reinventing the industry as Detroit, Munich, and Yokohama all get a Silicon Valley makeover, while Chinese manufacturers with little or no automotive experience only a decade ago are turning out affordable, mass-market "batteries on wheels," some with wired chill-out luxury. It has become much easier to build an EV today, as noted by Joost de Vries, CEO of the DeLorean Motor Company, which announced in 2022 that it too was jumping onto the EV bandwagon to give a new look to an old brand: "15 years ago there was no supply chain, there was no contract manufacturing, so everything had to be vertically integrated. . . . The barriers to entry from a technology perspective and a supply-chain perspective are much lower today."[63]

Next-generation electric vehicles will not only change how we drive but transform how we *think* about travel. With the advent of 200-mile EV range, self-driving (reduced accidents and efficient routing), and car- and ride-sharing schemes, General Motor's CEO Mary Barra stated in 2016 that mobility would change more in 5 years than in the past 50,[64] the economics of which are still up for grabs. Barra noted that 80% of commutes are less than 25 miles, 50% of all trips are under 5 miles, and that new ideas about how we drive are needed to reduce costs and waste.

According to the US Department of Transportation, the average American driver travels less than 37 miles (60 km) per day,[65] easily covered by an EV on one hour of charge. The changeover is taking longer than Barra first thought, but GM is lining up with other industry stalwarts to catch the next big automotive wave, investing in autonomous ride-sharing vehicles as well as planning for the day when a futuristic fleet of self-driving cars overhauls inner-city travel. However long the transition takes, an entirely new market is opening up.

Bloomberg estimated that "By 2040, more than half of all new car sales and a third of the planet's automobile fleet – equal to 559 million vehicles – will be electric."[66] If the estimates prove right the makeover will be astounding, from roughly one million EVs in 2016 (including less than 100,000 in the USA) to more than 500 million in a quarter of a century, a 20% increase per year. By comparison, it took until the 1930s for more than half of American families to own a gasoline-powered car, starting from 1% in 1910 in a similar growth pattern.[67] Fueling the increased uptake in EV sales, the EU has also set ambitious targets to reduce transport emissions by 60% by 2050 (measured against 1990 levels).[68] Soon, there will be no choice other than electric.

The founder of Singulato Motors – a young Chinese entrepreneur whiz-kid named Tiger Shen – sussed the transformative power of EVs for China after seeing the launch of Tesla's Model S in 2012, and thinks that software-controlled cars and 4- to 5-times more battery range in the next 20 years will spell the end of the gasmobile altogether.[69] Singulato's first production model, the iS6, was purposefully designed to help clean up smog-filled cities and reduce congestion in China, which has 8 of the world's 10 most-congested urban centers,[70] especially Beijing where the average driving speed is only 7.5 mph.[71] Selling his brand of the future as a car with a conscience, Shen noted, "It is our duty to get the blue sky back in Beijing and other cities in China."[72] On top of adding 5 million EVs by 2020, China is also aiming for 20% highly autonomous vehicles by 2025 and 10% fully autonomous vehicles by 2030 (a.k.a. smart EVs or SEVs).

To kick-start the transition to all-electric road transport, some countries have started banning gas-guzzlers, which contribute about 30% to global GHG emissions. With two-thirds of all oil consumption from cars and trucks, the stakes couldn't be higher.[73] In Norway, EVs already outsell gasmobiles with expectations of 100% EV sales by 2025, while Germany is aiming for 6 million EVs before 2030, by which time all new cars must be emissions-free. France announced it would phase out petrol-/diesel-fueled cars by 2040 in a bid to fight climate change, its ecology minister stating that the move was "a way to fight against air pollution" and a "veritable revolution," as did the UK soon after. In 2022, California announced it was banning the sale of gasmobiles by 2035, followed by the EU as a whole (with a German-led carve-out for so-called carbon-neutral e-fuels), while China announced it may ban all ICE vehicles even sooner, pushing anyone serious about the industry to ramp up EV production or risk losing out on the world's largest market.

The carmakers are gearing up in response, some more so than others. In 2017, Volvo announced it would manufacture only hybrid or all-electric vehicles from 2019, the first major car manufacturer to offer a complete electric retool in conjunction with the Swedish battery maker Northvolt. Volvo's Polestar 1 (hybrid) and Polestar 2 (all-electric) are now available, although still on the high-end of the market. Volvo's chief executive also signaled a likely end to the diesel engine, citing loss of sales. In 2018, Volkswagen announced plans to spend a "staggering $50 billion over the next five years in an 'electric offensive,'"[74] including $25 billion on batteries. VW's CEO predicted that Volkswagen would produce 2–3 million EVs per year by 2030 or 25% of company sales, and is already reaping the rewards with increased market share.[75] If VW can make good its ambitious goals, the EV market will no longer be limited by cost. China is also expecting to sell over 5 million EVs *per year* by 2025, including an "anti-Tesla" – the ultra-low-priced $9,000 Chery eQ[76] – while India will sell only electric cars by 2030.[77]

Pickups are particularly important to the American market. In 2019, Tesla's angular-shaped and *Blade Runner*-inspired electric Cybertruck was unveiled in typical flashy style with CEO Elon Musk throwing a giant ball-bearing through the window and waging a tug-of-war with a Ford F-150, netting 200,000 pre-orders in 2 days. Without having built one truck, the late-to-the-game challenger Nikola saw its share price more than double in 2020 after announcing plans for a commercial EV pickup, subsequently wiped out after a fierce short-selling attack. Finally recognizing the peril of Tesla's growing market share, Ford followed in 2021, announcing its all-electric F-150 – the most popular vehicle in the USA with about 1 million in annual sales – nabbing more than 150,000 reservations. A year later, GM announced its 2022, all-electric Silverado pickup, albeit with a hefty $60,000 price tag.

Underscoring the importance of the transition and growing competition, US electric truck maker Rivian went public in November 2021 in one of the largest-ever IPO evaluations at over $100 billion, despite having only made a few units. Setting up operations in an abandoned Mitsubishi plant in Normal, Illinois, Rivian is planning to build a competitively priced all-electric SUV (R1S) and truck (R1T). One of the so-called "Tesla wannabes," Rivian also hopes to start delivering on an order of 100,000 delivery vans from Amazon, which owns a 22% stake in the company.[78]

Increased EV uptake will naturally have its share of winners and losers. In part because of "Dieselgate," diesel cars are in fatal decline, and will soon be banned in most if not all European cities, its largest market, drastically cutting pollution in urban centers. Some form of buyback will be needed to help those who were encouraged to buy cheaper diesels only to have their resell value destroyed with a stroke of a government pen. Witness the regular disruptions in Paris by the so-called *gilet jaunes*, protesting increased diesel-fuel taxes imposed in 2018 to fund France's COP21 GHG-reduction plan. Highly urbanized countries have much to gain by going green and removing transportation pollution from their streets, but the transition must be fairly managed. A proposed end to a 50-year-old billion-dollar fuel subsidy in 2019 in Ecuador also sparked clashes between the government and indigenous rural communities. In early 2023, the EU as a whole banned the sale of all petrol/diesel cars and vans from 2035, giving sellers and buyers plenty of lead time to get ready, although one should expect problems and pushbacks.

Thanks to the Silicon Valley model, anyone with deep-enough pockets can try to break into the notoriously hard-to-crack industry. Tucker, Bricklin, DeLorean, and Fisker are among the most famous auto-industry roadkill, although Fisker resurfaced in 2020 with new financing and a competitively priced, outsourced SUV with flexible-length leasing, while the infamous gull-winged DeLorean reappeared from the past under new ownership (maximum speed 88 mph).

There is still much opposition, however, especially in the USA, the nay-sayers citing lack of demand, limited roadside recharging, and excessive regulations that decree how many EVs a company must make, a standard chicken-and-egg debate over whether infrastructure or demand comes first. The traditional carmakers have had their way for decades, but are begrudgingly being forced to adapt. Everything is going electric – bikes, scooters, unicycles, three-wheeled delivery vehicles, even rickshaws – while the legendary motor-cycle company Harley-Davidson announced it too would start making electric bikes. No one can afford to ignore the fundamental changes rippling through the transport industry.

***

Electric refits are also on the rise, which swap out an internal combustion engine for electric propulsion. HyPer 9 and Tesla are common EV conversion motors, adapted to the existing gearbox locked to one gear, no longer needed but cheaper to leave in. Series-wired batteries provide the juice, distributed front and back for balance, for example, eight 6.3-kW Teslas each made of 516 Li-ion cells. The overall weight is typically higher because of the extra battery weight, so the suspension may need to be adjusted, but the lower mount handles better while the horsepower is much higher. Older drum brakes may also need to be upgraded to disk brakes, essential with more power under the hood. A potentially massive future market, a refit changes more than just compression ratio to amps and fuel level to state of charge, but improves the whole feel of the drive.

Mike Brewer, co-host of the popular car show *Wheeler Dealers*, was instantly hooked after driving his first all-electric conversion – a VW van with a three-phase brushless AC motor that doubled the factory horsepower – chuffed by the increased performance and instant acceleration as he blew past the petrol competition in silence. Co-host and mechanic Edd China added to the allure for doubting petrol heads – "It is such a different and exciting experi-ence" getting "maximum power from the first rpm." Swapping out an old motor may soon become more than a hobbyist's weekend tinkering, although billions of makeovers will be needed to convert the existing rolling stock of gasmobiles, most of which are too valuable to junk (the average car on the road is 7–8 years old and lasts about 15 years[79]).

Mike Jackson, CEO of AutoNation, the largest automotive retailer in the USA, acknowledged the challenges over the next two decades with electrification, autonomous driving, and ride sharing, but expects 20% of vehicles sold in the USA to be fully electric by 2030, noting that "The industry is entering an epic new era, where for the first time through new technologies the industry can comprehensibly address the social cost of fatalities associated with our products." OPEC and the EIA both think oil demand will rise for another 15 years in Asia, while world demand will peak around 2030.[80]

Traffic is also reduced because of car- and ride-sharing schemes and better coordination between vehicles, drastically lowering car sales, perhaps by as much as 90%.[81] Sold to eager consumers for over a century, the dream of owning one's wheels is not as compelling in an age of shrinking urban space and expanding virtual worlds, while we are already seeing a decline in car sales as Millennials discard an outdated lifestyle that sold rugged individualism behind one's own set of wheels. Some wonder if we are already at "peak car" in the West as more drivers prefer access over ownership and the number of licensed young Americans continues to decline, down 25% from 1970 to 2010.[82]

In the end, economics will rule. It's no easy job to make cars in a labor-friendly way – an assembly line is inhuman by nature, green or otherwise – and with more companies vying for market share, the competition is stiff and the margins slim. Tesla is already experiencing growing pains with lower workers' pay and non-union contracts, while trying to ramp up production levels to that of VW, Toyota, GM, and Nissan, who all make about 25,000 cars per day. Nonetheless, a diverse mix of technologies will be with us for a while yet – gasoline/petrol, hybrid, hydrogen, propane, natural gas, all-electric – just as steam-powered manufacturing continued well into the early years of electricity.

China has benefitted most by slow-footed policies in the West, its controlled economy a boon to manufacturing. Blaming ICE vehicles for one-third of air pollution, China is implementing a rollout of increasing quotas that required 8% of new cars to be NEV in 2019 and petrol cars banned by 2040. Not wanting to be left behind, Volkswagen, Tesla, Renault, Nissan, and Ford are all making EVs in China now, the world's largest car market, having surpassed the USA during the 2008 financial crisis. China may be the next gold rush as secure and sustainable supply chains are built up and managed in Beijing rather than Washington.

Whether the new tech stars (Tesla, Singulato, Fisker, Nio, etc.), traditional automakers (GM, Nissan, BMW, VW/Porsche, etc.), a combination thereof (via joint venture or merger), or an as-yet-unknown startup with the deepest pockets of all can win the EV world by providing a popular, inexpensive, mass-market car, electrification is changing how we drive and reshaping an industry that has been too slow to change its dirty ways. To be sure, universal adoption takes time, especially in a business with over 1 billion units and 100 million new sales each year. As Hamish McKenzie notes in *Insane Mode*, "In 1900, less than 10 percent of US households had access to electricity. In 1960, less than 10 percent of US households owned a color TV. In 1990, less than 10 percent of US households had a cell phone."[83] Same as for the washing machine (1907), vacuum cleaner (1908), home refrigerator (1912), radio (1920s), television (1940s), microwave (1970s), PC (1970s), the Internet (1990s), and now the EV. Once an EV can do the same (or more) as a gasmobile at the same (or cheaper) price and everyone can buy one, no one will want yesterday's goods (Figure 6.5).

In a sign of things to come, an onboard setup can even be updated on the fly, reducing costly recalls and production delays for stressed-out carmakers. Tesla's Model 3 was initially given a bad review by *Consumer Reports* because of poor braking time (152 feet at 60 mph), but after a remote, over-the-air, software update cut the stopping distance by 19 feet, the director of auto testing changed his recommendation, noting "I've been at CR for 19 years and tested more than 1,000 cars and I've never seen a car that could improve its track performance with an over-the-air update."[84] Clearly, we have entered a new automotive era.

Time will tell if EVs are a fad, a Tesla cult, or the real deal. So far, the prognosis is good, but whether Tesla is a success, goes belly up (finally making money for a record number of short sellers), is bought out by an established carmaker, or buys out one of the slow-to-the-game Big Three, Tesla's CEO did what he said he would in his Master Plans – "provide zero emission electric power generation options." The price may also become more affordable in the next few years, the final piece of the puzzle. In March 2023, Tesla announced it would reduce the cost of its next-generation vehicles by half, bringing its world-beating product into the range of most car buyers. Zero to $20,000 in 20 years – impressive manufacturing chops to go with an impressive all-electric car.

Turning luxury into everyday affordable utility is the biggest challenge to expand the EV market, from golf carts to runabouts, from self-driving mini-cars for hire to everyday A-to-B commuter wheels. When we look back at the birth of electric vehicles, Tesla will be acknowledged for its pivotal role in making the car world electric, along with Nissan, Toyota, GM, and the earlier electric pioneers. Soon, EVs will simply be called cars.

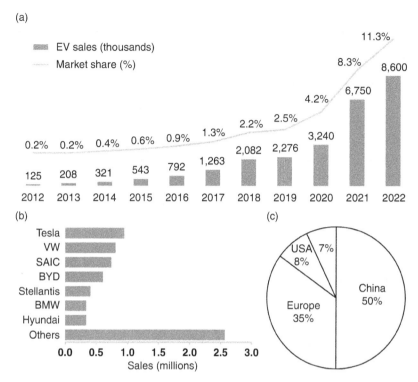

Figure 6.5 The EV future: (a) BEV/PHEV sales and market share (2012–2022) (*source*: Irle, R., "Global EV Sales for 2021," EV-volumes.com, 2022. www .ev-volumes.com/) and (b) 2021 EV car sales by company and region (*source*: Low, J. and Amberkar, A., "Global electric vehicle sales up 109% in 2021, with half in Mainland China," Canalys, February 14, 2022. https://canalys .com/newsroom/global-electric-vehicle-market–2021).

6.5 The New Machine: Autonomous and Semi-Autonomous Driving

Having rediscovered their early twentieth-century roots, electric delivery vehicles are once again appearing on our streets, increasingly more viable thanks to smarter and cheaper charging. In 2018, Atlanta-based UPS began converting 35,000 gasoline-powered vans in the USA and Europe to electric and hybrid, with plans to build an all-electric version from scratch. Announcing its own electric van in 2021, GM estimated that the delivery vehicle market for parcels and food will exceed $850 billion by 2025.[85] Costing the same as a conventional van with an easily manageable 100-mile (160-km) range, electric vans make more sense than gasmobiles and are cheaper to run, especially with staggered smart-charging software. Rental transport services are also going electric, beginning with Hermes in Europe and Ryder in North America. Ryder's director of advanced vehicle technology understands that the future of transportation is electric, noting that "Smart charging is key to creating scale."[86] Even the street cleaners are going green.

Although still in development, self-driving or "autonomous" vehicles (AV) have also begun creeping onto our roads in tandem with electric vehicles. Automated guided vehicle systems (AGVS) have been around for decades, such as driverless internal office mail delivery systems, guided either by underground signal wires, UV chemical paint strips, or imbedded magnets,[87] while autonomous, laser-guided stock-moving continues to revolutionize modern warehousing as various-sized electric transporters scamper to and fro in hectic precision (a.k.a Industry 4.0). Loading and unloading in the port of Rotterdam, Europe's largest seaport, is a highly coordinated operation as thousands of cargo containers are automatically directed to their destinations by built-in identification transponders, moving massive loads via remote control onto and off of waiting ships and trucks. The autonomous port vehicles even direct themselves to a warehouse to replace their own batteries.

Tampa Bay was the first to operate an automated guided transit (AGT) system at its airport in 1971, while other driverless mass transit systems or "people movers" operate in various cities around the world, including Kobe (the first in 1981), Vancouver's Sky Train (1985, the world's longest since 2016), and the Miami Metromover (1986, free since 2002). The Docklands Light Railway line in London has operated without drivers since opening in 1987, although a ticket checker is always onboard. Initially running on two lines from the City of London and Stratford to Island Gardens on the Isle of Dogs in East London, the system now comprises 45 stations over almost 40 km of track.

Driverless tractors are an established part of modern farming, while robot trains run in remote mining regions of Western Australia, where Rio Tinto now operates almost 200 trains on its Pilbara rail network via on-board cameras from an operations center in Perth. Now discontinued, a driverless, narrow-gauge, electric Mail Rail system scurried under the streets of London between Paddington and White Chapel for three-quarters of a century, conveniently bypassing the perennially aboveground congestion.

Tepidly taking to the roads (and skies) where safety is paramount, autonomous-driving technology (a.k.a. full-service driving) is improving by the day, despite the dangers and weirdness of human-less vehicles on our streets. One shouldn't be surprised if the future of travel includes something like the talking Johnny Cab in Paul Verhoeven's film *Total Recall* (based on the Philip K. Dick short novel

We Can Remember It for You Wholesale). The changes are indeed mind-blowing as chip-controlled sensors and self-steering software replace human drivers with a slew of cameras, sonars, radars, eye-safe NIR lidars, IR lasers, and real-time algorithmic route-mapping. Imagine an expanded Hawk-Eye tennis line-calling system that employs machine vision to provide instantaneous 360-degree-angle coverage with no blind spots, operating over a wide range of temperatures, illumination levels, and reflective surfaces (black tire bits are highly absorbing).

Uber (part funded by Google, Toyota, and Volvo), Lyft (GM, Ford, Jaguar), Hailo, and other ride-sharing companies are all planning to disrupt the urban-transit market, investing billions in driverless automated guided vehicles (AGVs), despite the occasional accident that horrifies a worried public. Remote-control systems ("teleoperation") are also in development, based partly on NASA's rover technology, where vehicles can even be delivered and returned by remote operator.

All the major players are involved – Ford, Toyota, Honda, Hyundai, Volkswagen, BMW, Mercedes-Benz, GM, Nissan, and Tesla – each with research centers in Silicon Valley to develop autonomous and connected driving.[88] Some are spending vast sums to win the human-less race for our streets, while others are taking a more measured approach. In 2017, Ford's head of autonomous vehicles noted that "Some view the opportunity with self-driving vehicles as a race to be first. We are focussing our efforts on building a service based around actual people's needs and wants."[89]

The first company to use real driverless cars on the road was Waymo – Google's original AV project – after practicing on a 91-acre test lot south of San Francisco. Waymo's autonomous Chevrolet minivans started for real in late 2017 in the Phoenix suburb of Chandler, the city's economic development director summing up the future of driving as "spectacularly unspectacular."[90] Pilot robotaxi projects have also started on the streets of Beijing and Seoul, hoping to get a head start on $2 trillion in annual business. The rollout has begun, the future as exciting as it is scary as if in a futuristic movie or indeed GM's own Futurama exhibit from the 1939 New York's World Fair, which displayed a streamlined and automated highway system with remote-controlled vehicles to combat growing transport congestion.

Started by two former Waymo engineers, the self-driving delivery service Nuro employs a one-third-sized AV minicar to deliver goods, such as groceries, dinner, and dry cleaning. Think of a scaled-up version of a *Star Wars* mouse droid without the annoyingly high-pitched screeching. While full licensing isn't yet available on the road, AV delivery systems are up and running for assembly lines, warehousing, and indoor errands, and will soon be appearing more frequently on our streets. Coming soon to a neighborhood near you: pizza delivered by autonomous minicar, scooter, or mouse droid (tipping optional). Humans need not apply.

Of course, change doesn't come without challenges and uncertainty, the public wary of a future world gone mad. An Autopilot-engaged Tesla was involved in the death of a 40-year-old former Navy Seal from Ohio in 2016, "the first known fatality in just over 130 million miles where Autopilot was activated," although the National Highway Traffic Safety Administration report found no fault in the automated system, while a 2018 California freeway crash killed another driver, although it was unclear if Autopilot was engaged. Deaths are not new in the transportation industry with accidents accounting for over 1 million fatalities per year across the globe. In the early days of steam locomotion, Charles Dickens was in a train crash on a viaduct in Kent in 1865 that killed 10 people and left 40 seriously injured, while in the same month that Tesla passed US car number 200,000 in 2018, 17 people died after a duck boat capsized on a Missouri lake. Safety is everyone's priority, but pointing to an isolated accident as a deal breaker for new technology is unfair in the history of industrial development.

To date, the occasional road accident has been attributed to human error – drivers are supposed to keep their hands on the wheel even in AV mode – but autonomous and semi-autonomous technology has already significantly reduced accidents with controlled constant distance and crash-avoidance braking. Road fatalities will be reduced further after the rules and regulations are worked out. Onboard sensors also detect driver alertness such as a nodding head, drooping eyes, and phone use while driving. More worrisome is the 200 people killed per day by human drivers on US and European roads with many more disabled.

Of course, automated decision-making can lead to tragic accidents if not properly implemented. Faulty programming was the cause of two fatal Boeing 737 Max crashes when the pilots couldn't override the Autopilot system, while "weaponized" vehicles brings a whole new meaning to the dangers of driving. One must be wary about ceding complete control to a machine as in the latest proposed full-service driving (FSD) systems, but the IEEE predicts that 75% of on-road vehicles could be autonomous by 2040, improving road safety and halving car use with increased ride-sharing, on-demand availability, and connectivity, while reducing GHG emissions by 20%.[91]

Started by two former Google Maps engineers, a new venture called Otto is developing autonomous driving technology for trucks, availing of the lower percentage costs of the expensive imaging equipment. Trucks still need a driver to respond to unpredictable events, such as sudden bad weather and other "edge cases" that the software can't anticipate, but autonomous and semi-autonomous trucking is greatly improving safety and reducing driver stress. Not legal everywhere, a form of autonomous driving is already allowed in some states such as Nevada and Texas, where drivers can sleep or take a live break on a safe stretch of road.

Autonomous EV trucks will be inherently safer with computer-directed guidance and crash-avoidance technology, along with a lower center of gravity thanks to bottom-mounted battery packs. In 2017, Tesla began developing a heavy-duty, all-electric, long-haul "Semi" truck with typical sleek design, quad-motor powertrain, and now-standard wow appeal, offering a range of 500 miles and $180,000 base price. Snagging orders from Wal-Mart, DHL, and Pepsi, Semi was unveiled 5 years later to instant approval. To remain competitive in the massive, trillion-dollar, road-transport industry, Daimler – the world's largest truck seller at about 500,000 per year – is building its own "eCascadia," while Thor Trucks is partnering with existing manufacturers to build the "ET-One" to meet California's new higher air-quality standards. All come with some form of advanced autonomy and improved safety.

We shouldn't crack open the champagne just yet, however, to celebrate a rapidly approaching, stranger-than-fiction, all-electric, AV future world. Consumer advocate Ralph Nader, who helped shape automobile safety in the 1960s with his seminal book *Unsafe at any Speed*, isn't convinced autonomous driving will come as advertised, calling driverless cars hype and hubris that distract "from addressing necessities of old 'mobilities' such as inadequate public transit and upgrading highway and rail infrastructure."[92] He commends robotic systems that don't get drunk, fall asleep at the wheel, or develop poor driving skills, but cautions against computers that fail and are susceptible to hacking, preferring more investment in clean energy and public transport. He is especially wary of allowing FSD on our roads, calling for federal regulators to ban "malfunctioning software which Tesla itself warns may do the 'wrong thing at the worst time' on the same streets where children walk to school."[93]

Driverless vehicles will also cost jobs as mechanical looms did in the 1800s and robotic manufacturing in the 1950s. In ports across the world, dockworkers are being replaced by unmanned EVs that grab, stack, and shuttle containers in coordinated, algorithmically designed maneuvers, reducing emissions but also impacting livelihoods. The port of Rotterdam has been fully automated since 2015. Same goes for public transit beyond the rails, where 5G autonomous minibuses will soon be rolled out after rigorous real-world testing. Stockholm and Montreal both began pilot projects in 2019.

Robotic automation and artificial intelligence (AI) is a concern for many workers with almost 50% of jobs at risk, although overall employment may increase in some larger companies and unsafe and boring repetitive jobs eliminated. We should also be careful with the motivation for replacing human decision-making as machines become more human-like. In his 1920 play R.U.R. (Rossum's Universal Robots) that introduced the word "robot" from the Czech word robota – meaning forced labor or compulsory work – Karel Čapek warned that "Those who think to master industry are themselves mastered by it." The "robot apocalypse" may not be coming anytime soon, but can no longer be considered beyond the realm of possibility or only the stuff of science fiction. Even Elon Musk is calling for caution, tweeting in his usual direct way, "We need to be super careful with AI. Potentially more dangerous than nukes."[94]

Some of the more ambitious ideas still in development will make our roads safer and easier to transit, such as an MIT-designed, ultrafast laser-reflection system that navigates through fog better than traditional visible-light detection. Thanks to AI and machine learning, traffic is also becoming more coordinated as machines optimize driving performance to calculate better routes, "seeing" bottlenecks before they happen, whether from heavy traffic, bad weather, or road closures. Alibaba's ET City Brain uses big data, machine intelligence, and real-time analysis to work out the best route for emergency vehicles, while CCTV images can be stitched together to predict better road options, providing drivers with real-time data to reduce traffic congestion and cut journey times. According to one AI-technology startup, London drivers spend 3 days a year just *stuck* in traffic, while congestion annually costs the typical LA driver $2,400.

Soon, Internet-connected cars (a.k.a. the Internet of vehicles or IoV) will communicate with each other via high-speed 5G links to improve traffic and safety in a coordinated network of machines, where real-time data become as important as electricity. Twenty-first-century "smart driving" will optimize group behavior on the road, although nothing beats smart citizens who can cut down, stagger, or share their rides. Tomorrow is coming faster than we can imagine, but we should welcome enhanced vision systems that can only increase safety for everyone. As Otto co-founder Anthony Levandowski noted, "There used to be elevator operators in New York City and there are not anymore."[95]

There are still plenty of road works ahead to realize a fully autonomous (level 5), all-EV world. At the very least, piloted semi-autonomous vehicles (levels 2–4) are making driving safer with crash control, auto parking, voice commands, and intelligent routing, while the next level of automation will reshape our roads via sophisticated learning algorithms, coordinated communications, and reduced car ownership. The coming automated future gives us all plenty to think about, not least whether humans or machines have a better future in the emergent behavior of our connected cities.

6.6 Twenty-First-Century Transit: E-Planes, E-Trains, and E-Automobiles

Akin to the early days of the internal combustion engine, every kind of vehicle is getting a makeover, from cars and buses to boats and planes, all powered by rechargeable batteries. In 2019, the first fire truck was converted in Menlo Park, San Francisco, the first Caterpillar excavator in Norway, and even the iconic *Maid of the Mist* ferry that has taken tourists to and from the edge of the Falls on the Niagara River since 1846. In one study, 50 million Americans indicated their next car will be electric, despite the US government cutting vehicle-emission standards and the ongoing promotion of gas guzzlers.

Many taxis are now hybrid, while buses are slowly making the conversion to hybrid or all-electric, charged via plug-in connection at an overnight depot (conductive charging) or fast wireless "top-up" at the end of or along a route (inductive charging).[96] In the commercial space, 300 miles is plenty of juice for the day, typically recharged overnight. For buses, short "opportunistic char-ging" is cheaper, however, as about one-sixth the cost of an electric bus is the battery. A one-hour-long bus route is also easily recharged at both termini using a smaller number of batteries rather than installing a full-blown system that tops up overnight for a full day's continuous operation, but full overnight depot charging could become the standard as batteries become cheaper, lighter, and longer-lasting.

Wireless, inductive charging can take as little as 6 minutes, employing coil-to-coil connectivity (a.k.a. magnetic resonance charging), where a vehicle parks on top of or under a charging pad. The same seamless simplicity is used to charge an electric toothbrush, where both the base and holder contain transformer coils (120 V AC for a 12-V DC toothbrush motor). Contactless proximity cards also transmit data by induction, such as credit cards, hotel room key cards, and reloadable transit fare cards, whenever the antenna coil in the card is placed or tapped near the embedded reader coil.[97]

Known as dynamic wireless charging (DWC), an inductive coupling system can even charge vehicles on the fly, for example, embedded in the road, albeit at high installation costs. South Korea has been inductively charging short-route electric buses in transit since 2010, a system almost as efficient as plug-in conductive charging.[98] A mile-long stretch of road is being trialed in Detroit, where buried inductive coils allow drivers to charge as they go or while stopped in traffic. Inductive charging makes for smaller batteries (reducing cost and weight), continuous supply (no range anxiety), and more efficient battery management.

Limited numbers of hydrogen, propane, and methane buses have also been rolled out, but suffer from poor filling infrastructure, while trams have been using overhead electrified catenaries for over a century across the globe, some reinstalled in the very same places they were ripped out after the rise of diesel-powered buses in the 1950s. In the center of Zaragoza, Spain, the trams run even without overhead wires via stored battery power and point charging prior to entering the city, sparing the need for unsightly overhead hook-ups in historic city centers and hard-to-install locations.

A similar point-charging system operates in Seville, where at each of five city-center tram stops an extendable, current-collecting pantograph is raised to contact the overhead power supply. Charging in about 2 minutes as passengers scurry off and on, the juice is sufficient to get to the next stop. Prior to installing the battery chargers in the heart of the famed Andalusian capital, the existing overhead lines were removed each spring to accommodate the annual Easter processions, underscoring the importance of a catenary-free power system.

As in other renewable sectors, China has led the way with an astonishing 99% of the world's electric buses, adding almost 2,000 zero-emission trans-porters per week. At the current rate of conversion, China's public transportation will be 100% electric by 2035, displacing more than 1 million barrels of diesel per day from the market. Aside from completely overhauling urban transit as we know it, EV buses are putting a noticeable dent on fuel demand – a diesel bus consumes about half a barrel of oil per day. If the rate of uptake continues, electric vehicles could displace almost 4% of the global oil supply by 2025, triggering a price crisis. The "inflection point" – the moment of price parity between gasmobiles and EVs – will signal the beginning of the end as battery prices continue to fall, coming sooner rather than later.

In the city of Shenzhen (population 12.5 million), a small fishing village prior to being designated as China's first special economic zone in 1980, all 16,359 public buses are electric with a 380-km range, helping to free com-muters from the vagaries of petroleum prices and the horrors of a smog-filled past – 20% of all vehicle fuel consumption in China is from buses and taxis.[99] As noted by Colin McKerracher, head of advanced transport at Bloomberg's London-based research unit, "This segment is approaching the tipping point. City governments all over the world are being taken to task over poor urban air quality. This pressure isn't going away, and electric bus sales are positioned to benefit."[100]

Made by the Shenzhen-based former Chinese battery-making company BYD, London added 11 shiny new all-electric buses to London's financial district, considered impressive at the time, although lagging well behind China. Although the fleet is expanding, thousands of dirty diesel buses still regularly pass through the center of London, including in a newly designated "ultra-low-energy zone" (ULEZ). In 2017, there were less than 1,000 electric buses in the five largest European countries, mostly hybrid. Today there are about 10,000 across the whole of Europe.[101] Operating 500 electric articulated buses since 2018 and aiming for 2,600 by 2024, the largest electric fleet is in Moscow.[102]

Cofounded in 1995 by chemist and engineer Wang Chuanfu and 10% owned by Warren Buffet's Berkshire Hathaway Energy, which invested $230 million, the former battery maker BYD Auto cut its electric teeth as a subcontractor manufacturing phone batteries for Motorola, Nokia, Samsung, and Sony Ericsson. With its hopeful corporate name "Build Your Dreams," the company seized on the opportunity to expand into a wider market and now makes all-electric cars, buses, and heavy-duty vehicles, including forklifts, garbage trucks, and cement mixers. Selling to 300 cities outside of China,[103] business is booming, helped by a $9 million grant in 2016 to buy 27 electric trucks for San Bernardino County, one of California's "most polluted air basins."[104] Within little more than a year, BYD was outselling Tesla. The second-largest battery maker in the world and largest electric bus maker in North America, BYD now employs 1,000 workers, mostly at its Lancaster, California plant that opened in 2013.

With more than 100 cities of over one million population, China is ripe for an electric makeover. Elsewhere, cities organize car-free days to reduce pollution and congestion, ban cars on certain days (for example, odd/even license plates), and restrict access with congestion charges and premium parking rates that unfairly disadvantage low-income citizens. The contrast between the old and new couldn't be starker than in the electrification of transit systems around the world.

In India, the government is focusing on shared mobility to combat inner-city gridlock and the horrendous levels of urban pollution, announcing subsidies in 2019 for 1 million electric scooters, 500,000 rickshaws, 7,000 buses, and 55,000 cars that must either be taxis or fleet vehicles.[105] In a developing country where 22 million two-wheelers were sold in 2018 compared to 2.2 million cars, the strategy is financially sound and eco-friendly, and will transform inner-city travel in the dizziest of urban environments.

***

The electric transport refit is not just confined to the road. Electric boats are an obvious solution to waterborne pollution since the first motorized boat put to sea over 150 years ago. Starting with small craft and expanding to larger boats and ships, going electric will make rivers, lakes, and oceans cleaner, both for drinking and long-suffering marine life. Accounting for 2.5% of GHG emissions as well as massive amounts of pollution, electric or biofuel marine travel will significantly improve water quality everywhere. Low-grade bunker fuel also contains 3,500 times as much sulfur as diesel fuel, making cruising an ecological disaster for beleaguered oceans and urban pit stops.[106] The leading European cruise ship company emits about 10 times more sulfur oxide than Europe's 260 million cars.[107]

Electric-boat propulsion is already revolutionizing ferry travel, especially where short crossings are an integral part of life. In 2015, the world's first all-electric ferry started up in Norway, providing passage across the Sognefjord between the villages of Oppedal and Lavik, a distance of 5.7 km or 3 nautical miles (NM). A crossing in the *Ampere* takes 20 minutes, has space for 120 cars and 360 people, while the propulsion system is powered by two 450-kW electric engines and a 10-ton, 1,000-kWh Li-ion battery array, recharged at each terminus during a 10-minute disembarking/embarking period.[108] Building on the success of the *Ampere*, which cut emissions by 95% and running costs by 80% in the first 2 years of operation, the Norwegian travel group Fjord1 ordered 75 battery-powered ferries from the Havyard shipyard in Fosnavåg, while low-emission operation and huge cost savings is fueling orders for more electric builds and diesel refits.[109]

The first medium-range, all-electric, car-and-passenger ferry began operations in 2019 on the Baltic Sea between the Danish mainland and Ærø Island at Søby with a single charging point at the Søby Havn terminus. Known as Ellen, the e-ferry can service 200 passengers as well as 31 cars (or five trucks) on two separate routes from Søby to Fynshav (10.7 NM) and Søby to Fåborg (9.6 NM). The longest maritime return distance of over 20 NM is powered by the largest-ever battery array (4.3 MWh), recharged in under 20 minutes by a portside, high-power, plug-in DC charger.[110] The Norwegian and Danish grids are also almost entirely green – hydropower in Norway and wind power in Denmark – and thus electric-powered ferries in the two environmentally conscious Scandinavian countries offer the cleanest marine transport option.

Fully electric, long-haul shipping is harder to implement, however, because the weight of a chemical battery is more than 30 times what is needed to generate the equivalent energy in a diesel engine, never mind the lack of mid-trip recharging at sea. Even with continued improvements in the energy density of Li-ion batteries, liquid fuel won't be replaced on long-haul ship travel anytime soon, although improved biofuels increase the overall efficiency as can hydrogen gas (as we'll see later).

Electric plane travel is also harder to implement because of the solid/liquid weight and refueling issues in the more extreme and dangerous operating environment of the sky, but development has begun on commercial, battery-powered, short-haul aircraft. Crucially, more power is needed for the roughly 3 minutes of elevation during takeoff than for in-flight cruising (up to twice for light aircraft, three times for airliners, and 20 times for short-range, vertical take-off urban air mobility vehicles[111]). A large commercial airliner would require the energy of 30,000 Teslas just to take off.[112]

In an electric Kitty Hawk moment, an all-electric seaplane stayed aloft for 3 minutes over the Fraser River east of Vancouver on the morning of December 10, 2019. The bright yellow Harbour Air DHC-2 de Havilland Beaver was powered by a standard electric motor and an NASA-approved Li-ion battery pack. Harbour Air's founder likened the prototype to a "Beaver on electric steroids," while one of the co-designers believes e-planes will eventually revolutionize short- to mid-haul flying.[113] By comparison, the famous gas-powered Wright Brothers Flyer biplane flew for only 12 seconds on its maiden flight.

The American aircraft design company Wright Electric is hoping to build a 120-seater with a 335-mile range for flights under 2 hours. Having already built a two-seater prototype, Wright Electric wants to roll out a commercial e-plane within a decade, although its co-founder Jeffrey Engler called the project "daunting from every perspective."[114] In 2021, United Airlines announced the purchase of 100 19-seater Heart Aerospace electric propeller planes, which it hopes to have running by 2026 for short-haul regional flights.

Such smaller electric planes are more affordable because they save on engine and maintenance costs as well as the huge liquid fuel costs. As Heart founder and CEO Anders Forslund noted, "Jet engines never made sense on small aircraft, it never made sense on short routes. Electric motors do."[115] The challenges of electric flight naturally increase with size and range, but starting with short-haul flight is good business and helps further the development of larger transport, potentially worth trillions of dollars in revenue. The first crewed electric helicopter was successfully tested at the Fiumicino Airport in Rome on October 6, 2022. The eVTOL or Volocopter 2X stayed aloft for 5 minutes with plans to roll out transport services from the airport's new "vertiport" to Rome in 20-minute trips by the start of 2025 and possibly for the Paris Olympics in 2024.

Solar power has already liberated the skies for the smallest of aircraft, but is primarily a toy for the hobbyist. Powered by a 3-kW system, the Solar Challenger plane flew across the English Channel in 1981, while Solar Impulse II made the first round-the-world, fixed-wing flight over 17 stages in 2015 and 2016, powered only by the Sun, a 66-kW$_p$ PV array, and 13 kW of Li-ion batteries. With a wing span of 72 m, just longer than that of a Boeing 747, Solar Impulse II circumnavigated the globe from Abu Dhabi to Abu Dhabi in

a tad under 24 flying days.[116] Weight is an obvious issue – there's not much
luggage room, let alone a co-pilot.

Lightweight drones, however, are easier to operate and keep aloft for longer
periods. Facebook is planning to run Internet and mobile networks in remote
regions with hundreds of airborne battery-powered drones that can fly for
months at 60,000 feet, well above the commercial-airplane routes that operate
at 35,000 feet. Although still hugely expensive, the founder of a company that
builds control systems for mobile networks commented on the reality of
incremental change: "It doesn't really matter if it is not inexpensive enough
today. We know it will be soon."[117]

Green commercial planes, suborbital flying, and low-altitude taxi drones may
eventually fill our skies with battery-powered aircraft as well as stratospheric
communication drones, but for now viable green ground transit may be the best
on offer, especially in densely populated areas. The prototypes are slowly getting
bigger as rechargeable lightweight batteries continue to revolutionize transportation, although the requisite charging infrastructure still lags behind.

In any transition, competing technologies continue to coexist as the novelty
of the new becomes known and the limitations of the old become impossible to
ignore. As one product replaces another, the losing technology touts its benefits: custom, familiarity, ease of use, even a presumed superiority. The telephone was a novelty to many before it supplanted the telegraph, taking until the
1970s to reach 90% penetration in the USA.[118] The first Philips light-bulb
factory in Eindhoven was initially lit by gas because electricity was not yet
available, while the electric bulb was considered by some to be too dim for
comfortable reading. Edison famously prattled on about DC versus AC for
long-distance power transmission, while as late as the 1950s 11% of US
households still had an ice box.[119]

There can be no denying that the world is on the cusp of another revolution
with the rise of EVs, storage batteries, and roadside charging, as well as the
coming V2G technology. As with any disruptive technology, one can expect
both an economic upswing and organized pushback.

6.7 Industrial Storage: The Ups and Downs of Managed Energy (Pumped Hydro, Thermal, Compressed Air, etc.)

When we generate electricity in a power plant, wind turbine, or solar panel, we
have to use it or lose it. Operating a plant below the nameplate capacity – for
example, using only 400 MW in a 500-MW power plant – allows some

tolerance to adjust for temporary increases in electrical demand to cover the ups and down of changing consumption patterns, as does a backup natural-gas peaker plant that can be fired up as needed (as we saw in the last chapter). Unfortunately, we cannot keep electricity in a jar and must convert any excess power to storable energy, whether the chemical energy of a battery, thermal energy of a heated material, or gravitational potential of water lifted to a height, three typical stored-energy systems employed to compliment variable grid power, where the energy is then available to be reconverted to electricity as needed. A battery-energy system (BES) or battery-management system (BMS) is like a squirrel gathering nuts for winter, the electrical energy stored for on-demand, easily dispatchable future use.

Electrical energy storage (EES) is essential to maintain our always-on power supply for a variety of needs, including backup power, load following, demand shifting (for example, day to night), seasonal storage, variable supply resource integration, real-time renewable integration, arbitrage, voltage support, frequency regulation, black start, and off-grid power, while smart technology is integrating more distributed energy resources and storage capacity into a growing bi-directional grid to improve efficiency and performance. Smart meters and grid software help manage the constantly changing demand to balance a variable real-time load (as we've already seen).

In Fairbanks, Alaska, where temperatures can fall to 40 degrees below zero, backup power is essential to keep the electricity on during regular winter outages. Since 2010, 13,600 NiCd batteries have been providing emergency backup power up to 40 MW for 7 minutes, until the diesel generators kick in. Tesla famously built a $50 million grid backup system near Adelaide to help stabilize a notoriously temperamental South Australia grid – the 129-MWh Hornsdale Power Reserve Li-ion battery array (later increased to 185 MWh) – while the city of Los Angeles plans to convert dirty natural-gas peaker plants to chemical-battery storage, employing lithium-ion batteries to cover the regular ups and downs of supply and demand.

Li-ion and other chemical battery storage have significantly advanced in the past decade to provide a viable alternative to natural-gas peaker plants (maintained at the ready as always-on spinning reserve). In 2021, a 100-MW Tesla Megapack battery system was installed on 3 acres of industrial land near Oxnard, Ventura County, enough to provide emergency juice to 80,000 homes and businesses for 4 hours. The Ventura Energy Storage system was built instead of a proposed fossil-fuel backup system and was championed by both environmentalists and residents throughout the coastal region. "It is a testament to what happens when communities say we want something better for the future, something better for our children," noted a local Oxnard councilwoman.[120]

One can even cannibalize the battery of an old car to provide stored juice, giving new meaning to the idea of a car as a battery on wheels. BMW made a battery farm from used i3 car batteries to help run its Leipzig factory, while old Nissan Leaf batteries were refashioned to help power the Johan Cruijff Arena in Amsterdam. Such "second-life" battery systems are ideal after a battery loses 20% of its rated capacity, transitioning from transport to stationary storage for industrial or off-grid use to recoup some of the original costs. An "end-of-life" (or EOL) battery is typically employed when 80% of the initial charging capacity remains.

Although chemical batteries are becoming cheaper and more viable, there are other ways to store energy besides separating charges in a lead–acid, alkaline, or lithium-ion battery, some essential for backup power and grid stability and others in various stages of development. We can raise water to a height (pumped storage hydropower), store heat in molten salt (at a concentrated solar plant), compress air in a confined space (large- or small-scale), or separate water into hydrogen gas and oxygen (to power a hydrogen fuel cell) to name a few energy-storage technologies available today, some of which we've already seen in limited action.

British environmentalist Jonathon Porritt once introduced a slide at a TED talk he feared was the most boring ever, showing a myriad of different storage technologies: "Sodium–sulfur batteries, rechargeable flow batteries, liquid metal batteries, molten salt thermal batteries, freshwater/saltwater batteries, pumped heat electricity storage, umpteen different kinds of flywheel, liquid air (or cryogen) systems, thin-film super capacitors integrated directly into appliances, pumped water storage, super-conducting magnetic storage, etc. etc."[121] Not so boring, perhaps, with hundreds of billions of dollars at stake for industrial and residential use, and growing every year.

***

The main, large-scale, energy-storage technology today is pumped storage hydropower (PSH), accounting for roughly 95% of a globally installed 170 GW of storage (9,000 GWh). PSH converts the kinetic energy of water into potential energy and vice versa as water is raised and lowered between two reservoirs for future, on-demand, electricity generation. We've already seen examples of storing water at height in El Hierro in the Canaries (to backup wind-generated power) and in La Muela II near Valencia (to augment hydropower).

About 60 km south of Dublin in the Wicklow Mountains, near the pictur-esque glacial valley of Glendalough, the Turlough Hill Power Station has been running since 1974, pumping water uphill from a lower reservoir to a higher reservoir during low demand – usually at night – and dumping it back via gravity as needed to run four hydraulic turbines during peak load (the same turbines are used to pump the water and generate electricity). Turlough Hill has a capacity of 292 MW, essentially on-demand with full generation achieved in 70 seconds.[122] Note that PSH doesn't create free energy – it costs to raise the water – but responds to a variable electric load in a fast and efficient way to help manage the grid, which is harder for fossil-fuel, nuclear, and renewable energy sources. The so-called "round-trip" efficiency for PSH is about 75% between pumping and generating.[123]

In a 2008 lecture entitled "Intelligent Energy Options for the Future," Igor Shvets, the Trinity College Dublin Chair of Applied Physics, outlined a plan to build hydro reservoirs along the west coast of Ireland by flooding mountain valleys from Donegal to Kerry. The stored energy would be released as needed to overcome the variable output of a proposed nearby wind farm, thus ensuring a steady supply of green energy and freeing Ireland from the vagaries of Middle Eastern oil and Russian natural gas. Alas, the ambitious project – called Spirit of Ireland – never received sufficient backing, lacking both the considerable funding and political will to disrupt private lands in the pristine Irish countryside.

Such projects are challenging, both environmentally and economically, even in centrally planned China, let alone the more politically fragile NIMBY West. Nonetheless, China is spending as much as $1 billion per gigawatt to install 60 *gigawatts* of pumped storage hydro, essential to manage their ever-evolving, state-run grid as more renewables-sourced, intermittent energy is added to the mix. A more than 100-fold global increase in stored energy is predicted by 2040[124] as more large-scale PSH plants are built, paired with renewables to lessen the impact of the inherent intermittency from changing weather patterns such as clouds and wind gusts.

Built in 1907, the world's oldest PSH installation is the 5-MW Engeweiher run-of-the-river hydroelectricity power plant in northern Switzerland near the town of Schaffhausen, where water is raised from the Rhine to create stored gravitational potential energy. Existing reservoir dams can also be repurposed to hold back water to provide dispatchable power as in the recently constructed 900-MW Nant de Drance PSH plant in the southwest Swiss canton of Valais that uses water from the existing Émosson Dam. Pumped 295 m to an upper reservoir, the 1,800-GWh plant can cover 10% of the Swiss grid for 20 hours, making money by gravitational arbitrage, pumping water up at low tariff and releasing water down at high tariff to generate electricity ("turbining").

A 440-MW PSH plant northwest of Glasgow is also capable of restoring power to the grid after a failure, a so-called "black start." Built inside the hollow of a mountain in the 1960s, the Cruachan Power Station pumps water 396 m off peak from the lower-lying Loch Awe that is then released during the day to generate electric power as needed, especially important in the event of an emergency restart. The pumping is done at night, initially using electrical output from the nearby Hunterston A nuclear plant before it was shut down in 2022. One of only four PSH plants in the UK, Cruachan can deliver up to 7.1 GWh or 440 MW for 16 hours and is ideally suited to tap the excess wind power available at night. Known as the "water battery" to locals, the 1.2-GW Northfield Mountain PSH facility in northern Massachusetts has been pumping water since 1972 from the Connecticut River to a man-made reservoir, producing power as needed to 1 million homes for up to 7.5 hours a day.

Although PSH is the main long-duration energy-storage (LDES) system for a real-time, fluctuating grid, not all require large expanses of land. The 19.9-MW Gemasolar CSP plant in southern Spain, 60 km northeast of Seville, occupies less than 2 km^2 and in 2011 was the first to use both a high-temperature solar receiving tower and molten salt for energy storage, providing electricity after sunset from heated sodium and potassium salt. Employing 2,650 heliostats to reflect the Sun onto a 140-m high solar tower (as we saw earlier), salt is pumped from a "cold" tank at 290°C and heated to 565°C in the collecting tower before being returned to a "hot" tank, where the hot salt is stored for later steam generation, producing up to 15 hours of electricity during the night or down-sun hours.[125]

Online since 2016, the 110-MW Crescent Dunes plant, located about 200 miles northwest of Las Vegas, also uses molten salt to store heat, generated by 10,000 115-m^2 heliostats to produce energy 24/7. As noted by the CEO of SolarReserve, the parent company of Crescent Dunes, "Storage in molten salt can stay hot for months. Normally, it isn't left there, of course, but cycled daily as needed, tapped by night for generating electricity, and replenished by day by the Sun."[126] Furthermore, a CSP plant is cleaner than a coal plant and doesn't require anywhere near the amount of water as a fossil-fuel or nuclear power plant.

On the edge of the Sahara desert in western Morocco, the first phase of the world's largest CSP plant, Noor 1, opened in 2015 with an installed capacity of 160 MW. When finished, Noor 1, 2, and 3 will generate 580 MW, enough to power 1 million homes. Unlike Gemasolar and Crescent Dunes, which are both power-tower CSP plants, Noor 1 is a parabolic-trough CSP plant that focuses the Sun from 500,000 12-m high parabolic mirrors onto liquid-containing steel troughs to create steam, but also uses molten salt to store heat after sunset for 3 hours. Noor 2 and Noor 3 will store enough energy for 8 hours, generating round-the-clock power to the Moroccan grid.[127]

Engineers at Ottawa's Carlton University have been investigating small-scale, passive heat-storage systems that collect heat in the summer that is then released during winter. In an experimental test house, heated water is piped from a south-facing, solar-collecting roof to a heavily insulated, 6 × 6 × 3 m deep, wet sand box, raising the temperature of the sand to 80°C, where the latent thermal energy is kept before being reradiated through the house as needed, greatly reducing electrical bills. Hoping to cover 90% of winter heating costs by the novel heat-transfer energy-storage system, the goal is to exploit thermal energy for seasonal storage in traditionally cold-weather climes.[128]

We can also generate heat directly from electricity, providing much more storage flexibility. A pumped thermal energy (PTE) system converts electricity to heat in an insulated tank, filled with gravel, sand, or water, converted back to electricity as needed via a heat engine. PTE can be employed with any source, is relatively cheap, easily scales, and can store more energy per volume than PSH. For example, 10 times more electricity is recovered from storing heat in 1 kg of water at 100°C than 1 kg of water raised to a height of 500 m, and is more than 90% efficient.[129]

The first commercial "hot-rocks" storage system started up as a demonstration plant in 2021 on the Danish island of Lolland, employing pea-sized crushed stones to store electrically generated heat in insulated steel tanks later reconverted to electricity. An already prosperous renewable-energy region that often produces 50% more wind power than local consumers can use, the excess generated power is now stored during times of surplus. As the CEO of Stiesdal Storage Technologies noted, "It is precisely in such a context that our storage technology can make a difference and contribute to a far more extensive integration of power from sun and wind than what has been feasible up till now."[130]

The first commercial thermal energy storage (TES) system employing sand was developed by a group of researchers in Finland, where 100 tonnes of low-grade builder's sand is heated in a dull gray steel silo to 500°C via cheap renewables for use as local district heating during winter, including heating the water of a nearby swimming pool.[131] Dubbed a "sand battery," the simple energy storage system can store heat for months at a time like an industrial-sized thermos that doesn't lose heat to the outside.

Underground thermal energy storage (UTES) and pit storage are also available for medium- to long-term seasonal storage, both heating and cooling (similar to how an ice house provides refrigeration). In 2019, Siemens

Gamesa installed a 130-MWh electro-thermal energy storage (ETES) system near Hamburg that electrically heats 1,000 tonnes of volcanic rock to 750°C, reconverting the heat to electricity via a steam turbine. Molecular absorption can also store sunlight released for future use as heat by the introduction of a catalyst, with the commercial potential to provide improved home and vehicle heating. A Chalmers University of Technology research team in Gothenburg, Sweden, has tested different liquids made up of carbon, hydrogen, and nitrogen to store solar energy.

Compressed air is also viable, the first compressed-air energy storage (CAES) system built in 1978 near Bremen, Germany, at the Huntorf CAES plant, a 290-MW system that augments baseload electricity from a nearby nuclear plant. The storage system compresses air and combustion gas by electrically driven compressors, stores the pressurized air–gas mixture at 50–70 MPa in two cylindrical salt caverns (200-m long by 30-m diameter) about 1 km underground, before the expanding air is returned to an on-site turbine to generate electrical power as needed. Only two large-scale CAES power plants are in operation – a similar 110-MW plant was built in 1991 in McIntosh, Alabama – but there are numerous salt domes and natural gas reservoirs available to store energy and level the load from an increasing number of nearby wind farms, good for 24-hour, dispatchable backup power.

Another novel compressed-air system pipes air into large underground or underwater balloons, released as needed to turn a turbine, while small-scale CAES uses pressure-resistant solid containers of ceramic, stone, cement, or cast iron, providing full capacity in minutes. CAES and TES can also be combined as in a hybrid system recently installed in a disused Polish coal mine, where the air is stored up to 80 times atmospheric pressure. As noted by the scientific research group at the Silesian University of Technology, "Mine shafts are usually located in proximity to power plants and/or distribution stations. This allows the use of existing grid connection infrastructure. In addition, proximity to highly industrialized areas reduces energy transmission losses."[132] The potential exists to extend the novel system to dozens of other decommissioned coal mines across the country. One can also make use of the temperature difference between air and liquid air by compressing air to a liquid at −196°C that expands back to a gas when released to turn a turbine.

Other storage systems are as simple as winding a watch, the hard part making the watch big enough without breaking. A mechanical flywheel is essentially a giant watch that stores and releases mechanical energy by coiling and uncoiling an internal spring. Massachusetts-based Beacon Power deployed a 1-MW grid-scale flywheel system in 2007 to "recycle" electricity in real time, storing energy from the grid when demand drops that is reinjected when demand rises. The system also regulates grid frequency by spinning slower or faster up to 1,600 rpm and can instantly change modes between charging and discharging.

Capacitors are another potential energy-storage device. Researchers at the University of Waterloo, Ontario, built a high-load supercapacitor from multi-layered, ultrathin, graphene sheets – 2D graphite with exceptionally high conductive surface area – sandwiched around a liquid-salt dielectric. Such "next-generation" electrochemical energy-storage devices are still under development, but graphene-based supercapacitors could ultimately charge a phone in seconds or an EV faster than filling a gas tank.[133]

Other storage systems are being designed to incorporate new thinking to old ways. A modern version of PSH employs a lifting mechanism to raise a solid weight rather than water, storing gravitational potential energy for future electric power generation as the weight returns to its starting location. Built in the small Swiss town of Arbedo-Castione, a prototype gravitational storage device uses a crane to lift two 35-tonne concrete blocks that produces 1 MW in 30 seconds when the blocks are returned to the ground, enough to power 1,000 homes. The blocks can even be made out of local dirt, gravel, or recycled materials, while a proposed scaled-up version could lift 7,000 blocks to power thousands of homes for 8 hours. "The greatest hurdle we have is getting low-cost storage," noted the CEO and co-founder of Energy Vault, the company that built the test site in the foothills of the Alps.[134]

Although still in its infancy, a major advantage to raising and lowering weights is that the system is not restricted to the proximity of water or mountains. The weights can also start below ground and be raised up in any suitably accessible location, whether in the air or underground to maximize space and safety, for example, in a disused mine shaft. Or even along a track as in a modified rollercoaster, skateboard half-pipe, or giant Newton's cradle. PSH is expensive and limited to geologically tailored sites, while weights are much cheaper and can be employed anywhere. All that is needed is a height difference to repeatedly raise and lower the weight in endless Sisyphean splendor to balance grid demand, especially useful to cover sudden spikes.

We have seen how energy can be readily stored for future power consumption using different processes to accommodate the intermittent nature of variable renewable energy (VRE): via gravitational difference (gravity cell), absorbed heat (heat cell), compressed air (pressure cell), in a spring (mechanical cell), or a capacitor (electrochemical cell), as well as various types of chemical batteries for large- and small-scale industrial and home use (chemical cell). Paramount is the power, lifetime, safety, and cost as measured per kWh

(a)   (b)

Figure 6.6 Pumped storage hydropower (a) upper and lower reservoirs at Turlough Hill, Wicklow (*source*: Archives ESB) and (b) diverted water from Lake Erie in conjunction with Sir Adam Beck power stations I and II down river from Niagara Falls, Ontario (*source*: Ontario Power Generation CC BY 2.0).

per life cycle. Instantaneous energy storage will be essential as the grid becomes greener.

Typically, PSH (Figure 6.6), thermal, CAES, and flywheels are best for large-scale storage systems, while batteries meet more small-scale needs, although larger-scale chemical batteries are now being increasingly deployed in commercial battery arrays, especially as cathode–anode–electrolyte chemistries improve. Supercapacitors, superconducting magnetic energy storage (SMES), and power-to-gas (P2G) systems are also available, each at different stages of development, from early R&D to mature commercialization. All kinds of reliable energy-storage systems – gravitational, thermal, mechanical, electrochemical – are available to meet our growing need for on-demand electricity to flatten a variable on-peak/off-peak load, save energy for future use, or as an emergency backup. We will need every one as renewables continue to provide more power to a revamped modern grid.

6.8 Hydrogen Gas: A New Old Pretender?

Employed in the past for motive power in limited applications, one particular energy-storage system has been kicking around for decades, with new ideas in the works for increased output, including dispatchable power, heating/cooking, steel and cement manufacturing, and transportation: hydrogen gas. Separating water (H_2O) into component parts is a well-developed storage technology, where the molecular hydrogen (H_2) and oxygen (O_2) is made via steam reformation of methane (so-called "gray" hydrogen), coal gasification ("black" hydrogen), or electrolysis in a renewable-energy electrolyzer ("green" hydrogen). The hydrogen gas is later recombined with oxygen in a fuel cell to generate electricity or combusted on its own for heat or power.[135]

As new facilities begin to produce more hydrogen to address the growing call to decarbonize energy, supply is not thought to be a problem, rather the pace of converting an existing fossil-fuel-based infrastructure to clean hydrogen, whether for home heating, stoking industrial blast furnaces, powering vehicles of all sizes, or energy storage. Hydrogen is also used to produce industrial ammonia, the primary component of artificial fertilizers, and is thus essential for farming. An energy carrier rather than an energy source, however, creating hydrogen is an inherently loss-making venture. Nonetheless, demand could increase five-fold from 90 million to 450 million tonnes by 2030, slashing 10% of global GHGs.[136]

For home heating and cooking, hydrogen gas can also be blended with natural gas (methane) using existing pipelines, especially important in countries such as the Netherlands, where 85% of homes are already connected to a natural-gas delivery infrastructure – hydrogen gas (2,660°C) and natural gas (2,770°C) conveniently burn at roughly the same temperature. There is some concern, however, about potential damage to the pipelines from standard T-junction blending techniques that can lead to pipe-wall embrittlement, especially near welds, because of poorly mixed hydrogen.[137] Retooling pipelines in other countries to carry 100% hydrogen is also possible and has been suggested for the vast US natural-gas pipeline system. Hydrogen-burning boilers have also been developed to replace household methane boilers, which are hurriedly being phased out in some countries and may soon even be outlawed.

Hydrogen gas can also be burned to produce the high temperatures needed to make steel rather than burning coking coal or natural gas, helping to decarbonize the high-emission manufacturing process. The hydrogen gas reacts with iron ore to make iron and water that is then turned into "green steel" via direct reduction, emitting 5% of the GHGs from a conventional coal-fired blast furnace. As electrolyzer costs decrease, low-carbon steel may soon become competitively priced, especially with increased investment. At a pilot plant in Lulea, northern Sweden, the HYBRIT coalition is already making green steel and aiming for large-scale production by 2025. Cement making is also ripe for a hydrogen makeover. Together, steel and cement manufacturing produce about 15% of global GHGs.

Some are even betting that hydrogen gas can replace gasoline and diesel to compete with lithium-ion batteries as a vehicle fuel. As seen in various test programs, hydrogen gas can power electric cars, trucks, and buses with plans in the works for trains, planes, and ships. A hydrogen fuel-cell EV (FCEV) is faster to refill and can provide more range than a battery-powered EV (BEV), although hydrogen fuel-cell or hydrogen combustion transportation may be more suited to long-haul shipping and aviation, where system size is less critical. About half as efficient as a BEV (45%/86%), FCEVs are nonetheless highly valued for heavier transport because of hydrogen's higher energy density.[138]

Touted as a viable green option to decarbonize the transportation sector, hydrogen fuel-cell electric vehicles are starting to compete with battery-electric vehicles, especially in a growing number of public bus systems. Fuel cells for EV propulsion are highly efficient, losing only about 20% as waste heat, while the by-product of combining hydrogen gas with oxygen is water, easily discarded from a vehicle's tailpipe with no emissions. Some corporate advocates believe that hydrogen is the perfect fuel to replace over 100 years of sludge-spewing, low-efficiency gasoline and diesel fuel. Considered low-carbon if the source is environmentally friendly and with 2.6 times the energy density of natural gas, hydrogen has even been hailed as the liquid fuel of the future, although ramping up to a global transportation network may be a road too far.

Unfortunately, the efficiency of a FCEV is half that of a BEV and while there are no emissions at the tailpipe, there are plenty in the manufacturing process if the hydrogen is made from methane, a.k.a. "gray" hydrogen, the primary source (95%) of today's industrial-scale hydrogen. One ton of gray hydrogen produces 10 tons of carbon dioxide, making any emissions savings in a FCEV much worse, although natural gas suppliers have started to promote "blue" hydrogen that captures waste carbon in a costly and still evolving technology (as we saw earlier with CCS). If green hydrogen can be shown to be economically viable, however, hydrogen fuel cells may turn out to be an industry savior.

∗∗∗

The hydrogen fuel cell was first demonstrated in 1842 by the Welsh scientist William Grove, who successfully combined hydrogen and oxygen to produce an electric current in a "gas battery." By the early 1940s, simple fuel-cell generators produced up to 5 kW, while a number of hydrogen storage systems have been operational ever since. NASA's Gemini and Apollo missions both used hydrogen fuel cells – the Apollo 11 command module *Columbia* was kitted out with three cells to provide electric power for the astronauts (and drinking water!) – while more modern, low-temperature, hydrogen-generating methods employ a proton exchange membrane electrolyzer (PEME). Submarines also generate oxygen for breathing while submerged via the electrolysis of seawater (the hydrogen is vented into the sea).

Primarily employed as a warehouse forklift, fuel-cell electric vehicles have successfully operated since the early 1960s, eliminating nasty tailpipe pollution in enclosed spaces while maintaining sufficient automotive power. Hydrogen gas has great performance, as good as or better than an internal combustion engine, with almost three times the energy per unit weight (hydrogen gas 120 MJ/kg versus gasoline 45.8 MJ/kg). To get an idea of the relative oomph, 1 kg of hydrogen gas equals about 1 gallon of gasoline.

Although most of today's hydrogen is gray, separating hydrogen gas from water is theoretically straightforward via electrolysis, where an electric current in an electrolyzer splits the water into component parts ($2H_2O$ + energy (electricity) $\rightarrow 2H_2 + O_2$). There are two half-reactions during electrolysis: one at the cathode (electro-accepting reduction) and one at the anode (electron-donating oxidation). Almost all hydrogen gas today (95%), however, is produced by heating methane (CH_4) and water at 1,000°C via steam reformation, where the methane is typically sourced by dirty extraction means ($CH_4 + H_2O$ + heat (1,000°C) $\rightarrow CO + 3H_2$).[139] Some hydrogen is also produced during flue gas recovery in petroleum refining and oil-sands extraction. Compared to steam methane reformation (SMR), green hydrogen gas is collected via a solar- or wind-powered electrolyzer, a low-carbon source already used in a number of clean-energy sites.

In Norway, forklifts, cars, trucks, and buses run on green hydrogen, separated from water via an electrolyzer in a solar farm in Trondheim and another at a wind farm in Berlevag at the northern tip of the country. Norway hopes to have 500,000 green-hydrogen-powered vehicles up and running after its mandated end to fossil-fuel-vehicle sales in 2025, part serviced by a hydrogen corridor of stations located between Oslo and Stavanger.[140] A 900-kW community wind turbine on Shapinsay, one of the 20 inhabited Orkney Islands off the north coast of Scotland, produces compressed green hydrogen via water electrolysis whenever wind-generated electricity rises above local grid demand. A veritable "energy island," Shapinsay's green hydrogen supply powers local domestic heating, commercial vehicles and ferries, and seasonal energy storage as excess wind power in winter is stored for calmer summer use.

Hydrogen gas is especially being touted as a grid-flattening savior, known as power-to-gas (P2G), where the stored chemical energy of the hydrogen gas is converted to electricity – for example, to accommodate low-sun winter needs, occasional summer doldrums, or because of intermittent sun and/or wind (as in Shapinsay). Utilized in a temporary energy-storage system, hydrogen buffers the grid at high demand, while the oxygen can be used to run an Allam Cycle generator. Chris Goodall, author of *The Switch*, thinks P2G will one day be cheaper than oil and is "a technology that the world urgently needs to make work on a large scale if it is to manage full decarbonisation of energy."[141]

As in any new technology, scale and cost is the limiting factor to make hydrogen a viable future eco-fuel or storage medium. A sample industrial green hydrogen electrolysis schematic is shown in Figure 6.7, powered either by solar or wind.

The European Union plans to increase electrolyzer capacity almost 1,000-fold to 40 GW by 2030 (10 million tonnes) at a cost of almost €40 billion, with another €340 billion earmarked for renewable sources and €65 billion in distribution and storage.[142] Globally, electrolyzer output is expected to rise from 3 GW in 2021 to 100 GW by 2030 with the cost per gigawatt more than halved, while at $1/kg the cost of green hydrogen could even be comparable to gray hydrogen within a few years.[143]

Some think blue hydrogen (SMR + CCS) can help enable the transition from gray "fossil" hydrogen to green hydrogen (GH2) as an intermediate step to scale up new technologies (for example, electrolyzers). Currently only twice as expensive as gray hydrogen to produce, blue hydrogen nonetheless suffers from minimal CCS infrastructure and a questionable technology that serves to keep the petroleum biz running and maintain fossil-fuel demand, akin to "clean coal." Others think new infrastructure funding should go straight to GH2 instead of wasting time with dirty fossil hydrogen (1 ton of gray H_2 produces 10 tons of CO_2), helping to reduce costs as electrolyzers become bigger and cheaper with improved membrane technology (for example, larger cells and higher current densities). GH2 is currently about 4 times the cost.

Although uncompetitive without including external costs, displacing harmful gray hydrogen with green hydrogen is certainly advantageous in the so-called

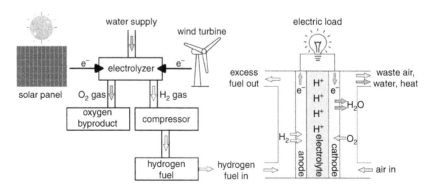

Figure 6.7 Hydrogen gas: (a) green P2G electrolysis system and (b) hydrogen fuel cell.

"hard-to-abate" industrial sectors such as steel, cement, and ammonia fertilizer manufacturing as well as for highly polluting long-haul ship and plane travel. Costs will drop rapidly as electrolyzers improve, especially if petroleum feedstock prices continue to rise. As noted in *New Scientist*, "The question today no longer seems to be if hydrogen will help us fight climate change, but a matter of whether it ends up as the star turn or just a bit player."[144]

New production plants are hastily being built as the hydrogen economy becomes more feasible. The world's largest green hydrogen production plant and storage hub is in the works for south Texas. Powered by wind, solar, and renewables from the ERCOT grid, the 60-GW Hydrogen City project will generate more than 2.5 billion kilograms of green hydrogen per year, stored in existing salt caverns in Duval County. Pipelines will transport the GH2 to hydrogen power plants as well as conversion facilities to make ammonia, aviation fuel, and rocket fuel.

Based in Spain, 30 European companies are collaborating to make 3.3 billion kilograms of hydrogen per year via 67 GW of electrolyzer capacity, distributed via a continent-wide transportation and storage network. The HyDeal Ambition project plans to provide green hydrogen at €1.5/kg, comparable to fossil fuels, even more so if oil prices rise above $100/barrel in strife-ridden times. The electrolyzers will eventually be fed by 90 GW of solar power. A company spokesperson noted that the project "constitutes a complete industrial ecosystem, covering the entire green hydrogen value chain (upstream, midstream, downstream and finance)."[145]

Understanding the importance of diversification in an energy-poor country, Japan has been a leader in hydrogen technology, both for storage and transportation. In 2020, a 10-MW solar-to-hydrogen pilot plant started up in the Fukushima prefecture, almost a decade after the Daiichi triple reactor meltdown. Importing liquefied hydrogen (LH2) has also begun, made from Australian coal, albeit increasing emissions if the CO_2 by-product isn't captured. Although hydrogen liquefies at a lower temperature ($-235°C$) than methane, existing LNG transportation technology is also being expanded to include hydrogen. The *Suiso Frontier* delivered a world-first cargo of pure LH2 in 2021 from Australia to Japan, where it was burnt for heat and power in a mixed-fuel, hydrogen–methane hybrid turbine at a Kobe pilot plant.

Hydrogen storage also scales more easily than lithium batteries to augment baseload grid power, doesn't degrade over time, and can be retrofitted to or blended with natural gas burners and turbines. The goal is to transition existing combined cycle gas-fired turbines (CCGT) from methane to 100% hydrogen, while the Japanese government also wants to increase the number of fuel-cell vehicles from under 4,000 in 2019 to 200,000 by 2025. With plans to transition from natural gas to hydrogen, the EU is also hurriedly building up infrastructure. The North Sea port cities of Hamburg, Wilhelmshaven, and Brunsbüttel (floating) are being turned into import hubs for LNG and LH2, piped from Scandinavia or shipped in from elsewhere.

                                    ***

Not all green hydrogen has to come from the wind or the Sun. In "biological microbiological methanation" (BHM), an electrolyzer creates hydrogen gas that can be combined with carbon dioxide – for example, generated from rotting biomass or in an anaerobic digester – and microbes to make methane, which is then burned in a CCGT to make electricity or injected into a natural-gas pipeline for consumer use, considered a low-carbon fuel if the CO_2 comes from a biological source. Fortunately, the microbes take what they want – that is, what we don't want! – while excreting the liquid fuel.

One such demonstration plant in Copenhagen uses a patented single-celled microorganism biocatalyst (biocat) called archaea to convert stranded electricity to pipeline-grade natural gas at low-demand times, which is then injected directly into the natural-gas supply. The two-step Electroarchea P2G process creates the hydrogen gas with an off-the-shelf electrolyzer fed into a separate bioreactor containing the archaea and mixed with carbon dioxide from a biogenic or industrial source.

Created by Jimmy Carter during the 1970s oil crisis, the Aquatic Species Program of the US Department of Energy (DOE) began research on algae in 1978 to create hydrogen as a possible alternative transportation fuel and to store CO_2. Although the program was eventually terminated, other potentially fruitful biofuel research projects are still in operation at the DOE's Bioenergy Technologies Office. Half oil by weight, algae includes a wide range of living organisms such as cyanobacteria, fungi, giant kelp, seaweed, and chlorella that like plants convert sunlight into energy via photosynthesis. Algae are very fast growing, can grow on marginal agricultural lands that require only water and sun, and have no pollutants. Wastewater or recycled water can also cultivate the algae.

Most systems maximize photosynthesis using a photo-bioreactor (PBR), a transparent container that slowly circulates algae with nutrients and CO_2. The Colorado School of Mines has been investigating photosynthetic processes and carbon fixation in algae with funding from ExxonMobil, one of the oil majors keen to develop alternatives to petroleum-based liquid fuels.[146] The interest in algae as an alternative to traditional transport fuels is growing, such that algae is being touted as a new darling by Big Oil under the dressed-up tag line: "It's tiny, green, and could be the future of biofuels."

Such carbon-negative, liquid-fuel production can also operate in a viable carbon capture system, especially as the technological limitations and economies of scale are worked out. A University of Waterloo project mimics photosynthesis to produce synthetic methanol gas (syngas) in the same way that plants absorb sunlight and CO_2 to make plant energy, a process likened to an artificial leaf. "A leaf produces glucose and oxygen. We produce methanol and oxygen," noted Yimin Wu, lead engineer on the project.[147] Another process, developed at the Carbon Recycling Institute in Iceland, combines hydrogen gas and waste geothermal CO_2 to produce methanol as a fuel additive (20%) or on its own in a methanol combustion engine (M100). The reaction is not entirely carbon negative, but produces no soot and can reduce carbon emissions by 90%. Although hydrogen is the more mature technology, methanol may eventually compete with hydrogen as a marine transportation fuel, for example, replacing bunker fuel, either by converting emissions to liquid or in a P2G system.

Natural hydrogen (a.k.a. "gold" or "white" hydrogen) is also being explored, tapped from subsurface underground stores. As reported in a 2023 *Science* article, "There might be enough natural hydrogen to meet burgeoning global demand for thousands of years, according to a US Geological Survey (USGS) model."[148] Still in its infancy, hydrogen boreholes are being drilled across the globe, including in Nebraska in 2019 and in the Lorraine region of northeastern France in 2023. Unlike petroleum, natural hydrogen is replenished underground as water reacts with iron minerals under high temperature and pressure, although the process is not completely understood or assured of commercial success. As noted by a geophysicist at Grenoble Alpes University, "Interest is growing fast, but the scientific facts are still lacking."[149]

The second Bush administration was keen on hydrogen fuel cells to the tune of $2 billion in research money, later shelved by Obama, but fuel cells are still expensive, as are hydrogen storage stations. Hydrogen gas also requires more space, up to six times that of gasoline, drastically restricting range. There is still much to do to hasten the timeline as stated in a 2004 *Scientific American* article: "Fuel-cell cars, in contrast [to hybrids], are expected on about the same schedule as NASA's manned trip to Mars and have about the same level of likelihood."[150] Hydrogen's biggest stumbling block is ramping up the scale of clean-energy electrolyzers, whether solar or wind.

There are always teething problems in any nascent industry, yet FCEV development has benefitted from a number of test programs, some discontinued and others still in operation. Iceland was the first to test hydrogen-powered vehicles for city transport, three DaimlerChrysler buses fueled by the world's first hydrogen refueling station in Reykjavik. Partly funded by the EU, the

€7 million ECTOS project was operational from 2003 to 2007 and worked through a number of technical issues involving safety, standards, and costs to integrate hydrogen technology into the Reykjavik public transport system. Only fresh water and electricity were used to produce the hydrogen in the compression, storage, and dispensing station.

The scenic resort town of Whistler, British Columbia, operated 20 buses with hydrogen fuel cells for the 2010 Winter Olympics, although the project was terminated after 5 years because of high costs, increased maintenance, and excessive fuel-importation distances. As noted by the program director, "[T]hese buses actually run, and they run well. It's a successful thing. But definitely things are not at a place where they can self-sustain."[151]

In 2015, Toyota launched its first fuel-cell electric car, the Mirai (未来), meaning "future," with a range of almost 300 miles. Japan is a country of 127 million people – about the size and shape of California – but has no natural petroleum reserves of its own and thus FCEV, PHEV, and BEV development are at the forefront of creating affordable transportation infrastructure. Parlaying extensive electronics expertise into large-scale, new-energy systems, Japan has championed EV technology for decades, once boasting over 10% of global sales (140,000/1.3 million in 2015[152]).

Toyota also sponsors *Energy Observer*, the first hydrogen-powered ship that makes hydrogen from seawater instead of fossil-fuel sources, employing a mix of Mirai saloon and Prius hybrid engines with Li-ion battery storage. The imposing 30.5-m maxi catamaran operates entirely carbon-free, augmented with solar cladding, two vertical-axis wind turbines, and wave power. Launched in 2017 from Saint-Malo, France, the zero-emissions vessel has visited more than 40 countries as it continues voyaging around the world.

The equivalent Kitty Hawk hydrogen-fuel-cell, e-plane moment occurred on September 29, 2016, when the four-seater DLR-HY4 glider took to the skies for a 15-minute test run at Stuttgart Airport, while 4 years later the first commercial-grade hydrogen fuel-cell plane – Zero Avia's Piper M-class six-seater – stayed aloft for a full pattern circuit at the company's R&D facility in Cranfield, England.

Developed by the French rail transportation company Alstom, the first hydrogen fuel-cell train, the Coradia iLint, began running in 2018 on a 100-km line in northern Germany. With a range similar to a diesel train, the project's manager noted, "buying a hydrogen train is somewhat more expensive than a diesel train, but it is cheaper to run."[153] Alstom is betting that the 5,000 diesel trains slated for retirement across Europe by 2035 could run on clean hydrogen, offering a cheaper alternative to grid-tied electrification. The port of Antwerp–Bruges began converting its tugboat fleet to hydrogen fuel in 2023, a potential bonanza with 200,000 tugs operating in ports worldwide. Dubbed Hydrotug, two V12 medium-speed engines can run either on hydrogen or traditional fuel.

In 2020, Airbus announced plans to fly the first commercial hydrogen-*combustion* plane by 2035, using a modified gas-turbine engine similar to Isaac de Rivaz's 1806 engine. Hydrogen combustion is not zero-emission, however, because NOx PM is possible and thus hydrogen can't qualify as a 100% green fuel when burned, although no nasty hydrocarbons are produced – water vapor is the main combustion product. Combusted hydrogen, however, is commercially attractive because existing internal combustion engines can be employed with little modification.

Potentially more realistic in the short term, the German truck maker Daimler is building a long-haul, liquid-hydrogen fuel-cell-powered 18-wheeler called GenH2. The essential refilling infrastructure will be supplied by Shell along a northern European hydrogen corridor between Rotterdam, Cologne, and Hamburg, managed from an import hub at Wilhelmshaven in northwest Germany. The first trucks off the blocks are expected to sport an impressive range of 1,000 km, easily competitive with current high-polluting diesels. Costs are typically the rate-determining step, but "diesel parity" is expected by 2027, while Hydrogen Europe predicts "10,000 hydrogen trucks on Europe's roads by 2025 and 100,000 by 2030."[154]

With improved engine technologies, powered either by fuel cells or chemical batteries, the roads are becoming cleaner, while shipping and aviation are still in the early stages of development. There is plenty of work ahead, but the prospects are good to challenge petroleum's monopoly on transportation fuel.

***

The battle lines are being drawn over the future powertrain to electrify the world's EVs: lithium versus hydrogen. Some think the market should decide, and what's important is to wean ourselves off of petroleum, while other proposed fuels include propane, ammonia (for example, combining hydrogen and nitrogen), methane (compressed natural gas), and methanol, but also suffer from high costs, infrastructure problems, and pollution. As a viable liquid fuel, hydrogen gas still has a number of drawbacks that have so far kept FCEVs from competing with the more successful BEVs: lack of infrastructure (although that can be built up), space problems (expensive, compressed, high-pressure storage needed), energy required to produce hydrogen (not a natural resource, but green if made from wind-, solar-, or biological-powered electrolyzers), and very few models available of either the fuel-cell or combustion variety. Nonetheless, as hydrogen technologies continue to mature, the competition is heating up.

Hydrogen's many detractors cite space, poor filling infrastructure, and safety (hydrogen is highly flammable when mixed with air) as deal breakers. Although much lighter and with 2.6 times the energy of gasoline, hydrogen gas nonetheless takes 6 times the volume to get the same bang for the buck because of a lower conversion efficiency.[155] Filling capacity is also limited and new stations expensive to build, but with roughly 3 times more energy per unit mass than diesel and 2.5 times natural gas, hydrogen is at least more efficient than gasoline to run. Fortunately, safety is not as worrisome as one might think. The Hindenburg airship famously caught fire while docking in New Jersey in 1937 – likely from a spark after a hydrogen gas leak, resulting in the death of 36 people – but fuel cells are much safer today, not least because of the smaller volume, solid metal containment, and fast dispersal of the lighter-than-air gas.[156]

The cost of powering a global fleet of over 1 billion vehicles, however, is an enormous task, replete with similar distribution issues to gasoline. Not nearly as problematic for electric vehicles that can easily plug into an existing, electric power grid, building up an entirely new infrastructure from scratch is a tall order. Furthermore, while some believe FCEVs will be a part of the future electric-transportation mix in some capacity, others worry that a limited number of suppliers will ultimately control production and distribution as with gasoline, the hype over hydrogen more about establishing a lucrative filling network than going green, similar to how natural gas was sold as a better cooking alternative to electricity. ExxonMobil, Chevron, BP, Shell, and TotalEnergies are all investing heavily in hydrogen hubs or clusters to take advantage of a potentially huge revenue stream. Conversely, liquid fuels don't need to rely on the grid, which is more susceptible to tampering, malfunction, and in some cases also subject to monopoly control.

Some think we should just burn the methane itself instead of using it to produce hydrogen because it is cheaper and more efficient, but both methane and hydrogen manufacturing (via natural gas reformation) are far from green, releasing vast amounts of carbon monoxide and carbon dioxide into the atmosphere. Hydrogen gas blended up to 20% with natural gas is also being studied to determine the burn properties for home delivery via existing natural gas pipelines, although substituting hydrogen made from methane is hardly a step in the clean direction. What's more, the engineering and delivery details are still murky.

Others are keen to note that hydrogen is not a *source* of energy, but only a means to *transfer* energy. Much like how Turlough Hill uses energy to raise water that is later lowered, energy is needed to create the hydrogen and oxygen gas in the first place, which is then later recombined in a fuel cell at our convenience (the round-trip efficiency of hydrogen is only 50% compared to pumped storage hydro at 75% and chemical batteries at 95%). Although large-scale, green hydrogen via renewable-energy electrolyzers is a viable production process that can easily slot into a P2G energy-storage system or as a transportation fuel to reduce greenhouse gas emissions and pollution, some are worried that hydrogen electrolyzers will reduce the amount of wind and solar energy available to decarbonize the grid.

Despite hydrogen's many problems, there are efficiencies to be made from the high-calorific value of hydrogen gas (~140 MJ/kg). But while fuel prices are comparable for hydrogen and gasoline (~$4/gallon), the sticker price is still daunting – a Toyota Mirai costs $58,385. Elon Musk thinks fuel cells are "a load of rubbish," but then FCEVs are in direct competition with Tesla and other BEV manufacturers. Engineer and author Saul Griffith thinks hydrogen vehicles are canonically silly when 50% is lost to make and then convert electricity to power, just "for the convenience of having a familiar fuel to fill a familiar tank."[157]

Nonetheless, some cities are testing hydrogen-powered buses (for example, eight on the London Covent Garden to Tower Gateway route) as part of green-transport initiatives, such as the Clean Hydrogen in European Cities (CHIC) project operating in Madrid, Hamburg, and London. In the USA, an NREL-funded, alternative-fuel evaluation program hopes to reduce dependence on imports, diversify energy sources, and decrease pollution and GHG emissions. We can expect to see more hydrogen buses competing with battery-powered buses on inner-city routes as well as vying for more government funding.

Berkeley physicist Richard Muller thinks "Hydrogen fuels are great for rockets, since for them weight is more critical than volume, and elaborate safety measures can be taken to allow the use of liquid hydrogen. Hydrogen might conceivably make sense for airplanes. But for autos, having super lightweight fuel doesn't add value."[158] Muller also believes FCEVs are more about establishing green credentials for manufacturers and wonders why we don't just run a car on methane in an existing engine, skipping the conversion process altogether. About the Mirai, he predicts "Like the Tesla Roadster, it will be left behind in history."[159] Stanford's Tony Seba thinks the argument is moot, citing a three-times higher efficiency of BEVs compared to FCEVs and the multi-*trillion*-dollar infrastructure needed to roll out hydrogen-gas filling stations: "If you look at it from the economics or the technology point of view, I think this is a no-brainer, I think hydrogen is a non-starter."[160] Standard conversion efficiency calculations peg fuel cells at around 19–23% comparable to 17–21% for petroleum, both far inferior to battery electric vehicles at 70–90%.[161]

Bill Gates also weighed in on how to solve the nasty problems of transportation emissions, basing his analysis on the importance of reducing GHGs to net zero by 2050 to avoid the "catastrophic" impact of man-made global warming: "It's rare that you can boil the solution for such a complex subject down into a single sentence. But with transportation, the zero-carbon future is basically this: Use electricity to run all the vehicles we can, and get cheap alternative fuels for the rest."[162] He cites cars, trucks, and buses in the first group that are indeed now being electrified everywhere and long-haul trucks, container ships, and airplanes in the second group that are slowly being adapted to run on more-efficient, lower-emission liquid fuels such as biofuels and hydrogen gas. Pioneering eco-explorer and pilot Raphaël Domjan, who captained the first solar-powered boat to circumnavigate the world and hopes to be the first to fly a solar plane beyond the stratosphere, also believes that short-haul planes will be electric within 30 years while long-haul flight will be powered by hydrogen.[163]

One should also note that the common knock about hydrogen not being a source but only a carrier of energy (or storage medium) – one that takes more energy to produce than is released in a fuel cell – also applies to gasoline and diesel. We shouldn't forget that petroleum was produced in the ground after a long geological process – for example, "diagenesis" that creates a physical and chemical change in compacted sediment and the formation of kerogen from degraded living matter – stored *in situ* and free to be dug up and used by us as needed. Hydrogen, however, is replenishable, clean, readily produced on demand, and essentially "free" if generated via wind-, solar-, or biological-powered electrolyzers. As fossil fuels continue to play havoc with global temperatures, the hydrogen economy can only grow and may indeed give lithium batteries a run for their money.

Not everyone is putting the green horse before the industry cart, however, as proponents and detractors weigh in with their analyses of an industry still very much in its infancy. Muller even argues that US automobiles haven't contributed that much to global warming ($1/40°C$ in the last 50 years and an estimated $1/25°C$ in the next 50 years),[164] presumably advocating for the status quo, neglecting that if developing countries consume as in the USA – 5% of world population, 25% of emissions, almost one car per person – the fractions will add up to greater than $1/2°C$ ($20/25°C$ according to my simple analysis). With rising average global temperatures of around $1°C$ already causing major climate problems, between one half and one degree Celsius is nothing to sneeze at.

Predicting winners and losers is notoriously difficult. Muller declared EVs to be a fad, primarily because of the expensive Li-ion battery pack and presumed high-replacement costs, calculating that the Tesla Roadster Li-ion battery cost $44,000 (40% of a $110,000 sticker price), the Chevy Volt 375-lb battery $15,000 (37.5% of $40,000), and the Nissan Leaf 400-lb battery $16,000 (46% of $34,700). Clearly, the battery is a hefty percentage of an EV's overall cost, more so if replacements are needed, but Muller also indicated little hope for any type of electric vehicle, oddly citing the limited range of a lead–acid battery car and GM's dumping of the fabled money-losing EV1 because of its expensive NiMH battery.[165]

But as the weight and cost of Li-ion batteries continues to decrease while energy density improves – perhaps doubling in the next decade – lower sticker prices will follow even without government incentives (as we will see next). Battery power is also becoming more efficient, lowering future replacement costs. Add in substantially lower fuel costs – $500 per year of electricity for an EV versus $2,500 for gas/diesel/hydrogen fill-ups – less engine maintenance (no oil changes needed), and huge green savings, more drivers are choosing electricity as the better, or indeed only, option.

Hydrogen will always have a future in limited transportation applications, low-temperature physics, high-temperature steel and cement manufacturing, and modern power-to-gas storage systems, but has a long and perhaps winding road ahead to become a major transportation fuel. Although not as daunting for localized use, a complete infrastructure overhaul is no small undertaking. One imagines Gottlieb Daimler, Karl Benz, Ransom E. Olds, and Henry Ford had similar concerns at the beginning of the petroleum era.

6.9 Industrial Charge Storage: The Future of Lithium

Charge storage is a central challenge to modernizing the grid, requiring more and larger-capacity batteries, while the future of both the electric vehicle and backup-storage market depends primarily on the availability and cost of lithium, the main ingredient in a Li-ion battery (LIB). China, Korea, and Japan lead the way in LIB manufacturing, although lithium is mostly extracted from hard-rock mining operations in Australia (47%), Chile (30%), and China (15%). Mainly found in high-altitude salt flats buried under deep sediment in the Andes, about half of the world's identified lithium resources are located in the so-called "lithium triangle" of South America, shared by Bolivia (21%), Chile (20%), and Argentina (11%), while Australia kicks in at 8% and China 7%.[166]

Because of its extreme reactivity, lithium does not exist in nature as a free metal, but is primarily found in rocks, soil, and water as either lithium carbonate (Li_2CO_3) or lithium hydroxide (LiOH), for example, where lithium salt has washed out from the mountains by erosion and separated by evaporation. In southwest Bolivia at a height of almost 4 km above sea level, Salar de Uyuni is the world's largest lithium lake at over 10,000 km^2. Created millions of years ago by subduction when the ocean floor was pushed up during the movement of two tectonic plates to form the Andes (salar means "salt flat" in Spanish), the lithium resides underneath a meter-deep crust of sodium chloride brine and requires pumping to remove. Holding about one-fifth of the world's identified lithium resources, a former Bolivian vice president called Salar de Uyuni an "infinite table of snowy white," and wished lithium could become "the engine of our economy."[167]

A component of the hydrogen bomb, the bottom fell out of the lithium market at the end of the Cold War in the early 1990s when the US Department of Energy started selling off stocks, but the price rebounded when Sony introduced its Li-ion battery and spiked again with the construction of Tesla's Nevada Gigafactory. A major component in pharmaceuticals and ceramics, lithium is now at the forefront of an evolving electric-charge storage industry. Trading at about \$5,000/metric ton (tonne), lithium-containing rock is known as spodumene (6% lithium oxide) and is mostly converted to battery-grade lithium in China.[168]

The competition for "white petroleum" has never been stiffer. Known deposits are on the order of about 100 million tonnes, 130,000 (~0.1%) of which were mined in 2022.[169] Three companies have historically controlled most of the production market – Chile's Sociedad Química y Minera (SQM) and two American corporations, FMC and Albemarle.[170] Reflecting its new-found worth, lithium carbonate (and lithium hydroxide) prices are highly volatile, tripling from 2015 to 2018 to over \$17,000 per tonne, while the Chinese spot price tripled in a single year to more than \$20,000 per tonne.[171] In 2022, lithium carbonate prices doubled from \$35,000 to \$67,000 per tonne, but may ultimately decrease as the main players ramp up capacity to discourage new investors.

Depending on the model, a single Tesla requires about 5–65 kg of lithium compounds or 1,000–10,000 cell phones worth, and thus a simple analysis shows that Tesla needs on average about 35,000 tonnes of lithium carbonate per year for an annual production of one million cars, roughly a quarter of current global production (there are 1,000 kg in a tonne). A Powerwall uses less than one-tenth the juice of a car, but will also put a dent in supplies as backup electrical storage grows. With a current estimated supply of recoverable lithium at almost 100 million tons, there is plenty to fuel the revolution.

In the meantime, Bolivia is aiming to become the world's "salt hub," producing its first major shipment of 10 tonnes of lithium carbonate in 2016 and hoping to one day set the global price.[172] Commercial-scale production, however, has been slow to develop, not least because of the lower evaporation rate and higher production costs for high-magnesium lithium deposits found in Bolivian salt flats. As more lithium is extracted, concerns have also been raised over salt-flat destruction and the depletion of water resources – 2,000 tons of water is evaporated in a couple months to make 1 ton of lithium.[173] Careless lithium extraction also causes pollution, harms native animal species (especially birds), and impacts local communities.

Production in a politically unstable region also comes with its own risks, heightened by former president Evo Morales stating that Bolivia wanted to make its own batteries instead of sending raw materials downstream. Preferring "partners, not owners" to ensure that all Bolivians share in the development of an estimated 20-million-ton supply, Morales was keen to create a domestic value chain with local factories (a.k.a. 100% *estatal*), but was ousted in a 2019 coup that may have had more to do with securing battery-grade lithium than politics.

With the election of another leftist leader, Gabriel Boric, in 2021, Chile is also seeking "new governance" for its 10 million tonnes of lithium, primarily located at Salar de Atacama in the north of the country, comprising 1,200 km^2 of salt flats. A proposed constitutional change would have nationalized all natural resources in Chile, including some of the world's largest lithium mines, but failed in a 2022 referendum. In 2022, the Mexican government created a state body to manage its lithium production, effectively nationalizing the industry, although exemptions were made for existing mining companies. Not as advanced as in Chile, a lithium mine in the northwest state of Sonora near the American border could potentially be the world's largest, pegged by some at over 200 million tonnes. A Chinese firm that has supplied batteries to Tesla plans to extract 35,000 tonnes per year.[174]

Around the world, old lithium mines are fast being upgraded to supply an expanding market. Lithium mining has increased in Cinovec, Czech Republic (European Metals Holdings), Tras os Montes, Portugal (Dakota Minerals), and Sonora, Mexico (Bacanora Minerals). The entire economy of Afghanistan was re-evaluated after lithium was found there in 2010, potentially worth $1 trillion, dwarfing its current $12 billion GDP. Based on data developed by American geologists from old Soviet mining maps made during the 1979–1989 Soviet occupation, a Pentagon-led team of geologists found battery-grade lithium in ground surveys of dry salt lakes in western Afghanistan. The then head of US Central Command, General David H. Petraeus, stated "There is stunning potential here. There are a lot of ifs, of course, but I think potentially it is hugely significant."[175] Underscoring the importance of protecting global supplies, talks have started over the creation of an OPEC-like lithium cartel.

Despite some of the world's largest lithium reserves, Silver Peak in Nevada is the only large-scale mine in the USA. Run by Albemarle and first opened in the 1960s, the heavy-brine site produces 5,000 tons per year or under 4% of current world production.[176] US production is set to increase, however, with two new sites employing different extraction methods. Lithium brine will be extracted from a 400-foot aquifer at Salton Sea, California, a large lake about 50 km north of the Mexican border, possibly in connection with a proposed geothermal power station. Less environmentally friendly, open-pit mining is being developed at McDermitt Caldera, an extinct super volcano near Thacker Pass, 100 km north of Winnemucca in northern Nevada. Electric propulsion is much cleaner than an internal combustion engine, but mining comes with its own set of red flags. The worry is that large amounts of water will be polluted or depleted, lowering the water table and impacting farming.

Salton Sea is particularly attractive, potentially holding enough lithium to power the entire American battery market. California's "Lithium Valley" is hurriedly being mapped to learn more about where the lithium comes from, the rate of decline after extraction from the brine, and whether geothermal lithium is replenished from the underlying rock. Mining practices for geothermal lithium are also considered more environmentally friendly than either salt-flat evaporation as in Bolivia and Chile or open-pit mining as in Australia and China. Possibly containing as much as 6 million tonnes, good for 100 years, California's governor Gavin Newsom called California "the Saudi Arabia of lithium."[177] Others are attempting to separate lithium from heated brines at existing geothermal plants.

Wherever the lithium comes from, the race is on to develop a stable, long-lasting, and cost-effective battery, both for electric vehicles and the expanding grid-storage industry to help manage a growing and continuously fluctuating consumer demand (see Table 6.2). Almost all EV batteries (96%) consist of a lithium cobalt-oxide (LCO) cathode, a graphite-based anode, and an organic liquid $LiFeP_6$ salt electrolyte that provides the lithium ions (Li^+) and electrons (e^-). Current research is directed toward improving the cathode chemistry, either lithium nickel–manganese–cobalt oxide (NMC) or lithium nickel–cobalt–aluminum oxide (NCA) found in most EVs today. Tesla uses NCA, but roughly 60% of Li-ion batteries are NMC, in particular NMC622 (that is, $LiNi_{0.6}Mn_{0.2}Co_{0.2}O_2$ in the $Ni_xMn_yCo_z$, $x + y + z = 1$ nomenclature).[178] The same is true of lithium batteries for grid storage technology (GST).

Table 6.2 *Lithium-ion (Li-ion) and lithium-ion polymer (Li-Po) battery chemistries*

Battery	Abbreviation	Formula	Energy (Wh/kg)	Main uses
Lithium cobalt oxide	LCO	$LiCoO_2$	200	Phones, laptops, cameras
Lithium NMC oxide	NMC	$LiNiMnCoO_2$	220	
Lithium manganese oxide	LMO	$LiMn_2O_4$	150	EVs, e-bikes, power tools, medical
Lithium iron phosphate	LFP	$LiFePO_4$	120	
Lithium NCA oxide	NCA	$LiNiCoAlO_2$	260	EVs and grid storage
Lithium titanate oxide	LTO	$Li_4Ti_5O_{12}$	80	

Source: "BU-205: Types of lithium-ion," Learn About Batteries, Battery University, October 22, 2021. https://batteryuniversity.com/article/bu-205-types-of-lithium-ion.

Reducing the cobalt content, however, is becoming a growing issue because of unethical mining practices in the Democratic Republic of Congo, where 70% of the world's supply of cobalt is found, much of it recovered by children and other cheap laborers using shovels and flashlights in unsafe working conditions. The Congo is also rife with corruption and insider deals. In 2019, the London Metal Exchange (LME) banned the trading of irresponsibly mined cobalt from 2022. Today, battery research is focused on switching to a non-cobalt cathode chemistry to avoid unsafe mining, in particular lithium–iron phosphate (LFP), which is also more thermally stable but has a lower energy density. The flammable liquid electrolyte is always a danger as battery sizes shrink and thus solid electrolytes are also being explored.

John B. Goodenough, co-inventor of the Li-ion battery, noted that "Cost, safety, energy density, rates of charge and discharge and cycle life are critical for battery-driven cars to be more widely adopted."[179] Acknowledging that a Tesla was still not affordable to the masses, CEO Elon Musk stated that the battery was the main issue. He hoped to make more powerful, longer-lasting batteries in house at half the cost, via automation, larger cells, and recycling, while also reducing the amount of cobalt. In 2020, Tesla began producing batteries with reduced cobalt as well as LFP batteries that shipped with new Model 3s in China. Tesla also announced that a 400 Wh/kg battery was doable, increasing EV range and facilitating the transition to electric flight. Three years was proposed as a time frame.

The control of lithium supplies could determine the ultimate winners and losers in a future energy market as the Battle of the Batteries begins. Musk speculated that 100 Gigafactories are needed to free the world of fossil fuels, well beyond the capability of his company, while significantly reducing global lithium reserves. With six Gigafactories already in operation, the location of Tesla's first Gigafactory outside Reno, Nevada, was strategically important, within easy delivery distance to its EV plant in Fremont, California, and 3 hours from Clayton Valley, Nevada, site of the only working American lithium mine (Albermarle's Silver Peak). The Gigafactories are a start, but much more is needed to fuel the revolution.

Headquartered in Ningde, southeast China, in the mining-rich province of Fujian, CATL – short for Contemporary Amperex Technology Co. Limited – also sells batteries to most major carmakers, including GM, VW, and Tesla. The Chinese government's policy of subsidizing EVs only for those vehicles powered by Chinese-made batteries helped CATL to become the #1 car-battery maker in the world with almost one-third of the global market (South Korea's LG Energy is #2 at 20%).[180]

China already has more than 10 times the EV-battery capacity of the USA, while CATL is planning a battery-making plant in Hungary three times the size of Tesla's Nevada Gigafactory (its second in Europe after a first in the works in Germany).[181] According to one analysis, 282 GFs could be up and running within a decade, increasing global capacity to about 6,000 GWh, 10 times current supply. Joint operations are slated for Tennessee and Kentucky by Ford and SK Innovation and for Michigan, Ohio, and Tennessee by GM and LG Energy.[182] All the major players are scrambling to catch up.

Having lagged behind the USA, Germany, and Japan, China is leapfrogging ICE technology with LIBs, turning the next big thing into a competition between petroleum-based propulsion and a LIB-powered future. Whoever can safely increase the energy density beyond 240 Wh/kg, reduce charging times, and lower prices will reap the enormous spoils of an industry expected to power half of all cars by 2030, a more than five-fold increase in the next decade.

***

As battery chemistry improves, the quest for the "million-mile" battery intensifies. Goodenough and a team of engineers at the University of Texas developed the first all-solid-state battery (ASSB), increasing energy density more than three times with an alkali–metal anode made of lithium, sodium, or potassium and an inorganic solid-glass electrolyte to reduce lithium dendrites. Likened to "metal whiskers," dendrites are the uneven build-up of Li-ions at the anode during recharging that can cross the interface and short circuit, leading to an explosive fire. The battery is non-combustible, fast-charging, long-lasting (1,200 cycles), and operates at –20°C.[183] Because of a solid rather than liquid electrolyte, the battery is lighter, smaller, and doesn't degrade as much, allowing for faster recharging and a longer life.

Another LIB chemistry being developed at Rice University uses porous carbon in the anode to mitigate dendrite formation and thermal overload. Showing a similar capacity to previous batteries, the charging rate was 24 times faster, allowing for a full recharge in 5 minutes instead of 2 hours, ideal for high-capacity batteries.[184]

In 2021, the Israeli company Storedot announced it was developing an "extreme fast-charging" production-line LIB that could recharge in 5 minutes to overcome range anxiety, comparable to liquid refueling. StoreDot CEO Doron Myersdorf stated "You're either afraid that you're going to get stuck on the highway or you're going to need to sit in a charging station for two hours. But if the experience of the driver is exactly like fuelling, this whole anxiety goes away."[185] The battery anode is made of germanium-based nanoparticles to store the Li^+ ions rather than graphite that turns into metal and can short circuit when rapidly charged. Further plans include transitioning to silicon, a technology also under development by Tesla and other major battery manufacturers. Rollout could begin soon, while the cost is expected to be the same as existing Li-ion batteries. As Myersdorf succinctly noted, "batteries are the new oil."[186]

Sodium and magnesium batteries could also be important for inorganic electrode chemistries as could sulfur in an improved lithium–sulfur (Li-S) liquid battery. Liquid sulfur is cheaper, more plentiful, and lighter, and thus able to store more charge, but manufacturing a lasting Li-S battery is not easy because the liquid cathode gradually dissolves into the electrolyte and becomes metallic, forming dangerous dendrites. Current research is based on making the Li-S cathode solid to prevent dissolution and adding carbon to reduce dendrite formation.

Sodium-ion batteries may provide the ideal solution for large-scale backup batteries. Sodium ($_{11}Na^{23}$) is heavier than lithium ($_3Li^7$) with a relative atomic mass of 23/7, but is much more plentiful as in common sodium chloride sea salt (NaCl). Sodium may not work for EV battery packs that need to be as light as possible, but is fine for stationary, charge-storage applications where energy density is not as important. A redesign is required for the cathode, anode, and electrolyte, but no cobalt or nickel is used in the metal-oxide cathode, while more research is needed to perfect the larger-pore anode and electrolyte concentration. The battery is also not as susceptible to flammable discharge and thus safer.

Aluminum–air (Al-air) "batteries" may also play a role in future electric propulsion via a lighter battery pack, higher energy density, and a cheaper, more plentiful supply that doesn't rely on critical components such as cobalt. Technically a power source as in a fuel cell (for example, in a solid rocket booster), aluminum powder reacts with oxygen (air) to provide the juice. Swapped rather than charged, Al-air puts less strain on the grid, while the lighter, higher-power engine scales more easily to marine and aviation transport. At the very least, Al-air could be used for onboard backup power, extending range whenever the main Li-ion battery runs low. Iron–air batteries are also in development.

Not all storage needs are the same. As we've seen, 100 miles of electric road travel requires about 30 kWh, found at the low-end of today's EV (the smallest Tesla battery is 60 kWh). For home electrical consumption, a family of four might use 30 kWh per day (covered by 10 kW of panels at $20,000 with a $3,500 backup charge-storage system), whereas an off-grid couple might need only 4 kWh (easily covered by a lead–acid or aqueous-ion battery system). Other electrochemical technologies such as flow batteries are also competing in our new battery-run world. In contrast to regularly recharged lithium-ion batteries that produce a large current fast (perfect for cars and other electric transport), flow batteries hold a lot more charge and last longer, but don't operate as easily or quickly (better for home and industrial use[187]). Metal-free, organic batteries are also being developed that could reduce the environmental dangers and unsafe practices of mineral mining.

The technology of fast, affordable charge storage continues to evolve. The former US energy secretary and 1997 Physics Nobel laureate Steven Chu stated in 2010 that a rechargeable battery would compete with the internal combustion engine if it could last 15 years (5,000 deep discharges), had a 5-times higher storage capacity (3.6 MJ/kg = 100 Wh), and was reduced three-fold in price,[188] while in 2013 the Joint Center for Energy Storage Research (JCESR) received a $120 million DOE grant to find better ways to store energy "to make batteries five times more powerful and five times cheaper in five years."[189]

The DOE's Battery 500 program hopes to increase the useable energy density in EV batteries from 300 to 500 Wh/kg, while high-storage-capacity lithium metal anodes and improved conversion–reaction cathodes are needed to exceed 1,000 Wh/kg or even 2,000 Wh/kg for viable commercial aircraft.[190] The head of the Battery and Storage Technology (BEST) Center at Penn State University, whose group is also developing faster charging by carefully heating a LIB to 60°C to speed up the flow of lithium ions, predicted that cheaper electric vehicles are on the way.[191]

As reported by Moody's, battery prices halved from 2010 to 2015,[192] while according to Bloomberg, LIB costs decreased almost 10-fold over a decade from $1,100/kWh in 2010 to $156/kWh in 2019.[193] As LIB costs drop below $100/kWh, EV–gasmobile parity may come as soon as 2026, while a further reduction of more than 50% is expected over the next decade,[194] ultimately reducing the cost of a 60-kWh Li-ion EV battery pack to under $3,000.

As battery technology changes the electric-storage landscape, prices and performance will continue to improve, akin to the exponential growth of integrated circuits (Moore's Law) and solar cells (Swanson's Law). Even just tweaking existing technology can have a major impact. Sustainability expert Chris Goodall noted that "Small increases in the energy density of cells will, for example, mean lower packaging costs per unit of energy capacity. The amount of electrode material per cell will fall, the speed of coating will rise and the assembly rates will increase."[195] Indeed, success breeds success and lower costs, a.k.a. the "experience curve" or "learning curve" that sees higher output and slashed costs over time in a virtuous circle with improved manufacturing, quality control, and automation, according to classical economic theory where increased supply produces increased demand.

In a ramping EV and stationary battery market that continues to improve each year, the competition for raw materials is enormous, as is the concern over unethical practices. Currently, Asia produces more than half of all minerals (59% or 10.5 billion of a total 17.9 billion tonnes), followed by North America (16%) and Europe (7%).[196] A 2021 IEA report, entitled "The Role of Critical Minerals in Clean Energy Transitions," noted that the demand for lithium in renewable energy technologies could increase by 40 times (160 GWh in 2020 to 6,200 GWh in 2040[197]) over the next two decades to supply the EV and charge-storage revolution, putting increased pressure on supplies and raising prices.

As we've seen, metals other than lithium are important, especially cathode materials made of lithium plus one of either cobalt, phosphate, manganese, or a combination of metal polymers (known as a lithium-ion polymer or Li-Po chemistry). Cobalt is essential for the lithium–cobalt-oxide cathodes still found in most mobile phones, laptops, and camera batteries. The demand for metal battery components will continue to rise with increased electrification, such that more minerals will be mined in the next three decades than throughout all of human history, requiring hundreds of mines and doubling supplies by 2050. China is at the forefront, manufacturing up to 90% of EV batteries and over half the magnets found in wind turbines and electric motors, while controlling 95% of the rare-earths market (the lanthanide series of metals such as neodymium used to make magnets).[198]

An EV is not considered "zero-emission," however, without accounting for all "imbedded emissions" in the supply chain. Cobalt mined in the Congo can be refined in Finland, processed in China, and turned into production packs in Nevada before being installed in an EV in Fremont, California (20,000 air miles[199]), while invasive deep-sea mining has also been proposed to make up a predicted future shortfall. The challenge is to ensure that growth is based on ethically mined sources and environmentally friendly practices to furnish tomorrow's clean energy supply.

Providing $369 billion of funding for new energy in the next decade, the 2022 Inflation Reduction Act will produce a run on mineral stocks in old and new locations, but particularly in North America (or US trading partners), where 50% of a US EV battery must be made by 2024 to qualify for the full $7,500 US tax credit (100% by 2028). Canada will especially benefit with numerous critical mineral exploration projects in development, including high-grade nickel. Even the recalcitrant US senator Joe Manchin is on board with the push to use local materials rather than rely on potentially unfriendly or competitive sources. Canada holds about 3% of the global reserves of cobalt and lithium, much of it in Ontario and Quebec, and is the sixth largest producer of nickel at 130,000 tons extracted in 2021.

Recycling is also becoming a bigger issue as batteries lose their ability to recharge, but one expects a similar practice to lead–acid batteries, 90% of which are now recycled. So-called "unmanufacturing" of phones, computers, and EVs, whose batteries have not yet reached an end-of-life state, can replace manufacturing in a rigorous circular economy that regulates all minerals and metals, and is cheaper than mining. Cathode sustainability must also be included for less plentiful cobalt along with lithium, nickel, etc., where batteries are melted down or ground up and upcycled to recover their elemental components. Newer batteries are also being designed to facilitate easier recycling. So-called "urban mining" is also on the upswing to reuse valuable demolition waste such as copper, aluminum, and steel.

As J. B. Straubel, founder of the battery recycling start-up Redwood Materials and a Tesla co-founder, noted, "By recycling a higher percentage of those materials that go into the battery, we can reduce dramatically the embedded emissions. . . . The more times you reuse them, the less emissions they'll have from when they were first created."[200] Emma Nehrenheim, chief environmental officer at Swedish battery maker Northvolt, noted that used batteries are an invaluable resource, where 95% of the minerals (especially cobalt, nickel, and lithium) can be recovered by shredding and hydrometallurgy – separating out reusable material from the "black mass" – rather than smelting, and restored to the same levels of purity in a fully functioning, off-the-shelf battery.[201] Northvolt is aiming to produce the first 100% recycled Li-ion battery, essential to service hundreds of millions of EVs expected in the next decade.

In 2019, Tesla began recycling its own batteries on site at GF1, after previously using third-party recyclers. Other manufacturers are following suit, reducing the environmental impact of digging up minerals. As of the start of 2023, 91 companies in the USA, Canada, and Europe were recycling lithium batteries or planning to do so.[202] Most used EV batteries still have 80% of their charging capacity and can either be recycled or repurposed as stationary, second-use, grid-storage backup, helping to keep toxic chemicals and valuable minerals from landfills.

"Battery passporting" technology will also help to ensure that batteries are made without cobalt or contain only ethically sourced components, while the tracking of past use will facilitate recycling, repurposing, and EOL management. As Amrit Chandan, the CEO and co-founder of Aceleron Energy, noted, "End-of-life battery waste is fast becoming one of the biggest issues the industry needs to tackle," adding that manufacturers will soon be "exploring alternatives to spot-welding technology and other permanent-assembly techniques, which will enable batteries to be recycled more easily."[203] Optimizing charging protocols and improved software management also extends battery life.

Rechargeable batteries didn't change much after Gaston Planté's 1859 lead–acid battery or the nickel–cadmium battery in 1910, but are now rapidly advancing since the advent of lithium-ion cells in 1991, increasing in energy density and shrinking in size. The market for higher-performance batteries will continue to grow with increased infrastructure, while clean-energy distribution makes life better for both the environment and the pocket book. As James Stafford of OilPrice.com succinctly noted, "Electric vehicles will be the rule rather than the exception; and lithium the number one commodity of our time."[204]

6.10 An Energy Technology Revolution: You Ain't Seen Nothin' Yet

Invented at the end of the nineteenth century, the incandescent light bulb was safer, cleaner, and less cumbersome than earlier forms of arc or gas lighting, yet required an entirely new infrastructure to connect multiple sites via a central station dynamo that eventually lead to the building of massive national grids and electrical power transmitted across thousands of miles. From simple beginnings in Menlo Park, Holborn Viaduct, and Lower Manhattan, a whole new industry grew, financed by enormous investment, both private and public. Today, green energy, electric vehicles, and power storage require similar backing to build up a viable distributed power and charging infrastructure.

It seems we are ever reinventing the electric wheel. After the 1876 Centennial Exhibition in Philadelphia – the first US world fair that had only one crude electric machine for electroplating and another that ran electric current to a single arc lamp – early electrification exploded. By the time of the 1893 Chicago Columbian Exposition, where the largest-ever switchboard operated 40 circuits and powered 180,000 incandescent lamps to light the whole of White City, the world was changing beyond recognition.[205] In under two decades, the incandescent light bulb, copper-wire electric transmission system, poly-phase AC motor, AC transformer, electric railway (the novel third rail first demonstrated at the Chicago fair), and electric elevator (rising 185 feet to a breathtaking promenade, also first demonstrated at the same fair) had all been developed.

As Elihu Thomson, founder of the Thomson-Houston Electric Company (which had merged the previous year with Edison Electric), said upon seeing the marvels at White City, "No similar period in the world's history has in any art shown so rapid development, so extensive and refined scientific study and experiment, so active invention, so varied application, such care and perfection in manufacture, as has taken place within the electrical field."[206] In his biography of Nikola Tesla, one of the chief architects of the electric world, history of science professor Iwan Rhys Morus noted, "The success of the Columbian Exposition – and the key role of its electrical exhibits in that success – cemented electricity in the public mind as the technology of the future."[207] There was no turning back the hands of progress. After electricity's triumphant Chicago display, even more extraordinary changes were on the way, ringing in a modernity we now take for granted.

But getting from A to B is never easy, especially when the entrenched technology is rooted to ever-familiar and already functional ways. The tried and true of a complacent past can seem impossible to nudge forward. Historian Maury Klein observed two kinds of technological change: "The first type improves or upgrades existing technologies in various ways but maintains the basic format of the machine or system. The second and more radical type simply replaces the existing technology with one based on altogether different principles."[208] Today, we are embarking on a third that upgrades *and* replaces past technology, using what works and discarding what doesn't. As such, we maintain our modernity but transform the meaning. As long as we get the mix right, expanding public infrastructure with the same gusto that enhances private gain, our energy future will be safe. If we're smart, more of us will share in the spoils.

In *Industrial Archaeology in Britain*, social historian R. A. Buchanan cites three conditions for industrial revolution that applies to both the proliferation of machinery in the Middle Ages and the Industrial Revolution of the nineteenth century, and is still valid today: (1) "key groups of people who are prepared to consider innovations seriously and sympathetically," (2) "technological innovation is being encouraged to match social needs," and (3) social resources such as "capital, materials and skilled personnel."[209] To turn the revolutionary into the mundane, however, takes more work. In *Ten Technologies to Save the Planet*, Chris Goodall lists four phases for universal adoption of new technologies: too expensive, waning enthusiasm over the slow progress, gradual acceptance by skeptics, and finally a dawning sense that we can do without what came before, in this case fossil fuels.[210]

Since the beginning of the industrial transformation of Western civilization, energy has been at the forefront of change. In *The Third Industrial Revolution: How lateral power is transforming energy, the economy, and the world*, Jeremy Rifkin notes that "Every great economic era is marked by the introduction of a new energy regime," describing how in a first Industrial Revolution steam-powered technology merged with the printing press to create a print-literate workforce (1830s–1890s), while in a second Industrial Revolution the oil-powered internal combustion engine merged with electronic communications to create electrified factories and mass-produced books in the 1900s.[211] But to continue to transition to the next fully fledged revolution, we must change from a wasteful, thermodynamically inefficient, high-carbon and high-polluting, fossil-fuel-dependent system that underpins the entire global economy. To make the change to a low-carbon, distributed, clean-energy economy Rifkin itemizes five pillars: (1) renewables, (2) micro-generators/grids, (3) storage technology (especially hydrogen), (4) an efficient, energy-sharing "intergrid," and (5) electric vehicles and power-to-grid technology.

There is still much to do as we grudgingly begin the change from dirty to clean. Indeed, peer-to-peer, distributed energy resources not only undermines the corporate structure of petroleum, but also threatens government administration over every facet of human life – a veritable political, economic, and social revolution. We may be stuck in the routine of a comfortable past, but our reticence is being pushed not for the sake of change or to build a better mousetrap, but because we have no choice. We must clean up our act, reducing pollution and GHG emissions. Fortunately, the past can serve as a guide rather than restrict our options.

The road from invention to product is never easy. All early adopters struggle with use, facility, and even need. Some technologies also have unintended uses that can derail the original intention or create new markets, such as paying passengers riding on coal-hauling trains, home appliances from power-station lighting, social media from number-crunching computers, grid stability with electric-car charging, all of which become accessible to the wealthy before being made available to all. But whether Mathew Bolton's original Soho Manufactory – the world's first great machine-making plant in Birmingham that churned out James Watt's new engines – Edison's Menlo Park invention factory where he cobbled together investors for a slate of new ideas, Westinghouse and Tesla's banker-backed Niagara Falls power plant, or the hothouse incubation of cutting-edge technology in today's Silicon Valley, innovation can only be nurtured with old-fashioned money. Smart capital makes technology work, and is now needed to turn the page on our dark and dirty industrial past.

In *From Luxury to Necessity*, Sjoerd Bakker lists a number of steps for any new technology to succeed – invention, rivalries, enthusiasm, unexpected uses, value recognition, adoption by the middle class, economies of scale, devaluation of infrastructure, and mass adoption – citing examples from birth to maturity of three now commonplace technologies: steam power (Watt's condenser in 1776 to Stephenson's locomotive in 1829), electricity (Siemens's electric generator in 1867 to Edison's power plant in 1882), and the internal combustion engine (Benz's "vehicle with gas-engine drive" patent in 1886 to Ford's mass-produced, assembly-line Model T in 1908).[212] All had various uses before becoming fully practical.

We are seeing similar changes today as maturing energy sources (especially solar panels and wind turbines), energy storage (especially lithium batteries), and electric vehicles spur on increased investment and wider consumer adoption. Peter Diamandis, founder of the XPRIZE and author of *Abundance: The future is better than you think*, believes "The convergence of solar, batteries and EVs will democratize energy production and offer billions of people access to cheap, carbon-neutral energy." Citing a UBS Investment Bank study that estimated a 6–8-year payback time for unsubsidized investment for battery storage, rooftop solar, and EVs, Diamandis predicts the smart grid will soon be in place:

> By 2025, everybody will be able to produce and store power. And it will be green and cost competitive, that is, not more expensive or even cheaper than buying power from utilities. It is also the most efficient way to produce power where it is consumed, because transmission losses will be minimized. Power will no longer be something that is consumed in a 'dumb' way. Homes and grids will be smart, aligning the demand profile with supply from (volatile) renewables.[213]

Some of the changes are already here, limited only by access, imagination, and investment. In Vermont, Green Mountain Power (GMP) initiated a pilot project to install grid-connected "behind-the-meter" batteries in customer's homes. Owning and operating the equipment as a capital expense like poles, wires, and transformers, GMP runs the batteries at peak times to store excess energy and as emergency backup, while advanced software monitors the weather and electric load to optimize resources (batteries, thermostats, water heaters, solar panels) to provide real-time "power conditioning." Operating as a test run across the UC San Diego campus, DERConnect is integrating various distributed energy resources (DERs) into the wider grid. More than 2,500 DERs – solar panels, wind turbines, smart buildings, electric-vehicle batteries, and fuel cells – will be connected campus-wide.[214] Other community-operated micro-grids are being built wherever consumers want to save by exploiting their habits to share power across a cooperative platform (as we saw in the last chapter).

Electric vehicles can also sell on dormant power using vehicle-to-grid (V2G) technology. V2G helps smooth out a constantly fluctuating load by providing intermediate dispatchable storage, putting money into consumers' pockets as they upload their own localized "stranded" power as well as saving money for utilities that must rapidly switch on costly backup systems during high demand. With V2G, connected users can earn as much as $200 a month selling to the grid, netting $24,000 over a 10-year expected lifetime of the car. All by just hooking up overnight or on demand as needed.

EVs are already being employed as flexible grid resources in a joint BMW/PG&E pilot project called "i ChargeForward" that manages EV charging at cheaper, off-peak rates in the San Francisco Bay Area as well as providing a "second-life" solar-charged lithium battery microgrid for on-road charging. Still in the early stages of development, V2G trials have begun in Denmark, the UK, and the Netherlands, where EV adoption in Europe has been greatest. A Nissan Leaf was connected to the grid for the first time in October 2018, as part of the Amsterdam Smart City initiative, while in 2019, Renault began a bi-directional charging project in Utrecht to connect 15 Renault Zoes. When grid power is higher than demand the plugged-in cars charge; when grid power is lower than demand the cars start discharging power back to the grid.[215]

Announced in 2021 and scheduled to hit the market by 2023, Ford's all-electric F-150 Lightning pickup includes a battery system big enough to power an average-sized house for 3 days (even longer if no heat or air-con is needed), perfect to cover downtimes, storms, or to sell on to the grid. Ford's chief engineer, Linda Zhang, noted, "It's your own personal power plant, automatically powering your house for three days during an outage."[216] Some owners are already planning their own electric-powered tailgating parties and off-grid camping. Aside from a handy, vehicle-to-load (V2L) capability, the F-150 also comes with a range of 230 miles (standard) to 320 miles (extended) in case you want to drive it too.

Burnishing its credentials as an energy company, Tesla revealed in 2023 that it would roll out bi-directional charging capability over 2 years by tweaking existing onboard power electronics. CEO Elon Musk suggested, however, that Tesla drivers would be unlikely to use the technology without a home-installed Powerwall, keen to add value to his growing energy company. For those who can, bi-directional charging could net owners up to $1,500/year, "discharging power back to the grid during peak hours and charging back up during off-peak hours," according to one energy company at the forefront of optimizing EV charging.[217] Soon, all EVs will offer standard bi-directional V2G technology following on the heels of Nissan, Renault, Ford, Tesla, and Lucid (now majority owned by Saudi Arabia's Public Investment Fund), subsidizing the original outlay.

After numerous summer blackouts, South Australia is building the world's largest virtual grid with 50,000 prosumer-battery homes, starting in Adelaide with 100 free Powerwalls. As Melbourne's first residential Powerwall user noted, he sells excess power to the grid during high-peak times worked out by a smart meter and software interface, a notification message sent directly to his iPad: "GridCredits Event. Your battery has been requested to discharge for 44 minutes starting at 05:15 PM. You will receive $2.47."[218] Such negotiated contracts recall a time when prices were bartered, before department stores fixed a set price on goods, not least to make life easier for inexperienced salesmen and wary customers.

The smart grid will also be able to vary demand via statistical models that calculate expected use in real time, while the EV charging load is flattened via territory-specific data and demand-side algorithms, optimizing overall system-charging needs. For example, wind power is typically strongest at night when conventional demand is low, and is ideally suited to charge an idle EV battery while we sleep. The so-called "duck curve" that measures output versus time will be statistically managed from the bottom up to accommodate a widening network of distributed energy resources, including a growing number of V2G batteries on wheels.

As the number of distributed energy resources increases, however, backup power storage must also be optimized to account for the real-time "dirtiness" or carbon intensity of the grid. Despite saving money and helping to stabilize the grid by charging a battery or raising water, energy storage generated by coal or natural gas rather than solar or wind won't decrease GHG emissions and can in fact increase emissions, in part because the grid is still mostly dirty, the round-trip efficiency of charging and discharging not 100%. As Vox writer David Roberts notes, "In terms of emissions, the when-and-where matter," both for grid storage and other flexible distributed energy resources.[219] In California, the Self-Generation Incentive Program (SGIP), designed in 2001 to reduce demand peaks, must now also reduce emissions via smart accounting to charge batteries at the right time, thus optimizing net emissions and improving overall green performance. To make the green grid work, everyone must mind their peaks and troughs.

Despite the continued dirtiness from legacy production, more customers will be paid to upload electricity on demand, while "load shedding" will play a part as customers voluntarily cut consumption at peak times in exchange for discounted rates. Current load shedding covers about 2% of the grid at peak times, but could rise to 10% with more contracted consumers.[220] Roadside charging stations with large-capacity battery systems could also become essential to managing real-time demand across an entire network of interconnected resources. All one needs to do is fill the electric tanks at the optimal time.

Battery energy storage systems are essential to integrate more users into the wider smart grid, condition power to satisfy peak/off-peak/super-off-peak demand, and aid peer-to-peer trading such as Blockchain (as we saw in the previous chapter). Anyone connected to the grid will be able to buy and sell power, but the goal is to shift peak load from optimum generation to optimum use, whatever type of energy storage from larger, pumped-storage-hydro commercial plants to smaller, integrated, battery-powered home or V2G-enabled installations.

***

The era of Internet-integrated energy has begun, cooperation maxed by real-time technology. Similar to the grid in the 1900s, the IoT is evolving in conjunction with connected devices, integrating more resources to maximize efficiency via mobile-operated apps and a wider Internet platform. Begun after the basics of electrical generation had been worked out in the twentieth century, cooperative power transmission created the grid, although the uptake was delayed by luxury pricing, the urban–rural (and suburban–urban) divide, uneven demand (not enough and variable loads), and inefficient power stations, all the same problems cited today when integrating modern renewables.

Installed at a nominal cost, smart meters relay real-time energy data to better monitor use, manage energy patterns, and encourage savings, while ending money-losing meter-reading services and estimated billing. Replacing monolithic utility companies reticent to share real-time statistics, the coming digitized networks require smart meters and sensors to detect and analyze data from the thousands and eventually millions of integrated household appliances and energy devices. For those not wanting to be grid-tied or who are fearful about loss of privacy, the meters can also be operated in "dumb" mode (for now at least).

Edison's English-born former secretary and Schenectady Machine Works manager, Samuel Insull, who left Edison's employ after the 1892 Thomson-Houston merger to run Chicago Edison, solved many of the growing pains of early electrification, in particular demand management. Writing the manual on how to share electricity in the early days of the grid – via novel rate plans, separate suburban and rural electrification schemes, and pioneering methods that balanced an uneven load to keep the machines and meters running 24/7 – Insull gobbled up competitors by modernizing and expanding power plants to enable universal electrification. For Insull, cooperation aided the bottom line of the monopoly utility, providing always-on power to the people.

Paradoxically, Insull noted that a monopoly reduces consumer bills, as high, fixed power-plant costs deterred competitors. To establish control over an uncertain emerging market, he devised the original two-tier rate, taking a loss on low-use residential consumers yet making a profit on high-use commercial and industrial consumers via the "demand meter," offering at-cost installation and discounts to secure more customers as well as undercutting the competition such as natural gas companies that supplied cooking, heating, and lighting. Following in the footsteps of Rockefeller's Standard Oil and Morgan's US Steel, electricity became Insull's monopoly, albeit with a significant twist. Because it cannot be stored, electricity has to be immediately consumed, which led to the careful managing of supply – flattening or spreading out the load, accomplished at first by balancing daytime streetcars with night-time industries, before hitting on the perfect solution to an uneven demand.

Based on his "load factor" analysis, Insull showed that large *average* use was preferred to high *maximum* use, flattening the load and keeping generators running at the highest possible efficiency, which he made possible by continuing to widen his customer base. Early utility companies didn't even generate power during the day because of low demand, instead operating only for a few hours, typically to provide night lighting.[221] Insull and GE also owned most of the streetcar systems, which at first were operated to smooth out uneven consumer demand.[222] But as demand increased, skyscrapers, department stores, and apartments all became ripe for an electric makeover, providing instantaneous lighting and lifting power, followed by meat-packing, ice-making, and other cold-storage industries, while streetcar networks would consume three times the electricity of all other users.[223] The advent of regular night shifts also helped to manage an unknown, intermittent load during the day.

At a time when flat charges and fluctuating loads were a major problem for underused power plants, Insull's changes spelled the end of the isolated power plant, residential prices dropping by 32% as consumption increased twice as fast as generating capacity in just 4 years,[224] cementing the era of the monopoly power provider. *How* one consumed electricity no longer mattered; only *when* and *how much*. By 1926, according to one survey, "85 percent of central station customers had irons, nearly 71 percent had vacuum cleaners, 42 percent had washing machines, and 31 percent had toasters."[225] Thanks to Insull, the total load was managed without worry about how the electricity was consumed.

Insull's genius was to make power the commodity rather than what power made, be it lighting, manufacturing, or running an elevator, or, in time, plugging in a kettle whenever one wanted. Where once we built a power plant to run a dedicated manufacturing operation and then to power a *changing* load, all that was needed was a way to spread the load and manage the ups and downs of an *averaged* consumer. Sharing was cheaper, as not all customers used power at the same time, spreading the risk to the utility company. Of course, regulations would be needed to keep the new monopolies in check and ensure fair services, lest the consumer-customer get lost in the numbers.

The problems of the early electricity market are comparable today as we manage an increasing supply of intermittent renewables, but unlike how monopoly utilities gobbled up smaller players, there are millions more local generators at the bottom level, distributing electricity on their own rather than following the dictates of a central power station. Yesterday's top-down monopoly is being turned into a seamless cooperative, where sharing between customers is not only encouraged but required. Perhaps we are returning to a more natural way of consuming electricity: creating energy as needed, storing the generated excess, and sharing what we don't use. In a powerful act of democratization, power is trickling *up*.

In the ultimate democratization of an interconnected market, today's energy-technology infrastructure is based on how an *individual* and not a mythical *average person* acts or how grouped workers behave, remaking the world via actual users' needs. Our modern virtual village is mimicking the original country village, where people came together to exchange goods and services, meeting along shared wires instead of in a local market or commons. As with other online commerce that rejigged markets via internet connectivity – selling everything from books to bookshops and cookies to cars – power is the latest traded commodity to be remade in a wired world. In this case, billions of customers will buy and sell to each other across a shared information and electrical conduit.

What's more, the old, centralized servo-mechanical, supply and transmission system is highly wasteful, operating at only 32% efficiency, unchanged since 1960.[226] Distributed sharing is a more efficient and reliable way to use resources, not unlike routing small-sized data packets along distributed nodes to send information along the Internet. The self-interested and politically beholden economic trading system with its presumed "invisible hand" – endlessly touted as the only natural model of efficient economic interaction – is being shown up as a wasteful and ecological failure. At the same time, the output from wind- and solar-powered plants can be shared to avail of differing weather conditions across larger regions via a more dynamic grid and networked transmission lines, connecting solar peaks with wind troughs and vice versa (all the more so with added storage, including idle EV batteries).

The implementation may not be the same for all, but no longer requires an outdated, one-directional, monopoly-run source. The fight over who owns what will continue between put-upon consumers and those eager to meter every electron, but with the rise of distributed energy resources and grid-independent electricity, customers will be able to compete with the established elite, while the needs of those who strive to be a part or apart can be met, including those least able to afford wired luxury. Autonomous, DER-enabled micro-grids in shared housing complexes are ideal for low- and middle-income households. One needn't rely on untrustworthy utility companies for power – a 2013 study found that 76% of residential customers don't trust utility companies anyway[227] – although we may need to depend on our neighbors.

Similar to how personal computers replaced centralized mainframes, an outdated, unidirectional, electrical grid is being upgraded to an Internet-powered, bi-directional smart grid, creating new economic activity via peer-to-peer nodal communications. What's more, the smart grid encourages "distributed capitalism" via localized energy exchanges, turning utility companies into net-work managers (if needed at all). We are on the cusp of another revolution, where prosumers interact to create, distribute, and price their own energy. Cooperation, not central control, becomes the administration, where consumers, prosumers, power providers, financers, governments, and regulators all must agree to meet individual *and* shared needs.

As noted by author Jeremy Rifkin, "If the First Industrial Revolution gave rise to dense urban cores, tenements, row housing, skyscrapers, and multilevel factories, and the Second Industrial Revolution spawned flat suburban tracts and industrial parks, the Third Industrial Revolution transforms every existing building into a dual-purpose dwelling – a habitat and a micro-power plant."[228]

Communities everywhere are being transformed from inefficient, centrally managed, top-down constructs to collaborative, interactive, and ecologically sustainable networks.

We must be cautious, however, as more microgrids lower the demand for utility power (the return of the "one-off"), increasing prices for those unable to hook up as richer users tune out and hoard their own juice (the smart divide again), as well as destabilizing a largely public grid, while a different set of raw materials stresses new supplies in environmentally and politically fragile areas. Balancing the social load is as important as matching the ups and downs of supply and demand as we bring more connected power to more people. Nor can we forget conservation, that is, not turning on at all when simple living is a better option. Improving passive energy – better windows, insulation, and efficient building design – is even more important now, addition by subtraction, a.k.a. "negawatts" (which we'll look at in Chapter 7).

The future is as exciting as it is scary, presumably no different than in the time of Edison, Westinghouse, Tesla, and Insull, when more people were able to consume electricity for the first time, freeing themselves from the drudgery of repetitive labor as they began a fascinating journey into modernity. Today, the very idea of "neighbor" is being redefined, which now includes borrowing a cup of charge. With more large-scale battery technology coming online, EV filling stations can even download large amounts of charge on site at low-demand times (typically overnight), helping to flatten the load and in effect acting as an efficient backup/peaker plant, remotely filling the filling stations. Coupled with rapid charging, roadside amenities will operate as before, except without the pollution or the mechanics.

Soon, we will all have a charge-storage unit mounted in our home, just the same as a hot-water tank, available on demand through a flexible smart grid that manages ongoing fluctuations and emergencies. We may not even need to press any buttons depending on the software. The car of the future will also add to overall capacity, powering homes and buying and selling charge in a bartered, shared-energy economy. Ready or not, tomorrow's energy is already here. As in any deeply interconnected multi-body system, however, the rules of attraction and engagement are only just being developed and fretted over.

6.11 The Cost of Revolution: How and When Who Gets What

To get more from the well, the smart city will employ smart energy and data for real-time energy trading, resource management, and environmental monitoring (via cameras, sensors, meters, and advanced algorithms). Based on a still-evolving emergent group behavior of individual users – call it Urban Renewal 2000 – today's energy generation, substation transmission, and distribution are all getting a modern makeover. A 2017 study examined how consumers feel about smart technologies, noting that customers are willing to pay a small premium for the coming integrated clean-energy world. Responses ranged from "We're okay; you can leave us alone" (the status quo, 16%) to "Smart technologies fit our environmentally aware, high-tech lifestyles" (green champions, 44%).[229]

A revolution can't advance, however, without a complete financial sizing. Economics dictates the speed of change, the spectrum of choice for tomorrow's grid as ever determined by money, where a "levelized cost of energy" (LCOE) compares different technologies, citing cost per delivered energy ($/kWh) and all installation variables: construction, financing, operation, maintenance, and repair over the lifetime of a power plant (as we saw briefly in Chapter 3). Global energy use over the last five decades is shown in Figure 6.8 for the main power technologies – coal, oil, gas, nuclear, hydro, and renewables (PV, CSP, wind, geothermal, biomass) – followed by a breakdown of LCOE ($US/kWh) with pertinent operational data – capacity factor, operational time, and dispatch-ability – in Table 6.3. Although power generation is still dominated by oil, coal, and gas (82%), how we produce and consume energy is changing as the fossil-fuel industry is undercut by cheaper integrated renewables. Renewable energy is no longer an awkward, younger sibling, but a vibrant, mature adult, albeit one still being bullied.

On its own, however, LCOE may be too narrow a metric to measure the green efficiency of an evolving electricity grid. A levelized cost is good to compare different sources, but should not prioritize generation when all assets

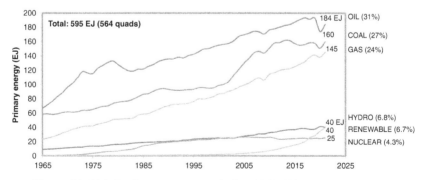

Figure 6.8 Global primary energy (exajoules; 1965–2021) (*source*: BP, *BP Statistical Review of World Energy*, 71st edition, BP, London, 2022).

Table 6.3 *Power technology comparison: levelized cost of electricity (low-to high-end), capacity factor, operation, and dispatchability*

Technology	LCOE ($US/kWh)	Min/Base/Max capacity factor (%)	Operation (hours)	Dispatch time
Coal (pulverized hard)	0.07–0.15	62	24	12 h
Biomass (wood pellets)	0.06–0.11	62	24	12 h
Natural gas (combined)	0.05–0.07	62	24	30 m
Natural gas (peaking)	0.15–0.20	8	24	5 m
Nuclear (EPR gen III)	0.13–0.20	40/74/95	24	24 h
Solar (p-Si PV farm)	0.03–0.04	14/28/39	Daylight	–
Solar (CSP PT)	0.13–0.16	15/27/35+	Daylight	–
Wind (onshore)	0.03–0.05	14/22/60	24	–
Wind (offshore)	0.07–0.10	20/39/55	24	–
Hydro (dam)	0.01–0.28	11/46/95	24	Instant
Geothermal	0.06–0.10	41/74/95	24	30 m

Sources: Electricity: "Lazard's Levelized Cost of Energy Analysis 2023," Lazard, Version 16.0, April 2023. https://www.lazard.com/media/typdgxmm/lazards-lcoeplus-april-2023.pdf.; capacity factor: Kis, Z., Pandya, N., and Koppelaar, R. H. E. M., "Electricity generation technologies: Comparison of materials use, energy return on investment, jobs creation and CO_2 emissions reduction," Energy Policy, May 2018).

should be equally valued, both generation and demand. Demand reduction, demand shifting, and efficiency – such as EVs, heat pumps, and home conservation – are just as valuable and should also be incentivized. As former UK parliamentarian Laura Sandys noted, "Fundamentally the market changes from 'supplying' as much as it can of the commodity to creating a new market in which optimised demand and optimised supply 'compete' with one another."[230] We are in uncharted lands, economic prowess beholden to our environmental survival.

Seeing how energy flows is also important to understand overall supply and demand. If we can see what we use energy on and where the energy comes from over time (~77% fossil fuels in 2021), we can concentrate on which areas are easier to decarbonize – in the USA, transportation (~27%), industrial (~26%), residential (~12%), and commercial (~9%). One such analysis is shown in a Sankey diagram as in Figure 6.9, first published in the 1970s by Lawrence

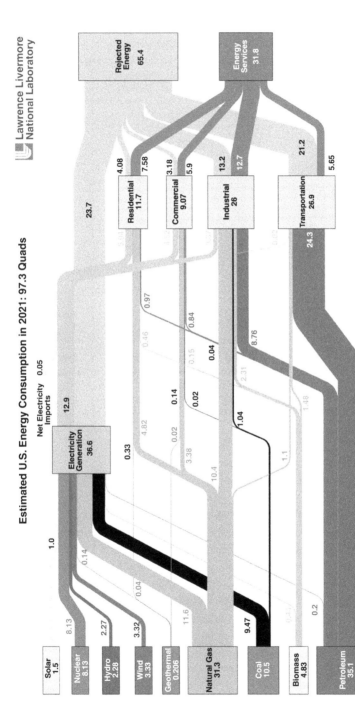

Figure 6.9 Energy flow chart: estimated US energy consumption in 2021 (97.3 quads) (*source*: "Energy Flow Charts: Charting the Complex Relationships among Energy, Water, and Carbon," Lawrence Livermore National Laboratory, Livermore, CA, March 2022. https://flowcharts.llnl.gov/).

Livermore National Laboratory to estimate US energy consumption. Saul Griffith, author of *Electrify: An optimist's playbook for our clean energy future*, calls it a "spaghetti chart" and notes that "this data summarizes all of the manifestations of our society in terms of its energy use and gives us quite a view into our collective human desires."[231] (Note, the data are in quads (~10^{18} J), but because the total is almost 100 quads, the percentages are similar.)

Admittedly, the data are difficult to follow in all their detail, but one gets the idea of what goes where and how we can cut down on emissions. For example, most of the petroleum supply (24.3 quads) goes to transportation and thus electrifying vehicles is critical to lowering GHG emissions, a goal easily supported by incentivized government policies. If we want to increase the amount of clean electricity generation, we see how much coal (9.5 quads) and natural gas (11.6 quads) must be converted to renewables. For example, to reduce natural gas in our homes (4.82 quads), we can convert gas cookers to induction or conduction and gas boilers to heat pumps (ground or air source). The amount of heat loss again shows the inherent inefficiency of burning fossil fuels (65.4 quads Rejected Energy).

Is history repeating itself? In 1923, Insull built a model electric house to show off the future of electricity to uncertain consumers. But despite the heroic efforts of early system builders, many Americans and others around the world still didn't have access to basic modern amenities almost 60 years after electricity first lit the financial district of Lower Manhattan. As Maury Klein noted in *The Power Makers: Steam, electricity, and the men who invented modern America*, "As late as 1941 seven out of ten rural inhabitants still lacked electricity. The lack of electric power, and all the amenities it brought, widened the already yawning chasm between urban and rural Americans."[232] Today, as in the past, progress is swift and uneven, especially when the cost of change is beyond the means of most.

To create new infrastructure requires not just technical and managerial skill, but sufficient investment to develop, capitalize, and build as well as important political cover. As the brains behind Chicago Edison, Commonwealth Edison, PSNCI, Middle West, and later Insull Utility Investments – the holding company he created to compete with the financial wizards vying for control of the nascent utility industry – Samuel Insull learned the ropes on the go as he built the American grid, creating a model that was emulated the world over.

In the end, Insull would be bankrupted, victim of a Depression-era bear campaign orchestrated by New York bankers wanting to break control of his web of companies and over 60 subsidiaries. Insull was even indicted for embezzlement and larceny when his over-leveraged stocks collapsed, although he was eventually acquitted.[233] The irony is that single-mindedness is often needed to will into being what does not exist, yet that same single-mindedness makes powerful opponents keen to maintain their own financial leverage. Those with sufficient assets to ride the ups and downs will always have their hands on the levers. Who pays for the future and who profits are rarely the same.

After Insull was ousted his companies continued without him as industrial innovation turned into corporate control, a lasting legacy to electrical power and modernity. As Klein noted, "The age of electricity had arrived in earnest, and no man had done more to make it so than Samuel Insull. What Edison, Tesla, and Westinghouse had done for the technology of electricity, Insull did for its distribution."[234] By the end of World War II, less than a decade after his death, grid electricity was an established way of life across the globe, without which none of us could power our lives and we assume is ever there at the flick of a switch.

Not unlike in Insull's time, today's energy revolution is being slowed by those who would divvy the spoils for themselves. Are the naysayers stalling progress so established businesses can jump aboard and position themselves to control the new distribution model as they learn the ropes? Are they casting doubt on proven technologies to ensure management of the products that emerge, while the rest of us worry about how to keep the heat on, fill our cars, and stay connected in an increasingly confused global village? Such is the way of expensive investment and indeed capitalism. Whatever the motives and however long it takes to implement a revamped, renewables-backed, cooperative grid, at least it will be harder this time to place a tariff on sunlight and wind.

There is still much to do to ensure a reliable and affordable energy system, but as with all new technologies one size does not fit all. Although Tesla CEO Elon Musk has gone on record as saying that "Most houses in the US have enough roof area to power all the needs of the house," he also predicted a 50-year time frame before solar roofs are widely integrated into the standard makeup of a home, longer than the life of a roof.[235] It will take time to replace the old, especially without sufficient investment and incentives to help.

Highlighting the importance of decarbonizing transportation, if every car in the USA were to run on electricity, the grid would be increased by only 20%, most of which can be accommodated at night or via on-off smart charging, and hence no new power plants needed other than to replace closed coal, oil, and gas plants to ensure a clean supply. But the electric vehicle is still only at single-figure-percent usage, hampered by high costs, poor infrastructure, and uncertain supply chains, while vehicle-to-grid service is seen as part of an attractive, yet complex, future. At the forefront of the fundamental changes to how we make, use, and share energy via distributed energy resources and backup storage, we are still figuring out how to implement even the basics of the latest technologies.

***

Invention is typically thought of as connected to an inventor: Einstein relativity, Faraday the electric generator, Copernicus the heliocentric solar system. In fact, all invention is tied to place and time. Without high North Sea winds blowing Dutch sails, large sperm whales off the coast of New England, or oil seeps near the Allegheny River in Pennsylvania, the world would never have profited from windmills, lamp oil, or petroleum in those locations, or as soon. It was no accident that the Industrial Revolution began in Britain, thanks to a plentiful and easily accessible supply of coal and a thriving Enlightenment curiosity that valued scientific enquiry. Steven Johnson notes in *The Invention of Air*, without its "unusually large stockpile of Carboniferous fuel, and the stockpile's shallow location . . . , it's entirely likely that the Industrial Revolution would have originated somewhere else."[236] If not for a 1,000-mile-plus distance between the coal mines of the Chinese interior and the coastal markets in Shanghai, the Industrial Revolution may well have begun centuries earlier in China.

Nor is one person able to press their genius or inventiveness on a global scale as easily as in earlier times. Sand was a necessary ingredient to make glass and lenses for the telescope and microscope, inventions both made in the Dutch Republic, thanks in part to plentiful dunes along the North Sea coast and an opportunistic lens maker presenting his work to the king in a giddy display of science conquering magic atop a palace roof. Four centuries later across an ocean of time and space, that same sand would be reinvented as a p–n junction, perfected by teams of dedicated workers united again by the necessities of war.

Investment is also vital to expansionary commercial exploits, such as the Dutch East Indies Company enlarging a young nation's exploring reach, aided by increased precision in measuring time and a growing collaboration among scientists backed by wealthy benefactors. Without the spread of discovery and increased scientific discourse, led by the Dutch physicist Christiaan Huygens, considered by some to be the first modern scientist, the Copernican revolution might well have been slowed for another generation. Without Matthew Bolton, James Watt may have languished in obscurity amid his aimless Sunday walks in Glasgow Green. Without Westinghouse, Tesla could never have fulfilled his electric dreams.

Today, invention is more complicated and requires both political will and deep pockets along with the usual brains and brawn. Without the intellectual hothouse of a war-tested Bell Labs, the forward-thinking energy policies of Germany's green-inspired government, and the manufacturing expertise of a centrally managed China (not to mention the plentiful sun of southern California), we would not be on the cusp of today's solar, charge storage, and EV revolutions. Indeed, the same as at the beginning of all great change, whether the Industrial Revolution of nineteenth-century Great Britain or the silicon-led information makeover of the twentieth century as exponential growth follows on from innovation. In the midst of change, all revolutions seem slow, new ideas grudgingly assimilated by wary investors and worried consumers. But eventually more of us begin to adopt the new, spurring on others to see the light. Just as Ptolemy gave way to Copernicus in the calculation of planetary orbits, Aristotle to Newton in the analysis of motion (and then to Einstein), and Priestley to Lavoisier in the role of oxygen in combustion, most change is incremental, indeed, until it isn't.

As the philosopher and science historian Thomas S. Kuhn noted in his seminal 1962 book, *The Structure of Scientific Revolutions*, "Paradigms gain their status because they are more successful than their competitors in solving a few problems that the group of practitioners has come to recognize as acute."[237] This was the case with the internal combustion engine, which blew away its peanut-oil competition to provide a reliable means of transportation in World War I, and is now *shifting* (to use his phrase) because of another acute need. Horse manure bunged up the works in the congested streets of the early twentieth century, while carbon-spewing, global-warming greenhouse gases and particulate-matter pollution have bunged up the air ever since, causing the latest set of acute problems.

Unfortunately, an already-paid-for fossil-fuel infrastructure is hampering the transition, managers compelled to continue their business-as-usual, dirty practices, locking in old power plants until the end of their current life cycles. All the more reason to get it right from the start, and not license more brown builds when renewable-energy options work better and are less expensive. Investment is needed now to scale up already-proven, renewable-energy technologies, providing more jobs, lower prices, and a greener world. Change is as much about will and want as it is about ways and means. Not for nothing have the oil and gas, coal, and nuclear industries been given precedence over wind, water, and sun (WWS). But change is here, focused more on the environment and how humans live and work.

As we ponder the newest of technologies, many challenges remain. This time, as we swap brown for green, we must change more than just the fuel. Interesting that such bounty should come just when we need to change as never before.

Part II Renewable Energy (Solar, Wind, . . .) and Electric Vehicles/Battery Milestones

Table II-1 *Milestones in the solar age*

~240 BCE	Eratosthenes shows that the Earth is spherical and calculates its radius
1666	Isaac Newton splits white light into component colors and recombines them with a prism
1839	Edmond Becquerel first observes the photoelectric effect
1864	John Tyndall shows that CO_2, CH_4, and other atmospheric gases absorb EM radiation (the greenhouse effect)
1905	Albert Einstein explains the photoelectric effect in the first of four *Annus Mirabilis* papers ("On a Heuristic Viewpoint Concerning the Production and Transformation of Light")
1913	Frank Shuman makes the first large-scale functioning CSP plant on the bank of the Nile near Cairo
1939	Russell Ohl produces a photovoltaic (PV) effect in silicon at Bell Labs
1954	Daryl Chapin, Calvin Fuller, and Gerald Pearson demonstrate the first "solar battery" at Bell Labs, with 6% efficiency (solar to electricity)
1976	Sharp makes the first solar calculator
1980s	NREL director Roland Hulstrom calculates that a 100×100 square mile PV farm could completely power US electrical needs
1982	ARCO's Lugo solar farm in Hesperia, California generates 1 MW
1984	The first modern CSP plant is built as part of the SEGS facility in the Mojave Desert
1999	Solar breaks the $1/watt barrier with a thin-film Cd-Te solar cell
2000	EEG (Erneuerbare-Energien-Gesetz or Renewable Energies Law) is passed in the German parliament
2016	Huawei's Yanchi solar farm in Qinghai, China, generates 1 GW
2020	The Bhadla Solar Park in Rajasthan, India, generates 2 GW

Table II-2 *Milestones in the renewables age*

1500	The first turret mills appear (free-swinging top), possibly invented by Leonardo da Vinci
1888	Charles F. Brush invents the first electric-generating wind turbine, powering his Cleveland home (12 kW)
1919	The first grid-connected wind turbine starts up in Denmark
1991	The first offshore wind farm is operational in Denmark (450 kW) located in the South Baltic Sea
2014	The most westerly Canary Island (El Hierro) goes 100% electric with the first wind-turbine/pumped storage hydro (11.5 MW: 5×2.3 MW)

Table II-2 *(cont.)*

2015	Denmark generates a one-day world record of 140% of its grid electricity from wind power
2017	The first floating wind farm (Hywind) begins operating off the coast of Aberdeen (30 MW: 5 × 6 MW)
2019	GE introduces the Haliade-X prototype WT with 107-m blades to the Port of Rotterdam, generating 12 MW
1911	Geothermal steam runs an electrical generator to power 5 lights in Larderello, Italy
1960	The world's largest geothermal power plant, the Geysers Complex, west of Sacramento, California, starts generating electricity
2014	The first three-technology hybrid power plant (PV, geothermal, CSP) begins operation in Stillwater, Nevada
1870s	The Pelton wheel and Francis turbine improve the conversion efficiency of falling water
1884	Charles Parson designs an efficient steam turbine for the Holborn Viaduct power station in London
1889	Nikola Tesla works as a Westinghouse consultant in Pittsburgh to make split-phase AC motors
1895	Long-distance power is sent from Niagara Falls to Buffalo, 22 miles away, establishing grid electricity
1935	The 2-GW Hoover Dam is built with a 220-m head at Lake Mead on the Colorado River employing 17 turbines
1942	The 6.8-GW Grand Coulee Dam is built over 29 stages on the Columbia River, which drops 2 feet per mile
2012	The 22.5-GW Three Gorges Dam is built with a 175-m head on the Yangtze River with 32, 700-MW turbo generators (the world's largest Francis turbines)
2022	After 7 years of construction, the $5-billion, 6-GW Grand Ethiopian Renaissance Dam (GERD) started tapping the falling waters of the Blue Nile
1966	The world's first tidal power station is built on the Rance River in northern France

Table II-3 *Milestones for electric vehicles and batteries*

1799	Alessandro Volta creates the first electrical battery using zinc and copper metal electrodes
1831	Michael Faraday demonstrates induction at the Royal Society in London
1859	Gaston Planté makes the first ever rechargeable chemical storage battery (lead–acid (PbA) "wet-cell" battery)
1879	Werner Siemens tests the first electric tram system in Berlin
1884	Thomas Parker builds the first electric vehicle (lead–acid battery)
1892	The first US electric commuter train begins operation (between Portland and Oregon City)

Table II-3 *(cont.)*

1951	Merle Williams invents the electric golf cart (lead–acid battery)
1970s	Solid-state physicist John B. Goodenough produces rechargeable lithium-ion (Li-ion) batteries at Oxford University
1996	General Motors launches the world's first modern all-electric car, the Chevy EV1, recalled in 2003 (lead–acid battery)
1997	Hybrid Toyota Prius launched in Japan (USA in 2000) (NiMH battery)
2003	Tesla Motors founded, goes public as Tesla Inc. in 2010
2006	Elon Musk releases Master Plan 1, citing path from expensive low-volume electric car to affordable high-volume electric car
2008	Tesla Roadster launched with chained Panasonic 18-650 Li-ion batteries
2010	Nissan launches first modern mass-produced all-electric vehicle, the Leaf (Li-ion battery)
2014	Guinness World Record for "shortest charging time to cross the United States in an electric vehicle" (Tesla Model S P85D, 12 h, 48 m, 19 s)
2015	Toyota launches the Mirai (hydrogen fuel-cell) electric car with 3 times the energy per unit mass than diesel and 2.5 times natural gas
2015	The first all-electric ferry begins operating across the Sognefjord between Oppedal and Lavik, a distance of 5.7 km
2016	Former battery company BYD makes all-electric buses and trucks for Chinese public transport and for export
2017	First "Gigafactory" (GF1) begins making Li-ion batteries in Sparks, Nevada, cutting battery costs by about one-third
2017	First driverless cars employed on real roads in a Phoenix suburb (Waymo's autonomous Chevrolet minivans)
2017	Prosumers buy and sell distributed electricity via virtual peer-to-peer trading in the Brooklyn Microgrid solar-powered community network
2019	Tesla's Model 3 racks up an estimated $10 billion in sales in 2 days, ultimately becoming the world's best-selling electric vehicle
2019	The first all-electric seaplane stayed aloft for 3 minutes east of Vancouver, powered by electric motor and Li-ion battery
2022	The US Congress passes the Inflation Reduction Act, earmarking $369 billion over 10 years for green initiatives, especially solar power, wind power, and electric vehicles

PART III

Less Is More

We came all this way to explore the moon, and the most important thing is that we discovered the Earth.

William Anders, Apollo 8 Astronaut

This universe is bigger than all of us. That earthrise photo gave us the sense that we live on a fragile planet, that we have limited resources, and that we better learn to take care of it.

Apollo 8 commander Frank Borman, remarking on seeing the Earth for the first time from the moon, a moment captured by crewmate Bill Anders in his famous earthrise *picture*

We are stardust / We are golden / And we've got to get ourselves / Back to the garden

Joni Mitchell, Woodstock

If you see a threat it's your responsibility to sound the alarm.

Greta Thunberg

7

Rethink, Rebuild, Rewire

7.1 Think Globally and Act Locally: We Can All Make a Difference

Raised by parents who grew up in post-war times, many of us learned to conserve: turn off lights, keep outside doors closed, and eat all your food (or no dessert). Conservation can be just as important as increasing our green-energy mix – every watt we don't consume is a watt we don't produce, a.k.a. *negawatts*, a concept introduced in 1985 by Rocky Mountain Institute scientist Amory Lovins. Alas, that ideal seems to have become lost in a modern race to consume without thought, 24/7 advertising singing the joys of excess at every turn.

But one can cut down in simple ways. The first step is to calculate how much energy you use to see the amount of money spent at home, work, and in transit. I have compiled a list of electrical energy consumption in our home, a three-bedroom terraced house in northern Spain (Table 7.1). Yours will be different in the detail depending on location, although likely similar in the bottom line: roughly 15 kWh/day per person (~5,000 kWh/year at €500/year per person).

Here, one sees how an electric bill is calculated from everyday use – about 15 kilowatt-hours per day per person on average – and the cost of each item, helping to identify ways to cut down. For example, what is the cost of leaving the lights on in empty room, keeping a printer permanently connected, or taking a 5-minute hot shower every morning? How bad is leaving a computer on overnight, running the clothes dryer for a small load, or half-filling a dishwasher? If one doesn't know the cost, it's hard to figure out where to save.

So, how does one calculate the cost of running an electric device or appliance? First, take a look at the power, P, in watts, usually indicated on the back or plug. For example, our 40-inch LED TV consumes 110 W as marked on a label on the back. If a device or appliance shows only current and voltage, you can

Table 7.1 *Sample household electricity usage ~30 kWh/day (*average,*
†winter, ‡night)[1]

Item	Output power (watts)	Use (hours or minutes per day)	Total energy (kWh/day)	Total cost (€/day)	Yearly cost (€)	Percent cost
Fridge*	160	24 h	3.84	0.58	211	12.7
Microwave	1,450	10 min	0.24	0.04	13	0.8
Oven/stove	1,200	30 min	0.60	0.09	33	2.0
Kettle	1,800	5 min	0.15	0.02	8	0.5
Toaster	1,200	–				
Dishwasher	2,400	3 min	0.12	0.02	7	0.4
Lighting*	240	6 h	1.44	0.22	79	4.8
Computer (2)	65	14 h	0.91	0.14	50	3.0
Printer	0	2 min	0.00	0.00	0	0.0
TV	110	3 h	0.33	0.05	18	1.1
DVD player	15	1 h	0.02	0.00	1	0.1
Stereo	40	1 h	0.04	0.01	2	0.1
Radio/CD player	14	1 h	0.01	0.00	1	0.1
Router	10	24 h	0.24	0.04	13	0.8
Phone (charger)	4	30 min	0.00	0.00	0	0.0
Washing machine	700	4 h (/week)	0.70	0.05	17	2.3
Clothes dryer	3,000	–				
Vacuum	1,400	10 min	0.23	0.04	13	0.8
Lawnmower	1,000	2 h (/season)	0.05	0.01	3	0.2
Hot water heater‡	2,000	2.5 h	5.00	0.34	123	16.5
Dehumidifier	240	12 (/week)	1.20	0.18	66	4.0
Hair dryer	1,200	5 min	0.10	0.02	5	0.3
Iron	2,000	30 min (/week)	0.17	0.03	9	0.6
Heating (storage)†‡	8,800	4 h	14.67	0.99	360	48.4
Portable heater†	2,000	30 min	0.25	0.04	14	0.8
Total			**30.32**	**2.86**	**1,045**	100

work out the power by multiplying the current (I) in amps by the voltage (V) in volts ($P = IV$). Next, the total energy consumed, E, in watt-hours (Wh) is the power (P) multiplied by the time (in hours). If we watch our 110-W television for 3 hours on average a day, we use 330 watt-hours or 0.33 kWh (110 W × 3 hours) each day. If electricity costs 15 cents per kilowatt-hour, the total cost is then 5 cents/day to run our TV (0.33 kWh × 15 cents/kWh). By calculating the power consumed, one can see that although a TV has a much lower power rating (110 W) than a washing machine (700 W), overall consumption in kWh is comparable. Note, a lesser rate/cost may apply after midnight.

As seen, the big-ticket items in our home are electric heating (49%), hot water (17%), refrigeration (13%), lighting (5%), and computers (3%). Household lighting isn't very expensive at 22 cents/day, but is certainly an area where we can easily save. Turning off one light won't save much, but converting to LED bulbs will. In 2004, the City of Chicago converted to LED traffic lights, cutting its annual energy consumption by 85%, CO_2 emissions by 23,000 tonnes, and costs by more than $2.5 million.[2] If all of the UK were to convert to LED lighting, the savings could equal the power generated at the 3.2-GW Hinkley C nuclear power station after it's up and running. Such savings are not lightly dismissed. Interestingly, one area I found easy to cut down is our printer, which I used to keep plugged in 24/7. As to the specifications, the deskjet printer is rated:

> Power consumption: 10 watts maximum, 10 watts (Active), 0.2 watts (Off), 1.6 watts (Standby), 0.8 watt (Sleep)

Okay, 10 watts is minimal and my 2 minutes of occasional printing doesn't even show up at 2-decimal accuracy in our household electrical usage table, either for total energy or cost (0.00 kWh and €0.00/day), but, amazingly, we spend more when the printer is supposedly *off* than when we use it – 14 times as much – all because of a green LED whose only function is to tell the world that the printer is connected. To be sure, 26 cents a year permanently plugged in compared to 1.8 cents a year for occasional printing won't break the bank, but why waste a quarter on nothing? If I multiply that quarter by other printer users on our street, we could share a fine wine at the summer block party. Multiply that quarter by the entire printer population and we could stop the construction of a new power plant.

A clothes dryer is a high-cost item, but one can easily save by drying clothes outdoors in warmer months or at the top of a stairwell in winter to avail of rising household heat (the clothes won't shrink either). In our home, we have plenty of south-facing windows so drying clothes without electricity is easy for us. Not everyone can, but passive heating significantly reduces the monthly bill. In our home, we never use the dryer and rarely the dishwasher (only after a large family dinner or party). A dishwasher is more efficient but only if full. A typical dryer or dishwasher load costs about 50 cents, worth it to cut down whenever one can.

Changing one's overall energy picture by changing habits is not as hard as we think. Some fixes are easier than others, such as installing a thermal blanket on a water heater, a low-flow showerhead in the bathroom, or a digital thermostat programmed to household wake–sleep patterns, although each change takes time and effort. Some require more work and/or money (for example, insulation, double-glazing, high-efficiency appliances, heat pumps), but every bit helps, while government incentives encourage better choices. Change is never easy, but managing energy should be as important as managing finances. At the very least, we should all examine our everyday consumption habits.

According to Environment California, the greatest household energy savings are (in order): solar hot water system, horizontal access clothes washer, condensing furnace, wall insulation, water heater blanket, ceiling insulation, dishwasher, high efficiency water heater, floor insulation, HVAC (heating, ventilation, and air-conditioning) testing and repair, duct repair, low-flow showerhead, faucet aerators, pipe wrap, programmable thermostat, duct insulation, and high-efficiency clothes dryer.[3] The top two – solar hot water system and horizontal access clothes washer that uses less water and spins faster than top loaders – equal as much as the rest combined. One needn't retrofit all at once, but one should consider efficiency ratings when renovating or buying new devices or appliances. The "Energy Star" rating system ranks appliances and is highly recommended to cut down waste. For example, if every computer in the USA was Energy Star certified, more than $1 billion in energy costs per year could be saved just via low-power "sleep-mode" during periods of inactivity. Similar certified efficiency rankings are available in Europe and elsewhere.

Other easy-to-make upgrades include window and door sealing and thicker curtains. It takes more work to replace single-pane windows with modern double glazing, but can reduce a heating bill by as much as 30%. Triple glazing is even better, while windows filled with transparent argon gas cut heat loss by about 85%.[4] A change to high-efficiency radiators will also net substantial savings. As winter approaches, many homeowners think of ways to save on heating, such as attic insulation, turning down the heat (especially at night), and closing off lesser-used spaces (aiding family togetherness if so desired). A wood-burning stove or cast-iron fireplace insert can heat an entire well-sealed home, reaping big savings on the heating bill, although a ventilation system may be needed to exchange indoor air.

The North American Insulation Manufacturers Association (NAIMA) calculated that only about 10% of homes are properly insulated. Its CEO noted that "insulation has a three times greater impact on the average home's energy and comfort than windows or doors do," while Dr. Jonathan Levy of Boston University's School of Public Health stated that "If all US homes were fitted with insulation based on the 2012 International Energy Conservation Code (IECC), residential electricity use nationwide would drop by about 5 percent and natural gas use by more than 10 percent."[5] Heating and air-conditioning (AC) costs are obviously lower with better insulation, considered "the single most important energy efficiency improvement the world can make"[6] by sustainability expert Chris Goodall, while retrofitting commercial buildings can reduce energy use by 60% (almost 90% with rooftop PV).[7] It's high time to start making the changes and get our governments to help with the makeover.

Some changes don't require money, just a change in thinking. Lower the heat on a stove element after water starts to boil because less energy is needed to keep the water boiling, fill a kettle only as much as needed, and wash clothes at a lower temperature, although hot water is recommended once a month to keep a washing machine free of bacteria and to kill lingering mites. Operate appliances/devices outside peak times. We can all make simple changes to save energy and money.

Of course, not all changes amount to similar savings. A turned-off yet plugged-in device on standby or "vampire" power may add only a few dollars a year to the electricity bill, but cutting down on the clothes dryer can save 10%. A microwave provides savings over an oven (higher wattage but less cooking time), while an induction stove is much more efficient than conduction heating or gas as well as keeping surroundings cooler and eliminating toxic indoor fumes. Leaving a computer on at night will cost about 5 cents a day, but reducing the heating or AC makes a big difference. Ceiling fans are better than AC, but you can also add awnings and outdoor greenery or even paint a roof white to reflect summer sunlight to cool a building, especially in tropical climates. Ultra-white "cool roofs" reflect 98% of incident radiation.[8] It's easy to install a programmable or smart-app temperature-controlled thermostat.

As always, if we multiply our own use by an entire population, conservation means big savings, although lower-income home owners and renters are less likely to avail of energy savings without government grants or subsidies. For example, excessive summer heat impacts poorer, inner-city neighborhoods more, which typically have fewer trees than leafy middle- and upper-class suburbs. Planting trees is a simple solution to help control temperatures in urban hotspots, while "nature-based" solutions have existed for ages, such as creating city parks that include more trees, capture carbon dioxide, and mitigate wet weather and potential flooding, as well as beautifying the environment and increasing biodiversity.

The Eden Project ran a "21st-Century Living" experiment to prepare the UK to reach its ambitious target of reducing GHG emissions by 80% by 2050 (27% of which comes from households), giving 100 families an introductory home audit, some initial guidance, and £500 to spend as they wished to reduce their environmental footprints.[9] The year-long study found that effective home-energy improvements are easily made with a little encouragement, resulting in not just increased savings (one family noted a 20% decrease in their energy bill), but further green spending (roughly £500 or £1 more for each £1

invested). The changes included installing low-energy light bulbs, loft insulation, cavity-wall insulation, a new boiler or upgraded heating system, double
glazing, compost bins, water butts, and an energy monitor for real-time feedback. Andy Jasper of the Eden Project noted that people don't need to be green
thinkers to change their ways: "If you give people money, the incentive, and the
information then they'll actually do something."[10]

The main conclusions of the experiment were: (1) change is easier than we think
(such as using less energy in the home and traveling more by public transport), (2)
simple ideas help change attitudes and behavior (thermal audit and energy monitors), (3) a radical change to utility bills is needed (gobbledygook for most,
including energy experts), and, perhaps most importantly, (4) green investment
stimulates more green buying. As stated in the findings, "This should make
interesting reading for policy-makers, manufacturers and retailers: give people
a voucher for £500 off, and they may well be tempted into spending an extra
£1,000 – if the right products are on the shelves."[11]

We can all make a difference despite the dauntless task of changing hard-
worn habits. As one participant remarked at the end of the year: "I think the
thing that most affects people is the feeling that you can change your own local
life, . . . , the life around about you. I don't think there's much point in beating
people around the head with statistics about global trends and one airplane trip
too many per year or whatever. It's what you get up with every day: the warmth
of your house, the productivity of your garden." Given a bit of helpful advice
and a gentle push, most of us will seek to reduce waste in our homes, be it
energy, water, or general everyday garbage, while consumers will choose
greener options if they are available.

The hope is that manufacturers will adapt to an increased consumer demand
for more energy-saving and energy-efficient products, encouraging cleaner consumer options. But we can also do our part by making and demanding changes
based not on convenience but on conscience. An economy that relies on fossil
fuels without concern for the environment is not prepared for the changing
energy landscape of the twenty-first century. Energy efficiency should be considered as important or even more important than cost and convenience, inspiring
the reformative value of our everyday actions. Every little bit does help.

7.2 Rethinking the Future: Quick and Easy Solutions

Everyone enjoys a good shower, but many of us don't need to lather to excess.
A 6-minute shower uses about 50 liters (half the amount of a mid-sized bath).
Three minutes would save $100 a year – not much to the individual, but if the

office-working population of Europe and the USA showered for 4 instead of 8 minutes a day we could annually consume at least 30 billion liters less water and save more than $2 billion in electricity and water costs. No small peanuts. Better yet, a low-flow showerhead or atomized water system reduces water consumption by over half, letting us soak in luxury for less.

It is also thought that many of us wash our hair too much, adding to unnecessary tap time and increased chemical waste (water-softening phosphates also promote algae growth). If you want to go completely eco-warrior, cut out colored or scented soap – it's the fats and oils that do the cleaning. To cut down on plastics and packaging use a biodegradable shampoo bar or simple unscented soap. Supermarket packaging is a system that makes life easier for robot assembly lines and long-distance transportation, yet ignores environmental and labor concerns. Excess packaging and product paraphernalia is everyone's problem.

Many of us are uncertain about how individual actions impact the world. Tom Millest, a London police inspector who participated in a local carbon-counting experiment, sums up the concern: "One thing that weakens my willingness to take action is that there seems to be so many unknowns in terms of the effect of what we do. When you wash your hands in public, what wastes more energy: the paper towels or the hand-dryer? Is a train better than a bus? Should we buy local food or support farmers in the Third World?"[12] Paper towels versus the hand dryer may be debated forever (about the same), while the train and bus are both good options compared to driving a car (either is fine if electric). As for food, buying local is always better, both for consumers and farmers in developing countries despite the presumed tweak at one's global conscience.

A farmers' market (or growing one's own) is a great way to save and become better connected to the land and one's community, while buying local helps everyone, from the farmer to those who work in the community or are looking for work. Purely on economic terms, 95% of the price at a farmers' market goes to the farmer, whereas 75% at a big-chain supermarket goes into "advertising, packaging, long-distance transport, and storage."[13] The local market might cost a tad more, be a bit less convenient, and add time to a shopping run, but the food is fresher, tastes better, and doesn't spoil as fast. Importantly, local markets support the local economy of which we are all a part, essential in poorer countries when farmers are turned into process-only shippers.

Although meat is still a main GHG emission item, buying local food also helps reduce one's carbon footprint by cutting down on long-distance food miles, including non-seasonal fruits and vegetables. As long-time vegan and carbon analyst Danny Chivers noted, "Emissions from livestock agriculture – including the methane from animals' digestive systems, deforestation, land use change and energy use – make up around 15 per cent of global emissions ... more than all of the world's cars, ships, trains and planes put together."[14] Energy-intensive, cereal-based meat production also contributes to global food shortages where 10 kg of cereal equates to 1 kg of beef.[15] Compounding the problem, meat now provides two-thirds of protein in the American diet, double a century ago.

Some artificial meat is as tasty as real meat, although not everyone wants to eliminate animal protein from their diet. Plant-based foods that look, taste, and even bleed like meat are now attracting more than just vegetarians and vegans, becoming more popular with meat eaters and "flexitarians" conscious about health, animal rights, and climate change. If you do eat meat, use a local butcher and not a chain supermarket (much better for you and cuts down on packaging!). Buying local also means fewer animals dying in transit.

Front and backyard herb and vegetable gardens are simple options to enhance salads, and aren't nearly as hard as one might think. A rainwater tank provides free water for household and garden plants, easily enclosed by affixing to a downspout to keep out mosquitoes and other water-loving bugs – a 100-liter tank is more than enough for home use. Gardens are harder for apartment dwellers, but balcony and terrace plots are becoming more popular among city dwellers, while community gardens are popping up in previously unused spaces and on apartment roofs. The popular allotment is getting a modern makeover.

With the advent of efficient LED lighting and stackable layers, "vertical" farms have also become more viable, especially for salad crops such as leafy greens, tomatoes, and herbs. Compared to traditional farming, a vertical farm has a much smaller energy and environmental footprint, consuming 10 times less water and requiring little or no soil. Growth is sped up in less space via controlled lighting, temperature, and irrigation, while avoiding pesticides, herbicides, and fungicides. Located much closer to customers, a vertical farm cuts down on food miles and waste and can also run 24/7 in urban locations or in poor farming regions. It's particularly important to cut down on food waste from farm to fork because of the extraordinary amount of resources required to grow, select, refrigerate, transport, and display what we eat.

Shockingly, in the USA and Britain, half of all food is wasted in production, while one-quarter of what does make it into our homes is thrown away. According to a 2021 UN study, we throw out over 2 trillion pounds (~1 trillion kilograms) of food per year, mostly from homes (67%), food services (22%), and retail (11%)![16] In *Waste: Uncovering the Global Food Scandal*, Tristram Stuart notes that food waste alone could alleviate the hunger of 1.5 billion people, while "wasting food represents a waste of water far greater than what we use in baths, toilets and washing machines," equal to the household water needs of the entire globe, not counting water use in the supply chain.[17] What's more, about 20% of GHG emissions comes from food production and preparation, more than half completely unnecessary if just thrown away.

Stuart proposes a simple hierarchy to stop putting excess food into landfills: (1) stop creating surplus (anything more than a 130% safety net is unnecessary), (2) cut out selection loss (primarily aesthetics such as perfectly straight carrots or good-looking but overripe produce), (3) no need to throw away what people can still eat (redistribute to food banks and community centers), and (4) anything leftover is then fed to livestock as animal feed or broken down by industrial digesters for heat, power, and soil-enriching compost: "The current practice of sending food to a landfill is the worst possible way of dealing with it and constitutes one of the clearest instances of mismanagement of resources."[18]

Take a trip to a landfill to see the waste up close and then ask yourself how much could have been saved, reused, or converted to energy. As you hold your nose because of all the rotting food, think of the resources used just to create waste. The numbers are staggering as are our feeble attempts to recycle perfectly good organic matter in cities. In the USA, for example, only 2.6% of municipal waste is recycled.

Elsewhere, food waste is taken more seriously, where keeping food out of a landfill is prioritized. South Korea requires everyone to separate organics from other recyclables and has made sending food waste to a landfill illegal, thus almost all food (98%!) is separated from regular household and industrial waste. Austria and Germany both have laws requiring the collection of biodegradable waste, which is then converted to energy (anaerobic digestion) or composted (anaerobic or aerobic breakdown), substantially reducing the amount of food waste entering a landfill.[19]

Given that food consumption accounts for at least 20% of all GHG emissions, cutting out waste is essential to reduce our carbon footprint, such as not growing too much in the first place (reducing deforestation), improving supply management in fields, supermarkets, and catering (smart food!), and preparing only as much as needed (drastically reducing table waste). Best-before dates are also overly cautious – most of us can easily judge if something has gone off. Why not use the freezer for excess food instead of automatically tossing out everything at a later date?

In conjunction with a regenerative biofuel program, overall GHG emissions could be cut by as much as 40%, making how we eat and the land needed to provide our daily bread a necessary ingredient in energy efficiency. Purpose-built, industrial biofuel power-generation continues to increase around the world, but reducing the vast amount of land used for agriculture has the potential to save much more energy. It's up to producers and consumers to make smarter choices about the food we make, use, and reuse. More waste means more energy and more emissions.

One must be careful not to oversimplify, but if we extrapolate our habits to others who do more or less the same as we do, simple actions can make a difference in a world with a population of 8 billion and counting. Conservation is an attitude rather than a directive. In the face of a mechanized system that has little concern for individual lives, we have to believe we can make a difference. As former New York mayor Michael Bloomberg aptly noted, "Individuals are the only ones willing to make a difference," while Greta Thunberg, the Swedish teenager who spearheaded an increased interest in green activism with her #FridaysForFuture school strikes, is even more emphatic about the power of the individual: "We cannot fight against climate change without individual change. The fact that it's the way we behave, the way we consume, that is the problem." Indeed, we can all do more in our daily lives, whether buying green, counting food miles, or reducing our everyday carbon output.

The time for change is now: at home, in the office, in the garden, everywhere. Figure 7.1 presents my top 10 list of simple and inexpensive ways to save, slow down, and improve the quality of life. Every little bit does help.

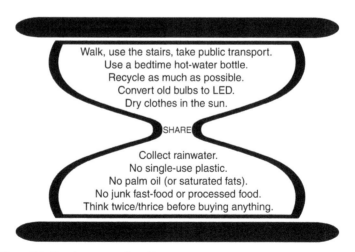

Figure 7.1 Simple green savings ideas.

7.3 Everyday Transit for Everyday People: Share and Share Alike

Simple transit efficiencies are a great way to reduce costs, fumes, and GHG emissions as more people turn to public transit, car-sharing, and bicycles, a.k.a. personal transporters, while EVs are fast becoming an everyday reality. Soon, the road more traveled will be a cleaner road and electric cars simply known as cars. In large cities, there is no alternative to make a city center fume-free and more accessible while reducing congestion. Take away the subway for a day as in the occasional strike or power failure and the big cities of the world become gridlocked, full of vehicles and people stranded everywhere. The horse and buggy of yesteryear would be faster.

"If you build it, he will come." That was the sage advice from a disembodied voice to a worried young farmer in *Shoeless Joe*, W. P. Kinsella's enchanting novel about building a baseball park on his family's mortgaged Iowa cornfield (Kevin Costner played the farmer in the movie version, *Field of Dreams*). When it comes to roads and inner-city highways, the reality is more like build it and they will come . . . and come and come and come.

Astonishingly, many city planners still think building more roads will somehow magically fix the weekday, bumper-to-bumper traffic crunch, made all the more horrific during morning and evening peaks, oddly called "rush hour." In fact, more cars bring more traffic and longer commutes. To try and gauge the insanity, more people are stuck in vehicles at this very moment in China than the entire population of Ireland, while a popular Beijing joke states that there is only one rush hour – *now*.

Some city planners are beginning to see the light, admitting that "induced demand" is the problem. As calculated by the transportation analytics firm Inrix, morning commute times worsened in both directions upon the completion of a $2.8 billion highway-expansion project in Houston, which added more lanes to the Katy Freeway – making it the widest stretch of highway in the world at 26 lanes! – while morning and evening commutes worsened in both directions after $1.6 billion was spent to widen the I-405 in Los Angeles.[20] Alarmed by the seemingly counterintuitive more-is-less logic, the city of Los Angeles voted unanimously to shelve further plans to spend another $6 billion for additional lanes in each direction of the 710 freeway in southeast LA County (a.k.a. "the diesel death zone") and will instead spend up to $120 billion on public transit projects and non-automotive mobility over the next 40 years via its Measure M fund.

The writing is on the wall, or the LED message is on the sign. Succinctly noted by *StreetsblogLA*, "More lanes means more traffic, more congestion, more pollution, more asthma, more cancer, more death."[21] After Highway 99 in Seattle – a transit route for 90,000 cars a day – was closed in 2019, the feared "Carmageddon" never materialized.[22] People just found other ways besides cars to get around.

In a telling scene from the 1996 movie *Swingers*, five friends living in Los Angeles each take separate cars from the Lava Lounge to "a party in the hills." Trent (Vince Vaughn) leads the way in his 1964 convertible Mercury Comet Caliente, followed by Mike (John Favreau) in his red Cavalier, and then Sue, Rob, and Charles each in their own cars. These young, would-be hipsters would never think to carpool or double up – living the dream is driving one's own ride, not sitting shotgun in another man's wheels. Neither would Carrie Bradshaw (Sarah Jessica Parker) or her suave New York girlfriends, who are presented with four separate, air-conditioned, half-million-dollar, chauffeured white Mercedes-Maybachs upon arrival in Abu Dhabi in *Sex and the City 2*. Theirs is the height of conspicuous consumption and waste, but one wonders why their baggage (emotional and otherwise) can't fit into one vehicle. "Living the dream" comes with costs none of us can afford.

Many cities now organize car-free days to reduce pollution and ease congestion, including Copenhagen, Brussels, Bogotá, Mexico City, and Jakarta, while World Car-Free Day is celebrated around the world on or about the September equinox. In 2016, Paris instigated a partial car-free day on the first Sunday of every month, restricting cars on 650 km of Parisian streets. On its first car-free day, nitrogen dioxide levels decreased by as much as 40% and sound levels by half.[23] During the 2018 running of the London Marathon, NOx emissions decreased by 89%, highlighting the effect of tailpipe pollution that directly contributes to the premature deaths of about 40,000 Britons each year, roughly the same number who ran the marathon.[24]

Congestion charges in London, Stockholm, and Milan have attempted to address inner-city emissions via so-called "cordon zones," but exclude lower-income drivers and those who don't want to pay, who then increase emissions by navigating longer detour routes.[25] Eventually, gasmobiles will be banned, starting with diesel cars in Rome in 2024. Sales are already falling as buyers fear the coming restrictions. Walking, cycling, public transit, and zero-emission vehicles are all becoming popular alternatives to the fume-spewing gasmobile.

If you do drive, there are simple ways to save. Diamond lanes, multi-occupant commuting (carpooling), and ride sharing (via various mobility apps) all help to connect drivers and passengers, while better and more accessible scheduling, kiss-and-go drop-offs, and transit station parking all reduce car use. Sadly, in the USA, only 5% of cars carry more than one occupant on their way to and from work.[26] One in 20!

Lowering your speed from 75 to 65 mph reduces fuel costs by 25%. As we saw in Chapter 2, Richard Nixon reduced the 60 mph speed limit to 50 during the 1970s fuel crisis, providing huge savings on imported oil, which Bill Clinton later revoked, while simple driving tips save money and emissions as we also saw earlier. Many new cars have automatic stop–start technology in traffic (although not advisable in freezing weather), while some show the fuel you saved, for example, 50 ml in 3 minutes of shutoffs. Natural Resources Canada estimated that fuel consumption was reduced by up to 10% from stopping a motor when stuck in traffic with annual fuel savings of $150.[27] App-found parking is also coming, happily decreasing the time to find a spot, although sadly not the cost. Increased efficiency helps, but driving less is best.

In Copenhagen, cycling is the preferred mode of transit, where almost two-thirds of inhabitants bike to work or school, even in winter, citing the convenience, cost, and flat terrain.[28] Why drive to work and go to a gym afterwards if you can fit both into the day by cycling to work (also reducing stress and saving on membership fees)? Not everyone can, but soon we may have no choice. We may be ecologically illiterate, but the Earth is not.

There is room for more bikes, scooters, and personal mobility vehicles (PMVs) in our cities, both people-powered and electric, lowering street pollution and liberating us from the jarring screeches and muffler-less revving as if stuck in a macho Fifties *Rebel Without a Cause* drag race. There are now more e-bikes than cars in China (as of 2020, over 200 million), because of lower costs and anti-pollution measures introduced in the 1990s, while charging is as easy as with a phone thanks to quick-release batteries. A 10-cent charge lasts for 100 km with top speeds that can reach 50 km/h. An e-kit can also be fitted to a pedal bike with wheel-hub motors that recharge on the go to help with the hills. Of course, muscle power is best wherever possible – good for the body and the grid.

Along with expanded public transit and more EVs to cut down on congestion and toxicity, sharing is on the rise thanks to mobility apps and improved connectivity. The sharing economy got a kick-start in Amsterdam in the mid-1960s with free white-painted bikes (Witte Fietsenplan) to counter rising inner-city car traffic and pollution, followed by a shopping-cart-style, coin-operated system in Copenhagen in the mid-1990s.[29] By the early 2000s, docking stations began appearing across Europe, where customers could rent bikes with a credit card or dedicated client card, dropping the bikes off at an open slot (sometimes not that easy). Paris rolled out Vélib' (Vélo liberté or "bicycle freedom"), London had its Boris Bikes (commonly called after the former mayor Boris

Johnson who oversaw their London arrival, although technically Santander Bikes), and Dublin dublinbikes (sponsored by Just Eat). The rates are reasonable, while journeys under 30 minutes are free in some cities. In the northern Spanish city of Gijón where we live, the municipal *Gijón-bici* bikes are completely free with a citizen card.

In most major cities, hundreds of docking stations with thousands of bikes are available, company hauliers busily moving stock around, refilling empty ranks, and repairing or replacing as needed. Today, millions of bikes are shared across the globe, helping to reduce congestion and toxic emissions, as many as 50 million rides a year recorded in China. So-called fourth-generation systems don't require a physical "hub-centric" docking station – the bikes are instead found via a GPS-tracked phone app and left anywhere, a.k.a. "free floating." With instant access, free-floating availability, and 25-cents-a-minute pricing, scooters have also become popular for short trips. Some automated e-scooters can even return themselves to a local station for recharging! Having bought out Segway – a self-balancing transporter invented in 1999 with computer-controlled gyroscope for stability – the Chinese scooter manufacturer Ninebot sold 1 million scooters in 2018 to claim 90% of the US "micro-mobility" market.[30]

Fewer emissions, less noise, and easy maintenance is good for one's health and sleep, although the rules of the road and etiquette between pedestrians and e-wheelers are still evolving, with limits placed on e-hooligans who free-wheel from road to lane to sidewalk to road, undeterred by hazard to themselves or others. Labeled "scooter anarchy" by a disgruntled mayor of Paris, e-scooters were barred from the pavement and then banned outright after a referendum, although dockless scooters still litter the streets in other cities such as Brussels, left helter-skelter without thought to others, making sidewalks difficult to navigate for pedestrians. In Amsterdam, which boasts more bikes than people, a lawless tyranny rules the road, unmonitored by authority.

Following the success of bike sharing, car sharing began in a few larger cities, both hub-centric and free-floating, and are mostly of the clean EV variety. Careco in Tokyo, Zity in Madrid, and Birò in Amsterdam are just a few rental companies operating at reasonable pay-as-you-go rates, while others offer on-demand EV sharing for a set monthly fee. The Daimler-owned car2go is the largest one-way car-sharing company in the world, providing cars to rent by the minute, although many are still of the petrol-fueled Mercedes kind with some Smart Fortwo EVs. Starting operations in 2010, car2go reached 1 million customers in 4 years and continues to grow by 30% a year. Future rentals may even incorporate autonomous technology – just buckle up and get your credit card ready.

Car sharing is becoming an important part of the growing disruption to urban mobility. Having grown up with easy access to computers, Gen Ys – the most influential market after the 2-billion strong Baby Boomers – desire "connectivity and convenience and can choose from an ever-increasing range of transportation types, alongside vehicle ownership, for getting from A to B" as noted in a 2017 Deloitte study.[31] Recasting the car from an individual or family possession to a "collective convenience," a shared car removes 20 others from circulation, optimizing generally idle resources (roughly 5% operating time).[32] Both the established car-rental companies and high-tech start-ups are vying for a larger piece of the action in an increasingly valuable stationary, peer-to-peer, and free-floating sharing market.

Fraught with financial peril, however, just staying afloat is hard enough. London's Bluecity is typical, providing an easy-access, hassle-free, all-EV service for £10 per hour. Started in Hammersmith and Fulham in 2017, fleet numbers quickly reached 200 cars across the greater London area, offering a fully charged range of 130 miles. Cars could be reserved in advance to ensure easy pickup and drop off, while charging points were managed by a mobile-phone/Internet app or onboard navigation. Drivers were expected to keep the vehicles clean in a community-spirited way. Alas, Bluecity (oddly colored red!) went bust after only 3 years.

Bluecity was based on the original car-sharing scheme in Paris (Le carsharing), begun in 2011 by French transportation giant Bolloré, which once boasted 4,000 EVs, 6,000 charge points, 100,000 subscribers, and half a million rentals a month before imploding in 2018 from high costs. Built to last, the little gray beasts were an iconic sight on the streets of Paris, darting in and out of traffic with characteristic French insouciance. Having expanded to other French cities – Bluecub in Bordeaux, Auto Bleu in Nice, and Bluely in Lyon – and further afoot to Indianapolis (BlueIndy), Los Angeles (BlueLA), and Singapore (BlueSG), where they were more successful, the Bolloré Group had hoped to rule the car-sharing world before crashing out of Paris. On-demand rental is certainly a smart idea to maximize idle resources, but comes with the usual problems in the sharing business – maintenance, round-the clock attention, and profitability. What's more, not everyone leaves a car as they found it, the golden rule of responsible community sharing.

Despite the growing pains, "Libre Comme L'air" (Free as the Air) is a laudable objective to rid our streets of stifling diesel-generated NOx contamination and burnt-carbon pollution. Something has to be done if Paris wants to meet its own 2015 COP21 goals and eliminate gasmobiles as mandated by 2030. Bolloré Group co-founder and director of development Cedric Bolloré noted that "Cleaning the air is not something new to us. We've been working for 25 years on that subject and spent an enormous amount of effort to develop batteries, cars and the most advanced car-sharing service in the world. . . . it works because it's simple."[33] Perhaps not so simple to make money as in any new venture, but break-even is worth the investment if inner-city transport becomes more accessible and fume-free.

In 2019, a shared mobility project began in Sacramento with 260 Chevy Bolts found via phone app, where drivers could buckle up and go for $2.50 per mile or $15 an hour,[34] while another test project made 140 VW e-Golfs available at $9/hour to low-income apartment dwellers, helping to democratize the EV transition across different-income zip codes. Funded by Electrify America, the first-of-its-kind, EV-charging platform was setup as part penalty for Volkswagen cheating on diesel emissions, making expensive EVs available to those who can't afford or don't want the extra hassles of ownership.

The road more traveled is changing with better-connected users and improved phone apps, especially among the young. Without public transit, however, congestion can actually increase if fewer commuters take the bus, tram, or train, as occurred in San Francisco after congestion rose 30% from 2010 to 2016 because of ride sharing.[35] Easy-access, integrated public transport is essential when it comes to inner-city travel.

***

Of course, modern infrastructure requires funding, the main argument against government-funded transit projects in debt-conscious cities. In truth, we can't afford not to build for the future, both for comfort and to kick-start investment to fund the green makeover of the auto industry and improve public transit systems around the world. A bold idea to cut congestion is to make public transport free as in parts of Calgary, Miami, and the entire country of Luxembourg, where the daily, cross-border influx of 200,000 foreign workers and high car ownership had turned the grand duchy into a giant parking lot. As of March 2020, all buses, trams, and trains are free, although one can still buy a first-class seat for those who don't want to mix with the hoi-polloi.

Austria tried a different tack, in 2021 offering unlimited nationwide travel for €3 a day, roughly €1,000 a year, following Vienna's €1 all-day travel-pass scheme, instigated a decade earlier. As noted in *New Scientist*, affordable public transit can start a virtuous circle: "Fewer cars on the road makes alternatives, including walking and cycling, more attractive too."[36] In 2022, air quality rose throughout Germany after a €9 all-you-can-travel nationwide train pass was introduced for the month of August. Spain followed suit, making train travel on commuter lines free from September to the end of the year (extended until the end of 2023). Incentivizing public transport is an obvious way to cut congestion, lower road emissions, and conserve fuel. In another novel project to wean people off their gas-guzzlers, car owners in Coventry can exchange their old clunkers for £3,000 in public transit credit – in the first pass, 150 drivers took up the offer.

Although the rollout will play out differently across the world – Americans prefer bigger cars and more space (the land of the SUV), Europeans better design (with easier parking on older, narrower streets), and the Chinese better connectivity (app-driven ease) – clean, connected transportation is improving by the day. Tomorrow's cities are all planning for a greener future with better vehicle design, low- and zero-emission EVs, electrified transport, connected and automated transport (CAT), data-driven network and traffic management, smart mobility and services, and a much-expanded and integrated public infrastructure. Of course, improved cyber security will be essential to protect against abuse in our increasingly wired world, including targeted cyberwarfare as commercial saboteurs and foreign agents hack into web systems to override car controls or try to shut down, hijack, and destroy power stations and industrial plants.

The opposition to public transit and vehicle electrification also remains strong, especially in the USA, where companies fight against anything that counters their core petroleum interests, even lying about the cause and effects of climate change. Paradoxically, in the land of the free and home of the brave, the USA is becoming the last stand against modernity as population densities approach those of Europe, where towns and cities founded in medieval times have been forced to rethink planning and mobility for an increasingly urbanized population. There will be no turning back, however, once the electric train leaves the station.

As cities grow and population densities increase, travel must become more efficient and less dependent on the car, a historically underused, difficult to store, and costly household item. As for the increasing urbanization in Asia, Harvard economist Edward Glaeser, who grew up in New York City, stated "It would be a lot better for the planet . . . [if they] built around the elevator, rather than in sprawling areas built around the car."[37] Alas, Le Corbusier's dream of building cities in the sky has come to pass, but we need to rethink the adjunct parking.

The most densely populated country in continental Europe is the Netherlands, averaging more than 500 people per square kilometer, many living in the two most populous provinces of North Holland and South Holland, where population densities are over 1,000 people per square kilo-meter – essentially one big city known as the Randstad marked by Amsterdam, The Hague, Rotterdam, and Utrecht. It's no surprise the Dutch have built an enviable world-class public-transport system with one transport card for the whole country and a 40% fare reduction outside peak times (now accessible with a credit card, mobile phone, or NFC-enabled smart watch).

Conversely, the average US population density is under 40 people per square kilometer, such that large public infrastructure is more difficult to construct in all but the most densely populated regions. New Jersey is the only state with a population density comparable to the Netherlands. By comparison, the whole of Japan has about 350 people per square kilometer, while Tokyo is a whopping 6,000, where individual car ownership for all would be crazy.

England is even more densely populated than the Netherlands – rural Scotland, Wales, and Northern Ireland skew the UK numbers – and is ever trying to improve its once-enviable railway system, regularly flip-flopping between public and private ownership. Trains make more sense than cars to combat urban congestion, while short-hop planes are especially bad given the amount of greenhouse gases emitted by flying. Taking a plane uses 10 times the energy of a train, while the jet fuel alone accounts for almost 40% of the cost of the ticket. Paul Hawken noted in *Drawdown* that electric high-speed rail is "the fastest way to travel between two points a few hundred miles apart and reduces carbon emissions up to 90 percent."[38]

Before COVID-19, airline emissions were becoming the "new coal" with 4 billion annual passengers, responsible for 2% of GHG emissions and expected to increase to 10% by 2050.[39] With characteristic cheek, Ryanair chief Michael O'Leary suggested a solution – ban first-class and business seating that takes up more space than regular seating, making flying more economical as well as less polluting/emitting per kilometer. O'Leary also noted that ships are more toxic than planes, although the amount of fuel alone in a low-cost flight makes short-haul flying a highly toxic alternative to any means of travel, typically untaxed at that. What's more, in Europe, rail tickets include per-kilometer infrastructure costs. Trains could certainly become more economical if the government planners redid the numbers. In 2021, the French government did ban short-haul domestic air travel if the same journey by train was available in less than two and half hours, a good start to curb inefficient travel, albeit currently applicable to less than 1% of flights. Even still, long-distance train travel is growing in Europe, including overnight journeys.

Despite the worrisome emissions, flying has nonetheless grown 10-fold in the last four decades, from roughly 1 million to over 10 million a day. For green travelers keen to lower their carbon footprint *flygskam* (Swedish for "flight shame") is catching on, the latest socially conscious practice, along with increased train travel or *tagskryt* (Swedish for "train bragging"). Germany has taken the challenge on board, Deutsche Bahn offering more intra-city train service, including at night, while slashing prices on all routes. Why take a train or metro to get to an airport to fly to another airport and then a train or metro into the city when city-to-city train travel is simpler, faster, and often cheaper? Reducing our carbon output means less flying, that is, if we are serious about our individual contributions to global warming. As economist Dieter Helm succinctly noted in *Net Zero: How we can stop climate change*, "Aviation and net zero just don't mix well."[40]

Decentralization also helps lessen travel, especially in an age of easy Internet connections. Forty percent of the Irish population doesn't need to live in Dublin because city planners can't think outside the urban box – or wherever a large capital city dominates the national interest – while urbanization increases uniformity as well as fueling a hostile "blue-red" divide, especially seen in the USA. Expanded teleworking during the COVID-19 pandemic showed that overcrowded urban hubs, in which hundreds of thousands of commuters pour into city centers and downtowns at the same time, is not the only work–life model. Decentralized power generation can help stimulate a reversal of hazardous urbanization trends and promote less wasteful lifestyles.

We have to do something to make our lives greener. Almost half the world is now urban, a percentage that continues to climb, while Latin America is already more than 70% urbanized. The urban/rural split is almost 75% in Western countries, the global average reduced by African and Asian countries excluding China (about 55%).[41] And while cities can operate efficiently, much of the energy produced is wasted. Today's cities occupy 2% of all land, contain half of all people, yet use 75% of the energy and create 70% of GHG emissions. As energy demand rises, cities will have to find ways to cut down and become more self-sufficient, including expanding localized green hubs rather than increasing traffic flow.

One can imagine a bustling yet tranquil time before the "puffing devils" took over. Perhaps we will see quieter days again in a less car-reliant or post-car urban future. After all, what is the point of a city – to prioritize movement at any cost or to provide convenience and safety for citizens? As off-grid capacity grows, the next decade may see the first trend away from city living since the start of the Industrial Revolution, for those who can afford to escape to the country and for others in smaller pocket-city developments where people walk to local amenities.

7.4 Reducing, Reusing, Recycling, etc.

I have been a city dweller almost my whole life (Toronto, Dublin, Amsterdam, and Gijón), where one can easily avail of public transport, recycling facilities, and shared activities. I have been "green" for more than 40 years and rarely buy stupid gadgets. I try to reduce as much as possible (less is definitely more!), reuse bags, save scrap pieces of paper for notes, and wear holey socks long past my wife's liking. Gone are the days of darning for all but the most ardent, although cotton items (including holey socks) and other clothing fibers can be ground into pulp to make beautifully textured paper.

Government policy has spurred on my contribution during a peripatetic life in Canada, Ireland, the Netherlands, and Spain. Like many of us, I now recycle paper, plastic, cans, glass, take old appliances and electronics back to the store or a second-hand outlet, and recycle food scraps/organic waste with a biodegradable bag that decomposes by moisture and heat. What I can't handle, I take to our local Punto Limpio (Clean Point).

Sidewalk recycling began in our Toronto neighborhood in the 1980s, reducing landfill waste (as well as letting one surreptitiously survey neighborhood eating and drinking habits). Ireland introduced a 15-cent plastic-bag levy at point of sale in 2002, increased to 22 cents in 2007, and a free waste electrical and electronic equipment (WEEE) return policy on a one-to-one basis when buying new electronic devices and appliances as part of a standard 2005 EU-wide directive that increased returns eight-fold. Spain began limited organic waste collection in 2014, while the Dutch have installed neighborhood composters in choice locations, amusingly called "worm hotels" where locals open a combination lock to deposit their food waste. Other cities have started up their own organic collection and drop off centers, such as GrowNYC and the NYC Compost Project.

Paper is the easiest to recycle, fibers converted into pulp feedstock to create more paper or other paper products. Metals are separated at source, primarily tin and aluminum, while glass is melted to make new glass products. Theoretically, organic waste is the easiest to recycle, but also the messiest as the biological material is converted into biofuels, mulch, and compost. Plastics is a whole other deal that is now being considered as a chemical feedstock and for fuel conversion, which we'll look at later in more detail.

It wasn't always easy to recycle. A 1955 *LIFE* article "Throwaway Living" sang the praises of our ultra-modern, disposable lifestyle, although a few people questioned the massive amounts of waste being generated, such that basic recycling is now standard across much of the globe. Each item not discarded in a dump/landfill saves on the environment and processing costs to handle all the used material, packaging, and garbage we discard to the tune of about 500 kg per person per year, roughly our own weight every 2 months.

"Waste diversion" makes sense, highlighted by the three-arrow Mobius Loop symbol created in the 1960s for a Chicago recycling-container company to encourage a "circular" economy. But there is still much to do. In Spain, only

30% of household waste is recycled, below the EU average of 45%, while in the USA almost half of all food is thrown away, roughly $170 billion or $500 per person per year.[42] Some of us barely think about what we dump. In the EU, the aim is to recycle 70% of all packaging waste by 2030 and 65% of all municipal waste by 2035 (paper, ferrous materials, aluminum, glass, plastics, and wood).[43]

Curbside recycling continues to evolve, cities providing different services with their own systems and colored bins. In Spain, paper is blue, plastic yellow, glass green, and food brown, while unsorted garbage is light green or gray. San Francisco provides a single-stream blue bin for all recycling material (no plastic/paper/glass/tin distinction), a green bin for composting (food and other organic waste), and a black bin for trash. Waste separation is mandatory, but full compliance isn't always possible, even though residents can lower their "pay as you throw" bill by recycling and composting more.

In the Netherlands, citizens put their refuse into dedicated collection containers located on street corners (blue paper, yellow glass, orange plastic) with limited organic recycling, although restaurants, supermarkets, and industrial producers are required by law to recycle organic waste and cooking oil, deposited in large plastic bins (food) or sealed kegs (oil), regularly collected by private handlers and converted into green electricity or biofuel, a.k.a. "second-generation" energy. In some cities in Spain, used cooking oil can be dropped off at recycling points or deposited in curbside boxes, the recycled oil converted to biofuel.

Care is required to put things in the right bin whatever the color code. Wrongly sorted materials and food scraps can literally clog up the system and ruin a perfectly good recycling batch. Paper soaks up food, so it's best to clean containers of any side-hugging yoghurt, tomato sauce, or peanut butter before binning (if it can't be eaten). Be extra careful with mixed plastic/paper (especially paper coffee cups with plastic lids), but small amounts of "contaminants" are okay such as plastic-lined tissue boxes and cellophane-window paper envelopes, although one can choose to cut out the windows and liners and add them to the plastics bin or better yet buy paper-only tissue boxes and envelopes.

In Toronto, more than 50,000 tonnes of non-recyclable items are incorrectly placed in the Blue Bin every year, costing millions of dollars as unexpected material clogs up the automated sorting machines.[44] The City of Toronto has an online Waste Wizard to determine what goes where to help with the non-obvious recyclables (Know Before You Throw!) as do many municipalities. In 2021, the city of Amsterdam suspended roadside plastic collection, claiming that managed waste identification machines can separate plastic better than humans. Instructing residents to put their plastics in with the regular waste, plastic recovery rose by 60% in the first year.[45] Perhaps in the future, we won't need to sort our garbage at all and can instead concentrate on reducing all types of waste.

Using less and reusing more is the simplest way to cut down. Conservation is about practice and policy. Many of us reuse containers and try creative ways to extend a product's life. My godparents saved small frozen orange juice cans to reuse as seedling protectors, wrapped their garden vegetables in plastic milk bags (an odd Canadian invention), and composted everything they could to make mulch and topsoil. Old clothes ("slow fashion"), books, and unneeded items can all be donated to a second-hand shop. Wherever possible, it is better to extend the lifecycle of an electronic device before returning them to the seller when it comes time to upgrade, now required by law in many countries.

Changing habits is the hardest as more than half of what is discarded can be diverted from our overflowing landfills. The goal is zero waste, or more realistically as-little-as-possible waste. Sadly, some "recycled" goods are being incinerated instead of reprocessed, while recycling is still low in many countries, including the USA (32%, 25 out of 32 major countries tracked), Canada (29%), and Japan (20%), behind number 1 recycler Slovenia (72%). In the USA, paper and paperboard is at 67% and metals 13%, but wood, plastics, glass, textiles, rubber, and leather are all under 5%.[46]

Real change comes from the bottom-up with improved habits and attitudes toward waste. Choose your own maxim: "think globally, act locally," "think globally, act neighborly," or as in the philosopher Immanuel Kant's categorical imperative, "What if everyone did as me?" Manufacturers will continue to assume we don't care unless shown otherwise, so it's up to us to be guided by more than whim. Fueled by an endless call for more, excess consumerism is a trap, a deepening hole from which no one can escape, unless one stops digging.

Charitable giving is also a way to cut down, typically increased with higher landfill costs, taxes, and fines, in particular food that feeds almost 10% of the American population.[47] Rather than encouraging poor working habits, donations help those in need, while reducing perfectly good surplus that would otherwise go straight to a landfill, helped out by tax breaks for company giving. Supermarkets offering free surplus food also help, say as a percentage of one's total bill in a loyalty scheme to deter purposeful freeloading, although only if the food isn't wasted in dubious "2-for-1" marketing schemes.

More than half of all plastics is packaging, but we can easily cut down by reducing the amount of excess wrapping. Do bananas need to be put in their own container and wrapped? They already have skins! Or individually sealed oranges with trademark twisted-paper to identify the source – no need to double-glaze fruit! It's easy to bring your own reusable fruit and vegetable bags when doing the shopping. A reusable drinking bottle pays for itself almost immediately and isn't any less convenient to lug around than a plastic throwaway bottle, nor is the bottled water guaranteed to be any better than tap water. It is easy to say "no" to excess packaging, unnecessary plastic, and useless frilly extras.

The single biggest use of plastics is packaging – about one-third of 400 million tons annually produced – most of which is tossed out in under 6 months and never recycled or incinerated. Of the 9.2 billion tons of plastic made in the last century (mostly since 1960), 6.9 billion tons is now waste (75%), a whopping 91% not recycled, while up to 14 million tons per year end up in the ocean from careless dumping practices in coastal regions.[48] Should containers be made of single-use plastic to make life easy for long-haul shipping instead of cardboard or paper as in the past?

Sadly, producers have not been tasked with providing alternatives to a plastic supply system that incentivizes wasteful packaging. Who can forget the picture of a seahorse riding a plastic cotton swab off the island of Sumbawa or dead whales found with huge amounts of plastic in their stomachs? One can never assume that garbage isn't potentially lethal – "out of sight, out of mind" does not mean out of existence. Substitution is always preferred, such as paper, organic material, or reusable packaging rather than single-use plastic, but short of doing away with plastics altogether, there are some things we can do to help (as we look at now).

7.5 The Plastics Problem: Reducing, Returning, Researching, etc.

Change won't happen without changing habits, while the problems will only get worse in the West now that China and other developing nations have banned the importation of "foreign garbage" since 2018, forcing the developed world to sort out its mess and acknowledge the failure of inadequate recycling systems. Made of long-chain carbon atoms, the most common plastic, polyethylene, was accidentally discovered in 1898 in Germany, and can literally take ages to biodegrade. Tossed into a landfill the polyethylene slowly turns into methane, while in water the polyethylene may never fully decompose without oxygen, breaking into smaller bits of harmful microplastics. Essentially, if it is plastic at some point it will end up in a landfill or, worse, the ocean and our bodies (even passed from mother to fetus). Although new biodegradable and compostable plastic bags are better, they are marginally more expensive and also release methane when broken down.

The annual global plastic output is roughly 400 million tons – 800 billion pounds or almost 5 kg per person per year – requiring 200,000 barrels of oil a day to manufacture and occupying a quarter of our landfills by volume.[49] Taking centuries to decompose, over one *trillion* pounds of plastic now litter American dumps. A completely new packaging system is needed that favors reusable, Earth-friendly materials in a comprehensive recycling process, while restricting or banning unnecessary single-use items. Wherever possible, we should all use reusable or recyclable metal, wood, glass, or paper.

To study how to "upgrade" or reduce the mounds of "post-consumer" plastic waste, various research projects have begun, such as plastic-eating bacteria and bugs, plant-based biodegradable plastic, and petroleum substitution. Researchers are also converting plastic waste into petrochemical feedstock (P2P) and fuel (P2F) via solvent-based liquefaction or thermal decomposition, that is, heating via pyrolysis (no oxygen) or gasification (low oxygen), despite environmental concerns and deleterious effects (Figure 7.2). According to Aachen University fuels engineer Peter Quicker, such chemical recycling is flawed: "The special value of plastic, the polymerized structure, is decomposed and transformed into an inferior product, such as a low quality oil that has to be treated with great effort in order to turn it back into plastic."[50]

New plastics are also being designed to "self-destruct" and exploit "depolymerization." As biochemist Robert M. Hazan, executive director of the Deep Carbon Observatory, noted, "As the polymer chain or network is fragmented into smaller, unconnected pieces, soluble molecular fragments can simply wash

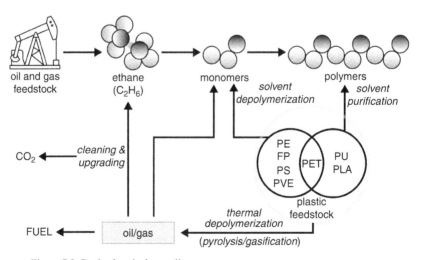

Figure 7.2 Basic chemical recycling processes.

away and return carbon atoms to their natural cycle. New breeds of plastics embrace this opportunity, with a special focus on polymers that can be broken down by hungry microbes."[51]

In 2016, Japanese scientists at the Kyoto Institute of Technology found a species of bacteria that "eats" polyethylene terephthalate (PET). The *Ideonella sakaiensis* bacterium breaks down and metabolizes PET with two hydrolyzing enzymes, a process that helps biodegrade long-lived plastics. We should be wary, however, of introducing a biological fix into the environment, such that we don't do more harm than good as in the disastrous DDT pesticides of the 1950s. Letting loose vast amounts of plastic-eating bacteria or bugs on a pile of polyethylene may seem like an intriguing solution to our ever-expanding plastic problem, but they can also eat through the good stuff.[52]

In an effort to cut down petroleum-based plastics, biodegradable plant-based plastics are becoming more popular. Polylactic acid (PLA) is made from fermented plant starch – mostly corn, although also potatoes, wheat, and beets – but is not easily separated from petroleum-based plastic in current recycling programs, causing problems to already challenging waste diversion systems. PLA is more environmentally friendly than petroleum-based plastic (65% less energy and 68% fewer GHG emissions[53]), decomposing into carbon dioxide and water, and is not as toxic when burned or broken down, but still takes from 100 to 1,000 years to biodegrade in a landfill. In a controlled industrial composter, PLA takes 3 months, alas not readily available to the average consumer.

Typically made from genetically modified organisms, PLA may ultimately cause more problems than it solves, although Eric Lombardi, president of the Grassroots Recycling Network, believes it is visionary even for us to think about using plant-based plastics and that the real problem is our attitude to waste: "We need a convenient, creative collection system with three bins: one for biodegradables, which we'll compost, one for recycling, and one for whatever's left."[54] Plastic composting additives also help break down bio-logical plastic in both anaerobic and aerobic environments, although the logis-tics of large-scale composting is still difficult.

Researchers in biological chemistry at University of Warwick have been investigating how to produce useful biodegradable plastic by mixing lignin – a natural glue found in trees discarded in paper making – and the genetically modified soil-living bacterium *Rhodococcus jostii*. As Warwick's lead researcher Timothy Bugg noted, "Because lignin is complex, as you break it down you get a complex mixture. But what's nice with these bacteria is they are able to funnel all this. We are hoping in five years that we will have something."[55] Previous research into wood-digesting enzymes is also helping biochemists understand how to create more sustainable biofuels.

At their core, plastics are polymers (macromolecules), but the long-chain hydrocarbons can theoretically be broken down for reuse as chemical feed-stocks and fuel, promoting circular use and reducing waste, but the process is messy, especially if the waste plastic is impure. Researchers at Chalmers University of Technology in Gothenburg, Sweden, are studying how to recon-vert waste plastic into reusable virgin plastic via steam cracking in a chemical reactor, while Utah engineering start-up Renewlogy is using pyrolysis (heating without oxygen) to create gas that can be condensed into oil, fuel (such as ultra-low-sulfur diesel), and other petroleum products. Despite the laudable goal of reducing plastic waste, the process is still technically incineration, however, as the reformed plastic fuel is ultimately burned. Furthermore, such processes are not yet scalable to real-world applications.

A more fruitful solution is to rethink plastic waste as a potential design material, for example, as a viable construction resource or even fashion item. Indian chemistry professor Rajagopalan Vasudevan prefers not to ban plastic because of its many uses and has focused instead on reusing plastic materials by shredding and mixing plastic with tar to increase the tensile strength of roads, making surfaces more durable and flexible. The bitumen-substituted plastic also helps prevent pothole formation by repelling water from seeping into the gravel and bitumen mix during heavy rains, while the road-laying process saves as much as 10% on bitumen by using the equivalent of 2 million plastic bags per kilometer. After a test road was constructed at his southern Indian university, almost 10,000 km of roads have been paved with Vasudevan's plastic-modified bitumen.[56] Another company has made plastic paving stones.

A New Brunswick-based Canadian start-up, Plaex Building Systems, con-verts shredded plastic waste into no-cut, interlocking, mortar-less bricks in a process that uses very little water and can significantly lower the building industry's large carbon footprint. The refashioned Lego-like plastic construc-tion bricks were initially designed to cut down on excess agricultural plastic waste, such as the immense amount of greenhouse caterpillar tunnels and ground cover, but can be applied to any industry, and is limited only by the imagination. The Dubai-based company DGrade is making textiles from shredded plastic PET bottles, weaving a patented polyester-like yarn into wearable products. The repurposed plastic t-shirts, caps, and bags can them-selves be further "down-cycled" into other recycled plastic goods.

Others are attempting to replace the many everyday products made of throwaway plastic with organic substitutes. Bagasse (or sugarcane pulp) is the leftover non-sugar part of pulverized sugarcane in the making of sugar, whose short fibers can be interwoven with long-fiber organic material such as bamboo to make biodegradable food containers, utensils, and coffee cups. Headed by mechanical engineer Hongli Zhu, a Northeastern University project in Boston has produced a compostable hybrid material twice as strong as standard cup plastic that completely rots away after just 6 months in the ground.[57] Another option is plant- and seaweed-based packaging materials that naturally biodegrade in weeks, a compostable product that won the UK company Notpla the 2022 Earthshot Prize.

Unfortunately, the sheer amount of plastic has become a crisis, despite the many novel schemes and research projects that aim to cut down on waste or break down chemical ingredients for reuse. One enterprising Dutch engineering student took it upon himself to recover ocean plastic, designing a system to vacuum floating plastic on the high seas with a self-orienting, U-shaped boom. Having seen more plastic than fish while swimming on a family holiday in Greece, a then 18-year-old Boyan Slat began The Ocean Cleanup to collect 5 trillion pieces of plastic in the five main ocean garbage patches of the world. He hopes to clean up 50% of the Great Pacific Garbage Patch – roughly the size of France – with a full-scale roll-out within 5 years. After working out a few design bugs, System 001 was ready to go in 2019, while the next version will autonomously navigate the sea. We can all learn from those who are valiantly taking on the ecologically disastrous packaging industry on their own.

The Great Bubble Barrier is another innovative device created in the Netherlands that emits air bubbles through a perforated tube, fixed diagonally across a canal bed to direct floating and submerged plastic to a bank-side container, stopping plastic waste from reaching the open seas. First installed in 2019 in Amsterdam's Westerdok, the 24/7 system lets fish, wildlife, and boats pass, but blocks waste from escaping. The goal is to reduce the 42,000 kg of plastic deposited annually in the inner waters of Amsterdam, while rolling out more systems across Europe and Asia. The bubbles also aid the aquatic environment by increasing oxygen levels in the canal water.

Sadly, we have been treating the Earth as a dump for too long, without thought to the damage. When a parcel arrives at our door, we rationalize the packaging as immaterial, a one-off in the grand scheme of global waste. But multiply that one parcel by 1,000 more in our neighborhood and again by hundreds more neighborhoods in thousands more cities – EVERY DAY – and the immaterial becomes too much for the Earth to handle. Yes, much more needs to be done to ensure recycling works and appropriate penalties levied for littering and dumping, while manufacturers must provide ecologically responsible packaging alternatives to plastic. But we too can cut down on our own waste and think of better ways to carry and package things, saving energy and reducing pollution. Carrying a simple scrunchable cotton bag in one's pocket is the least we can do.

Unfortunately, we can't wait for science and technology to solve our growing plastic problem, absolving of us of our duty to consume more responsibly and use less. As in the Plastic Bank slogan, we must "Be part of the solution not the pollution."

7.6 The Circular Economy: Say No to Waste

We live in abundance, often paying little attention to how we consume. Changing behavior isn't easy, but the idea is to cut down however one can and wherever one sees a better alternative. In *How to Live a Low-Carbon Life*, Chris Goodall shows us how to reduce CO_2 emissions from 14 to 3 tonnes per person, essential to beat global warming, including reduced air and car travel, installing better home heating and appliances, and eating local, unprocessed, and unpackaged organic food. Happily today, using affordable, carbon-neutral products is no longer a choice between cost and the environment.

On a planet of more than 8 billion people that increases by about 75 million a year (20 more by the time you finish reading this sentence), we have to cut down. As we've seen, one simple way to improve efficiency is to reduce food waste and intake (also good for the waistlines of an increasingly overweight public). Turning off computers at night or during extended breaks provides huge savings. Simple power-management software can dramatically cut down consumption as well as air conditioning costs – for example, Dell Computers using its own power-management system to reduce energy spending by 40%.[58] Wherever possible, use appliances outside of peak times.

If you play the stock market, invest in your favorite green company or companies that won't destroy the world. Sustainable investing is becoming increasingly important because of concerns about resources, global warming, and social responsibility, especially for Millennials who want to invest in ethically minded companies and funds. Called ESG (environmental, social, governance), the central issues beyond company fundamentals include GHG emissions, supply chains, working conditions, pay (including CEO), workforce diversity, fossil-fuel use, water quality, and energy efficiency as well as avoiding more traditional "sin stocks" such as alcohol, tobacco, and weapons. Although share price rules in the investment game, some ESG funds now outperform conventional stocks and are considered important enough to be included in rating agencies' measurement criteria. After centuries of ignoring green fundamentals, pure profit may not be the only guide to future spending.

Others stress community participation, such as productive giving (local green investments), practical acts (fun and easy ways to reduce emissions and energy bills, including rooftop solar panels), and shared learning (mentoring and training in one's community and elsewhere). Community beautification and local trash pickup days combine the best of neighborhood participation as in the annual Irish Tidy Towns competition. Group behavior educates us all with new ideas and levels of experience about what does and doesn't work.

In 2015, Sweden opened the world's first all-used-goods, low-carbon mall next to a recycling center in Eskilstuna, 60 km west of Stockholm, with 14 shops full of pre-loved stuff, featuring only items that have been recycled, repaired, or refurbished for resale. Called ReTuna Återbruksgalleria, thousands of customers regularly shop in mall-style comfort for perfectly good "upscaled" items such as clothes, furniture, computers, toys, gardening tools, and building materials.

In Santa Barbara, where support was galvanized for the environmental movement and first ever Earth Day (April 22, 1970) after a disastrous oil spill off the coast a year earlier, water use is now less than half the state average. Some changes were simple, even laughably obvious, such as banning pavement washing, fitting idle pools and hot tubs with covers, and serving water to restaurant patrons only on request.[59] Partly because of necessity in the midst of record droughts, Arizona and Nevada have reduced their overall water use by about a third (~300 down to 200 gallons per person per day).

Japan launched the Cool Biz campaign in the summer of 2005 to reduce air conditioning. Office temperatures were restricted to no lower than 28°C, while workers were encouraged to dress down, forgoing suits and ties for more comfortable, hot-weather casual attire. Government office lights are also dimmed for an hour each day to save energy and remind everyone of the stakes.[60] The campaign now runs annually from May to October.

In response to the energy crisis caused by the Ukraine war, temperatures in government buildings in Germany and Spain were limited (between 19 and 27°C for Spain), while as the 2022 winter approached France announced an "energy sobriety" to cut consumption by 10%. With the tag line "Every gesture counts," the French government asked citizens to turn down thermostats to 19°C, switch off lights, and wear more sweaters (including turtle necks), while putting a price cap on gas and increasing work-from-home allowance for public employees.[61]

Started by a local artist in 1982 in Vancouver, British Columbia, and promoted by the activist magazine *Adbusters*, Buy Nothing Day is a great way to cut down and can be extended to Buy Nothing Week (originally November 16, but now the last Saturday in November). Indeed, the best way to save is not to buy what you don't need in the first place. Buy Nothing groups are also sprouting up online to share and exchange unwanted items in a virtual

24/7 garage or yard sale. Social media is perfect for exchanging disused items with others, for example, via a local Facebook group. Just because one person no longer has a need for something doesn't mean it is waste – a baby chair, car seat, clothes, furniture . . . the list is endless.

In 2010, after becoming the fastest sailor to sail solo around the world 5 years earlier, Ellen MacArthur started a foundation in her name to aid the circular economy. Alone at sea for 71 days and dependent only on what was in her boat, MacArthur recognized that her world was limited, writing in her logbook, "What I have on this boat is all I have. There is no more." Seeking to change the wasteful ways in which we use things in our own everyday lives, MacArthur asks us to make a straight line into a circle, as if returning to the start after circumnavigating the globe. She encourages us to work with nature to pass on a healthier planet to the next generation, calling on us all to fix our problems: "Rather than taking, making and wasting, we design out waste and pollution, we keep products and materials in use for as long as possible and we regenerate nature. Then, it's an economy that doesn't run out."[62]

The 3 Rs we learned in childhood (reduce, reuse, recycle) are expanding everywhere: refuse (just say "no"), remove (picking up garbage), refill (so simple), rot (organic decomposition), replenish, repair, regenerate, recover, refurbish, repurpose, redistribute, resell, recharge, rewild, rebel, etc. Perhaps new to some, "refuse" may be the most important – if you don't let something you don't need into your life in the first place you don't need to get rid of it later! Just say "no." My 3 Rs – rethink, rebuild, rewire – invite us all to examine the world as it is and see how and where we can make life better.

It's fun to learn new ways and put them into practice. RRR is an attitude, a choice, a lifestyle. We can all make the change (Figure 7.3).

Figure 7.3 No junk mail, recycle logo, no plastics.

7.7 Shrinking Our Energy Impact: A Few Comparisons

I was surprised to read in an otherwise excellent book on renewable energy that the author didn't think every little bit helps. In fact, he wrote "every big helps." The self-professed "technologist" Bill Gates expressed a similar idea in his book *How to Avoid Climate Disaster: The solutions we have and the break-throughs we need*, stating that he wouldn't fund a project that captures less than 1% of carbon, while eschewing conservation as the most obvious way to reduce our carbon footprint. Highlighting the compartmentalized approach to solving problems after the fact, rather than looking at simple causes, Gates's technologist's answer to climate change doesn't even include an index entry for *conservation*.

To be sure, unplugging my printer when not in use won't save the world, but doing so is worth it because good behavior begets more good behavior. I also believe many hands make light work and that simple individual actions make a difference, spurring on new ideas and better habits all round. Our attitudes are what make us responsible, despite those who think that simple solutions are unimportant or only symbolic. "Less is more" is an ethos. A little can turn into a lot. Sure, we need the big boys to change their ways, but change the little and the big can happen.

In 1785, the German philosopher Immanuel Kant proposed a simple test in his *Groundwork of the Metaphysics of Morals* to determine to what extent individual actions matter. Essentially he asked "What would happen if every-one were to do as I do?" (a.k.a. the categorical imperative). Working at the University of Königsberg in Old Prussia (now the modern Russian enclave of Kaliningrad), it was said you could set your watch to Kant's daily stroll, and so, aside from being rather ordered in one's habits, what would happen if each individual action did have consequences? All we have to do is stop, take a look around, and refuse to participate in wasteful, ecologically illiterate, and inhu-mane acts.

We simply can't keep axing, extracting, and destroying without replacing. Cutting down forests to make animal grazing space for short-term gain today without thinking about the consequences tomorrow is crazy, especially com-pared to sustainably harvesting trees to produce a reusable supply of resources, the basis of a replenishable economy. Extracting more oil as if we will never run out is insane compared to utilizing more sun and wind, especially as generating solar and wind power is similar to how we farm crops – renewably and sustainably. Wind and sun also provide less chance for resource conflict over disputed sources compared to oil, gas, and water. Destroying precious water tables and aquifers with chemical poisons to coax out more fossil fuels from the ground and transporting fracked gas across thousands of miles of precious farmland and pristine waters is madness. And yet we continue.

Few people worried about depletion in frontier times, when resources were plentiful and we were free to exploit at will, but the logic is flawed as the global population rises beyond the capacity of the Earth to adapt and where basic ecosystems are excluded in the bottom line. The prevailing ethos of endless growth is inherently flawed, while strategies that tell us "more" is the only way forward and that GNP is the only measure of success are wrong. Now more than ever we need to level the playing field and pave the way for a "just transition" that supports green jobs, environmental justice, and social consensus. As author Jeremy Rifkin notes, "Our very approach to engineering has to be recalibrated to synchronize with the regenerative periodicities of nature rather than simply the productive rhythms of market efficiency."[63]

The transition from dirty to clean is not only about energy, but about replacing an objectified, exploitative world with a subjective, interactive world, where humans act as custodians rather than masters of the Earth. In a just, clean, and sustainable world, we buy time and space instead of material. We become guests not hosts, renting existence rather than trying to own the future, which as temporal beings is impossible. Using science and engineering as tools to exploit resources is no longer tenable, while single-use materials are of little value in a limited world. Caring for what we have is a far better path than growing beyond our means.

There are numerous ways to redress the imbalance and take more control of our lives. Buy local. Use less. Be choosy. Keep a personal carbon diary. In essence, a sustainable future reverses a failed system of consumerism and globalization. Transporting goods halfway across the world that can be made around the corner is a losing venture. We've got it backwards when we prioritize convenience and compartmentalized processes from factory to transportation to the shelf to our own overconsumption. If we don't change, we will end up as the final cog in the system (robot trucks coming soon).

What is the cost of our easy lifestyle that gobbles up petroleum at 100 million barrels a day and throws away 1 trillion pounds of food, 5 trillion plastic bags, and five hundred billion bottles every year? There are many online calculators that determine your carbon footprint or how many *Earths* you take up. I worked out that I generate 500 kg of waste and 10 tons of CO_2 a year, based on my home, travel, food, and shopping habits. In all, I use 2 Earths, that is, a world full of me would need twice the Earth's resources, better than typical Western practice, yet still well above the global average and 5 times more than needed to stop global warming.

Of course, there is no such thing as the "average person," especially on a global scale where the numbers split into rich and poor humps, but if you live in the developed world you are in the richer, narrower hump, energy-wise, and using more than average. If you live in the USA, Canada, or Europe, much more. Half of the world uses almost all the world's energy, while the 20 largest economies are responsible for 79% of emissions (Figure 7.4). Adding to the madness of our wasteful ways, at the top of the consumption scale a single billionaire emits *one million times* the GHGs as the average person (2.76 tons).[64]

The easiest change is lighting. A light-emitting diode is more efficient than either incandescent or fluorescent lighting, converting photons to electrons to make light, instead of producing wasteful heat in an inefficient and outdated process. A tungsten wire burns white hot to emit light thermionically in an evacuated bulb, essentially unchanged since Edison's time, while in a fluorescent tube electrons excite mercury vapor atoms to emit UV light that collides with a phosphor coating to make light. A modern compact fluorescent lamp is four times better than an incandescent bulb, but LEDs are even more efficient and longer-lasting, saving both money and the environment.

The 16-year-old Swedish activist Greta Thunberg highlighted the importance of low-carbon travel, taking trains instead of planes on a European speaking tour. Starting in Stockholm, Thunberg stopped in Basel, Rome, and London, before returning to Sweden, a distance of over 6,000 km by rail. She then crossed the Atlantic in a monohull sailing yacht to speak at the UN in New York, returning home to Europe by catamaran. Both crossings were

Figure 7.4 Percentage of CO_2 emissions by world population.

carbon neutral, although admittedly not for the squeamish or to everyone's taste. But low-impact travel is growing because of carbon awareness and movements such as Sweden's "flygskam" (flight shame), providing huge environmental savings because a train generates 15% the CO_2 per kilometer compared to flying.[65]

In the nineteenth century, we began to move beyond our birthplaces for the first time with relative ease thanks to the engineering wonders of steam-powered locomotion, while twentieth-century technology created interconnected modern cities allowing even more people to move as they wished for work or pleasure. The twenty-first-century city is striving to remove the grime and noise that comes with the freedom of such unfettered movement. We need to relearn the value of interdependency instead of compartmentalization as nature has shown us (circular not linear).

If we have any hope of reducing fossil-fuel production, the polluters must pay, which will of course increase the cost of our current lifestyle, as it should if the market is properly and efficiently accounted. All carbon-intensive goods and services that have had a free ride until now will become more expensive, making it easier to switch to cleaner alternatives. Putting a true cost on environmental pollution and GHG emissions will naturally incentivize green technologies that have been disadvantaged for too long by carbon's privileged status.

But it is not enough to reduce industrial production of carbon, we must also reduce consumption. Net zero locally doesn't mean net zero globally if all we do is outsource the damage to other countries with lower emission standards, replacing domestically made high-carbon products with equally bad or worse foreign imports. Getting someone else to do our dirty emissions is no savings, such as importing high-carbon steel from abroad.

In *Ending Fossil Fuels: Why net zero is not enough*, Holly Jean Buck writes that net zero is a "collective delusion" and that current thinking on decarbonization has been shaped by a fossil-fuel industry keen to maintain production.[66] Buck doesn't mince her words, stating that although climate scientists may be optimistic about a green transition and eventual end to fossil fuels, net zero is based on an impossible infrastructure, rationalized by complex academic accounting of fanciful, net-zero modeling and arcane metrification. She prefers a managed phase out starting now, however hard to imagine let alone implement by governments beholden to bottom lines and corporate performance.

In *The Carbon Crunch*, Dieter Helm called government reduction plans a "fairy tale" that promotes hype as reality. He notes that we are unlikely to limit global warming to "anything like 2°C" (3.6°F) – never mind the current target of 1.5°C (2.7°F) – because "the gap between what needs to be done and what is on offer is too big."[67] Sadly, the status quo remains because of cheap fossil fuels that aren't priced to include environmental damage. Helm does suggest a number of fixes, such as adding new technology, increased R&D and other "public goods," and redoubling efforts to turn a passive grid into an active grid (a.k.a. "managed demand"). He also calls for the polluter to pay via proper carbon pricing, regulated carbon offsets, and a border adjustment tax to discourage offshoring carbon pollution. An economy is not efficient if not all costs are included in the price of goods.

Although doubtful about the ability to change our ways in the next decade, Buck also lists a number of concrete measures to end fossil fuels rather than let Big Oil continue to fudge the ledger with its insufficient and dangerous "net zero" language, such as "more bans on exploration and extraction, subsidy reform, expanding international coalitions and networks to develop a global approach, planning the retirement of infrastructure piece by piece."[68] Availing of their vast engineering experience, Buck wryly suggests that the oil companies could end up as retooled carbon removal companies.

Some think we are in a repeat of the 1900s economic Wild West, when the world was remade with oil, automobiles, and shady banking, others the start of a catastrophic green failure and more government control. In fact, converting to a zero-emission economy would cost a fraction of current economic production when all factors are considered, while homegrown power isn't metered. However the change comes, the twenty-first-century makeover will at least transform me-first consumers into a collaborative group of we-first prosumers.

In *A Sand County Almanac*, American conservationist Aldo Leopold reported that annual patterns in a changing rural environment were brought on by increased human activity, noting that conservation was undervalued because economics is based on the "Abrahamic concept" of ownership rather than an ecologically responsible "land ethic." His Earth-based spirituality asks us to consider our privileged status and give back what we have so callously taken. The Earth and civilization needn't be in opposition, however, but must coexist. Conservation, ecological equilibrium, and everyday environmentalism should be everyone's concern, lest we kill our host. It is no longer a matter of slowing down. If we want to preserve our world we *have* to change. As Leopold stated, "We need a 'new kind of farmer, banker, voter, consumer.'"[69]

In 2004, two Princeton research scientists, biologist Stephen Pacala and physicist Robert Socolow, listed 15 strategies to limit carbon dioxide emissions to between 450 and 550 ppm by 2050. Bunched into seven "stabilization wedges," the main efforts were energy efficiency and conservation, fuel shifting (replacing coal plants), carbon capture and storage, nuclear power, renewable energy, and forest conservation and low-tillage farming, all of which were available at the time to varying degrees and could be scaled up "to meet the world's energy needs over the next 50 years and limit atmospheric carbon dioxide to a trajectory that avoids a doubling of the preindustrial concentration" (280 to 560 ppm).[70] If implemented, each wedge would remove 1 GtC/year rather than add 14 GtC/year by 2050 in an ongoing business-as-usual (BAU) trajectory. Carbon dioxide concentrations were already 375 ppm at the time and have steadily risen 2 ppm per year to more than 415 ppm over two decades. Alas, having missed horribly on the proposed mitigation targets, "business as usual" or "steady state" is no longer viable if we seriously want to confront the problem of global warming.

It is doubtful we can agree to reduce emissions via international agreements in an overly competitive, nationalistic world, but our economies can be retooled to replace dirty for clean, forging a bottom-up solution to the difficult challenges. We can embrace new ideas that seek to replace and repair old-world thinking, changing old and dirty for new and clean as we embark on another future. Just as we transformed a largely agrarian society in an earlier round of modernity with coal, oil, and gas, we must change again. We have no choice if we want to survive. The "race to zero" has begun.

7.8 We Are Not Alone: Less Is More

Like the Ptolemaic geocentric universe, the petroleum industry is complicated and convoluted. The heliocentric system is so much simpler. Putting the Sun at the center of a modern renewables world will help fix the damage caused by almost three centuries of burning fossil fuels and increase the "carbon efficiency" of the global economy.

But the multi-trillion-dollar petroleum industry won't give up its dominant market share without a fight. As Peder Holk Nielsen, CEO of Novozymes, a world-leading Danish enzyme manufacturer with interests in microorganisms and biofuels, noted, "Let's be very clear on this. The oil companies will not help the world to switch to renewable energy – that will never happen. They are part of a system that protects the business they have. The only way the world gets more renewables is if bold politicians step up to it and mandate."[71] And if each of us makes essential changes, learning about clean technologies and using them in our lives.

Some think we will be able to fix the damage before it's too late with large geo-engineering or climate-engineering projects, the last high hurdle to tame the untamed Earth as if inspired by a relentless Hollywood hero ethos, where good always champions bad in a morally uplifting finale. Unfortunately, real life is not so simple or glossy, each technological fix more complex than the next, such as running thousands of industrial-scale direct-air-capture CO_2 filters (as we saw in Chapter 2), strategically injecting sulfur dioxide aerosols in the atmospheric to mimic the reflective ash clouds of volcanic eruption (likely increasing respiratory diseases), or controlling the weather via cloud brightening or cloud seeding (banned since 1977 for good reason[72]).

Others think sea-level rise can be mitigated by digging up an amount of ocean-floor sediment equal in volume to all the melted glacial ice, the exact opposite solution proposed by those who want to dump dead plant matter at the bottom of the sea where it won't rot and thus continue to sequester carbon. Still others want to genetically modify trees to grow deeper roots to take in more CO_2. One of the craziest ideas being discussed is an atmospheric canopy of space mirrors to reflect incoming sunlight back into space, a.k.a. solar radiation management (SRM).

One especially bold SRM idea proposes to shade the Earth with a 1,200-mile diameter umbrella in space at the Lagrangian point (L1) – a neutral gravity point 1/100th the way to the Sun, where the gravitational pull of the Sun and Earth balance each other – keeping our Earth-saving heat screen in stable orbit without spiraling into a fiery death.[73] Hoping to prod interest, American astrophysicist Lowell Wood calculated in 2001 that a 1,600,000 km^2 wire-mesh mirror array would reflect 1% of incoming solar radiation back into space. Bill Gates also proposes to dim the Sun by 1%, either by spraying fine particles in the upper atmosphere to scatter more sunlight or by brightening 10% of the Earth's cloud cover by 10% that will also scatter more light and keep us cool.[74] Calling for preparations now rather than later, when there may be no choice, the self-professed "technologist" hopes such literal pie-in-the-sky thinking will lower temperatures without crippling the economy.

Of course, nobody wants to talk about who pays or what happens if something goes wrong. Blocking 1% of the Sun's radiation may counter global warming, but could also alter the Earth's ecosystem or irreparably damage animal species as we play with eons of natural evolution. And what happens to an already fragile climate as we flip on the civilization-saving sky switch? A Pandora's Box of uncertainty awaits. Such fantasist thinking is akin to trusting a Bond villain to save humanity – Blofeld, Drako, Spectre, take your pick.

In reality, such wishful geo-engineering is a failed technological response to bad planning that aims to extend the burning of unsustainable fossil fuels without addressing the root problems of pollution, global warming, and ecological disaster. While not dismissing all geo-engineering ideas out of hand, such as naturally increasing the carbon uptake in soil, plants, and trees, as well as underground CCS injection, Chris Goodall notes that "the idea that we could cleanly counteract the consequences of increasing greenhouse-gas levels by simple techniques such as reflecting the Sun's energy is hubris of the worst sort."[75] The German biogeochemist Meinrat Andreae is more scathing in his analysis, likening geo-engineering to "a junkie figuring out new ways of stealing from his children."[76] In effect, such "technical optimism" or "technological salvation" aims to geo-engineer nature solely to keep adding more carbon to an already damaged ecosystem.

Even if such fantasist projects were to get the green light, there are also important ethical issues to work out – one wouldn't want to make things worse between neighboring countries that may disagree about who can do what to the borderless sky. Chief among the concerns are financial disparities between governments that can afford to experiment versus those with little or no money to spend on expensive unproven fixes. One could even imagine poorer countries having to pay richer countries to remove their emissions in a future, for-pay, carbon-sucking world. One wonders too about the possibility of misuse and dangerous experimentation as the Global North tells the Global South what to do. What's more, such massive fixes are inordinately expensive, require international cooperation, and need consensus between hopelessly divided factions, some of whom refuse to agree on or even address the causes and effects of global warming, let alone how to act in unison.

Isn't it just simpler to work *with* the Earth and learn how the Earth operates, rather than trying to implement poorly scoped and unproven industrial-sized bandages after the fact? Haven't we done enough human geo-engineering to a fragile ecosystem that can easily survive without us? We are at a point where just leaving the Earth alone is a better solution than trying to fix it. Mother Nature cleans herself so much better than we do.

Michael Mann, the Penn State climatology professor, who brought us the famous "hockey stick" curve in 1999 that showed a rapid increase in global temperatures in the industrial era, called geo-engineering a "non-solution solution" along with natural gas and carbon capture. He further calls those opposing proper carbon pricing "inactivists" whose real goals are deniability, misinformation, and switching blame from fossil fuel companies to consumers. All to cast aspersions on how we live our lives so they can continue business as usual.

More dedicated government support for existing clean, green technologies that are already known to work is a more sensible goal. A proper costing of harmful emissions in all phases of fossil-fuel use – say a $50/ton carbon or

GHG tax – would also aid the transition to the tune of $2.5 trillion per year (50 billion tons/year × $50/ton), a paltry 3% of world GDP that would require a minimal increase to our electricity and fuel costs. A $100/ton tax would aid the transition fund to $5 trillion per year. A huge opportunity exists for those who can invest now in a clean future, while the consequences to the Earth and our own health can no longer be ignored. The argument is not one of backing winners and losers, but of leveling the playing field and including the real cost of fossil fuels in the economy after centuries of ignorance and neglect. Properly priced, there can be no return to brown.

Begun in the Fertile Crescent and elsewhere between 5,000 and 10,000 years ago during the Earth's current interglacial period, farming is the greatest geo-engineering experiment of all that has increased deforestation (3 trillion, or half, of all trees gone) and wet-rice cultivation, which has, in turn, increased carbon dioxide and methane levels enough to warm global temperatures and delay the onset of the next ice age.[77] People and domesticated animals now account for about 90% of the "mammal biomass," an almost 1,000-fold increase since the beginning of farming.[78] Without humans tipping the balance, the natural up-and-down temperature range from "orbital forcing" – periodic changes in the Earth's eccentricity, tilt, and precession – should have put us in an ice age by now, but instead the Earth is precipitously warming from a spike in carbon dioxide since the start of industrial burning of fossil fuels and increased methane from animal farming, thawing Arctic tundra, and leaky fracking.

Typical glacial/interglacial ranges are 180–280 ppm (CO_2) and 350–700 ppb (CH_4),[79] but CO_2 is now almost 420 ppm and CH_4 2,000 ppb, levels not seen for 3–5 million years when temperatures were 2–3 degrees higher and sea levels 10–20 m higher. As is, we may be hard-pressed to end the dominant geo-engineering of our own civilization that has seen the human population grow from 1 million at the start of farming to 500 million in 1500, 1 billion in 1800, and more than 8 billion today, putting us on the edge of what UCL earth scientists Simon Lewis and Mark Maslin think will become an "unusually long super-interglacial" with "much future suffering" if we don't stop burning fossil fuels.[80]

The term "sustainable development" was coined in 1987 in the UN-commissioned Brundtland Report to promote economic growth *with* environ-mental protection and social equality in an intricately connected global bio-sphere, made increasingly vulnerable since the rise of global travel and the impact of increasing world population. Alas, *growth* itself may be the problem. We can't keep growing and be sustainable, acting as if we have a natural right to exist without consequences to our actions or that humans as a species will live forever, which logically we cannot here on Earth because of increased heat from an eventually nuclear maturing Sun.

The problem may be in the DNA of the conquering man since the start of the Anthropocene, deemed by Lewis and Maslin to have begun in 1610 with the regrowth of lost forests in the newly discovered Americas after the death of 50 million indigenous people led to a significant drop in global temperatures that were caused by a return of low-carbon-storage farming into high-carbon-storage trees on a vast scale, called the Orbis Spike. According to Lewis and Maslin, the continental "rewilding" marked "the first global impacts of the Columbian Exchange,"[81] followed by the "world-wide hybridization of cultures through modern transport and industrialization"[82] as noted by Aldo Leopold. Others believe the Anthropocene began either in 1712 with Newcomen's atmospheric steam pump; 1784 with James Watt's general-purpose steam engine; 1804 when coal use increased the concentration of carbon dioxide beyond the Holocene maximum of 284 ppm; July 16, 1945, with the Trinity atomic bomb test that released traceable radionuclides into the atmosphere; or 1950 with the exponential growth in fossil-fuel carbon emissions at the start of the "Great Acceleration."[83]

The English scientist and engineer James Lovelock, who first measured the dangerous effect of CFCs on the ozone layer and designed instruments to measure the Martian atmosphere, puts the start at 1712 with Newcomen's atmospheric steam pump and has even come up with the epoch *after* the Anthropocene when technology gets out of hand. According to Lovelock, the installation of Newcomen's engine, first used in 1712 in a mine in Griff, Warwickshire, initiated "the domination of human power over the entirety of the planet" as "humans first began to convert stored solar energy into useful work."[84] Lovelock neatly divides Earth's existence into three solar-farming stages, paying homage to our life-giving Sun: (1) when organisms first began to convert sunlight into energy via chemical photosynthesis, (2) the present Anthropocene, and (3) the coming Novocene, when cyborgs replace photosynthesis with semiconductor solar harvesting into information (also possibly ending the production of atmospheric oxygen and terrestrial animal life as we know it).

Regardless the exact date or start of a human-less world that may one day come to pass, the Anthropocene is a geological bifurcation to which there is no return, and can be divided into three modern stages: (1) 1750–1945 (thermo-industrial), (2) 1945–2000 (the Great Acceleration, aided by the industrial mobilization of World War II and new post-war markets), and (3) 2000 onwards, with an increased public awareness of anthropogenic global warming.[85] Historians Christophe Bonneuil and Jean-Baptiste Fressoz also noted that the Anthropocene can otherwise be labeled to highlight the human causes, such as the Oliganthropocene (unequal use and abuse), the Anglocene (as of 1980, almost half of global carbon dioxide emissions were by the USA and Great Britain), the Thantocene (wars), or the Capitalocene (beginning with mercantilism in the sixteenth century).[86]

But whatever one wants to call our new age of human-induced, environmental change, science and technology have clearly failed if the ecology of our planet is dominated by an unforgiving economy or if the improved technology of manufacturing things to make life easier makes life impossible. An ideology of growth that came with Watt's steam engine continues to favor industry and monetary wealth while destroying the very system it claims to aid. We need another way to interact with the Earth, one that doesn't objectify its bounty, oddly life-giving and yet at the same time treated by us as inanimate.

As much as we think we can tame or control nature, Mother Earth will always be in charge. We can only adapt to the increasingly commonplace climate changes wrought by an ecocidal lifestyle, now more potent than in a previously stable, albeit often unpredictable, recent past. We might also think – thanks to our presumed mastery of science and technology – that we will one day live in a perfect, post-nature "technosphere" with flying cars, skyscraper farms, and embryo nurseries, but we are a long way from such curious utopias and limitless bounty. We should also be careful what we wish for or think we can achieve, as seen in any number of appealing yet sterile future worlds such as those depicted in *2001*, *Star Trek*, and *Star Wars*. We should instead heed what we do now under nature's watch. No matter the future, we can't continue to ignore our past mistakes.

In *Truth to Power*, Al Gore's 2017 sequel to *An Inconvenient Truth*, he wrote about the importance of children to help change thinking, "When I was a young child, my father taught me about soil conservation on our family farm. For me, that was the beginning of my awareness of and concern about the impact people have on the environment. If we can help children make sense of the environmental problems they're inheriting, they'll be better equipped to solve them. Start early with simple concepts, and make sure to nurture an ongoing conversation."[87] Children have always been agents of change. As Damasia Ezcurra, head of sustainability programs in Buenos Aires, noted after more than halving in five years the amount of trash going to a landfill: "If a child teaches a parent how to recycle, it is much more effective than a politician telling them what to do."[88]

Ever since her first Friday school strike in August 2018 outside the Swedish parliament, Greta Thunberg has energized the environmental movement with a plain-talking anger, armed with her intellect, enthusiasm, and handmade sign

("SKOLSTREJK FÖR KLIMATET"). Thunberg continued to spread the message in an inspiring TED talk and at COP24 before 40 heads of state and government (both later that same year) and in a speech to the 2019 UN climate summit in New York City, where she chastised the delegates, "The young people are starting to understand your betrayal. The eyes of all future generations are upon you, and if you choose to fail us, I say, we will never forgive you."[89]

Others are just as adamant that we change now before it's too late. As 350. org founder Bill McKibben notes in *Eaarth: Making a life on a tough new planet*, we have to mature as a civilization, minding what we use after "the global volume of trade grew fivefold in the quarter century that began with Ronald Reagan's inauguration."[90] The Earth can't handle much more expansion, certainly not fueled by Colonel Drake's "mad" venture.

In *Connections*, James Burke wonders if we should return to "intermediate" technologies that are more land-based, socially interactive, and easier to understand, "encouraging participation on the part of the electorate in decision-making which would relate to simpler, more fundamental matters."[91] In so doing, we use only environmentally sound, renewable resources such as wind, water, and sun, all viable, efficient, and affordable to a modern energy-hungry world. No more bigger-is-best, super-sized, centralized answers that trucks in or wires in resources from afar but conveniently forgets about the impact of our actions. We need decarbonized electrification, reduced consumption, and increased conservation. As Paul Allen, one of the authors of *Zero Carbon Britain 2030* stated, "this is not a technical challenge, it's social and political."[92] The endless us-and-them debate about public–private financing shouldn't keep us from making important changes today.

There is still much to learn and even more to question. McKibben notes that natural gas as a "bridge fuel" is more environmentally dangerous than coal and oil, likening it to "the equivalent of losing weight by cutting your hair."[93] Ramping up natural gas while reducing coal as in China's so-called "twin pillar" policy of renewables *and* natural gas may even make global warming worse because of increased methane leaks, much more dangerous than the coal it is meant to replace. Exporting fracked LNG from the USA to Europe to counter Russian military aggression is just as bad.

Bill Gates noted that natural gas cannot act as a bridge fuel if we are serious about net zero by 2050, stating that such "gradualism" is throwing good money after bad and locks us into a mistaken direction by providing short-term gain yet long-term failure. He prefers all-out zero-carbon electricity and wide-scale electrification, "everything from vehicles to industrial processes and heat pumps."[94] Naomi Oreskes thinks the pro-nukes lobby is "a new, strange form of denial" and although Al Gore doesn't have "a theological opposition to nuclear power," he states "It's simply not cost competitive." At the end of *Hell and High Water*, Joseph Romm succinctly states, "Get informed, get outraged, and then get political."[95]

Hopefully, we are becoming more informed and willing to act. We stand on the shoulders of those who came before us: Isaac Newton ("If I have seen further it is by standing on ye shoulders of Giants"), Alessandro Volta, Michael Faraday, Thomas Edison, George Westinghouse, Charles Parson, Nikola Tesla, William Stanley, Samuel Insull, Rachel Carson, Hermann Scheer. If we want to see even further, we have to make changes in *our* lives and demand that governments include the environment in theirs.

In a 1955 article entitled *Can we survive technology?*, the mathematician and theoretical physicist John von Neumann – who escaped Europe along with so many others during World War II to come to the USA – wrote, "Technologies are always constructive and beneficial, directly or indirectly. Yet their consequences tend to increase instability." That is, at first, until we figure out the changes. As he further noted, presaging current thinking about the modern energy–technology power grid, "Fundamentally, improvements in control are really improvements in communicating information within an organization or mechanism."[96] It's time for all of us to be smart as we usher in an era of smart energy and smart grids.

We have the means to move mountains. With informed thinking, we can do anything. Real power starts with each and every one of us doing what we know is right and joining together to increase the effect of our individual efforts. The logic is simple: $1 + 1 > 2$.

Afterword

Some have begun to wonder whether we will survive the next century as we continue to grow and pollute without concern for the Earth. We should all be worried about the difficult changes coming our way, but our Earth will ultimately survive in the long view of geological time. In North America, 12,000 years ago, the Mojave Desert was under water, the Great Lakes didn't exist, and Alaska was full of grassland, while the first human inhabitants were just crossing from Asia to North America via an exposed land bridge across the Bering Strait. Those same geological conditions will appear again in time as natural orbital forcing lowers the amount of incident radiation arriving from the Sun. That doesn't mean much to us today as humans, who measure a life not on the scale of a slowly flip-flopping hot and cold planet, but in fractions of a century.

Yet as a species who want to live beyond its short recorded history in the manner we do now, we must at some point adapt and redress our failing relationship with the Earth, begun the moment we first tamed fire and steam from the middle of the eighteenth century on. Because of our own accelerating contribution to global warming, that time is now. If we continue to extract and burn all the fossil fuels we have discovered until they are gone, we will fail in the simple guardianship of our planet. If we avoid asking the hard questions about burning coal, oil, and gas, or about how to make energy, we will have failed as humans. As we contemplate our future, we must begin to see the Earth as other than a static resource, but as something alive and as fragile as ourselves.

Although much harder to implement in non-autocratic ways, we must also consider "degrowth" to counter a prevailing economic system that sees the future only as an ever-expanding opportunity, which as we know cannot continue indefinitely. There are simply not enough resources to live as in frontier times. We must start to ask the harder question about our overly consumptive ways and become content with what we have. This book is an attempt to understand and learn from our mistakes, to seek a better way to live sustainably, and to foster more respect for our fragile, only, home. Fortunately, we can reinvent how we live *with* the Earth, rather than as lord and master over all we seek to subjugate in a never-ending global ownership.

If we can build a nuclear bomb, land humans on the moon, and ring the globe with satellites and phone masts to send wireless messages wherever we want in an instant, we can run a world on renewable energy. Certainly we can afford the cost when one considers the more than \$6 trillion spent on Middle East wars, \$5 trillion to maintain a nuclear arsenal (which can never be used without destroying civilization), and \$5 trillion on *annual* fossil-fuel subsidies. The question is no longer one of can but of will and how to catch up to the science.

Some think a green future is the start of world government with distant administrators telling us what to do and think. But such thinking couldn't be more wrong. Today, most of our energy is already controlled by distant powers, whether renewable or not, either in uncaring corporate boardrooms or by autocratic state players. If we want more control of our daily lives, smaller-sized, scalable renewable energy allows us to become self-sufficient, letting us make our own decisions about our own needs. With an off-grid power setup, no one can tell me what to do. Most importantly, no one has to go to war again in my name to secure my energy.

New discoveries in science, engineering, and technology have seen the creation of revolutions in steam power, electricity, the grid, the automobile, satellites, and the Internet. Renewable energy is one more revolution to transform our lives for the better and remake our world again.

We must become more than custodians of an increasingly ravaged Earth, sweeping up the debris of our own excess as we count down to a presumed end. We can take back from those who have willfully neglected our beautiful bounty and begin again to treat the Earth as we would treat ourselves – the ground, the air, and the water we all need to coexist with nature. Most urgently, the atmosphere can no longer be used as a CO_2 and CH_4 dump for our carelessness. The well-being of the Earth is our only future.

Appendix A Unit Abbreviations

m, km	meter, kilometer	length
h, m, s	hour, minute, second	time
m/s	meters per second	speed
mph, km/h	miles per hour, kilometers per hour	speed
rpm	rotations per minute	speed (angular)
l, m^3	liter, cubic meter	volume
bbl	barrel	volume
mbpd	million barrels per day	volume/day
kg, g	kilogram, gram	mass
kg/m^3	kilograms per cubic meter	density
°C, °F, K	degree Celsius, Fahrenheit, Kelvin	temperature
W	watt	power (kW, MW, GW, ...)
MWe, MWp	megawatt (electric), megawatt (peak)	power
Wh	watt-hour	energy (kWh, MWh, GWh, ...)
BTU	British thermal unit	energy
J	joule	energy
eV, MeV	electron volt, mega electron volt	energy
V	volt	electric potential
A	amp (ampere)	electric current
Ah, mAh	amp-hour (milliamp-hour)	electric charge
Wh/kg	watt-hour per kilogram	electric energy density
W/m^2	watts per square meter	power density (solar flux, irradiance)
Hz	hertz	frequency
Pa, psi	pascal, pounds per square inch	pressure
Bq, Ci	becquerel, curie	radiation activity
Sv	sievert	ionizing radiation
Gy	gray	ionizing radiation (absorbed)

Appendix B Metric Prefixes

f	femto	$1\,e^{-15}$		k	kilo	$1\,e^{3}$
p	pico	$1\,e^{-12}$		M	mega	$1\,e^{6}$
n	nano	$1\,e^{-9}$		G	giga	$1\,e^{9}$
μ	micro	$1\,e^{-6}$		T	tera	$1\,e^{12}$
m	milli	$1\,e^{-3}$		P	peta	$1\,e^{15}$
c	centi	$1\,e^{-2}$		E	exa	$1\,e^{18}$

Appendix C Useful Acronyms

ABWR	advanced boiling water reactor
AC/DC	alternating/direct current
AD	anaerobic digester
AGR	advanced gas-cooled reactor
AI	artificial intelligence
API	American Petroleum Institute
APR	advanced power reactor
ASHP	air source heat pump
ASSB	all-solid-state battery
AV	autonomous vehicle
BAU	business as usual
BESS	battery energy storage system
BIPV(T)	building-integrated photovoltaic(/thermal)
BMS	battery-management system
BOS(S)	balance of system (and services)
BWR	boiling-water reactor
CAFE	Corporate Average Fuel Economy
CANR	chemically assisted nuclear reaction
CANDU	Canada Deuterium Uranium
CARB	California Air Resources Board
CAT	connected and automated transport
CCS	carbon capture and storage/sequester
CDR	carbon dioxide removal
CEO	chief executive officer
CHP	combined heat and power
CNG	compressed natural gas
COP	Conference of the Parties
CP-1	Chicago Pile One
CPV	concentrator photovoltaic
CREZ	creative renewable energy zone
CSP	concentrated solar power
CTR	controlled thermonuclear reaction
DAC	direct-air capture

DAPL	Dakota Access Pipeline
DER	distributed energy resource
DGE	deep geothermal energy
DGR	deep geologic repository
DMG	Daimler-Motoren-Gesellschaft
DT	deuterium–tritium
DWC	dynamic wireless charging
ECCS	emergency core coolant system
EDF	Électricité de France
EEG	Erneuerbare-Energien-Gesetz
EES	electrical energy storage
EEZ	exclusive economic zone
EM	electromagnetic
EPA	Environmental Protection Agency
EPR	European pressurized reactor
(CA/ET/T/UT) ES	compressed-air/electro-thermal/thermal/underground thermal energy storage
ESG	environmental, social, governance
EU	European Union
EV	electric vehicle
(B/H/S/FC/PH) EV	(battery/hybrid/smart/fuel-cell/plug-in hybrid) electric vehicle
F/SCP	fast/slow charging protocol
FGD	flue-gas desulfurization
FIT	feed-in tariff
FOG	fats, oils, and greases
FSD	full-service driving
GD/NP	gross domestic/national product
GECF	Gas Exporting Countries Forum
GERD	Grand Ethiopian Renaissance Dam
GF	gigafactory
G/LH2	green/liquefied hydrogen
GHG	greenhouse gas
GSHP	ground source heat pump
GWP	global warming potential
HAWT	horizontal-axis wind turbine
HC	hydrocarbon
HCPV	high-concentration PV
HELE	high-efficiency low-emission
HLRW	high-level radioactive waste
HVAC	high-voltage alternating current
HVDC	high-voltage direct current
IAEA	International Atomic Energy Agency
ICE	internal combustion engine
ICF	inertial confinement fusion
IEA	International Energy Agency
INES	International Nuclear Event Scale
IoT	Internet of things

IPO	initial public offering
IR	infrared
IRA	Inflation Reduction Act
IRENA	International Renewable Energy Agency
ITC	investment tax credit
ITER	International Tokamak Experimental Reactor
JET	Joint European Torus
KXL	Keystone XL pipeline
LCOE	levelized cost of energy
LCPV	low-concentration PV
LED	light-emitting diode
LENR	low-energy nuclear reaction
LIB	lithium-ion battery
LDES	long-duration energy storage
LLNL	Lawrence Livermore National Laboratory
LLRW	low-level radioactive waste
LNG	liquefied natural gas
LOCA	loss of coolant accident
LPG	liquefied petroleum gas
LWR	light-water reactor
MAD	mutually assured destruction
MCF	magnetic confinement fusion
MENA	Middle East North Africa
MSAT	mobile source air toxics
MSR	molten salt reactor
MTF	magnetic target fusion
NDC	nationally determined contribution
NET	negative-emission technology
NIF	National Ignition Facility
NIMBY	not in my backyard
NLWR	non-light-water reactor
NRC	Nuclear Regulatory Commission
NREL	National Renewable Energy Laboratory
NS1/2	Nord Stream 1/2
O&G	oil and gas
OPEC	Organization of Petroleum Exporting Countries
OTEC	ocean thermal energy conversion
P2G	power to gas
P2P	peer to peer
PCC	pulverized coal combustion
PERC	passivated emitter and rear cell
PHMSA	Pipeline and Hazardous Materials Safety Administration
PHWR	pressurized heavy-water reactor
PM	particulate matter
PMV	personal mobility vehicle
PPA	power purchase agreement
ppm	parts per million

PSC	perovskite solar cell
PSH	pumped storage hydropower
PV	photovoltaic
PWR	pressurized-water reactor
QD	quantum dot
RBMK	reaktor bolshoy moshchnosty kanalny
R&D	research and development
REE	rare earth element
RES	renewable energy sector
RFS	renewable fuel standard
RON	research octane number
RPH	range per hour
RPS	renewable portfolio standard
RRR	reduce, reuse, recycle
SCCO2	social cost of carbon dioxide
SGE	shallow geothermal energy
SI	System International
SMR	small modular reactor
SMSL	Silicon Module Super League
SOC	state of charge
STC	standard test conditions
TEC	tidal energy conversion
TEPCO	Tokyo Electric Power Company
TIR	Third Industrial Revolution
TMI	Three Mile Island
TOU	time of use
UN	United Nations
UNFCCC	UN Framework Convention on Climate Change
UV	ultraviolet
V2G	vehicle to grid
VAWT	vertical-axis wind turbine
VGI	vehicle–grid integration
VOC	volatile organic compound
VPP	virtual power plant
VRE	variable renewable energy
VVER	water–water energetic reactor
WEC	wave energy conversion
WEEE	waste electrical and electronic equipment
WHO	World Health Organization
WIPP	waste isolation pilot plant
WNA	World Nuclear Association
WT	wind turbine
WTE	waste to energy
WTI	West Texas Intermediate
WWS	wind, water, sun

Notes

Introduction

1. "World Energy Balances: Database Documentation (2020 edition)," International Energy Agency, 2020. https://iea.blob.core.windows.net/assets/4f314df4-8c60-4e48-9f36-bfea3d2b7fd5/WorldBAL_2020_Documentation.pdf.
2. "Spain breezes into record books as wind power becomes main source of energy," *El País (in English)*, January 15, 2014.
3. Morison, R. and Warren, H., "British sun beats natural gas to provide most electricity," Bloomberg, May 8, 2018. www.bloomberg.com/news/articles/2018-05-08/british-sun-beats-natural-gas-to-provide-most-electricity.
4. Lewis, M., "California runs on 100% clean energy for the first time, with solar dominating," Electrek, May 2, 2022. https://electrek.co/2022/05/02/california-runs-on-100-clean-energy-for-the-first-time-with-solar-dominating/.
5. Okonjo-Iweala, N., "Green energy for the poor," *The New York Times*, September 10, 2015.
6. "Solar Power Revolution – Here Comes The Sun," *Backlight* [documentary], directed by Rob van Hattum, VPRO, The Netherlands, October 20, 2008.
7. Crane, D. and Kennedy Jr., R. F., "Solar panels for every home," *The New York Times*, December 13, 2012.
8. "Solar Power," Sacramento Municipal Utility District. www.smud.org/en/about-smud/environment/renewable-energy/solar.htm.
9. Zweibel, K., Mason, J., and Fthenakis, V., "A solar grand plan," *Scientific American*, pp. 48–57, January 2008.
10. Parfit, M., "Future power: Where will the world get its next energy fix?" *National Geographic*, pp. 2–35, August 2005.
11. Scheer, H., *The Solar Economy: Renewable energy for a sustainable global future*, p. XI, Earthscan, London, 2002.
12. Jewkes, S. and Piovaccari, G., "'30 years of blah blah blah': Thunberg questions Italy climate talks," Reuters, September 28, 2021. www.reuters.com/world/europe/protests-proposals-activists-face-climate-talks-test-2021-09-28/.
13. Hubbert, M. K., *Nuclear Energy and the Fossil Fuels*, American Petroleum Institute, San Antonio, TX, March 1956.
14. Plumer, B., "All of the world's power plants, in one handy map," *The Washington Post*, December 8, 2012.
15. Scheer, *The Solar Economy*, p. XVII.

16. Carson, R., *Silent Spring*, p. 177, Houghton Mifflin Company, New York, NY, 2002 (first published 1962).
17. "Executive Order on Tackling the Climate Crisis at Home and Abroad," US Presidential Action, January 27, 2021. www.whitehouse.gov/briefing-room/presi dential-actions/2021/01/27/executive-order-on-tackling-the-climate-crisis-at-home-and-abroad/.
18. "Pathway to critical and formidable goal of net-zero emissions by 2050 is narrow but brings huge benefits, according to IEA special report," Press Release, International Energy Agency (IEA), May 18, 2021. www.iea.org/news/pathway-to-critical-and-formidable-goal-of-net-zero-emissions-by-2050-is-narrow-but-brings-huge-benefits-according-to-iea-special-report.

1 Wood to Coal

1. Wedgwood also noted that blue was hotter than red, the first quantum understanding of energy where the frequency of light emitted is proportional to its energy.
2. Power = work/time; work = force × distance; force = mass × acceleration (g), that is, $P = W/t$; $W = Fd$; $F = mg$, so $P = mgd/t$. Thus, to convert horsepower to watts: 550 lb / 2.20462 lb/kg × 9.80665 m/s^2 × 12 inches × 2.54 cm/inch / 100 cm/m / 1 s = 745.7 watts. Thus 1 hp = 746 W. Watt himself defined 1 horsepower as a standard draft horse lifting 33,000 pounds 1 foot high per minute.
3. Rhodes, R., *Energy: A human history*, p. 32, Simon and Schuster, New York, NY, 2018.
4. The Devon inventor and engineer Thomas Savery was the first to pump water in a British mine via heat and steam, although his Miner's Friend was too primitive (and dangerous) to be considered as a working engine. Pumping solely by vacuum pressure, the device could also only raise water about 30 feet, hardly useful in deeper mines.
5. Rhodes, *Energy: A human history*, p. 55.
6. Rhodes, *Energy: A human history*, p. 58.
7. Crump, T., *The Age of Steam: The power that drove the Industrial Revolution*, p. 31, Robinson, London, 2007.
8. Rhodes, *Energy: A human history*, pp. 72–73.
9. Patented in 1769, No. 913: "New Invented Method of Lessening the Consumption of Steam and Fuel in Fire Engines."
10. Rhodes, *Energy: A human history*, pp. 75–76.
11. Klein, M., *The Power Makers: Steam, electricity, and the men who invented modern America*, pp. 48–49, Bloomsbury Press, New York, NY, 2008.
12. Crump, *The Age of Steam*, p. 81.
13. Klein, *The Power Makers*, p. 50.
14. Crump, *The Age of Steam*, p. 82.
15. Crump, *The Age of Steam*, p. 106.
16. The precise gauge was set by the width between the wheels of a horse-drawn wagon, which in turn was based on the width of two horses yoked side by side, set as far back as Roman times.
17. Williams, H., *Days That Changed the World*, pp. 135–139, Quercus, London, 2008.
18. Williams, *Days That Changed the World*, pp. 135–139.

19. Isambard Kingdom Brunel's iron-hulled SS *Great Britain* was the first ocean-going vessel to employ a screw propeller, making the 3,000-mile crossing from Bristol to New York in 1845 in 2 weeks.
20. Crump, *The Age of Steam*, pp. 301–304.
21. Woolcock, N., "Join voyage of discovery to track down ancestors," *The Times*, January 10, 2007.
22. "Why the Industrial Revolution happened here" [documentary], presented by Jeremy Black, BBC Two, January 22, 2013.
23. Williams, M., *On the Rails*, Episode 7, "Big Country," Discovery Channel, October 14, 2002.
24. McNaught, K. W., Saywell, J. T., and Ricker, J. C., *Manifest Destiny: A short history of the United States*, p. 95, Clarke, Irwin & Company Ltd., Toronto, 1980.
25. Klein, *The Power Makers*, p. 57.
26. Crump, *The Age of Steam*, p. 200.
27. Livesay, H. C., *Andrew Carnegie and the Rise of Big Business*, p. 8, HarperCollins, New York, NY, 1975.
28. McNaught, Saywell, and Ricker, *Manifest Destiny*, p. 95.
29. Crump, *The Age of Steam*, pp. 163–167.
30. Boyle's Law: The volume (V) of a gas is inversely proportional to its pressure (P); Charles' Law: The volume of a gas is proportional to its temperature (T); Gay-Lussac's Law: Gas pressure is proportional to its temperature. The three are combined in the Ideal Gas Law: $PV = nRT$. After examining the smoke from burning wood and determining it was not air, Boyle also overturned Aristotle's idea of four elements (five if you include "quintessence"), initiating the concept of irreducible chemical elements that he called "corpuscles."
31. "December 1840: Joule's abstract on converting mechanical power into heat," *APS News*, December 2009, 18(11), www.aps.org/publications/apsnews/200912/physicshistory.cfm.
32. Joule heating is a good thing in a toaster, hair dryer, or space heater, where heat is generated by passing a current through a wire, but detrimental in electrical power transmission or an electronic circuit.
33. Ridley, M., "It's a bio-mess. Burning wood is a disaster," *The Times*, June 20, 2013.
34. Bonneuil, C. and Fressoz, J.-B. (trans. Fernbach, D.), p. 233, *The Shock of the Anthropocene*, Verso, London, 2017.
35. Bonneuil and Fressoz, *The Shock of the Anthropocene*, p. 256.
36. Mytting, L., *Norwegian Wood: Chopping, stacking, and drying wood the Scandinavian way*, p. 20, MacLehose Press, London, 2021.
37. World Energy Council, "World Energy Resources, 2013 Survey," Chapter 6, 2013.
38. Carroll, R., "End of an era as Ireland closes its peat bogs 'to fight climate change,'" *The Guardian*, November 27, 2018. www.theguardian.com/world/2018/nov/27/ireland-closes-peat-bogs-climate-change.
39. Lane, A., "For peat's sake," *New Scientist*, January 1, 2022.
40. Restoration and rewetting helps protect and return degraded areas, including planting donor moss from healthy peatlands, which can be done by those who already work the land. Rewetting is essential to stop degraded peatlands from becoming a source of CO_2.
41. Smil, V., *Energy and Civilization: A history*, p. 230, The MIT Press, Cambridge, MA, 2017.
42. Dickens, C., *Hard Times*, p. 108, Penguin Books, Harmondsworth, 1983 (first published in 1854).

43. Bonneuil and Fressoz, *The Shock of the Anthropocene*, p. 263.
44. van Gogh, V., *Vincent van Gogh: The letters*, #151, Petit-Wasmes, between April 1 and April 16, 1879. http://vangoghletters.org/vg/letters/let151/letter.html.
45. Paxman, J., *Black Gold: The history of how coal made Britain*, p. 154, William Collins, London, 2022.
46. Orwell, G., *The Road to Wigan Pier*, p. 18, Penguin Books, London, 2020 (first published in 1937).
47. Orwell, *The Road to Wigan Pier*, pp. 31, 101.
48. Orwell, *The Road to Wigan Pier*, p. 18.
49. Smil, *Energy and Civilization*, p. 273.
50. Paxman, *Black Gold*, p. 173.
51. Williams, M., *On the Rails*, Episode 10, "Out of Steam," Discovery Channel, November 4, 2002.
52. "When British Coal Was King" [documentary], directed by Kate Thomas, BBC Four, November 4, 2013.
53. Paxman, *Black Gold*, pp. 32, 40.
54. Hand, M., "'Rape rooms': How West Virginia women paid off coal company debts," *CounterPunch*, October 2, 2015.
55. Sinclair, U., *King Coal: A novel of the Colorado coal country*, Section 6, published by the author, Station A, Pasadena, CA, 1917.
56. "Coal Mining: A Dangerous Way to Make Money," *Modern Marvels*, Season 10, Episode 13, April 2, 2003, The History Channel.
57. McClure, T. and Corlett, E., "'It hits you hard': shock as bodies of Pike River miners found 11 years after New Zealand disaster," *The Guardian*, November 17, 2021. www.theguardian.com/world/2021/nov/17/pike-river-bodies-found-of-miners-from-2010-new-zealand-disaster.
58. Helm, D., *Net Zero: How we can stop climate change*, pp. 27, 212, William Collins, London, 2020.
59. Griffith, S., *Electrify: An optimist's playbook for our clean energy future*, pp. 41, 58, The MIT Press, Cambridge, MA, 2021.
60. US Energy Information Administration, "Coal Data: By state and mine type," October 18, 2022.
61. Weir, B., "More coal-fired power plants have closed under Trump than in Obama's first term," CNN, January 7, 2019. https://edition.cnn.com/2019/01/07/politics/pennsylvania-coal-plants-weir-wxc/index.html.
62. Mann, C. C., *The Wizard and the Prophet: Science and the future of our planet*, pp. 333–334, Picador, London, 2019.
63. "Under the Dome" [documentary], directed by Ming Fan, People's Daily, China, February 28, 2015. www.youtube.com/watch?v=T6X2uwlQGQM.
64. Xuefei, T. and Huiying, Z., "From coal to clean: One man's journey," *China Daily*, January 2, 2019.
65. Wacket, M., "Germany to phase out coal by 2038 in move away from fossil fuels," Reuters, January 26, 2019. www.reuters.com/article/us-germany-energy-coal/germany-to-phase-out-coal-by-2038-in-move-away-from-fossil-fuels-idUSKCN1PK04L.
66. International Renewable Energy Agency, "Renewable Energy and Jobs – Annual Review 2020," Abu Dhabi, 2020. www.irena.org/-/media/Files/IRENA/Agency/Publication/2020/Sep/IRENA_RE_Jobs_2020.pdf.
67. Rowles, N., "King Coal is dead and McConnell knows it," *USA Today*, February 12, 2017. https://eu.cincinnati.com/story/news/local/kentoncounty/2017/02/12/column-king-coal-dead-mcconnell-knows/97827400/.

68. Bronowski, J., *The Ascent of Man*, Episode 8, The Drive for Power, June 23, 1973.
69. When a hydrocarbon (HC) is burned, the hydrogen (H) is converted to water and the carbon (C) is converted to carbon dioxide. For example, for methane (the simplest hydrocarbon), $CH_4 + 2O_2 \rightarrow 2H_2O + CO_2$. Other elements contained in the hydrocarbon fuel and atmosphere are also burnt, producing a host of by-products. For example, nitrogen is converted to nitric oxide ($2N + O_2 \rightarrow 2NO$) and nitrogen dioxide ($N + O_2 \rightarrow NO_2$) and sulfur is converted to sulfur dioxide ($S + O_2 \rightarrow SO_2$). Particulate matter, hydrocarbon gases, and trace metals are also generated in the exhaust. Gasoline, diesel, and kerosene contain many more higher-order hydrocarbons, such as pentane (C_5H_{12}), hexane (C_6H_{14}), heptane (C_7H_{16}), and octane (C_8H_{18}), all with the general formula C_nH_{2n+2}.
70. Paxman, *Black Gold*, pp. 255, 257.
71. Paxman, *Black Gold*, p. 256.
72. Schobert, H. H., *Energy and Society: An introduction*, p. 246, Taylor & Francis, New York, NY, 2002.
73. Rhodes, *Energy: A human history*, p. 298.
74. Dirty air is much worse on cold, windless nights as a temperature inversion traps the smoke at ground level under a blanket of warm air (the smoke can't rise), while winter is the worst as the cold air sits close to the ground and pollutants and particulates are trapped. At temperatures below zero, smoke can even come back down a chimney.
75. Barry, A., "'It creeps menacingly': When deadly smog choked Dublin's skies," *thejournal.ie*, January 17, 2013. www.thejournal.ie/smog-dublin-764941-Jan2013/.
76. Kelly, I. and Clancy, L., "Mortality in a general hospital and urban air pollution," *Irish Medical Journal* 77(10):322–4, November 1984.
77. Hincks, J., "The clouds over city life," *Time*, July 9, 2018.
78. Ruskin, L. and Holden, E., "Natural but deadly: Huge gaps in US rules for wood-stove smoke exposed," *The Guardian*, March 16, 2021. www.theguardian.com/environment/2021/mar/16/wood-smoke-alaska-state-regulators-air-quality.
79. Wallace-Wells, D., *The Uninhabitable Earth: A story of the future*, p. 102, Penguin Random House, London, 2019.
80. Wong, E. "China smog puts leaders to test," *The New York Times*, December 11, 2015.
81. Lelieveld, J., Klingmüller, K., Pozzer, A., et al., "Cardiovascular disease burden from ambient air pollution in Europe reassessed using novel hazard ratio functions," *European Heart Journal*, 40(20):1590–6, March 12, 2019. https://doi.org/10.1093/eurheartj/ehz135.
82. "EU says one in eight deaths is linked to pollution," BBC News, September 8, 2020. www.bbc.com/news/world-europe–54071380.
83. Buck, H. J., *Ending Fossil Fuels: Why net zero is not enough*, p. 59, Verso, London, 2021.
84. Kikstra, J. S., Waidelich, P., Rising, J., et al., "The social cost of carbon dioxide under climate-economy feedbacks and temperature variability," *Environmental Research Letters*, 16(9), 2021. https://doi.org/10.1088/1748-9326/ac1d0b.
85. "The Coal Resource: A Comprehensive Overview of Coal," Section 4, *How is Coal Used?*, pp. 19–25. https://is.muni.cz/el/fss/podzim2015/MEB412/um/World_Coal_Institute_2009.pdf
86. A more efficient yet also more expensive method to burn coal than in a PCC plant is to first heat the coal in a low-oxygen environment to create hydrogen (syngas) that is then burned, from which the hot exhaust creates the turbine steam to

generate electrical power, called an integrated gasification combined cycle (IGCC). Carbon monoxide is also produced in the gasification stage, where it can be separated prior to burning and is thus more easily captured (rather than turned into CO_2 during burning), although the cleaner IGCC process generates up to 20% less electrical power than in a PCC plant and is not as commercially viable.

87. "Coal's assault on human health," Physicians for Social Responsibility, Washington, DC, November 2009. https://psr.org/wp-content/uploads/2018/05/coals-assault-on-human-health.pdf.
88. US Energy Information Administration, "Coal plants without scrubbers account for a majority of US SO2 emissions," Today in Energy, December 21, 2011, www.eia.gov/todayinenergy/detail.cfm?id=4410.
89. US Energy Information Administration, "Coal plants without scrubbers account for a majority of US SO2 emissions," Today in Energy, December 21, 2011, www.eia.gov/todayinenergy/detail.cfm?id=4410
90. World Energy Council, "World Energy Resources, 2013 Survey," Chapter 1, 2013.
91. O'Keefe, E., Nakamura, D., and Mufson, S., "GOP congressional leaders denounce US–China deal on climate change," *The Washington Post*, November 12, 2014.
92. Shotter, J. and Hook, L., "Coal looms large over Katowice climate talks," *The Financial Times*, December 1/2, 2018.
93. Vaughan, A., "Battle over new UK coal mine puts climate commitments to the test," *New Scientist*, September 11, 2021.
94. Laville, S., "'Not the future we should be going for': The reopening of Wales's Aberpergwm coalmine," *The Guardian*, March 18, 2022. www.theguardian.com/environment/2022/mar/18/not-the-future-we-should-be-going-for-the-reopening-of-waless-aberpergwm-coalmine.
95. "Make coal history," *The Economist*, December 5, 2020.
96. Ambrose, J., "Most plans for new coal plants scrapped since Paris agreement," *The Guardian*, September 2021. www.theguardian.com/environment/2021/sep/14/most-plans-for-new-coal-plants-scrapped-since-paris-agreement.
97. Gislam, S., "Defiant Polish PM refuses to shut coal mine despite EU court fine," *Industry Europe*, September 22, 2021. https://industryeurope.com/sectors/politics-economics/defiant-polish-pm-refuses-to-shut-coal-mine-despite-eu-court-fine/.
98. The first telegraph system was invented 33 years earlier thanks to the invention of Alessandro Volta's chemical battery and William Sturgeon's electromagnet.
99. Friedrich Engels coined the term "Industrial Revolution" after comparing the political changes of the French Revolution to the mechanized changes occurring in Manchester and elsewhere (his father owned large textile factories in England and Germany). Starting roughly around the beginning of the eighteenth century, the Industrial Revolution was an uneven progression of industrialization, resulting from the mechanization of labor, ongoing metallurgical advances, the rise of steel and coking, coal-fired steam power, and electrification.
100. Rodgers, H.B., "Manchester, England, United Kingdom," *Encyclopaedia Britannica*. www.britannica.com/place/Manchester-England.
101. Brown, F., "Percentage of global population living in cities, by continent," *The Guardian*, August 24, 2009.
102. Bronowski, J., *The Ascent of Man*, Episode 8, "The Drive for Power," June 23, 1973.

103. Broadberry, S., et al., *British Economic Growth, 1270–1870*, Cambridge University Press, 2015. https://ourworldindata.org/grapher/total-GDP-in-the-UK-since-1270?time=1700..1913
104. Crump, *The Age of Steam*, p. 123.
105. Williams, M., *On The Rails*, Episode 5, "Carriage Kings," Discovery Channel, September 30, 2002.
106. Byron, G. G., "Speech of February 27, 1812," in T. C. Hansard (1812), *The Parliamentary Debates*, vol. 21, pp. 966–72.
107. Galbraith, J. K., The Prophets and Promise of Classical Capitalism [documentary], directed by Mike Jackson, Episode 1, "The Age of Uncertainty." London: BBC, 1977.
108. Marx, K. and Engels, F., *The Communist Manifesto*, p. 9, Oxford University Press, Oxford, 1998 (first published 1848).
109. Scheer, H., *The Solar Economy: Renewable energy for a sustainable global future*, p. 8, Earthscan, London, 2002.
110. "Wood: The fuel of the future," *The Economist*, April 6, 2013.
111. "Wood: The fuel of the future," *The Economist*, April 6, 2013.
112. "Wood: The fuel of the future," *The Economist*, April 6, 2013.
113. With higher temperatures from global warming, trees grow faster, though die sooner. Carbon take-up is also faster in trees in the tropics than at higher latitudes.
114. Mytting, L., *Norwegian Wood*, p. 62.
115. Sterman, J. D., Siegel, L., and Rooney-Varga, J. N., "Does replacing coal with wood lower CO_2 emissions? Dynamic lifecycle analysis of wood bioenergy," *Environmental Research Letters*, 13, 015007, January 18, 2018.
116. Gill, T., "A guide to biomass power plants," theecoexperts, November 30, 2022. www.theecoexperts.co.uk/blog/biomass-power-plant.
117. "Power from waste – the world's biggest biomass power plants," power-technology.com, April 1, 2014. www.power-technology.com/features/feature power-from-waste–the-worlds-biggest-biomass-power-plants–4205990/.
118. "US biomass power plants," *Biomass Magazine*, November 21, 2022. https://biomassmagazine.com/plants/listplants/biomass/US/page:1/sort:plant/direc tion:asc.
119. McGrath, M., "Most wood energy schemes are a 'disaster' for climate change," BBC, February 23, 2017. www.bbc.com/news/science-environment–39053678.
120. Wells, J., "Deforestation the next big corporate risk," *The Toronto Star*, September 5, 2019.
121. de Puy Kamp, M., "How marginalized communities in the South are paying the price for 'green energy' in Europe," CNN, July 9, 2021. https://edition.cnn.com/interactive/2021/07/us/american-south-biomass-energy-invs/.
122. Betts, M. G., Yang, Y., Hadley, A. S., et al., "Forest degradation drives widespread avian habitat and population declines," *Nature: Ecology & Evolution*, 6, 709–719, 2022. https://doi.org/10.1038/s41559-022-01737-8.
123. World Energy Council, "World Energy Resources, 2013 Survey," Chapter 7, Bioenergy, 2013.
124. Mytting, *Norwegian Wood*, pp. 31–32.
125. World Energy Council, "World Energy Resources, 2013 Survey," Chapter 7, Bioenergy, 2013.
126. "Electric Nation, America Revealed" [documentary], directed by Jack Youngelson, PBS, 2012.
127. Parfit, M., "Future power: Where will the world get its next energy fix?" *National Geographic*, p. 22, August 2005.

128. Rhodes, *Energy: A human history*, p. 215.
129. Rhodes, *Energy: A human history*, pp. 235–236.
130. The sugar is squeezed out as a juice from the pulverized sugarcane feedstock. The leftover nonsugar part, a.k.a. bagasse, is often burned onsite to power the sugar mill. Bagasse can also make a biodegradable plastic replacement when a small amount of long-fiber bamboo is blended with short-fiber bagasse.
131. UNICA, "UNICA's Comments on Brazilian Sugarcane Ethanol Availability for the LCFS (Low Carbon Fuel Standard)," October 16, 2014.
132. US Environmental Protection Agency, "Overview for Renewable Fuel Standard." www.epa.gov/renewable-fuel-standard-program/overview-renewable-fuel-standard.
133. Runge, C. F., "The case against more ethanol: It's simply bad for environment," *Yale Environment 360*, May 25, 2016. https://e360.yale.edu/features/the_case_against_ethanol_bad_for_environment.
134. Runge, C. F., "The case against more ethanol."
135. Meyer, G., "Why farmers are falling out of love with Trump," *The Financial Times*, November 22, 2019.
136. Conca, J. "It's final – Corn ethanol is of no use," *Forbes Energy*, April 20, 2014. www.forbes.com/sites/jamesconca/2014/04/20/its-final-corn-ethanol-is-of-no-use/.
137. Hazen, R. M., *Symphony in C: Carbon and the evolution of (almost) everything*, p. 215, William Collins, London, 2019.
138. We get biogas from cellulosic feedstock via anaerobic digestion. For example, yeast-catalyzed sugar fermentation produces the main by-products biogas (ethanol) and carbon dioxide: $C_6H_{12}O_6$ (glucose) $\rightarrow$ $2C_2H_5OH$ (ethanol) + $2CO_2$ or $C_{12}H_{24}O_{12}$ (sucrose) $\rightarrow$ $4C_2H_5OH$ (ethanol) + $4CO_2$. As in beer making, fermentation produces carbon dioxide.
139. Eller, D., "DuPont sells Iowa ethanol plant to German company; it will soon make renewable natural gas," *The Des Moines Register*, December 16, 2019. https://eu.desmoinesregister.com/story/money/agriculture/2018/11/08/dupont-cellulosic-ethanol-plant-nevada-sold-german-company-verbio-north-america-claus-sauter/1938321002/.
140. US Energy Information Administration, "Biofuels: Ethanol & Biodiesel." www.eia.gov/energyexplained/index.cfm?page=biofuel_home.
141. Conca, "It's final – Corn ethanol is of no use."
142. "Soybean as a biofuel feedstock," *Crop Watch*, University of Nebraska-Lincoln. https://cropwatch.unl.edu/bioenergy/soybeans.
143. McInnis, A., "The transformation of rapeseed into canola: A Cinderella story," *The Saskatchewan Story Exhibit*, North Battleford, Saskatchewan, May 21, 2004.
144. Walker, G. and King, D., *The Hot Topic: What we can do about global warming*, p. 162, Harcourt Inc., Orlando, FL, 2008.
145. Le Page, M., "Give solar the green light," *New Scientist*, November 5, 2022.
146. Romm, J., *Hell and High Water: The global warming solution*, p. 167, Harper Perennial, New York, NY, 2007.
147. Howard, A., *Farming and Gardening for Health or Disease*, Faber and Faber, London, 1945.
148. Burgen, S., "'A role model': How Seville is turning leftover oranges into electricity," *The Guardian*, February 23, 2021. www.theguardian.com/environment/2021/feb/23/how-seville-is-turning-leftover-oranges-into-electricity.

149. Yu, A., "Waste not, want not: Why aren't more farms putting poop to good use?," *npr*, April 23, 2017. www.npr.org/sections/thesalt/2017/04/23/524878531/waste-not-want-not-why-arent-more-farms-putting-poop-to-good-use.

150. Stuart, T., *Waste: Uncovering the global food scandal*, p. 233, Penguin Books, London, 2009.

151. Goodall, C., *Ten Technologies to Save the Planet*, p. 253, GreenProfile, London, 2008.

152. US Environmental Protection Agency, "Livestock Anaerobic Digester Database," May 2022. www.epa.gov/agstar/livestock-anaerobic-digester-database.

153. Daly, P., "Lowell biogas plant making electricity," *Grand Rapids Business Journal*, February 13, 2015. https://grbj.com/news/lowell-biogas-plant-making-electricity/.

154. On its own, a cow also emits about 500 liters of methane per day in burps and farts as it breaks down cellulose in grass feeds, contributing to almost 10% of global greenhouse gas emissions, although enriched-feed additives can reduce those emissions by almost half. Similar to humans, cows also belch less with a better diet! Once you get over the adolescent giggles (and smell), capturing waste emissions from livestock is a simple way to save.

155. Stuart, *Waste: Uncovering the global food scandal*, p. 241.

156. Bronowski, J., *The Ascent of Man*, Episode 2, "The Harvest of the Seasons," May 12, 1973.

157. Crump, *The Age of Steam*, p. 84.

158. Founded in 1662, the Royal Society of London was the world's first scientific society and evolved from an amateur congregation of neo-alchemist apothecaries and natural philosophers to an order of professional scientists studying and codifying natural laws through the able stewardship from 1703 to 1727 of its 12th president, Sir Isaac Newton. Sir Humphry Davy was president from 1820 to 1827.

159. Rhodes, *Energy: A human history*, p. 169.

160. Klein, *The Power Makers*, pp. 80–81.

161. Rhodes, *Energy: A human history*, pp. 174–176.

162. The terms *ion*, *electrode*, *anode*, and *cathode* were all coined by Michael Faraday, who also coined *electrolysis* from *electron*, Greek for amber, a material known to exhibit electric charge when rubbed. The Irish physicist George Stoney would later call the fundamental charge carrier of electricity an *electron*.

163. Suplee, C., *Milestones of Science*, p. 146, National Geographic, Washington, DC, 2000.

164. Ørsted's serendipitous classroom demonstration may be apocryphal, but he had certainly been experimenting with electric currents soon after Volta's battery. Ørsted was also the first to isolate aluminum, using electrolysis to separate aluminum from aluminum chloride.

165. To explain gravitational action at a distance, Faraday would introduce the concept of the "luminiferous ether," a mysterious invisible substance that also befuddled scientists hoping to explain how light travels in a vacuum. That is, until the idea of an all-encompassing ether was disproved.

166. Mann, C. C., *The Wizard and the Prophet*, p. 270.

167. Rhodes, *Energy: A human history*, p. 121.

168. Castaneda, C., "Manufactured and Natural Gas Industry," *EH.Net Encyclopedia*, ed. Robert Whaples, September 3, 2001. http://eh.net/encyclopedia/manufac tured-and-natural-gas-industry/.

169. Paxman, *Black Gold*, p. 142.

170. Joseph Swan also invented a short-life, low-resistance incandescent filament bulb, although Edison greatly improved the process and commercially exploited his own long-lasting, high-resistance, filament bulb in a distributed electrical system. Swan's low-resistance, high-current, filament light bulbs (~2 ohms, 7 amps) were configured to run in series like a string of old-style Christmas tree lights with the same all-or-nothing aggravation when one doesn't work compared to Edison's more functional high-resistance, low-current bulbs (140 ohms, 0.85 amps) in a parallel system.

171. Stross, R. *The Wizard of Menlo Park: How Thomas Alva Edison invented the modern world*, p. 133, Three Rivers Press, New York, NY, 2007.

172. "Edison's electric light," *The New York Times*, September 5, 1882.

173. In January 1882, a simple demonstration took place in the Edison Electric Light Station at the Holborn Viaduct in London, powered by a 125-hp steam engine. Electricity was distributed to street lamps in the area and in a few establishments such as the Old Bailey, but was not considered a working power station.

174. Stross, *The Wizard of Menlo Park*, p. 138.

175. Gold, R., *Superpower: One man's quest to transform American energy*, p. 21, Simon & Schuster, New York, NY, 2019.

176. Klein, *The Power Makers*, p. 220.

177. *Power* [documentary] Episode 1, "The Genius of Invention," directed by Victoria Bell, BBC Two, January 24, 2013.

178. Hirsh, R. F., "Powering the past: A look back," Virginia Polytechnic Institute and State University. http://americanhistory.si.edu/powering/past/history2.htm.

179. Bronowski, J., *The Ascent of Man*, Episode 8, "The Drive for Power," June 23, 1973.

2 Oil and Gas: Twentieth-Century Prosperity

1. Rhodes, R., *Energy: A human history*, p. 141, Simon and Schuster, New York, NY, 2018.

2. Mann, A. N., "Some petroleum pioneers of Pittsburgh," *Western Pennsylvania History*, Summer 2009.

3. Mann, "Some petroleum pioneers of Pittsburgh."

4. The analysis was completed on April 16, 1855, by Yale science professor Benjamin Silliman Jr., who by a roundabout way received a seep sample from the nearby McClintock farm. According to Peake, Silliman "showed that by ignition over carbon the crude could be converted to a good gas, but decided that it was too valuable for other purposes to consider making gas from it. He attempted to distill the oil at higher temperatures, reaching 750°C, and obtaining a thick dark oil which, on cooling, produced crystallized paraffin. He recognized that this would be very valuable for candles, but supposed, erroneously, that the paraffin was formed by some chemical cracking process during the heating."

5. Pees, S. T., "The Drake chapters" in *Oil History*. www.petroleumhistory.org/OilHistory/pages/drake/drake.html.

6. Rhodes, *Energy: A human history*, p. 134.

7. Melville, H., *Moby-Dick*, p. 117, Signet, New York, NY, 1980 (first published in 1851).

8. Coleman Jr., J. L. "The American whale oil industry: A look back to the future of the American Petroleum Industry? (Part 1)," pp. 13–16, *Houston Geological Society Bulletin*, October 1994.

9. Coleman Jr., J. L. "The American whale oil industry: A look back to the future of the American Petroleum Industry? (Part 2)," pp. 15–19, *Houston Geological Society Bulletin*, November 1994.

10. Coleman Jr., "The American whale oil industry (Part 2)."

11. Coleman Jr., "The American whale oil industry (Part 1)."

12. Coleman Jr., "The American whale oil industry (Part 2)."

13. Bowling, B., "Boom and bust for first oil well driller Edwin Drake," July 6, 2008, *Pittsburgh Tribune-Review*. http://triblive.com/x/pittsburghtrib/news/pittsburgh/s_576222.html.

14. Dickey, P. A. "The first oil well," Oil Industry Centennial, *The Journal of Petroleum Technology*, 1959.

15. Pees, "The Drake chapters."

16. Scruggs, M. H., "The first oil well fire," Penn States University Libraries, Spring 2010. www.pabook.libraries.psu.edu/literary-cultural-heritage-map-pa/feature-articles/first-oil-well-fire.

17. Scruggs, "The first oil well fire."

18. Sinclair, U., *Oil!*, p. 157, published by the author, Station A, Pasadena, CA, 1926, 1927.

19. Rhodes, *Energy: A human history*, pp. 158–159.

20. Davé, U., "Edwin Drake and the oil well drill pipe," Penn States University Libraries, Summer 2008. www.pabook.libraries.psu.edu/literary-cultural-heritage-map-pa/feature-articles/edwin-drake-and-oil-well-drill-pipe.

21. Mann, "Some petroleum pioneers of Pittsburgh."

22. Pees, "The Drake chapters."

23. Mann, "Some petroleum pioneers of Pittsburgh."

24. Dickey, "The first oil well."

25. Rhodes, *Energy: A human history*, p. 143.

26. Rhodes, *Energy: A human history*, p. 143.

27. Rhodes, *Energy: A human history*, p. 160.

28. In the 1820s, naphtha was extracted from coal-tar waste, which dissolved rubber, ideal for shaping waterproof clothing such as the first rubberized gloves, shoes, and Mackintosh coat (mac). As noted by James Burke in *Connections*, a few decades later aniline dyes were derived from coal-tar naphtha to make the brightly colored mauves, crimson fuscines, violets, and greens of Victorian fashion.

29. Mann, "Some petroleum pioneers of Pittsburgh."

30. Wright, W., *The Oil Regions of Pennsylvania*, Harper, New York, NY, 1865.

31. Frehner, B., "From creekology to geology; Finding and conserving oil on the southern plains, 1859–1930." PhD dissertation, University of Oklahoma, Norman, OK, 2004.

32. Schwartz, J., "Rockefellers, heirs to an oil fortune, will divest charity of fossil fuels," *The New York Times*, September 21, 2014.

33. Tarbell, I., *The History of the Standard Oil Company*, Vol. 2, p. 31, McClure, Philips & Co., New York, NY, 1905.

34. Tarbell, I., *The History of the Standard Oil Company*, Vol. 1, p. 43, McClure, Philips & Co., New York, NY, 1905.

35. Tarbell, *The History of the Standard Oil Company*, Vol. 2, p. 241.

36. Tarbell, *The History of the Standard Oil Company*, Vol. 2, p. 252.

37. Lloyd, H. D., "Story of a great monopoly," *Atlantic Monthly*, March 1881.

38. Lloyd, "Story of a great monopoly."
39. Chernow, R., "Book Discussion on Titan: The Life of John D. Rockefeller, Sr," *C-SPAN*, May 7, 1998. www.c-span.org/video/?105430-1/book-discussion-titan-life-john-d-rockefeller-sr.
40. "The Men Who Built America" [documentary], Episode 2, directed by Ruan Magan, The History Channel, 2012.
41. Mauldin, J., "There are alarming economic similarities between now and 1873," *Business Insider*, November 4, 2014. www.businessinsider.com/mauldin-on-panic-of-1873-2014–11.
42. Lloyd, "Story of a great monopoly."
43. Lloyd, "Story of a great monopoly."
44. Tarbell, *The History of the Standard Oil Company*, Vol. 2, p. 66.
45. Tarbell, *The History of the Standard Oil Company*, Vol. 2, p. 112.
46. Lloyd, "Story of a great monopoly."
47. McNaught, K. W., Saywell, J. T., and Ricker, J. C., *Manifest Destiny: A short history of the United States*, p. 159, Clarke, Irwin & Company Ltd., Toronto, 1980.
48. Tarbell, *The History of the Standard Oil Company*, Vol. 2, p. 193.
49. Tarbell, *The History of the Standard Oil Company*, Vol. 2, p. 198.
50. Tarbell, *The History of the Standard Oil Company*, Vol. 2, p. 205.
51. Lloyd, "Story of a great monopoly."
52. Lloyd, "Story of a great monopoly."
53. Crude Oil Production" (1859–2015), US Energy Information Agency. www.eia.gov/dnav/pet/pet_crd_crpdn_adc_mbblpd_a.htm.
54. Frehner, "From creekology to geology."
55. "Oil and petroleum products explained: Refinery Rankings," US Energy Information Agency, January 1, 2022. www.eia.gov/energyexplained/oil-and-petroleum-products/refining-crude-oil-refinery-rankings.php.
56. Petroleum acts as a good lubricant or grease because hydrocarbon molecules have full outer electron shells and don't attract other materials, easily sliding past lubricated substances as well as being difficult to wash off.
57. "Crude Oil Production," US Energy Information Agency, October 31, 2021. www.eia.gov/dnav/pet/pet_crd_crpdn_adc_mbblpd_a.htm. Note, Gulf of Mexico is Federal Offshore (PADD 3).
58. Boorstin, D. J., *The Americans: The democracy experience*, p. 567, Phoenix Press, London, 1973.
59. "The Men Who Built America" [documentary], Episode 2, directed by Ruan Magan, The History Channel, 2012.
60. Kolbert, E. "Hosed: Is there a quick fix for the climate?" *The New Yorker*, November 16, 2009.
61. Fair, G. "Otto-Volk: Building an Otto-Langen atmospheric engine," *Gas Engine Magazine*, December/January 2002.
62. "The Internal Combustion Engine," The Secret Life of Machines [documentary], directed by Nigel Maslin, Channel 4 (Antifax), January 15, 1991. www.youtube.com/watch?v=gfr3_AwuO9Y.
63. Bellis, M., "The history of steam-powered cars," *ThoughtCo*, August 26, 2020. www.thoughtco.com/history-of-steam-powered-cars-4066248.
64. Ernst, J., "The history of the gasoline engine at Mercedes-Benz," Daimler, Stuttgart, June 2008.
65. "August 1888: Bertha Benz takes world's first long-distance trip in an automobile," Daimler, Stuttgart, June 27, 2008. https://group-media.mercedes-

benz.com/marsMediaSite/en/instance/ko/August-1888-Bertha-Benz-takes-worlds-first-long-distance-trip-in-an-automobile.xhtml?oid=9361401.

66. Ernst, J., "The history of the gasoline engine at Mercedes-Benz."

67. "The history behind the Mercedes-Benz brand and the three-pointed star," Daimler, Stuttgart, April 17, 2008. https://group-media.mercedes-benz.com/marsMediaSite/en/instance/ko/The-history-behind-the-Mercedes-Benz-brand-and-the-three-pointed-star.xhtml?oid=9912871.

68. "The history behind the Mercedes-Benz brand and the three-pointed star."

69. "The Diesel Story" [documentary], directed by Lionel Cole, Shell Oil Company, 1952. www.youtube.com/watch?v=Ee8Do7bF3gE.

70. Brake hp is the amount of horse power after friction losses are subtracted.

71. *The Diesel Engine*, p. 31, Busch-Sulzer Bros., Diesel Engine Co., Saint Louis, MO, 1913.

72. *The Diesel Engine*, p. 20.

73. *The Diesel Engine*, pp. 44–45.

74. Launched in 1920, 300 men serviced the oil-fired battlecruiser HMS *Hood*, half that needed for the coal-fired battlecruiser HMS *Lion* launched a decade earlier.

75. Usher, B., *Renewable Energy: A primer for the twenty-first century*, p. 78, Columbia University Press, New York, NY, 2019.

76. Crump, T., *The Age of Steam: The power that drove the Industrial Revolution*, p. 305, Robinson, London, 2007.

77. "Occupational Exposures in Petroleum Refining; Crude Oil and Major Petroleum Fuels," p. 39, *IARC Monographs on the Evaluation of Carcinogenic Risks to Humans*, Volume 45, IARC, Lyon, France, 1989.

78. Smil, V., *Energy and Civilization: A history*, p. 290, The MIT Press, Cambridge, MA, 2017.

79. Agco also acquired Massey Ferguson, which started out in the 1850s making mechanical threshers in rural Ontario to separate edible grain from inedible hulls, stalks, and chaff, before becoming a world leader with the "combine" – cutting, threshing, and cleaning in one machine – forever changing how we farm.

80. "About clean Diesel: Agriculture," Diesel Technology Forum. https://dieselforum.org/about-clean-diesel/agriculture.

81. McGlothlin, M., "11 Diesel myths busted!: Clearing-up diesel misinformation," *Truck* Trend, March 1, 2013. www.motortrend.com/reviews/1303dp-11-diesel-myths-busted/.

82. A new geometry was developed in 1956 by the German engineer Felix Wankel, thought to be the next best thing before succumbing to high fuel costs. Wankel's unique "rotary" engine employed an asymmetric rotating wheel within a wheel (essentially a three-pointed piston in an oval cylinder) to create the large and small volumes required to achieve a sufficient compression ratio instead of an up-and-down piston in a cylinder, greatly reducing vibrations. Primarily manufactured by Mazda, its last rotary-engine model, the RX-8, was discontinued in 2012.

83. An electric motor is much more efficient than an internal combustion engine because it doesn't need to discard waste heat at a lower temperature and thus almost all the energy can be converted into mechanical work.

84. Rhodes, R., *Energy: A human history*, p. 299.

85. "History of the automobile," GM Canada.

86. Ernst, "The history of the gasoline engine at Mercedes-Benz."

87. "History of the automobile," GM Canada.

88. The decline of horses inside the city also led to table scraps and organic waste thrown out as garbage instead of recycled as animal feed, which, along with flush

toilets and artificial fertilizers, ended the need to collect human excrement for farming.

89. Stross, R. *The Wizard of Menlo Park: How Thomas Alva Edison invented the modern world*, pp. 234, 252, Three Rivers Press, New York, NY, 2007.
90. "Before the Model T: Henry Ford's letter cars," *Past Forward: Activating The Henry Ford Archive of Innovation*, September 4, 2013. www.thehenryford.org/explore/blog/before-the-model-t-henry-fords-letter-cars.
91. "Ford Historic Model T" [documentary], narrated by Kelly, Car Data Video, August 13, 2008. www.youtube.com/watch?v=S4KrIMZpwCY.
92. Casey, J., Dodge, J., and Dodge, H., "Henry Ford and Innovation, 'From the Curators'." thehenryford.org. 2010. www.thehenryford.org/docs/default-source/default-document-library/default-document-library/henryfordandinnovation.pdf?sfvrsn=0.
93. "History of the automobile," GM Canada.
94. Casey et al., "Henry Ford and Innovation."
95. "History of the automobile," GM Canada.
96. Casey et al. "Henry Ford and Innovation."
97. Boorstin, *The Americans*, p. 426.
98. Hamper, B., *Rivethead: Tales from the assembly line*, p. 41, Warner Books, New York, NY, 1991.
99. "History of the automobile," GM Canada.
100. Yergin, D., *The Prize: The epic quest for oil, money, & power*, p. 207, Simon & Schuster, New York, NY, 1991.
101. Bonneuil, C. and Fressoz, J.-B. (trans. Fernbach, D.), *The Shock of the Anthropocene*, pp. 138, 139, Verso, London, 2017.
102. Rifkin, J., *The Third Industrial Revolution: How lateral power is transforming energy, the economy, and the world*, p. 19, Palgrave Macmillan, New York, NY, 2011.
103. McGrath, R., "1949 Hudson Commodore – The car that started the modern road movies," *Autoweek*, November 29, 2012. www.autoweek.com/car-life/a1976266/1949-hudson-commodore-car-started-modern-road-movies/.
104. Kerouac, J., *On the Road*, pp. 133–134, Penguin Books, London, 1955.
105. Schobert, H. H., *Energy and Society: An introduction*, p. 323, Taylor & Francis, New York, NY, 2002.
106. Bonneuil and Fressoz, *The Shock of the Anthropocene*, p. 164.
107. Chrysler is now part of Stellantis, formed in 2020 when Groupe PSA (makers of Citroën, Opel, and Peugeot) merged with Fiat/Chrysler (formed in an earlier 2014 merger).
108. "Registrations or Sales of New Vehicles – All Types, OICA (2017)," OICA, 2017. www.oica.net/wp-content/uploads/total-sales-2016.pdf.
109. Smil, *Energy and Civilization*, p. 329.
110. Nader, R., "Federal regulation saves millions of lives," *CounterPunch*, September 12, 2016. www.counterpunch.org/2016/09/12/federal-regulation-saves-millions-of-lives/.
111. Nader, "Federal regulation saves millions of lives."
112. Yergin, *The Prize*, p. 111.
113. National Research Council, "Oil in the sea: Inputs, fates, and effects," p. 17. The National Academies Press, Washington, DC, 1985. https://nap.nationalacademies.org/read/314/chapter/4.
114. Recalling its origins, petrol/gasoline is still called benzin in German and Danish, benzine in Dutch, and bensin in Swedish.

115. No two oils or gases are the same and depend on their region of origin and means of extraction and refining. As calculated by Rocky Mountain Institute's OCI+, life-cycle emissions also vary widely – from dry gas and coal-bed methane at the low end to extra-heavy oil at the high end – each having a unique climate impact.

116. In the petroleum lexicon, alkanes are also called "paraffins," cycloalkanes "naphthenes," and benzene and its derivatives "aromatics." The famous C_6 benzene ring is also the parent compound of a branch of chemistry known as aromatics.

117. Complete combustion (a.k.a. stoichiometric combustion) occurs when enough oxygen is present to completely burn the carbon-containing fuel (coal, natural gas, gasoline, etc.), producing carbon dioxide (CO_2) and water (H_2O) and the maximum amount of heat possible, extracting all the energy from the fuel. If insufficient oxygen is present, soot (an oxygen-starved reaction) and carbon monoxide (CO) (incomplete combustion) is produced instead of CO_2. One hundred percent complete combustion is not possible, however, and heat loss is inevitable along with unburned combustibles.

118. Yergin, *The Prize*, p. 211.

119. Yergin, *The Prize*, p. 211.

120. A "cetane" number is used instead to measure burn as a function of ignition delay, expressed as a percentage between 0 for heptamethylnonane and 100 for cetane ($C_{16}H_{34}$). A typical diesel fuel has a cetane rating of 50 and an octane rating of 25.

121. Schobert, *Energy and Society*, p. 304.

122. Rhodes, *Energy: A human history*, pp. 241–243.

123. Rhodes, *Energy: A human history*, p. 247.

124. Schobert, *Energy and Society*, p. 305.

125. US Environmental Protection Agency, "EPA Sets New Limits on Lead in Gasoline," Press Release, March 4, 1985. www.epa.gov/archive/epa/aboutepa/epa-sets-new-limits-lead-gasoline.html.

126. Kitman, J. L., "The secret history of lead," *The Nation*, March 2, 2000. www.thenation.com/article/archive/secret-history-lead/.

127. Rhodes, *Energy: A human history*, p. 299.

128. Haagen-Smit, A. J., "A lesson from the smog capital of the world," *Proceedings of the National Academy of Sciences of the United States of America*, 67(2), pp. 887–897, October 1970.

129. Gardner, S., "LA smog: The battle against air pollution," *MarketPlace*, American Public Media, July 14, 2014. www.marketplace.org/2014/07/14/sustainability/we-used-be-china/la-smog-battle-against-air-pollution/.

130. Valavanidis, A., Fiotakis, K., and Vlachogianni, T., "Airborne particulate matter and human health: Toxicological assessment and importance of size and composition of particles for oxidative damage and carcinogenic mechanisms," *Journal of Environmental Science and Health, Part C*, 26: 4, pp. 339–362, DOI: 10.1080/10590500802494538, November 26, 2008.

131. "Tailpipe emissions," Greenercars.org. https://greenercars.org/greenercars-ratings/tailpipe-emissions.

132. "Fuel-efficient driving," *Eartheasy*. https://learn.eartheasy.com/guides/fuel-efficient-driving/.

133. "Fuel-efficient driving," *Eartheasy*.

134. Topham, G., Clarke, S., Levett, C., Scruton, P., and Fidler, M., "The Volkswagen emissions scandal explained," *The Guardian*, September 23, 2015. www.theguardian.com/business/ng-interactive/2015/sep/23/volkswagen-emissions-scandal-explained-diesel-cars.

135. Carrington, D., "38,000 people a year die early because of diesel emissions testing failures," *The Guardian*, May 15, 2017. www.theguardian.com/environment/2017/may/15/diesel-emissions-test-scandal-causes-38000-early-deaths-year-study.
136. "Deputy attorney general Sally Q. Yates delivers remarks at press conference announcing $14.7 billion Volkswagen settlements," *Justice News*, US Department of Justice, Washington, DC, June 28, 2016. www.justice.gov/opa/speech/deputy-attorney-general-sally-q-yates-delivers-remarks-press-conference-announcing–147.
137. McLean, B., "Emissions of guilt," *The New York Times*, June 7, 2017.
138. Although the regulatory bodies have differing priorities in Europe (reduced CO_2 global warming emissions) and the USA (reduced NOx smog emissions), which helps explain the higher market penetration of diesel in Europe and gasoline in the USA, one wonders what the extent of the environmental disregard would be *without* regulation.
139. Benzie, R., "Ontario's 'climate change action plan'," *The Toronto Star*, June 7, 2016.
140. Voelcker, J., "1.2 billion vehicles on world's roads now, 2 billion by 2035," *Green Car Reports*, July 29, 2014. www.greencarreports.com/news/1093560_1-2-billion-vehicles-on-worlds-roads-now-2-billion-by-2035-report.
141. McGlothlin, "11 Diesel myths busted!"
142. "Unsafe at any speed: 50th anniversary," *Ralph Nader Radio Hour*, KPFK Studios, November 28, 2015, California. www.ralphnaderradiohour.com/p/unsafe-at-any-speed-50th-anniversary–823.
143. Yergin, *The Prize*, p. 59.
144. A liquid sloshes around in any unfilled container space during transport (called the "free surface" effect), which can capsize a vessel unless the container is entirely full or empty (or nearly so), or is divided into baffled compartments thus reducing the sloshing forces.
145. Lichtman, M. A., "Alfred Nobel and his prizes: From dynamite to DNA," *Rambam Maimonides Medical Journal*, Vol 8(3), July 2017.
146. Yergin, *The Prize*, p. 70.
147. "Occupational Exposures in Petroleum Refining; Crude Oil and Major Petroleum Fuels," p. 40, in *IARC Monographs on the Evaluation of Carcinogenic Risks to Humans*, Volume 45, IARC, Lyon, France, 1989.
148. Dennett, C., *The Crash of Flight 3804: A lost spy, a daughter's quest, and the deadly politics of the great game for oil*, p. 147, Chelsea Green, Vermont, 2020.
149. Rhodes, *Energy: A human history*, pp. 256–257.
150. Rhodes, *Energy: A human history*, pp. 256–257.
151. Barr, J., *Lords of the Desert: Britain's struggle with America to dominate the Middle East*, p. 36, Simon & Schuster, London, 2018.
152. Yergin, *The Prize*, p. 393.
153. Yergin, *The Prize*, pp. 399–400.
154. Yergin, *The Prize*, p. 395.
155. Yergin, *The Prize*, p. 404.
156. Yergin, *The Prize*, p. 404.
157. Bonneuil and Fressoz, *The Shock of the Anthropocene*, pp. 243–244.
158. Barr, *Lords of the Desert*, p. 222.
159. Yergin, *The Prize*, p. 697.
160. Dennett, *The Crash of Flight 3804*, pp. 13–14.
161. Barr, *Lords of the Desert*, p. 96.

162. Dennett, *The Crash of Flight 3804*, p. 93.
163. Maass, P., *Crude World*, pp. 14–15, Penguin Books, London, 2009.
164. "Natural gas – production (cubic meters) > TOP 100 – World," IndexMundi (from *The CIA World Factbook*), 2020. www.indexmundi.com/map/?t=100&v=136&r=xx&l=en.
165. "Natural gas – exports (cubic meters) > TOP 100 – World," IndexMundi (from *The CIA World Factbook*), 2020. www.indexmundi.com/map/?t=100&v=136&r=xx&l=en.
166. Weaver, A. M., "Revolution from the top down," p. 94, *National Geographic*, March 2003.
167. The six ruling sons of Ibn Saud are Saud (1953–1964), Faisal (1964–1975), Khalid (1975–1982), Fahd (1982–2005), Abdullah (2005–2015), and Salman (2015–).
168. "Fight for Oil: 100 Years in the Middle East," Bavarian Public TV, directed by Dieter Schroeder, Falls Church, VA, Landmark Media, 2007.
169. "Fight for Oil: 100 Years in the Middle East."
170. Sorkhabi, R., "The road to OPEC 1960," *GeoExPro*, 7(5), 2016.
171. "House of Saud: A Family at War," Episode 2 [documentary], directed by Leo Telling, BBC2, London, January 16, 2018.
172. "House of Saud: A Family at War," Episode 2 [documentary].
173. "Obituary: King Fahd," BBC News, August 1, 2005. http://news.bbc.co.uk/2/hi/middle_east/255097.stm.
174. "Interview with deputy crown prince Mohammed bin Salman," *Al Arabiya*, April 25, 2016. https://english.alarabiya.net/webtv/programs/special-interview/2016/04/25/Deputy-Crown-Prince-This-is-the-Saudi-vision–2030.
175. The long-awaited IPO was eventually offered in late 2019, posted to the domestic Riyadh exchange and limited only to Saudi nationals, avoiding the need to prove or publicly disclose reserves to international market regulators.
176. Daoudy, M., "Scorched Earth: Climate and conflict in the Middle East," p. 55, *Foreign Affairs*, March/April 2022.
177. Perkins, J., *Confessions of an Economic Hitman*, p. 95, Plume, New York, NY, 2006.
178. Bonneuil and Fressoz, *The Shock of the Anthropocene*, p. 249.
179. Nixon, R., "Address to the nation about policies to deal with the energy shortages," The American Presidency Project, *UC Santa Barbara*, November 7, 1973. www.presidency.ucsb.edu/documents/address-the-nation-about-national-energy-policy.
180. Nixon, R., "Address to the nation about national energy policy," The American Presidency Project, *UC Santa Barbara*, November 25, 1973. www.presidency.ucsb.edu/documents/address-the-nation-about-national-energy-policy.
181. Nixon, "Address to the nation about national energy policy."
182. Schobert, *Energy and Society*, p. 513.
183. Paris, A. J., "Harvard study urges conservation and solar use over synthetic fuel," *The New York Times*, July 12, 1979.
184. Schobert, *Energy and Society*, pp. 513–514.
185. Yergin, *The Prize*, p. 685.
186. Schobert, *Energy and Society*, p. 513.
187. "The Secret of the Seven Sisters: Desert Storm" [documentary], Sunset Presse, Al Jazeera, April 26, 2013. www.aljazeera.com/program/featured-documentaries/2013/4/26/the-secret-of-the-seven-sisters.
188. Paxman, J., *Black Gold: The history of how coal made Britain*, p. 290, William Collins, London, 2022.

189. *Foreign Relations of the United States*, p.45, vol. 8, Government Printing Office, Washington, DC, 1945.

190. One hundred million barrels per day (100 mbpd) overestimates current use, but is a good round number to quantify global petroleum production. Note that in the arcane nomenclature of petroleum measures, "bbl" is a barrel (perhaps for "blue barrel," the original color), while MBPD (all caps) is also sometimes used for thousand barrels per day and MMBPD for billion barrels per day (you will also occasionally see mbd). I will stick to the non-standard but much less confusing mbpd to mean million barrels per day.

191. Ambrose, J., "Supertankers drafted in to store glut of crude oil," *The Guardian*, April 19, 2020. www.theguardian.com/business/2020/apr/19/supertankers-drafted-in-to-store-glut-of-crude-oil-coronavirus.

192. Note that specific gravity gives the relative density to water, which has a specific gravity of 1.00, a density of 1,000 kg/m^3, and an API gravity of 10.

193. "Sweet vs. sour crude oil," Petroleum.co.uk. www.petroleum.co.uk/sweet-vs-sour?p=sweet-vs-sour.

194. Freudenburg, W. R. and Gramling, R., *Blowout in the Gulf: The BP oil spill disaster and the future of energy in America*, p. 92, The MIT Press, Cambridge, MA, 2011.

195. Freudenburg and Gramling, *Blowout in the Gulf*, p. 93.

196. Freudenburg and Gramling, *Blowout in the Gulf*, p. 148.

197. Hudson, M., *Global Fracture*, p. 152, Pluto Press, London, 2005.

198. "1965: Sea Gem oil rig collapses," On this Day, 27 December, BBC. http://news.bbc.co.uk/onthisday/hi/dates/stories/december/27/newsid_4630000/4630741.stm.

199. "A day in history: This rig gave birth to Norwegian oil industry," *Offshore Energy Today*. www.offshoreenergytoday.com/a-day-in-history-this-rig-gave-birth-to-norwegian-oil-industry/.

200. Lavelle, M., "Coast guard blames Shell risk-taking in Kulluk rig accident," *National Geographic*, April 4, 2014. www.nationalgeographic.com/news/energy/2014/04/140404-coast-guard-blames-shell-in-kulluk-rig-accident/.

201. Federman, A., "Leaked memo: Government scientists warned Trump's oil plan would threaten Alaska's polar bears," *Mother Jones*, December 11, 2018. www.motherjones.com/environment/2018/12/anwr-polar-bears-trump/.

202. A coming, year-long, ice-free Arctic passage will provide shorter container ship routes – certainly attractive to transportation – and easier access to difficult-to-reach oil reserves. But the accompanying increased albedo (decreased solar reflectivity) and increased methane will vastly alter the previously stable and reasonably predictable weather cycle that built the market system.

203. Staalesen, A., "Moscow outlines €210 billion plan for Arctic oil," *The Barents Observer*, February 4, 2020. https://thebarentsobserver.com/en/arctic-industry-and-energy/2020/02/moscow-outlines-eu208-billion-plan-arctic-oil.

204. Blackman, S., "Risky business: Challenges of deepwater drilling in the North Sea," June 21, 2012, *Offshore Technology*. www.offshore-technology.com/features/featurerisky-business-deepwater-drilling-north-sea/.

205. Katona, V., "The world's largest offshore oil field is back in action," OilPrice.com, September 18, 2017. http://oilprice.com/Energy/Crude-Oil/The-Worlds-Largest-Offshore-Oil-Field-Is-Back-In-Action.html.

206. Yergin, D., *The Quest: Energy, security, and the remaking of the modern world*, p. 248, Penguin Books, New York, NY, 2012.

207. "Gulf of Mexico," *Drain the Oceans*, Season 1, Episode 3 [documentary], directed by Anthony Barwel, National Geographic, June 4, 2018.
208. "Worldwide rig count," Baker Hughes, December 30, 2022. https://rigcount .bakerhughes.com/static-files/cd074597-60a1-4cc60-8f6b-0718f75abe0e.
209. Yergin, *The Quest*, p. 340.
210. Rutte, M., "Churchill Lecture," Europa Institut, University of Zurich, February 13, 2019. www.government.nl/documents/speeches/2019/02/13/churchill-lecture-by-prime-minister-mark-rutte-europa-institut-at-the-university-of-zurich.
211. White, J. K., "Petroleum wars in the age of climate disaster: A bridge fuel too far," *CounterPunch*, June 3, 2022. www.counterpunch.org/2022/06/03/petroleum-wars-in-the-age-of-climate-disaster-a-bridge-fuel-too-far/.
212. "Putin's energy weapon," *The Economist*, January 29, 2022.
213. Dempsey, J., "Europe's energy strategy and South Stream's demise," *Carnegie Europe*, December 4, 2014. http://carnegieeurope.eu/strategiceurope/?fa=57386.
214. Klare, M. T., *Resource Wars: The new landscape of global conflict*, p. 104, Henry Holt and Company, New York, NY, 2001.
215. Dennett, *The Crash of Flight 3804*, pp. 13–14.
216. Yergin, *The Quest*, p. 336.
217. Yergin, *The Quest*, p. 331.
218. "Drilling Productivity Report," US Energy Information Administration, January 17, 2023. www.eia.gov/petroleum/drilling/.
219. Kuchment, A., "Methane hunters," *Scientific American*, September 2021.
220. Spiller, H. A., Hale, J. R., and de Boer, J. Z., "The Delphic Oracle: A multidisciplinary defense of the gaseous vent theory," *Clinical Toxicology*, 40(2), 189–196 (2002).
221. Leahy, S., "Fracking boom tied to methane spike in earth's atmosphere," *National Geographic*, August 15, 2019. www.nationalgeographic.com/environment/art icle/fracking-boom-tied-to-methane-spike-in-earths-atmosphere.
222. Horn, S., "Obama administration approved over 1,500 offshore fracking permits," *CounterPunch*, June 30, 2016. www.counterpunch.org/2016/06/30/obama-admin-approved-over-1500-offshore-fracking-permits/.
223. "Shattered Ground," *The Nature of Things* [documentary], directed by Leif Kaldo, Zoot Pictures, CBC, February 7, 2013. www.cbc.ca/natureofthings/episodes/shat tered-ground.
224. US Environmental Protection Agency, "EPA Publishes 21st Annual US Greenhouse Gas Inventory," Press Release, April 15, 2016. https://yosemite.epa .gov/opa/admpress.nsf/0/80447044E54D9F7385257F96005F5278.
225. Durkee, A., "EPA rescinds Obama-era methane rules as White House speeds environmental rollbacks ahead of election," *Forbes*, August 13, 2020. www .forbes.com/sites/alisondurkee/2020/08/10/epa-reportedly-set-to-rescind-obama-era-methane-rules-as-white-house-speeds-environmental-rollbacks-ahead-of-election/#2d2a26b443fb.
226. Yu, J., Hmiel, B., Lyon, D. R., et al., "Methane emissions from natural gas gathering pipelines in the Permian Basin," *Environmental Science and Technology Letters*, October 4, 2022. https://doi.org/10.1021/acs.estlett.2c00380.
227. Cribb, R., Sonntag, P., Elliott, P. W., and McSheffrey, E., "That rotten stench in the air? It's the smell of deadly gas and secrecy," *The Toronto Star*, October 1, 2017.
228. McKenzie, L. M., Blair, B., Hughes, J., et al., "Ambient nonmethane hydrocarbon levels along Colorado's northern front range: Acute and chronic health risks," *Environmental Science and Technology*, March 27, 2018. https://doi.org/10.1021/acs.est.7b05983.

229. Clark, C. J., Johnson, N. P., Soriano Jr., M., et al., "Unconventional oil and gas development exposure and risk of childhood acute lymphoblastic leukemia: A case-control study in Pennsylvania, 2009–2017," *Environmental Health Perspectives*, 130(8), August 17, 2022. https://doi.org/10.1289/EHP11092.
230. Kuchment, "Methane hunters."
231. Rhodes, *Energy: A human history*, p. 263.
232. Schulz, R., McGlade, C., and Zeniewski, P., "Flaring emissions," Tracking report, International Energy Agency, November 2021. www.iea.org/reports/flaring-emissions.
233. Meyer, G., "Surge in gas flaring sparks controversy for US shale sector," *The Financial Times*, January 5, 2019.
234. Lauvaux, T., Giron, C., Mazzolini, M., et al., "Global assessment of oil and gas methane ultra-emitters," *Science*, 375(6580), 557–561, February 3, 2022. www.science.org/doi/10.1126/science.abj4351.
235. "The Methane Hunters," Storylines [documentary], produced by Alan Jeffries, Bloomberg Originals, September 14, 2021. www.youtube.com/watch?v=62rkNvfuTlg.
236. Kuchment, "Methane hunters."
237. "Low-hanging fruit," *The Economist*, June 25, 2022.
238. Fountain, H., "In Alberta, fracking is directly tied to quakes," *The New York Times*, November 19, 2016.
239. Frank, J., "Frack, rattle and roll," *CounterPunch*, 21(4), 2014.
240. Frank, "Frack, rattle and roll."
241. "Places Near Oklahoma, United States," Earthquake Track, 2022. https://earthquaketrack.com/p/united-states/oklahoma/recent.
242. "Commission recommendation on minimum principles for the exploration and production of hydrocarbons (such as shale gas) using high-volume hydraulic fracturing" (2014/70/EU), Section 5.2, *Official Journal of the European Union*, February 2, 2014.
243. Cartwright, J., "Settling the fracking question," *Physics World*, March 2012.
244. Conley, J., "After overwhelming public opposition, Scotland announces fracking ban," *CommonDreams*, October 3, 2017. www.commondreams.org/news/2017/10/03/after-overwhelming-public-opposition-scotland-announces-fracking-ban.
245. "Groningen is being 'choked to death' by government neglect," Dutch News, February 19, 2023. www.dutchnews.nl/features/2023/02/groningen-is-being-choked-to-death-by-government-neglect/.
246. "Green light for more gas extraction in Drenthe, Groningen," DutchNews, November 15, 2018. www.dutchnews.nl/news/2018/11/green-light-for-more-gas-extraction-in-drenthe-groningen/.
247. Felcenloben, P., "Groningen gas: 80% of the homes in this village are being demolished," Dutch News, June 14, 2021. www.dutchnews.nl/features/2021/06/between-rubble-and-rebirth-overschild-residents-greet-unequal-rebuild-with-frustration/.
248. US Energy Information Administration, "Shale Gas Production," December 30, 2022. www.eia.gov/dnav/ng/ng_prod_shalegas_s1_a.htm.
249. Rapier, R., "The irony of President Obama's oil legacy," *Forbes*, January 15, 2016. www.forbes.com/sites/rrapier/2016/01/15/president-obamas-petroleum-legacy/#24d6c012c10f.
250. Nussbaum, A. and Steel, A., "Shale billionaire Hamm slams 'exaggerated' US oil projections," BloombergMarkets, September 21, 2017. www.bloomberg.com/

news/articles/2017-09-21/shale-billionaire-hamm-slams-exaggerated-u-s-oil-projections.

251. *Age of Paradox: Exploring the uncertain world of energy 2000–2020*, Clingendael International Energy Programme, The Hague, 2011.
252. "Shattered Ground."
253. "Shattered Ground."
254. Lebel, E. C., Finnegan, C. J., Ouyang, Z., and Jackson, R. B., "Methane and NOx emissions from natural gas stoves, cooktops, and ovens in residential homes," *Environmental Science and Technology*, 56(4), 2529–2539, January 27, 2022. https://doi.org/10.1021/acs.est.1c04707.
255. Myers, A., "Methane leaks are far worse than estimates, at least in New Mexico, but there's hope," *Stanford Earth Matters*, March 24, 2022. https://earth.stanford.edu/news/methane-leaks-are-far-worse-estimates-least-new-mexico-theres-hope.
256. Krauss, C., "Industry grapples with gas leaks," *The New York Times*, July 13, 2016.
257. Helm, D., *The Carbon Crunch: Revised and updated*, p. 212, Yale University Press, New Haven, CT, 2015.
258. "Quantum-enabled camera detects methane leaks," *Photonics Spectra*, November 2021.
259. Harvey, F., "'Golden age of gas' threatens renewable energy, IEA warns," *The Guardian*, May 29, 2012. www.theguardian.com/environment/2012/may/29/gas-boom-renewables-agency-warns.
260. Gerken, J., "Exxon CEO Rex Tillerson sues to block water tower that might supply fracking operations," February 26, 2014, *Huffington Post*. www.huffingtonpost.com/2014/02/21/exxon-ceo-rex-tillerson-lawsuit_n_4833185.html.
261. Jamail, D., "Could COVID-19 spell the end of the fracking industry as we know it?," *Truth Out*, March 20, 2020. https://truthout.org/articles/could-covid-19-spell-the-end-of-the-fracking-industry-as-we-know-it/.
262. Yergin, *The Quest*, p. 325.
263. Reed, S., "Liquid gas makes Qatar an energy giant," *The New York Times*, August 6, 2015.
264. Smyth, J., "Shell takes $14bn gamble on floating LNG," *The Financial Times*, August 18, 2017.
265. Schobert, *Energy and Society*, p. 528.
266. Lee, M., "'Hubris': LNG plant officials saw trouble days before blast," *E&E News*, November 1, 2022. www.eenews.net/articles/hubris-lng-plant-officials-saw-trouble-days-before-blast/.
267. Horn, S. and Frank, J., "Trump admin quietly pushing 'small scale' LNG exports that avoid environmental reviews," *CounterPunch*, September 8, 2017. www.counterpunch.org/2017/09/08/trump-admin-quietly-pushing-small-scale-lng-exports-that-avoid-environmental-reviews/.
268. "Interview with Tellurian CEO Meg Gentle," *Oil & Gas 360*, March 16, 2017. www.oilandgas360.com/oil-gas-360-exclusive-interview-tellurian-ceo-meg-gentle/.
269. Meyer, G., "Climate fears stoke natural gas opposition," *The Financial Times*, February 7, 2020.
270. Rhodes, *Energy: A human history*, p. 262.
271. Rhodes, *Energy: A human history*, pp. 266–271.
272. Paterson, L. and Wirfs-Brock, J., "Protesters say pipelines are dangerous. Are they?" *Inside Energy*, November 18, 2016. http://insideenergy.org/2016/11/18/protesters-say-pipelines-are-dangerous-are-they/.

273. Renshaw, J. and Kumar, D. K., "Technology designed to detect US energy pipeline leaks often fails," Reuters, September 30, 2016. www.reuters.com/article/us-usa-pipelines-colonial-analysis/technology-designed-to-detect-u-s-energy-pipeline-leaks-often-fails-idUSKCN1200FQ.

274. Stevenson, M., "Death toll climbs to at least 85 in Mexico fuel pipeline explosion," *USA Today*, January 20, 2019. https://eu.usatoday.com/story/news/world/2019/01/20/mexico-pipeline-explosion-death-toll-rises/2634734002/.

275. McGowan, E. and Song, L., "The dilbit disaster: Inside the biggest oil spill you've never heard of, Part 1," *InsideClimate News*. https://insideclimatenews.org/news/20120626/dilbit-diluted-bitumen-enbridge-kalamazoo-river-marshall-michigan-oil-spill-6b-pipeline-epa.

276. Parrish, W., "The fires of Standing Rock: How a new resistance movement was ignited," *CounterPunch*, December 30, 2016. www.counterpunch.org/2016/12/30/the-fires-of-standing-rock-how-a-new-resistance-movement-was-ignited/.

277. Joseph, G., "30 years of oil and gas pipeline accidents, mapped," *CityLab*, November 30, 2016. www.citylab.com/environment/2016/11/30-years-of-pipe line-accidents-mapped/509066/.

278. Cohen, M., "Portion of Keystone Pipeline shut down after 380,000-gallon oil leak in North Dakota," *USA Today*, November 1, 2019. https://eu.usatoday.com/story/news/nation/2019/11/01/keystone-pipeline-leak-oil-spilled-north-dakota/4121954002/.

279. Cardwell, D., "Oil and gas glut creates transport headache," *The New York Times*, October 10, 2013.

280. Mikulka, J., "What have we learned from the Lac-Megantic oil train disaster?" *Desmog*, December 21, 2016. www.desmogblog.com/2016/12/21/what-have-we-learned-lac-megantic-oil-train-disaster.

281. "Lessons learned in the Lac-Mégantic railway disaster," 2015 Conference and Expo, National Fire Protection Association, Chicago, June 24, 2015.

282. Quenneville, G., Seglins, D., and Loiero, J., "Why crude oil trains keep derailing and exploding in Canada – even after the Lac-Mégantic disaster," June 15, 2020, CBC News. www.cbc.ca/news/canada/saskatoon/lac-megantic-crude-oil-train-canada-guernsey-saskatchewan-rail-1.5608769.

283. Vigo, J., "Tar-ma is a bitch!: The real tragedy of Fort McMurray," *CounterPunch*, May 9, 2016. www.counterpunch.org/2016/05/09/tar-ma-is-a-bitch-the-real-tra gedy-of-fort-mcmurray/.

284. Muru, T., "Canada's 'dirty oil' climate change dilemma," *BBC HARDtalk*, August 22, 2016. www.bbc.com/news/world-us-canada–37094763.

285. "Pipe Dreams" [documentary], directed by Leslie Iwerks, Leslie Iwerks Productions, USA, 2011.

286. "Pipe Dreams."

287. Nikiforuk, A., "David Schindler, the scientific giant who defended fresh water," *The Tyee*, March 9, 2021. https://thetyee.ca/News/2021/03/09/David-Schindler-Scientific-Giant-Fresh-Water-Defender/.

288. Langenbrunner, B., "Crude awakening: Oil industry pursuing massive build-out of new pipelines, led by projects in US, India, China, Russia," Global Energy Monitor, September 2022. https://globalenergymonitor.org/report/crude-awaken ing-oil-industry-pursuing-massive-build-out-of-new-pipelines-led-by-projects-in-u-s-india-china-russia/.

289. Volume is often given in tonnes, but barrels and gallons may be more intuitive. Recall there are 42 gallons of oil in a barrel and 7.15 barrels in a metric ton (tonne).

290. Freudenburg and Gramling, *Blowout in the Gulf*, p. 171.

291. Pokharel, S. S., Bishop, G. A., and Stedman, D. H., "An on-road motor vehicle emissions inventory for Denver: An efficient alternative to modelling," *Atmospheric Environment*, 36(33), 5177–5184, 2002.

292. World Health Organization, "WHO global urban ambient air pollution database (update 2016)," 2018. www.who.int/phe/health_topics/outdoorair/databases/cit ies/en/.

293. Vidal, J., "Air pollution rising at an 'alarming rate' in world's cities," *The Guardian*, May 12, 2016. www.theguardian.com/environment/2016/may/12/air-pollution-rising-at-an-alarming-rate-in-worlds-cities.

294. Wallace-Wells, D., *The Uninhabitable Earth: A story of the future*, p. 102, Penguin Random House, London, 2019.

295. US Environmental Protection Agency, "EPA Sets Tier 3 Motor Vehicle Emission and Fuel Standards," EPA-420-F-14-009, Office of Transportation and Air Quality, March 2014.

296. Carrington, D., "Health effects of diesel 'cost European taxpayers billions'," *The Guardian*, November 27, 2018. www.theguardian.com/environment/2018/nov/27/health-effects-of-diesel-cost-european-taxpayers-billions.

297. Carson, R., *Silent Spring*, p. 85, Houghton Mifflin Company, New York, NY, 2002 (first published 1962).

298. Leopold, A., *A Sand County Almanac and Sketched Here and There*, p. 172, Penguin Books, London, 2020 (first published 1949).

299. Weart, S. R., *The Discovery of Global Warming*, 2nd edition, Harvard University Press, Harvard, MA, 2008.

300. Williams, D. R., "Earth fact sheet: Terrestrial atmosphere," NASA Goddard Space Flight Center, December 21, 2021. http://nssdc.gsfc.nasa.gov/planetary/factsheet/earthfact.html.

301. "Carbon Cycle Greenhouse Gases," Global Monitoring Laboratory, US National Oceanic & Atmospheric Administration. https://gml.noaa.gov/ccgg/trends/.

302. Segrè, G., *A Matter of Degrees*, p. 116, Penguin Books, New York, NY, 2002.

303. Walker, G. and King, D., *The hot Topic: What we can do about global warming*, p. 11, Harcourt Inc., Orlando, FL, 2008.

304. Loeb, N. G., Johnson, G. C., Thorsen, T. J., et al., "Satellite and ocean data reveal marked increase in earth's heating rate," *Geophysical Research Letters*, June 15, 2021. https://doi.org/10.1029/2021GL093047.

305. Walker and King, *The Hot Topic*, pp. 10–11.

306. Molena, F., "Remarkable weather of 1911," *Popular Mechanics*, pp. 339–342, March 1912.

307. US Environmental Protection Agency, "Global Greenhouse Gas Emissions Data," 2014. www.epa.gov/ghgemissions/global-greenhouse-gas-emissions-data (based on the IPCC Fifth Assessment Report, 2014).

308. Carbon dioxide is stored across the Earth in reservoirs of "biogenic carbon," for example, in chalk cliffs (compressed calcium carbonate), surface limestone, and of course in oceans, soil, and plants and trees, without which the temperature of the Earth would be too high for human habitation (the current average terrestrial temperature is 15°C).

309. Gillis, J., "Beyond conference, an ambitious path to cutting emissions," *The New York Times*, December 1, 2015.

310. Page, B., Turan, G., and Zapantis, A., "Global Status of CCS 2020," Global CCS Institute, November 2020.

311. If the carbon is generated from biomass, the process is a bioenergy carbon capture system (BECCS).

312. Goodall, C., *Ten Technologies to Save the Planet*, p. 193, *Green*Profile, London, 2008.
313. Goodall, *Ten Technologies to Save the Planet*, p. 193.
314. Boffey, D. "Empty North Sea gas fields to be used to bury 10m tonnes of CO_2," *The Guardian*, May 9, 2019. www.theguardian.com/environment/2019/may/09/empty-north-sea-gas-fields-bury-10m-tonnes-c02-eu-ports.
315. Friends of the Earth Scotland, "Report: Fossil fuel carbon capture & storage," January 11, 2021. https://foe.scot/resource/report-carbon-capture-storage-energy-role/.
316. Gayle, D., "Carbon capture is not a solution to net zero emissions plans, report says," *The Guardian*, September 1, 2022. www.theguardian.com/environment/2022/sep/01/carbon-capture-is-not-a-solution-to-net-zero-emissions-plans-report-says.
317. Shell Global, "Shell to construct world's first oil sands carbon capture and storage (CCS) project," September 5, 2012. www.shell.com/media/news-and-media-releases/2012/quest-first-oil-sands-ccs-project-05092012.html.
318. Bakx, K., "Shell unveils new carbon capture project amid wave of new CCS proposals in Alberta," CBC News, July 13, 2021. www.cbc.ca/news/business/shell-carbon-capture-alberta-government-1.6099797.
319. Bakx, "Shell unveils new carbon capture project."
320. US Energy Information Administration, "Petra Nova is one of two carbon capture and sequestration power plants in the world," October 31, 2017. www.eia.gov/todayinenergy/detail.php?id=33552#.
321. Griffith, S., *Electrify: An optimist's playbook for our clean energy future*, p. 192, The MIT Press, Cambridge, MA, 2021.
322. Kusnetz, N., "In a bid to save its coal industry, Wyoming has become a test case for carbon capture, but utilities are balking at the pricetag," *Inside Climate News*, May 29, 2022. https://insideclimatenews.org/news/29052022/coal-carbon-capture-wyoming/.
323. Beuttler, C., Charles, L., and Wurzbacher, J., "The role of direct air capture in mitigation of anthropogenic greenhouse gas emissions," *Frontiers in Climate*, 1 (Article 1), November 7, 2019.
324. Mundy, S., "Journey to the climate frontlines," p. 30, *The Financial Times*, October 30, 2021.
325. Gates, B., *How to Avoid a Climate Disaster: The solutions we have and the breakthroughs we need*, pp. 63–64, Allen Lane, London, 2021.
326. Hunziker, R., "Direct air capture and big oil," *CounterPunch*, March 12, 2021. www.counterpunch.org/2021/03/12/direct-air-capture-and-big-oil/.
327. Hazen, R. M., *Symphony in C: Carbon and the evolution of (almost) everything*, p. 126, William Collins, London, 2019.
328. Fox, D., "The carbon rocks of Oman," *Scientific American*, July 2021.
329. Fox, "The carbon rocks of Oman."
330. Mundy, "Journey to the climate frontlines."
331. Blain, L., "MIT team makes a case for direct carbon capture from seawater, not air," *New Atlas*, February 17, 2023. https://newatlas.com/environment/mit-carbon-capture-seawater/.
332. "Coal in Kentucky" [documentary], directed by Stephen Bailey, University of Kentucky, 2009.
333. "State of Climate Action 2022," Climate Action Tracker, October 26, 2022. https://climateactiontracker.org/publications/state-of-climate-action–2022/.

334. "Pittsburgh to Paris" [documentary], directed by Sidney Beaumont and Michael Bonfiglio, National Geographic, 2018.
335. Harvey, C. and House, K., "Every dollar spent on this climate technology is a waste," *The New York Times*, August 16, 2022. www.nytimes.com/2022/08/16/opinion/climate-inflation-reduction-act.html.
336. Freudenburg and Gramling, *Blowout in the Gulf*, p. 70.
337. Freudenburg and Gramling, *Blowout in the Gulf*, p. 83.
338. Hubbert, M. K., "Nuclear energy and the fossil fuels," p. 8, presented at the spring meeting of the Southern District Division of Production, American Petroleum Institute, San Antonio, TX, March 1956.
339. Hubbert, "Nuclear energy and the fossil fuels," p. 26.
340. Fanchi, J. R., *Energy in the 21st Century*, p. 221, World Scientific Publishing Co., Singapore, 2005.
341. Others have used the reserve-to-production ratio (R/P or RPR), but consumption determines how fast we deplete something.
342. Maass, *Crude World*, p. 17.
343. Hubbert, "Nuclear energy and the fossil fuels," p. 36.
344. Mann, C. C., *The Wizard and the Prophet: Science and the future of our planet*, p. 281, Picador, London, 2019.
345. Goodstein, D., *Out of Gas: The end of the age of oil*, pp. 126–129, W. W. Norton & Company, New York, NY, 2004.
346. Goodstein, *Out of Gas*, p. 130.
347. Goodstein, *Out of Gas*, pp. 130–131.
348. Yergin, *The Quest*, pp. 231–235.
349. Freudenburg and Gramling, *Blowout in the Gulf*, p. 173.
350. "Mexico oil production falls again, raising questions about Pemex's plans," *Bnamericas*, November 26, 2020. www.bnamericas.com/en/news/mexico-oil-production-falls-again-raising-questions-about-pemexs-plans.
351. Sage, A., "Taxpayer at risk of rise in North Sea clear-up bill," *The Times*, January 25, 2019.
352. Wallace-Wells, D., "Going deeper into the unknown," *The New York Times*, January 15, 2011.
353. Frank, J. "Will the frackers go bust?" *CounterPunch*, 22(1), 2015.
354. Bonneuil and Fressoz, *The Shock of the Anthropocene*, pp. 123–124.
355. Sheppard, D., "Field discovery lifts hope of revival in Gulf state's output," *The Financial Times*, October 4, 2018.
356. Smil, *Energy and Civilization*, p. 278.
357. Maass, *Crude World*, p. 53.

3 The Nuclear World: Atoms for Peace

1. In the past, the percentage U-235 in natural uranium was greater because of the different radioactive half-lives: U-235 ($t_{1/2}$ = 700 million years) and U-238 ($t_{1/2}$ = 4.5 billion years). Two billion years ago, U-235 comprised about 3.5% of natural uranium.
2. Richtmyer, F. K., Kennard, E. H., and Cooper, J. N., *Introduction to Modern Physics*, p. 748, 6th edition, Tata McGraw-Hill, New Delhi, 1997.

3. The nuclear "cross-section" relates to the probability of a reaction and is measured in barns (1×10^{-28} m^2). One barn (b) is about the size of a uranium nucleus, the extent of the nuclear force.
4. Pitchblende comes from the German for *Pech* and *Blende*, meaning *bad luck mineral*.
5. Schobert, H. H., *Energy and Society: An introduction*, p. 387, Taylor & Francis, New York, NY, 2002.
6. Thompson initially called electrons "corpuscles" before they were shown to be the same as the charge carriers in an electric current (that propagate as a traveling EM wave) and high-energy beta rays in nuclear decay.
7. The smashing produced unstable beryllium that then broke apart into two helium atoms plus a lot of energy: $_3\text{Li}^7 + _1\text{H}^1 \rightarrow _4\text{Be}^{8}* \rightarrow _2\text{He}^4 + _2\text{He}^4 + 17.2$ MeV.
8. Richtmyer et al., *Introduction to Modern Physics*, p. 693.
9. Invented in 1908 by the German physicist Hans Geiger, the initial Geiger counter only detected alpha particles. From 1928, Geiger counters fitted with a Geiger-Muller tube detected alpha, beta, and gamma radiation. Geiger worked in Rutherford's Cambridge lab and conducted the famous gold-foil experiment with Ernest Marsden, from which Rutherford formulated the structure of the atom.
10. Pais, A., *Inward Bound: Of matter and forces in the physical world*, p. 436, Clarendon Press, Oxford, 1986.
11. Isaacson, W., *Einstein: His life and universe*, p. 469, Simon & Schuster, London, 2007.
12. Originally called ausonium and hesperium by the Italian researchers, elements 93 and 94 were eventually named neptunium in 1940 and plutonium in 1942, following the order of planets out from uranium (element 92).
13. Kragh, H., *Quantum Generations: The history of physics in the twentieth century*, p. 260, Princeton University Press, Princeton, NJ, 1999.
14. Kragh, *Quantum Generations*, p. 261.
15. Primary neutrons are generated in the original splitting, while secondary neutrons are created after the original fission by decaying fission products.
16. The famous letter was initiated by Szilard during two hurried meetings at Einstein's summer rental home on Long Island. Based on two drafts dictated in German by Einstein, the first was translated by Wigner and the second by Teller. Because of delays by the proposed courier – a friend of Roosevelt's – the letter wasn't delivered until October 11. The Manhattan Project started up in earnest two years later.
17. Bizony, P., *Atom*, pp. 115–116, Icon Books, London, 2017.
18. Isaacson, *Einstein*, pp. 480–481.
19. Clynes, T., *The Boy Who Played with Fusion*, p. 144, Faber and Faber, London, 2015.
20. Designing a plutonium bomb (via Pu-239) was deemed important because only enough U-235 could be produced for one uranium bomb (separating out weapons-grade U-235 at Oak Ridge was difficult). The implosion detonation method was then considered to avoid predetonation in a plutonium gun-design because critical mass could not be achieved fast enough without removing the high spontaneous-fission Pu-240 generated along with Pu-239 in the Hanford reactor (a difficult and time-consuming process). G Division worked on the U-235 gun Gadget and X Division on the Pu-239 eXplosive-lens.
21. Baggott, J., *Atomic: The first war of physics and the secret history of the atomic bomb*, p. 285, Icon Books, London, 2019.

22. According to Oppenheimer, the name Trinity was inspired by the English poet John Donne's Holy Sonnet XIV ("Batter My Heart, Three-Person'd God"), whose force could "break, blow, burn."

23. Kragh, *Quantum Generations*, p. 270.

24. Oppenheimer, J. R., "NBC white paper. The decision to drop the bomb," *National Broadcasting Company*, New York, 1965.

25. Baggott, *Atomic*, pp. 249–250.

26. Baggott, *Atomic*, p. 230.

27. Segrè, E., *Enrico Fermi: Physicist*, p. 159, The University of Chicago Press, Chicago, IL, 1970.

28. Isaacson, *Einstein*, p. 485.

29. Baggott, *Atomic*, p. 295.

30. Rhodes, R., *Energy: A human history*, p. 319, Simon and Schuster, New York, NY, 2018.

31. Baggott, *Atomic*, p. 456.

32. Baggott, *Atomic*, p. 460.

33. Smil, V., *Energy and Civilization: A history*, p. 374, The MIT Press, Cambridge, MA, 2017.

34. McNaught, K. W., Saywell, J. T., and Ricker, J. C., *Manifest Destiny: A short history of the United States*, p. 313, Clarke, Irwin & Company Ltd., Toronto, 1980.

35. After the 33,000-foot, mid-air collision between the nuclear-armed B-52 and KC-135 refueling plane, two bombs remained largely intact, while two broke apart after their TNT detonators exploded on ground impact, spreading plutonium dust to the northeast. The search for the four bombs in early 1966 occupied a worrying two and half months in the region.

36. Faus, J. and González, M., "Washington and Madrid to seal deal over 1966 Palomares nuclear accident," *El País*, October 13, 2015. https://english.elpais.com/elpais/2015/10/12/inenglish/1444641825_987132.html.

37. Clynes, *The Boy Who Played with Fusion*, pp. 215–216.

38. Smil, *Energy and Civilization*, p. 378.

39. Baggott, *Atomic*, p. 472.

40. Smil, *Energy and Civilization*, p. 378.

41. Haynes, R. D., *From Faust to Strangelove: Representations of the scientist in Western literature*, p. 303, The Johns Hopkins University Press, Baltimore, MD, 1994.

42. The Pu-239 cores for the Trinity and Nagasaki Fat Man bombs would be produced at the Hanford reactor from the decay of U-239, separated out from the spent fuel reaction products: $_0n^1 + _{92}U^{238} \rightarrow _{92}U^{239} \rightarrow _{93}Np^{239} + _{-1}\beta^0 \rightarrow _{94}Pu^{239} + _{-1}\beta^0$.

43. At Columbia, Fermi had constructed an intermediate pile that did not reach a critical dimension to sustain a chain reaction. The Chicago pile was improved by reducing the parasitic absorption impurities in all materials. Fermi also attained criticality in CP-1 prior to the December 2 event officially witnessed by the upper brass.

44. Fermi, L., *Atoms in the Family*, University of Chicago Press, Chicago, IL, 1954.

45. Segrè, *Enrico Fermi*, p. 129.

46. Kragh, *Quantum Generations*, p. 266.

47. Baggott, *Atomic*, pp. 157, 282.

48. Fermi, E., "Discovery of fission," *American Institute of Physics*, 1952. https://history.aip.org/history/exhibits/mod/fission/fission1/10.html.

49. Kragh, *Quantum Generations*, p. 285.

50. Rhodes, *Energy: A human history*, p. 281.

51. Rhodes, *Energy: A human history*, p. 282.

52. Rhodes, *Energy: A human history*, p. 290.
53. Jacobs, R., "Nuclear Stockholm Syndrome," *CounterPunch*, July 9, 2021. www .counterpunch.org/2021/07/09/nuclear-stockholm-syndrome/.
54. Rhodes, *Energy: A human history*, p. 287.
55. Rhodes, *Energy: A human history*, p. 290.
56. Schobert, H. H., *Energy and Society: An introduction*, p. 396, Taylor & Francis, New York, NY, 2002.
57. Yergin, D., *The Prize: The epic quest for oil, money, & power*, p. 378, Simon & Schuster, New York, NY, 1991.
58. Schobert, *Energy and Society*, p. 395.
59. Hubbert, M. K., "Nuclear energy and the fossil fuels," p. 31, presented at the spring meeting of the Southern District Division of Production, *American Petroleum Institute*, San Antonio, TX, March 1956.
60. "World Nuclear Power Reactors & Uranium Requirements," *World Nuclear Association*, November 2022. www.world-nuclear.org/information-library/facts-and-figures/world-nuclear-power-reactors-and-uranium-requireme.aspx.
61. St. Clair, J., "Fukushima mon amour," Roaming Charges, *CounterPunch*, 22(9): 5, 2015.
62. Bonneuil, C. and Fressoz, J.-B. (trans. Fernbach, D.), *The Shock of the Anthropocene*, p. 132, Verso, London, 2017.
63. Johnstone, B., *Switching to Solar: What we can learn from Germany's success in harnessing clean energy*, p. 33, Prometheus Books, New York, NY, 2011.
64. Bizony, *Atom*, p. 115.
65. Boese, A., *Electrified Sheep: And other bizarre experiments*, pp. 137–139, Pan Books, London, 2017.
66. Boese, *Electrified Sheep*, p. 129.
67. "World Nuclear Power Reactors & Uranium Requirements."
68. "Plans For New Reactors Worldwide," *World Nuclear Association*, December 2022. www.world-nuclear.org/information-library/current-and-future-generation/plans-for-new-reactors-worldwide.aspx.
69. LaForge, J., "Nuclear power: Dead in the water it poisoned," *CounterPunch*, 21(2), 2014.
70. Hernandez, A., "Europe's sputtering nuclear renaissance," *Politico*, March 11, 2021. www.politico.eu/article/europes-sputtering-nuclear-renaissance/amp/.
71. Gundersen, A., "An open letter to Bill Gates about his Wyoming atomic reactor," *CounterPunch*, August 20, 2021. www.counterpunch.org/2021/08/20/an-open-letter-to-bill-gates-about-his-wyoming-atomic-reactor/.
72. Mallapaty, S., "China prepares to test thorium-fuelled nuclear reactor," *Nature*, September 9, 2021. www.nature.com/articles/d41586-021-02459-w.
73. McCombs, B. and Gruver, M., "In tiny Wyoming town, Bill Gates bets big on nuclear power," *AP News*, January 18, 2022. https://apnews.com/article/climate-technology-business-wyoming-bill-gates-19a36eb0bd65e0999d26c0cc122f6158.
74. Kramer, A. E., "Russia tests a climate innovation: Nuclear-powered showers," *New York Times*, November 5, 2021.
75. Green, J., "Small modular reactor rhetoric hits a hurdle," *Renew Economy*, June 23, 2020. https://reneweconomy.com.au/small-modular-reactor-rhetoric-hits-a-hurdle–62196.
76. Rhodes, *Energy: A human history*, p. 289.
77. Hubbert, M. K., "Nuclear energy and the fossil fuels," p. 36.
78. Schobert, *Energy and Society*, p. 376.

79. Radioactive carbon (C-14) has a half-life of 5,730 years and can be used to date organic material from time of death by comparing the ratio of C-14 to C-12. Carbon-14 is created by neutron capture in atmospheric nitrogen (N-14 + n → C-14 + p) and comprises roughly 1 part in 1 trillion in continuously replenished living material (increased somewhat by atomic testing). After death, the uptake of carbon ceases and the amount of still decaying C-14 can be compared to the amount of regular carbon (C-12) in a sample to determine its age since death. The time range of "carbon-dating" is limited to about 40,000 years, because there are no longer enough C-14 atoms left after seven half-lives. To go back further in time, other radioactive dating methods are needed, such as potassium–argon and uranium–lead dating that can date materials back billions of years in the past.

80. To break a chemical bond in most materials takes about 3 eV of energy, roughly the violet edge of the visible spectrum (380 nm = 3.26 eV), and so EM radiation that is more energetic than the visible range is considered "ionizing." To convert from nanometers (nm) to electron-volts (eV): 1 eV = (hc/e)/λ, where h is Planck's constant, c is the speed of light, e is the fundamental charge on the electron, and λ is the wavelength in nm. Plugging in the numbers, 1 eV = 1240/λ [nm], thus 1 eV = 1,240 nm (in the IR) and 3 eV = 413 nm (violet).

81. Schobert, *Energy and Society*, p. 405.

82. Schobert, *Energy and Society*, p. 406.

83. Curie, P., "Radioactive substances, especially radium," *Nobel Lecture*, June 6, 1905. www.nobelprize.org/uploads/2018/06/pierre-curie-lecture.pdf.

84. Sanger, P., *Blind Faith: The nuclear industry in one small town*, p. 26, McGraw-Hill Ryerson, Toronto, 1981.

85. Sanger, *Blind Faith*, pp. 109, 138.

86. Marra, J. F., *Hot Carbon: Carbon-14 and a revolution in science*, pp. 42–44, Columbia University Press, New York, NY, 2019.

87. Wellerstein, A., "Counting the dead at Hiroshima and Nagasaki," *Bulletin of the Atomic Scientists*, August 4, 2020. https://thebulletin.org/2020/08/counting-the-dead-at-hiroshima-and-nagasaki/.

88. Sutou, S., "Black rain in Hiroshima: A critique to the life span study of A-bomb survivors, basis of the linear no-threshold model," *Genes and Environment*, 42(1), 2020. https://doi.org/10.1186/s41021-019-0141-8.

89. "US Congress, Senate, Special Committee on Atomic Energy," 79th Congress, Washington, DC, November 1945.

90. The ratio of strontium isotopes ^{87}Sr/^{86}Sr in bones and teeth can indicate the location of one's diet and thus the migratory patterns of early humans.

91. Schobert, *Energy and Society*, p. 412.

92. LaForge, "Chernobyl, and cesium, at 30," *CounterPunch*, April 22, 2016.

93. Clynes, *The Boy Who Played with Fusion*, pp. 246, 248.

94. Berman, B., *Zapped: From infrared to X-rays, the curious history of invisible light*, p. 130, Oneworld Publications, London, 2017.

95. Sanger, *Blind Faith*, p. 106.

96. Sanger, *Blind Faith*, p. 77.

97. Rhodes, *Energy: A human history*, pp. 322–323.

98. Rhodes, *Energy: A human history*, p. 320.

99. The amount of radioactive decay is inversely related to its half-life: Ra-226 is about 16 times more active than Pu-239, and so its half-life is about 16 times shorter (Ra-226 $t_{1/2}$ = 1,600 years, Pu-239 $t_{1/2}$ = 24,000 years).

100. "Personal Annual Radiation Dose Calculator," *US Nuclear Regulatory Commission*, October 2, 2017. www.nrc.gov/about-nrc/radiation/around-us/calculator.html.
101. "Green Warriors: South Africa, Toxic Townships" [documentary], directed by Martin Boudot, *Premières Lignes Télévision 3*, 2018.
102. "Leak First, Fix Later: Uncontrolled and Unmonitored Radioactive Releases from Nuclear Power Plants," *Beyond Nuclear*, 2010. https://archive.beyondnuclear.org/reports/.
103. Gunter, P., "Atomic power's silent spills," *CounterPunch*, 22(6): 10–13, 2015.
104. Schobert, *Energy and Society*, p. 414.
105. Biello, D., "Spent nuclear fuel: A trash heap deadly for 250,000 years or a renewable energy source?" *Scientific American*, January 28, 2009.
106. Sanger, *Blind Faith*, pp. 20, 23.
107. The uranium mines were in the Ore Mountains of Bohemia that border Germany and the Czech Republic near the spa town of Joachimsthal (renamed Jáchymov after World War II). The site was also a major silver mine in the sixteen century where *joachimsthaler* coins were minted, later called a *thaler* from which we get the word *dollar*.
108. Pitkanen, L. L., *"A hot commodity: Uranium and containment in the nuclear state,"* p. 301, PhD thesis, Department of Geography, University of Toronto, 2014.
109. "Old reactors," *No2NuclearPower*, November 3, 2012. www.no2nuclearpower.org.uk/old-reactors/.
110. Lawson, T., *Crazy Caverns: How one small community challenged a technocrat juggernaut and won*, pp. 12, 29, Port Hope, Ontario, 2013.
111. Lawson, *Crazy Caverns*, p. 55.
112. Lawson, *Crazy Caverns*, pp. 54, 61.
113. Aulakh, R., "Port Hope's nuclear past pits economic interests against health," *The Toronto Star*, April 1, 2011.
114. Aulakh, "Port Hope's nuclear past pits economic interests against health."
115. Lane, R. S. D, Frost, S. E., Howe, G. R., and Zablotska, L. B., "Mortality (1950–1999) and cancer incidence (1969–1999) in the cohort of Eldorado uranium workers," *Radiation Research*, 174(6): 773–785, 2010.
116. Chen, J., Moir, D., Lane, R., and Thompson, P., "An ecological study of cancer incidence in Port Hope, Ontario from 1992 to 2007," *Journal of Radiological Protection*, 33(1): 227–242, 2013.
117. Nelson, J., "Radioactive folly at Lake Huron," *CounterPunch*, 22(8), 2015.
118. Nelson, J., "Saugeen Ojibway Nation has saved Lake Huron from a nuclear waste dump," *CounterPunch*, February 4, 2020. www.counterpunch.org/2020/02/04/saugeen-ojibway-nation-has-saved-lake-huron-from-a-nuclear-waste-dump/.
119. Butler, C., "30,000 shipments of nuclear waste would move through Ontario cities, farmland under draft plan," *CBC News*, January 18, 2022. www.cbc.ca/news/canada/london/nuclear-waste-transport-ontario-1.6309561.
120. Lunau, K. and Oberhaus, D., "The plan to build a million-year nuclear waste dump on the Great Lakes," *Motherboard*, May 2, 2017. https://motherboard.vice.com/en_us/article/ez3k5e/kincardine-deep-geological-repository-opg-wipp-new-mexico-nuclear-waste-bruce-power.
121. Black, R., "Finland buries its nuclear past," *BBC News*, April 27, 2006. http://news.bbc.co.uk/2/hi/science/nature/4948378.stm.
122. Helm, D., *The Carbon Crunch: Revised and updated*, p. 127, Yale University Press, New Haven, CT, 2015.

123. Pearce, F., "Shocking state of world's riskiest nuclear waste site," *New Scientist*, January 21, 2015. www.newscientist.com/article/mg22530053-800-shocking-state-of-worlds-riskiest-nuclear-waste-site/.

124. Pearce, "Shocking state of world's riskiest nuclear waste site."

125. Vaughan, A., "Price tag rises for UK's planned nuclear waste facility," *New Scientist*, March 5, 2022.

126. Walker, S., "Russia begins cleaning up the Soviets' top-secret nuclear waste dump," *The Guardian*, July 2, 2017.

127. "US Nuclear Power Policy," *World Nuclear Association*, November 2017. www.world-nuclear.org/information-library/country-profiles/countries-t-z/usa-nuclear-power-policy.aspx.

128. Lunau and Oberhaus, "The plan to build a million-year nuclear waste dump on the Great Lakes."

129. Hamblin, J. D., "Nuclear power and promise" (Review of *Confessions of a Rogue Nuclear Regulator*)," *Science*, January 19, 2019.

130. Flowers, B., "Nuclear power and the environment," p. 131, *Sixth Report of the Royal Commission on Environmental Pollution*, London, 1976.

131. Rhodes, R., "Living with the bomb," *National Geographic*, August 2005.

132. Wakeford, R., "A double diamond anniversary – Kyshtym and Windscale: The nuclear accidents of 1957," *Journal of Radiological Protection*, 37: E7, 2017.

133. "Safety of Nuclear Power Reactors," *World Nuclear Association*, May 2016. www.world-nuclear.org/information-library/safety-and-security/safety-of-plants/safety-of-nuclear-power-reactors.aspx.

134. Wakeford, "A double diamond anniversary."

135. Schobert, *Energy and Society*, p. 419.

136. The generally accepted Ukrainian spelling of the Ukraine capital Kyiv is used rather than the russified Kiev, as are the English spellings of Chernobyl and Pripyat. Kyiv, Chornobyl, and Prypiat are Ukrainian.

137. Schobert, *Energy and Society*, pp. 422–423.

138. "Chapter II, The release, dispersion and deposition of radionuclides" in Chernobyl: Assessment of Radiological and Health Impact, *Nuclear Energy Agency*, 2002. www.oecd-nea.org/rp/chernobyl/c02.html.

139. Fairlie, I. and Sumner, D., *"The other report on Chernobyl (TORCH),"* Commissioned by the Greens/EFA in the European Parliament, Berlin, Brussels, Kyiv, April 2006.

140. Plokhy, S., *Chernobyl: History of a tragedy*, p. 162, Penguin, London, 2019.

141. Fairlie and Sumner, *"The other report on Chernobyl (TORCH)."*

142. Fairlie and Sumner, *"The other report on Chernobyl (TORCH)."*

143. Plokhy, *Chernobyl*, p. xii.

144. LaForge, "Chernobyl, and cesium, at 30."

145. Plokhy, *Chernobyl*, p. 345.

146. Fairlie and Sumner, *"The other report on Chernobyl (TORCH)."*

147. LaForge, "Chernobyl, and cesium, at 30," *CounterPunch*, April 22, 2016. www.counterpunch.org/2016/04/22/chernobyl-and-cesium-at–30/.

148. Plokhy, *Chernobyl*, p. 339.

149. Hjelmgaard, K., "Chernobyl's legacy: Kids with bodies ravaged by disaster," *USA Today*, April 17, 2016. www.usatoday.com/story/news/world/2016/04/17/clinic-ukraine-chernobyl-30th-anniversary-health-impact/82892592/.

150. Plokhy, *Chernobyl*, pp. xii–xii.

151. Plokhy, *Chernobyl*, p. xv.

152. "The Fukushima Daiichi Accident: Technical Volume 1/5 (Description and Context of the Accident)," pp. 152–154, International Atomic Energy Agency, Vienna, 2015.
153. "A decade after Fukushima, Japanese towns find it difficult to bounce back, despite incentives," *Global Times*, March 11, 2021. www.globaltimes.cn/page/202103/1218076.shtml.
154. Denyer, S., "A decade after Fukushima nuclear disaster, contaminated water symbolizes Japan's struggles," *The Washington Post*, March 6, 2021.
155. Hunziker, R., "Fukushima takes a turn for the worse," *CounterPunch*, January 10, 2022. www.counterpunch.org/2022/01/10/fukushima-takes-a-turn-for-the-worse/.
156. Hunziker, R., "Dumping Fukushima's water into the ocean: What could possibly go wrong?" *CounterPunch*, November 1, 2020. www.counterpunch.org/2020/11/01/dumping-fukushimas-water-into-the-ocean/.
157. McCurry, J., "Dying robots and failing hope: Fukushima clean-up falters six years after tsunami," *The Guardian*, March 9, 2017. www.theguardian.com/world/2017/mar/09/fukushima-nuclear-cleanup-falters-six-years-after-tsunami.
158. Yamaguchi, M., "Robot photos appear to show melted fuel at Fukushima reactor," *AP News*, February 10, 2012. https://apnews.com/article/science-technology-business-japan-tsunamis-9e906166eab4f838abb1251af1880a5f.
159. St. Clair, "Fukushima mon amour."
160. Muller, R. A., *Energy for Future Presidents: The science behind the headlines*, pp. 19, 23–24, W. W. Norton and Company, New York, NY, 2012.
161. Fairlie and Sumner, "*The other report on Chernobyl (TORCH).*"
162. "Punk science," *The Economist*, December 23, 2017.
163. Fackler, M., "Residents torn on return to Fukushima, 4 years after disaster," *The New York Times*, August 10, 2015.
164. McCurry, J., "Dying robots and failing hope."
165. Caldicott, H., "The Dangers of Nuclear War" [press conference], *Centre for Research on Globalization*, GRTV, Montreal, March 18, 2011. www.youtube.com/watch?v=eMmaduq-5bw.
166. Ryall, J., "US sailors who 'fell sick from Fukushima radiation' allowed to sue Japan, nuclear plant operator," *The Telegraph*, June 23, 2017.
167. Johnson, R., "Fukushima: A foreseeable consequence of nuclear dependency," *openDemocracy*, March 16, 2012. www.opendemocracy.net/en/5050/fukushima-foreseeable-consequence-of-nuclear-dependency/.
168. Yergin, *The Prize*, p. 416.
169. Acton, J. M. and Hibbs, M., "Why Fukushima was preventable," The Carnegie Endowment for International Peace, Washington, DC, March 6, 2012.
170. Acton and Hibbs, "Why Fukushima was preventable."
171. McCurry, J., "Fukushima nuclear disaster: ex-bosses of owner Tepco ordered to pay ¥13tn," *The Guardian*, July 13, 2022. www.theguardian.com/world/2022/jul/13/fukushima-nuclear-disaster-ex-bosses-of-owner-tepco-ordered-to-pay-yen-13tn-tokyo-court.
172. "Japan's Nuclear Nightmare," *Four Corners*, Australian Broadcasting Company, 2011.
173. Yergin, *The Prize*, pp. 376–378.
174. Yergin, *The Prize*, p. 412.
175. Lavelle, M. and Garthwaite, J., "Is Armenia's nuclear plant the world's most dangerous?" *National Geographic*, April 14, 2011. www.nationalgeographic.com/science/article/110412-most-dangerous-nuclear-plant-armenia.

176. Mangano, J., "Tritium: Toxic tip of the nuclear iceberg," *CounterPunch*, June 26, 2017. www.counterpunch.org/2017/06/26/tritium-toxic-tip-of-the-nuclear-ice berg/.
177. Hunziker, R., "Nuclear fuel buried 108 feet from the sea," March 19, 2021, *CounterPunch*. www.counterpunch.org/2021/03/19/nuclear-fuel-buried-108-feet-from-the-sea/.
178. Amy, J., "Georgia nuclear plant cost tops $27B as more delays unveiled," *AP News*, July 30, 2021. https://apnews.com/article/business-environment-and-nature-georgia-90bbe5cc8e3a1a6077b9e4318e2bbf7e.
179. Goodall, C., "Nugen's AP1000 nuclear reactor – Is it any better than the EPR?" *The Ecologist*, July 17, 2015. https://theecologist.org/2015/jul/17/nugens-ap1000-nuclear-reactor-it-any-better-epr.
180. LaForge, J., "Nuclear reactors, bankrupting their owners, closing early," *CounterPunch*, February 6, 2018. www.counterpunch.org/2018/02/06/nuclear-reactors-bankrupting-their-owners-closing-early/.
181. "World Nuclear Power Reactors & Uranium Requirements."
182. "Nuclear family," *The Economist*, June 25, 2022.
183. Macalister, T., "EU approves Hinkley Point nuclear power station as costs raise by £8bn," *The Guardian*, October 8, 2014. www.theguardian.com/world/2014/oct/08/hinkley-point-european-commission-nuclear-power-station-somerset.
184. Vaughan, A., "How serious is the nuclear power plant radiation leak in China?" *New Scientist*, June 15, 2021. www.newscientist.com/article/2280903-how-ser ious-is-the-nuclear-power-plant-radiation-leak-in-china/.
185. Watt, H., "Hinkley Point: The 'dreadful deal' behind the world's most expensive power plant," *The Guardian*, December 21, 2017. www.theguardian.com/news/2017/dec/21/hinkley-point-c-dreadful-deal-behind-worlds-most-expensive-power-plant.
186. Elliott, D., "False promise: Nuclear power: past, present and (no) future," *The Ecologist*, April 12, 2017. https://theecologist.org/2017/apr/12/false-promise-nuclear-power-past-present-and-no-future.
187. Mallet, B., "EDF hopeful end in sight for long-delayed, budget-busting nuclear plant," Reuters, June 16, 2022. www.reuters.com/business/energy/edf-hopeful-end-sight-long-delayed-budget-busting-nuclear-plant-2022-06-16/.
188. "World Nuclear Power Reactors & Uranium Requirements."
189. Goodall, C., *Ten Technologies to Save the Planet*, pp. 269–270, *Green*Profile, London, 2008.
190. "World Nuclear Power Reactors & Uranium Requirements."
191. "Nuclear Power in India," *World Nuclear Association*, May 2022. www.world-nuclear.org/information-library/country-profiles/countries-g-n/india.aspx.
192. Langton, J., "Power for the people," *The National*, July 28, 2017. https://nuclear power.thenational.ae/.
193. Muller, *Energy for future Presidents*, p. 84.
194. "Nuclear Power in China," *World Nuclear Association*, January 2018. www .world-nuclear.org/information-library/economic-aspects/economics-of-nuclear-power.aspx.
195. Dorfman, P., "EDF facing bankruptcy as decommissioning time for France's ageing nuclear fleet nears," *The Ecologist*, March 16, 2017. https://theecologist .org/2017/mar/16/edf-facing-bankruptcy-decommissioning-time-frances-ageing-nuclear-fleet-nears.

196. Drogan, M., "The nuclear imperative: Atoms for peace and the development of US policy on exporting nuclear power, 1953–1955," *Diplomatic History*, 40(5): 948–974, September 18, 2015.

197. Rhodes, *Energy: A human history*, p. 314.

198. Grubler, A., "The costs of the French nuclear scale-up: A case of negative learning by doing," *Energy Policy*, 38(9): 5174–5188, 2010. https://doi.org/10.1016/j.enpol.2010.05.003.

199. Elliott, "False promise."

200. Green, J., "Nuclear power's annus horribilus," *CounterPunch*, July 17, 2017. www.counterpunch.org/2017/07/17/nuclear-powers-annus-horribilus/.

201. Wealer, B., Bauer, S., Göke, L, von HIrschhausen, C., and Kemfert, C., "High-priced and dangerous: Nuclear power is not an option for the climate-friendly energy mix," *DIW Weekly Report*, 30, pp. 235–243, 2019. www.diw.de/sixcms/detail.php?id=diw_01.c.670590.de.

202. LaForge, "Nuclear power: Dead in the water it poisoned."

203. Pfund, N. and Healey, B., "What would Jefferson do?" *DBL Investors*, September 2011.

204. Lawson, *Crazy Caverns*, p. 53.

205. Aldred, J. and Stoddard, K., "Timeline: Nuclear power in the United Kingdom," *The Guardian*, May 27, 2008. www.theguardian.com/environment/2008/jan/10/nuclearpower.energy.

206. Adams, L., "Why a national day of remembrance for downwinders is not enough," *Union of Concerned Scientists*, January 26, 2021. https://allthingsnuclear.org/guest-commentary/why-a-national-day-of-remembrance-for-downwinders-is-not-enough.

207. Geranios, N. K., "US government works to 'cocoon' old nuclear reactors," *AP News*, November 4, 2021. https://apnews.com/article/science-lifestyle-business-washington-world-war-ii-bf6f8f675a3d75628e1cf600ac8d1ea2.

208. Jacobson, M. and Delucchi, M., "A path to sustainable energy by 2030," *Scientific American*, November 2009.

209. Jacobson, M. Z., *100% Clean, Renewable Energy and Storage for Everything*, Cambridge University Press, New York, NY, 2020.

210. Abnett, K. and Jessop, S., "Impact of nuclear energy needs more study before getting green label, EU told," Reuters, July 2, 2021. www.reuters.com/business/energy/impact-nuclear-energy-needs-more-study-before-getting-green-label-eu-told-2021-07-02/.

211. Lovins, A. B., "Nuclear reactors make climate change worse," *Beyond Nuclear*, September 13, 2020. https://beyondnuclearinternational.org/2020/09/13/nuclear-reactors-make-climate-change-worse/.

212. Lovins, A. B., "'Low-carbon' misses the point: Arguments favoring nuclear power as a climate 'solution' are fundamentally misframed," *CounterPunch*, October 6, 2021. www.counterpunch.org/2021/10/06/low-carbon-misses-the-point-arguments-favoring-nuclear-power-as-a-climate-solution-are-fundamentally-misframed/.

213. Pearce, F., "Industry meltdown: Is the era of nuclear power coming to an end?" *E360*, Yale School of Forestry & Environmental Studies, May 15, 2017. http://e360.yale.edu/features/industry-meltdown-is-era-of-nuclear-power-coming-to-an-end.

214. "Global Citizen Reaction to the Fukushima Nuclear Plant Disaster," *IPSOS*, June 2011.

215. Using atomic mass units (amu), $1 - m_{He}/4m_{H} = 1 - 4.00260$ amu$/(4 \times 1.00797$ amu$) = 0.00726 \approx 1\%$.

216. Berman, *Zapped*, pp. 75–76.
217. Harvey, F., "Iter nuclear fusion project reaches key halfway milestone," *The Guardian*, December 6, 2017. www.theguardian.com/environment/2017/dec/06/iter-nuclear-fusion-project-reaches-key-halfway-milestone.
218. "Wendelstein 7-X fusion device produces its first hydrogen plasma," *Max-Planck-Institut für Plasmaphysik*, February 3, 2016. www.ipp.mpg.de/4010154/02_16.
219. Amos, J., "Major breakthrough on nuclear fusion energy," *BBC News*, February 9, 2022. www.bbc.com/news/science-environment–60312633.
220. Nuckolls, J., Wood, L., Thiessen, A., and Zimmerman, G., "Laser compression of matter to super-high densities: Thermonuclear (CTR) applications," *Nature*, September 1972.
221. Dunne, M., "Fusion's bright new dawn," *Physics World*, 23(5), May 2010.
222. Dunne, "Fusion's bright new dawn."
223. Clery, D., "Laser fusion reactor approaches 'burning plasma' milestone," *Science*, November 23, 2020, www.sciencemag.org/news/2020/11/laser-fusion-reactor-approaches-burning-plasma-milestone.
224. Tollefson, J., "Exclusive: Laser-fusion facility heads back to the drawing board," *Nature*, Volume 608, August 4, 2022.
225. Tennenbaum, J., "Hydrogen-boron fusion could be a dream come true," *Asia Times*, April 19, 2020. https://asiatimes.com/2020/04/hydrogen-boron-fusion-could-be-a-dream-come-true/.
226. Austin-Morgan, T., "An Oxford-based fusion project is closing on producing gain using a cheaper method than the mainstream," *Eureka!*, October 10, 2019. www.eurekamagazine.co.uk/design-engineering-features/technology/project-ile-fusion-goes-for-gain/219979/.
227. Pressurized high-temperature or even room-temperature superconductors lower transmission-wire friction (electric resistance), greatly reducing power loss in electric grids, as well as improving magnets for fusion and maglev trains.
228. Ball, P., "The chase for fusion energy," *Nature*, November 17, 2021. www.nature.com/immersive/d41586-021-03401-w/index.html.
229. Cho, A., "Scientists rally around plan for fusion power plant," *Science*, p. 1258, December 11, 2020.
230. Ball, "The chase for fusion energy."
231. Clynes, *The Boy Who Played with Fusion*, p. 131.
232. Tutt, K., *The Scientist, the Madman, the Thief and Their Lightbulb*, p. 149, Pocket Books, London, 2003.
233. Tutt, *The Scientist, the Madman, the Thief and Their Lightbulb*, p. 150.
234. Clery, D., "Out of gas," *Science*, June 24, 2022.

4 Old to New: The Sun and All Its Glory

1. If we count by number of atoms, the composition is hydrogen 91.2% and helium 8.7%. The rest of the Sun is composed of other higher elements, mostly oxygen and carbon (2% by mass or 0.1% by number of atoms).
2. One nanometer (nm) is equal to 1 billionth of a meter (1×10^{-9} m). There is no exact border between UV and the viz and between the viz and IR. The ICNIRP defines the visible range as between 380 and 780 nm.
3. In 1676, the Danish astronomer Ole Rømer calculated the speed of light by measuring eclipses of the moons of Jupiter at different distances from the Earth: when the Earth

is closest to Jupiter (that is, on the same side of the Sun at ~4.3 AU) and when the Earth is furthest from Jupiter (that is, on the opposite sides of the Sun at ~6.3 AU). In 1850, Hippolyte Fizeau and Léon Foucault reflected light from a rotating mirror to measure the speed. The symbol c is from *celeritas*, Latin for speed. Proxima Centauri (in the Alpha Centauri multiple star system) is the next nearest star to Earth after the Sun, taking 4.23 light years for its radiated energy to reach earth from a distance of almost 40 trillion km (270,000 AU).

4. For example, the four main colors emitted from heated hydrogen gas have wavelengths of 656 nm (red), 486 nm (blue–green), 434 nm (blue–violet), and 410 nm (violet), representing excited transitions from the nth to the 2nd atomic shell, $n = 3$–6 (a.k.a. the Balmer series). The energy, E, equals hc/λ, where h is Planck's constant and c is the speed of light.

5. Atmospheric scattering (S) is the ability of light to bounce off atoms and molecules in the air and depends on the wavelength of the light, called Rayleigh scattering after its discoverer ($S \propto 1/\lambda^4$). If the thickness of the atmosphere is taken as 100 km, sunlight must then travel through 1100 km of atmosphere at sunset or sunrise, that is, the relative thickness of air at zenith (when the sky is seen as red) to horizon (when the sky is seen as blue) is 11 times.

6. The idea of six or seven colors is a human construct based on perceived bands in the visible spectrum: violet, (sometimes indigo), blue, green, yellow, orange, and red. In fact there are millions of colors as seen in the hexadecimal color code #RRGGBB, which ranges from #000000 (black) to #FFFFFF (white) and produces more than 16 million different colors (FF^3 or $256 \times 256 \times 256$). Newton worked out his theory of colors at home aged 21–22 while the University of Cambridge was closed during the outbreak of bubonic plague from 1665 to 1666.

7. The material property of light was much debated from the mid-seventeenth century on, especially with the advent of practical ground lenses and the birth of the telescope. Newton (and Nicolas Fatio) favored the corpuscular theory, while Huygens (and Gottfried Leibniz) the wave theory. We know now that both were right as explained by wave-particle duality proposed in the twentieth century. As noted by Aldersey-Williams, the many correspondences between this "international quadriga" and other natural philosophers-cum-scientists also helped to establish collaborative discourse as "the norm within the scientific community of Europe."

8. Aldersey-Williams, H., *Dutch Light: Christiaan Huygens and the making of science in Europe*, p. 306, Picador, London, 2020.

9. Stoddard, M. C., Eyster, H. N., Hogan, B. G., et al., "Wild hummingbirds discriminate nonspectral colors," *Proceedings of the National Academy of Science*, 117(26): 15112–15122, 2020. https://doi.org/10.1073/pnas.1919377117.

10. As ground-dwelling terrestrial creatures, our bodies have also adapted to the temperature, pressure, and elemental composition of the atmosphere.

11. Wien's Displacement Law relates the temperature of a black body to its maximum wavelength: $T = b/\lambda$, where $b = 2.8977729 \times 10^{-3}$ m K.

12. In a "black body" no light is reflected and thus all emitted light is because of its temperature. Any difference between the "idealized" black-body radiation and the measured radiation indicates absorption at certain wavelengths.

13. The radiated power of the Sun is $P = 4\pi R^2 \sigma T^4 = 4\pi (695,700 \times 10^3$ m$)^2 \times 5.67 \times 10^{-8}$ W/m^2/K$^4 \times (5,778$ K$)^4 = 3.84 \times 10^{26}$ W. P to Earth $= 3.84 \times 10^{26}$ W $\times 4.6 \times 10^{-10} = 1.73 \times 10^{17}$ W. The net radiated power of the human body, $P = \sigma e A(Ts^4 - Tr^4)$, is 5.67×10^{-8} W/m^2/K$^4 \times 0.97 \times 1.5$ m$^2 \times (306^4 - 293^4) = 115$ W. The result is calculated for an average human body with surface area (A) of 1.5 m^2, skin

temperature (T_s) of 33°C (306 K), in a room with an ambient temperature (T_r) of 20°C (293 K) (needed because our skin also absorbs heat), and an emissivity, e, of 0.97 (the human body is not a total black body and reflects back a tiny bit of incident energy).

14. P to Earth per square meter = P to Earth / (πR_E^2) = 1.73×10^{17} W / $\pi(6,400 \times 10^3$ m$)^2$ = 1,374 W/m^2 (the solar constant). We can also integrate the measured irradiance (I) in Figure 4.1 to get the actual power per square meter: $P = \int I \, d\lambda$.

15. The sunspot cycle is actually a 22-year cycle as the Sun's magnetic field flips every 11 years due to its varying latitudinal rotations. Increased sunspot activity may also account for increased temperatures on earth and decreased sunspot activity for decreased temperatures (as during the Little Ice Age of 1645–1715).

16. Zweibel, K., Mason, J., and Fthenakis, V., "A solar grand plan," *Scientific American*, p. 48, January 2008.

17. Johnstone, B., *Switching to Solar: What we can learn from Germany's success in harnessing clean energy*, p. 126, Prometheus Books, New York, NY, 2011.

18. Despite the decreased insolation, summer temperatures in the northern hemisphere (at aphelion) are higher than summer temperatures in the southern hemisphere (at perihelion) because of the much greater land mass north of the equator – land has a lower heat capacity than water and thus heats (and cools) much quicker.

19. Some think the asteroid that struck Earth was an early proto-planet called Theia, a Mars-sized object competing for orbital space with earth. In the collision, almost half of the magma covering the molten Earth was ejected, which solidified to form the moon.

20. You can calculate local minimum and maximum solar angles based on one's latitude (L). The minimum angle is 90 – (L + 23.4)° at the winter solstice and the maximum 90 – (L – 23.4)° at the summer solstice. For example, in Los Angeles (latitude 34.0 °N), sun angles through the year range from 32.6° in the dead of winter to 79.4° at the height of summer, providing almost twice the strength from solstice to solstice (relative sun strengths are the ratio of the sines of the maximum/ minimum angles).

21. Relative sun strength at two angles is the ratio of the sine of the two angles; thus, Los Angeles at 34.0° latitude = 1.8 (sin(79.4°)/sin(32.6°) and Gijón at 43.5° latitude = 2.4 (sin(70°)/sin(23°). In general, the relative sun strength between peak summer and peak winter is sin(90° – (L – 23.4)° / sin(90° – (L + 23.4)°), where L is the latitude.

22. In the northern hemisphere above the tropics, look south. In the southern hemisphere below the tropics, look north. In the tropics, look north, south, or directly above your head depending on the time of year.

23. To calculate total panel or array energy for a single day, integrate the instantaneous power over time for 24 hours ($E = \int P \, dt$ in kWh, where E is total energy, P is the incremental power, and dt is the time increment). The summer-to-winter ratio is the ratio of energies (areas under each curve). If the time increments are the same, one can simply sum the power data. The data were from our apartment roof in Amsterdam on three consecutive days in July and December, where the output was recorded in 15-minute increments.

24. Muller, R. A., *Energy for Future Presidents: The science behind the headlines*, p. 146, W. W. Norton and Company, New York, NY, 2012.

25. Muller, *Energy for Future Presidents*, p. 146.

26. There are differing ideas about the end of the Earth's atmosphere and the start of space. The Kármán line is a good marker at 100 km above mean sea level,

indicating where the atmosphere is too thin to provide conventional aircraft with enough lift to fly. The Kármán line is also where radio blackouts occur in descending spacecraft, caused by interference from ionized air created by aerodynamic heating during atmospheric compression at re-entry. The Arabic scholar Al-Haytham (Alhazen) (965–1039) calculated the thickness of the Earth's atmosphere at roughly 90 km (and thus the edge of space) by measuring the Sun's elevation below the horizon at the end of twilight and the duration of twilight (between sunset and last glow on the horizon). NASA uses 80 km, above which a traveler is considered an "astronaut." At roughly 1,000 km, there is hardly any atmosphere at all, while the exosphere extends to 10,000 km. If the Earth were the size of a basketball, the atmosphere would be roughly 2 mm thick.

27. Tyndall, J., "On radiation through the earth's atmosphere," Chapter XIII, *In the domain of radiant heat*, pp. 421–424, Longmans, Green, and Co., London, 1872.

28. In a 1975 *Science* article "Climactic change: Are we on the brink of a pronounced global warming?" Columbia University geophysicist Wallace Broecker coined the term "global warming," noting that man-made carbon dioxide (and now methane) would soon contribute to an exponential rise in global temperatures.

29. The yellow helium line at 587.56 nm is now known as one of the Fraunhofer lines – spectral features observed as dark absorption bands in the Sun's optical spectrum – and lies just beside the famous yellow sodium doublet, seen by heating salt in a flame. Joseph von Fraunhofer magnified Isaac Newton's earlier prism-dispersed sunlight to see about 600 of the 30,000 known lines.

30. McEvoy, J. P., *A Brief History of the Universe: From ancient Babylon to the Big Bang*, p. 156, Robison, London, 2010.

31. Utrecht's Sonnenborgh Observatory created the first detailed photometric atlas of the solar spectrum, listing spectral lines from 333.2 to 887.1 nm, consisting of 120 m of pages.

32. The solar constant is measured at the edge of space before sunlight enters the Earth's atmosphere (called AM0 for air mass zero). At the Earth's surface, the solar constant is reduced after being absorbed by the atmosphere (a.k.a. atmospheric extinction), differentially depending on scattering, absorption, and latitude. In the tropics, where the Sun is almost directly overhead, the incident sunlight is reduced by 1 atmosphere of thickness (called AM1). Outside the tropics, AM is greater than 1 as more atmosphere is traversed and thus more solar radiation absorbed. AM1.5 is the standard air mass used for solar cell testing, indicating 1.5 atmospheres traversed at a solar zenith angle of 48.2°. AM38 is essentially horizontal. If the absorption is 33%, AM1 = 0.67 × AM0 = 0.67 × 1,374 W/m^2 = 920 W/m^2.

33. Del Chiaro, B. and Telleen-Lawton, T., "Solar water heating: How California can reduce its dependence on natural gas," Environment California Research & Policy Center, April 2007.

34. Hawken, P. (ed.), *Drawdown: The most comprehensive plan ever proposed to reverse global warming*, p. 36, Penguin Books, London, 2017.

35. Weiss, W. and Spörk-Dür, M., "Solar Heat Worldwide, Global Market Development and Trends 2021," AEE – Institute for Sustainable Technologies, Gleisdorf, Austria, May 2022. www.iea-shc.org/Data/Sites/1/publications/Solar-Heat-Worldwide-2022.pdf.

36. Del Chiaro and Telleen-Lawton, "Solar Water Heating."

37. Bonneuil, C. and Fressoz, J.-B. (trans. Fernbach, D.), *The Shock of the Anthropocene*, p. 112, Verso, London, 2017.

38. In 2020, on the winter "solstice," which means "sun stands still" in Latin, the Sun rose over the Boyne Valley and entered Newgrange at 8:58 a.m. on December 21, while sunlight was observed in the inner chamber from December 18 to 23. The times and dates are similar for other years.
39. Note that Eratosthenes's calculation doesn't prove the Earth is spherical. His elegant demonstration calculated a circular polar distance along a line of longitude, but it doesn't follow that the equatorial circumference is necessarily the same. For example, from his analysis the Earth could be watermelon-shaped and the polar distance on a line through Alexandria and Syene longer or shorter than the equatorial distance. We know now that the Earth bulges at the equator and is more correctly called an "oblate spheroid." The polar and equatorial difference is minor, however, and thus for most purposes the Earth is considered a sphere.
40. Syene/Aswan is ideally located at 23.5°, practically the Tropic of Cancer (one of the five major parallels of latitude) and where the Sun is exactly overhead on the summer solstice.
41. Xenophon, *Memorabilia of Socrates*, Book 3, Chapter 8, Section 9, ed. E. C. Marchant, Harvard University Press, Cambridge, MA, and William Heinemann, Ltd, London, 1923.
42. "Death Ray MiniMyth," *MythBusters*, The Discovery Channel, 2004. www.discovery.com/tv-shows/mythbusters/videos/death-ray-minimyth/.
43. "Archimedes Death Ray," 2.009 Product Engineering Processes, MIT, October 2005. http://designed.mit.edu/gallery/data/2011/homepage/experiments/deathray/10_ArchimedesResult.html.
44. Kemp, C. M., "Apparatus for utilizing the Sun's rays for heating water," US Patent No. 451,384, April 28, 1891.
45. Madrigal, A., *Powering the Dream. The history and promise of green technology*, pp. 84–86, Da Capo, Cambridge MA, 2011.
46. Kapstein, E. B., "A solar energy heyday," *The Washington Post*, September 16, 1979. www.washingtonpost.com/archive/opinions/1979/09/16/a-solar-energy-heyday/1515beee-f27e-4bb9-9583-ff35029ad0da/.
47. Ragheb, M., "Solar thermal power and energy storage historical perspective," October 9, 2014. www.solarthermalworld.org/sites/default/files/story/2015-04-18/solar_thermal_power_and_energy_storage_historical_perspective.pdf.
48. Shuman, F., "Power from sunshine: A pioneer solar power plant," *Scientific American*, September 30, 1911.
49. Shuman, "Power from sunshine."
50. Strathern, P., *Mendeleyev's Dream: The quest for the elements*, p. 248, Penguin Books, London, 2001.
51. The purest silicon comes from quartz that is ground up into dust and separated out. Interestingly, silicon was originally discarded in the mining of silica (silicon dioxide) from quartz. Some of the best quartz is found in the Blue Ridge Mountains in the Spruce Pine mining region of North Carolina.
52. Mann, C. C., *The Wizard and the Prophet: Science and the future of our planet*, p. 283, Picador, London, 2019.
53. Smith, W., "Effect of light on selenium during the passage of an electric current," *Nature*, p. 303, February 20, 1873.
54. Hawken (ed.), *Drawdown*, p. 11.
55. Perlin, J., "The invention of the solar cell," *Popular Science*, April 22, 2014. www.popsci.com/article/science/invention-solar-cell.

56. De Decker, K., "How to build a low-tech solar panel?" *Low-Tech Magazine*, October 2021. www.lowtechmagazine.com/2021/10/how-to-build-a-low-tech-solar-panel.html.

57. Planck's constant, h, converts the energy of light to the wavelength (or frequency) from the relation $E = hc/\lambda$ (or $E = hcf$), where E is the energy of the light, λ is the wavelength, and f is the frequency. In the usual first-year experiment, Planck's constant is calculated from the slope of the line in a plot of stopping voltage versus wavelength (or frequency) of three different incident light energies. The typical wavelengths are 578 nm (yellow), 546 nm (green), and 436 nm (blue). One can also work out the work function of the material from the intercept. The value of h is 6.626×10^{-34} Js.

58. The diode is considered a "non-ohmic" device because the current is not directly proportional to the applied voltage as in Ohm's law, $I = V/R$.

59. Suplee, C., *Physics in the 20th Century*, p. 53, Harry N. Abrams Inc., New York, NY, 1999.

60. Riordan, M. and Hoddeson, L., *Crystal Fire: The invention of the transistor and the birth of the Information Age*, pp. 59–60, W. W. Norton and Company, New York, NY, 1997.

61. Riordan, M. and Hoddeson, L., "The origins of the *pn* junction," *IEEE Spectrum*, June 1997.

62. Riordan and Hoddeson, "The origins of the *pn* junction."

63. Hemour and Wu, "Radio-frequency rectifier for electromagnetic energy harvesting."

64. Pearson, G. L. and Brattain, W. H., "History of semiconductor research," *Proceedings of the IRE*, 43(12): 1794–1806, December 1955.

65. Pearson and Brattain, "History of semiconductor research."

66. Ohl, R. S., "Light-sensitive electric device," US Patent 2402662 A, May 27, 1941. https://patents.google.com/patent/US2402662.

67. Riordan and Hoddeson, *Crystal Fire*, p. 88.

68. "Conductors and insulators," Hyperphysics, Georgia State University. http://hyperphysics.phy-astr.gsu.edu/hbase/electric/conins.html.

69. Riordan and Hoddeson, "The origins of the pn junction."

70. Gertner, J., *The Idea Factory: Bell Labs and the great age of American innovation*, pp. 82–83, Penguin, London, 2012.

71. Brattain and Bardeen were so annoyed at Shockley's end-run around their work that Brattain never worked with him again and Bardeen left Bell Labs in 1951 for the University of Illinois, Urbana, where he would later win another Nobel for his work on superconductors.

72. The middle, *p*-type "meat" acts as the grid, sandwiched by two thicker, outer pieces of *n*-type "bread" that are like the cathode (source) and anode (plate) in a vacuum tube. Shockley later perfected the *pnp* junction transistor that used electrons as the main charge carrier rather than holes. The new transistor terms "base," "emitter," and "collector" were coined soon after. About a decade later the planar field-effect transistor (FET) was perfected, and is the transistor type used in most applications today.

73. Gertner, *The Idea Factory*, p. 102.

74. Gertner, *The Idea Factory*, p. 169.

75. Gertner, *The Idea Factory*, p. 170.

76. Perlin, "The invention of the solar cell."

77. "This month in physics history: April 25, 1954: Bell Labs demonstrates the first practical silicon solar cell," *APS News*, 18(4), April 2009. www.aps.org/publica tions/apsnews/200904/physicshistory.cfm.

78. "Vast power of the Sun is tapped by battery using sand ingredient," *The New York Times*, April 26, 1954.

79. Gertner, *The Idea Factory*, p. 203.

80. Riordan and Hoddeson, *Crystal Fire*, pp. 210–212.

81. Kolker, R., "The original germaniac," in The Elements (Special Issue), *Bloomberg Businessweek*, September 2, 2019.

82. In a current-controlled, bipolar junction transistor (BJT), the three electrodes are the base, emitter, and collector. In a voltage-controlled, field effect transistor (FET), the three electrodes are the gate, source, and drain.

83. Moore, G. E., "Cramming more components onto integrated circuits," *Electronics*, 38(8), 1965.

84. The original Silicon Valley semiconductor company, Shockley Semiconductor Laboratory, was started by transistor co-inventor William Shockley after he left Bell Labs in New Jersey to start up a business in his home town of Palo Alto. Shockley is considered the man who put the "silicon" in Silicon Valley or Moses because he led the industry to the promised land but never quite got there himself. Noyce, Moore, and Andy Grove would go on to form Intel, known as the "Intel trinity."

85. Wolfe, T., "Two young men who went west," in *Hooking Up: Essays and fiction*, p. 35, Johnathan Cape, London, 2000.

86. Technically, a conduction electron recombines with a valence hole, giving off light with energy equal to the band gap.

87. To convert between energy and wavelength, use Planck's equation, $E = hc/\lambda$. Thus, E [eV] = $1240/\lambda$ [nm].

88. Giancoli, D. C., *Physics: Principles with applications*, 6th ed., p. 830, Pearson Education International, Englewood Cliffs, NJ, 2006.

89. Webb, J., "Invention of blue LEDs wins physics Nobel," BBC News, October 7, 2014. www.bbc.com/news/science-environment–29518521.

90. Smil, V., *Energy and Civilization: A history*, pp. 402, 406, The MIT Press, Cambridge, MA, 2017.

91. "LED efficacy: What America stands to gain," Building Technologies Office, US Department of Energy, November 2015. www.energy.gov/sites/prod/files/2015/11/f27/LED%20Efficacy-What%20America%20Stands%20to%20Gain_%20November%202015.pdf.

92. "LED efficacy: What America stands to gain."

93. Keller, M., "World's largest LED retrofit will cut Chase Bank's lighting bill in half," *GE Reports*, February 18, 2016. www.ge.com/news/reports/worlds-largest-led-retrofit-will-cut-chase-banks-lighting-bill-in-half.

94. Gertner, *The Idea Factory*, p. 336.

95. CRAY-3 Supercomputer Systems, Cray Computer Corporation, Colorado Springs, 1993.

96. In a hurry to match the Soviets after *Sputnik*, the first *Vanguard* exploded on the launch pad, watched live on TV by millions of horrified American viewers and adding to the worry of the Cold War.

97. Green, C. M. and Lomask, M., Chapter 13, The National Academy of Sciences and the Scientific Harvest, 1957–1959, in *Vanguard, a history*, NASA SP-4202, The NASA Historical Series, NASA, Washington, DC, 1970.

98. Segrè, G., *A Matter of Degrees*, p. 87, Penguin Books, New York, NY, 2002.

99. Cho, A., "After botched launch, orbiting atomic clocks confirm Einstein's theory of relativity," *Science*, December 7, 2018. www.science.org/content/article/after-botched-launch-orbiting-atomic-clocks-confirm-einsteins-theory-relativity.

100. Gertner, *The Idea Factory*, pp. 221–224, Penguin, London, 2012.

101. "Solar timeline," The history of solar, US Department of Energy. www1.eere .energy.gov/solar/pdfs/solar_timeline.pdf.

102. "Solar arrays," International Space Station, NASA. www.nasa.gov/mission_ pages/station/structure/elements/solar_arrays.html#.WIcc_LmGO_o.

103. Johnstone, *Switching to Solar*, p. 31.

104. Johnstone, *Switching to Solar*, p. 35.

105. Gertner, *The Idea Factory*, p. 221.

106. "Monocrystalline vs. polycrystalline solar panels" [video], presented by Amy Beaudet, *AltE*, October 23, 2015. www.youtube.com/watch?v=TCq0K3DlFdc.

107. Solar cell sizes continue to grow, allowing for larger assembled modules, higher rated power, and lower costs. Prior to 2010, cell dimensions were 125 mm × 125 mm (labeled 125) and assembled in a rectangular lattice. As of 1984, cells have been classified from M0 (156), M2 (156.75, 5.5 W), M4 (161.7, 5.81 W), M6 (166, 6.17 W), M9 (192, 8.29 W), M10 (200, 9.00 W), and M12 (9.92 W).

108. Johnstone, *Switching to Solar*, pp. 307–308.

109. Boxwell, M., *Solar Electricity Handbook*: *2019 Edition*, p. 61, Greenstream Publishing, London, 2019.

110. "UNSW's Green awarded global energy prize," *Photonics Spectra*, September 2018.

111. Philipps, S. P. and Bett, A. W., "Current status of concentrator photovoltaic (CPV) technology," Version 1.2, Fraunhofer Institute for Solar Energy Systems ISE, Freiburg, Germany, February 2016.

112. Knier, G., "How do photovoltaics work?" *Science Beta*, NASA. https://science .nasa.gov/science-news/science-at-nasa/2002/solarcells.

113. STC assumes a module temperature of 25°C, 1-Sun irradiance (1,000 W/m^2), and typical luminosity at the Earth's surface (AM1.5) for "single-junction" devices (that is, not concentrated, multi-junction, or textured), thus measuring material properties only. Note, AM is the air mass (relative atmospheric thickness), where AM1.0 is the zenith sun. The theoretical maximum efficiency of a single-junction silicon cell is 32% (matched to its 1.1-eV band gap, called the Shockley–Queisser limit), while higher efficiencies can be obtained using other materials or adding layers in a "multi-junction" cell, theoretically as high as 87%.

114. "Best research-cell efficiency chart," *NREL*, November 22, 2021. www.nrel.gov/ pv/interactive-cell-efficiency.html.

115. Muller, *Energy for Future Presidents*, p. 151.

116. Roney, J. M., "China leads world to solar power record in 2013," Earth Policy Institute, June 18, 2014. www.earth-policy.org/indicators/C47.

117. Johnstone, *Switching to Solar*, p. 236.

118. Johnstone, *Switching to Solar*, p. 237.

119. Johnstone, *Switching to Solar*, p. 251.

120. White, J. K., "Let the Sun shine: Making solar power work," *CounterPunch*, November 12, 2021. www.counterpunch.org/2021/11/12/let-the-sun-shine-mak ing-solar-power-work/.

121. Johnstone, *Switching to Solar*, pp. 247–248.

122. Johnstone, *Switching to Solar*, p. 243.

123. Bonneuil and Fressoz, *The Shock of the Anthropocene*, p. 113.

124. Bonneuil and Fressoz, *The Shock of the Anthropocene*, p. 113.

125. Swets, A., Jäger, K., Isabella, O., van Swaaij, R., and Zeman, M. *Solar Energy: The physics and engineering of photovoltaic conversion, technologies and systems*, UIT Cambridge Ltd, Cambridge, 2016.

126. Johnstone, *Switching to Solar*, p. 301.

127. "The Breakthrough in Renewable Energy," Backlight, [documentary], directed by Martijn Kieft, VPRO, The Netherlands, 2016.

128. Milman, O., "Elon Musk leads Tesla effort to build house roofs entirely out of solar panels," *The Guardian*, August 19, 2016. www.theguardian.com/technol ogy/2016/aug/19/elon-musk-tesla-solar-panel-roofs-solarcity.

129. Carr, A. and Eckhouse, B., "Did Elon Musk forget about Buffalo?" Bloomberg, November 20, 2018. www.bloomberg.com/news/features/2018-11-20/inside-elon-musk-s-forgotten-gigafactory-2-in-buffalo.

130. Walker, A., "James Cameron's plan to fix solar panels," *Gizmodo*, June 30, 2015. https://gizmodo.com/james-cameron-s-plan-to-fix-solar-panels–1714720243.

131. Mossalgue, J., "In France, all large parking lots now have to be covered by solar panels," *Electrek*, November 8, 2022. https://electrek.co/2022/11/08/france-require-parking-lots-be-covered-in-solar-panels/.

132. "Power generation you can see through," *MSU Today*, August 25, 2021. https://msutoday.msu.edu/news/2021/solar-glass-panels-installed.

133. "Carport solar array receives 2018 Innovative Project Award," *MSU Today*, November 9, 2018. https://msutoday.msu.edu/news/2018/carport-solar-array-receives-2018-innovative-project-award.

134. Enman, S., "Coney Island station's pricy solar panels have been dark for 7 years. No one knows when they're coming back," *Brooklyn Daily Eagle*, November 22, 2019. https://brooklyneagle.com/articles/2019/11/22/coneys-subway-solar-panels-went-dark-after-only-7-years-theyre-still-off-the-grid-today/.

135. d'Estries, M., "5 solar-powered buildings that will forever change architecture," *Treehugger*, May 31, 2017. www.treehugger.com/solar-powered-buildings-will-forever-change-architecture–4868157.

136. Lambert, F., "Toyota brings back the solar panel on the plug-in Prius Prime – but now it powers the car," *Electrek*, June 20, 2016. https://electrek.co/2016/06/20/toyota-prius-plug-prime-solar-panel/.

137. "BIPV combined energy rooftop system opened in Australia," *Solar + Power Management*, June 17, 2014. https://solarpowermanagement.net/article/94023/BIPV_combined_energy_rooftop_system_opened_in_Australia.

138. "Turlock Irrigation District selected to pilot first-in-the-nation water-energy Nexus Project involving solar panels over canals," Turlock Irrigation District, February 8, 2022. www.prnewswire.com/news-releases/turlock-irrigation-dis trict-selected-to-pilot-first-in-the-nation-water-energy-nexus-project-involving-solar-panels-over-canals-301478069.html.

139. An organic photovoltaic (OPV) cell consists of very high energy-absorbing materials such as small molecules, polymers, and dyes rather than silicon, deposited between two electrodes (the electron donor and electron acceptor). Made from inexpensive carbon-rich compounds, organic photovoltaics are thinner, lighter, and highly flexible, offering unique ways to provide energy in solar applications such as "stick-on" wall panels. Organic photovoltaics can be bent when applied to curved plastic substrates, woven into fabrics, and even used as a paint coating (semitransparent in any color), thus facilitating mass production. The world's largest OPV façade at time of installation in 2016 is on a warehouse in the inland port of Duisburg, Germany, 185 m^2 of stick-on strips glued directly to the building with self-adhesive backing.

140. Of particular interest are "dye-sensitized" solar cells (DSSC) that employ small-molecule organic absorber dyes as dopants, found in cheap "sensitizer" substances that absorb the light, including blackcurrents, raspberries, and spinach! Adsorbed onto electron-accepting materials, such as titanium dioxide (TiO_2) and zinc oxide (ZnO), a dye-sensitized solar cell mimics photosynthesis through electron transfer. The highest yield is with polypyridyl ruthenium derivatives in a Grätzel cell, although thermal instability limits longevity (dyes contract and expand with changing temperature). In an attractive and elegant design, the SwissTech Convention Center in Lausanne incorporates vertical-strip DSSCs into a 300-m^2 multi-colored glass façade that optimally tracks the Sun to maximize each panel's output.

141. As the name implies, quantum dots (QDs) are small – 2 to 10 nm thick (10–50 atoms!) with a high surface to volume ratio – and can produce different colors by changing the energy difference between the highest valence band and lowest conduction band (typically II–VI or III–V heterojunctions, such as InAs and PbS). Discovered in 1980, quantum dots are still in their infancy, but are highly prized because they tune light by changing their dot size and are particularly useful in medical applications to light up specific cells in a nanoscopic light bulb.

142. Goodall, C., *The Switch: How solar, storage and new tech means cheap power for all*, p. 73, Profile Books, London, 2016.

143. Hogan, H., "Photonic technologies energize sustainability," *Photonics*, pp. 48–51, July 2020.

144. Watanabe, C., "Researchers think they're getting closer to making spray-on solar cells a reality," Bloomberg, March 21, 2017. www.bloomberg.com/news/articles/2017-03-21/the-wonder-material-that-may-make-spray-on-solar-cells-a-reality.

145. Foley, M., "Printed solar panels a shining light for saving energy," *The Sydney Morning Herald*, July 11, 2020. www.smh.com.au/environment/sustainability/printed-solar-panels-a-shining-light-for-saving-energy-20200707-p559po.html.

146. Savage, N., "Solar cells that make use of wasted light," *Nature*, June 24, 2021. www.nature.com/articles/d41586-021-01673-w.

147. Yun, Y., Mühlenbein, L., Knoche, D. S., Lotnyk, A., and Bhatnagar, A., "Strongly enhanced and tunable photovoltaic effect in ferroelectric-paraelectric superlattices," *Science Advances*, June 2, 2021, 7(23), DOI: 10.1126/sciadv.abe4206.

148. Sakharkar, A., "New efficiency record set for ultrathin solar cells," *InceptiveMind*, February 16, 2022. www.inceptivemind.com/new-efficiency-record-set-ultrathin-solar-cells/23346/.

149. Buttgenbach, T., "Why Solar PV Power Plants Will Fundamentally Change the Way We Power the Planet" [seminar], Urban Growth Seminars, USC Price, March 21, 2016. www.youtube.com/watch?v=jFZ8g0W7jV0.

150. Johnstone, *Switching to Solar*, p. 71.

151. Fraas, L. and Partain, L., *Solar Cells and Their Applications*, 2nd ed., p. 208, John Wiley and Sons, Hoboken, NJ, 2010.

152. Cimochowski, A. B., "Solar in-depth: Sun 'tracker' arrays," Solar Cell Central. http://solarcellcentral.com/solar_page.html.

153. Goodall, *The Switch*, p. 95.

154. "The Wiki-Solar Database," Wiki-Solar, 2015. https://wiki-solar.org/data/index.html.

155. Clover, I., "China's Three Gorges connects part of 150 MW floating solar plant," *PV Magazine*, December 12, 2017. www.pv-magazine.com/2017/12/12/chinas-three-gorges-connects-part-of-150-mw-floating-solar-plant/.

156. "Dutch start trials of flexible floating solar panels which move with the waves," *Dutch News*, November 29, 2021. www.dutchnews.nl/news/2021/11/dutch-start-trials-of-flexible-floating-solar-panels-which-move-with-the-waves/.

157. Egan, N., "Secretive energy startup backed by Bill Gates achieves solar breakthrough," *CNN Business*, November 19, 2019. https://edition.cnn.com/2019/11/19/business/heliogen-solar-energy-bill-gates/index.html.

158. Swets et al., *Solar Energy*.

159. "Concentrating Solar Power Projects," The National Renewable Energy Laboratory, July 1, 2022. www.nrel.gov/csp/solarpaces/.

160. Mosiori, C. O., *The Solar Cell: Solar cell technology*, p. 161, Rift Valley Institute of Science and Technology, Nakuru, Kenya, 2016.

161. Goodall, C., *Ten Technologies to Save the Planet*, p. 69, *Green*Profile, London, 2008.

162. Rifkin, J., *The Third Industrial Revolution: How lateral power is transforming energy, the economy, and the world*, pp. 156–157, Palgrave Macmillan, New York, NY, 2011.

163. Phiddian, E., "Sun Cable in voluntary administration: Why is the energy project in trouble?" *Cosmos*, January 19, 2023. https://cosmosmagazine.com/technology/sun-cable-administration-explainer/.

164. Turner, J. A., "A realizable renewable energy future," *Science*, 285(5428): 687–689, July 30, 1999. DOI: 10.1126/science.285.5428.687.

165. Zweibel et al., "A solar grand plan."

166. "Schwarzenegger's Green Challenge," 60 Minutes, CBS News, December 19, 2008.

167. "US Solar Market Insight Executive Summary," Q3 2022.

168. Grossman, K., "Our solar bonanza!" *CounterPunch*, November 27, 2015. www.counterpunch.org/2015/11/27/our-solar-bonanza/.

169. Warrick, J., "Utilities wage campaign against rooftop solar," *The Washington Post*, March 7, 2015.

170. " International Technology Roadmap for Photovoltaic," in *ITRPV*, 13th ed., pp. 59–61, VDMA, Frankfurt, March 2022. www.vdma.org/international-technology-roadmap-photovoltaic.

171. Cardwell, D., "Homes with their own personal power grid," *The New York Times*, July 31, 2017.

172. Bruggink, A. and van der Hoeven, D., *More with Less: Welcome to the precision economy*, pp. 29–31, Biobased Press, Amsterdam, 2017.

173. Osborne, S., "Sweden phases out fossil fuels in attempt to run completely off renewable energy," *The Independent*, May 24, 2016.

174. "Net Zero Scorecard," The Energy and Climate Intelligence Unit (ECIU), UK, 2021. https://eciu.net/netzerotracker.

175. "SolarCity and Tesla: Ta'u microgrid" [video], SolarCity, November 22, 2016. www.youtube.com/watch?v=VZjEvwrDXn0.

176. Roy, E. A., "South Pacific island ditches fossil fuels to run entirely on solar power," *The Guardian*, November 28, 2016. www.theguardian.com/environment/2016/nov/28/south-pacific-island-ditches-fossil-fuels-to-run-entirely-on-solar-power.

177. Fialka, J., "As Hawaii aims for 100% renewable energy, other states watching closely," *E&E News* (reprinted in *Scientific American*), April 27, 2018. www.scientificamerican.com/article/as-hawaii-aims-for-100-renewable-energy-other-states-watching-closely/.

178. Colthorpe, A., "Hawaiian Electric wants all solar in next procurement round to be paired with battery storage," *Energy News*, October 21, 2021. www.energy-stor age.news/hawaiian-electric-wants-all-solar-in-next-procurement-round-to-be-paired-with-battery-storage/.

179. Baidawi, A., "Tesla founder puts world's biggest battery in rural Australia," *The New York Times*, December 1, 2017.

180. Matich, B., "Neoen completes expansion of Tesla big battery in Australia," *PV Magazine*, September 3, 2020. www.pv-magazine.com/2020/09/03/neoen-com pletes-expansion-of-tesla-big-battery-in-australia/.

181. "Costa Rica continues to run on 100% renewable energy," The Costa Rica News, April 23, 2015. https://thecostaricanews.com/costa-rica-runs-on-renewable-energy/.

182. Brown, L. R. (with Larsen, J., Roney, J. M., and Adams, E. E.), *The Great Transition: Shifting from fossil fuels to solar and wind energy*, p. 81, Earth Policy Institute, Washington, DC, 2015.

183. Mahtani, N., "Chile, the land of mines, leads the way in solar energy," *El País*, November 7, 2022. https://english.elpais.com/society/2022-11-07/chile-the-land-of-mines-leads-the-way-in-solar-energy.html.

184. Dipaola, A., "Saudis seek up to $50 billion in renewable-energy expansion," Bloomberg Business, January 16, 2017. www.bloomberg.com/news/articles/2017-01-16/saudis-seek-up-to-50-billion-for-first-phase-of-renewables-plan.

185. "DEWA's adoption and development of storage CSP systems ensure Dubai's energy supply security at the lowest cost," Dubai Electricity & Water Authority, November 9, 2022. www.dewa.gov.ae/en/about-us/media-publications/latest-news/2022/11/dewas-adoption-and-development-of-storage-csp-systems-ensure.

186. Usher, B., *Renewable Energy: A primer for the twenty-first century*, p. 106, Columbia University Press, New York, NY, 2019.

187. Vaughan, A., "Solar panels in space could help power the UK by 2039," *New Scientist*, August 28, 2021.

188. Rifkin, *The Third Industrial Revolution*, p. 158.

189. Brown, *The Great Transition*, p. 77.

190. Brown, *The Great Transition*, p. 77.

191. Smil, *Energy and Civilization*, pp. 389, 392.

192. Bonneuil and Fressoz, *The Shock of the Anthropocene*, p. 101.

193. Clynes, T., *The Boy Who Played with Fusion*, p. 106, Faber and Faber, London, 2015.

194. Pielke, R and Byerly, R. "Shuttle programme lifetime cost," *Nature*, 472(38), April 6, 2011. https://doi.org/10.1038/472038d.

195. Borenstein, S., Daly, M., and Phillis, M., "Sweeping climate bill pushes American energy to go green," *AP News*, August 12, 2022. https://apnews.com/article/technology-science-congress-climate-and-environment-f084d23d61ebb068068d4aa92c82fdbb.

196. Fortuna, C., "Paul Krugman: The Inflation Reduction Act is 'A very big deal' for the climate," *CleanTechnica*, August 12, 2022. https://cleantechnica.com/2022/08/12/paul-krugman-the-inflation-reduction-act-is-a-very-big-deal-for-the-cli mate/.

197. Sachs, E., "Testimony to Select Committee on Energy Independence and Global Warming," US House of Representatives, July 28, 2009.

5 The Old Becomes New Again: More Sustainable Energy

1. Gimpel, J., *The Medieval Machine: The Industrial Revolution of the Middle Ages*, p. 12, Holt, Rinehart and Winston, New York, NY, 1976.
2. Cervantes, M., *Don Qixote*, p. 42, Woodsworth Editions, London, 1993 (first published in 1605).
3. Smil, V., *Energy and Civilization: A history*, p. 391, The MIT Press, Cambridge, MA, 2017.
4. Burke, J., *Connections*, p. 24, MacMillan, London, 1978.
5. A Flemish polymath, Stevin also designed a "sand-yacht" that he tested on a beach near The Hague. As noted by Aldersey-Williams, Stevin's infamous *zeilwagen* could carry six people at about 30 mph.
6. Schobert, H. H., *Energy and Society: An introduction*, p. 108, Taylor & Francis, New York, NY, 2002.
7. Smil, *Energy and Civilization*, p. 158.
8. When the wind moves away from the blades, the fantail catches the wind and turns the blades back into the wind, one of the earliest negative feedback systems in a self-regulating machine.
9. Smil, *Energy and Civilization*, p. 160.
10. Having added about 20% to their original land mass, the Dutch are proud to say "God created the world, but the Dutch created the Netherlands."
11. Smil, *Energy and Civilization*, p. 161.
12. Rhodes, R., *Energy: A human history*, p. 328, Simon and Schuster, New York, NY, 2018.
13. $P \propto Av^3$. If you double the blade length, say from 50 to 100 m, the swept area and hence power output increases fourfold ($P \propto A$, $A = \pi r^2$, thus $P \propto r^2$). Double the wind speed, say from 5 to 10 m/s, and the power output increases eightfold ($P \propto v^3$).
14. Rohrig, K., Berkhout, V., Callies, D., et al., "Powering the 21st century by wind energy – Options, facts, figures," *Applied Physics Review*, 6, 031303, August 13, 2019.
15. You can work out the rpm by counting the time for successive blades to pass the central tower, a simple, mesmerizing task: if 2 blades take 1 second to transit the tower, the rotational speed is 20 revolutions per minute (1 second × 1/3 revolution × 60 seconds/minute). The corresponding tip speed is about 100 m/s.
16. Durakovic, A., "World's largest, most powerful wind turbine stands complete," OffshoreWIND.biz, November 12, 2021. www.offshorewind.biz/2021/11/12/worlds-largest-most-powerful-wind-turbine-stands-complete/.
17. Durakovic, A., "18 MW offshore wind turbine launches in China," OffshoreWind.biz, January 6, 2023. www.offshorewind.biz/2023/01/06/18-mw-offshore-wind-turbine-launches-in-china/.
18. Lee, J. and Zhao, F., "Global Wind Report 2022," Global Wind Energy Council, Brussels, April 4, 2022.
19. Gronholt-Pedersen, J., "Denmark sources record 47% of power from wind in 2019," Reuters, January 2, 2020. www.reuters.com/article/us-climate-change-denmark-windpower-idUSKBN1Z10KE.
20. Selby, W. G., "Barack Obama says wind power cheaper in Texas than power from 'dirty fossil fuels'," *PolitiFact*, April 12, 2016. www.politifact.com/factchecks/2016/apr/12/barack-obama/barack-obama-says-wind-power-cheaper-texas-power-d/.

21. Wiser, R. and Bolinger, M., *2015 Wind Technologies Market Report*, pp. v, 7, US Department of Energy, August 2016.
22. Webber, T., "Wind blows by coal to become Iowa's largest source of electricity," *USA Today*, April 16, 2020. https://eu.desmoinesregister.com/story/tech/science/environment/2020/04/16/wind-energy-iowa-largest-source-electricity/5146483002/.
23. Brown, L. R. (with Larsen, J., Roney, J. M., and Adams, E. E.), *The Great Transition: Shifting from fossil fuels to solar and wind energy*, p. 87, Earth Policy Institute, Washington, DC, 2015.
24. Brown et al., *The Great Transition*, p. 86.
25. Roth, S., "In Wyoming wind, a conservative billionaire sees California's future," *The Desert Sun*, February 1, 2017. https://eu.desertsun.com/story/tech/science/energy/2017/02/01/wyoming-wind-philip-anschutz/95452488/.
26. Hernández, J. C., "It can power a small nation. But this wind farm in China is mostly idle," *The New York Times*, January 15, 2017.
27. Fanchi, J. R., *Energy in the 21st Century*, pp. 98–99, World Scientific Publishing Co, Singapore, 2005.
28. Archer, C. L. and Jacobson, M. Z., "Evaluation of global wind power," *Journal of Geophysical Research*, 110, D12110, June 30, 2005. doi:10.1029/2004JD005462.
29. A wind turbine should be built sufficiently far from residential sites. Dutch regulations require a 400-m distance for a 2-MW, 80-m HH turbine and 2.5 km for an 8-MW, 130-m HH turbine to minimize sight and sound impact (HH = hub height). Some countries stipulate 4 HH, a.k.a. the setback distance, while at least 2 km is best for health.
30. MacDonald, M., "'Rare' collapse of 80-metre wind turbine in Cape Breton believed to be a Canadian first," *The National Post*, August 25, 2016.
31. "BrightSource seeks to set 'the record straight' on bird deaths at CSP plant," *Renewable Energy Magazine, August* 20, 2014. www.renewableenergymagazine.com/solar_thermal_electric/brightsource-seeks-to-set-the-record-straight–20140820.
32. Drouin, R., "For the birds (and the bats): 8 ways wind power companies are trying to prevent deadly collisions," *grist*, January 3, 2014. https://grist.org/climate-energy/for-the-birds-and-the-bats-8-ways-wind-power-companies-are-trying-to-prevent-deadly-collisions/.
33. Stross, R., *The Wizard of Menlo Park: How Thomas Alva Edison invented the modern world*, p. 127, Three Rivers Press, New York, NY, 2007.
34. Rostock Business and Technology Development GmbH, "Wind energy regions: Denmark." www.southbaltic-offshore.eu/regions-denmark.html.
35. Brown et al., *The Great Transition*, p. 95.
36. Goodall, C., *Ten Technologies to Save the Planet*, p. 32. *Green*Profile, London, 2008.
37. Harrabin, R., "Boris Johnson: Wind farms could power every home by 2030," BBC, October 6, 2020. www.bbc.com/news/uk-politics–54421489.
38. The Maritime Executive, "European offshore wind capacity jumped in 2018," February 6, 2019. www.maritime-executive.com/article/european-offshore-wind-capacity-jumped-in–2018.
39. Royte, E., "This historic community is pushing the nation toward a wind power revolution," *Smithsonian Magazine*, April 2022. www.smithsonianmag.com/science-nature/historic-community-blockisland-pushing-nation-toward-wind-power-revolution–180979789/.

40. Berwick, A., "Offshore wind may finally take off with big projects, none named Cape Wind," *The Boston Globe*, March 23, 2016. www.bostonglobe.com/maga zine/2016/03/23/offshore-wind-may-finally-take-off-with-big-projects-none-named-cape-wind/FZ9Ng715HYkgKFFoNtlZHN/story.html.

41. Usher, B., *Renewable Energy: A primer for the twenty-first century*, pp. 38–39, Columbia University Press, New York, NY, 2019.

42. Gold, R., *Superpower: One man's quest to transform American energy*, p. 203, Simon & Schuster, New York, NY, 2019.

43. Jacobs, J., "Wind power gains force in oil-rich Gulf of Mexico," p. 8, *The Financial Times*, November 5, 2021.

44. Moskvitch, K., "Will falling oil prices kill wind and solar power?" *Scientific American*, January 22, 2015. www.scientificamerican.com/article/will-falling-oil-prices-kill-wind-and-solar-power/.

45. Eirgrid Group, "Electricity grid to run on 75% variable renewable generation following successful trial," April 7, 2022. www.eirgridgroup.com/newsroom/ electricity-grid-to-run-o/index.xml.

46. Frayer, L., "Tiny Spanish island nears its goal: 100 percent renewable energy," *NPR*, September 28, 2014. www.npr.org/sections/parallels/2014/09/17/ 349223674/tiny-spanish-island-nears-its-goal-100-percent-renewable-energy.

47. Rummelhoff, I., "World's first floating wind farm has started production," *Equinor*, October 18, 2017. www.equinor.com/en/news/worlds-first-floating-wind-farm-started-production.html.

48. Hill, J. S., "World's second floating offshore wind farm takes next big step forward," *CleanTechnica*, October 23, 2018. https://cleantechnica.com/2018/10/ 23/worlds-second-floating-offshore-wind-farm-takes-next-big-step-forward/.

49. Ramirez, V. B., "A Swedish company wants to transform offshore wind with vertical-axis turbines," *Singularity Hub*, September 14, 2022. https://singularity hub.com/2022/09/14/a-swedish-company-wants-to-transform-offshore-wind-with-vertical-axis-turbines/.

50. Griffith, S., *Electrify: An optimist's playbook for our clean energy future*, p. 106, The MIT Press, Cambridge, MA, 2021.

51. Dart, T. and Milman, O., "The wild west of wind: Republicans push Texas as unlikely green energy leader," *The Guardian*, February 20, 2017. www.theguar dian.com/us-news/2017/feb/20/texas-wind-energy-green-turbines-republicans-environment.

52. The upward movement of molten rock is also the mechanism for continental drift because of plate movements in the upper mantle (as well as earthquakes).

53. An air-source heat pump (ASHP) is also available to maintain a constant household temperature year round, operating essentially like a refrigerator or air conditioner in reverse. ASHPs are more efficient, require less maintenance, and are cheaper to run than traditional home-heating systems, but are more expensive and harder to install without assistance.

54. "Enel Green Power sets new geothermal power generation record 2015," ThinkGeoEnergy, February 11, 2016. www.thinkgeoenergy.com/enel-green-power-sets-new-geothermal-power-generation-record–2015/.

55. Enel Green Power, "Stillwater Solar Geothermal Hybrid Project, Fallon, Nevada" [video], March 21, 2016. www.youtube.com/watch?v=U_kayrbkkus.

56. Jones, B. and McKibben, M., "How a few geothermal plants could solve America's lithium supply crunch and boost the EV battery industry," *The Conversation*, March 21, 2022. https://theconversation.com/how-a-few-geother

mal-plants-could-solve-americas-lithium-supply-crunch-and-boost-the-ev-bat
tery-industry–179465.

57. Aldred, J., "Iceland's energy answer comes naturally," *The Guardian*, April 22, 2008. www.theguardian.com/environment/2008/apr/22/renewableenergy .alternativeenergy.

58. "Unique geothermal technology innovation laboratory," *Ocean Energy Resources*, pp. 32–33, 3/4, 2019.

59. Keating, D., "European cities share heating decarbonisation tips," EURACTIV, October 4, 2019. www.euractiv.com/section/energy/news/european-cities-share-heating-decarbonisation-tips/.

60. Slavin, T., "An incinerator with a view: Copenhagen waste plant gets ski slope and picnic area," *The Guardian*, October 26, 2016. www.theguardian.com/cities/2016/oct/26/incinerator-copenhagen-waste-plant-bjarke-ingels-ski-slope.

61. Edelman, L., "Facebook's hyperscale data center warms Odense," *Tech at Meta*, July 7, 2020. https://tech.fb.com/engineering/2020/07/odense-data-center–2/.

62. Winn, Z., "Tapping into the million-year energy source below our feet," *MIT News*, June 28, 2022. https://news.mit.edu/2022/quaise-energy-geothermal–0628.

63. "DOE announces $20 million to lower costs of geothermal drilling," Energy.gov, February 4, 2022. www.energy.gov/articles/doe-announces-20-million-lower-costs-geothermal-drilling.

64. World Energy Council, "World Energy Resources (Geothermal)," 2016.

65. Oppenheimer, D., "How does geothermal drilling trigger earthquakes?" Ask the Experts, *Scientific American*, p. 81, September 2010.

66. Carlson, W. B., *Tesla: Inventor of the electrical age*, pp. 113–115, Princeton University Press, Princeton, NJ, 2013.

67. Tesla would later be acknowledged instead of Marconi as the original patent holder for wireless radio.

68. The world's first water-powered factory was built for cotton spinning in 1771 in Cromford, England (between Nottingham and Sheffield), using Richard Arkwright's spinning frame to turn raw cotton into thread.

69. Jonnes, J., *Empires of Light: Edison, Tesla, Westinghouse, and the race to electrify the world*, p. 281, Random House, New York, NY, 2004.

70. Jonnes, *Empires of Light*, p. 286.

71. Lake Erie lies at 173 m mean elevation, while the mouth of the Niagara River, which empties into Lake Ontario, is at 74 m mean elevation, providing an overall drop of 99 m. The Falls is about 15 km from Lake Ontario.

72. Jonnes, *Empires of Light*, pp. 319–320, 328.

73. Klein, M., *The Power Makers: Steam, electricity, and the men who invented modern America*, p. 352, Bloomsbury Press, New York, NY, 2008.

74. Carborundum (silicon carbide) is an artificial abrasive used to polish gems, second hardest on the Mohs scale at 9 after diamond the hardest at 10.

75. Klein, *The Power Makers*, p. 351.

76. Berton, P., *Niagara: A history of the Falls*, p. 166, Penguin Books, New York, NY, 1998.

77. Berton, *Niagara*, pp. 169–170.

78. Carlson, *Tesla*, p. 163.

79. Cooper, J., "Niagara Falls," *Oscar Wilde in America*, March 7, 2016. https://oscarwildeinamerica.wordpress.com/2016/03/07/niagara-falls/.

80. Klein, *The Power Makers*, p. 346.

81. Shortridge, R. W., "Some early history of hydroelectric power," *Hydro Review*, June 1988.

82. "Run by falls power," reprinted as "A Tesla short story: Tesla's dream – 1893," *Buffalo Gazette*, Saturday, August 28, 2010. www.buffalohistorygazette.net/2010/08/a-tesla-short-story.html.

83. Morus, I. R., *Nikola Tesla and the Electrical Future*, p. 108, Icon Books Ltd., London, 2019.

84. A transformer consists of a pair of wire-wound metal induction coils: a small number of windings in the primary coil beside (or inside) a large number of windings in the secondary coil to increase or "step up" the voltage. At the other end, the process is reversed to decrease or "step down" the voltage. The step-up/step-down voltage is equal to the ratio of the number of windings in the secondary/primary coils. Power is equal to the current times the voltage ($P = IV$), while power loss is proportional to the current squared ($P = I^2R$). Voltage is thus increased to minimize I^2 line-loss during transmission, which correspondingly decreases the current yet keeps the power the same. For example, doubling the voltage to half the current reduces line-loss by a factor of 4.

85. Rhodes, *Energy*, p. 196.

86. Klein, *The Power Makers*, pp. 365, 346.

87. Klein, *The Power Makers*, p. 216.

88. Long-distance power-line losses are about 3.5% per kilometer for HVDC and 6.5% per kilometer for HVAC, although AC is more economical for short distances because of the simplicity of step-up/step-down transformers. Beyond a "break-even distance" of about 600–800 km, however, HVDC is more effective because of zero "skin effect" (all of a wire's cross-sectional area conducts electrical power with DC, but only the central core can for AC) and reduced "coronal losses" (power leakage to the surrounding air, often seen as a bluish discharge arc). HVDC is also better for voltage regulation (no inductance) and for mismatched transnational interconnections (AC is non-standard at 50 Hz and 60 Hz). HVDC is also less disruptive, needing less material (two thin wires for DC, but three thicker three-phase wires for AC) and thus less right-of-way corridor space. Primarily because of parallel-line interference, HVDC is also more economical for subsea transmission above about 50 km.

89. Berton, *Niagara*, p. 217.

90. Berton, *Niagara*, p. 235.

91. *Preservation and Enhancement of the American Falls at Niagara, Appendix D – Hydraulics*, p. D9 (Article 4), American Falls International Board, June 1974.

92. Berton, *Niagara*, p. 314.

93. Bakke, G., *The Grid: The fraying wires between Americans and our energy future*, p. 37, Bloomsbury, New York, NY, 2017.

94. Shortridge, "Some early history of hydroelectric power."

95. Klein, *The Power Makers*, p. 375.

96. Shortridge, "Some early history of hydroelectric power."

97. Montaigne, F., "A river dammed," p. 17, *National Geographic*, April 2001.

98. Walker, G. and King, D., *The Hot Topic: What we can do about global warming*, p. 128, Harcourt Inc., Orlando, 2008.

99. Michaels, A., *The Winter Vault*, p. 233, McClelland & Stewart, Toronto, 2009.

100. Kumagai, J., "The Grand Ethiopian Renaissance Dam gets set to open," *Spectrum*, IEEE, December 30, 2016. https://spectrum.ieee.org/the-grand-ethiopian-renaissance-dam-gets-set-to-open.

101. A "gravity" dam holds the water weight by its own structure, while an "arch" dam transmits the weight of the reservoir water to each riverside.

102. International Hydropower Association, "Hydropower Status Report 2022," London, June 2022. www.hydropower.org/status-report.

103. "Jirau and Santo Antonio dams on Madeira River, Brazil," *Environmental Justice Atlas*, June 23, 2016. https://ejatlas.org/conflict/jirau-and-santo-antonio-dams-on-madeira-river-brazil.

104. Today, short-distance, low-voltage transmission is primarily AC while long-distance, high-voltage and underwater transmission is more commonly DC (HVDC) because DC costs more to step up and step down, while AC has higher capacitive losses during transmission (electric motor matching is also considered in engineering applications).

105. Handwerk, B., "China's Three Gorges Dam, by the numbers," *National Geographic*, June 9, 2006. www.nationalgeographic.com/science/2006/06/china-three-gorges-dam-how-big/.

106. Bache, R., "Niagara's power from the tides," *Popular Science Monthly*, May 1924.

107. Interreg North-West Europe Marine Energy Alliance, "Discover the potential of marine energy," p. 4, 2019.

108. Sofge, E., "The energy fix: Engineering triumphs over wave and tidal forces," *Popular Science*, June 12, 2013. www.popsci.com/science/article/2013-05/future-energy-water.

109. Goodall, *Ten Technologies to Save the Planet*, p. 92.

110. Gimpel, *The Medieval Machine*, p. 23.

111. Kemeny, R., "The Spanish town powered by waves," BBC, January 23, 2023. www.bbc.com/travel/article/20230122-the-spanish-town-powered-by-waves.

112. Goodall, *Ten Technologies to Save the Planet*, p. 91.

113. Sofge, "The energy fix."

114. Fanchi, *Energy in the 21st Century*, p. 106.

115. The European Marine Energy Centre Limited, "Orbital Marine Power: SR250," 2023, www.emec.org.uk/about-us/our-tidal-clients/orbital-marine-power/.

116. Nova Innovation, "Europe Case Study – Shetland Tidal Array," 2016. www.novainnovation.com/markets/scotland-shetland-tidal-array/.

117. Hanley, S., "Harnessing the power of the tides in Scotland," *CleanTechnica*, November 15, 2021. https://cleantechnica.com/2021/11/15/hoarnessing-the-power-of-the-tides-in-scotland/.

118. Hopper, T., "Nova Scotia just fulfilled its longstanding dream to harness the tides with an underwater windmill," *The Financial Post*, November 22, 2016. http://business.financialpost.com/news/energy/nova-scotia-just-fulfilled-its-longstanding-dream-to-harness-the-tides-with-an-underwater-windmill?__lsa=2ae3-6a29.

119. Obermann, E., "Marine renewable energy: Contributing to Canada's low-carbon future," p. 8, *Ocean News & Technology*, September 2019.

120. Klein, *The Power Makers*, p. 374.

121. Klein, *The Power Makers*, p. 415.

122. Carlson, *Tesla*, p. 402.

123. Rhodes, *Energy*, p. 189.

124. Jonnes, *Empires of Light*, p. 70.

125. Edison's Pearl Street power station ran from September 4, 1882, until January 2, 1890, with only a single 3-hour disruption before being destroyed in a fire.

126. Klein, *The Power Makers*, p. 188.

127. Smil, *Energy and Civilization*, p. 304.

128. Numerous energy dashboards are available online to see the real-time grid supply, such as the UK national grid: https://grid.iamkate.com/.

129. Berton, *Niagara*, p. 306.

130. Bakke, *The Grid*, pp. 124–125.
131. Bakke, *The Grid*, pp. 119–122.
132. Roth, S., "Texas blackouts show the power grid isn't ready for climate change," *Los Angeles Times*, February 16, 2021. www.latimes.com/environment/story/2021-02-16/texas-blackouts-california-climate-change.
133. Hackett, S., "New net, new grid," *Future in Review (FiRe)*, Utah, USA, October 6–9, 2015.
134. Bakke, *The Grid*, p. xxii.
135. "Renewable energy: A world turned upside down," *The Economist*, February 25, 2017.
136. Keay, M., "Electricity markets are broken – Can they be fixed?" Oxford Institute for Energy Studies, January 2016. www.oxfordenergy.org/wpcms/wp-content/uploads/2016/01/Electricity-markets-are-broken-can-they-be-fixed-EL-17.pdf.
137. Klein, *The Power Makers*, pp. 280, 382.
138. Cardwell, D., "Solar project lets neighbors exchange energy," *The New York Times*, March 17, 2017.
139. Klein, *The Power Makers*, p. 382.
140. "The green continent: Powering Africa," *The Economist*, November 5, 2022.
141. "Microgrids: An idea whose time has come?" CBC, January 24, 2020. www.cbc.ca/news/technology/what-on-earth-newsletter-microgrids-green-energy-1.5437568.
142. Bakke, *The Grid*, p. 201.
143. Bakke, *The Grid*, p. 199.
144. Bakke, *The Grid*, pp. 207–209.
145. Crowley, K., Steel, A., and Gilblom, K., "Shell says it can be world's top power producer and profit," Bloomberg, March 11, 2019. www.bloomberg.com/news/articles/2019-03-11/shell-says-it-can-be-world-s-top-power-producer-and-make-money.
146. Beresford, M., "Game changers come to Dubai," Blockchain in Dubai, p. 12, *Time: Buzz Business*, April 19, 2018.
147. Reed, S., "Energy plan: Sell less power," *The New York Times*, August 16, 2017.
148. Bakke, *The Grid*, p. 152.
149. Goodall, C., *The Switch: How solar, storage and new tech means cheap power for all*, p. 161, Profile Books, London, 2016.
150. Because of their heavy-metal composition, the secondary batteries NiCd and NiMH need to be recycled and should never be discarded as household waste or put in a fire. Dead-battery receptacles are found in most supermarkets and also now in the base of some city advertising signs.
151. Mars Perseverance continues the line of Li-ion-powered Martian vehicles that must function at minus 60°C.
152. Lee, S.-H. and Mozur, P., "Samsung stumbles in Galaxy Note 7 recall," *The New York Times*, September 17, 2016.
153. Jeong, E.-Y., "Samsung's massive Galaxy Note 7 recall brings battery-maker into focus," *The Wall Street Journal*, September 5, 2016. www.wsj.com/articles/samsungs-massive-galaxy-note-7-recall-brings-battery-maker-into-focus–1473082175.
154. Gibbs, S., "Charged issue: How phone batteries work – And why some explode," *The Guardian*, October 10, 2016. www.theguardian.com/technology/2016/oct/10/samsung-how-batteries-work-smartphones-explode.
155. To lengthen battery lifetime and get more charge cycles, the state of charge (SOC) should not go below 20% (that is, the depth of discharge (DOD) should not exceed

80%). Conversely, when recharging, the SOC should not exceed 80% (that is, DOD should remain below 20%, unless of course you need more charge for a trip). Typically, an upper and lower SOC buffer stops a battery recharging at 95% or discharging at 5%. At full discharge, internal battery resistance increases because of oxidation at the anode from excess Li^+ ions.

156. Sometimes referred to as "capacity," a unit of electric charge is technically measured in coulombs (C), but amp-hours is more common (1 mAh = 3.6 C).
157. McKenzie, H., *Insane Mode: How Elon Musk's Tesla sparked an electric revolution to end the age of oil*, pp. 40–43, Faber and Faber, London, 2018.
158. Hawken, P. (ed.), *Drawdown: The most comprehensive plan ever proposed to reverse global warming*, p. 34, Penguin Books, London, 2017.
159. "US Energy Storage Monitor: Q3 2016 Executive Summary," GTM Research/ ESA, September 2016.
160. "Coal and gas to stay cheap, but renewables still win race on costs," *Bloomberg New Energy Finance*, June 12, 2016. https://about.bnef.com/press-releases/coal-and-gas-to-stay-cheap-but-renewables-still-win-race-on-costs/.
161. Scheer, H., *The Solar Economy: Renewable energy for a sustainable global future*, p. 30, Earthscan, London, 2002.
162. Lambert, F., "Tesla is finally going to expand Gigafactory Nevada," *Electrek*, October 3, 2022. https://electrek.co/2022/10/03/tesla-expand-gigafactory-nevada/.

6 Driving the Revolution Revolution: From Volta to Tesla and Back

1. Morus, I. R., *Nikola Tesla and the Electrical Future*, pp. 72–73, Icon Books Ltd., London, 2019.
2. "On track: Siemens presents the world's first electric railway," Siemens, https:// new.siemens.com/global/en/company/about/history/stories/on-track.html.
3. Klein, M., *The Power Makers: Steam, electricity, and the men who invented modern America*, p. 243, Bloomsbury Press, New York, 2008.
4. Morus, *Nikola Tesla and the Electrical Future*, p. 73.
5. Klein, *The Power Makers*, p. 249.
6. Klein, *The Power Makers*, p. 251.
7. "Batteries 1," Modern Marvels [documentary], Season 13, Episode 30, produced by Don Cambou, USA, August 3, 2006.
8. Stross, R. *The Wizard of Menlo Park: How Thomas Alva Edison invented the modern world*, p. 216, Three Rivers Press, New York, NY, 2007.
9. Stross, *The Wizard of Menlo Park*, p. 236.
10. Stross, *The Wizard of Menlo Park*, p. 237.
11. McKenzie, H., *Insane Mode: How Elon Musk's Tesla sparked an electric revolution to end the age of oil*, p. 29, Faber and Faber, London, 2018.
12. Stross, *The Wizard of Menlo Park*, p. 250.
13. Jonnes, J., *Empires of Light: Edison, Tesla, Westinghouse, and the race to electrify the world*, p. 352, Random House, New York, NY, 2004.
14. Karsten, J. and West, D. M., "Five emerging battery technologies for electric vehicles," TechTank, The Brookings Institution, September 15, 2015. www .brookings.edu/blog/techtank/2015/09/15/five-emerging-battery-technologies-for-electric-vehicles/.

15. As the first major users of distributed electric power, electrified transit systems stimulated the building of neighborhood lighting, commercial machine power, and household electrification along their routes.

16. "People in transport: Golf gets wheels," *People Today,* p. 47, August 11, 1954.

17. "People in transport: Golf gets wheels."

18. Captured disk-brake heat is typically stored as chemical energy in the electric battery, but can also be stored in a capacitor (or supercapacitor) or in a flywheel.

19. Hawken, P. (ed.), *Drawdown: The most comprehensive plan ever proposed to reverse global warming,* p. 149, Penguin Books, London, 2017.

20. A low-speed acoustic vehicle alerting system (AVAS) is now required for electric vehicles.

21. Neil, D., "The 50 worst cars of all time: 1997 GM EV1," #42, *Time,* September 7, 2007. https://content.time.com/time/specials/2007/article/0,28804,1658545_1658544_1658535,00.html.

22. "Revenge of the Electric Car" [documentary], directed by Chris Paine, West Midwest Productions, USA, 2011.

23. "Bu-204: How do lithium batteries work?" Learn About Batteries, Battery University, February 22, 2022. https://batteryuniversity.com/article/bu-204-how-do-lithium-batteries-work.

24. Other high-tech spinoffs financed by former PayPal developers include LinkedIn, YouTube, and Yelp.

25. McKenzie, *Insane Mode,* p. 78.

26. McKenzie, *Insane Mode,* p. 121.

27. "Revenge of the Electric Car."

28. McKenzie, *Insane Mode,* p. 166.

29. Romm, J., *Hell and High Water: The global warming solution,* pp. 143–144, Harper Perennial, New York, NY, 2007.

30. McKenzie, *Insane Mode,* p. 165.

31. "Revenge of the Electric Car."

32. "The future of mobility – Auto China 2020 special," *Biz Talk,* CGTN, October 10, 2020. https://news.cgtn.com/news/784d544d7930575a306c5562684a335a764a4855/index.html.

33. "Revenge of the Electric Car."

34. McKenzie, *Insane Mode,* p. 175.

35. Parkinson, G., "Paris COP21: Climate talks may not matter, because coal and oil will be redundant anyway," *Renew Economy,* December 4, 2015. http://reneweconomy.com.au/paris-cop21-climate-talks-may-not-matter-because-coal-and-oil-will-be-redundant-anyway–15990/.

36. Gifford, D., "The GM strike is really about the switch to electric cars," *Market Watch,* October 5, 2019. www.marketwatch.com/story/the-gm-strike-is-really-about-the-switch-to-electric-cars-2019-10–05.

37. To highlight the basic induction principle, one can make a simple, hand-held "homopolar" motor by placing a metal screw between a disc magnet and a battery and touching a copper wire to the ends (use a 1.5-V AA battery and a neodymium magnet). As you complete the circuit the screw turns.

38. McKenzie, *Insane Mode,* p. 67.

39. Sivak, M. and Schoettle, B., "Relative costs of driving electric and gasoline vehicles in the individual US states," SWT-2018-1, The University of Michigan, Sustainable Worldwide Transportation, January 2018. http://umich.edu/~umtriswt/PDF/SWT-2018-1_Abstract_English.pdf.

40. "US, Europe, & China electric car sales in 2016," *EVObsession*. https://evobses
sion.com/us-europe-china-electric-car-sales–2016/.
41. "Tesla unveils Model 3" [video], Tesla, March 31, 2016. www.youtube.com/
watch?v=Q4VGQPk2Dl8.
42. Randall, T. and Halford, D., "Tesla Model 3 tracker," Bloomberg, January 1, 2020.
www.bloomberg.com/graphics/2018-tesla-tracker/.
43. Recall that power is in kilowatts (kW) and energy in kilowatt-hours (kWh), while
battery capacity is the amount of total stored energy in kWh.
44. Vance, A., *Elon Musk: How the billionaire CEO of SpaceX and Tesla is shaping
our future*, pp. 301, 305, 362, Virgin Books, London, 2015.
45. McKenzie, *Insane Mode*, p. 58.
46. The nomenclature is still changing, but for now: L1 = 120 V, L2 = 240 V, and L3 =
rapid.
47. Of course, total cost will vary with the cost of electricity in one's area and the type
of electric vehicle. A more efficient vehicle will go further on one hour of charge
and thus cost less to operate.
48. "Untethered" cables are portable, while "tethered" cables are permanently
attached to a charging station.
49. "Electric car subsidies largely benefit 'rich' Tesla and Jaguar drivers," Dutch
News, January 30, 2019. www.dutchnews.nl/news/2019/01/electric-car-subsid
ies-largely-benefit-rich-tesla-and-jaguar-drivers/.
50. Caulfield, B., Furszyfer, D., Stefaniec, A., and Foley, A., "Measuring the equity
impacts of government subsidies for electric vehicles," *Energy* 248, 123588,
February 24, 2022.
51. IEA, "A rapid rise in battery innovation is playing a key role in clean energy
transitions," Sustainable Development Scenario, September 22, 2020. www.iea
.org/news/a-rapid-rise-in-battery-innovation-is-playing-a-key-role-in-clean-
energy-transitions.
52. To get an idea of the power we consume, a flat-screen TV is rated at 0.1 kW, a laptop
half that, while the main-hog household appliances are the refrigerator (1.6 kWh
per day), clothes washer (2.3 kWh per load), and clothes dryer (3.3 kWh per load).
53. Musk and other board members would be sued by Tesla shareholders for the
$2.6 billion SolarCity deal, citing a breach of fiduciary duty in acquiring the highly
indebted SolarCity. The suit alleged that as CEO and board chairman of both Tesla
and SolarCity, Musk used controlling pressure to push through the acquisition.
Although the other board members settled the "derivative action," Musk took his
chances at court and was cleared.
54. McKenzie, *Insane Mode*, p. 201.
55. Miller, J., "VW chief's plan to catch Tesla hit by union pledge to protect jobs,"
p. 6, *The Financial Times*, November 5, 2021.
56. Musk, E., *Master Plan, Part Deux*, July 20, 2016. www.tesla.com/blog/master-
plan-part-deux.
57. Cardwell, D., "Tesla ventures into solar power storage," *The New York Times*,
May 2, 2015.
58. Campbell, P., "Tesla's rivals cruise towards the starting line," *The Financial
Times*, September 22, 2018.
59. McKenzie, *Insane Mode*, p. 194.
60. Field, K., "BloombergNEF: Lithium-ion battery cell densities have almost tripled
since 2010," *CleanTechnica*, February 19, 2020. https://cleantechnica.com/2020/
02/19/bloombergnef-lithium-ion-battery-cell-densities-have-almost-tripled-
since–2010/.

61. McKenzie, *Insane Mode*, p. 112.
62. Bradshaw, T., "Apple's car ambitions cloaked in mystery," *The Financial Times*, September 23, 2016.
63. De Vries, J., "First Move with Julia Chatterley," CNN, July 12, 2022.
64. "The First Electric Car for the Masses: Mary Barra Talks Bolt EV and Future of Mobility" [conference], Wired Business, June 29, 2016. www.youtube.com/watch?v=_7aW84SXru4.
65. McKenzie, *Insane Mode*, p. 53.
66. "Gas guzzlers set to fade as China sparks surge for electric cars," *Bloomberg News*, May 21, 2018. www.bloomberg.com/news/articles/2018-05-21/gas-guzzlers-set-to-fade-as-china-sparks-surge-for-electric-cars.
67. Bakker, S., *From Luxury to Necessity: What the railways, electricity and the automobile teach us about the IT revolution*, p. 133, Boom, Amsterdam, 2017.
68. Neslen, A., "VW and Shell accused of trying to block EU push for electric cars," *The Guardian*, April 28, 2016. www.theguardian.com/environment/2016/apr/28/vw-and-shell-try-to-block-eu-push-for-cleaner-cars.
69. McKenzie, *Insane Mode*, p. 147.
70. "Great wheels of China," *The Economist*, April 6, 2019.
71. McKenzie, *Insane Mode*, p. 257.
72. Walz, E., "Singulato Motors CEO Tiger Shen hopes its iS6 EV will help bring blue skies back to Beijing," *Future Car*, October 2, 2017. www.futurecar.com/1497/Singulato-Motors-CEO-Tiger-Shen-Hopes-its-iS6-EV-Will-help-Bring-Blue-Skies-Back-to-Beijing.
73. Hawken (ed.), *Drawdown*, p. 142.
74. Vega, N., "Volkswagen plans 'electric offensive' against Tesla," *The New York Post*, November 16, 2018. https://nypost.com/2018/11/16/volkswagen-plans-electric-offensive-against-tesla/.
75. McKenzie, *Insane Mode*, p. 169.
76. Spring, J., "China's anti-Teslas: cheap models drive electric car boom," Reuters, January 12, 2017. www.reuters.com/article/us-usa-autoshow-china-electric-idUSKBN14V1H3.
77. Brodie, C., "India will sell only electric cars within the next 13 years," *World Economic Forum*, May 23, 2017. www.weforum.org/agenda/2017/05/india-electric-car-sales-only–2030/.
78. Waters, R., "Rivian picked up speed in Tesla's slipstream but it will hit traffic," p. 6, *The Financial Times*, November 5, 2021.
79. Buck, H. J., *Ending Fossil Fuels: Why net zero is not enough*, p. 39, Verso, London, 2021.
80. Ward, E., "Oil groups warned over failure to embrace renewable energy," *The Financial Times*, November 19, 2016.
81. "Roadkill," *The Economist*, August 12, 2017.
82. "Gloom and boom," Special Report: Cars, *The Economist*, April 20, 2013.
83. McKenzie, *Insane Mode*, p. 218.
84. Olsen, P., "Tesla Model 3 gets CR recommendation after braking update," *Consumer Reports*, May 30, 2018. www.consumerreports.org/car-safety/tesla-model-3-gets-cr-recommendation-after-braking-update/.
85. Klayman, B., "GM turns to supplier to build initial EV vans while it readies plant in Canada," Reuters, July 12, 2021. www.reuters.com/business/sustainable-business/exclusive-gm-turns-supplier-build-initial-ev-vans-while-it-readies-plant-canada-2021-07-12/.

86. Carey, N., "UPS-workhorse electric van deal shows progress on charging costs," *Reuters*, February 22, 2018. www.reuters.com/article/ups-workhorse-group-elec tric-vehicles/ups-workhorse-electric-van-deal-shows-progress-on-charging-costs-idUSL2N1Q6117.

87. "Egemin Automation Inc.: Guiding automation," *Industry Today*, July 26, 2016. https://industrytoday.com/guiding-automation/.

88. McKenzie, *Insane Mode*, p. 110.

89. Campbell, P., "Ford and Lyft to develop self-driving taxis by 2021," *The Financial Times*, September 28, 2017.

90. Hook, L. and Water, R., "Waymo puts driverless car project in the fast lane," *The Financial Times*, November 10, 2017.

91. Hawken (ed.), *Drawdown*, pp. 185–186.

92. Nader, R., "Driverless cars: Hype, hubris and distractions," *CounterPunch*, June 27, 2017. www.counterpunch.org/2017/06/27/driverless-cars-hype-hubris-and-distractions/.

93. Nader, R., "Statement by Ralph Nader on Tesla Full Self-Driving (FSD) technology," Nader.org, August 10, 2022. https://nader.org/2022/08/10/statement-by-ralph-nader-on-tesla-full-self-driving-fsd-technology/.

94. Musk, E., Twitter [Internet], August 3, 2014. https://twitter.com/elonmusk/status/495759307346952192?lang=en.

95. Markoff, J., "Want to buy a self-driving car? Trucks may come first," *The New York Times*, May 16, 2016.

96. Prior to the success of the gasoline-powered car, many taxis in the USA were electric, powered by lead–acid batteries and fueling the growth of cab companies in the late 1890s.

97. The market for easy wireless charging of household products is also expanding, but remains in flux. For example, the Samsung Galaxy offers wireless inductive charging with its Power Store, although Apple discarded its AirPower wireless charging mat, citing problems with heat management.

98. Le Page, M., "Parking charge," *New Scientist*, February 16, 2019.

99. McKenzie, *Insane Mode*, p. 132.

100. Hodges, J., "Electric buses are hurting the oil industry," Bloomberg, April 24, 2018. www.bloomberg.com/news/articles/2018-04-23/electric-buses-are-hurting-the-oil-industry.

101. "EU electric bus market grew 50% in 2021 (over 3,000 registrations). Mercedes debuts in the leaders' group," *Sustainable Bus*, February 10, 2022. www.sustain able-bus.com/news/eu-electric-bus-market–2021/.

102. Morris, C., "Europe's largest electric bus fleet adds its 500th vehicle," *EV Engineering News*, December 15, 2020. https://chargedevs.com/newswire/eur opes-largest-electric-bus-fleet-adds-its-500th-vehicle/.

103. "Great wheels of China."

104. McKenzie, *Insane Mode*, p. 132.

105. Mundy, S., "Subsidies set to power up India's electric vehicles," *The Financial Times*, p. 14, March 22, 2019.

106. Hawken (ed.), *Drawdown*, p. 140.

107. "Luxury cruise giant emits 10 times more air pollution (SOx) than all of Europe's cars – study," *Transport & Environment*, June 4, 2019. www.transportenviron ment.org/discover/luxury-cruise-giant-emits-10-times-more-air-pollution-sox-all-europes-cars-study/.

108. "Ampere electric-powered ferry," *Ship Technology*. www.ship-technology.com/projects/norled-zerocat-electric-powered-ferry/.

109. Lambert, F., "A new fleet of all-electric ferries with massive battery packs is going into production," Electrek, March 5, 2018. https://electrek.co/2018/03/05/all-electric-ferries-battery-packs/.
110. "E-ferry final conference: About Ellen," E-ferry. www.conf.eferry.eu/.
111. Viswanathan, V., Epstein, A. H., Chiang, Y.-M., et al., "The challenges and opportunities of battery-powered flight," Nature, 601, pp. 519–525, January 27, 2022. https://doi.org/10.1038/s41586-021-04139-1.
112. Viswanathan, et al., "The challenges and opportunities of battery-powered flight."
113. Larsen, K., "All systems go: 1st all-electric seaplane takes flight in B.C.," CBC News, December 10, 2019, www.cbc.ca/news/canada/british-columbia/electric-seaplane-float-plane-test-flight-harbour-air-1.5390816.
114. Rovnick, N., "EasyJet joins forces to create electric commercial line," The Financial Times, September 28, 2017.
115. "Quest Means Business," CNN, July 21, 2021.
116. The Quiet Achiever made the first solar-powered trans-continental road trip in 1982, taking 20 days to trek the more than 4,000 km across Australia from Perth to Sidney, 10 days fewer than the first gasoline-powered trip and prompting the now biennial World Solar Challenge.
117. Goodall, C., The Switch: How solar, storage and new tech means cheap power for all, p. 187, Profile Books, London, 2016.
118. Lack of telephones in rural areas contributed to a famously bad poll prior to the 1936 presidential election, which projected that Republican Alf Landon would win in a landslide over Democrat Franklin Roosevelt. At the time, two-thirds of Americans didn't have telephones, most of whom were less well-off and ultimately voted Democrat. Roosevelt would win in the most lopsided Electoral College result ever at 98.5%.
119. Bakker, From Luxury to Necessity, p. 95.
120. Leung, W., "Battery storage project slated near Oxnard will be among the nation's biggest," Ventura County Star, May 15, 2020. https://eu.vcstar.com/story/news/2020/05/15/battery-storage-project-oxnard-ventura-energy-storage-tesla-strata-solar/3110101001/.
121. Porritt, J., "Sustainability for all," TEDxExeter, April 2013. www.jonathonporritt.com/talks.
122. "Generation Asset Map, Turlough Hill, Co Wicklow," Irish Electricity Supply Board. www.esb.ie/our-businesses/generation-energy-trading-new/generation-asset-map#turlough-hill.
123. A novel variant of PSH (or PHES for Pumped Hydro Energy Storage) is PHCAES (Pumped Hydro Compressed Air Energy Storage), where the lower reservoir is a depleted underground gas, oil, or water well that uses compressed air to elevate the water to the upper reservoir rather than pumps. Still in the early stages of development, a PHCAES system would cost less by using an existing reservoir with a much smaller upper reservoir, saving space compared to conventional PSH. The potential is enormous, especially in the USA with almost 3 million abandoned oil and gas wells.
124. Reynolds, M., "Depth charge," Wired, January/February 2022.
125. National Renewable Energy Laboratory, "Gemasolar Thermosolar Plant / Solar TRES CSP Project," October 21, 2022. https://solarpaces.nrel.gov/project/gemasolar-thermosolar-plant-solar-tres.
126. Kraemer, S., "Crescent Dunes 24-hour solar tower is online," CleanTechnica, February 22, 2016. https://cleantechnica.com/2016/02/22/crescent-dunes-24-hour-solar-tower-online/.

127. Neslen, A., "Morocco poised to become a solar superpower with launch of desert mega-project," *The Guardian*, October 26, 2015. www.theguardian.com/environment/2015/oct/26/morocco-poised-to-become-a-solar-superpower-with-launch-of-desert-mega-project.
128. "Urbandale Centre for Home Energy Research," Sustainable Building Energy Systems, Carleton University. https://carleton.ca/sbes/research-facilities/urbandale-centre-for-home-energy-research/.
129. Koen, A. and Farres Antunez, P., "How heat can be used to store renewable energy," *The Conversation*, February 25, 2020. https://theconversation.com/how-heat-can-be-used-to-store-renewable-energy–130549.
130. Snieckus, D., "Stiesdal 'hot rocks' energy storage flagship to power up on Danish island of Lolland," *Recharge*, September 2, 2021. www.rechargenews.com/energy-transition/stiesdal-hot-rocks-energy-storage-flagship-to-power-up-on-danish-island-of-lolland/2-1–1061093.
131. McGrath, M., "Climate change: 'Sand battery' could solve green energy's big problem," BBC, July 5, 2022. www.bbc.com/news/science-environment–61996520.
132. Bellini, E., "Hybridizing compressed air, thermal energy storage in post mining infrastructure," *PV Magazine*, July 15, 2022. www.pv-magazine.com/2022/07/15/hybridizing-compressed-air-thermal-energy-storage-in-post-mining-infrastructure/.
133. Clegg, B., *The Graphene Revolution: The weird science of the ultrathin*, p. 138, Icon Books, London, 2018.
134. Reynolds, "Depth charge."
135. Hydrogen gas (H_2) is designated by its own rainbow of colors, depending on the manufacturing process: (1) *green* hydrogen is made via electrolysis of water, where the electricity is generated from a renewable-energy source (for example, wind or solar) and no CO_2 produced, (2) *gray* hydrogen via steam reformation of natural gas that releases CO_2 into the atmosphere, (3) *black* hydrogen via coal gasification that also releases CO_2 into the atmosphere, (4) *blue* hydrogen via coal or natural gas where the CO_2 is sequestered (essentially gray or black hydrogen + CCS), (5) *turquoise* hydrogen from the pyrolysis of methane (with CCS), (6) *purple* or *pink* hydrogen, where the electricity is generated via nuclear power. One may also hear of *brown* hydrogen produced from brown coal (lignite) or *black* hydrogen from black coal (bituminous). Natural hydrogen is degassed from shallow subsurface stores (also during fracking) (*gold* or *white*) or pumped out by water from deep underground stores (*orange*). As of 2022, 95% of hydrogen gas production is gray, generated by steam reformation of methane. (Because of a much lower material density than natural gas, hydrogen gas is pumped at higher speeds to achieve the same energy output.)
136. "Briefing the hydrogen economy: A very big balancing act," *The Economist*, October 9, 2021.
137. Eames, I., Austin, M., and Wojcik, A., "Injection of gaseous hydrogen into a natural gas pipeline," *International Journal of Hydrogen Energy*, May 31, 2022. https://doi.org/10.1016/j.ijhydene.2022.05.300.
138. Kluth, A., "How hydrogen is and isn't the future of energy," Bloomberg, November 9, 2020. www.bloomberg.com/opinion/articles/2020-11-09/how-hydrogen-is-and-isn-t-the-future-of-energy-for-the-eu-china-japan.
139. Seba, T. "Clean Energy & Transportation: Market & Investment Opportunities" [presentation], CSP BUS227, Stanford, May 12, 2014. www.youtube.com/watch?v=23lz9ercqvA.

140. Muskus, J., "Getting on the hydrogen highway," in The Elements (Special Issue), *Bloomberg Businessweek,* September 2, 2019.
141. Goodall, *The Switch*, pp. 253–254, 232.
142. Liebreich, M., "Separating hype from hydrogen – Part one: The supply side," *BloombergNEF*, October 8, 2020. https://about.bnef.com/blog/liebreich-separat ing-hype-from-hydrogen-part-one-the-supply-side/.
143. "Briefing the hydrogen economy."
144. Vaughn, A., "Hope or hype?" *New Scientist*, February 6, 2021.
145. Sánchez Molina, P., "New European project to drive 95 GW of solar and 67 GW of hydrogen," *PV Magazine*, February 12, 2021. www.pv-magazine.com/2021/02/ 12/new-european-project-to-drive-95-gw-of-solar-and-67-gw-of-hydrogen/.
146. "ExxonMobil funds algal biofuels research collaboration at Colorado School of Mines," Colorado School of Mines, May 19, 2015. www.minesnewsroom.com/ press-releases/exxonmobil-funds-algal-biofuels-research-collaboration-color ado-school-mines.
147. "Scientists create 'artificial leaf' that turns carbon dioxide into fuel," *Waterloo News,* November 4, 2019. https://uwaterloo.ca/news/news/scientists-create-artifi cial-leaf-turns-carbon-dioxide-fuel.
148. Hand, E., "Hidden hydrogen: Does earth hold vast stores of a renewable, carbon-free fuel?" *Science*, February 16, 2023. www.science.org/content/article/hidden-hydrogen-earth-may-hold-vast-stores-renewable-carbon-free-fuel.
149. Hand, "Hidden hydrogen."
150. Wald, M. L., "Questions about a hydrogen economy," *Scientific American*, May 2004.
151. Pyper, J., "Hydrogen buses struggle with expense," *Scientific American*, December 16, 2013.
152. Block, D. and Brooker, P., "2015 electric vehicle market summary and barriers," University of Central Florida, June 2016. www.fsec.ucf.edu/en/publications/pdf/ FSEC-CR-2027-16.pdf
153. "Germany launches world's first hydrogen-powered train," France 24, September 17, 2018. www.france24.com/en/20180917-germany-rolls-out-worlds-first-hydrogen-train.
154. Parnell, J., "Shell, Volvo and Daimler back hydrogen as Europe sets its sights on truck emissions," gtm, December 15, 2020. www.greentechmedia.com/articles/ read/shell-volvo-daimler-back-hydrogen-as-europe-turns-sights-on-truck-emissions.
155. Muller, R. A., *Energy for Future Presidents: The science behind the headlines*, p. 237, W. W. Norton and Company, New York, NY, 2012.
156. Hydrogen is combustible when mixed with oxygen: 10% is flammable and 20% explosive. A modern gas-filled airship – a.k.a. dirigible, blimp, or zeppelin – uses helium gas, while a hot-air balloon rises by heated air, both of which are safer than hydrogen gas.
157. Griffith, S., *Electrify: An optimist's playbook for our clean energy future*, p. 52, The MIT Press, Cambridge, MA, 2021.
158. Muller, *Energy for Future Presidents*, p. 238.
159. Muller, *Energy for Future Presidents*, p. 272.
160. Seba, T. "Clean Energy & Transportation: Market & Investment Opportunities" [presentation], CSP BUS227, Stanford, May 12, 2014. www.youtube.com/watch? v=23lz9ercqvA.
161. The efficiency of a fuel cell is limited by the Gibbs free energy, while the efficiency of an internal combustion engine is limited by the Carnot efficiency.

Typically, a fuel cell is much more efficient than an internal combustion engine. Depending on the drive cycle and regenerative braking, an electric engine is much more efficient than either a fuel cell or combustion engine.

162. Gates, B., *How to Avoid a Climate Disaster: The solutions we have and the breakthroughs we need*, p. 147, Allen Lane, London, 2021.
163. Somara, S. and Harries, D., "Flying to the edge of space using only solar power," CGTN: RAZOR, December 11, 2020. https://newseu.cgtn.com/news/2020-12-11/Flying-to-the-edge-of-space-using-only-solar-power-RAZOR-VXiAFUPwVG/index.html.
164. Muller, *Energy for Future Presidents*, p. 256.
165. Muller, *Energy for Future Presidents*, p. 255.
166. Jaskula, B., "Lithium," US Geological Survey, Mineral Commodity Summaries, January 2023. https://pubs.usgs.gov/periodicals/mcs2023/mcs2023-lithium.pdf.
167. Draper, R., "This metal is powering today's technology – At what price?" *National Geographic*, February 2019.
168. Sanderson, H., "Lithium miners pin hopes on 'unstoppable' shift to electric cards," *The Financial Times*, January 17, 2020.
169. Jaskula, "Lithium."
170. West, J. and Paes-Braga, B., "Podcast: Lithium X Energy CEO on why Argentina is the place to be," MidasLetter, *Financial Post*, September 20, 2016. https://financialpost.com/midas-letter/podcast-lithium-x-energy-ceo-on-why-argentina-is-the-place-to-be.
171. Home, A., "What price lithium, the metal of the future?" Reuters, June 6, 2016. www.reuters.com/article/us-lithium-batteries-ahome-idUSKCN0YS19D.
172. Sanderson, H. and Schipani, A., "Bolivia makes first shipment of lithium to China," *The Financial Times*, August 17, 2016.
173. Prashad, V. and Zúñiga Silva, T., "Chile's lithium provides profit to the billionaires but exhausts the land and the people," *CounterPunch*, August 2, 2022. www.counterpunch.org/2022/08/02/chiles-lithium-provides-profit-to-the-billionaires-but-exhausts-the-land-and-the-people/.
174. Morland, S. and Torres, N., "Analysis: Lithium experts skeptical on success of Mexico's state-run miner," Reuters, August 26, 2022. www.reuters.com/markets/commodities/lithium-experts-skeptical-success-mexicos-state-run-miner-2022-08-26/.
175. Risen, J., "US identifies vast mineral riches in Afghanistan," *The New York Times*, June 13, 2010.
176. Penn, I. and Lipton, E., "The lithium gold rush: Inside the race to power electric vehicles," *The New York Times*, May 6, 2021. www.nytimes.com/2021/05/06/business/lithium-mining-race.html.
177. Chao, J., "Quantifying California's Lithium Valley: Can it power our EV revolution?" Berkeley Lab, February 16, 2022. https://newscenter.lbl.gov/2022/02/16/quantifying-californias-lithium-valley-can-it-power-our-ev-revolution/.
178. Salgado, R. M., Danzi, F., Oliveira, J. E., et al., "The latest trends in electric vehicles batteries," *Molecules*, 26, 3188, 2021. https://doi.org/10.3390/molecules26113188.
179. "Lithium-ion battery inventor introduces new technology for fast-charging, noncombustible batteries," *UT News*, February 28, 2017. https://news.utexas.edu/2017/02/28/goodenough-introduces-new-battery-technology.
180. Jung, A.-S., "LG Energy Solution readies $11bn IPO to take on Chinese EV battery rivals," *Financial Times*, January 11, 2022.

181. Bradsher, K. and Forsythe, M., "Why a Chinese company dominates electric car batteries," *New York Times*, December 22, 2021.

182. "The EV revolution: Cell-side analysis," *The Economist*, August 20, 2022.

183. "Lithium-ion battery inventor introduces new technology for fast-charging, noncombustible batteries," *UT News*, February 28, 2017. https://news.utexas.edu/2017/02/28/goodenough-introduces-new-battery-technology/.

184. Williams, M., "Asphalt helps lithium batteries charge faster," Rice University, October 2, 2017. https://news2.rice.edu/2017/10/02/asphalt-helps-lithium-batteries-charge-faster/.

185. Carrington, D., "Electric car batteries with five-minute charging times produced," *The Guardian*, January 19, 2021. www.theguardian.com/environment/2021/jan/19/electric-car-batteries-race-ahead-with-five-minute-charging-times.

186. Carrington, "Electric car batteries."

187. Flow batteries are like an inside-out battery, where the electrolyte is outside the electrodes. The vanadium redox (V-flow) battery uses multiple valence states of liquid vanadium to exchange ions through a fuel-cell/battery membrane (V^{2+}, V^{3+}, V^{4+}, V^{5+}) rather than metals of differing electronegativities. V-flow batteries are longer lasting than Li-ion and PbA batteries (10,000 cycles and 20 years), stay completely discharged for long periods (easier to maintain), and easily scale using larger tanks. Goodall notes that 1 MWh of backup power could fit in a standard stackable 40-foot-long shipping container, making V-flow ideal for large grid storage.

188. Chu, S., "Building a Sustainable Energy Future" [presentation], United Nations Climate Change Conference, COP-16, Cancún, Mexico, December 6, 2010. www.youtube.com/watch?v=c980D0VSZsg.

189. "The future of energy: Batteries included?" *The Economist*, February 2, 2013.

190. Viswanathan, et al., "The challenges and opportunities of battery-powered flight."

191. Carrington, "Electric car batteries."

192. Goodall, *The Switch*, p. 173.

193. Henze, V., "Battery pack prices fall as market ramps up with market average at $156/kWh in 2019," *Bloomberg NEF*, December 19, 2019. https://about.bnef.com/blog/battery-pack-prices-fall-as-market-ramps-up-with-market-average-at-156-kwh-in-2019/.

194. Partridge, J., "Electric cars 'will be cheaper to produce than fossil fuel vehicles by 2027'," *The Guardian*, May 9, 2021. www.theguardian.com/business/2021/may/09/electric-cars-will-be-cheaper-to-produce-than-fossil-fuel-vehicles-by-2027.

195. Goodall, *The Switch*, p. 179.

196. Reichl, C. and Schatz, M., *World Mining Data 2021*, Volume 36, Federal Ministry of Agriculture, Regions and Tourism, Vienna, April 27, 2021. www.world-mining-data.info/wmd/downloads/PDF/WMD2021.pdf.

197. International Energy Agency, "The role of critical minerals in clean energy transitions," International Energy Agency, Paris, 2021. www.iea.org/reports/the-role-of-critical-minerals-in-clean-energy-transitions.

198. Ings, S., "Zero carbon, high costs" (review of *The Rare Metals War* by Guillaume Pitron), p. 31, *New Scientist*, January 30, 2021.

199. McGee, P., "This Tesla co-founder has a plan to recycle your EV batteries," *Financial Times*, September 15, 2021. www.ft.com/video/d59734d9-5745-46d1-afbe-622c200b3783.

200. McGee, "This Tesla co-founder."

201. Nehrenheim, E., "Old batteries get new life," p. 70, *Wired Annual*, 2022.

202. Jaskula, "Lithium."

203. Chandan, A., "Batteries will get greener," p. 32, *Wired Annual*, 2022.
204. Stafford, J., "Move over oil – Lithium is the future of transportation, OilPrice.com, June 21, 2016. https://oilprice.com/Energy/Energy-General/Move-Over-Oil-Lithium-Is-The-Future-Of-Transportation.html.
205. Klein, *The Power Makers*, p. 314.
206. Klein, *The Power Makers*, p. 320.
207. Morus, *Nikola Tesla and the Electrical Future*, p. 124.
208. Klein, *The Power Makers*, p. 319.
209. Gimpel, J., *The Medieval Machine: The Industrial Revolution of the Middle Ages*, pp. 229–230, Holt, Rinehart and Winston, New York, NY, 1976.
210. Goodall, C., *Ten Technologies to Save the Planet*, p. 5, GreenProfile, London, 2008.
211. Rifkin, J., *The Third Industrial Revolution: How lateral power is transforming energy, the economy, and the world*, pp. 208, 3538, Palgrave Macmillan, New York, NY, 2011.
212. Bakker, *From Luxury to Necessity*, pp. 222–223.
213. Diamandis, P., "Solar energy revolution: A massive opportunity," *Forbes Magazine*, September 2, 2014. www.forbes.com/sites/peterdiamandis/2014/09/02/solar-energy-revolution-a-massive-opportunity.
214. Patringenaru, I., "$39 million grant to better integrate renewables into power grid," thisweek@ucsandiego, October 29, 2020. https://ucsdnews.ucsd.edu/feature/39-million-grant-to-better-integrate-renewables-into-power-grid.
215. "Renault rolling-out V2G trials across Europe," electrive.com, March 24, 2019. www.electrive.com/2019/03/24/renault-kicks-off-v2g-projects-in-utrecht-porto-santos/.
216. "Ford unveils electric version of F-150 pickup truck – And it can power your house for 3 days," CBC News, May 19, 2021. www.cbc.ca/news/business/ford-electric-f150-truck-1.6033607.
217. Halvorson, B., "Tesla lineup to get bidirectional charging capability by 2025," *Green Car Reports*, March 2, 2023. www.greencarreports.com/news/1138917_tesla-lineup-to-get-bidirectional-charging-capability-by–2025.
218. Barber, G., "Melbourne's first Tesla Powerwall home battery system – One year on," *One Step Off the Grid*, February 15, 2017. https://onestepoffthegrid.com.au/melbournes-first-tesla-powerwall-home-battery-system-one-year/.
219. Roberts, D., "California solves batteries' embarrassing climate problem," *Vox*, December 2, 2019. www.vox.com/energy-and-environment/2019/12/2/20983341/climate-change-california-batteries-emissions-watttime.
220. Goodall, *Ten Technologies to Save the Planet*, p. 41.
221. Bakker, *From Luxury to Necessity*, pp. 78, 88.
222. Bonneuil, C. and Fressoz, J.-B. (trans. Fernbach, D.), *The Shock of the Anthropocene*, p. 115, Verso, London, 2017.
223. Klein, *The Power Makers*, pp. 418, 432.
224. Klein, *The Power Makers*, pp. 411, 413.
225. Klein, *The Power Makers*, p. 432.
226. Rifkin, *The Third Industrial Revolution*, p. 211.
227. Tweed, K., "Survey: 76% of consumers don't trust their utility," Greentech Media, July 8, 2013. www.greentechmedia.com/articles/read/consumer-trust-in-utilities-continues-to-nosedive.
228. Rifkin, *The Third Industrial Revolution*, p. 45.

229. Pyper, J. "SGCC's latest consumer-pulse study finds high levels of interest in grid edge tech, but also hesitancy," Greentech Media, June 9, 2017. www.greentech media.com/articles/read/survey-what-electricity-customers-really-want.
230. Sandys, L., "We're not costing energy correctly: Reward clean energy optimisation, not maximum generation," *Energy Post*, July 23, 2021. https:// energypost.eu/were-not-costing-energy-correctly-reward-clean-energy-optimisa tion-not-maximum-generation/.
231. Griffith, *Electrify*, p. 36.
232. Klein, *The Power Makers*, p. 433.
233. Klein, *The Power Makers*, pp. 439–440.
234. Klein, *The Power Makers*, p. 443.
235. Musk, E., "The future we're building – and boring," TED Talk, April 2017. www .ted.com/talks/elon_musk_the_future_we_re_building_and_boring/transcript.
236. Johnson, S., *The Invention of Air*, p. 126, Penguin Books, London, 2009.
237. Kuhn, T. S., *The Structure of Scientific Revolutions*, 2nd ed., p. 23, The University of Chicago, Chicago, IL, 1970.

7 Rethink, Rebuild, Rewire

1. Day and night rates are often different: where we live, roughly 15 cents/kWh during the day and 6.7 cents at night (midnight to 8 a.m.). Some appliances have higher start-up draws, such as air conditioners and induction appliances, while refrigerator condensers run intermittently – average rates are used. Overall daily consumption of 30 kWh and average daily cost is based on a 5-month winter period from November to March.
2. "What is the rest of the world doing?" p. 63, *The Sunday Times Magazine*.
3. Del Chiaro, B. and Telleen-Lawton, T., "Solar water heating: How California can reduce its dependence on natural gas," Environment California, April 2007.
4. Smil, V., "How to save ourselves," *New Scientist*, January 7, 2023.
5. Fitzgerald-Redd, S., "Ninety percent of US homes are under insulated," *PR Newswire*, September 30, 2015. www.prnewswire.com/news-releases/ninety-per cent-of-us-homes-are-under-insulated-300151277.html.
6. Goodall, C., *Ten Technologies to Save the Planet*, p. 271, *Green*Profile, London, 2008.
7. Rifkin, J., *The Third Industrial Revolution: How lateral power is transforming energy, the economy, and the world*, p. 211, Palgrave Macmillan, New York, NY, 2011.
8. Steffen, L., "Ultra-white paint reflects 98% of heat from the Sun," *Intelligent Living*, July 13, 2020. www.intelligentliving.co/ultra-white-paint-reflects-heat-from-sun/.
9. "The 21st Century Living Project" [documentary], directed by Anna Meneer, Eden Project, United Kingdom, January 11, 2011. www.youtube.com/watch? v=hTLwY_1_yD0.
10. "The 21st Century Living Project."
11. "21st Century Living Project," The Eden Project. www.edenproject.com/eden-story/our-ethos/21st-century-living-project.
12. Flintoff, J.-P., "After lunch we'll save the planet," p. 18, *The Sunday Times Magazine*, November 29, 2009.
13. McKibben, B., *Earth: Making a life on a tough new planet*, p. 176, Times Books, New York, NY, 2010.

14. Chivers, D., "Cowspiracy: Stampeding in the wrong direction?" *New Internationalist*, February 10, 2016. https://newint.org/blog/2016/02/10/cowspiracy-stampeding-in-the-wrong-direction.
15. Stuart, T., *Waste: Uncovering the global food scandal*, pp. 139, 142, Penguin Books, London, 2009.
16. "Food Waste Index Report 2021," United Nations Environment Programme, Nairobi, 2021. https://wedocs.unep.org/bitstream/handle/20.500.11822/35280/FoodWaste.pdf.
17. Stuart, *Waste*, p. 90.
18. Stuart, *Waste*, p. 202.
19. Stuart, *Waste*, pp. 215, 217.
20. Gillis, J. and Harvey, H., "Cars are ruining our cities," *The New York Times*, April 25, 2018. www.nytimes.com/2018/04/25/opinion/cars-ruining-cities.html.
21. Wick, J., "L.A. taps the brakes on freeway expansion," *CityLab*, March 12, 2018. www.citylab.com/transportation/2018/03/la-says-no-to-freeway-expansion/555353/.
22. Schmitt, A., "Seattle's viadoom: The 'Carmageddon' that wasn't," Streetsblog USA, January 24, 2019. https://usa.streetsblog.org/2019/01/24/seattles-viadoom-the-carmageddon-that-wasnt/.
23. Van Mead, N., "Paris is banning traffic from half the city. Why can't London have a car-free day?" *The Guardian*, September 22, 2016. www.theguardian.com/cities/2016/sep/22/paris-ban-traffic-london-world-car-free-day.
24. Hull, R., "Air pollution in central London dropped 89% on Sunday, as cars were barred from streets for the marathon," ThisisMoney.co.uk, April 23, 2018. www.thisismoney.co.uk/money/cars/article-5646509/Air-pollution-London-fell-89-marathon-driving-restrictions.html.
25. Sallee, J., "Vehicle charge zones: Too high, and driver detours can increase emissions," energypost.eu, November 25, 2021. https://energypost.eu/vehicle-charge-zones-too-high-and-driver-detours-can-increase-emissions/.
26. Hawken, P. (ed.), *Drawdown: The most comprehensive plan ever proposed to reverse global warming*, p. 145, Penguin Books, London, 2017.
27. "Idle stop–start technology," Natural Resources Canada, April 9, 2018. www.nrcan.gc.ca/energy/efficiency/energy-efficiency-transportation-and-alternative-fuels/choosing-right-vehicle/tips-buying-fuel-efficient-vehicle/factors-affect-fuel-efficiency/idle-stop-start-technology/21020.
28. Todhunter, C., "I want to ride my bicycle: Urban planning and the Danish concept of 'hygge,'" *CounterPunch*, April 30, 2018. www.counterpunch.org/2018/04/30/i-want-to-ride-my-bicycle-urban-planning-and-the-danish-concept-of-hygge/.
29. van der Zee, R., "Story of cities #30: How this Amsterdam inventor gave bike-sharing to the world," *The Guardian*, April 26, 2016. www.theguardian.com/cities/2016/apr/26/story-cities-amsterdam-bike-share-scheme.
30. "Segway-Ninebot: Torque of the town," *The Economist*, November 24, 2018.
31. Schiller, T., Scheidl, J., and Pottebaum, T., "Car sharing in Europe business models, national variations and upcoming disruptions," Monitor Deloitte, June 2017. www2.deloitte.com/content/dam/Deloitte/de/Documents/consumer-industrial-products/CIP-Automotive-Car-Sharing-in-Europe.pdf.
32. Rifkin, *The Third Industrial Revolution*, pp. 123–124.
33. Musk, J., "Bolloré's Bluecity electric car sharing launched in Hammersmith & Fulham," *Autovolt*, June 12, 2017. www.autovolt-magazine.com/bollores-bluecity-electric-car-sharing-launched-in-hammersmith-fulham/.

34. Berman, B., "Are electric cars only for the rich? Sacramento is challenging that notion," *The New York Times*, January 24, 2019.
35. "Stop and go," *The Economist*, August 3, 2019.
36. Webb, R., "A transport revolution," *New Scientist*, January 9, 2021.
37. Kunzig, R., "The city solution," *National Geographic*, December 2011.
38. Hawken (ed.), *Drawdown*, p. 138.
39. The less-than-ambitious CORSIA agreement will try to limit the growth of airline emissions by "carbon offsets," such as tree planting, investing in a wind farm, solar farm, geothermal plant, or other green energy projects such as installing solar panels, helping purchase efficient stoves, heating, or clean-water systems in developing countries. Some consider carbon offsets the latest corporate "greenwash," especially because airlines could help by stopping the industry-wide practice of "tinkering," that is, loading up on cheap fuel in distant locations to save money and then flying with the excess fuel. Sometimes the savings are miniscule.
40. Helm, D., *Net Zero: How we can stop climate change*, p. 140, William Collins, London, 2020.
41. Smil, V., *Energy and Civilization: A history*, p. 354, The MIT Press, Cambridge, MA, 2017.
42. Cosgrove, B., "'Throwaway living': When tossing out everything was all the rage," *Time*, May 15, 2014.
43. "EU Circular Economy Action Plan," European Commission, 2015. https://ec.europa.eu/environment/circular-economy/first_circular_economy_action_plan.html.
44. "What goes in my blue bin?" City of Toronto. www.toronto.ca/services-payments/recycling-organics-garbage/houses/what-goes-in-my-blue-bin/.
45. "Amsterdam rounds up more plastic waste, after abolishing collection points," Dutch News, November 5, 2022. www.dutchnews.nl/news/2022/11/amsterdam-rounds-up-more-plastic-waste-after-abolishing-collection-points/.
46. Behm, J., "All the US recycling facts you need to know," Dumpsters.com, September 7, 2022. www.dumpsters.com/blog/us-recycling-statistics.
47. Stuart, *Waste*, p. 225.
48. Parker, L., "We made plastic. We depend on it. Now we're drowning in it," *National Geographic*, June 2018. www.nationalgeographic.com/magazine/2018/06/plastic-planet-waste-pollution-trash-crisis/.
49. Royte, E., "Corn plastic to the rescue," *Smithsonian Magazine*, August 2006. www.smithsonianmag.com/science-nature/corn-plastic-to-the-rescue–126404720/.
50. Quicker, P., "Evaluation of recent developments regarding alternative thermal waste treatment with a focus on depolymerisation processes," in F. Winter et al. (eds.), *Waste Management, Volume 9, Waste-to-Energy*, Thomé-Kozmiensky, Neurippen, 2019.
51. Hazen, R. M., *Symphony in C: Carbon and the evolution of (almost) everything*, p. 159, William Collins, London, 2019.
52. Ball, P., "Plastic-eating bugs? It's a great story – But there's a sting in the tail," *The Guardian*, April 25, 2017. www.theguardian.com/commentisfree/2017/apr/25/plastic-eating-bugs-wax-moth-caterpillars-bee.
53. Royte, "Corn plastic to the rescue."
54. Royte, "Corn plastic to the rescue."
55. Knapton, S., "Scientists create eco-friendly plastic from tree molecules," *The Daily Telegraph*, July 5, 2018.
56. Thiagarajan, K., "The man who paves India's roads with old plastic," *The Guardian*, July 9, 2018. www.theguardian.com/world/2018/jul/09/the-man-who-paves-indias-roads-with-old-plastic.

57. "Would you like sugar cane in that?" *The Economist*, November 14, 2020.
58. Harris, J., "Powering down the PC would be big cost-saver," *The Financial Post*, August 14, 2009.
59. Daly, S., "When it rains here in California, it is front-page news on the '*Los Angeles Times*,'" *The Irish Times*, December 2, 2015. www.irishtimes.com/life-and-style/generation-emigration/when-it-rains-here-in-california-it-is-front-page-news-on-the-los-angeles-times-1.2450779.
60. Schlossberg, T., "Japan's worries on climate change," *The New York Times*, December 7, 2016.
61. "France launches 'sobriety' energy savings drive to avoid winter power cuts," *Euronews*, October 6, 2022. www.euronews.com/2022/10/06/france-launches-sobriety-energy-savings-drive-to-avoid-winter-power-cuts.
62. MacArthur, E., "Princess of Asturias Award for International Cooperation 2022" (speech), Oviedo, Spain, October 28, 2022. www.fpa.es/en/princess-of-asturias-awards/laureates/2022-ellen-macarthur.html?texto=discurso&especifica=0.
63. Rifkin, *The Third Industrial Revolution*, pp. 224, 225.
64. Oxfam, "A billionaire emits a million times more greenhouse gases than the average person," November 7, 2022. https://westafrica.oxfam.org/en/latest/press-release/billionaire-emits-million-times-more-greenhouse-gases-average-person.
65. Orange, R., "Greta Thunberg's train journey through Europe highlights no-fly movement," *The Guardian*, April 26, 2019. www.theguardian.com/environment/2019/apr/26/greta-thunberg-train-journey-through-europe-flygskam-no-fly.
66. Buck, H. J., *Ending Fossil Fuels: Why net zero is not enough*, p. 54, Verso, London, 2021.
67. Helm, D., *The Carbon Crunch: Revised and updated*, p. 217, Yale University Press, New Haven, CT, 2015.
68. Buck, *Ending Fossil Fuels*, p. 180.
69. Leopold, A., *A Sand County Almanac and Sketched Here and There*, p. xxii, Penguin Books, London, 2020 (first published 1949).
70. Pacala, S. and Socolow, R., "Stabilization wedges: Solving the climate problem for the next 50 years with current technologies," *Science*, 305(5686), 968–972, August 13, 2004. DOI: 10.1126/science.1100103.
71. Harvey, F., "François Hollande calls for 'miracle' climate agreement at Paris talks," *The Guardian*, May 20, 2015.
72. The USA tried to weaponize cloud seeding in more than 2,600 aerial missions during the Vietnam War to induce artificial rain on the Ho Chi Minh trail, leading to the 1977 UN convention against climate manipulation. The Soviets may have tried to produce artificial rain via cloud seeding at Chernobyl in 1986 to stop a radioactive plume reaching Moscow.
73. Goodstein, D., *Out of Gas: The end of the age of oil*, p. 102, W. W. Norton & Company, New York, NY, 2004.
74. Gates, B., *How to Avoid a Climate Disaster: The solutions we have and the breakthroughs we need*, pp. 176–177, Allen Lane, London, 2021.
75. Goodall, *Ten Technologies to Save the Planet*, p. 275.
76. Walker, G. and King, D., *The Hot Topic: What we can do about global warming*, p. 96, Harcourt Inc., Orlando, FL, 2008.
77. After the last ice age about 12,000 years ago, *Homo sapiens* migrated north across the Earth to begin their dominance over the planet, especially with the advent of agriculture.
78. Crane, N., *Why Geography Matters: A brief guide to the planet*, p. 110, Weidenfeld and Nicolson, London, 2020.

79. Lewis, S. J. and Maslin, M. A., *The Human Planet: How we created the Anthropocene*, p. 141, Pelican Books, London, 2018.
80. Lewis and Maslin, *The Human Planet*, p. 272.
81. Lewis and Maslin, *The Human Panet*, p. 315.
82. Leopold, *A Sand County Almanac*, p. 144.
83. Bonneuil, C. and Fressoz, J.-B. (trans. Fernbach, D.), *The Shock of the Anthropocene*, pp. 3, 16, 53, Verso, London, 2017.
84. Lovelock, J., *Novacene: The coming age of hyperintelligence*, pp. 37, 39, Penguin Books, London, 2020.
85. Bonneuil and Fressoz, *The Shock of the Anthropocene*, pp. 50, 51, 145–146.
86. Bonneuil and Fressoz, *The Shock of the Anthropocene*, pp. 71, 116, 122, 222.
87. Gore, A., *Truth to Power*, p. 218, Rodale Inc., New York, NY, 2017.
88. Newbery, C., "Green campaign looks to youth and the altered habits of adults," *The Financial Times*, April 4, 2017.
89. Greta Thunberg tells world leaders "you are failing us," as nations announce fresh climate action, Climate Action Summit, UN News, September 24, 2019. www.un .org/development/desa/youth/news/2019/09/greta-thunberg/.
90. McKibben, B., *Earth: Making a life on a tough new planet*, p. 88, Times Books, New York, NY, 2010.
91. Burke, J., *Connections*, p. 293, MacMillan London Limited, London, 1978.
92. Worth, J., "Powering up to zero," *New Internationalist*, December 2010.
93. Schwartz, J., "Differences on how to fight climate change," *The New York Times*, July 13, 2016.
94. Gates, *How to Avoid a Climate Disaster*, p. 197.
95. Romm, J., *Hell and High Water: The global warming solution*, p. 237, Harper Perennial, New York, NY, 2007.
96. von Neumann, J., "Can we survive technology?" *Fortune Magazine*, June 1955.

Select Bibliography

Baggott, J., *Atomic: The first war of physics and the secret history of the atomic bomb*, Icon Books, London, 2019.

Bakke, G., *The Grid: The fraying wires between Americans and our energy future*, Bloomsbury, New York, 2017.

Bakker, S., *From Luxury to Necessity: What the railways, electricity and the automobile teach us about the IT revolution*, Boom, Amsterdam, 2017.

Barr, J., *Lords of the Desert: Britain's struggle with America to dominate the Middle East*, Simon & Schuster, London, 2018.

Berman, B., *Zapped: From infrared to X-rays, the curious history of invisible light*, Oneworld Publications, London, 2017.

Berton, P., *Niagara: A history of the Falls*, Penguin Books, New York, 1998.

Bizony, P., *Atom*, Icon Books, London, 2017.

Bonneuil, C. and Fressoz, J.-B. (trans. Fernbach, D.), *The Shock of the Anthropocene*, Verso, London, 2017.

Boorstin, D. J., *The Americans: The democracy experience*, Phoenix Press, London, 1973.

Brown, L. R. (with Larsen, J., Roney, J. M., and Adams, E. E.), *The Great Transition: Shifting from fossil fuels to solar and wind energy*, Earth Policy Institute, Washington, DC, 2015.

Bruggink, A. and van der Hoeven, D., *More with Less: Welcome to the precision economy*, Biobased Press, Amsterdam, 2017.

Buck, H. J., *Ending Fossil Fuels: Why net zero is not enough*, Verso, London, 2021.

Burke, J., *Connections*, MacMillan London Limited, London, 1978.

Carlson, W. B., *Tesla: Inventor of the electrical age*, Princeton University Press, Princeton, NJ, 2013.

Carson, R., *Silent Spring*, Houghton Mifflin Company, New York, NY, 2002.

Clynes, T., *The Boy who Played with Fusion*, Faber and Faber, London, 2015.

Crump, T., *The Age of Steam: The power that drove the Industrial Revolution*, Robinson, London, 2007.

Fanchi, J. R., *Energy in the 21st Century*, World Scientific Publishing Co., Singapore, 2005.

Freudenburg, W. R. and Gramling, R., *Blowout in the Gulf: The BP oil spill disaster and the future of energy in America*, The MIT Press, Cambridge, MA, 2011.

Gates, B., *How to Avoid a Climate Disaster: The solutions we have and the breakthroughs we need*, Allen Lane, London, 2021.

Gertner, J., *The Idea Factory: Bell Labs and the great age of American innovation*, Penguin, London, 2012.

Giancoli, D. C., *Physics: Principles with applications*, 6th ed., Pearson Education International, Englewood Cliffs, NJ, 2006.

Gimpel, J., *The Medieval Machine: The Industrial Revolution of the Middle Ages*, Holt, Rinehart and Winston, New York, NY, 1976.

Goodall, C., *Ten Technologies to Save the Planet*, Green Profile, London, 2008.

Goodall, C., *The Switch: How solar, storage and new tech means cheap power for all*, Profile Books, London, 2016.

Goodstein, D., *Out of Gas: The end of the age of oil*, W. W. Norton & Company, New York, NY, 2004.

Gore, A., *Truth to Power*, Rodale Inc., New York, NY, 2017.

Griffith, S., *Electrify: An optimist's playbook for our clean energy future*, The MIT Press, Cambridge, MA, 2021.

Hawken, P. (Ed.), *Drawdown: The most comprehensive plan ever proposed to reverse global warming*, Penguin Books, London, 2017.

Haynes, R. D., *From Faust to Strangelove: Representations of the scientist in western literature*, The Johns Hopkins University Press, Baltimore, MD, 1994.

Helm, D., *The Carbon Crunch: Revised and updated*, Yale University Press, New Haven, CT, 2015.

Helm, D., *Net Zero: How we can stop climate change*, William Collins, London, 2020.

Hudson, M., *Global Fracture*, Pluto Press, London, 2005.

Isaacson, W., *Einstein: His life and universe*, Simon & Schuster, London, 2007.

Jacobson, M. Z., *100% Clean, Renewable Energy and Storage for Everything*, Cambridge University Press, Cambridge, 2020.

Johnson, S., *The Invention of Air*, Penguin Books, London, 2009.

Johnstone, B., *Switching to Solar: What we can learn from Germany's success in harnessing clean energy*, Prometheus Books, New York, NY, 2011.

Jonnes, J., *Empires of Light: Edison, Tesla, Westinghouse, and the race to electrify the world*, Random House, New York, NY, 2004.

Klein, M., *The Power Makers: Steam, electricity, and the men who invented modern America*, Bloomsbury Press, New York, NY, 2008.

Kragh, H., *Quantum Generations: The history of physics in the twentieth century*, Princeton University Press, Princeton, NJ, 1999.

Kuhn, T. S., *The Structure of Scientific Revolutions*, 2nd ed., The University of Chicago Press, Chicago, IL, 1970.

Lawson, T., *Crazy Caverns: How one small community challenged a technocrat juggernaut and won*, Port Hope, Ontario, 2013.

Lewis, S. J. and Maslin, M. A., *The Human Planet: How we created the Anthropocene*, Pelican Books, London, 2018.

Livesay, H. C., *Andrew Carnegie and the Rise of Big Business*, HarperCollins, New York, NY, 1975.

Maass, P., *Crude World*, Penguin Books, London, 2009.

Mann, C. C., *The Wizard and the Prophet: Science and the future of our planet*, Picador, London, 2019.

McKenzie, H., *Insane Mode: How Elon Musk's Tesla sparked an electric revolution to end the age of oil*, Faber and Faber, London, 2018.

McKibben, B., *Eaarth: Making a life on a tough new planet*, Times Books, New York, NY, 2010.

McNaught, K. W., Saywell, J. T., and Ricker, J. C., *Manifest Destiny: A short history of the United States*, Clarke, Irwin & Company Ltd., Toronto, 1980.

Morus, I. R., *Nikola Tesla and the Electrical Future*, Icon Books, London, 2019.

Muller, R. A., *Energy for Future Presidents: The science behind the headlines*, W. W. Norton and Company, New York, NY, 2012.

Orwell, G., *The Road to Wigan Pier*, Penguin Books, London, 2020.

Pais, A., *Inward Bound: Of matter and forces in the physical world*, Clarendon Press, Oxford, 1986.

Paxman, J., *Black Gold: The history of how coal made Britain*, William Collins, London, 2022.

Plokhy, S., *Chernobyl: History of a tragedy*, Penguin, London, 2019.

Rhodes, R., *Energy: A human history*, Simon and Schuster, New York, NY, 2018.

Richtmyer, F. K., Kennard, E. H., and Cooper, J. N., *Introduction to Modern Physics*, 6th ed., Tata McGraw-Hill, New Delhi, 1997.

Rifkin, J., *The Third Industrial Revolution: How lateral power is transforming energy, the economy, and the world*, Palgrave Macmillan, New York, NY, 2011.

Riordan, M. and Hoddeson, L., *Crystal Fire: The invention of the transistor and the birth of the Information Age*, W. W. Norton and Company, New York, NY, 1997.

Romm, J., *Hell and High Water: The global warming solution*, Harper Perennial, New York, NY, 2007.

Sanger, P., *Blind Faith: The nuclear industry in one small town*, McGraw-Hill Ryerson, Toronto, 1981.

Scheer, H., *The Solar Economy: Renewable energy for a sustainable global future*, Earthscan, London, 2002.

Segrè, E., *Enrico Fermi: Physicist*, The University of Chicago Press, Chicago, IL, 1970.

Segrè, G., *A Matter of Degrees*, Penguin Books, New York, NY, 2002.

Sinclair, U., *King Coal: A novel of the Colorado coal country*, published by the author, Station A, Pasadena, CA, 1917.

Sinclair, U., *Oil!: A novel*, published by the author, Station A, Pasadena, CA, 1926, 1927.

Smil, V., *Energy and Civilization: A history*, The MIT Press, Cambridge, MA, 2017.

Strathern, P., *Mendeleyev's Dream: The quest for the elements*, Penguin Books, London, 2001.

Stross, R., *The Wizard of Menlo Park: How Thomas Alva Edison invented the modern world*, Three Rivers Press, New York, NY, 2007.

Stuart, T., *Waste: Uncovering the global food scandal*, Penguin Books, London, 2009.

Tarbell, I. M., *The History of the Standard Oil Company*, McLure, Phillips and Co., New York, NY, 1905.

Tutt, K., *The Scientist, the Madman, the Thief and their Lightbulb*, Pocket Books, London, 2003.

Usher, B., *Renewable Energy: A primer for the twenty-first century*, Columbia University Press, New York, NY, 2019.

Vance, A., *Elon Musk: How the billionaire CEO of SpaceX and Tesla is shaping our future*, Virgin Books, London, 2015.

Walker, G. and King, D., *The Hot Topic: What we can do about global warming*, Harcourt Inc., Orlando, FL, 2008.

Wallace-Wells, D., *The Uninhabitable Earth: A story of the future*, Penguin Random House, London, 2019.

Yergin, D., *The Prize: The epic quest for oil, money, & power*, Simon & Schuster, New York, NY, 1991.

Yergin, D., *The Quest: Energy, security, and the remaking of the modern world*, Penguin Books, New York, NY, 2012.

Index

Page numbers in **"bold"** refer to tables, *"Italics"* refer to figures, and "n" refer to "notes".

"21st-Century Living" experiment, 555
1851 Crystal Palace Exhibition, 18, 320, 456
1893 Chicago World's Fair, 412
60-GW Hydrogen City project, 509

AAPowerLink, 367
Abe, Shinzō, 263
Aberpergwm Colliery (south Wales), 40
AC induction motor, 410
AC lighting, 410
AC to DC conversion, 324
Act of Parliament (1826), 17
Adams, Edward Dean, 411
advanced boiling water reactor (ABWR), 269
advanced gas-cooled (AGR) reactors, 243
advanced pressurized reactors (AP1000), 268
Africa, northern, sunshine in, 307
Age of Steam, The (Crump), 16
airline emissions, 568–569
Akademik Lomonosov power station, 222
Akasaki, Isamu, 337
Al Rekayyat, 158
Alaskan fuels, hydrocarbon concentration in, **102**
Albemarle, 518, 520
Alexander, Lamar, 395
Al-Falih, Khalid, 373
Allen, Paul, Zero Carbon Britain 2030, 592
Alliance to Protect Nantucket Sound, 395
all-solid-state battery (ASSB), 522
alternating current system, 416
    induction motors, Tesla's, 423
    power station. *See also* AC power
    the transformer and, 415

Aluminum Company of America (Alcoa), 414
aluminum-air (Al-air) batteries, 524
Al-Yamamah project, 123
Amano, Hiroshi, 337
Ambient Nonmethane Hydrocarbon Levels
    Along Colorado's Northern Front Range …
    Report (Colorado University), 149
American Association for the Advancement of
    Science, 233
American Petroleum Institute (API), 5, 131,
    184, 215
American Revolution, 20
Amsterdam Smart City initiative, 531
anaerobic digester (AD), 53
Andasol power station, 366
Anderson, Paul Thomas, 75
Andreae, Meinrat, views on geo-
    engineering, 588
Annapolis Royal Generating Station, 430
Annus Mirabilis papers (Einstein), 322
Anschutz, Philip, 389
Anthropocene, 590–591
anthropogenic global warming (AGW), 172
anti-American Iranian revolution, 126
Apollo 11, 506
Appleton power station, 418
Aquatic Species Program, 510
Arabian Peninsula, 114
Arab-Israeli Six-Day War, 117
ArcelorMittal, 175
Archimedes, 315, 330, 364
    screw, 380, 431
Arctic National Wildlife Region (ANWR), 138

Milton Keynes UK
Ingram Content Group UK Ltd.
UKHW021854021224
3319UKWH00035B/415